Studies in Classification, Data Analysis, and Knowledge Organization

Titles in the Series

W. Gaul and D. Pfeifer (Eds.)
From Data to Knowledge. 1995

H.-H. Bock and W. Polasek (Eds.)
Data Analysis and Information Systems. 1996

E. Diday, Y. Lechevallier, and O. Opitz (Eds.) Ordinal and Symbolic Data Analysis. 1996

R. Klar and O. Opitz (Eds.)
Classification and Knowledge Organization. 1997

C. Hayashi, N.Ohsumi, K.Yajima, Y. Tanaka, H.-H. Bock, and Y. Baba (Eds.)
Data Science, Classifaction, and Related Methods. 1998

I. Balderjahn, R. Mathar, and M. Schader (Eds.)
Classification, Data Analysis, and Data Highways. 1998

A. Rizzi, M. Vichi, and H.-H. Bock (Eds.)
Advances in Data Science and Classification 1998.

M. Vichi and O. Optiz (Eds.)
Classification and Data Analysis. 1999

W. Gaul and H. Locarek-Junge (Eds.)
Classification in the Information Age. 1999

H.-H. Bock and E. Diday (Eds.)
Analysis of Symbolic Data. 2000

H. A. L. Kiers, J.-P. Rasson, P. J. F. Groenen, and M. Schader (Eds.)
Data Analysis, Classification, and Related Methods. 2000

W. Gaul, O. Opitz, M. Schader (Eds.)
Data Analysis. 2000

R. Decker and W. Gaul (Eds.)
Classification and Information Processing at the Turn of the Millenium. 2000

S. Borra, R. Rocci, M. Vichi, and M. Schader (Eds.)
Advances in Classification and Data Analysis. 2000

W. Gaul and G. Ritter (Eds.)
Classification, Automation, and New Media. 2002

K. Jajuga, A. Sokołowski, and H.-H. Bock (Eds.)
Classification, Clustering and Data Analysis. 2002

M. Schwaiger and O. Opitz (Eds.)
Exploratory Data Analysis in Empirical Research. 2003

M. Schader, W. Gaul, and M. Vichi (Eds.)
Between Data Science and Applied Data Analysis. 2003

H.-H. Bock, M. Chiodi, and A. Mineo (Eds.)
Advances in Multivariate Data Analysis. 2004

D. Banks, L. House, F.R. McMorris, P. Arabie, and W. Gaul (Eds.)
Classification, Clustering, and Data Minig Applications. 2004

D. Baier and K.-D. Wernecke (Eds.)
Innovations in Classification, Data Science, and Information Systems. 2005

M. Vichi, P. Monari, S. Mignani, and A. Montanari (Eds.)
New Developments in Classification and Data Analysis. 2005

D. Baier, R. Decker, and L. Schmidt-Thieme (Eds.)
Data Analysis and Decision Support. 2005

C. Weihs and W. Gaul (Eds.)
Classification - the Ubiquitous Challenge. 2005

M. Spiliopoulou, R. Kruse, C. Borgelt, A. Nürnberger, and W. Gaul (Eds.)
From Data and Information Analysis to Knowledge Engineering. 2006

V. Batagelj, H.-H. Bock, A. Ferligoj, and A. Žiberna (Eds.)
Data Science and Classification. 2006

S. Zani, A. Cerioli, M. Riani, M. Vichi (Eds.)
Data Analysis, Classification and the Forward Search. 2006

Reinhold Decker
Hans-J. Lenz
Editors

Advances in Data Analysis

Proceedings of the 30th Annual Conference of the Gesellschaft für Klassifikation e.V., Freie Universität Berlin, March 8-10, 2006

With 202 Figures and 92 Tables

Springer

Professor Dr. Reinhold Decker
Department of Business Administration and Economics
Bielefeld University
Universitätsstr. 25
33501 Bielefeld
Germany
rdecker@wiwi.uni-bielefeld.de

Professor Dr. Hans - J. Lenz
Department of Economics
Freie Universität Berlin
Garystraße 21
14195 Berlin
Germany
hjlenz@wiwiss.fu-berlin.de

Library of Congress Control Number: 2007920573

ISSN 1431-8814

ISBN 978-3-540-70980-0 Springer Berlin Heidelberg New York

Springer is a part of Springer Science+Business Media

springer.com

Production: LE-TEX Jelonek, Schmidt & Vockler GbR, Leipzig
Cover-design: WMX Design GmbH, Heidelberg

SPIN 12022755 43/3100YL - 5 4 3 2 1 0 Printed on acid-free paper

Preface

This volume contains the revised versions of selected papers presented during the 30th Annual Conference of the German Classification Society (Gesellschaft für Klassifikation – GfKl) on "Advances in Data Analysis". The conference was held at the Freie Universität Berlin, Germany, in March 2006. The scientific program featured 7 parallel tracks with more than 200 contributed talks in 63 sessions. Additionally, thanks to the support of the DFG (German Research Foundation), 18 plenary and semi-plenary speakers from Europe and overseas could be invited to talk about their current research in classification and data analysis. With 325 participants from 24 countries in Europe and overseas this GfKl Conference, once again, provided an international forum for discussions and mutual exchange of knowledge with colleagues from different fields of interest. From altogether 115 full papers that had been submitted for this volume 77 were finally accepted.

The scientific program included a broad range of topics from classification and data analysis. Interdisciplinary research and the interaction between theory and practice were particularly emphasized. The following sections (with chairs in alphabetical order) were established:

I. Theory and Methods

Clustering and Classification (H.-H. Bock and T. Imaizumi); Exploratory Data Analysis and Data Mining (M. Meyer and M. Schwaiger); Pattern Recognition and Discrimination (G. Ritter); Visualization and Scaling Methods (P. Groenen and A. Okada); Bayesian, Neural, and Fuzzy Clustering (R. Kruse and A. Ultsch); Graphs, Trees, and Hierarchies (E. Godehardt and J. Hansohm); Evaluation of Clustering Algorithms and Data Structures (C. Hennig); Data Analysis and Time Series Analysis (S. Lang); Data Cleaning and Pre-Processing (H.-J. Lenz); Text and Web Mining (A. Nürnberger and M. Spiliopoulou); Personalization and Intelligent Agents (A. Geyer-Schulz); Tools for Intelligent Data Analysis (M. Hahsler and K. Hornik).

II. Applications

Subject Indexing and Library Science (H.-J. Hermes and B. Lorenz); Marketing, Management Science, and OR (D. Baier and O. Opitz); E-commerce, Recommender Systems, and Business Intelligence (L. Schmidt-Thieme); Banking and Finance (K. Jajuga and H. Locarek-Junge); Economics (G. Kauermann and W. Polasek); Biostatistics and Bioinformatics (B. Lausen and U. Mansmann); Genome and DNA Analysis (A. Schliep); Medical and Health Sciences (K.-D. Wernecke and S. Willich); Archaeology (I. Herzog, T. Kerig, and A. Posluschny); Statistical Musicology (C. Weihs); Image and Signal Processing (J. Buhmann); Linguistics (H. Goebl and P. Grzybek); Psychology (S. Krolak-Schwerdt); Technology and Production (M. Feldmann).

Additionally, the following invited sessions were organized by colleagues from associated societies: Classification with Complex Data Structures (A. Cerioli); Machine Learning (D.A. Zighed); Classification and Dimensionality Reduction (M. Vichi).

The editors would like to emphatically thank the section chairs for doing such a great job regarding the organization of their sections and the associated paper reviews. The same applies to W. Esswein for organizing the Doctoral Workshop and to H.-H. Hermes and B. Lorenz for organizing the Librarians Workshop. Cordial thanks also go to the members of the scientific program committee for their conceptual and practical support (in alphabetical order): D. Baier (Cottbus), H.-H. Bock (Aachen), H.W. Brachinger (Fribourg), R. Decker (Bielefeld, Chair), D. Dubois (Toulouse), A. Gammerman (London), W. Gaul (Karlsruhe), A. Geyer-Schulz (Karlsruhe), B. Goldfarb (Paris), P. Groenen (Rotterdam), D. Hand (London), T. Imaizumi (Tokyo), K. Jajuga (Wroclaw), G. Kauermann (Bielefeld), R. Kruse (Magdeburg), S. Lang (Innsbruck), B. Lausen (Erlangen-Nürnberg), H.-J. Lenz (Berlin), F. Murtagh (London), A. Okada (Tokyo), L. Schmidt-Thieme (Hildesheim) M. Spiliopoulou (Magdeburg), W. Stützle (Washington), and C. Weihs (Dortmund). The review process was additionally supported by the following colleagues: A. Cerioli, E. Gatnar, T. Kneib, V. Köppen, M. Meißner, I. Michalarias, F. Mörchen, W. Steiner, and M. Walesiak.

The great success of this conference would not have been possible without the support of many people mainly working in the backstage. Representative for the whole team we would like to particularly thank M. Darkow (Bielefeld) and A. Wnuk (Berlin) for their exceptional efforts and great commitment with respect to the preparation, organization and post-processing of the conference. We thank very much our web masters I. Michalarias (Berlin) and A. Omelchenko (Berlin). Furthermore, we would cordially thank V. Köppen (Berlin) and M. Meißner (Bielefeld) for providing an excellent support regarding the management of the reviewing process and the final editing of the papers printed in this volume.

The GfKl Conference 2006 would not have been possible in the way it took place without the financial and/or material support of the following institutions and companies (in alphabetical order): Deutsche Forschungsgemeinschaft, Freie Universität Berlin, Gesellschaft für Klassifikation e.V., Land Software-Entwicklung, Microsoft München, SAS Deutschland, Springer-

Verlag, SPSS München, Universität Bielefeld, and Westfälisch-Lippische Universitätsgesellschaft. We express our gratitude to all of them.

Finally, we would like to thank Dr. Martina Bihn of Springer-Verlag, Heidelberg, for her support and dedication to the production of this volume.

Berlin and Bielefeld, January 2007

Hans-J. Lenz
Reinhold Decker

Contents

Part III Data and Time Series Analysis

Part VI Marketing, Management Science and Economics

Part XI Archaeology

Part I

Clustering

Mixture Models for Classification

Gilles Celeux

Inria Futurs, Orsay, France; Gilles.Celeux@inria.fr

Abstract. Finite mixture distributions provide efficient approaches of model-based clustering and classification. The advantages of mixture models for unsupervised classification are reviewed. Then, the article is focusing on the model selection problem. The usefulness of taking into account the modeling purpose when selecting a model is advocated in the unsupervised and supervised classification contexts. This point of view had lead to the definition of two penalized likelihood criteria, ICL and BEC, which are presented and discussed. Criterion ICL is the approximation of the integrated completed likelihood and is concerned with model-based cluster analysis. Criterion BEC is the approximation of the integrated conditional likelihood and is concerned with generative models of classification. The behavior of ICL for choosing the number of components in a mixture model and of BEC to choose a model minimizing the expected error rate are analyzed in contrast with standard model selection criteria.

1 Introduction

Finite mixtures models has been extensively studied for decades and provide a fruitful framework for classification (McLachlan and Peel (2000)). In this article some of the main features and advantages of finite mixture analysis for model-based clustering are reviewed in Section 2. An important interest of finite mixture model is to provide a rigorous setting to assess the number of clusters in an unsupervised classification context or to assess the stability of a classification function. It is focused on those two questions in Section 3.

Model-based clustering (MBC) consists of assuming that the data come from a source with several subpopulations. Each subpopulation is modeled separately and the overall population is a mixture of these subpopulations. The resulting model is a finite mixture model. Observations $\mathbf{x} = (\mathbf{x}_1, \ldots, \mathbf{x}_n)$ in $\mathbf{R}^{nd}$ are assumed to be a sample from a probability distribution with density

$$p(\mathbf{x}_i \mid K, \theta_K) = \sum_{k=1}^{K} p_k \phi(\mathbf{x}_i \mid \mathbf{a}_k) \tag{1}$$

where the p_k's are the mixing proportions ($0 < p_k < 1$ for all $k = 1, \ldots, K$ and $\sum_k p_k = 1$), $\phi(. \mid \mathbf{a}_k)$ denotes a parameterized density and $\theta_K = (p_1, \ldots, p_{K-1}, a_1, \ldots, a_K)$. When data are multivariate continuous observations, the component parameterized density is usually the d-dimensional Gaussian density and parameter $a_k = (\mu_k, \Sigma_k)$, μ_k being the mean and Σ_k the variance matrix of component k. When data are discrete, the component parameterized density is usually the multivariate multinomial density which is assuming conditional Independence of the observations knowing their component mixture and the $a_k = (a_k^j, j = 1, \ldots, d)$'s are the multinomial probabilities for variable j and mixture component k. The resulting model is the so-called Latent Class Model (see for instance Goodman (1974)).

The mixture model is an incomplete data structure model: The complete data are

$$\mathbf{y} = (\mathbf{y}_1, \ldots, \mathbf{y}_n) = ((\mathbf{x}_1, \mathbf{z}_1), \ldots, (\mathbf{x}_n, \mathbf{z}_n))$$

where the missing data are $\mathbf{z} = (\mathbf{z}_1, \ldots, \mathbf{z}_n)$, with $\mathbf{z}_i = (z_{i1}, \ldots, z_{iK})$ are binary vectors such that $z_{ik} = 1$ iff $\mathbf{x}_i$ arises from group k. The $\mathbf{z}$'s define a partition $P = (P_1, \ldots, P_K)$ of the observed data $\mathbf{x}$ with $P_k = \{\mathbf{x}_i \mid z_{ik} = 1\}$.

In this article, it is considered that the mixture models at hand are estimated through maximum likelihood (ml) or related methods. Despite it has received a lot of attention, since the seminal article of Diebolt and Robert (1994), Bayesian inference is not considered here. Bayesian analysis of univariate mixtures has became the standard Bayesian tool for density estimation. But, especially in the multivariate setting a lot of problems (possible slow convergence of MCMC algorithms, definition of subjective weakly informative priors, identifiability, ...) remain and it cannot be regarded as a standard tool for Bayesian clustering of multivariate data (see Aitkin (2001)). The reader is referred to the survey article of Marin et al. (2005) for a readable state of the art of Bayesian inference for mixture models.

2 Some advantages of model-based clustering

In this section, some important and nice features of finite mixture analysis are sketched. The advantages of finite mixture analysis in a clustering context, highlighted here, are: Many versatile or parsimonious models are available, many algorithms to estimate the mixture parameters are available, special questions can be tackled in a proper way in the MBC context, and, last but not least, finite mixture models can be compared and assessed in an objective way. It allows in particular to assess the number of clusters properly. The discussion on this important point is postponed to Section 3.

Many versatile or parsimonious models are available.

In the multivariate Gaussian mixture context, the variance matrix eigenvalue decomposition

$$\Sigma_k = V_k D_k^t A_k D_k$$

where $V_k = |\Sigma_k)|^{1/d}$ defines the component volume, D_k the matrix of eigenvectors of Σ defines the component orientation, and A_k the diagonal matrix of normalized eigenvalues defines the component shape, leads to get different and easily interpreted models by allowing some of these quantities to vary between components. Following Banfield and Raftery (1993) or Celeux and Govaert (1995), a large range of fourteen versatile (from the most complex to the simplest one) models derived from this eigenvalue decomposition can be considered. Assuming equal or free volumes, orientations and shapes leads to eight different models. Assuming in addition that the component variance matrices are diagonal leads to four models. And, finally, assuming in addition that the component variance matrices are proportional to the identity matrix leads to two other models.

In the Latent Class Model, a re-parameterization is possible to lead to various models taking account of the scattering around centers of the clusters in different ways (Celeux and Govaert (1991)). This re-parameterization is as follows. The multinomial probabilities $\mathbf{a}_k$ are decomposed in $(\mathbf{m}_k, \varepsilon_k)$ where binary vector $\mathbf{m}_k = (\mathbf{m}_k^1, \ldots, \mathbf{m}_k^d)$ provides the mode levels in cluster k for variable j

$$(m_k^{jh}) = \begin{cases} 1 & \text{if } h = \arg\max_h a_k^{jh} \\ 0 & \text{otherwise,} \end{cases}$$

and the ε_k^j can be regarded as scattering values.

$$(\varepsilon^{jh}) = \begin{cases} 1 - \alpha^{jh} & \text{if } a^{jh} = 1 \\ \alpha^{jh} & \text{if } a^{jh} = 0. \end{cases}$$

For instance, if $\mathbf{a}_k^j = (0.7, 0.2, 0.1)$, the new parameters are $\mathbf{m}_k^j = (1, 0, 0)$ and $\varepsilon_k^j = (0.3, 0.2, 0.1)$. This parameterization can lead to five latent class models. Denoting $h(jk)$ the mode level for variable j and cluster k and $h(ij)$ the level of object i for the variable j, the model can be written

$$f(\mathbf{x}_i; \theta) = \sum_k p_k \prod_j \left((1 - \varepsilon_k^{jh(jk)})^{x_i^{jh(jk)}} (\varepsilon_k^{jh(ij)})^{x_i^{jh(ij)} - x_k^{jh(jk)}} \right).$$

Using this form, it is possible to impose various constraints to the scattering parameters ε_k^{jh}. The models of interest are the following:

- The standard latent class model $[\varepsilon_k^{jh}]$: The scattering is depending upon clusters, variables and levels.
- $[\varepsilon_k^j]$: The scattering is depending upon clusters and variables but not upon levels.
- $[\varepsilon_k]$: The scattering is depending upon clusters, but not upon variables.
- $[\varepsilon^j]$: The scattering is depending upon variables, but not upon clusters.
- $[\varepsilon]$: The scattering is constant over variables and clusters.

Many algorithms available from different points of view

The EM algorithm of Dempster et al. (1977) is the reference tool to derive the ml estimates in a mixture model. An iteration of EM is as follows:

- *E step:* Compute the conditional probabilities t_{ik}, $i = 1, \ldots, n$, $k = 1, \ldots, K$ that x_i arises from the kth component for the current value of the mixture parameters.
- *M step:* Update the mixture parameter estimates maximizing the expected value of the completed likelihood. It leads to use standard formulas where the observation i for group k is weighted with the conditional probability t_{ik}.

Others algorithms are taking profit of the missing data structure of the mixture model. For instance, the classification EM (CEM), see Celeux and Govaert (1992) is directly concerned with the estimation of the missing labels $\mathbf{z}$. An iteration of CEM is as follows:

- *E step:* As in EM.
- *C step:* Assign each point x_i to the component maximizing the conditional probability t_{ik} using a maximum a posteriori (MAP) principle.
- *M step:* Update the mixture parameter estimates maximizing the completed likelihood.

CEM aims to maximize the completed likelihood where the component label of each sample point is included in the data set. CEM is a K-means-like algorithm and, contrary to EM, it converges in a finite number of iterations. But CEM provides biased estimates of the mixture parameters. This algorithm is interesting in a clustering context when the clusters are well separated (see Celeux and Govaert (1993)).

From an other point of view, the Stochastic EM (SEM) algorithm can be useful. It is as follows:

- *E step:* As in EM.
- *S step:* Assign each point x_i at random to one of the component according to the distribution defined by the $(t_{ik}, k = 1, \ldots, K)$.
- *M step:* Update the mixture parameter estimates maximizing the completed likelihood.

SEM generates a Markov chain whose stationary distribution is (more or less) concentrated around the ML parameter estimator. Thus a natural parameter estimate from a SEM sequence is the mean of the iterates values obtain after a burn-in period. An alternative estimate is to consider the parameter value leading to the largest likelihood in a SEM sequence. In any cases, SEM is expected to avoid insensible maxima of the likelihood that EM cannot avoid, but SEM can be jeopardized by spurious maxima (see Celeux et al. (1996) or McLachlan and Peel (2000) for details). Note that different variants (Monte Carlo EM, Simulated Annealing EM) are possible (see, for instance, Celeux et al. (1996)). Note also that Biernacki et al. (2003) proposed simple strategies for getting sensible ml estimates. Those strategies are acting in two ways to deal with this problem. They choose particular starting values from CEM or SEM and they run several times EM or algorithms combining CEM and EM.

Special questions can be tackled in a proper way in the MBC context

Robust Cluster Analysis can be obtained by making use of multivariate Student distributions instead of Multivariate Gaussian distributions. It lead to attenuate the influence of outliers (McLachlan and Peel (2000)). On an other hand, including in the mixture a group from a uniform distribution allows to take into account noisy data (DasGupta and Raftery (1998)).

To avoid spurious maxima of likelihood, shrinking the group variance matrix toward a matrix proportional to the identity matrix can be quite efficient. On of the most achieved work in this domain is Ciuperca et al. (2003).

Taking profit of the probabilistic framework, it is possible to deal with missing data at random in a proper way with mixture models (Hunt and Basford (2001)). Also, simple, natural and efficient methods of semi-supervised classification can be derived in the mixture framework (an example of pioneer article on this subject, recently followed by many others, is Ganesalingam and McLachlan (1978)). Finally, it can be noted that promising variable selection procedures for Model-Based Clustering begin to appear (Raftery and Dean (2006)).

3 Choosing a model in a classification purpose

In statistical inference from data selecting a parsimonious model among a collection of models is an important but difficult task. This general problem receives much attention since the seminal articles of Akaike (1974) and Schwarz (1978). A model selection problem consists essentially of solving the bias-variance dilemma. A classical approach to the model assessing problem consists of penalizing the fit of a model by a measure of its complexity. Criterion AIC of Akaike (1974) is an asymptotic approximation of the expectation of the deviance. It is

$$\mathrm{AIC}(m) = 2 \log \mathbf{p}(\mathbf{x}|m, \hat{\theta}_m) - 2\nu_m. \tag{2}$$

where $\hat{\theta}_m$ is the ml estimate of parameter θ_m and ν_m is the number of free parameters of model m.

An other point of view consists of basing the model selection on the integrated likelihood of the data in a Bayesian perspective (see Kass and Raftery (1995)). This integrated likelihood is

$$\mathbf{p}(\mathbf{x}|m) = \int \mathbf{p}(\mathbf{x}|m, \theta_m)\pi(\theta_m)d\theta_m, \tag{3}$$

$\pi(\theta_m)$ being a prior distribution for parameter θ_m. The essential technical problem is to approximate this integrated likelihood in a right way. A classical asymptotic approximation of the logarithm of the integrated likelihood is the BIC criterion of Schwarz (1978). It is

$$\text{BIC}(m) = \log \mathbf{p}(\mathbf{x}|m, \hat{\theta}_m) - \frac{\nu_m}{2} \log(n). \tag{4}$$

Beyond technical difficulties, the scope of this section is to show how it can be fruitful to take into account the purpose of the model user to get reliable and useful models for statistical description or decision tasks. Two situations are considered to support this idea: Choosing the number of components in a mixture model in a cluster analysis perspective, and choosing a generative probabilistic model in a supervised classification context.

3.1 Choosing the number of clusters

Assessing the number K of components in a mixture model is a difficult question, from both theoretical and practical points of view, which had received much attention in the past two decades. This section does not propose a state of the art of this problem which has not been completely solved. The reader is referred to the chapter 6 of the book of McLachlan and Peel (2000) for an excellent overview on this subject. This section is essentially aiming to discuss elements of practical interest regarding the problem of choosing the number of mixture components when concerned with cluster analysis.

From the theoretical point of view, even when K^* the right number of component is assumed to exist, if $K^* < K_0$ then K^* is not identifiable in the parameter space Θ^{K_0} (see for instance McLachlan and Peel (2000), chapter 6).

But, here, we want to stress the importance of taking into account the modeling context to select a reasonable number of mixture components. Our opinion is that, behind the theoretical difficulties, assessing the number of components in a mixture model from data is a weakly identifiable statistical problem. Mixture densities with different number of components can lead to quite similar resulting probability distributions. For instance, the galaxy velocities data of Roeder (1990) has became a benchmark data set and is used by many authors to illustrate procedures for choosing the number of mixture components. Now, according to those authors the answer lies from $K = 2$ to $K = 10$, and it is not exaggerating a lot to say that all the answers between 2 and 10 have been proposed as a good answer, at least one time, in the articles considering this particular data set. (An interesting and illuminating comparative study on this data set can be found in Aitkin (2001).) Thus, we consider that it is highly desirable to choose K by keeping in mind what is expected from the mixture modeling to get a relevant answer to this question. Actually, mixture modeling can be used in quite different purposes. It can be regarded as a semi parametric tool for density estimation purpose or as a tool for cluster analysis.

In the first perspective, much considered by Bayesian statisticians, numerical experiments (see Fraley and Raftery (1998)) show that the BIC approximation of the integrated likelihood works well at a practical level. And, under regularity conditions including the fact that the component densities are finite, Keribin (2000) proved that BIC provides a consistent estimator of K.

But, in the second perspective, the integrated likelihood does not take into account the clustering purpose for selecting a mixture model in a model-based clustering setting. As a consequence, in the most current situations where the distribution from which the data arose is not in the collection of considered mixture models, BIC criterion will tend to overestimate the correct size regardless of the separation of the clusters (see Biernacki et al. (2000)).

To overcome this limitation, it can be advantageous to choose K in order to get the mixture giving rise to partitioning data with the greatest evidence. With that purpose in mind, Biernacki et al. (2000) considered the integrated likelihood of the complete data $(\mathbf{x}, \mathbf{z})$ (or integrated completed likelihood). (Recall that $\mathbf{z} = (\mathbf{z}_1, \ldots, \mathbf{z}_n)$ is denoting the missing data such that $\mathbf{z}_i = (z_{i1}, \ldots, z_{iK})$ are binary K-dimensional vectors with $z_{ik} = 1$ if and only if $\mathbf{x}_i$ arises from component k.) Then, the integrated complete likelihood is

$$\mathbf{p}(\mathbf{x}, \mathbf{z} \mid K) = \int_{\Theta_K} \mathbf{p}(\mathbf{x}, \mathbf{z} \mid K, \theta) \pi(\theta \mid K) d\theta, \tag{5}$$

where

$$\mathbf{p}(\mathbf{x}, \mathbf{z} \mid K, \theta) = \prod_{i=1}^{n} p(\mathbf{x}_i, \mathbf{z}_i \mid K, \theta)$$

with

$$p(\mathbf{x}_i, \mathbf{z}_i \mid K, \theta) = \prod_{k=1}^{K} p_k^{z_{ik}} \left[\phi(\mathbf{x}_i \mid \mathbf{a}_k)\right]^{z_{ik}}.$$

To approximate this integrated complete likelihood, those authors propose to use a BIC-like approximation leading to the criterion

$$\mathrm{ICL}(K) = \log \mathbf{p}(\mathbf{x}, \hat{\mathbf{z}} \mid K, \hat{\theta}) - \frac{\nu_K}{2} \log n, \tag{6}$$

where the missing data have been replaced by their most probable value for parameter estimate $\hat{\theta}$. (Details can be found in Biernacki et al. (2000)). Roughly speaking criterion ICL is the criterion BIC penalized by the mean entropy

$$E(K) = -\sum_{k=1}^{K} \sum_{i=1}^{n} t_{ik} \log t_{ik} \geq 0,$$

t_{ik} denoting the conditional probability that $\mathbf{x}_i$ arises from the kth mixture component ($1 \leq i \leq n$ and $1 \leq k \leq K$).

As a consequence, ICL favors K values giving rise to partitioning the data with the greatest evidence, as highlighted in the numerical experiments in Biernacki et al. (2000), because of this additional entropy term. More generally, ICL appears to provide a stable and reliable estimate of K for real data sets and also for simulated data sets from mixtures when the components are not too much overlapping (see McLachlan and Peel (2000)). But ICL, which is not aiming to discover the true number of mixture components, can underestimate the number of components for simulated data arising from mixture with poorly separated components as illustrated in Figueiredo and Jain (2002).

On the contrary, BIC performs remarkably well to assess the true number of components from simulated data (see Biernacki et al. (2000), Fraley and Raftery (1998) for instance). But, for real world data sets, BIC has a marked tendency to overestimate the numbers of components. The reason is that real data sets do not arise from the mixture densities at hand, and the penalty term of BIC is not strong enough to balance the tendency of the loglikelihood to increase with K in order to improve the fit of the mixture model.

3.2 Model selection in classification

Supervised classification is about guessing the unknown group among K groups from the knowledge of d variables entering in a vector $\mathbf{x}_i$ for unit i. This group for unit i is defined by $\mathbf{z}_i = (z_{i1}, \ldots, z_{iK})$ a binary K-dimensional vector with $z_{ik} = 1$ if and only if $\mathbf{x}_i$ arises from group k. For that purpose, a decision function, called a classifier, $\delta(\mathbf{x}) : \mathbf{R}^d \to \{1, \ldots, K\}$ is designed from a learning sample $(\mathbf{x}_i, \mathbf{z}_i), i = 1, \ldots, n$. A classical approach to design a classifier is to represent the group conditional densities with a parametric model $\mathbf{p}(\mathbf{x}|m, z_k = 1, \theta_m)$ for $k = 1, \ldots, K$. Then the classifier is assigning an observation $\mathbf{x}$ to the group k maximizing the conditional probability $p(z_k = 1|m, \mathbf{x}, \theta_m)$. Using the Bayes rule, it leads to set $\delta(\mathbf{x}) = j$ if and only if $j = \arg\max_k p_k \mathbf{p}(\mathbf{x}|m, z_k = 1, \hat{\theta}_m)$, $\hat{\theta}_m$ being the ml estimate of the group conditional parameters θ_m and p_k being the prior probability of group k. This approach is known as the generative discriminant analysis in the Machine Learning community.

In this context, it could be expected to improve the actual error rate by selecting a generative model m among a large collection of models $\mathcal{M}$ (see for instance Friedman (1989) or Bensmail and Celeux (1996)). For instance Hastie and Tibshirani (1996) proposed to model each group density with a mixture of Gaussian distributions. In this approach the number of mixture components *per* group are sensitive tuning parameters. They can be supplied by the user, as in Hastie and Tibshirani (1996), but it is clearly a sub-optimal solution. They can be chosen to minimize the v-fold cross-validated error rate, as done in Friedman (1989) or Bensmail and Celeux (1996) for other tuning parameters. Despite the fact the choice of v can be sensitive, it can be regarded as a nearly optimal solution. But it is highly CPU time consuming and choosing tuning parameters with a penalized loglikelihood criterion, as BIC, can be expected to be much more efficient in many situations. But, BIC measures the fit of the model m to the data $(\mathbf{x}, \mathbf{z})$ rather than its ability to produce a reliable classifier. Thus, in many situations, BIC can have a tendency to overestimate the complexity of the generative classification model to be chosen. In order to counter this tendency, a penalized likelihood criterion taking into account the classification task when evaluating the performance of a model has been proposed by Bouchard and Celeux (2006). It the so-called Bayesian Entropy Criterion (BEC) that it is now presented.

As stated above, a classifier deduced from model m is assigning an observation $\mathbf{x}$ to the group k maximizing $p(z_k = 1|m, \mathbf{x}, \hat{\theta}_m)$. Thus, from the classification point of view, the conditional likelihood $\mathbf{p}(\mathbf{z}|m, \mathbf{x}, \theta_m)$ has a central position. For this very reason, Bouchard and Celeux (2006) proposed to make use of the integrated conditional likelihood

$$\mathbf{p}(\mathbf{z}|m, \mathbf{x}) = \int \mathbf{p}(\mathbf{z}|m, \mathbf{x}, \theta_m)\pi(\theta_m|\mathbf{x})d\theta_m, \tag{7}$$

where

$$\pi(\theta_m|\mathbf{x}) \propto \pi(\theta_m)\mathbf{p}(\mathbf{x}|m, \theta_m)$$

is the posterior distribution of θ_m knowing $\mathbf{x}$, to select a relevant model m. As for the integrated likelihood, this integral is generally difficult to calculate and has to be approximated. We have

$$\mathbf{p}(\mathbf{z}|m, \mathbf{x}) = \frac{\mathbf{p}(\mathbf{x}, \mathbf{z}|m)}{\mathbf{p}(\mathbf{x}|m)} \tag{8}$$

with

$$\mathbf{p}(\mathbf{x}, \mathbf{z}|m) = \int \mathbf{p}(\mathbf{x}, \mathbf{z}|m, \theta_m)\pi(\theta_m)d\theta_m, \tag{9}$$

and

$$\mathbf{p}(\mathbf{x}|m) = \int \mathbf{p}(\mathbf{x}|m, \theta_m)\pi(\theta_m)d\theta_m. \tag{10}$$

Denoting

$$\hat{\theta}_m = \arg\max_{\theta_m} \mathbf{p}(\mathbf{x}, \mathbf{z}|m, \theta_m),$$

$$\tilde{\theta}_m = \arg\max_{\theta_m} \mathbf{p}(\mathbf{x}|m, \theta_m)$$

and

$$\theta^\star_m = \arg\max_{\theta_m} \mathbf{p}(\mathbf{z}|m, \mathbf{x}, \theta_m),$$

BIC-like approximations of the numerator and denominator of equation (8) leads to

$$\log \mathbf{p}(\mathbf{z}|m, \mathbf{x}) = \log \mathbf{p}(\mathbf{x}, \mathbf{z}|m, \hat{\theta}_m) - \log \mathbf{p}(\mathbf{x}|m, \tilde{\theta}_m) + \mathrm{O}(1). \tag{11}$$

Thus the approximation of $\log \mathbf{p}(\mathbf{z}|m, \mathbf{x})$ that Bouchard and Celeux (2006) proposed is

$$\text{BEC} = \log \mathbf{p}(\mathbf{x}, \mathbf{z}|m, \hat{\theta}_m) - \log \mathbf{p}(\mathbf{x}|m, \tilde{\theta}_m). \tag{12}$$

The criterion BEC needs to compute $\tilde{\theta}_m = \arg\max_{\theta_m} \mathbf{p}(\mathbf{x}|m, \theta_m)$. Since, for $i = 1, \ldots, n$,

$$p(\mathbf{x}_i|m, \theta_m) = \sum_{k=1}^{K} p(z_{ik} = 1|m, \theta_m) p(\mathbf{x}_i|z_{ik} = 1, m, \theta_m),$$

$\tilde{\theta}$ is the ml estimate of a finite mixture distribution. It can be derived from the EM algorithm. And, the EM algorithm can be initiated in a quite natural and unique way with $\hat{\theta}$. Thus the calculation of $\tilde{\theta}$ avoids all the possible difficulties which can be encountered with the EM algorithm. Despite the need to use the EM algorithm to estimate this parameter, it would be estimated in a stable and reliable way. It can also be noted that when the learning data set has been obtained through the diagnosis paradigm, the proportions in the mixture distribution are fixed: $p_k = \text{card}\{i \text{ such that } z_{ik} = 1\}/n$ for $k = 1, \ldots, K$.

Numerical experiments reported in Bouchard and Celeux (2006) show that BEC and cross validated error rate criteria select most of the times the same models contrary to BIC which often selects suboptimal models.

4 Discussion

As sketched in the Section 2 of this article, finite mixture analysis is definitively a powerful framework for model-based cluster analysis. Many free and valuable softwares for mixture analysis are available: C.A.Man, Emmix, Flemix, MClust, MIXMOD, Multimix, Sob ... We want to insist on the software MIXMOD on which we are working for years (Biernacki et al. (2006)). It is a mixture software for cluster analysis and classification which contains most of the features described here and which last version is quite rapid. It is available at url http://www-math.univ-fcomte.fr/mixmod.

In the second part of this article, we highlighted how it could be useful to take into account the model purpose to select a relevant and useful model. This point of view can lead to define different selection criteria than the classical BIC criterion. It has been illustrated in two situations: modeling in a clustering purpose and modeling in a supervised classification purpose. This leads to two penalized likelihood criteria ICL and BEC for which the the penalty is data driven and is expected to choose a useful, if not true, model.

Now, it can be noticed that we did not consider the modeling purpose when estimating the model parameters. In both situations, we simply considered the maximum likelihood estimator. Taking into account the modeling purpose in the estimation process could be regarded as an interesting point of view. However we do not think that this point of view is fruitful and, moreover, we think it can jeopardize the statistical analysis. For instance, in the cluster analysis context, it could be thought of as more natural to compute the parameter value maximizing the complete loglikelihood $\log \mathbf{p}(\mathbf{x}, \mathbf{z}|\theta)$

rather than the observed loglikelihood $\log \mathbf{p}(\mathbf{x}|\theta)$. But as proved in Bryant and Williamson (1978), this strategy leads to asymptotically biased estimates of the mixture parameters. In the same manner, in the supervised classification context, considering the parameter value θ^* maximizing directly the conditional likelihood $\log \mathbf{p}(\mathbf{z}|\mathbf{x}, \theta)$ could be regarded as an alternative to the classical maximum likelihood estimation. But this would lead to difficult optimization problems and would provide unstable estimated values. Finally, we do not recommend taking into account the modeling purpose when estimating the model parameters because it could lead to cumbersome algorithms or provoke undesirable biases in the estimation. On the contrary, we think that taking into account the model purpose when assessing a model could lead to choose reliable and stable models especially in unsupervised and supervised classification context.

References

AITKIN, M. (2001): Likelihood and Bayesian Analysis of Mixtures. *Statistical Modeling, 1, 287–304.*

AKAIKE, H. (1974): A New Look at Statistical Model Identification. *IEEE Transactions on Automatic Control, 19, 716–723.*

BANFIELD and RAFTERY, A.E. (1993): Model-based Gaussian and Non-Gaussian Clustering. *Biometrics, 49, 803–821.*

BENSMAIL, H. and CELEUX, G. (1996): Regularized Gaussian Discriminant Analysis Through Eigenvalue Decomposition. *Journal of the American Statistical Association, 91, 1743–48.*

BIERNACKI, C., CELEUX., G. and GOVAERT, G. (2000): Assessing a Mixture Model for Clustering with the Integrated Completed Likelihood. *IEEE Trans. on PAMI, 22, 719–725.*

BIERNACKI, C., CELEUX., G. and GOVAERT, G. (2003): Choosing Starting Values for the EM Algorithm for Getting the Highest Likelihood in Multivariate Gaussian Mixture Models. *Computational Statistics and Data Analysis, 41, 561–575.*

BIERNACKI, C., CELEUX, G., GOVAERT G. and LANGROGNET F. (2006): Model-based Cluster Analysis and Discriminant Analysis With the MIXMOD Software, *Computational Statistics and Data Analysis* (to appear).

BOUCHARD, G. and CELEUX, G. (2006): Selection of Generative Models in Classification. *IEEE Trans. on PAMI, 28, 544–554.*

BRYANT, P. and WILLIAMSON, J. (1978): Asymptotic Behavior of Classification Maximum Likelihood Estimates. *Biometrika, 65, 273–281.*

CELEUX, G., CHAUVEAU, D. and DIEBOLT, J. (1996): Some Stochastic Versions of the EM Algorithm. *Journal of Statistical Computation and Simulation, 55, 287–314.*

CELEUX, G. and GOVAERT, G. (1991): Clustering Criteria for Discrete Data and Latent Class Model. *Journal of Classification, 8, 157–176.*

CELEUX, G. and GOVAERT, G. (1992): A Classification EM Algorithm for Clustering and Two Stochastic Versions. *Computational Statistics and Data Analysis, 14, 315–332.*

CELEUX, G. and GOVAERT, G. (1993): Comparison of the Mixture and the Classification Maximum Likelihood in Cluster Analysis. *Journal of Computational and Simulated Statistics, 14, 315–332.*

CIUPERCA, G., IDIER, J. and RIDOLFI, A.(2003): Penalized Maximum Likelihood Estimator for Normal Mixtures. *Scandinavian Journal of Statistics, 30, 45–59.*

DEMPSTER, A.P., LAIRD, N.M. and RUBIN, D.B. (1977): Maximum Likelihood From Incomplete Data Via the EM Algorithm (With Discussion). *Journal of the Royal Statistical Society, Series B, 39, 1–38.*

DIEBOLT, J. and ROBERT, C. P. (1994): Estimation of Finite Mixture Distributions by Bayesian Sampling. *Journal of the Royal Statistical Society, Series B, 56, 363–375.*

FIGUEIREDO, M. and JAIN, A.K. (2002): Unsupervised Learning of Finite Mixture Models. *IEEE Trans. on PAMI, 24, 381–396.*

FRALEY, C. and RAFTERY, A.E. (1998): How Many Clusters? Answers via Model-based Cluster Analysis. *The Computer Journal, 41, 578–588.*

FRIEDMAN, J. (1989): Regularized Discriminant Analysis. *Journal of the American Statistical Association, 84, 165–175.*

GANESALINGAM, S. and MCLACHLAN, G. J. (1978): The Efficiency of a Linear Discriminant Function Based on Unclassified Initial Samples. *Biometrika, 65, 658–662.*

GOODMAN, L.A. (1974): Exploratory Latent Structure Analysis Using Both Identifiable and Unidentifiable Models. *Biometrika, 61, 215–231.*

HASTIE, T. and TIBSHIRANI, R. (1996): Discriminant Analysis By Gaussian Mixtures. *Journal of the Royal Statistical Society, Series B, 58, 158–176.*

HUNT, L.A. and BASFORD K.E. (2001): Fitting a Mixture Model to Three-mode Three-way Data With Missing Information. *Journal of Classification, 18, 209–226.*

KASS, R.E. and RAFTERY, A.E. (1995): Bayes Factors. *Journal of the American Statistical Association, 90, 773–795.*

KERIBIN, C. (2000): Consistent Estimation of the Order of Mixture. *Sankhya, 62, 49–66.*

MARIN, J.-M., MENGERSEN, K. and ROBERT, C.P. (2005): Bayesian Analysis of Finite mixtures. *Handbook of Statistics*, Vol. 25, Chapter 16. Elsevier B.V.

MCLACHLAN, G.J. and PEEL, D. (2000): *Finite Mixture Models.* Wiley, New York.

RAFTERY, A.E. (1995): Bayesian Model Selection in Social Research (With Discussion). In: P.V. Marsden (Ed.): *Sociological Methodology 1995*, Oxford, U.K.: Blackwells, 111–196.

RAFTERY, A.E. and DEAN, N. (2006): *Journal of the American Statistical Association, 101, 168–78.*

ROEDER, K. (1990): Density Estimation with Confidence Sets Exemplified by Superclusters and Voids in Galaxies. *Journal of the American Statistical Association, 85, 617–624.*

SCHWARZ, G. (1978): Estimating the Dimension of a Model. *The Annals of Statistics, 6, 461–464.*

How to Choose the Number of Clusters: The Cramer Multiplicity Solution

Adriana Climescu-Haulica

Institut d'Informatique et Mathématiques Appliquées, 51 rue des Mathématiques, Laboratoire Biologie, Informatique, Mathématiques, CEA 17 rue des Martyrs, Grenoble, France; adriana.climescu@imag.fr

Abstract. We propose a method for estimating the number of clusters in data sets based only on the knowledge of the similarity measure between the data points to be clustered. The technique is inspired from spectral learning algorithms and Cramer multiplicity and is tested on synthetic and real data sets. This approach has the advantage to be computationally inexpensive while being an *a priori* method, independent from the clustering technique.

1 Introduction

Clustering is a technique used in the analysis of microarray gene expression data as a preprocessing step, in functional genomic for example, or as the main discriminating tool in the tumor classification study (Dudoit et al. (2002)). While in recent years many clustering methods were developed, it is acknowledged that the reliability of allocation of units to a cluster and the computation of the number of clusters are questions waiting for a joint theoretical and practical validation (Dudoit et al. (2002)).

Two main approaches are used in data analysis practice to determine the number of clusters. The most common procedure is to use the number of clusters as a parameter of the clustering method and to select it from a maximum reliability criteria. This approach is strongly dependent on the clustering method. The second approach uses statistical procedures (for example the sampling with respect to a reference distribution) and are less dependent on the clustering method. Examples of methods in this category are the Gap statistic (Tibshirani et al. (2001)) and the Clest procedure (Fridlyand and Dudoit (2002)). All of the methods reviewed and compared by Fridlyand and Dudoit (2002) are *a posteriori* methods, in the sense that they include clustering algorithms as a preprocessing step. In this paper we propose a method for choosing the number of clusters based only on the knowledge of the similarity measure between the data points to be clustered. This criterion is inspired from spectral learning algorithms and Cramer multiplicity. The novelty of the method is given by the direct extraction of the number of clusters from data, with no assumption about effective clusters. The procedure is the following: the clustering problem is mapped to the framework of spectral graph theory by means of the "min-cut" problem. This induces the passage from the discrete domain to the continuous one, by the definition of the time-continous Markov process associated with the graph. It is the analysis on the continuous domain which allows the screening of the Cramer multiplicity, otherwise set to 1 for the discrete case. We evaluate the method on artificial data obtained as samples from different gaussian distributions and on yeast cell data for which the similarity metric is well established in the literature. Compared to methods evaluated in Fridlyand and Dudoit (2002) our algorithm is computationally less expensive.

2 Clustering by min-cut

The clustering framework is given by the min-cut problem on graphs. Let $G = \langle V, E \rangle$ be a weighted graph where to each vertex in the set V we assign one of the m points to be clustered. The weights on the edges E of the graph represent how "similar" one data point is to another and are assigned by a similarity measure $S : V \times V \to I\!R_+$. In the min-cut problem a partition on subsets $\{V_1, V_2, \ldots V_k\}$ of V is searched such that the sum of the weights corresponding to the edges going from one subset to another is minimized. This is a NP-hard optimization problem. The Laplacian operator associated with the graph G is defined on the space of functions $f : V(G) \to I\!R$ by

$$\mathcal{L} = I - D^{-1}S \tag{1}$$

Therefore, the min-cut problem corresponds to the following optimization problem

$$C(S) = min_g \frac{\langle g, \mathcal{L}g \rangle}{\langle g, g \rangle}. \tag{2}$$

An heuristic approximation to this problem is given by the spectral clustering method described in Ng et al. (2002) and uses eigenvectors of a normalized similarity matrix. A special case of the spectral clustering is obtained when the normalized similarity matrix is a stochastic matrix, i.e. the sum of the rows equals 1. Then, on the corresponding graph, a probability distribution is well defined around each vertex. In Meila and Shi (2001) the normalized similarity measure between the vertices i and j is interpreted as the probability that a random walk moves from vertex i to vertex j in one step. We associate with the graph G a time-continuous Markov process with the state space given by the vertex set V and the transition matrix given by the normalized similarity matrix. Assuming the graph is without loops, the paths of the time-continuous Markov process are Hilbert space valuated with respect to the norm given by the quadratic mean.

3 Cramer multiplicity

As a stochastic process, the time-continuous Markov process defined by means of the similarity matrix is associated with a unique, up to an equivalence class, sequence of spectral measures, by means of its Cramer decomposition (Cramer (1964)). The length of the sequence of spectral measures is named the Cramer multiplicity of the stochastic process. More precisely, let $X : \Omega \times [0, \infty) \to V$ be the time continuous Markov process associated with the graph G and let N be its Cramer multiplicity. Then, by the Cramer decomposition theorem (Cramer (1964)) there exist N mutually orthogonal stochastic processes with orthogonal increments $Z_i : \Omega \times [0, \infty) \to V$, $1 \leq n \leq N$ such that

$$X(t) = \sum_{n=1}^{N} \int_{0}^{t} g_n(t,u) dZ_n(u), \tag{3}$$

$g_n(t,u), 1 \leq n \leq N$ are complex valuated deterministic functions such that $\sum_{n=1}^{N} \int_0^t \mid g_n(t,u) \mid^2 dF_{Z_n}(u) < \infty$ where $F_{Z_n}(t) = \mathbf{E} \mid Z_n(t) \mid^2$, $1 \leq n \leq N$ are the spectral measures associated with the stochastic process X, forming a decreasing sequence of measures with respect to the absolute continuity relationship

$$F_{Z_1} \gg F_{Z_2} \ldots \gg F_{Z_N}. \tag{4}$$

No representation of the form 3 with these properties exists for any smaller value of N. If the time index set of a process is discrete then its Cramer multiplicity is 1. It is easily seen that Cramer decomposition is a generalization of the Fourier representation, applying to stochastic processes and allowing a more general class of orthogonal bases. The processes $Z_n, 1 \leq n \leq N$ are interpreted as innovation processes associated with X. The idea of considering the Cramer multiplicity as the number of clusters resides in this interpretation.

4 Envelope intensity algorithm

We present here an algorithm derived heuristically from the observations above. Nevertheless the relationship between the Laplacian of the graph G and the Kolmogorov equations associated with the time continuous Markov process X and hence its Cramer representation is an open problem to be addressed in the future. The input of the algorithm is the similarity matrix and the output is a function we called the envelope intensity associated with the similarity matrix. This is a piecewise "continuous" increasing function whose number of jumps contributes to the approximation of the Cramer multiplicity.

1. Construct the normalized similarity matrix $W = D^{-1}S$ where D is the diagonal matrix with elements the sum of the corresponding rows from the matrix S.
2. Compute the matrix $L = I - W$ corresponding to the Laplacian operator.

3. Find $y_1, y_2, \ldots y_m$, the eigenvectors of L, chosen to be orthogonal to each other in the case of repeated eigenvalues and form the matrix $Y = [y_1\ y_2 \ldots y_m] \in I\!R^{m \times m}$ by stacking the eigenvectors in columns.
4. Compute the Fourier transform of Y column by column and construct W the matrix corresponding to the absolute values of matrix elements.
5. Assign, in increasing order, the maximal value of each column from W to the vector $U \in I\!R_+$ called envelope intensity.

Steps 1 to 3 are derived from the spectral clustering and steps 3 to 5 are parts of the heuristic program to approximate the Cramer multiplicity. Step 3, corresponding to the spectral decomposition of the graph's Laplacian, is their junction part. Two spectral decompositions are applied: the Laplacian decomposition on eigenvectors and the eigenvectors decomposition on their Hilbert space, exemplified by the Fourier transform.

5 Data analysis

In order to check the potential of our approach on obtaining *a priori* information about the number of clusters, based only on a similarity matrix, primary we have to apply the envelope intensity algorithm to classes of sets whose number of clusters is well established. We choose two cate-

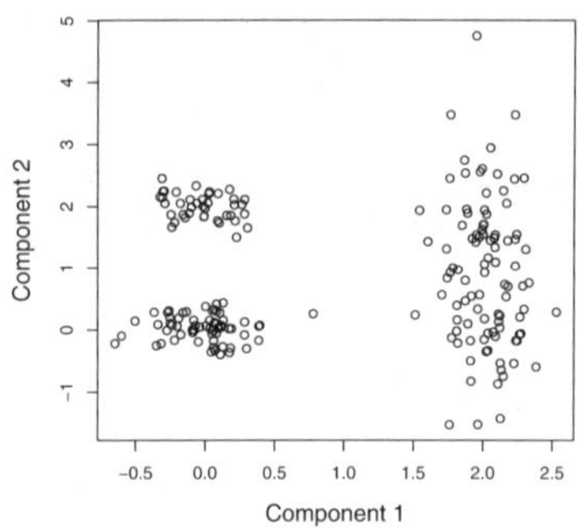

Fig. 1. Example of data sampled from three different gaussian distributions.

gories of sets. The first category is classical for the clustering analysis, we give two sets as examples. The first set is represented in Figure 1 and is given by 200 points from a mixture of three Gaussian distributions (from *http://iew3.technion.ac.il/CE/data/3gauss.mat*). The plot in Figure 3 corresponds to a mixture of five Gaussian distributions generated by $x = m_x + R \cos U$ and $y = m_y + R \sin U$ where (m_x, m_y) is the local mean point chosen from the set $\{(3,18), (3,9), (9,3), (18,9), (18,18)\}$. R and U are random variables distributed $Normal(0,1)$ and $Uniform(0,\pi)$ respectively. The

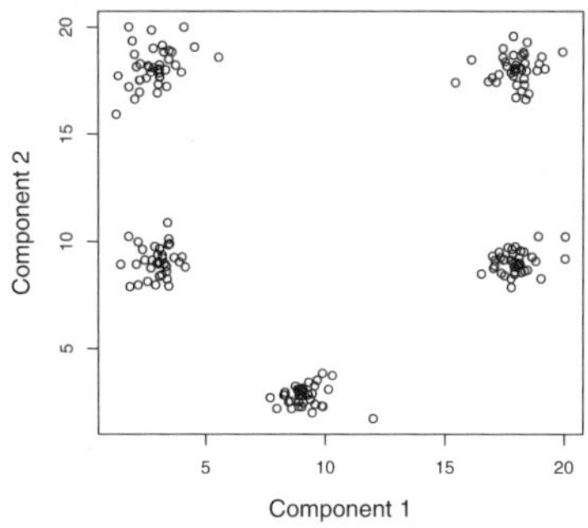

Fig. 2. Example of data sampled from five different Gaussian distributions.

similarity measure used for those data is given by

$$S_{ij} = \exp\left(-d(a(i,*), a(j,*))^2/2\sigma^2\right) \tag{5}$$

where $a(i,*)$ are the coordinates of the data points i and $d()$ is the euclidian distance. σ is a parameter chosen to be equal to the data standard deviation. The envelope intensities obtained from the algorithm presented in

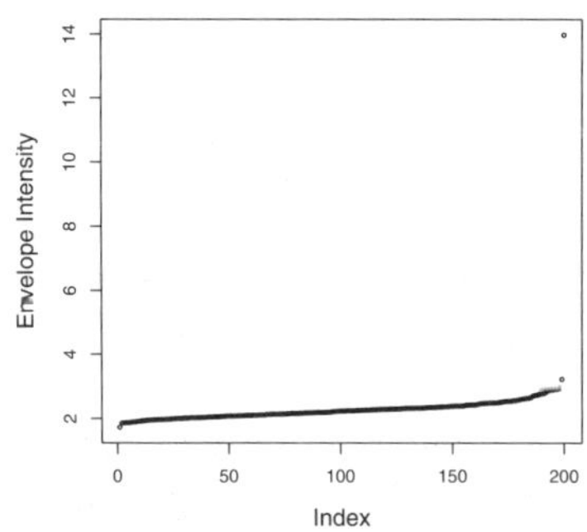

Fig. 3. Three Gaussian data: The envelope intensity for Fourier transformed eigenvalues has three regions - one continuous region and two single points.

the previous section are shown in Figure 1 and Figure 4, respectively. Those pictures show the separation by jumps of three and five regions of the envelope intensity respectively, the upper regions being mostly single points. For comparison we computed directly the equivalent of the envelope intensity on non Fourier transformed eigenvectors. The result for the three Gaussian mixture distribution is shown as example in Figure 2. It follows clearly that no information about the number of clusters could be retrieved from this computation. We use as second category of test for our approach the yeast cell

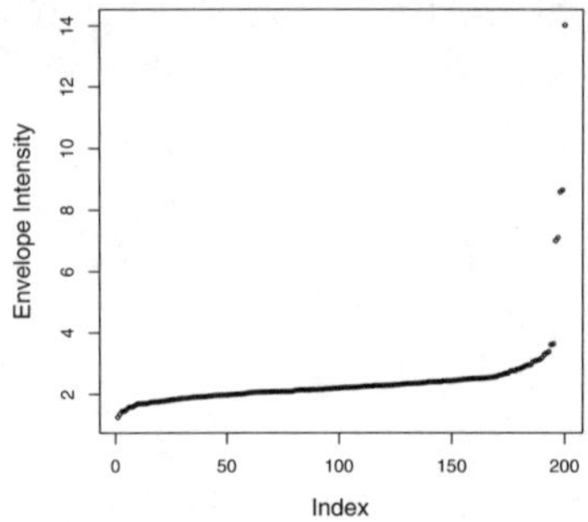

Fig. 4. Five Gaussian distributions: The envelope intensity for Fourier transformed eigenvalues has five well separated regions.

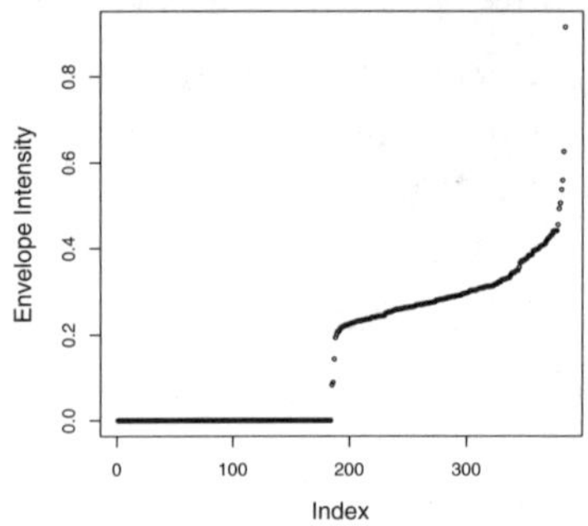

Fig. 5. Three Gaussian distribution data: The envelope intensity of the non Fourier transformed eigenvectors do not give information about the number of clusters.

data from *http://faculty.washington.edu/kayee/model/* already analyzed by a model based approach and by spectral clustering in Meila and Verma (2001). This data set has the advantage to come from real experiments and meanwhile, the number of clusters to be intrinsically determined, given by the five phases of the cell cycle. We applied the envelope intensity algorithm for two sets. The first yeast cell set contains 384 genes selected from general data bases such that each gene has one and only one phase associated to it. To each gene corresponds a vector of intensity points measured at 17 distinct time points. The raw data is normalized by a Z score transformation. The result of the envelope intensity computation, with respect to the similarity measure given by the correlation plus 1, as in Meila and Verma (2001) is shown in Figure 6. The five regions appear distinctly separated by jumps. The second yeast cell set is selected from the first one, corresponding to some functional categories and only four phases. It contains 237 genes, it is log normalized and the simi-

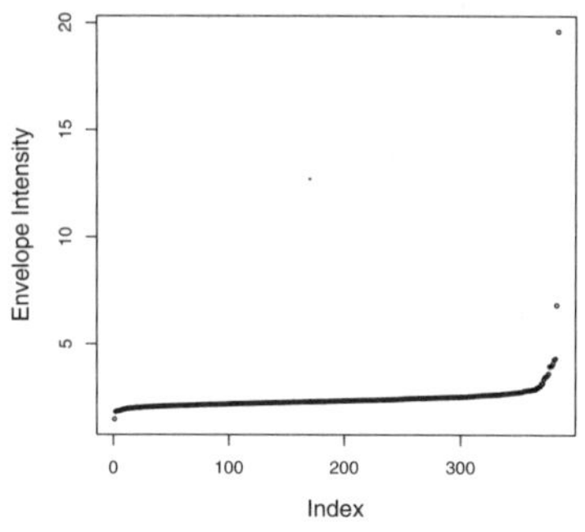

Fig. 6. Full yeast cycle: five regions on the envelope intensity of the Fourier transformed eigenvectors.

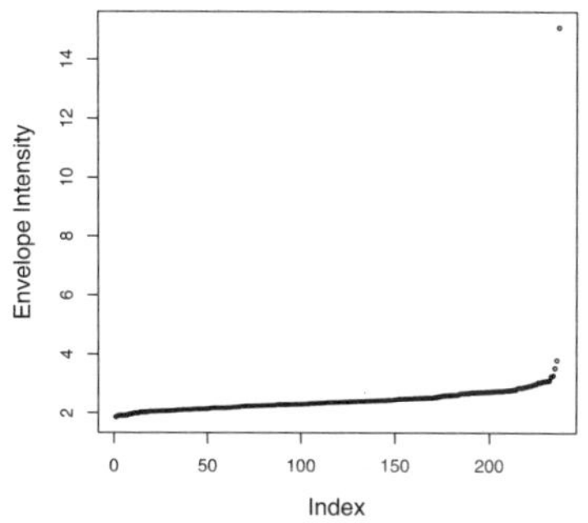

Fig. 7. Partial yeast cycle: the envelope intensity of the Fourier transformed eigenvalues contains four regions well separated.

larity matrix is given by the covariance. The corresponding envelope intensity computed from our algorithm is given in Figure 7. Between the four regions, three are single points and are clearly separated, while the "continuous" region is a well connected curve. For both sets the computation of the envelope intensity of non Fourier transformed eigenvectors gives no information about the number of clusters.

6 Conclusions

We propose an algorithm which is able to indicate the number of clusters based only on the data similarity matrix. This algorithm is inspired from ideas on spectral clustering, stochastic processes on graphs and Cramer decomposition theory. It combines two types of spectral decomposition: the matrix spectral decomposition and the spectral decomposition on Hilbert spaces. The algorithm is easy to implement as it is resumed to the computation of the envelope intensity of the Fourier transformed eigenvectors of the Laplacian associated with the similarity matrix. The data analysis we performed shows that the envelope intensity computed by the algorithm is separated by jumps in connected or single point regions, whose number coincides with the number of clusters. Still more theoretical results have to be developed, this algorithm is an exploratory tool on clustering analysis.

References

CRAMER, H. (1964): Stochastic Processes as Curves on Hilbert Space. *Theory Probab. Appl. 9, 193–204*

DUDOIT, S., FRIDLYAND, J. and SPEED, T. (2002): Comparison of Discrimination Methods for the Classification of Tumors Using Gene Expression Data. *Journal of the American Statistical Association, 97, 77–87.*

FRIDLYAND, J. and DUDOIT, S. (2002): A Prediction-based Resampling Method to Estimate the Number of Clusters in a Dataset. *Genome Biology, 3, 7.*

MEILA, M. and SHI, J. (2001): A Random Walks View of Spectral Segmentation. *Proceedings of the International Workshop of Artificial Intelegence and Statistics.*

MEILA, M. and VERMA, D. (2001): A Comparison of Spectral Clustering Algorithms. *UW CSE Technical Report.*

NG, A., JORDAN, M. and WEISS, Y. (2002): Spectral Clustering: Analysis and an Algorithm. In: T. Dietterich, S. Becker, and Z. Ghahramani (Eds.): *Advances in Neural Information Processing Systems (NIPS).*

TIBSHIRANI, R., GUENTHER, W.G. and HASTIE, T. (2001): Estimating the Number of Clusters in a Dataset Via the Gap Statistic. *Journal of the Royal Statistical Society, B, 63, 411-423.*

Model Selection Criteria for Model-Based Clustering of Categorical Time Series Data: A Monte Carlo Study

José G. Dias

Department of Quantitative Methods – GIESTA/UNIDE,
ISCTE – Higher Institute of Social Sciences and Business Studies,
Av. das Forças Armadas, 1649–026 Lisboa, Portugal; jose.dias@iscte.pt

Abstract. An open issue in the statistical literature is the selection of the number of components for model-based clustering of time series data with a finite number of states (categories) that are observed several times. We set a finite mixture of Markov chains for which the performance of selection methods that use different information criteria is compared across a large experimental design. The results show that the performance of the information criteria vary across the design. Overall, AIC3 outperforms more widespread information criteria such as AIC and BIC for these finite mixture models.

1 Introduction

Time series or longitudinal data have played an important role in the understanding of the dynamics of the human behavior in most of the social sciences. Despite extensive analyses for continuous data time series, little research has been conducted on the unobserved heterogeneity for categorical time series data. Exceptions are the application of finite mixtures of Markov chains, e.g., in marketing (Poulsen (1990)), machine learning (Cadez et al. (2003)) or demography (Dias and Willekens (2005)). Despite the increasing use of these mixtures little is known about model selection of the number of components.

Information criteria have become popular as a useful approach to model selection. Some of them such as Akaike Information Criterion (AIC) and Bayesian Information Criterion (BIC) have been widely used. The performance of information criteria has been studied extensively in the finite mixture literature, mostly focused on finite mixtures of Gaussian distributions (McLachlan and Peel (2000)). Therefore, in this article a Monte Carlo experiment is designed to assess the ability of the different information criteria to retrieve the true model and to measure the effect of the design factors for finite mixtures of Markov chains. The results reported in this paper extend the conclusions in Dias (2006) from the zero-order Markov model (latent class model) to the one-order Markov model (finite mixture of Markov chains).

This paper is organized as follows. Section 2 describes the finite mixture of Markov chains. In Section 3, we review the literature on model selection criteria. In Section 4, we describe the design of the Monte Carlo study. In Section 5, we present and discuss the results. The paper concludes with a summary of main findings, implications, and suggestions for further research.

2 Finite mixture of Markov chains

Let X_{it} be the random variable denoting the category (state) of the individual i at time t, and x_{it} a particular realization. We will assume discrete time from 0 to T ($t = 0, 1, \ldots, T$). Thus, the vectors $\mathbf{X}_i$ and $\mathbf{x}_i$ denote the consecutive observations (time series) – respectively X_{it} and x_{it} –, with $t = 0, \ldots, T$. The probability density $P(\mathbf{X}_i = \mathbf{x}_i) = P(X_{i0} = x_{i0}, X_{i1} = x_{i1}, ..., X_{iT} = x_{iT})$ can be extremely difficult to characterize, due to its possibly huge dimension ($T+1$). A common procedure to simplify $P(\mathbf{X}_i = \mathbf{x}_i)$ is by assuming the Markov property stating that the occurrence of event $X_t = x_t$ only depends on the previous state $X_{t-1} = x_{t-1}$; that is, conditional on X_{t-1}, X_t is independent of the states at the other time points. From the Markov property, it follows that

$$P(\mathbf{X} = \mathbf{x}_i) = P(X_{i0} = x_{i0}) \prod_{t=1}^{T} P(X_{it} = x_{it} | X_{i,t-1} = x_{i,t-1}), \qquad (1)$$

where $P(X_{i0} = x_{i0})$ is the initial distribution and $P(X_{it} = x_{it} | X_{i,t-1} = x_{i,t-1})$ is the probability that individual i is in state x_{it} at t, given that he is in state $x_{i,t-1}$ at time $t-1$. A first-order Markov chain is specified by its transition probabilities and initial distribution. Hereafter, we denote the initial and the transition probabilities as $\lambda_j = P(X_{i0} = j)$ and $a_{jk} = P(X_t = k | X_{t-1} = j)$, respectively. Note that we assume that transition probabilities are time homogeneous, which means that our model is a stationary first-order Markov model.

The finite mixture of Markov chains assumes discrete heterogeneity. Individuals are clustered into S segments, each denoted by s ($s = 1, \ldots, S$). The clusters, including its number, are not known *a priori*. Thus, in advance one does not know how the sample will be partitioned into clusters. The component that individual i belongs to is denoted by the latent discrete variable $Z_i \in \{1, 2, ..., S\}$. Let $\mathbf{z} = (z_1, ..., z_n)$. Because $\mathbf{z}$ is not observed, the inference problem is to estimate the parameters of the model, say φ, using only information on $\mathbf{x} = (\mathbf{x}_1, \ldots, \mathbf{x}_n)$. More precisely, the estimation procedure has to be based on the marginal distribution of $\mathbf{x}_i$ which is obtained as follows:

$$P(\mathbf{X}_i = \mathbf{x}_i; \varphi) = \sum_{s=1}^{S} \pi_s P(\mathbf{X}_i = \mathbf{x}_i | Z_i = s). \qquad (2)$$

This equation defines a finite mixture model with S components. The component proportions, $\pi_s = P(Z_i = s; \varphi)$, correspond to the *a priori* probability that individual i belongs to the segment s, and gives the segment relative size. Moreover, π_s satisfies $\pi_s > 0$ and $\sum_{s=1}^{S} \pi_s = 1$.

Within each latent segment s, observation $\mathbf{x}_i$ is characterized by $P(\mathbf{X}_i = \mathbf{x}_i | Z_i = s) = P(\mathbf{X}_i = \mathbf{x}_i | Z_i = s; \theta_s)$ which implies that all individuals in segment s have the same probability distribution defined by the segment-specific parameters θ_s. The parameters of the model are $\varphi = (\pi_1, ..., \pi_{S-1}, \theta_1, ..., \theta_S)$. The θ_s includes the transition and initial probabilities $a_{sjk} = P(X_{it} = k | X_{i,t-1} = j, Z_i = s)$ and $\lambda_{sk} = P(X_{i0} = k | Z_i = s)$, respectively. A finite mixture of Markov chains is not a Markov chain, which enables the modeling of very complex patterns (see Cadez et al. (2003), Dias and Willekens (2005)). The independent parameters of the model are $S-1$ prior probabilities, $S(K-1)$ initial probabilities, and $SK(K-1)$ transition probabilities, where K is the number of categories or states. Thus, the total number of independent parameters is $SK^2 - 1$. The log-likelihood function for φ, given that $\mathbf{x}_i$ are independent, is $\ell_S(\varphi; \mathbf{x}) = \sum_{i=1}^{n} \log P(\mathbf{X}_i = \mathbf{x}_i; \varphi)$ and the maximum likelihood estimator (MLE) is $\hat{\varphi} = \arg\max_{\varphi} \ell_S(\varphi; \mathbf{x})$. For estimating this model by the EM algorithm, we refer to Dias and Willekens (2005).

3 Information criteria for model selection

The traditional approach to the selection of the best among different models is using a likelihood ratio test, which under regularity conditions has a simple asymptotic theory (Wilks (1938)). However, in the context of finite mixture models this approach is problematic. The null hypothesis under test is defined on the boundary of the parameter space, and consequently the regularity condition of Cramer on the asymptotic properties of the MLE is not valid. Some recent results have been achieved (see, e.g., Lo et al. (2001)). However, most of these results are difficult to implement and usually derived for finite mixtures of Gaussian distributions.

As an alternative information statistics have received much attention recently in finite mixture modeling. These statistics are based on the value of $-2\ell_S(\hat{\varphi}; \mathbf{x})$ of the model adjusted for the number of free parameters in the model (and other factors such as the sample size). The basic principle under these information criteria is the parsimony: all other things being the same (log-likelihood), we choose the simplest model (with fewer parameters). Thus, we select the number S which minimizes the criterion $C_S = -2\ell_S(\hat{\varphi}; \mathbf{x}) + dN_S$, where N_S is the number of free parameters of the model. For different values of d, we have the Akaike Information Criterion (AIC: Akaike (1974)) ($d = 2$), the Bayesian Information Criterion (BIC: Schwarz (1978)) ($d = \log n$), and the Consistent Akaike Information Criterion (CAIC: Bozdogan (1987)) ($d = \log n + 1$).

Bozdogan (1993) argues that the marginal cost per free parameter, the so-called magic number 2 in AIC's equation above, is not correct for finite mixture models. Based on Wolfe (1970), he conjectures that the likelihood ratio for comparing mixture models with p_1 and p_2 free parameters is asymptotically distributed as a noncentral chi-square with noncentrality parameter δ and $2(p_1 - p_2)$ degrees of freedom instead of the usual $p_1 - p_2$ degrees of freedom as assumed in AIC. Therefore, AIC3 uses $d = 3$ as penalizing factor.

The Average Weight of Evidence (AWE) criterion adds a third dimension to the information criteria described above. It weights fit, parsimony, and the performance of the classification (Banfield and Raftery (1993)). This measure uses the so-called classification log-likelihood ($\log L^c$) and is defined as $AWE = -2 \log L^c + 2N_S(\frac{2}{3} + \log n)$.

Apart from the five information criteria reported above, we also investigated a modified definition of the BIC, CAIC and AWE. Some researchers (e.g., Ramaswamy et al. (1993), DeSarbo et al. (2004)) have considered as *sample size* the repeated measurements from each observation. Therefore, the penalization would be function of $n(T + 1)$ instead of n.

4 Experimental design

A Monte Carlo (MC) study was conducted for the assessment of the performance of these criteria. The experimental design controls the number of time points $(T+1)$ and number of categories (K), the sample size (n), the balance of component sizes (whether component sizes are equal or unequal), and the level of separation of components. The number of time points $(T+1)$ was set at levels 10, 20, and 40; and the number of categories (K) at levels 2 and 4. The sample size (n) assumes the levels: 300, 600, and 1200. For simulation of data we use $(S = 2)$, and models with one, two, and three components are estimated. The size of components can be equal or unequal: $\pi = (0.5, 0.5)$ and $\pi = (0.4, 0.6)$, respectively.

Controlling the level of separation of component distributions is more challenging. The (true) parameter values are shown in Table 1. These *ad hoc* values try to cover different situations in empirical data sets. In particular, there is an attempt to include persistent patterns usually observed in empirical data sets with heavy retention probabilities (states almost absorbent). The distance between a_{1kk} and a_{2kk}, $|a_{1kk} - a_{2kk}| = |P(X_{it} = k|X_{i,t-1} = k, Z_i = 1) - P(X_{it} = k|X_{i,t-1} = k, Z_i = 2)|$, and between λ_{s1} and λ_{s2}, $|\lambda_{1k} - \lambda_{2k}| = |P(X_{i0} = k|Z_i = 1) - P(X_{i0} = k|Z_i = 2)|$, sets the level of separation.

To avoid local optima, for each number of components (2 and 3) the EM algorithm was repeated 5 times with random starting centers, and the best solution (maximum likelihood value out of those 5 runs) and model selection results were kept. The EM algorithm ran until the difference between log-likelihoods being smaller than a given tolerance. We include the tolerance as

Table 1. The (true) initial and conditional probabilities λ_{sk} and a_{sjk} for the simulated data

Parameters	$K = 2$		$K = 4$			
	$k = 1$	$k = 2$	$k = 1$	$k = 2$	$k = 3$	$k = 4$
Moderate-separated components						
λ_{1k}	0.60	0.40	0.30	0.20	0.30	0.20
λ_{2k}	0.40	0.60	0.20	0.30	0.20	0.30
a_{11k}	0.75	0.25	0.45	0.15	0.25	0.15
a_{12k}	0.33	0.67	0.20	0.40	0.30	0.10
a_{13k}	-	-	0.40	0.40	0.10	0.10
a_{14k}	-	-	0.30	0.30	0.20	0.20
a_{21k}	0.25	0.75	0.15	0.45	0.15	0.25
a_{22k}	0.67	0.33	0.40	0.20	0.10	0.30
a_{23k}	-	-	0.10	0.10	0.40	0.40
a_{24k}	-	-	0.20	0.20	0.30	0.30
Ill-separated components						
λ_{1k}	0.40	0.60	0.20	0.30	0.10	0.40
λ_{2k}	0.50	0.50	0.30	0.30	0.30	0.10
a_{11k}	0.94	0.06	0.85	0.05	0.05	0.05
a_{12k}	0.12	0.88	0.10	0.70	0.10	0.10
a_{13k}	-	-	0.10	0.01	0.80	0.09
a_{14k}	-	-	0.01	0.01	0.13	0.85
a_{21k}	0.95	0.05	0.95	0.05	0.00	0.00
a_{22k}	0.05	0.95	0.05	0.90	0.00	0.05
a_{23k}	-	-	0.03	0.00	0.90	0.07
a_{24k}	-	-	0.04	0.02	0.04	0.90

a factor being controlled, with levels: 10^{-2}, 10^{-3}, and 10^{-4}. It is important to understand and identify possible interplays between model selection and the EM stopping rule.

This MC study sets a $2^3 \times 3^3$ factorial design with 216 cells. Special care needs to be taken before arriving at conclusions based on MC results. In this study, we performed 25 replications within each cell to obtain the frequency of obtaining the true model, resulting in a total of 5400 data sets. The programs were written in MATLAB. The main performance measure used is the frequency with which each criterion picks the correct model. For each data set, each criterion is classified as *underfitting*, *fitting*, or *overfitting*, based on the relation between S and the estimated S by those criteria. Underfitting means that the estimated number of components is smaller than the true value; and overfitting happens whenever the estimated number of components is larger than the true value.

5 Results

The key feature of the results is the overall remarkable performance of AIC3 (Table 2). While most of the criteria perform satisfactory, AIC3 identifies the true model 82.9% of the times. The AIC3 presents minor overfitting (0.6%), and therefore outperforms other traditional criteria such as AIC, BIC, and CAIC. AIC presents unacceptable overfitting (17.2%). For CAIC, BIC, and AWE the penalization $n(T + 1)$ decreases their performance and it is not considered hereafter.

Table 2. Results of the Monte Carlo study

Factors		Criteria							
		AIC	AIC3	CAIC		BIC		AWE	
				n	$n(T+1)$	n	$n(T+1)$	n	$n(T+1)$
Sample size (n)									
	Underfit	0.161	0.223	0.332	0.403	0.322	0.379	0.491	0.552
300	Fit	0.697	0.771	0.668	0.597	0.678	0.621	0.509	0.448
	Overfit	0.142	0.006	0.000	0.000	0.000	0.000	0.000	0.000
	Underfit	0.111	0.155	0.313	0.330	0.282	0.330	0.417	0.443
600	Fit	0.706	0.839	0.687	0.670	0.718	0.670	0.583	0.557
	Overfit	0.183	0.006	0.000	0.000	0.000	0.000	0.000	0.000
	Underfit	0.091	0.117	0.196	0.274	0.172	0.246	0.417	0.417
1200	Fit	0.719	0.876	0.804	0.726	0.827	0.754	0.583	0.583
	Overfit	0.190	0.007	0.000	0.000	0.001	0.000	0.000	0.000
Number of variables ($T+1$)									
	Underfit	0.218	0.267	0.431	0.473	0.404	0.455	0.574	0.612
10	Fit	0.637	0.728	0.569	0.527	0.596	0.545	0.426	0.388
	Overfit	0.145	0.005	0.000	0.000	0.000	0.000	0.000	0.000
	Underfit	0.118	0.179	0.252	0.319	0.245	0.294	0.500	0.500
20	Fit	0.690	0.815	0.748	0.681	0.755	0.706	0.500	0.500
	Overfit	0.192	0.006	0.000	0.000	0.000	0.000	0.000	0.000
	Underfit	0.027	0.049	0.157	0.215	0.127	0.206	0.251	0.300
40	Fit	0.794	0.942	0.842	0.785	0.873	0.794	0.749	0.700
	Overfit	0.179	0.009	0.001	0.000	0.000	0.000	0.000	0.000
Number of states (K)									
	Underfit	0.240	0.310	0.436	0.477	0.414	0.470	0.500	0.501
2	Fit	0.662	0.678	0.563	0.523	0.585	0.530	0.500	0.499
	Overfit	0.098	0.012	0.001	0.000	0.001	0.000	0.000	0.000
	Underfit	0.001	0.020	0.124	0.195	0.103	0.167	0.383	0.440
4	Fit	0.753	0.979	0.876	0.805	0.897	0.833	0.617	0.560
	Overfit	0.246	0.001	0.000	0.000	0.000	0.000	0.000	0.000
Proportions									
	Underfit	0.000	0.000	0.000	0.000	0.000	0.000	0.049	0.075
Equal	Fit	0.748	0.990	1.000	1.000	1.000	1.000	0.950	0.925
	Overfit	0.252	0.010	0.000	0.000	0.000	0.000	0.001	0.000
	Underfit	0.241	0.330	0.560	0.671	0.517	0.637	0.834	0.867
Unequal	Fit	0.666	0.667	0.437	0.329	0.483	0.363	0.166	0.133
	Overfit	0.093	0.003	0.003	0.000	0.000	0.000	0.000	0.000
Level of separation									
	Underfit	0.121	0.169	0.278	0.334	0.257	0.311	0.441	0.476
Moderate-separated	Fit	0.704	0.824	0.721	0.666	0.743	0.689	0.559	0.524
	Overfit	0.175	0.007	0.001	0.000	0.000	0.000	0.000	0.000
	Underfit	0.121	0.161	0.282	0.338	0.260	0.326	0.442	0.465
Ill-separated	Fit	0.710	0.833	0.718	0.662	0.740	0.674	0.558	0.535
	Overfit	0.169	0.006	0.000	0.000	0.000	0.000	0.000	0.000
Tolerance									
	Underfit	0.138	0.179	0.283	0.338	0.263	0.317	0.441	0.472
10^{-2}	Fit	0.712	0.812	0.717	0.662	0.736	0.683	0.559	0.528
	Overfit	0.150	0.009	0.000	0.000	0.001	0.000	0.000	0.000
	Underfit	0.116	0.161	0.278	0.337	0.255	0.320	0.441	0.471
10^{-3}	Fit	0.713	0.834	0.722	0.663	0.745	0.680	0.559	0.529
	Overfit	0.171	0.005	0.000	0.000	0.000	0.000	0.000	0.000
	Underfit	0.108	0.154	0.279	0.332	0.257	0.318	0.443	0.469
10^{-4}	Fit	0.697	0.839	0.721	0.668	0.743	0.682	0.557	0.531
	Overfit	0.195	0.007	0.000	0.000	0.000	0.000	0.000	0.000
Overall	Underfit	0.121	0.165	0.280	0.336	0.259	0.318	0.442	0.471
	Fit	0.707	0.829	0.720	0.664	0.741	0.682	0.558	0.529
	Overfit	0.172	0.006	0.000	0.000	0.000	0.000	0.000	0.000

A second objective of the study was the comparison of these criteria across the design factors. Increasing the sample size always improves the performance of the information criteria, and reduces underfitting. However, for AIC increasing the sample size tends to increase the overfitting. Increasing the number of time points $(T + 1)$ improves the performance of the information criteria and reduces the underfitting. Increasing the state space (K) reduces the underfitting, and improves the performance of the information criteria. The balance of component sizes has a dramatic effect on the information criteria. For unequal proportions the performance of the information criteria has a substantial reduction (comparing to the equal size level) and increase in underfitting. Moreover, we observe that despite different levels of separation of components considered, this factor has no effect on the results. This may suggest that more structured methods for controlling the level of separation of components are required. On the other hand, the tolerance level (at least 10^{-2}) seems to have a small impact on the performance of the information criteria.

6 Conclusion

The AIC3 dominated remaining criteria over most of the experimental conditions, which is in agreement with results reported in Dias (2006) for the latent class model. It is also shown that information criteria which are function of sample size work better with n than with $n(T + 1)$ as penalization. Applications of model-based clustering have to take into account these new results, whenever they are set to time series data. For the first time a Monte Carlo study analyzes the behavior of information criteria for finite mixture of Markov chains. Because most of the information criteria are derived from asymptotics, this Monte Carlo study allowed their assessment for realistic sample sizes. We included the most commonly used information criteria in applied research for this type of longitudinal model-based clustering approach. This first attempt to understand the behavior of information criteria for the mixture of Markov chains model points out the need for detailed replications of these results, which are restricted to $S = 2$ and have to be extended to a larger number of components. On the other hand for a larger number of components the level of separation of components is even harder to set. Our results suggest that a more structured procedure for controlling the level of separation of components is needed for this model. Recently Dias (2004) has introduced such a method for latent class models. Further research is needed on the way that this method can be extended to mixtures of Markov chains. Moreover, this study suggests that the shape of the log-likelihood function prevents the tolerance level to have influence on the performance of information criteria, assuming a maximum of 10^{-2}. Future research could extend our findings to more general finite mixtures for time series data.

References

AKAIKE, H. (1974): A New Look at Statistical Model Identification. *IEEE Transactions on Automatic Control, AC-19, 716–723.*

BANFIELD, J.D. and Raftery, A.E. (1993): Model-based Gaussian and Non-Gaussian Grupoing. *Biometrics, 49, 803–821.*

BOZDOGAN, H. (1987): Model Selection and Akaike's Information Criterion (AIC): The General Theory and Its Analytical Extensions. *Psychometrika, 52, 345–370.*

BOZDOGAN, H. (1993): Choosing the Number of Component Clusters in the Mixture-Model Using a New Informational Complexity Criterion of the Inverse-Fisher Information Matrix. In: O. Opitz, B. Lausen and R. Klar (Eds.): *Information and Classification, Concepts, Methods and Applications.* Springer, Berlin, 40–54.

CADEZ, I., HECKERMAN, D., MEEK, C., SMYTH, P. and WHITE, S. (2003): Visualization of Navigation Patterns on a Web Site Using Model-Based Clustering. *Data Mining and Knowledge Discovery, 7, 399–424.*

DESARBO, W.S., LEHMANN, D.R. and HOLLMAN, F.G. (2004): Modeling Dynamic Effects in Repeated-measures Experiments Involving Preference/Choice: An Illustration Involving Stated Preference Analysis. *Applied Psychological Measurement, 28, 186–209.*

DIAS, J.G. (2004): Controlling the Level of Separation of Components in Monte Carlo Studies of Latent Class Models. In: D. Banks, L. House, F.R. McMorris, P. Arabie and W. Gaul (Eds.): *Classification, Clustering, and Data Mining Applications.* Springer, Berlin, 77–84.

DIAS, J.G. (2006): Model Selection for the Binary Latent Class Model. A Monte Carlo Simulation. In: V. Batagelj, H.-H. Bock, A. Ferligoj and A. Ziberna (Eds.): *Data Science and Classification.* Springer, Berlin, 91–99.

DIAS, J.G. and WILLEKENS, F. (2005): Model-based Clustering of Sequential Data with an Application to Contraceptive Use Dynamics. *Mathematical Population Studies, 12, 135–157.*

LO, Y., MENDELL, N.R. and RUBIN, D.B. (2001): Testing the Number of Components in a Normal Mixture. *Biometrika, 88, 767-778.*

MCLACHLAN, G.J. and PEEL, D. (2000): *Finite Mixture Models.* John Wiley & Sons, New York.

POULSEN, C.S. (1990): Mixed Markov and Latent Markov Modelling Applied to Brand Choice Behavior. *International Journal of Research in Marketing, 7, 5–19.*

RAMASWAMY, V., DESARBO, W.S., REIBSTEIN, D.J. and ROBINSON, W.T. (1993): An Empirical Pooling Approach for Estimating Marketing Mix Elasticities with PIMS Data. *Marketing Science, 12, 103–124.*

SCHWARZ, G. (1978): Estimating the Dimension of a Model. *Annals of Statistics, 6, 461–464.*

WILKS, S.S. (1938): The Large Sample Distribution of the Likelihood Ratio for Testing Composite Hypotheses. *Annals of Mathematical Statistics, 9, 60–62.*

WOLFE, J.H. (1970): Pattern Clustering by Multivariate Mixture Analysis. *Multivariate Behavioral Research, 5, 329-350.*

Cluster Quality Indexes for Symbolic Classification – An Examination

Andrzej Dudek

Wrocław University of Economics, Department of Econometrics and Computer Science, Nowowiejska 3, 58-500 Jelenia Góra, Poland; andrzej.dudek@ae.jgora.pl

Abstract. The paper presents difficulties of measuring clustering quality for symbolic data (such as lack of a "traditional" data matrix). Some hints concerning the usage of known indexes for such kind of data are given and indexes designed exclusively for symbolic data are described. Finally, after the presentation of simulation results, some proposals for choosing the most adequate indexes for popular classification algorithms are given.

1 Introduction

In a typical classification procedure, cluster validation is one of the crucial steps. Typically, in the validation step an internal cluster quality index is used. There is a variety of such kind of indexes with over fifty measures (Milligan and Cooper (1985), Weingessel et al. (1999))

The problem of choosing the most adequate cluster quality index for data measured on different scales and classified by various clustering methods is well described in literature. Milligan and Cooper (1985) suggest to use Caliński and Harabasz, Hubert and Levine, Baker and Hubert indexes, and also the Silhouette index and the Krzanowski and Lai index are quite commonly used.

The situation differs in case of symbolic data. There are no hints in literature which indexes are most appropriate for that kind of data. This paper describes cluster quality indexes that can be used in this case.

In the first part clustering methods that can be used for symbolic data and methods designed exclusively for this kind of data are described. The second part presents main groups of cluster quality indexes along with examples of indexes from each group (due to the lack of space only the most frequently used indexes are described). The third part describes the classification process for symbolic data. In the next part cluster quality indexes are compared on 100 sets of symbolic data with known structures and for three clustering methods. Furthermore, there is a short summary which of them most accurately represents the structure of the clusters. Finally some conclusions and remarks are given.

2 Clustering methods for symbolic data

Symbolic data, unlike classical data, are more complex than tables of numeric values. Bock and Diday (2000) define five types of symbolic variables:

- single quantitative value,
- categorical value,
- interval,
- multi-valued variable,
- multi-valued variable with weights.

Variables in a symbolic object can also be, regardless of theirs type (Diday (2002)):

- taxonomic representing hierarchical structure,
- hierarchically dependent,
- logically dependent.

A common problem with the usage of symbolic data in classification algorithms is the fact, that for this kind of data, due to their structure, operations of addition, subtraction, multiplication, squaring, calculation of means or calculation of variance are not defined. Thus, methods based on data matrices cannot be used. Only methods based on distance matrices are applicable. Among them the most popular ones are:
Hierarchical agglomerative clustering methods (Gordon (1999, p. 79)):

- Ward,
- single linkage,
- complete linkage,
- average linkage,
- McQuitty (1966),
- centroid,
- median.

Optimization methods:

- Partitioning around medoids, also called k-medoids method (Kaufman and Rousseeuw (1990)).

Algorithms developed for symbolic data (Chavent et al. (2003), Verde (2004)):

- divisive clustering of symbolic objects (DIV),
- clustering of symbolic objects based on distance tables (DCLUST),
- dynamic clustering of symbolic objects (SCLUST),
- hierarchical and pyramidal clustering of symbolic objects (HiPYR).

Popular methods like k-means and related ones like hard competitive learning, soft competitive learning, Isodata and others cannot be used for symbolic data.

3 Cluster quality indexes

Over fifty internal cluster quality indexes are described in the literature of subject. They can be arranged in three main groups (Weingessel et al. (2003)), for each group a few well-known representatives are enumerated:

Indexes based on inertia (Sum of squares):

- Caliński and Harabasz (1974) index (pseudo F-statistics),
- Hartigan(1975) index,
- Ratkovski index (Ratkovski and Lance (1978)),
- Ball (1965) index,
- Krzanowski and Lai (1988) index.

Indexes based on scatter matrices:

- Scott index (Scott and Symons (1971)),
- Marriot (1971) index,
- Friedman index (Friedman and Rubin (1967)),
- Rubin index (Friedman and Rubin (1967)).

Indexes based on distance matrices:

- Silhouette (Rousseeuw (1987), Kaufman and Rousseeuw (1990)),
- Baker and Hubert (Hubert (1974), Baker and Hubert (1975)),
- Hubert and Levine (1976).

A different, relatively small group is defined by indexes dedicated only for symbolic data. Those indexes are (Verde (2004)):

- Inertia for symbolic objects,
- homogeneity based quality index.

4 Clustering quality indexes – symbolic objects case

Figure 1 summarizes the usage of clustering quality indexes for symbolic objects. For symbolic objects, clustering methods based on data matrices cannot be used. If the clustering algorithm is based on a distance matrix then, in validation, indexes based on the inertia and indexes based on a distance matrix are allowed. If an algorithm designed strictly for symbolic data is used then for validation indexes based on inertia and "symbolic" indexes are most appropriate. Thus, four paths of classification procedure may be distinguished:

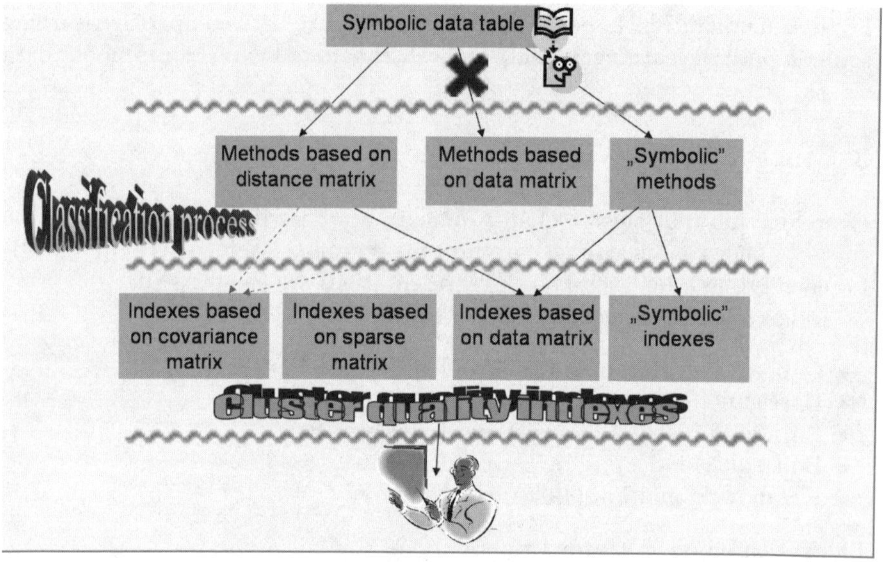

Fig. 1. Clustering method and cluster quality indexes for symbolic data. (Source: Own research based on Verde (2004), Chavent at al. (2003), Weingessel (1999))

- Clustering procedure based on dissimilarity matrix, validation with cluster quality index based on inertia;
- Clustering procedure based on dissimilarity matrix, validation with cluster quality index based on dissimilarity/distance matrix;
- "Symbolic" clustering procedure, validation with cluster quality index based on inertia;
- "Symbolic" clustering procedure, validation with cluster quality index designed for symbolic data.

5 Comparison of cluster quality indexes in symbolic objects case – computational results

Many authors like Milligan and Copper (1985) have compared cluster quality indexes so far, and some hints at the usage of them can be found in the literature. But no such comparison has been done for symbolic data yet (as far as the author knows).

100 symbolic data sets with known class structure have been clustered, and a compatibility measure for each index has been calculated according to the condition: "If the best value of the index is achieved for a number of clusters corresponding to the real structure of the data set then the compatibility measure is incremented, if it is second in the row 0.5 is added and for third in the row 0.25 is added".

Three clustering algorithms have been used: Ward hierarchical clustering method, partitioning around medoids method and dynamical clustering for symbolic objects methods. For each algorithm the compatibility measure has been calculated separately. Calculations have been made with the use of the symbolicDA library (written in R and C language by the author).

The data for the experiment has been generated artificially. The main reason for this is lack of real symbolic datasets with known data structure. There are only a few datasets shipped with the SODAS Software. But we can assume that switching from artificial to real data wouldn't change the results of the simulation, as far as the real cluster sizes are approximately equal. For datasets with one "large" and few "small" clusters the situation probably differs. Each data set contained a fixed number of objects (150), a random number (from 2 to 5) of single numerical variables, a random number (from 2 to 10) of variables in form of intervals and a random number (from 2 to 10) of multi-nominal variables. 20 data sets have 3 clusters structure, 25 have 4 clusters structure, 30 contain 5 clusters and 25 have 7 clusters structure, which is indicated in 1-4 headings.

The following indexes have been compared:

- S – Silhouette index,
- G2 – Baker and Hubert index,
- G3 – Hubert and Levine index,
- F – Caliński and Harabasz index,
- H – Hartigan index,
- KL – Krzanowski and Lai index,
- SI – inertia for symbolic objects,
- SH – homogeneity based quality index.

The Ichino and Yaguchi distance measure has been used to calculate the distance matrix. The results of the experiment are presented in Tables 1-4. Calculations were made in the R environment using the symbolicDA library.

Table 1. Comparison of cluster quality indexes for symbolic data – Ward hierarchical clustering.

Index	3 clusters (20 sets)	4 clusters (25 sets)	5 clusters (30 sets)	7 clusters (25 sets)	Total
S	6.5	4.5	0.25	0.5	11.75
G2	17	20.25	5.75	21.5	64.5
G3	18.25	17.25	15.5	7	58
F	6.5	7.25	0.5	0.75	15
H	6.75	26.25	5.25	0.25	38.5
KL	8	5.25	1.5	2.25	17
SI	6.5	3.5	8.75	0.75	19.5
SH	7.25	4.75	10.25	2	24.25

Table 2. Comparison of cluster quality indexes for symbolic data – k-medoids algorithm.

Index	3 clusters (20 sets)	4 clusters (25 sets)	5 clusters (30 sets)	7 clusters (25 sets)	Total
S	17.25	4.5	5.75	1	28.5
G2	5.25	17.5	18.75	11	52.5
G3	20	5.75	27.25	6.75	59.75
F	1.5	4.75	0.25	5.75	12.25
H	11.5	0.25	0.5	0	12.25
KL	13.75	5.75	6.25	3	28.75
SI	20	0.25	26.5	0.75	47.5
SH	20	1.25	28	4.25	53.5

For Ward hierarchical clustering of symbolic objects Hubert and Levine (G3) and Baker and Hubert (G2) indexes most adequately represent the real structure of the data. Only the Hartigan index provides significantly good results and the correlation between other indexes values and the real class structure is at a very low level. For the k-medoids algorithm for symbolic objects Hubert and Levine (G2), Baker and Hubert (G3), symbolic inertia (SI) and homogeneity based quality index (SH) may be used to validate classification results. And again for dynamical clustering of symbolic objects the Hubert and Levine (G2) and the Baker and Hubert (G3) indexes most adequately represent the real structure of data. Table 4 summarizes the results of the experiments. The G2 and G3 indexes are significantly better than the other indexes. It can be explained by the fact, that these indexes, are based on distance matrices, however the third index from this group (Silhouette index) is not as good as the two others. Indexes designed for symbolic data: symbolic inertia and homogeneity based quality index can also be used for symbolic cluster validation but the results may be worse than those achieved by using the Hubert and Levine or the Baker and Hubert index.

6 Final remarks

In this paper several cluster quality indexes were compared on 100 artificially generated symbolic data sets. The experiment showed that the most adequate ones for this kind of data are the Hubert and Levine and the Baker and Hubert indexes. We can assume that the usage of these indexes in case of real symbolic data validation should also give good results. The preliminary experiments with real symbolic data sets, done by the author, also confirm the quality of these indexes in the symbolic data case.

The results can be explained by the fact that Hubert and Levine and the Baker and Hubert indexes are based on distance matrices and for them, limitations of symbolic methods, described in section 2, do not exist.

Table 3. Comparison of cluster quality indexes for symbolic data – Dynamical clustering.

Index	3 clusters (max 20)	4 clusters (max 25)	5 clusters (max 30)	7 clusters (max 25)	Total
S	17.25	4	3.75	4.5	29.5
G2	5.25	10.25	17	16.75	49.25
G3	20	5.25	28	4.5	57.75
F	4.5	0	2.75	0	7.25
H	10	0	0	0.5	10.5
KL	12.25	3.5	3.25	4	23
SI	0	0.25	9.25	0	9.5
SH	2	2.75	17.5	0.5	22.75

Table 4. Comparison of cluster quality indexes for symbolic data – Aggregated results.

Index	3 clusters (60 sets)	4 clusters (75 sets)	5 clusters (90 sets)	7 clusters (75 sets)	Total
S	41	13	9.75	6	69.75
G2	27.5	48	41.5	49.25	166.25
G3	58.25	28.25	70.75	18.25	175.5
F	12.5	12	3.5	6.5	34.5
H	28.25	26.5	5.75	0.75	61.25
KL	34	14.5	11	9.25	68.75
SI	26.5	4	44.5	1.5	76.5
SH	29.25	8.75	55.75	6.75	100.5

Note that only two strictly "symbolic" indexes (symbolic inertia and homogeneity based quality index) have been taken into consideration. Currently new proposals are available (see for example Hardy (2005) for a description of the nbstat procedure), so this comparison should be repeated when more indexes and cluster number determination procedures are introduced.

References

BAKER, F.B. and HUBERT, L.J. (1975): Measuring the Power of Hierarchical Cluster Analysis. *Journal of the American Statistical Association, 70, 349, 31-38.*

BALL, F.B. and HALL, D.J. (1965): *ISODATA, A Novel Method of Data Analysis and Pattern Classification.* Tech. Rep. NTIS No.AD 699616, Stanford Research Institute, Menlo Park.

BOCK, H.-H. and DIDAY, E. (2000): *Analysis of Symbolic Data. Explanatory Methods for Extracting Statistical Information from Complex Data.* Springer, Berlin.

CALIŃSKI, R.B. and HARABASZ, J. (1974): A Dendrite Method for Cluster Analysis. *Communications in Statistics, 3, 1-27.*

CHAVENT, M., DE CARVALHO, F.A.T., VERDE, R. and LECHEVALLIER, Y. (2003): Trois Nouvelle Méthodes de Classification Automatique de Données Symboliques de Type Intervalle. *Revue de Statistique Appliquée , LI 4, 5-29.*

DIDAY, E. (2002): An Introduction to Symbolic Data Analysis and the SODAS Software. *J.S.D.A., International EJournal.*

FRIEDMAN, H.P. and RUBIN, J. (1967): On Some Invariant Criteria for Grouping Data. *Journal of the American Statistical Association, 62, 1159-1178.*

GORDON, A.D. (1999): *Classification*, Chapman & Hall/CRC, London.

HARDY, A. (2005): Validation of Unsupervised Symbolic Classification. *Proceedings of ASMDA 2005 Conference.* Available at URL: `http://asmda2005.enst-bretagne.fr/IMG/pdf/proceedings/379.pdf`.

HARTIGAN, J.A. (1975): *Clustering Algorithms.* New York, Wiley.

HUBERT, L.J. (1974): Approximate Evaluation Technique for the Single-link and Complete-link Hierarchical Clustering Procedures. *Journal of the American Statistical Association, 69, 347, 698-704.*

HUBERT, L.J. and LEVINE, J.R. (1976): Evaluating Object Set Partitions: Free Sort Analysis and Some Generalizations. *Journal of Verbal Learning and Verbal Behaviour, 15, 549-570.*

KAUFMAN, L. and ROUSSEEUW, P.J. (1990): *Finding Groups in Data: An Introduction to Cluster Analysis.* Wiley, New York.

KRZANOWSKI, W.J. and LAI, Y.T. (1988): A Criterion for Determining the Number of Groups in a Data Set Using Sum of Squares Clustering. *Biometrics, 44, 23-34.*

MARRIOT, F.H. (1971). Practical Problems in a Method of Cluster Analysis. *Biometrics, 27, 501-514.*

MCQUITTY, L.L. (1966): Similarity Analysis by Reciprocal Pairs for Discrete and Continuous Data. *Educational and Psychological Measurement, 26, 825-831.*

MILLIGAN, G.W. and COOPER, M.C. (1985): An Examination of Procedures for Determining the Number of Clusters in a Data Set. *Psychometrika, 2, 159-179.*

RATKOVSKI, D.A. and LANCE, G.N. (1978) A Criterion for Determining a Number of Groups in a Classification. *Australian Computer Journal, 10, 115-117.*

ROUSSEEUW, P.J. (1987): Silhouettes: A Graphical Aid to the Interpretation and Validation of Cluster Analysis. *Journal of Computational and Applied Mathematics, 20, 53-65.*

SCOTT, A.J. and SYMONS, M.J. (1971) Clustering Methods Based on Likelihood Ratio Criteria. *Biometrics, 27, 387-397.*

VERDE, R. (2004): Clustering Methods in Symbolic Data Analysis. In: D. Banks et al. (Eds.): *Classification, Clustering and Data Mining Applications*, Springer, Berlin, 299-318.

WEINGESSEL, A., DIMITRIADOU, A. and DOLNICAR, S. (1999): An Examination Of Indexes For Determining The Number Of Clusters In Binary Data Sets. Available at URL: `http://www.wu-wien.ac.at/am/wp99.htm#29`.

Semi-Supervised Clustering: Application to Image Segmentation

Mário A.T. Figueiredo

Instituto de Telecomunicações and Instituto Superior Técnico, Technical University of Lisbon, 1049-001 Lisboa, Portugal; mario.figueiredo@lx.it.pt

Abstract. This paper describes a new approach to semi-supervised model-based clustering. The problem is formulated as penalized logistic regression, where the labels are only indirectly observed (via the component densities). This formulation allows deriving a generalized EM algorithm with closed-form update equations, which is in contrast with other related approaches which require expensive Gibbs sampling or suboptimal algorithms. We show how this approach can be naturally used for image segmentation under spatial priors, avoiding the usual hard combinatorial optimization required by classical Markov random fields; this opens the door to the use of sophisticated spatial priors (such as those based on wavelet representations) in a simple and computationally very efficient way.

1 Introduction

In recent years there has been a considerable amount of interest in semi-supervised learning problems (see Zhu (2006)). Most formulations of semi-supervised learning approach the problem from one of the two ends of the unsupervised-supervised spectrum: either supervised learning in the presence of unlabelled data (see, e.g., Belkin and Niyogi (2003), Krishnapuram et al. (2004), Seeger (2001), Zhu et al. (2003)) or unsupervised learning with additional information (see, e.g., Basu et al. (2004), Law et al. (2005), Lu and Leen (2005), Shental et al. (2003), Wagstaff et al. (2001)). The second perspective, known as semi-supervised clustering (SSC), is usually adopted when labels are completely absent from the training data, but there are (say, pair-wise) relations that one wishes to enforce or simply encourage.

Most methods for SSC work by incorporating the desired relations (or constraints) into classical algorithms such as the expectation-maximization (EM) algorithm for mixture-based clustering or the K-means algorithm. These relations may be imposed in a hard way, as constraints (Shental et al. (2003),

Wagstaff et al. (2001)), or used to build priors under which probabilistic clustering is performed (Basu et al. (2004), Lu and Leen (2005)). This last approach has been shown to yield good results and is the most natural for applications where one knows that the relations should be encouraged, but not enforced (e.g., in image segmentation, neighboring pixels should be encouraged, but obviously not enforced, to belong to the same class). However, the resulting EM-type algorithms have a considerable drawback: because of the presence of the prior on the grouping relations, the E-step no longer has a simple closed form, requiring the use of expensive stochastic (e.g., Gibbs) sampling schemes (Lu and Leen (2005)) or suboptimal methods such as the *iterated conditional modes* (ICM) algorithm (Basu et al. (2004)).

In this paper, we describe a new approach to semi-supervised mixture-based clustering for which we derive a simple, fully deterministic *generalized EM* (GEM) algorithm. The keystone of our approach is the formulation of semi-supervised mixture-based clustering as a penalized logistic regression problem, where the labels are only indirectly observed. The linearity of the resulting complete log-likelihood, with respect to the missing group labels, will allow deriving a simple GEM algorithm.

We show how the proposed formulation is used for image segmentation under spatial priors which, until now, were only used for real-valued fields (e.g., image restoration/denoising): Gaussian fields and wavelet-based priors. Under these priors, our GEM algorithm can be implemented very efficiently by resorting to fast Fourier or fast wavelet transforms. Our approach completely avoids the combinatorial nature of standard segmentation methods, which are based on Markov random fields of discrete labels (see Li (2001)).

Although we focus on image segmentation, SSC has been recently used in other areas, such as clustering of image databases (see Grira et al. (2005)), clustering of documents (see Zhong (2006) for a survey), and bioinformatics (see, e.g., Nikkilä et al. (2001), Cebron and Berthold (2006)). Our approach will thus also be potentially useful in those application areas.

2 Formulation

We build on the standard formulation of finite mixtures: let $\mathcal{X} = \{\mathbf{x}_1, ..., \mathbf{x}_n\}$ be an observed data set, with each $\mathbf{x}_i \in I\!R^d$ assumed to have been generated (independently) according to one of a set of K probability densities $\{p(\cdot|\phi^{(1)}), ..., p(\cdot|\phi^{(K)})\}$. Associated with $\mathcal{X}$, there's a hidden/missing label set $\mathcal{Y} = \{\mathbf{y}_1, ..., \mathbf{y}_n\}$, where $\mathbf{y}_i = [y_i^{(1)}, ..., y_i^{(K)}]^T \in \{0,1\}^K$, with $y_i^{(k)} = 1$ if and only if $\mathbf{x}_i$ was generated by source k ("1-of-K" binary encoding). Thus,

$$p\left(\mathcal{X} \,\middle|\, \mathcal{Y}, \phi^{(1)}, ..., \phi^{(K)}\right) = \prod_{i=1}^{n} \prod_{k=1}^{K} \left[p(\mathbf{x}_i|\phi^{(k)})\right]^{y_i^{(k)}}. \tag{1}$$

In standard mixture models, the hidden labels $\mathbf{y}_i$ are assumed to be (independent) samples from a multinomial variable with probabilities $\{\eta^{(1)}, ..., \eta^{(K)}\}$, i.e., $P(\mathcal{Y}) = \prod_i \prod_k (\eta^{(k)})^{y_i^{(k)}}$. This independence assumption will clearly have to be abandoned in order to insert grouping constraints or prior preference for some grouping relations. In Basu et al. (2004) and Lu and Leen (2005), this is done by defining a prior $P(\mathcal{Y})$, in which the $\mathbf{y}_1, ..., \mathbf{y}_n$ are not independent. However, any such prior destroys the simple structure of EM for standard finite mixtures, which is critically supported on the independence assumption. Here, we follow a different route which in which the $\mathbf{y}_1, ..., \mathbf{y}_n$ are not modelled as independent, but for which we can still derive a simple GEM algorithm.

Let the set of hidden labels $\mathcal{Y} = \{\mathbf{y}_1, ..., \mathbf{y}_n\}$ depend on a new set of variables $\mathcal{Z} = \{\mathbf{z}_1, ..., \mathbf{z}_n\}$, where each $\mathbf{z}_i = [z_i^{(1)}, ..., z_i^{(K)}]^T \in I\!\!R^K$, according to a multinomial logistic model (see Böhning (1992)):

$$P(\mathcal{Y}|\mathcal{Z}) = \prod_{i=1}^{n} \prod_{k=1}^{K} \left(P[y_i^{(k)} = 1|\mathbf{z}_i] \right)^{y_i^{(k)}} = \prod_{i=1}^{n} \prod_{k=1}^{K} \left(\frac{e^{z_i^{(k)}}}{\sum_{l=1}^{K} e^{z_i^{(l)}}} \right)^{y_i^{(k)}}. \quad (2)$$

Due to the normalization constraint $\sum_{k=1}^{K} P[y_i^{(k)} = 1|\mathbf{z}_i] = 1$, we set (without loss of generality) $z_i^{(K)} = 0$, for $i = 1, ..., n$ (see Böhning (1992)). We are thus left with a total of $(n(K-1))$ real variables, i.e., $\mathcal{Z} = \{\mathbf{z}^{(1)}, ..., \mathbf{z}^{(K-1)}\}$, where $\mathbf{z}^{(k)} = [z_1^{(k)}, ..., z_n^{(k)}]^T$. Since the variables $z_i^{(k)}$ are real-valued and totally unconstrained, it's formally simple to define a prior $p(\mathcal{Z})$ and to perform optimization w.r.t. $\mathcal{Z}$. This contrasts with the direct manipulation of $\mathcal{Y}$, which, due to its discrete nature, brings a combinatorial nature to the problems.

The prior grouping relations are now expressed by a prior $p(\mathcal{Z})$; in particular, preferred pair-wise relations are encoded in a Gaussian prior

$$p(\mathcal{Z}|\mathbf{W}, \boldsymbol{\alpha}) \propto \prod_{k=1}^{K-1} \exp\left[-\frac{\|\mathbf{z}^{(k)} - \alpha^{(k)}\mathbf{1}\|^2}{2} - \frac{1}{4} \sum_{i,j=1}^{n} W_{i,j}(z_i^{(k)} - z_j^{(k)})^2 \right], \quad (3)$$

where $\mathbf{1} = [1, ..., 1]^T$ is a vector of n ones, $\boldsymbol{\alpha} = [\alpha^{(1)}, ..., \alpha^{(K-1)}]^T$, where $\alpha^{(k)}$ is a global mean for $\mathbf{z}^{(k)}$, and $\mathbf{W}$ is a matrix (with zeros in the diagonal) encoding the pair-wise preferences: $W_{i,j} > 0$ expresses a preference (with strength proportional to $W_{i,j}$) for having points i and j in the same cluster; $W_{i,j} = 0$ expresses the absence of any preference concerning the pair (i, j). The first term pulls the variables in $\mathbf{z}^{(k)}$ towards a common mean $\alpha^{(k)}$. If all $W_{i,j} = 0$, we have a standard mixture model in which each probability $\eta^{(k)}$ is a function of the corresponding $\alpha^{(k)}$. Defining

$$\mathbf{z} = [z_1^{(1)}, ..., z_n^{(1)}, z_1^{(2)}, ..., z_n^{(2)}, ..., z_1^{(K-1)}, ..., z_n^{(K-1)}]^T = \left[(\mathbf{z}^{(1)})^T, ..., (\mathbf{z}^{(K-1)})^T \right]^T$$

and matrix $\boldsymbol{\Delta}$ (the well-known graph-Laplacian; see Belkin and Niyogi (2003)),

$$\boldsymbol{\Delta} = \mathrm{diag}\left\{\sum_{j=1}^{n} W_{1,j}, ..., \sum_{j=1}^{n} W_{n,j}\right\} - \mathbf{W}, \tag{4}$$

allows writing the prior (3) in the more standard Gaussian form

$$\log p(\mathbf{z}|\mathbf{W}, \boldsymbol{\alpha}) = \log \mathcal{N}(\mathbf{z}|\boldsymbol{\beta}, \boldsymbol{\Psi}^{-1}) = -\frac{1}{2}(\mathbf{z}-\boldsymbol{\beta})^T \boldsymbol{\Psi}(\mathbf{z}-\boldsymbol{\beta}) + \frac{1}{2}\log\left(|\boldsymbol{\Psi}|\,(2\pi)^{-n}\right), \tag{5}$$

where the mean $\boldsymbol{\beta}$ and inverse covariance $\boldsymbol{\Psi}$ are given by

$$\boldsymbol{\beta} = \boldsymbol{\alpha} \otimes ((\mathbf{I}_n + \boldsymbol{\Delta})^{-1}\mathbf{1}_n) \qquad \text{and} \qquad \boldsymbol{\Psi} = \mathbf{I}_{K-1} \otimes (\mathbf{I}_n + \boldsymbol{\Delta}). \tag{6}$$

In (6), $\otimes$ is the Kronecker matrix product, $\mathbf{I}_a$ stands for an $a \times a$ identity matrix, and $\mathbf{1}_a = [1, 1, ..., 1]^T$ is a vector of a ones. From this point on, we consider $\mathbf{W}$ (but not $\boldsymbol{\alpha}$) as fixed, thus we omit it and write simply $p(\mathbf{z}|\boldsymbol{\alpha})$.

3 Model estimation

3.1 Marginal maximum a posteriori and the GEM algorithm

Based on the above formulation, semi-supervised clustering consists in estimating the unknown parameters of the model, $\boldsymbol{\alpha}$, $\mathbf{z}$, and $\boldsymbol{\phi} = \{\boldsymbol{\phi}^{(1)}, ..., \boldsymbol{\phi}^{(K)}\}$, taking into account that $\mathcal{Y}$ is missing. For this purpose, we adopt the marginal *maximum a posteriori* criterion, obtained by marginalizing out the hidden labels; thus, since by Bayes law $p(\mathcal{X}, \mathcal{Y}, \mathbf{z}|\boldsymbol{\phi}, \boldsymbol{\alpha}) = p(\mathcal{X}|\mathcal{Y}, \boldsymbol{\phi})\, P(\mathcal{Y}|\mathbf{z})\, p(\mathbf{z}|\boldsymbol{\alpha})$,

$$\left(\widehat{\mathbf{z}}, \widehat{\boldsymbol{\phi}}, \widehat{\boldsymbol{\alpha}}\right) = \arg\max_{\mathbf{z}, \boldsymbol{\phi}, \boldsymbol{\alpha}} \sum_{\mathcal{Y}} p(\mathcal{X}|\mathcal{Y}, \boldsymbol{\phi})\, P(\mathcal{Y}|\mathbf{z})\, p(\mathbf{z}|\boldsymbol{\alpha}),$$

where the sum is over all the possible label configurations, and we are assuming flat priors for $\boldsymbol{\phi}$ and $\boldsymbol{\alpha}$. We address this estimation problem using a generalized EM (GEM) algorithm (see, e.g., McLachlan and Krishnan (1997)), that is, by iterating the following two steps (until some convergence criterion is met):

E-step: Compute the conditional expectation of the complete log-posterior, given the current estimates $(\widehat{\mathbf{z}}, \widehat{\boldsymbol{\phi}}, \widehat{\boldsymbol{\alpha}})$ and the observations $\mathcal{X}$:

$$Q(\mathbf{z}, \boldsymbol{\phi}, \boldsymbol{\alpha}|\widehat{\mathbf{z}}, \widehat{\boldsymbol{\phi}}, \widehat{\boldsymbol{\alpha}}) = E_{\mathcal{Y}}[\log p(\mathcal{X}, \mathcal{Y}, \mathbf{z}|\boldsymbol{\phi}, \boldsymbol{\alpha})|\widehat{\mathbf{z}}, \widehat{\boldsymbol{\phi}}, \widehat{\boldsymbol{\alpha}}, \mathcal{X}]. \tag{7}$$

M-step: Update the estimate, that is, compute $(\widehat{\mathbf{z}}_{\text{new}}, \widehat{\boldsymbol{\phi}}_{\text{new}}, \widehat{\boldsymbol{\alpha}}_{\text{new}})$, such that these new values are guaranteed to improve the Q function, *i.e.*,

$$Q(\widehat{\mathbf{z}}_{\text{new}}, \widehat{\boldsymbol{\phi}}_{\text{new}}, \widehat{\boldsymbol{\alpha}}_{\text{new}}|\widehat{\mathbf{z}}, \widehat{\boldsymbol{\phi}}, \widehat{\boldsymbol{\alpha}}) \geq Q(\widehat{\mathbf{z}}, \widehat{\boldsymbol{\phi}}, \widehat{\boldsymbol{\alpha}}|\widehat{\mathbf{z}}, \widehat{\boldsymbol{\phi}}, \widehat{\boldsymbol{\alpha}}). \tag{8}$$

It is well known that, under mild conditions, a GEM algorithm converges to a local maximum of the marginal log-posterior $p(\mathcal{X}, \mathbf{z}|\boldsymbol{\phi}, \boldsymbol{\alpha})$ (see Wu (1983)).

3.2 E-step

Using equation (1) for $p(\mathcal{X}|\mathcal{Y},\boldsymbol{\phi})$, equation (2) for $P(\mathcal{Y}|\mathbf{z})$ (notice that $\mathbf{z}$ and $\mathcal{Z}$ are the same), and equation (5) for $p(\mathbf{z}|\boldsymbol{\alpha})$, leads to

$$\log p(\mathcal{X},\mathcal{Y},\mathbf{z}|\boldsymbol{\phi},\boldsymbol{\alpha}) \doteq \sum_{i=1}^{n}\sum_{k=1}^{K} y_i^{(k)} \log p(\mathbf{x}_i|\boldsymbol{\phi}^{(k)}) - \frac{(\mathbf{z}-\boldsymbol{\beta})^T\boldsymbol{\Psi}(\mathbf{z}-\boldsymbol{\beta})}{2} + \sum_{i=1}^{n}\left[\sum_{k=1}^{K} y_i^{(k)} z_i^{(k)} - \log \sum_{k=1}^{K} e^{z_i^{(k)}}\right], \tag{9}$$

where $\doteq$ stands for "equal apart from an additive constant". The important thing to notice here is that this function is linear w.r.t. the hidden variables $y_i^{(k)}$. Thus, the E-step reduces to the computation of their conditional expectations, which are then plugged into $p(\mathcal{X},\mathcal{Y},\mathbf{z}|\boldsymbol{\phi},\boldsymbol{\alpha})$.

As in standard mixtures, the missing $y_i^{(k)}$ are binary, thus their expectation (denoted as $\widehat{y}_i^{(k)}$) are equal to their probabilities of being equal to one, which can be obtained via Bayes law:

$$\widehat{y}_i^{(k)} \equiv E[y_i^{(k)}|\widehat{\mathbf{z}},\widehat{\boldsymbol{\phi}},\widehat{\boldsymbol{\alpha}},\mathcal{X}] = P(y_i^{(k)}=1|\widehat{\mathbf{z}}_i,\widehat{\boldsymbol{\phi}},\mathbf{x}_i) = \frac{p(\mathbf{x}_i|\widehat{\boldsymbol{\phi}}^{(k)})\,P(y_i^{(k)}=1|\widehat{\mathbf{z}}_i)}{\sum_{j=1}^{K} p(\mathbf{x}_i|\widehat{\boldsymbol{\phi}}^{(j)})\,P(y_i^{(j)}=1|\widehat{\mathbf{z}}_i)}. \tag{10}$$

Notice that (10) is similar to the E-step for a standard finite mixture (see, e.g., McLachlan and Krishnan (1997)), with the probabilities $P(y_i^{(k)}=1|\widehat{\mathbf{z}}_i) = \exp(z_i^{(k)})/\sum_j \exp(z_i^{(j)})$ playing the role of the class probabilities. Finally, the Q function is obtained by plugging the $\widehat{y}_i^{(k)}$ into (9).

3.3 M-Step: Density parameters $\boldsymbol{\phi}$

It's clear from (9) that the maximization w.r.t. $\boldsymbol{\phi}$ can be decoupled into

$$\widehat{\boldsymbol{\phi}}^{(k)}_{\text{new}} = \arg\max_{\boldsymbol{\phi}^{(k)}} \sum_{i=1}^{n}\sum_{k=1}^{K} \widehat{y}_i^{(k)} \log p(\mathbf{x}_i|\boldsymbol{\phi}^{(k)}). \tag{11}$$

This is the well-known weighted maximum likelihood criterion, exactly as it appears in the M-step for standard mixtures. The specific form of this update depends on the choice of $p(\cdot|\boldsymbol{\phi}^{(k)})$; *e.g.*, this step can be easily applied to any finite mixture of exponential family densities (see Banerjee et al. (2004)), of which the Gaussian is by far the one most often adopted.

3.4 M-step: $\mathbf{z}$ and $\boldsymbol{\alpha}$

The $\mathbf{z}$ and $\boldsymbol{\alpha}$ estimates are updated by maximizing (or increasing, see (8))

$$L(\mathbf{z},\boldsymbol{\alpha}) \equiv \sum_{i=1}^{n}\left[\sum_{k=1}^{K}\widehat{y}_i^{(k)} z_i^{(k)} - \log\sum_{k=1}^{K} e^{z_i^{(k)}}\right] - \frac{1}{2}(\mathbf{z}-\boldsymbol{\beta})^T\boldsymbol{\Psi}(\mathbf{z}-\boldsymbol{\beta}). \tag{12}$$

Ignoring the second term (the log-prior), this would correspond to a standard *logistic regression* (LR) problem, with an identity design matrix (Böhning (1992)), but where instead of the usual hard labels $y_i^{(k)} \in \{0,1\}$ we have soft labels $\widehat{y}_i^{(k)} \in [0,1]$.

The standard approaches to maximum likelihood LR are the Newton-Raphson algorithm (also known as *iteratively reweighted least squares* – IRLS; see Hastie et al. (2001)) and the *bound optimization approach* (BOA) (see Böhning (1992) and Lange et al. (2000)). In the presence of the log-prior, with a fixed $\boldsymbol{\alpha}$ (thus fixed $\boldsymbol{\beta}$), we have a quadratically penalized LR problem, and it's easy to modify either the IRLS or the BOA (see below) for this case. However, since we assume that $\boldsymbol{\alpha}$ is unknown, we adopt a scheme in which we maximize w.r.t. $\mathbf{z}$ and $\boldsymbol{\alpha}$ in an iterative fashion, by cycling through

$$\widehat{\boldsymbol{\alpha}}_{\text{new}} = \arg\max_{\boldsymbol{\alpha}} L(\widehat{\mathbf{z}},\boldsymbol{\alpha}) \tag{13}$$

$$\widehat{\mathbf{z}}_{\text{new}} = \arg\max_{\mathbf{z}} L(\mathbf{z},\widehat{\boldsymbol{\alpha}}_{\text{new}}). \tag{14}$$

It turns out that although (13) has a very simple closed form solution,

$$\widehat{\alpha}_{\text{new}}^{(k)} = \tfrac{1}{n}\textstyle\sum_{i=1}^{n}\widehat{z}_i^{(k)}, \quad \text{for} \quad k = 1,..,K-1, \tag{15}$$

the maximization (14) can only be solved iteratively. Adopting the BOA will lead to a simple update equation which, for certain choices of $\mathbf{W}$, can be implemented very efficiently.

We now briefly recall the BOA for maximizing a concave function with bounded Hessian (see Böhning (1992)). Let $G(\theta)$ be a concave differentiable function, such that its Hessian $\mathcal{H}(\theta)$ is bounded below by $-\mathbf{B}$ (i.e. $\mathcal{H}(\theta) \succeq -\mathbf{B}$ in the matrix sense, meaning that $\mathcal{H}(\theta)+\mathbf{B}$ is semi-definite positive), where $\mathbf{B}$ is a positive definite matrix. Then, it's easy to show that the iteration

$$\widehat{\theta}_{\text{new}} = \arg\max_{\theta}\left\{\theta^T\mathbf{g}(\widehat{\theta}) - \frac{1}{2}(\theta-\widehat{\theta})^T\,\mathbf{B}\,(\theta-\widehat{\theta})\right\} = \widehat{\theta} + \mathbf{B}^{-1}\mathbf{g}(\widehat{\theta}),$$

where $\mathbf{g}(\widehat{\theta})$ denotes the gradient of $G(\theta)$ at $\widehat{\theta}$, monotonically improves $G(\theta)$, i.e., $G(\widehat{\theta}_{\text{new}}) \geq G(\widehat{\theta})$. In our specific problem, the gradient of the logistic log-likelihood function (*i.e.*, (12) without the log-prior) is $\mathbf{g}(\widehat{\mathbf{z}}) = \mathbf{y}-\mathbf{p}$ and the Hessian verifies

$$\mathcal{H}(\mathbf{z}) \succeq -\frac{1}{2}\left[\left(\mathbf{I}_{K-1} - \frac{\mathbf{1}_{K-1}\,\mathbf{1}_{K-1}^T}{K}\right)\otimes\mathbf{I}_n\right] \equiv -\mathbf{B}, \tag{16}$$

where $\mathbf{y} = [y_1^{(1)},...,y_n^{(1)},y_1^{(2)},...,y_n^{(2)},y_1^{(K-1)}...,y_n^{(K-1)}]^T$ is a vector arrangement of $\mathcal{Y}$, and $\mathbf{p} = [p_1^{(1)},...,p_n^{(1)},p_1^{(2)},...,p_n^{(2)},p_1^{(K-1)},...,p_n^{(K-1)}]^T$ with $p_i^{(k)} = e^{z_i^{(k)}}/\sum_j e^{z_i^{(j)}}$.

The update equation for solving (14) via a BOA is thus

$$\begin{aligned}
\widehat{\mathbf{z}}_{\text{new}} &= \arg\max_{\mathbf{z}} \left\{ 2\,\mathbf{z}^T \mathbf{g}(\widehat{\mathbf{z}}) - (\mathbf{z}-\widehat{\mathbf{z}})^T \mathbf{B}\,(\mathbf{z}-\widehat{\mathbf{z}}) - (\mathbf{z}-\widehat{\boldsymbol{\beta}}_{\text{new}})^T \boldsymbol{\Psi} (\mathbf{z}-\widehat{\boldsymbol{\beta}}_{\text{new}}) \right\} \\
&= (\mathbf{B}+\boldsymbol{\Psi})^{-1} \left(\mathbf{g}(\widehat{\mathbf{z}}) + \mathbf{B}\,\widehat{\mathbf{z}} + \boldsymbol{\Psi}\,\widehat{\boldsymbol{\beta}}_{\text{new}} \right) \\
&= (\mathbf{B}+\boldsymbol{\Psi})^{-1} \left[\mathbf{g}(\widehat{\mathbf{z}}) + \mathbf{B}\,\widehat{\mathbf{z}} + \widehat{\boldsymbol{\alpha}}_{\text{new}} \otimes \mathbf{1}_n \right], \qquad (17)
\end{aligned}$$

where, according to the definition of $\boldsymbol{\beta}$ in (6), we write $\widehat{\boldsymbol{\beta}}_{\text{new}} = \widehat{\boldsymbol{\alpha}}_{\text{new}} \otimes \left[(\mathbf{I}_n + \boldsymbol{\Delta})^{-1}\mathbf{1}_n\right]$. The equality $\boldsymbol{\Psi}\,\widehat{\boldsymbol{\beta}}_{\text{new}} = \widehat{\boldsymbol{\alpha}}_{\text{new}} \otimes \mathbf{1}_n$ is show in the appendix.

Notice that $(\mathbf{B}+\boldsymbol{\Psi})^{-1}$ needs only be computed once. This is the fundamental advantage of the BOA over IRLS, which would require the inversion of a new matrix at each iteration (see Böhning (1992)).

Summarizing our GEM algorithm: the E-step is the application of (10), for all i and k; the M-step consists in (11) followed by (one or more) applications of (13)-(14). Eq. (13) is implemented by (15), and (14) is implemented by (one or more) applications of (17).

3.5 Speeding up the algorithm

The inversion $(\mathbf{B}+\boldsymbol{\Psi})^{-1}$, although it can be performed off-line, may be costly because $(\mathbf{B}+\boldsymbol{\Psi})$ is a $(n(K-1)) \times (n(K-1))$ matrix. We can alleviate this problem at the cost of using a less tight bound in (16), as shown by the following lemma (proved in the appendix):

Lemma 1. *Let $\xi_K = 1/2$, if $K > 2$, and $\xi_K = 1/4$, if $K = 2$; let $\mathbf{B}$ be defined as in (16). Then, $\mathbf{B} \preceq \xi_K\, \mathbf{I}_{n(K-1)}$.*

This lemma allows replacing $\mathbf{B}$ by $\xi_K\, \mathbf{I}_{n(K-1)}$ in the BOA (because, obviously $\mathcal{H}(\mathbf{z}) \succeq -\mathbf{B} \succeq -\xi_K\, \mathbf{I}_{n(K-1)}$). The matrix inversion in (17) becomes (see proof in appendix):

$$(\xi_K\, \mathbf{I}_{n(K-1)} + \boldsymbol{\Psi})^{-1} = \mathbf{I}_{K-1} \otimes ((\xi_K + 1)\mathbf{I}_n + \boldsymbol{\Delta})^{-1}, \qquad (18)$$

which means that we avoid the $(n(K-1)) \times (n(K-1))$ inversion and are left with a single $n \times n$ inversion; for a general matrix (assuming the standard inversion cost of $O(n^3)$), this yields a computational saving of roughly $(K-1)^3$, which for large K can be very meaningful. Finally, careful observation of

$$\widehat{\mathbf{z}}_{\text{new}} = \left[\mathbf{I}_{K-1} \otimes [(\xi_K+1)\mathbf{I}_n + \boldsymbol{\Delta}]^{-1} \right] (\mathbf{g}(\widehat{\mathbf{z}}) + \xi_K\, \widehat{\mathbf{z}} + \widehat{\boldsymbol{\alpha}} \otimes \mathbf{1}_n)$$

reveals that it can be decoupled among the several $\mathbf{z}^{(k)}$, yielding

$$\widehat{\mathbf{z}}^{(k)}_{\text{new}} = [(\xi_K+1)\mathbf{I}_n + \boldsymbol{\Delta}]^{-1} \left(\mathbf{y}^{(k)} - \mathbf{p}^{(k)} + \xi_K\, \widehat{\mathbf{z}}^{(k)} + \widehat{\alpha}^{(k)} \mathbf{1}_n \right). \qquad (19)$$

4 Application to image segmentation

Let $\mathcal{L} = \{i = (r,c),\ r = 1,...,N,\ c = 1,...,M\}$ be a $2D$ lattice of $n = |\mathcal{L}| = MN$ sites/pixels. A K-segmentation $\mathcal{R} = \{R_k \subseteq \mathcal{L},\ k = 1,...,K\}$ is a partition of $\mathcal{L}$ into K regions. As above, $\mathbf{y}$ is a "1-of-K" encoding of $\mathcal{R}$, i.e., $(y_i^{(k)} = 1) \Leftrightarrow (i \in R_k)$. The observation model (1) covers intensity-based or texture-based segmentation (each $\mathbf{x}_i$ can be a d-dimensional vector of texture features) and segmentation of multi-spectral images (e.g. color or remote sensing images, with d the number of spectral bands).

4.1 Stationary Gaussian field priors

If $W_{i,j}$ only depends on the relative position of i and j the Gaussian field prior is stationary. If, in addition, the neighborhood system defined by $\mathbf{W}$ has periodic boundary conditions, both $\mathbf{W}$ and $\boldsymbol{\Delta}$ are block-circulant, with circulant blocks (see Balram and Moura (1993)), thus are diagonalized by a 2D discrete Fourier transform (2D-DFT): $\boldsymbol{\Delta} = \mathbf{U}^H\mathbf{D}\mathbf{U}$, where $\mathbf{D}$ is diagonal, $\mathbf{U}$ is the matrix representation of the 2D-DFT, and $(\cdot)^H$ denotes conjugate transpose. Since $\mathbf{U}$ is orthogonal ($\mathbf{U}^H\mathbf{U} = \mathbf{U}\mathbf{U}^H = \mathbf{I}$), (19) can be written as

$$\widehat{\mathbf{z}}^{(k)}_{\text{new}} = \mathbf{U}^H\left[(\xi_K + 1)\mathbf{I}_n + \mathbf{D}\right]^{-1}\mathbf{U}\left(\mathbf{y}^{(k)} - \mathbf{p}^{(k)} + \xi_K\,\widehat{\mathbf{z}}^{(k)} + \widehat{\alpha}^{(k)}\mathbf{1}_n\right). \quad (20)$$

We now have a trivial diagonal inversion, and the matrix-vector products by $\mathbf{U}$ and $\mathbf{U}^H$ are not carried out explicitly but rather (very efficiently) via the fast Fourier transform (FFT) algorithm.

4.2 Wavelet-based priors

It is known that piece-wise smooth images have sparse wavelet-based representations (see Mallat (1998)); this fact underlies the excellent performance of wavelet-based denoising and compression techniques. Piece-wise smoothness of the $\mathbf{z}^{(k)}$ translates into segmentations in which pixels in each class tend to form connected regions. Consider a wavelet expansion of each $\mathbf{z}^{(k)}$

$$\mathbf{z}^{(k)} = \mathbf{L}\theta^{(k)}, \quad k = 1,...,K-1, \quad (21)$$

where the $\theta^{(k)}$ are sets of coefficients and $\mathbf{L}$ is now a matrix where each column is a wavelet basis function; $\mathbf{L}$ may be orthogonal or have more columns than lines (e.g., in over-complete, shift-invariant, representations) (see Mallat (1998)). The goal now is to estimate $\theta = \{\theta^{(1)},...,\theta^{(K-1)}\}$, under a sparseness prior $p(\theta)$. Classical choices for $p(\theta)$ are independent generalized Gaussians (see Moulin and Liu (1999)); a particular well-known case is the Laplacian,

$$p(\theta) = \textstyle\prod_{k=1}^{K-1}\prod_j (\lambda/2)\exp\{-\lambda\,|\theta_j^{(k)}|\}, \quad (22)$$

which induces a strongly non-Gaussian, non-Markovian prior $p(\mathbf{z})$, via (21).

The impact of adopting this wavelet-based prior is that the logistic regression equations (see (12)) now have $\mathbf{L}$ as the design matrix, instead of identity; thus, matrix $\mathbf{I}_n$ in (16) must be replaced by $\mathbf{L}^T\mathbf{L}$ and all occurrences of $z_i^{(k)}$ replaced by $(\mathbf{L}\theta^{(k)})_i$. Finally, in Lemma 1, ξ_K becomes $C/2$ and $C/4$, for $K > 2$ and $K = 2$, respectively, where C is the maximum eigenvalue of $\mathbf{L}^T\mathbf{L}$. Propagating all these changes through the derivations of the GEM algorithm leads to a simple closed-form update equation which involves the well-known soft-threshold non-linearity (see Figueiredo (2005) for details).

5 Experiments

5.1 Semi-supervised clustering

We show a simple toy experiment illustrating the behavior of the algorithm. The 900 data points in Fig. 1 (a) were generated by 6 circular Gaussians; the desired grouping is the one shown by the symbols: stars, circles, and dots. In Fig. 1(b), we show the result of estimating a 3-component Gaussian mixture using standard EM, which is of course totally unaware of the desired grouping. Prior information about the desired grouping is then embodied in the following $\mathbf{W}$ matrix: we randomly choose 300 pairs of points (out of $900 \times 899/2 \simeq 4 \times 10^5$ possible pairs) such that both points belong to the same desired group, and set the corresponding $W_{i,j}$ to one. The remaining elements of $\mathbf{W}$ are zero; notice that this is a highly sparse matrix. The mixture components produced by the proposed algorithm, under this prior knowledge, is shown in Fig. 1 (d). Convergence is obtained in about $30 \sim 50$ GEM iterations.

5.2 Image segmentation

We only have space to show a simple example (see Fig. 2). The observed image contains 4 regions, following Gaussian distributions with standard deviation 0.6 and means 1, 2, 3, and 4. Matrix $\mathbf{W}$ for the Gaussian field prior has $W_{i,j} = 1$ if i and j are nearest neighbors, zero otherwise. More details and results, namely with real images, can be found in Figueiredo (2005).

6 Conclusions

We have introduced a new formulation for semi-supervised clustering and shown how it leads to a simple GEM algorithm. We have demonstrated how our formulation can be applied to image segmentation with spatial priors, and have illustrated this using Gaussian field priors and wavelet-based priors. Future work includes a thorough experimental evaluation of the method, application to other problems, and extension to the unknown K case.

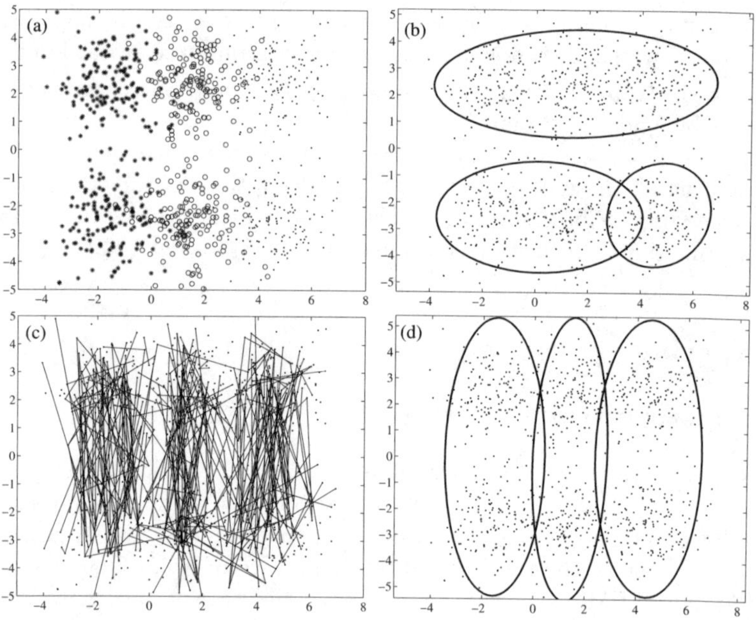

Fig. 1. Toy experiment with semi-supervised clustering; see text for details.

Appendix: Some proofs

Proof of $\boldsymbol{\Psi}\,\widehat{\boldsymbol{\beta}}_{\mathbf{new}} = \widehat{\boldsymbol{\alpha}}_{\mathbf{new}} \otimes \mathbf{1}_n$ (used in (17)): Since $(\mathbf{M} \otimes \mathbf{P})(\mathbf{Q} \otimes \mathbf{R}) = \mathbf{MQ} \otimes \mathbf{PR}$, we have $\boldsymbol{\Psi}\,\widehat{\boldsymbol{\beta}}_{\text{new}} = [\mathbf{I}_{K-1} \otimes (\mathbf{I}_n + \boldsymbol{\Delta})]\left[\widehat{\boldsymbol{\alpha}}_{\text{new}} \otimes ((\mathbf{I}_n + \boldsymbol{\Delta})^{-1}\mathbf{1}_n)\right] = \mathbf{I}_{K-1}\,\widehat{\boldsymbol{\alpha}}_{\text{new}} \otimes \left[(\mathbf{I}_n + \boldsymbol{\Delta})(\mathbf{I}_n + \boldsymbol{\Delta})^{-1}\mathbf{1}_n\right] = \widehat{\boldsymbol{\alpha}}_{\text{new}} \otimes \mathbf{1}_n$. □ **Proof Lemma 1:** Inserting $K = 2$ in (16) yields $\mathbf{B} = \mathbf{I}/4$. For $K > 2$, the inequality $\mathbf{I}/2 \succeq \mathbf{B}$ is equivalent to $\lambda_{\min}(\mathbf{I}/2 - \mathbf{B}) \geq 0$, which is equivalent to $\lambda_{\max}(\mathbf{B}) \leq (1/2)$. Since the eigenvalues of the Kronecker product are the products of the eigenvalues of the matrices, $\lambda_{\max}(\mathbf{B}) = \lambda_{\max}(\mathbf{I} - (1/K)\,\mathbf{1}\,\mathbf{1}^T)/2$. Since $\mathbf{1}\,\mathbf{1}^T$ is a rank-1 matrix with eigenvalues $\{0, ..., 0, K-1\}$, the eigenvalues of $(\mathbf{I} - (1/K)\,\mathbf{1}\,\mathbf{1}^T)$ are $\{1, ..., 1, 1/K\}$, thus $\lambda_{\max}(\mathbf{I} - (1/K)\,\mathbf{1}\,\mathbf{1}^T) = 1$, and $\lambda_{\max}(\mathbf{B}) = 1/2$. □ **Proof of equality (18):** Using $(\mathbf{M} \otimes \mathbf{P})^{-1} = \mathbf{M}^{-1} \otimes \mathbf{P}^{-1}$ and $\mathbf{I}_a \otimes \mathbf{I}_b = \mathbf{I}_{ab}$, and the definition of $\boldsymbol{\Psi}$ in (6), we can write

$$
\begin{aligned}
(\xi_K\,\mathbf{I}_{n(K-1)} + \boldsymbol{\Psi})^{-1} &= \left(\xi_K\,\mathbf{I}_{n(K-1)} + \mathbf{I}_{K-1} \otimes (\mathbf{I}_n + \boldsymbol{\Delta})\right)^{-1} \\
&= \left(\xi_K\,\mathbf{I}_{K-1} \otimes \mathbf{I}_n + \mathbf{I}_{K-1} \otimes (\mathbf{I}_n + \boldsymbol{\Delta})\right)^{-1} \\
&= \left(\mathbf{I}_{K-1} \otimes ((\xi_K + 1)\mathbf{I}_n + \boldsymbol{\Delta})\right)^{-1} \\
&= \mathbf{I}_{K-1} \otimes ((\xi_K + 1)\mathbf{I}_n + \boldsymbol{\Delta})^{-1}.
\end{aligned}
$$

□

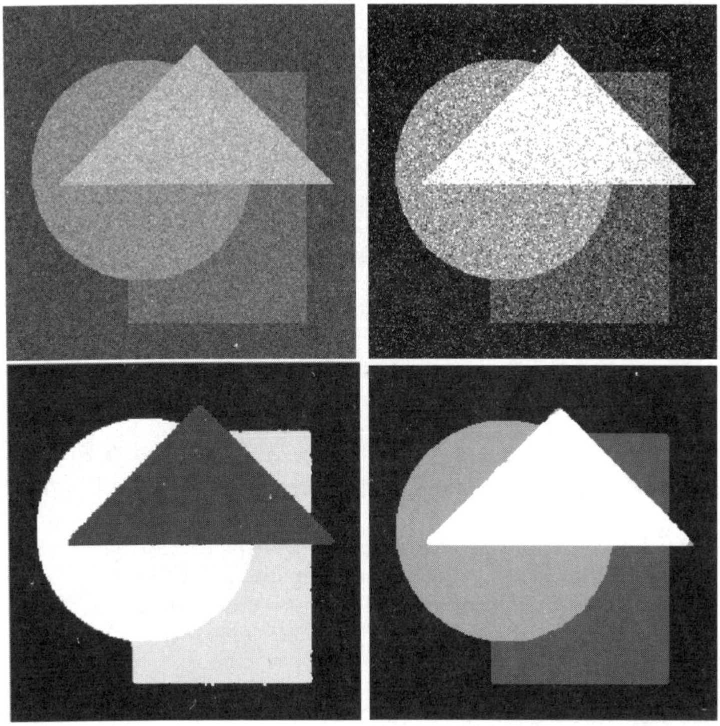

Fig. 2. Image segmentation example: upper left, observed image; upper right, maximum likelihood segmentation; lower left, SSC-based segmentation under Gaussian field prior; lower right, SSC-based segmentation under wavelet-based prior.

References

BALRAM, N. and MOURA, J. (1993): Noncausal Gauss-Markov Random Fields: Parameter Structure and Estimation. *IEEE Transactions on Information Theory, 39, 1333–1355.*

BANERJEE, A., MERUGU. S., DHILLON, I. and GHOSH, J. (2004): Clustering With Bregman Divergences. *Proc. SIAM International Conference on Data Mining,* Lake Buena Vista.

BASU, S., BILENKO, M. and MOONEY, R. (2004): A Probabilistic Framework for Semi-supervised Clustering. *Proc. International Conference on Knowledge Discovery and Data Mining,* Seattle.

BELKIN, M. and NIYOGI, P. (2003): Using Manifold Structure for Partially Labelled Classification. *Proc. Neural Information Processing Systems 15,* MIT Press, Cambridge.

BÖHNING, D. (1992): Multinomial Logistic Regression Algorithm. *Annals of the Institute of Statistical Mathematics, 44, 197–200.*

CEBRON, N. and BERTHOLD, M. (2006): Mining of Cell Assay Images Using Active Semi-supervised Clustering. *Proc. Workshop on Computational Intelligence in Data Mining,* Houston.

FIGUEIREDO, M. (2005): Bayesian Image Segmentation Using Wavelet-based Priors. *Proc. IEEE Conference on Computer Vision and Pattern Recognition*, San Diego.

GRIRA, N., CRUCIANU, M. and BOUJEMAA, N. (2005): Active and Semi-supervised Clustering for Image Database Categorization. *Proc. IEEE/EURASIP Workshop on Content Based Multimedia Indexing*, Riga, Latvia.

HASTIE, T., TIBSHIRANI, R. and FRIEDMAN, J. (2001): *The Elements of Statistical Learning.* Springer, New York.

KRISHNAPURAM, B., WILLIAMS, D., XUE, Y., HARTEMINK, A., CARIN, L. and FIGUEIREDO, M. (2005): On Semi-supervised Classification. *Proc. Neural Information Processing Systems 17*, MIT Press, Cambridge.

LANGE, K., HUNTER, D. and YANG, I. (2000): Optimization Transfer Using Surrogate Objective Functions. *Jour. Computational and Graphical Statistics, 9, 1–59.*

LAW, M., TOPCHY, A. and JAIN, A. K. (2005): Model-based Clustering With Probabilistic Constraints. *Proc. SIAM Conference on Data Mining*, Newport Beach.

LI, S. (2001): *Markov Random Field Modelling in Computer Vision*, Springer, Tokyo.

LU, Z. and LEEN, T. (2005): Probabilistic Penalized Clustering. *Proc. Neural Information Processing Systems 17*, MIT Press, Cambridge.

MALLAT, S. (1998): *A Wavelet Tour of Signal Processing.* Academic Press, San Diego, USA.

MCLACHLAN, G. and KRISHNAN, T. (1997): *The EM Algorithm and Extensions.* Wiley, New York.

MOULIN, P. and LIU, J. (1999): Analysis of Multiresolution Image Denoising Schemes Using Generalized-Gaussian and Ccomplexity Priors. *IEEE Transactions on Information Theory, 45, 909–919.*

NIKKILÄ, J., TÖRÖNEN, P., SINKKONEN, J. and KASKI, S. (2001): Analysis of Gene Expression Data Using Semi-supervised Clustering. *Proc. Bioinformatics 2001*, Skövde.

SEEGER, M. (2001): *Learning With Labelled and Unlabelled Data.* Technical Report, Institute for Adaptive and Neural Computation, University of Edinburgh.

SHENTAL, N., BAR-HILLEL, A., HERTZ, T. and WEINSHALL, D. (2003): Computing Gaussian Mixture Models With EM Using Equivalence Constraints. *Proc. Neural Information Processing Systems 15*, MIT Press, Cambridge.

WAGSTAFF, K., CARDIE, C., ROGERS, S. and SCHRÖDL, S. (2001): Constrained K-means Clustering With Background Knowledge. *Proc. International Conference on Machine Learning*, Williamstown.

WU, C. (1983): On the Convergence Properties of the EM Algorithm. *Annals of Statistics, 11, 95–103.*

ZHONG, S. (2006): Semi-supervised Model-based Document Clustering: A Comparative Study. *Machine Lerning*, 2006 (in press).

ZHU, X. (2006): *Semi-Supervised Learning Literature Survey.* Technical Report, Computer Sciences Department, University of Wisconsin, Madison.

ZHU, X., GHAHRAMANI, Z. and LAFFERTY, J. (2003): Semi-supervised Learning Using Gaussian Fields and Harmonic Functions. *Proc. International Conference on Machine Learning*, Washington DC.

A Method for Analyzing the Asymptotic Behavior of the Walk Process in Restricted Random Walk Cluster Algorithm

Markus Franke and Andreas Geyer-Schulz

Institute for Information Systems and Management, Universität Karlsruhe (TH), D-76128 Karlsruhe, Germany; {maf, ags}@em.uni-karlsruhe.de

Abstract. The Restricted Random Walk clustering algorithm is based on the execution of a series of random walks on a similarity or distance graph such that in the course of the walk only edges with growing similarity are visited. Cluster construction follows the idea that the later a pair of nodes occurs in a walk, the higher their similarity and thus their tendency to belong to the same cluster.

The resulting clusters show a grade of stochastic variation that depends on the number of walks. In this research paper, we scrutinize the asymptotic behavior of this stochastic process for an infinite number of walks. Thus, we are able to establish a starting point for the analysis of the influence of stochastics on the clusters.

To this end, we construct a cycle-free graph based on the transition matrix of the walk process. The edges of the similarity graph form the nodes of the new graph. Its edges are determined by the predecessor-successor relation of the similarity graph's edges. We then use a combination of shortest and longest path algorithms to calculate the highest possible position of each node pair in the walks that determines the cluster composition. In order to give an idea of the potential results of such an analysis, we show an exemplary comparison with single linkage clustering. On a local view, the clusters are very similar, however, a global view reveals differences in the order in which linkage of clusters takes place. This is due to the fact that restricted random walk clustering only has a local perspective of the similarity matrix while single linkage takes into account the whole matrix.

1 Introduction

Clustering with Restricted Random Walks (RRW) is a stochastic algorithm that is based on sampling the distances or similarities between the objects to be clustered with a specific form of random walks. The character of the resulting clusters depends strongly on the number of walks used to explore the data set. In this paper we will show how to perform an analysis for the sampling process with an infinite number of walks. In that case every conceivable path through the data set is taken at least once almost surely, i.e. with a probability of one.

The global goal of the analysis is to determine the method's behavior when different numbers of walks m are executed on a data set: with few walks, there is a high stochastic influence on the outcome of the clusters. On the one hand, this is favorable since it reduces the chances of bridging and since the computation is relatively cheap. On the other hand, results obtained with few walks are not stable, they may change between consecutive runs of the algorithm, and the risk that the walks miss important links is high. Many walks, on the contrary, are computationally expensive and increase the risk of bridging, but the results are stable and the data set is thoroughly explored.

The considerations will finally allow us to determine the number of walks that need to be executed on a given similarity matrix in order to obtain the best compromise between the two extremes. Currently, a fixed number of walks, usually five to ten, is started from each object as suggested by Schöll and Schöll-Paschinger (2003). Clearly, this is not satisfying.

Before presenting the asymptotic analysis in section 3, we give a short introduction into RRW clustering in section 2. A more detailed description was e.g. given by Franke and Geyer-Schulz (2006). Section 4 summarizes the results of the paper and gives an outlook to further research.

2 Clustering with restricted random walks

RRW clustering as presented by Schöll and Schöll-Paschinger (2003) is a stochastic process that makes use of a special form of random walks on object sets with a distance or similarity measure. With growing walk length, only more and more similar objects are selected. Thus the later in a walk a pair of objects occurs, the higher is the tendency that they belong to the same cluster. This relation is used for the construction of clusters from the accumulated walk data in the cluster construction stage.

2.1 Walk stage

For the analysis we use a modified notation that leads to a infinite, irreducible first-order Markov chain that can be analyzed by the techniques presented for instance in Karr's book (1993). As discussed by Franke and Thede (2005), the two formulations of the problem are equivalent in the asymptotic view.

Consider a similarity graph $G = (V, E, \omega)$ where V is the (finite) set of nodes, containing the objects to be clustered. E contains an edge for each pair of nodes that has a positive similarity, and the weight ω_{ij} on the edge between node i and j denotes that similarity. The similarity measure used here only has to comply to nonnegativity and symmetry, but not necessarily to the triangle inequality. Based on G, we construct a stochastic process with a state set $S = E \cup \{\Omega\}$ containing the edges of the graph as well as a transition state Ω. Ω is also the start state of the process. From there, the transition probability to the first step, i.e. the first edge $e_0 = (i_0, i_1)$ of the first walk is

$$P(i_0 i_1|\Omega) = \frac{1}{|V|\deg(i_0)} \,. \tag{1}$$

This probability comes from the original formulation: We pick one of the nodes with equal probability ($\frac{1}{|V|}$) and from there, take one of the incident edges in order to find a successor ($\frac{1}{\deg(i_0)}$). For the edge $e_{k-1} = (i_{k-1}, i_k)$ visited in the k-th step of the walk, the set of possible successors is defined as

$$T_{i_{k-1}i_k} = \{(i_k, j) \in E|\omega_{i_k j} > \omega_{i_{k-1}i_k}\} \tag{2}$$

containing only those edges with a higher weight than e_{k-1}. The next edge, e_k, is picked from a uniform distribution over $T_{i_{k-1}i_k}$. We thus obtain a sequence of states or edges such that the weight associated with the edges increases with each step. Given that E is finite, the walk will end in finite time, because at some point, $T_{i_{k-1}i_k}$ is empty. In that case, the process returns to Ω, from where another walk is started. For the scope of this paper, we assume that the process continues infinitely, i.e. the number of walks m is infinite.

2.2 Cluster construction

Two methods have been developed for the generation of clusters from the walk data: component clusters and walk context clusters. The former produces disjunctive clusters, while walk context clusters may be overlapping. Both produce hierarchical clusterings; disjunctive component clusters can be represented for instance by a dendrogram like the one in Fig. 2. In order to obtain a partitional clustering from the hierarchy, a cut is made through the dendrogram at a given level l. l is chosen in accordance with the requirements of the application: at the lowest level, there is only one cluster if the graph is connected or a few large ones otherwise. When increasing l, the clusters are successively split until, at the highest level, only singleton clusters exist.

We will detail here the analysis for component clusters. This method was proposed by Schöll and Schöll-Paschinger (2003). In order to obtain clusters, connected subgraphs are sought among the graph's edges visited during all executed walks. First, a series of graphs

$$G_k = (V, E_k) \tag{3}$$

is constructed. The set of vertices is identical to that of the original similarity graph, while E_k contains all edges that were visited during the k-th step of any of the m walks. For the cluster construction at level l, the G_k are aggregated:

$$H_l = \cup_{k=l}^{\infty} G_k \tag{4}$$

where the union of graphs consists of the common node set and the union of the edge sets. Clusters are defined as connected subgraphs in H_l.

As an extension, we introduce additional level measures as an alternative to the step concept that serve to further improve the quality of the resulting clusters. This is especially important if the graph's density differs among its different parts. The simplest of the three is the relative level

$$l = \frac{\text{step number}}{\text{total walk length}} \tag{5}$$

where the step number refers to the number of the step in which the node pair is visited in the respective walk. This normalization serves to also take into account last, i.e. important steps from shorter walks. Furthermore, contrary to the step number that is only bounded by the number of edges of the graph, the relative level has the range $[0, 1]$. On the other hand, l completely removes any influence of the walk length by contemplating only the relative position. This is not favorable since long walks tend to carry more information and are more reliable because stochastic influences present at the beginning of each walk are better reduced in longer than in shorter walks. The two measures

$$l^+ = \frac{\text{step number}}{\text{total walk length} + 1} \tag{6}$$

and

$$l^- = \frac{\text{step number} - 1}{\text{total walk length}} . \tag{7}$$

also give some weight to the walk length. For practical purposes, l^+ and l^- can be considered equivalent in terms of the resulting cluster quality as discussed by Franke and Geyer-Schulz (2006).

As shown by Franke and Thede (2005) the algorithm has an overall complexity of $O(n \log^2 n)$ for a data set containing n objects. For the general performance of the algorithm compared to some standard algorithms we refer the reader to the original contribution by Schöll and Schöll-Paschinger (2003).

3 Asymptotic behavior

The behavior of the algorithm can best be characterized by the clusters that result from the positions of node pairs in the walks. We are especially interested in the maximum position of each node pair since it is decisive for the pair's inclusion in the H_l graphs. Consequently, we will present a method for the computation of the maximum position for each pair of nodes. Since each possible walk is almost surely present for $m \to \infty$, it is safe to assume that this maximum position is sufficient for the description of the clustering with an infinite number of walks.

The general structure is given by Fig. 1 that contains the schematic of the walk process. Each walk starts at state A (capital α), chooses a successor state in each step according to the rules given in section 2, and ends in Ω. The maximum position of all node pairs can be computed with Dijkstra's (1959) shortest path algorithm in order to construct the asymptotic clusters.

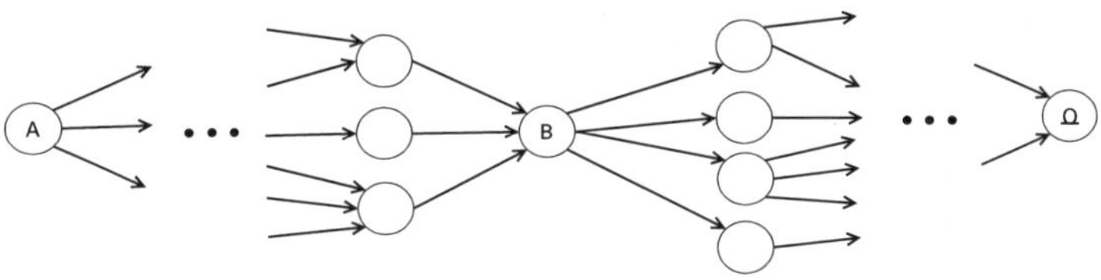

Fig. 1. The transition graph for the stochastic process

Consider state B in Fig. 1. To compute the highest level that this state – and thus the associated pair of objects – can achieve using one of the relative level measures l, l^+, and l^-, we have to identify the walk in which B has both a minimal distance to Ω and a maximal distance to A, where distance is defined as the number of intermediary steps. Per definition, this walk contains the state in question at the maximum position: Let a be the number of steps from A to B, and o the number of steps from B to Ω. The total length of the underlying walk is then $a + o - 1$, the maximum relative level for l is given as

$$l^* = \frac{a}{a + o - 1} = \frac{1}{1 + \frac{o-1}{a}} . \tag{8}$$

Clearly, if a is maximal for the state, and o is minimal, then l^* is maximal. This argument applies analogously to l^+ and l^-. The levels thus computed can then be used for the cluster construction as described in the last section. We use the following algorithm to determine o; $\mathcal{S}_B$ is the set of successors of B. Since the graph is loop-free per construction, the algorithm will terminate, which is not generally the case when searching for maximum path lengths.

1. $\mathcal{F} = \{\Omega\}$
2. $o_\Omega = 0$
3. while $\mathcal{F} \neq E \cup \{A, \Omega\}$ do
4. select state B with $B \notin \mathcal{F}$ and $\mathcal{S}_B \subseteq \mathcal{F}$
5. $o_B = min_{C \in \mathcal{S}_B}\{o_C\} + 1, \ \mathcal{F} = \mathcal{F} \cup \{B\}$
6. end

a can be computed analogously, with $\mathcal{P}_B$ the set of predecessors of B:

1. $\mathcal{F} = \{A\}$, $o_\Omega = 0$
2. $\forall e \in E \cup \{\Omega\} : a_e = 1$
3. while $\mathcal{F} \neq E \cup \{A, \Omega\}$ do
4. select state B with $B \notin F$ and $\mathcal{P}_B \subseteq \mathcal{F}$
5. $a_B = max_{C \in \mathcal{P}_B}\{a_C\} + 1, \ \mathcal{F} = \mathcal{F} \cup \{B\}$
6. end

3.1 Results

In this section we are going to present two aspects that differentiate the asymptotic RRW clustering e.g. from single linkage (SL). In the following, the term

RRW is always used with reference to the asymptotic view, using component clusters and the level measure l^+. The effects discussed here will be exemplarily shown on the cluster hierarchy for the Deep South data set by Davis et al. (1948) that contains data on 18 women and their attendance of 14 social events. Their presence or absence at a particular event is coded in a binary matrix, and the similarity between two women is defined by the number of events they have attended together; the corresponding matrix is given in Tab. 1, and the dendrograms created by SL and RRW are depicted in Fig. 2 and 3.

In general, SL and RRW produce quite similar results. The first obvious difference pertains to the perspective or horizon of the algorithms: the SL algorithm operates on the whole matrix at once while RRW samples only local similarities. The second difference is that, in consequence of the context provided by the walks, RRW uses transitive information from the neighborhood of a node pair, whereas for SL only the direct similarity between the nodes is decisive for a possible join. This can lead to the following effects: first, pairs of objects that have the same similarity may be merged at different levels. Using SL, this is not possible since the global comparison operator employed by the algorithm marks two object pairs with the same similarity for a merge on the same level. In the case of RRW, this is possible since the evaluation is based on the local neighborhood of each of the object pairs. For instance, an object pair located in a region with high density will in general be merged on a lower

Table 1. The similarity matrix for the Deep South data set

ids	1	2	3	4	5	6	7	8	9	10	11	12	13	14	15	16	17	18
1	-	6	7	6	3	4	3	3	3	2	2	2	2	2	1	2	1	1
2	6	-	6	6	3	4	4	2	3	2	1	1	2	2	2	1	0	0
3	7	6	-	6	4	4	4	3	4	3	2	2	3	3	2	2	1	1
4	6	6	6	-	4	4	4	2	3	2	1	1	2	2	2	1	0	0
5	3	3	4	4	-	2	2	0	2	1	0	0	1	1	1	0	0	0
6	4	4	4	4	2	-	3	2	2	1	1	1	1	1	1	1	0	0
7	3	4	4	4	2	3	-	2	3	2	1	1	2	2	2	1	0	0
8	3	2	3	2	0	2	2	-	2	2	2	2	2	2	1	2	1	1
9	3	3	4	3	2	2	3	2	-	3	2	2	3	2	2	2	1	1
10	2	2	3	2	1	1	2	2	3	-	3	3	4	3	3	2	1	1
11	2	1	2	1	0	1	1	2	2	3	-	4	4	3	3	2	1	1
12	2	1	2	1	0	1	1	2	2	3	4	-	6	5	3	2	1	1
13	2	2	3	2	1	1	2	2	3	4	4	6	-	6	4	2	1	1
14	2	2	3	2	1	1	2	2	2	3	3	5	6	-	4	1	2	2
15	1	2	2	2	1	1	2	1	2	3	3	3	4	4	-	1	1	1
16	2	1	2	1	0	1	1	2	2	2	2	2	2	1	1	-	1	1
17	1	0	1	0	0	0	0	1	1	1	1	1	1	2	1	1	-	2
18	1	0	1	0	0	0	0	1	1	1	1	1	1	2	1	1	2	-

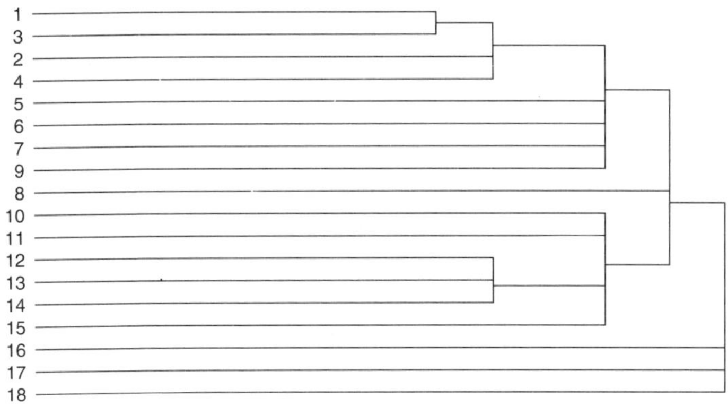

Fig. 2. Single linkage dendrogram of the Deep South data set

level than one in a low-density region where it might constitute even a local similarity maximum, thus receiving the last step of walks in the region. An example for this effect are the pairs 12/13 and 2/4. Both have a similarity of six, but are merged on different levels. This is due to the fact that the walks heading for the pair 12/13 have a maximal length of six whereas the walks ending in 2/4 have no more than five steps. Thus, the algorithm is able to find an additional differentiation between nodes based on the local neighborhood for nodes that the SL method treats as equivalent.

The other effect is that object pairs with different similarities are merged on the same level because they are embedded in a similar neighborhood. This behavior of the algorithm can be explained by the shortsightedness of the algorithm: it is not able to "see" that the two pairs occurring at the same level l^+ have in fact different similarities since it lacks the global perspective. In the Deep South data set, such a configuration is observable for the pairs 1/3 and 12/13. This behavior is highly desirable for applications where it is more important to find local maxima in the similarity matrix and to group them than to identify the global ones.

4 Conclusion and outlook

We have presented an analysis technique for the asymptotic behavior of the walk stage of the RRW clustering method. First results suggest that the resulting clusterings have a more local perspective than their SL counterparts. For applications that rather rely on local groups, for instance recommender services, this is an advantage over SL. On the other hand, for applications requiring a global classification SL is advantageous.

As a next step, we will theoretically analyze the cluster construction stage both for component and walk context clusters. With this, one extremal point for the analysis of the process at different walk numbers is established. Consequently, it will be interesting to identify the tipping points at which the be-

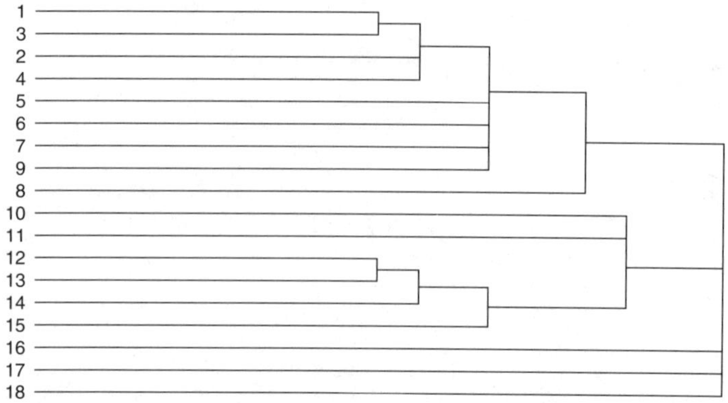

Fig. 3. RRW dendrogram of the Deep South data set with l^+

havior of the method changes significantly when varying the number of walks. For this the convergence of the graph-construction process to the steady-state graph (Fig. 1) must be investigated. As a final goal, we should be able to derive recommendations for the number of walks needed in order to obtain certain characteristics of the resulting clusters.

Acknowledgment. We gratefully acknowledge the funding of the project "SESAM" by the Bundesministerium für Bildung und Forschung (BMBF). Also, we would like to thank Peter Kleiweg for the software den.c used for the dendrograms (`http://www.let.rug.nl/~kleiweg/clustering/clustering`).

References

DAVIS, A., GARDNER, B.B. and GARDNER, M.R. (1948): *Deep South. A Social Anthropological Study of Caste and Class.* University of Chicago Press, Chicago.

DIJKSTRA, E.W. (1959): A Note on Two Problems in Connexion With Graphs. *Numerische Mathematik, 1, 269–271.*

FRANKE, M. and THEDE, A. (2005): Clustering of Large Document Sets with Restricted Random Walks on Usage Histories. In: C. Weihs and W. Gaul (Eds.): *Classification – the Ubiquitous Challenge.* Springer, Heidelberg, 402–409.

FRANKE, M. and GEYER-SCHULZ, A. (2006): Using Restricted Random Walks for Library Recommendations and Knowledge Space Exploration. *International Journal of Pattern Recognition and Artificial Intelligence, Special Issue on Personalization Techniques for Recommender Systems and Intelligent User Interfaces* (to appear).

KARR, A.F. (1993): *Probability.* Springer, Heidelberg.

SCHÖLL, J. and SCHÖLL-PASCHINGER, E. (2003): Classification by Restricted Random Walks. *Pattern Recognition*, 36, 6, 1279–1290.

Cluster and Select Approach to Classifier Fusion

Eugeniusz Gatnar

Institute of Statistics, Katowice University of Economics,
Bogucicka 14, 40-226 Katowice, Poland; egatnar@ae.katowice.pl

Abstract. The key issue in classifier fusion is diversity of the component models. In order to obtain the most diverse candidate models we generate a large number of classifiers and divide the set into K disjoint subsets. Classifiers with similar outputs are in the same cluster and classifiers with different predicted class labels are assigned to different clusters. In the next step one member of each cluster is selected, e.g. the one that exhibits the minimum average distance from the cluster center. Finally the selected classifiers are combined using majority voting.

Results from several experiments have shown that the candidate classifiers are diverse and their fusion improves classification accuracy.

1 Introduction

Tumer and Ghosh (1996) have shown that the key issue in classifier fusion is diversity of the component models, i.e. error of the fuser decreases with the reduction in correlation between base classifiers.

In order to obtain the most diverse candidate models we have proposed the "cluster and select" method, being a variant of the so-called "overproduce and choose" general framework.

In this method we have generated a large number of classifiers and clustered them into disjoint subsets. Classifiers with similar outputs were in the same cluster, and classifiers with different predicted class labels were assigned to different clusters. Then, one member of each cluster has been selected, e.g. the one with the highest accuracy. Finally, the selected classifiers were combined by the majority voting.

In this paper we compare the performance of six measures of dissmilarity of the base classifiers and four clustering methods used in the proposed approach.

The paper is organised as follows. In Section 2 we review existing work within the "overproduce and choose" framework and propose our "cluster and select" approach. Section 3 gives a short description of six measures of diversity between pairs of base classifiers. In Section 4 we recall four hierarchical clustering methods used in our experiments and the silhouette index applied to find the optimal number of clusters. The selection phase is explained in Section 5. Section 6 gives a brief description of our experiments and the obtained results. The last section contains a short summary.

2 Cluster and select approach

Direct creation of accurate and diverse classifiers is a very difficult task. Therefore, Partridge and Yates (1996) proposed a method based on the "overproduce and choose" general framework. Also, Sharkey et al. (2000) developed a method that follows the framework.

According to the "overproduce and choose" framework, an initial large set of candidate classifiers is created on the basis of a training set. Then a subset of the most error independent classifiers is selected. Partridge and Yates (1996) also introduced the measure (6) from Section 3 to guide the selection of the most diverse classifiers.

Giacinto et al. (2000) proposed to use the measure (5) from Section 3 as the distance measure and complete linkage clustering method. They used neural networks and k-nearest neighbors as the component classifiers. Kuncheva (2000) developed a simple clustering and selection algorithm to combine classifier outputs. She divided the training set into clusters using the k-means method and found the most accurate classifier for each cluster. Then the selected classifier was nominated to label the observations in the Voronoi cell of the cluster centroid. The proposed method performed slightly better than the majority vote and the other combining methods.

Giacinto and Roli (2001) proposed a hierarchical method for ensemble creation. At the first step, each of M base classifiers is a cluster. Next, the two least diverse classifiers (clusters) are joined in a cluster, and the more accurate of them is selected as the cluster representative. Then the representatives of all clusters form an ensemble. The procedure is repeated until all the classifiers are joined together.

They produced ensembles of $1, 2, \ldots, M-1$ classifiers, and the final ensemble size is the size of the one with the lowest classification error estimated on a test set.

In our method, a set of M candidate classifiers is divided into K disjoint subsets $\{L_1, L_2, \ldots, L_K\}$, using a clustering method. Then a member of each cluster is selected, e.g. the one with the highest accuracy or the one that exhibits the maximum average distance from all other cluster centers. Finally, the selected classifiers are combined using the majority voting.

The algorithm of the proposed method is as follows:

1. Produce M candidate classifiers using the training set.

2. Build the dissimilarity matrix $\mathbf{D} = [d_{ij}]$ that contains the pairwise diversities between classifiers.
3. Divide the classifiers into $K \in \{2, \ldots, M-1\}$ clusters using a clustering method and find the optimal value of K.
4. Select one classifier C_k from each cluster.
5. Combine the K selected classifiers using the majority voting:

$$C^*(\mathbf{x}) = argmax_{y \in Y} \left\{ \sum_{k=1}^{K} I(C_k(\mathbf{x}) = y) \right\}. \tag{1}$$

Considering the above algorithm, some questions arise: how to build the dissimilarity matrix $\mathbf{D}$, i.e. what measure should be used as the dissimilarity measure between classifiers? Which clustering method should be applied to the candidate classifiers? How to detect the optimal number of clusters, i.e. the number of ensemble members? In order to answer the questions, we performed several comparisons, described in the Section 6.

3 Pairwise diversity measures

In the "cluster" phase of the algorithm the candidate classifiers are clustered on the basis of a dissimilarity or distance matrix $\mathbf{D} = [d_{ij}]$ between all pairs of them. In order to find the most appropriate dissmilarity measure, we have examined six different pairwise diversity measures[1], presented in Kuncheva and Whitaker (2003) and Gatnar (2005).

Let $[\hat{C}(\mathbf{x}_1), \hat{C}(\mathbf{x}_2), \ldots, \hat{C}(\mathbf{x}_N)]$ be a vector of predictions of the classifier C for the set of examples $V = \{(\mathbf{x}_1, y_1), (\mathbf{x}_2, y_2), \ldots, (\mathbf{x}_N, y_N)\}$. The relationship between a pair of classifiers C_i and C_j can be shown in the form of the 2×2 contingency table (Table 1). In order to use this table for any number of classes, the "oracle" labels are applied. We define the "oracle" output (R_i) of the classifier C_i as:

$$R_i(\mathbf{x}_n) = \begin{cases} 1 \; if \; \hat{C}_i(\mathbf{x}_n) = y_n \\ 0 \; if \; \hat{C}_i(\mathbf{x}_n) \neq y_n \end{cases}. \tag{2}$$

In other words, the value of $R_i = 1$ means that the classifier C_i is correct, i.e. it recognizes the true class (y_n) of the example $\mathbf{x}_n$, and $R_i = 0$ means that the classifier is wrong.

The binary version of the Pearson's correlation coefficient:

$$d_{ij} = \frac{ad - bc}{\sqrt{(a+b)(c+d)(a+c)(b+d)}}. \tag{3}$$

[1] In fact, some of them are similarity measures and have been transformed into dissimilarity ones.

Table 1. A 2×2 contingency table for the "oracle" outputs.

Classifiers	$R_j = 1$	$R_j = 0$
$R_i = 1$	a	b
$R_i = 0$	c	d

can be used, after tranformation, as simple diversity measure. Its values range from -1 to 1, with 0 indicating independence of the two classifers.

Kuncheva et al. (2000) proposed the Yule's Q statistics to evaluate the diversity of all possible component classifier pairs. The Yule's Q statistics is the original measure of dichotomous agreement, designed to be analogous to the Pearson's correlation:

$$d_{ij} = \frac{ad - bc}{ad + bc}. \tag{4}$$

This measure is pairwise and symmetric, and varies between -1 and 1. A value of 0 indicates statistical independence of classifiers, positive values mean that the classifiers have recognized the same examples correctly and negative values – that the classifiers commit errors on different examples:

Giacinto and Roli (2000) have introduced a measure based on the compound error probability for the two classifiers, and named *compound diversity*:

$$d_{ij} = \frac{d}{a + b + c + d}. \tag{5}$$

This measure is also named "double-fault measure" because it is the proportion of the examples that have been misclassified by both classifiers. Partridge and Yates (1996) and Margineantu and Dietterich (1997) have used a measure named within-set generalization diversity. This measure is simply the κ statistics, and it measures the level of agreement between two classifiers with the correction for chance. Its pairwise version is calculated as:

$$d_{ij} = \frac{2(ac - bd)}{(a + b)(c + d) + (a + c)(b + d)}. \tag{6}$$

Skalak (1996) reported the use of the disagreement measure to characterize the diversity between base classifiers:

$$d_{ij} = \frac{b + c}{a + b + c + d}. \tag{7}$$

This is the ratio between the number of examples on which one classifier is correct and the other is wrong to the total number of examples. Gatnar (2005) has proposed the diversity measure based on the Hamann's coefficient that is simply the difference between the matches and mismatches as a proportion of the total number of entries:

$$d_{ij} = \frac{(a+d)-(b+c)}{a+b+c+d}. \tag{8}$$

It ranges from -1 to 1. A value of 0 indicates an equal number of matches to mismatches, -1 represents perfect disagreement, and 1 – perfect agreement.

4 Clustering methods

Having a set of M candidate classifiers $\{C_1, C_2, \ldots, C_M\}$, we divide them into K clusters: $\{L_1, L_2, \ldots, L_K\}$ on the basis of a dissimilarity matrix $\mathbf{D}$. Classifiers with similar errors are in the same cluster, and classifiers with different errors are assigned to different clusters. In our experiments we have applied four hierarchical clustering methods that use the dissimilarity matrix $\mathbf{D}$:

- *single linkage* (nearest neighbor method), where we treat the smallest dissimilarity between an observation in the first cluster and an observation in the second cluster as the distance between two clusters,
- *complete linkage* (furthest neighbor method), where we use the largest dissimilarity between a point in the first cluster and a point in the second cluster as the distance between the two clusters,
- *average method*, where the distance between two clusters is the average of the dissimilarities between the observations in one cluster and the observations in the other cluster.
- *Ward's method*, where the dissimilarity between two clusters is the Euclidean distance between their centroids.

In order to determine the optimal number of clusters K we have used the silhouette index developed by Rousseeuw (1987) and implemented in the `cluster` package described in Kaufman and Rousseeuw (1990).

For each observation x_i, the silhouette width $s(x_i)$ that ranges from 0 to 1 is computed. Observations with a high value of $s(x_i)$ are well clustered, while small value of $s(x_i)$ means that the observation x_i lies between two clusters. Observations with a negative $s(x_i)$ belong to the wrong cluster. We find the optimal clustering as the one that maximizes the sum of silhouette widths for all observations in the learning set:

$$\max\left\{\sum_{i=1}^{N} s(x_i)\right\}. \tag{9}$$

5 Selecting a representative of the cluster

Having an optimal clustering $\{L_1, L_2, \ldots, L_K\}$, we choose one classifier C_k from each cluster L_k. Several different selection strategies can be considered in the "select" phase, but in our experiment we selected the classifier that has the lowest classification error estimated on the appropriate test set.

6 Results of experiments

In order to find the optimal diversity measure and the clustering method for the cluster and select approach, we have applied it to 9 benchmark datasets from the Machine Learning Repository at the UCI (Blake et al. (1998)).

Some of them were already divided into learning and test part, but in some cases we divided them randomly.

For each dataset we have generated 14 sets of candidate classifiers of different sizes: $M = 10, 20, 30, 40, 50, 60, 70, 80, 90, 100, 150, 200, 250, 300$, and we used classification trees[2] as the base models.

For example, average classification errors for different diversity measures and different clustering methods for the DNA dataset are shown in Figures 1 and 2.

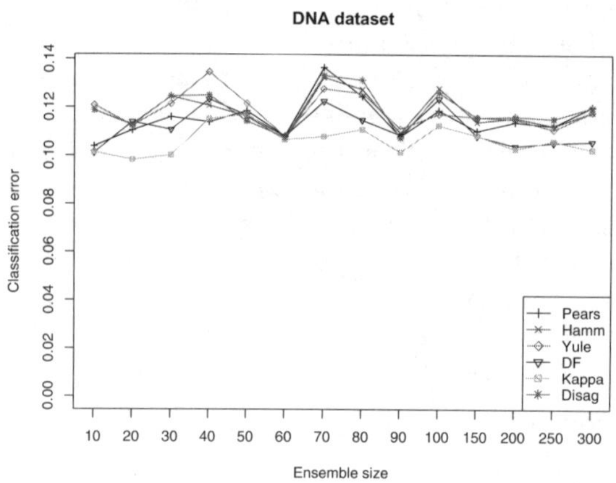

Fig. 1. Average classification error for different diversity measures for the DNA dataset

We have also computed the number of lowest error combinations of diversity measure and clustering method for each ensemble size. For example, Table 2 shows the distribution of winning combinations for the DNA data set.

7 Summary

In this paper we have proposed a modification of the cluster-and-select approach to classifier fusion. In order to find the most appropriate dissimilarity measure and clustering method, we performed several comparisons.

---

[2] In order to grow trees, we have used the Rpart procedure written by Therneau and Atkinson (1997) for the R environment.

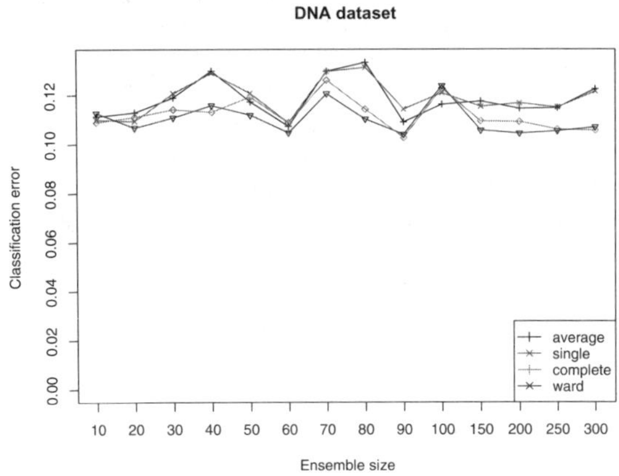

Fig. 2. Average classification error for different clustering methods for the DNA dataset

Table 2. Winning combinations for the DNA dataset

	average	single	complete	ward
Pearson	1	0	0	0
Hamann	0	0	1	1
Yule	0	0	0	0
Double fault	0	0	5	4
Kappa	3	1	0	3
Disagreement	0	0	1	0

Comparing different diversity measures, we have observed that using double fault measure or Kappa statistics lead to the most accurate ensembles, but Hamman's coefficient also works quite well. Comparing clustering methods, we conclude that the complete linkage and Wards method outperformed the other clustering methods.

References

BLAKE, C., KEOGH, E. and MERZ, C.J. (1998): UCI Repository of Machine Learning Databases. Department of Information and Computer Science, University of California, Irvine.

GATNAR, E. (2005): A Diversity Measure for Tree-Based Classifier Ensembles. In: D. Baier, R. Decker and L. Schmidt-Thieme (Eds.): *Data Analysis and Decision Support.* Springer, Heidelberg.

GIACINTO, G. and ROLI, F. (2001): Design of Effective Neural Network Ensembles for Image Classification Processes. *Image Vision and Computing Journal, 19, 699–707.*

GIACINTO, G., ROLI, F. and FUMERA, G. (2000): Design of Effective Multiple Classifier Systems by Clustering of Classifiers. *Proc. of the Int. Conference on Pattern Recognition*, ICPR'00, IEEE.

HANSEN, L.K. and SALAMON, P. (1990): Neural Network Ensembles. *IEEE Transactions on Pattern Analysis and Machine Intelligence 12, 993–1001.*

KAUFMAN, L. and ROUSSEEUW, P.J. (1990): *Finding Groups in Data: An Introduction to Cluster Analysis.* Wiley, New York.

KUNCHEVA, L. and WHITAKER, C. (2003): Measures of Diversity in Classifier Ensembles and Their Relationship with the Ensemble Accuracy. *Machine Learning 51, 181–207.*

KUNCHEVA, L. (2000): Clustering-and-Selection Model for Classifier Combination. *Proceedings of the 15th International Conference on Pattern Recognition*, Barcelona, Spain.

KUNCHEVA, L., WHITAKER, C., SHIPP, D. and DUIN, R. (2000): Is Independence Good for Combining Classifiers? In: J. Kittler,and F. Roli (Eds.): *Proceedings of the First International Workshop on Multiple Classifier Systems.* LNCS 1857, Springer, Berlin.

MARGINEANTU, M.M. and DIETTERICH, T.G. (1997): Pruning Adaptive Boosting. *Proceedings of the 14th International Conference on Machine Learning*, Morgan Kaufmann, San Mateo.

PARTRIDGE, D. and YATES, W.B. (1996): Engineering Multiversion Neural-net Systems. *Neural Computation 8, 869–893.*

ROUSSEEUW P.J. (1987): Silhouettes: A Graphical Aid to the Interpretation and Validation of Cluster Analysis. *J. Comput. Appl. Math. 20, 53–65.*

SHARKEY, A.J.C., SHARKEY, N.E., GERECKE, U. and CHANDROTH, G. (2000): The Test and Select Approach to Ensemble Combination. In: J. Kittler and F. Roli (Eds.): *Proceedings of the First International Workshop on Multiple Classifier Systems.* LNCS 1857, Springer, Berlin.

SKALAK, D.B. (1996): The Sources of Increased Accuracy for two Proposed Boosting Algorithms. *Proceedeings of the American Association for Artificial Intelligence AAAI-96*, Morgan Kaufmann, San Mateo.

THERNEAU, T.M. and ATKINSON, E.J. (1997): *An Introduction to Recursive Partitioning Using the RPART Routines*, Mayo Foundation, Rochester.

TUMER, K. and GHOSH, J. (1996): Analysis of Decision Boundaries in Linearly Combined Neural Classifiers. *Pattern Recognition 29, 341–348.*

Random Intersection Graphs and Classification

Erhard Godehardt[1], Jerzy Jaworski[2] and Katarzyna Rybarczyk[2]

[1] Clinic of Thoracic and Cardiovascular Surgery, Heinrich Heine University, 40225 Düsseldorf, Germany; godehard@uni-duesseldorf.de

[2] Faculty of Mathematics and Computer Science, Adam Mickiewicz University, 60769 Poznań, Poland; jaworski@amu.edu.pl, kryba@amu.edu.pl

Abstract. We study properties of random intersection graphs generated by a random bipartite graph. We focus on the connectedness of these random intersection graphs and give threshold functions for this property and results for the size of the largest components in such graphs. The application of intersection graphs to find clusters and to test their randomness in sets of non-metric data is shortly discussed.

1 Introduction

It is sometimes necessary to divide a set of objects into clusters based on a given pattern of properties chosen by (or related to) each object. Quite often, relations between objects and their properties can be described by a bipartite graph with the 2-partition $(\mathcal{V}, \mathcal{W})$ of the vertex set $\mathcal{V} \cup \mathcal{W}$, where the n-element subset $\mathcal{V}$ represents the objects and the m-element subset $\mathcal{W}$ represents the properties. In such a bipartite graph, edges would connect objects (the elements of $\mathcal{V}$) with their properties (the elements of $\mathcal{W}$). Two objects from $\mathcal{V}$ are called "similar", if they have at least one property in common (or more generally if they share at least s properties for a given integer s). Two properties of $\mathcal{W}$ are similar if there exists at least one object (or generally, if there exist at least s different objects) having both properties. We are interested in clusters of similar objects and in those of similar properties. Every two similar elements will be included in the same cluster.

A useful concept to describe connections between similar objects or between similar properties is that of intersection graphs generated by the bipartite graph describing the original relations between objects and properties. The first intersection graph would have the vertex set $\mathcal{V}$ of objects where two objects are joined by an edge if and only if the sets of properties of the corresponding objects have a non-empty intersection (or, more generally, if these two objects share at least s properties). This graph will be called the active intersection graph $\mathcal{IG}^{active}$, since its vertices represent the objects which have (actively) chosen their properties. Similarly, we get an intersection graph with the properties as vertices. This intersection graph is called a passive intersection graph $\mathcal{IG}^{passive}$, since the properties have been chosen. In the context of cluster analysis, certain subgraphs of the active intersection graph on $\mathcal{V}$ correspond to object clusters, and specific subgraphs of the passive intersection graph on $\mathcal{W}$ correspond to property clusters. In the simplest case, the similarity clusters will be the connected components of the intersection graphs.

A key feature of this approach is that this model of bipartite graphs and related intersection graphs enables us to find cluster structures in non-metric data sets. Under the hypothesis of homogeneity in a data set, the bipartite graph and the related intersection graphs can be interpreted as the realization of a random bipartite graph together with its two random intersection graphs. A probability model for these graphs has been introduced in Godehardt and Jaworski (2002) (for more information on how graph-theoretical concepts can be used in defining cluster models, revealing hidden clusters in a data set, and testing the randomness of such clusters see Bock (1996), Godehardt (1990), Godehardt and Jaworski (1996) for metric data, and Godehardt and Jaworski (2002) for non-metric data).

2 Definitions

We will start with the formal definition of a random bipartite graph which serves to introduce the two models of active and passive intersection graphs. This bipartite model was introduced in Godehardt and Jaworski (2002), and it is an analogue of a general model of random digraphs (see Jaworski and Palka (2002), Jaworski and Smit (1987)).

Definition 1 (Random bipartite graph). *Let $\mathcal{P}_{(m)} = (P_0, P_1, \ldots, P_m)$ be a discrete probability distribution, i.e., an $(m+1)$-tuple of non-negative real numbers satisfying $P_0 + P_1 + \cdots + P_m = 1$. Denote by $\mathcal{BG}_{n,m}\left(n, \mathcal{P}_{(m)}\right)$ a random bipartite graph on a vertex set $\mathcal{V} \cup \mathcal{W} = \{v_1, v_2, \ldots, v_n\} \cup \{w_1, w_2, \ldots, w_m\}$, such that*

1. *each $v \in \mathcal{V}$ chooses its degree $X^*(v)$ independently of all other vertices from $\mathcal{V}$ according to the probability distribution $\mathcal{P}_{(m)} = (P_0, P_1, \ldots, P_m)$:*
$$\Pr\{X^*(v) = k\} = P_k, \quad k = 0, 1, \ldots, m;$$
2. *each $v \in \mathcal{V}$ with degree k chooses (independently of all other vertices from $\mathcal{V}$) its set of neighbors $\Gamma(v)$ (properties) uniformly at random from all k-element subsets of $\mathcal{W}$.*

If $\mathcal{P}_{(m)}$ is a binomial distribution $B(m,p)$, then the model $\mathcal{BG}_{n,m}\left(n, B(m,p)\right)$ is equivalent to the standard random bipartite graph $\mathcal{G}_{n,m,p}$ on $n+m$ labeled vertices where each of the $n\,m$ possible edges between the sets $\mathcal{V}$ and $\mathcal{W}$ appears independently with a given probability p (see Janson et al. (2001)). The idea of using this model to define random intersection graphs was proposed in Karoński et al. (1999). Let $\mathcal{BG}_{n,m}(n, \mathcal{P}_{(m)})$ be defined as above. The active and passive random intersection graphs are defined in the following way.

Definition 2 (Active intersection graph). *A graph with the vertex set $\mathcal{V} = \{v_1, v_2, \ldots, v_n\}$, in which (v_i, v_j) is an edge if and only if v_i and v_j have at least s common neighbors in $\mathcal{BG}_{n,m}(n, \mathcal{P}_{(m)})$, i.e., $|\Gamma(v_i) \cap \Gamma(v_j)| \geq s$ in $\mathcal{W}$ holds, is called an active intersection graph $\mathcal{IG}^{act}(n, \mathcal{P}_{(m)})$.*

Definition 3 (Passive intersection graph). *A graph with the vertex set $\mathcal{W} = \{w_1, w_2, \ldots, w_m\}$, in which (w_i, w_j) is an edge if and only if w_i and w_j have at least s common neighbors in $\mathcal{BG}_{n,m}(n, \mathcal{P}_{(m)})$, i.e., $|\Gamma(w_i) \cap \Gamma(w_j)| \geq s$ in $\mathcal{V}$ holds, is called a passive intersection graph $\mathcal{IG}^{pas}(m, \mathcal{P}_{(m)})$.*

In this article, results will concern intersection graphs for $s = 1$. In this case every edge of an intersection graph connects vertices whose sets of neighbors are intersecting in the original bipartite graph.

3 Connectivity of intersection graphs

In Godehardt and Jaworski (2002), basic properties of $\mathcal{BG}_{n,m}\left(n, \mathcal{P}_{(m)}\right)$ have been proved, and the distribution of the vertex degree and of the number of isolated vertices for the corresponding random intersection graphs have been studied (see also Jaworski et al. (2006)). In this paper, we are going to give results concerning the connectivity of random intersection graphs, generated by $\mathcal{BG}_{n,m}\left(n, \mathcal{P}_{(m)}\right)$. In particular, under some constraints on the distribution $\mathcal{P}_{(m)}$, we will give the relation between m and $n = n(m)$, for which these random graphs will be connected asymptotically almost surely (a.a.s.), i.e., with probability tending to 1 as m tends to infinity.

3.1 Main results

All results below are given for m tending to infinity and $n = n(m)$. We also assume that the distribution $\mathcal{P}_{(m)} = (P_0, P_1, \ldots, P_m)$ is such that $P_k = 0$ for $k = r+1, r+2, \ldots, m$, where r is fixed (independent of m).

Theorem 1. *Let $\mathcal{P}_{(m)} = (P_0, P_1, \ldots, P_r)$, where r is a fixed integer; and let $\omega(m) = \frac{n}{m}\mathrm{E}(X^*) - \frac{nP_1}{m} - \ln m = \frac{n}{m}\sum_{i=2}^{r} i\, P_i - \ln m$. Then*

$$\lim_{m\to\infty} \Pr\left\{\mathcal{IG}^{pas}(m, \mathcal{P}_{(m)}) \text{ is connected}\right\} = \begin{cases} 0, & \text{if } \omega(m) \to -\infty \\ e^{-e^{-c}}, & \text{if } \omega(m) \to c \\ 1, & \text{if } \omega(m) \to \infty \end{cases}$$

Theorem 2. *Let $\mathcal{P}_{(m)} = (P_0, P_1, \ldots, P_r)$, where r is fixed, and $P_0 = 0$. Furthermore, let $\frac{n}{m}\mathrm{E}(X^*) - \frac{nP_1}{m} - \ln m = \frac{n}{m}\sum_{i=2}^{r} i\, P_i - \ln m \to \infty$ as $m \to \infty$. Then $\mathcal{IG}^{act}(n, \mathcal{P}_{(m)})$ is connected a.a.s.*

On the other hand, the results for isolated vertices in $\mathcal{IG}^{act}(n, \mathcal{P}_{(m)})$ (see Godehardt and Jaworski (2002)), imply the following result.

Corollary 1. *For a given $r \geq 2$, let $\mathcal{P}_{(m)} = (P_r)$ be the one-point distribution. Furthermore let $n = m\,(\ln m + o(\ln\ln m))/r^2$. Then $\mathcal{IG}^{act}(n, \mathcal{P}_{(m)})$ is not connected a.a.s.*

Since connectivity is a monotone property it follows from the fact above that under the same assumptions on m and n, the probability that $\mathcal{IG}^{act}(n, \mathcal{P}_{(m)})$ is connected also tends to 0 if $\mathcal{P}_{(m)} = (P_0, P_1, \ldots, P_r)$ for r fixed.

To get a better insight into the structure of "typical" random intersection graphs, especially at the "time" just before they become connected, we give a theorem on the size of their largest component (by $\mathrm{E}_2(X^*)$ we denote the second moment of the random variable "degree").

Theorem 3. *Let $\mathcal{P}_{(m)} = (P_0, P_1, \ldots, P_r)$, r fixed, and let*

$$\alpha(m) = \frac{n}{m}\sum_{i=2}^{r} i(i-1)P_i = \frac{n}{m}\,\mathrm{E}_2(X^*).$$

If $\alpha(m)$ tends to a finite limit $\alpha > 0$ as $m \to \infty$, then

(i) *for $\alpha < 1$, a.a.s. the size of the largest connected component in both $\mathcal{IG}^{pas}(m, \mathcal{P}_{(m)})$ and $\mathcal{IG}^{act}(n, \mathcal{P}_{(m)})$ is of order $\ln m$;*
(ii) *for $\alpha > 1$, a.a.s. the size of the largest connected component in both $\mathcal{IG}^{pas}(m, \mathcal{P}_{(m)})$ and $\mathcal{IG}^{act}(n, \mathcal{P}_{(m)})$ is of order m.*

In fact, a more precise version of the theorem above can be given (see Rybarczyk (2005)). In the case of a degenerate distribution $\mathcal{P}_m = (P_r)$ for fixed $r \leq 3$ we can prove, e.g., that if $n = \lfloor m/(r\,(r-1))\rfloor$, then the largest connected component in $\mathcal{IG}^{pas}(m, \mathcal{P}_{(m)})$ as well as in $\mathcal{IG}^{act}(n, \mathcal{P}_{(m)})$ has less than $\omega(m)m^{2/3}\ln m$ and more than $m^{2/3}/\omega(m)$ vertices, where $\omega(m)$ is a function tending to infinity arbitrarily slowly with m (see Rybarczyk (2005)).

3.2 Sketch of the proofs

We will give a sketch of the proofs of the main results stated above. The detailed proofs can be found in Rybarczyk (2005). The key idea of our approach is to study hypergraphs generated by bipartite graphs and to use known results for models of random hypergraphs. A hypergraph is a pair $\langle V, E\rangle$, in which V is a non-empty set of vertices and E is a family of non-empty subsets of V which we call edges (in fact these elements are a generalization of the edges in graphs). For a given bipartite graph $\mathcal{BG}_{n,m}$ on the vertex set $\mathcal{V} \cup \mathcal{W}$, $\mathcal{H}(\mathcal{BG}_{n,m})$ is a hypergraph with the vertex set $\mathcal{W}$ and an edge set such that $S \subseteq \mathcal{W}$ is an edge of the hypergraph $\mathcal{H}(\mathcal{BG}_{n,m})$ if and only if there exists $v \in \mathcal{V}$ such that S is a set of neighbors of v ($\Gamma(v) = S$). Therefore each edge of $\mathcal{H}(\mathcal{BG}_{n,m})$ is related to at least one vertex from $\mathcal{V}$. In fact each edge is a set of all the properties of some object. An edge may be related to more than one vertex if there exist two vertices $v_1, v_2 \in \mathcal{V}$ in $\mathcal{BG}_{n,m}$ having the same sets of neighbors, $\Gamma(v_1) = \Gamma(v_2)$, i.e., if some objects have identical properties.

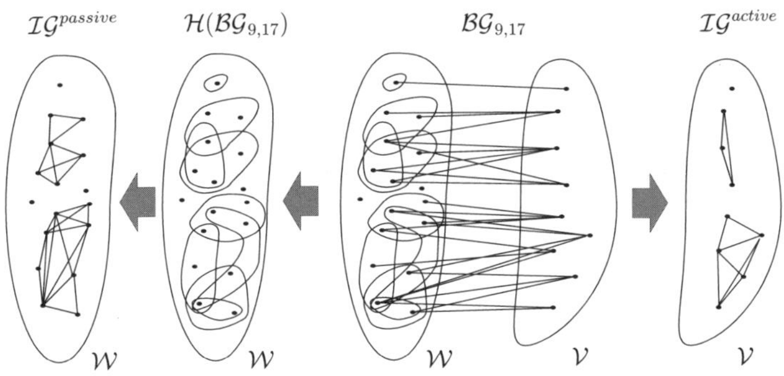

Fig. 1. The hypergraph $\mathcal{H}(\mathcal{BG}_{9,17})$ has five connected components, thus $\mathcal{IG}^{passive}$ has five connected components. Only three of the components in $\mathcal{H}(\mathcal{BG}_{9,17})$ contain edges and they relate to all three connected components in $\mathcal{IG}^{active}$.

In $\mathcal{IG}^{passive}$, two vertices are linked by an edge if and only if there exists an edge in $\mathcal{H}(\mathcal{BG}_{n,m})$ containing both these vertices (an edge of $\mathcal{H}(\mathcal{BG}_{n,m})$ corresponds to a complete subgraph in $\mathcal{IG}^{passive}$). Each pair of vertices joined by an edge in $\mathcal{IG}^{active}$ is related to intersecting edges in $\mathcal{H}(\mathcal{BG}_{n,m})$. Thus by making some observations concerning walks and paths in $\mathcal{H}(\mathcal{BG}_{n,m})$, $\mathcal{IG}^{active}$ and $\mathcal{IG}^{passive}$, we can prove the following facts about relations between connected components of $\mathcal{H}(\mathcal{BG}_{n,m})$ and components of random intersection graphs.

Lemma 1. *The graph $\mathcal{IG}^{passive}$ is connected if and only if $\mathcal{H}(\mathcal{BG}_{n,m})$ is connected. Furthermore a subset $S \subseteq \mathcal{W}$ is a set of vertices of a connected component of $\mathcal{H}(\mathcal{BG}_{n,m})$ if and only if it forms a connected component of $\mathcal{IG}^{passive}$.*

Lemma 2. *The graph $\mathcal{IG}^{active}$ is connected if and only if none of the vertices in $\mathcal{V}$ in $\mathcal{BG}_{n,m}$ is of degree 0 and if $\mathcal{H}(\mathcal{BG}_{n,m})$ outside the largest connected component has only isolated vertices which are not one-element edges. For any subset $S \subseteq \mathcal{W}$ with $|S| > 1$, which is a set of vertices of a connected component of $\mathcal{H}(\mathcal{BG}_{n,m})$, there exists exactly one connected component of $\mathcal{IG}^{active}$ whose vertices relate to edges of the connected component with the vertex set S.*

Figure 1 gives examples of a hypergraph $\mathcal{H}(\mathcal{BG}_{n,m})$, and of the intersection graphs $\mathcal{IG}^{active}$ and $\mathcal{IG}^{passive}$, generated by the bipartite graph $\mathcal{BG}_{n,m}$. The facts given above imply that the conditions for the connectedness of $\mathcal{H}(\mathcal{BG}_{n,m})$ and $\mathcal{IG}^{passive}$, generated by the same $\mathcal{BG}_{n,m}$, are the same and also that, under the assumption that there are no isolated vertices in $\mathcal{BG}_{n,m}$, the connectedness of $\mathcal{H}(\mathcal{BG}_{n,m})$ will imply directly the connectedness of $\mathcal{IG}^{active}$.

Using Chebyshev's inequality, we may prove that a.a.s. the number of vertices $v \in \mathcal{V}$ choosing degree i in $\mathcal{BG}_{n,m}(n, \mathcal{P}_{(m)})$ is equal to $n_i(m) = P_i\, n(m) + \omega_i'(m)\sqrt{n(m)}$ where $\omega_i'(m)$ are functions which tend to infinity arbitrary slowly as $m \to \infty$. Also, if we consider all the vertices with degree i in $\mathcal{BG}_{n,m}(n, \mathcal{P}_{(m)})$, only few of them would choose the same set of neighbors (properties). More precisely, we can show that if $n_i(m)$ vertices have degree i in $\mathcal{BG}_{n,m}(n, \mathcal{P}_{(m)})$, then the number of edges related to them in the random hypergraph $\mathcal{H}\left(\mathcal{BG}_{n,m}(n, \mathcal{P}_{(m)})\right)$ would be equal to $n_i(m) - \omega_i''(m)\, n_i(m)^2/\binom{m}{i}$ a.a.s. where $\omega_i''(m)$ are functions which tend to infinity arbitrary slowly as $m \to \infty$. Hence, a.a.s. the number of edges in $\mathcal{H}\left(\mathcal{BG}_{n,m}(n, \mathcal{P}_{(m)})\right)$ containing exactly i vertices is $P_i\; n(m) + \omega_i(m)\, \max\{\sqrt{n(m)}\,,\, (P_i\, n(m))^2/\binom{m}{i}\}$, where $\omega_i(m)$ are functions which tend to infinity arbitrary slowly as $m \to \infty$. In fact, the structure of this random hypergraph does not differ much from the structure of a random hypergraph having exactly $n_i(m) = P_i\, n(m)$ edges with i vertices (for more details see Lemma 5.11 in Rybarczyk (2005)). This leads to the standard random hypergraph model $\mathcal{H}(\leq r, m, \underline{n})$ (where $\underline{n} = (n_1, n_2, \ldots, n_r)$ is a given vector of non-negative integers), a hypergraph that is chosen uniformly at random from all hypergraphs on m vertices containing exactly n_i edges of size i for all $1 \leq i \leq r$. Thus, to finish the proof of Theorem 1 it is enough to determine conditions for the connectedness of $\mathcal{H}(\leq r, m, \underline{n})$. More precisely, we need generalized results concerning the connectedness of the second standard random hypergraph model $\mathcal{H}(\leq r, m, \underline{p})$ (where $\underline{p} = (p_1, p_2, \ldots, p_r)$ is a vector of probabilities for the existence of edges), a hypergraph on m vertices such that for all $1 \leq i \leq r$ each edge with i vertices is included independently of all others in $\mathcal{H}(\leq r, m, \underline{p})$ with probability p_i, together with theorems giving conditions for the equivalence of random hypergraphs $\mathcal{H}(\leq r, m, \underline{p})$ and $\mathcal{H}(\leq r, m, \underline{n})$.

A theorem about the connectedness of r-uniform hypergraphs was proved by Kordecki (see Kordecki (1985)). A similar approach leads to a more general result on the connectedness of $\mathcal{H}(\leq r, m, \underline{p})$ (see Rybarczyk (2005)). We get for any hypergraph $\mathcal{H}(\leq r, m, \underline{p})$, for which $\underline{p} = (p_1, p_2, p_3, \ldots, p_r)$ and $p_j(m) = a_j(m)\,(j-1)!\,(\ln m + c_j(m) + o(1))\, n^{-(j-1)}$ holds, where $0 \leq a_j(m) \leq 1$ for $j = 2, \ldots, r$ and $\sum_{j=2}^{r} a_j(m) = 1$, $\sum_{j=2}^{r} a_j(m) c_j(m) = c + o(1)$, that

$$\lim_{m\to\infty} \Pr\{\mathcal{H}(\leq r, m, \underline{p}) \text{ is connected}\} = e^{-e^{-c}}.$$

Connectedness is a monotone hypergraph property. Thus, if $n_i = p_i\binom{m}{i}$ for all $1 \leq i \leq r$, then the probability of connectedness of both $\mathcal{H}(\leq r, m, \underline{p})$ and $\mathcal{H}(\leq r, m, \underline{n})$ tends to the same limit as $m \to \infty$ (for precise conditions of the equivalence of random r-uniform hypergraphs see Janson et al. (2001), for the equivalence of $\mathcal{H}(\leq r, m, \underline{p})$ and $\mathcal{H}(\leq r, m, \underline{n})$ see Rybarczyk (2005)). The equivalence of random hypergraphs together with the equivalence of $\mathcal{H}(\leq r, m, \underline{n})$ and $\mathcal{H}\left(\mathcal{BG}_{n,m}(n, \mathcal{P}_{(m)})\right)$ gives us conditions for the connectedness of random intersection graphs $\mathcal{IG}^{pas}(m, \mathcal{P}_{(m)})$. Because $\mathcal{IG}^{pas}(m, \mathcal{P}_{(m)})$ is connected if and only if $\mathcal{H}\left(\mathcal{BG}_{n,m}(n, \mathcal{P}_{(m)})\right)$ is connected it finishes the proof of Theorem 1. Theorem 2 is a simple corollary from Theorem 1. If $\mathcal{IG}^{pas}(m, \mathcal{P}_{(m)})$ is connected and $P_0 = 0$, then $\mathcal{IG}^{act}(n, \mathcal{P}_{(m)})$ is connected.

To prove Theorem 3, we also use Lemmas 1 and 2 and equivalence theorems. In Schmidt-Pruzan and Shamir (1985), a theorem concerning the size of the largest component in $\mathcal{H}(\leq r, m, \underline{p})$ was proved. Because "having at least k vertices in the largest component" and "having at most k vertices in the largest component" both are monotone properties, we can use all equivalence theorems mentioned above together with results from Schmidt-Pruzan and Shamir (1985) to estimate the sizes of the components in $\mathcal{H}\left(\mathcal{BG}_{n,m}(n, \mathcal{P}_{(m)})\right)$. From Facts 1 and 2 we know that the sizes of the largest components in $\mathcal{H}\left(\mathcal{BG}_{n,m}(n, \mathcal{P}_{(m)})\right)$ and $\mathcal{IG}^{pas}(m, \mathcal{P}_{(m)})$ are the same. The size of the largest component in $\mathcal{IG}^{act}(n, \mathcal{P}_{(m)})$ would be at most r times smaller (r is a constant, bounding from above the number of vertices included in an edge) than the size of the corresponding component of $\mathcal{H}\left(\mathcal{BG}_{n,m}(n, \mathcal{P}_{(m)})\right)$ since each vertex $v \in \mathcal{V}$ of $\mathcal{IG}^{act}(n, \mathcal{P}_{(m)})$ relates to $|\Gamma(v)| \leq r$ vertices in the component of $\mathcal{H}\left(\mathcal{BG}_{n,m}(n, \mathcal{P}_{(m)})\right)$. This leads to the assertion of the theorem.

4 Conclusions

In this paper, we studied the probability of connectedness of active and passive random intersection graphs under the assumption that an edge arises if the respective sets of neighbors have a non-empty intersection in the generating random bipartite graph $\mathcal{BG}_{n,m}\left(n, \mathcal{P}_{(m)}\right)$ (the case $s = 1$). Assuming independent choices of vertex degrees in the set $\mathcal{V}$ of objects and of neighbors from the set $\mathcal{W}$ of properties, we gave threshold functions for the connectivity of active and passive random intersection graphs $\mathcal{IG}^{act}(n, \mathcal{P}_{(m)})$ and $\mathcal{IG}^{pas}(m, \mathcal{P}_{(m)})$ for $m \to \infty$ and $n = n(m)$ and for specific distributions $\mathcal{P}_{(m)} = (P_0, P_1, \ldots, P_m)$ of the vertex degrees in $\mathcal{V}$, namely for those with $P_k = 0$ for $k = r+1, r+2, \ldots, m$ (for some fixed r). We also included results on the size of the largest component in such random graphs (under the same assumptions). For the proofs, relations between the generating random bipartite graphs and hypergraphs were used.

Bipartite graphs provide the user with a procedure to detect clusters in sets of non-metric data where we have two distinct sets of vertices, a set $\mathcal{V}$ of active objects and a set $\mathcal{W}$ of passive properties. We can define object clusters as the components of the active intersection graph derived from the data. Similarly, if we want to check a set of non-metric data for a structure between the properties, then we can define property clusters as subgraphs of the passive intersection graph. If we are looking for a "real" cluster structure in a data set, then we can use the corresponding probability models for active or passive intersection graphs to derive test statistics to test the hypothesis of randomness of the outlined clusters for both, clusters of objects and clusters of properties. Connectedness is an important property in the "evolution" of random graphs. For finding clusters and for testing a cluster structure against the hypothesis of homogeneity, this property and the (minimal) number of edges which is necessary to get a connected graph with positive probability can be used. This number should be big enough so that we can expect not too many components (clusters) if the data set is not homogeneous; and for homogeneous data, we expect only one single connected component. Thus, the distribution of the number of components or of isolated vertices can be used as a test statistic in such a case (see Godehardt and Jaworski (2002)).

Acknowledgement. J. Jaworski acknowledges the support by the Marie Curie Intra-European Fellowship No. 501863 (RANDIGRAPH) within the 6th European Community Framework Programme. This work also had been supported by the Deutsche Forschungsgesellschaft (grant no. 436 POL 17/3/06).

References

BOCK, H.-H. (1996): Probabilistic Models in Cluster Analysis. *Computational Statistics and Data Analysis, 23, 5–28.*

GODEHARDT, E. (1990): *Graphs as Structural Models.* Vieweg, Braunschweig.

GODEHARDT, E. and JAWORSKI, J. (1996): On the Connectivity of a Random Interval Graph. *Random Structures and Algorithms, 9, 137–161.*

GODEHARDT, E. and JAWORSKI, J. (2002): Two Models of Random Intersection Graphs for Classification. In: M. Schwaiger and O. Opitz (Eds.): *Exploratory Data Analysis in Empirical Research.* Springer, Berlin, 68–81.

JANSON, S., ŁUCZAK, T. and RUCIŃSKI, A. (2001): *Random Graphs.* Wiley, New York.

JAWORSKI, J., KAROŃSKI, M. and STARK, D. (2006): The Degree of a Typical Vertex in Generalized Random Intersection Graph Models. *Discrete Mathematics, 306, 18, 2152–2165.*

JAWORSKI, J. and PALKA, Z. (2002): Remarks on a general model of a random digraph. *Ars Combinatoria, 65, 135–144.*

JAWORSKI, J. and SMIT, I. (1987): On a Random Digraph. *Annals of Discrete Mathematics, 33, 111–127.*

KAROŃSKI, M., SCHEINERMAN, E.R. and SINGER-COHEN, K.B. (1999): On Random Intersection Graphs: The Subgraph Problem. *Combinatorics, Probability and Computing, 8, 131–159.*

KORDECKI, W. (1985): On the Connectedness of Random Hypergraphs. *Commentaciones Mathematicae, 25, 265–283.*

RYBARCZYK, K. (2005): *O pewnych zastosowaniach hipergrafów losowych.* MA Thesis, Faculty of Mathematics, Adam Mickiewicz University, Poznań.

SCHMIDT-PRUZAN, J. and SHAMIR, E. (1985): Component Structure in the Evolution of Random Hypergraphs. *Combinatorica, 5.1, 81–94.*

Optimized Alignment and Visualization of Clustering Results

Martin Hoffmann, Dörte Radke and Ulrich Möller

Junior Research Group Bioinformatics / Pattern Recognition, Leibniz Institute for Natural Products Research and Infection Biology - Hans Knöll Institute, Beutenbergstr. 11a, D-07745 Jena, Germany; martin.hoffmann@hki-jena.de

Abstract. The grouping of data by clustering generally depends on the clustering method used and its specific parameter settings. Therefore, the comparison of results obtained from different clusterings is generally recommended (ensemble clustering). The present study presents a simple and an optimized method for visualizing such results by drawing a two-dimensional color map that associates data with cluster memberships. The methodology is applicable to any unsupervised and supervised classification results.

1 Introduction

The clustering of objects is a frequent task in data analysis. The resulting object classification generally depends on the clustering method used and its specific parameter settings (e.g. the number of clusters). Different clusterings may capture different aspects of the data and the comparison of different clustering results (ensemble clustering) is generally recommended in order to arrive at an object grouping that best reflects the underlying data structure.

Torrente et al. (2005) have presented a method for comparing and visualizing two clusterings, one flat (prototype) and one hierarchical clustering. The commercially available tool ArrayMiner (Optimal Design) offers the comparison of up to nine clusterings. Each class is displayed as a rectangle and two classes of successive clusterings are connected by lines representing the objects in their intersection. Associated lists of objects can be displayed on mouse click, but there is no general view with reference to the individual objects. The overall match between two clusterings can be assessed by summary statistics like the adjusted Rand index used in Monti et al. (2003). However, such measures are averages over all objects and are not informative about possibly consistently grouped subsets.

This study presents two methods for generating a complete two-dimensional color map that associates objects with cluster memberships maintaining full reference to the individual data objects. This enables an intuitive and detailed analysis of stable and unstable object subgroups. Both methods sort the original class labels in order to simplify the graphical representation. The proposed optimized method improves over the simple method by further defragmenting the color map. This can considerably facilitate result interpretation.

2 Methods

The color map generation is designed as a two step process. First, the matrix of class labels $\boldsymbol{c} = \{c_{ij}\}$ obtained by clustering data objects D_i $(i = 1, \ldots, n)$ using clustering methods C_j $(j = 1, \ldots, m)$ is reordered and denoted by $\hat{\boldsymbol{c}}$ (clusterings resulting from different parameter settings of a single method are treated as different clustering methods). Second, the class labels $\hat{c}_{ij}$ are assigned to specific colors $\hat{M}_{ij}$ that constitute the color map $\hat{\boldsymbol{M}}$. The overall procedure thus reads: $\boldsymbol{c} \rightarrow$ reordering $\rightarrow \hat{\boldsymbol{c}} \rightarrow$ coloring $\rightarrow \hat{\boldsymbol{M}}$.

The present study assumes a fixed column order. This is appropriate in applications with a natural order of the clusterings (e.g. with respect to an increasing number of classes) or a given priority ranking. The coloring problem is straightforward and different methods were implemented to fit the user's needs. The specific method used for generating the figures of this manuscript aims at using contrasting colors for neighboring classes in one column and assigning the same color to maximally overlapping classes in successive columns.

The simple and optimized sorting methods both determine the row order by successively including columns from the right, i.e. the rows are first ordered with respect to column 1, then with respect to column 2, etc. This implies a perfect ordering for the first column and usually a decreasing order towards the last column.

Simple sorting. This ordering method successively sorts the data with respect to their class label vectors $\boldsymbol{c}_{i\cdot} = (c_{i1}, \ldots, c_{im})$. An illustrating example of a class label matrix $\boldsymbol{c} = \{c_{ij}\}$ before and $\hat{\boldsymbol{c}}_s = \{c_{s(i)j}\}$ after simple sorting is shown in Fig. 1 ($\boldsymbol{s} = \{s(i)\}$ denotes the permutation vector of simple sorting).

Optimized sorting. Simple sorting is suboptimal in that the class labels are usually not as contiguous as they could be. The proposed optimized sorting method rearranges the data in order to account for this potential improvement. The class label matrix $\hat{\boldsymbol{c}}_o = \{c_{o(i)j}\}$ resulting from optimized sorting is shown on the right hand side of Fig. 1 ($\boldsymbol{o} = \{o(i)\}$ denotes the associated data permutation vector).

Permutations. The contiguity of the class labels for clustering B achieved by optimized sorting is a result of i) rearranging the classes of clustering A and ii) rearranging the intersections of each class in A with classes in B. As an example, in Fig. 1 the A-ordering of $\hat{\boldsymbol{c}}_{\boldsymbol{o}}$ is $A2$ $A3$ $A1$. $A2$ intersects with $B3$ and $B5$. $A3$ intersects with $B4$ and $B1$. And $A1$ intersects with $B1, B2$, and $B4$. The $B1$ and $B4$ intersections of $A3$ are rearranged ($B4$ to top, $B1$ to bottom) in order to match the top $B1$ end of $A1$. Thus, optimized sorting involves permutations of the A-classes and, for each such A-permutation, the

$$
\begin{array}{cc} A & B \end{array} \qquad \begin{array}{cc} A & B \end{array} \qquad \begin{array}{cc} A & B \end{array} \quad \begin{array}{ccc} \bar{b} & \underline{b} & I \end{array}
$$

$$
\boldsymbol{c} = \begin{pmatrix} 2 & 5 \\ 3 & 1 \\ 2 & 3 \\ 1 & 2 \\ 3 & 4 \\ 3 & 1 \\ 2 & 3 \\ 1 & 4 \\ 1 & 1 \\ 1 & 4 \end{pmatrix} \quad \hat{\boldsymbol{c}}_S = \begin{pmatrix} 1 & 1 \\ 1 & 2 \\ 1 & 4 \\ 1 & 4 \\ 2 & 3 \\ 2 & 3 \\ 2 & 5 \\ 3 & 1 \\ 3 & 1 \\ 3 & 4 \end{pmatrix} \quad \hat{\boldsymbol{c}}_O = \begin{pmatrix} 2 & 3 \\ 2 & 3 \\ 2 & 5 \\ 3 & 4 \\ 3 & 1 \\ 3 & 1 \\ 1 & 1 \\ 1 & 2 \\ 1 & 4 \\ 1 & 4 \end{pmatrix} \begin{array}{ccc} 1 & 1 & 1 \\ 1 & 1 & 1 \\ 1 & 1 & 2 \\ 2 & 2 & 3 \\ 2 & 3 & 4 \\ 2 & 3 & 4 \\ 2 & 4 & 5 \\ 2 & 5 & 6 \\ 2 & 5 & 7 \\ 2 & 5 & 7 \end{array}
$$

Fig. 1. Class labels for $n = 10$ objects and $m = 2$ clusterings A and B with 3 and 5 classes, respectively. Class label matrix $\boldsymbol{c}$ before sorting, $\hat{\boldsymbol{c}}_s$ after simple sorting, and $\hat{\boldsymbol{c}}_o$ after optimized sorting. The class labels are almost contiguous after optimized sorting, except for the fourth label 4 of the second classification B. The permutation vector of simple sorting is $\boldsymbol{s} = (9, 4, 8, 10, 3, 7, 1, 2, 6, 5)$, that of optimized sorting $\boldsymbol{o} = (3, 7, 1, 5, 2, 6, 9, 4, 8, 10)$. The rightmost columns $\bar{b}$, $\underline{b}$, and I indicate the 2 superblocks, 5 subblocks, and the intersection clustering, respectively, created by optimized sorting.

permutations of the respective B-class intersections. For each combined AB permutation the result is evaluated by an objective function.

Blockings. The first alignment step is to sort the data with respect to the class labels of the first clustering. The contiguous A-classes are then preserved during future alignments, i.e. permutations are allowed within each A-class only. The A-classes are said to be blocked. The new contiguous $B1$-segment formed in the AB alignment spans across $A3$ and $A1$ (Fig. 1). Therefore, $A3$ and $A1$ constitute a superblock ($\bar{b}_2$) that fixes their order. $A2$ is a superblock on its own and both superblocks can still be permuted to improve future alignments. In addition to superblockings, subblockings must be introduced that prevent e.g. the top part of $B1$ to be permuted with its bottom part, which would destroy the A-blocking. The purpose of blockings is thus to preserve contiguous segments of class labels during the alignment of the respective next clustering. They provide a bookkeeping for the available degrees of permutational freedom. The intersection clustering I of A and B, i.e. the clustering that contains the serially numbered intersections between classes in A and classes in B (Fig. 1), becomes the A-clustering of the next step and the newly to be aligned classification is denoted by B. Permutations of the A-classes, which were unrestricted in the first alignment, must now conform to the blocking structure given by $\bar{b}$ and $\underline{b}$.

Objective function. Optimized sorting ranks candidate alignments according to five criteria (applied to the current B clustering): 1) the number of class label fragments (i.e. the number segments with contiguous class labels), 2) the overall number of data in the respective main fragments (one main fragment for each class label), 3) the number of superblocks, 4) the number of subblocks, and 5) the cumulative weighted distance of the scattered fragments to their respective main fragments (using the scattered fragment size for weighting). These criteria were used to rank candidates by first sorting with respect to criterion 1, then with respect to criterion 2, etc. The best candidates were selected for the next alignment. This approach implements Pareto optimality with lexicographic order for the best compromise solution. The criteria 1, 2, and 5 were also used to assess the improvement of optimized over simple sorting in the results section.

3 Data

The present study used publicly available gene expression microarray data and some simulated data. Most data were taken from the data collection provided by Monti et al. (2003) (supplementary material): Uniform1, Gaussian1, Gaussian5, Leukemia (Golub et al. (2002)), Novartis multi-tissue (Su et al. (2002)), St. Jude leukemia (Yeoh et al. (2002)), Lung cancer (Bhattacharjee et al. (2001)), CNS tumors (Pomeroy et al. (2002)), Normal tissues (Ramaswamy et al. (2001)) (full references to the original articles can be found in Monti et al. (2003)). The cell cycle data were published by Cho et al. (2003), the infection data by Boldrick et al. (2002), and the cancer cell line data by Scherf et al. (2000). Five data sets were simulated by the authors (point distributions from Gaussian mixture models in two dimensions).

4 Results

The alignment and visualization methods of the present study are demonstrated using the experimental and simulated data sets described in the previous section. Depending on the intention of the respective experiments either genes or samples (patients) were taken as data objects.

The simple and optimized row sorting methods described in Section 2 are first demonstrated for the cell cycle data of Cho et al. (2003). The data (genes) were clustered using the fuzzy c-means clustering algorithm with an increasing number of clusters. Figure 2 shows the resulting color maps after sorting with respect to the first column only, after simple sorting, and after optimized sorting. Clearly, the best representation is achieved by using optimized sorting. An application to the classification of samples (patients) is shown in Fig. 3 for the leukemia data of Yeoh et al. (2002) (taken from Monti et al. (2003)). Again, optimized sorting improves over simple sorting.

The improvement of optimized sorting over simple sorting depends on the data set and the clustering method used. It was quantified using 21 ensembles of clusterings with an increasing number of clusters (like those shown in Figs. 2 and 3) obtained by clustering 17 different data sets (Section 3). Figure 4 shows the mean and standard deviation of the relative improvement as a function of the number of clusters. The decrease in the number of fragments and the

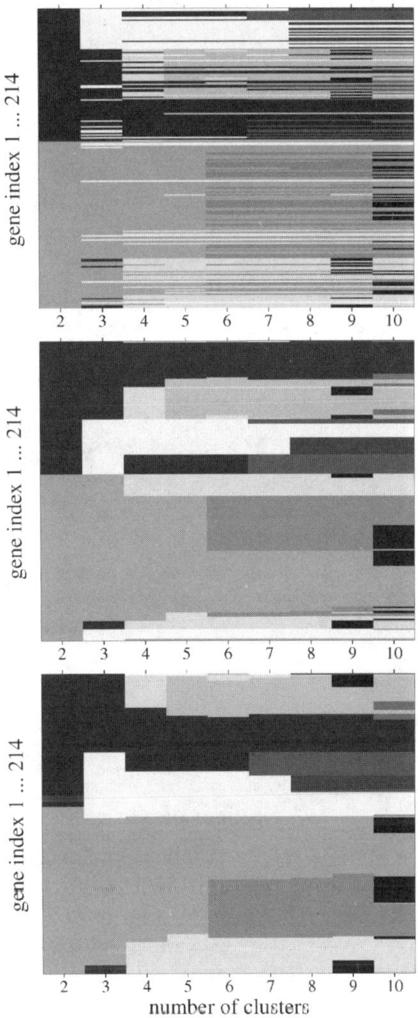

Fig. 2. Color maps for the class labels resulting from clustering the yeast cell cycle data of Cho et al. (2003) (214 selected genes) for an increasing number of clusters. The color maps were obtained by sorting the rows with respect to the class labels of the first clustering only (upper panel) or by applying the simple sorting (mid-panel), or optimized sorting (lower panel) methods. Simple sorting recursively sorts class labels for each clustering from left to right. Optimized sorting improves over simple sorting by grouping class label fragments together. This is best seen for the white (lower panel: middle), light gray (lower panel: bottom), and black (lower panel: above white) class labels. For the cell cycle data a class number of 4 or 5 is expected depending on whether in addition to the G_1, S, G_2, and M cell cycle phases a separate M/G_1 phase is assumed or not. The clarity of the color maps is increasing from top to bottom and analyzing the results is clearly easiest using optimized sorting. Colors were transformed to grayscales for printing.

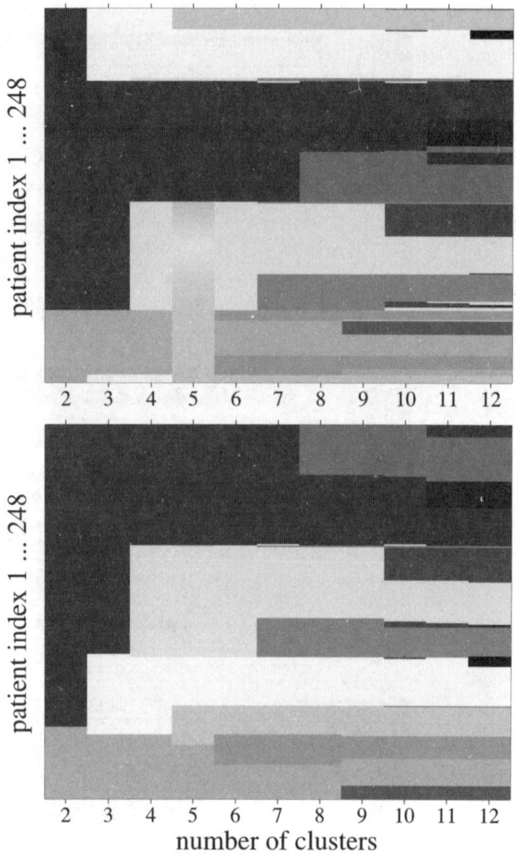

Fig. 3. Color maps for the class labels resulting from clustering the pediatric acute lymphoblastic leukemia (ALL) data of Yeoh et al. (2002) (taken from Monti et al. (2003)) (248 pediatric ALL patients). The structure of the data is clearer for optimized sorting (lower panel) compared to simple sorting (upper panel). Mainly, this is due to the defragmentation of the white (lower panel: lower middle), light gray (lower panel: below white), and gray (lower panel: below light gray) class labels. The number of expected classes is 6 for the leukemia data corresponding to the ALL subtypes T-ALL, E2A-PBX1, BCR-ABL, TEL-AML1, MLL, and hyperdiploid>50. Clearly, the 5-cluster result does not fit to the neighboring clusterings and from 6 clusters onwards the clusters are basically subdivided (hierarchical cluster structure) supporting the 6 ALL subtypes. This is also best recognized using the optimized sorting method (lower panel). Colors were transformed to grayscales for printing.

increase in the number of data in the main fragments were strictly non-negative in all cases. The decrease in weighted distance was negative in some cases. Thus, the application of optimized sorting always improved the visual representation of the results with respect to the first two criteria, the number of fragments and the main fragment size.

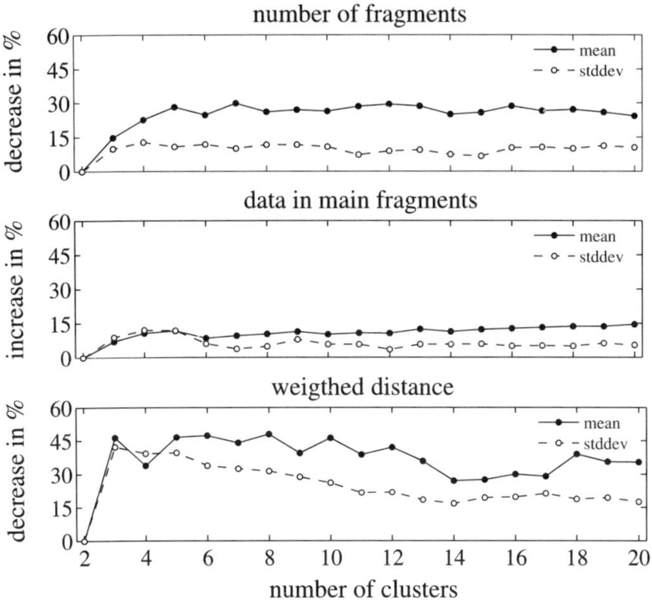

Fig. 4. Improvement of optimized over simple sorting as a function of the number of clusters. The changes are measured relative to simple sorting. The curves show the means (solid) and standard deviations (dashed) across 21 clustering ensembles obtained by clustering 17 different data sets (Section 3). The reduction in the number of fragments is 25-30% for most cluster numbers. The increase in the data in the main fragments is 10-15%. The decrease in the weighted distance is most effective for 5 to 12 clusters (40-45%) and otherwise between 25 and 40%.

5 Conclusions and future research

This study presented a simple and an optimized method for drawing an object-versus-class-label color map that facilitates the comparison of different clustering results. Apart from clustering, the proposed methodology is applicable to any classification problem for which a comparison of the results obtained from different classifiers is of importance. The results of this study showed that the proposed optimized sorting method improves over the simple method by performing an effective defragmentation of the color map. The improvement relative to the simple sorting method was 10 - 45% depending on the fragmentation measure and the data set. Indeed, the results of both methods are identical if the classifications are hierarchically organized, as e.g. in hierarchical linkage clustering.

The proposed visualization method may become a useful application for a broad range of classification problems. It provides easy and informative visualization of results obtained from classifier ensembles by giving a detailed view of the classification results and maintaining reference to the individual data objects. This study demonstrated the working of the proposed methodology using clustering with an increasing number of clusters. Instead, different clustering algorithms, distance measures, and parameter settings can be used. Evaluating the results of these methods is very important especially for classification in post-genomics. This is work in progress. Also, different approaches for determining an optimized column order are currently investigated in our group.

Acknowledgments

This work was supported by the German Federal Ministry of Education and Research (BMBF) through the Jena Centre for Bioinformatics (JCB, grant FKZ 0312704D to R. Guthke).

References

BOLDRICK, J.C., ALIZADEH, A.A., DIEHN, M., DUDOIT, S., LIU, C.L., BELCHER, C.E., BOTSTEIN, D., STAUDT, L.M., BROWN, P.O. and RELMAN, D.A. (2002): Stereotyped and Specific Gene Expression Programs in Human Innate Immune Responses to Bacteria, *Proc. Natl. Acad. Sci. U.S.A., 99, 2, 972–977.*

CHO, R.J., CAMPBELL, M.J., WINZELER, E.A., STEINMETZ, L., CONWAY, A., WODICKA, L., WOLFSBERG, T.G., GABRIELIAN, A.E., LANDSMAN, D., LOCKHART, D.J. and DAVIS R.W. (1998): A Genome-Wide Transcriptional Analysis of the Mitotic Cell Cycle, *Molecular Cell, 2, 1, 65–73.*

MONTI, S., TAMAYO, P., MESIROV, J. and GOLUB, T. (2003): Consensus Clustering: A Resampling-Based Method for Class Discovery and Visualization of Gene Expression Microarray Data, *Machine Learning, 52, 91–118.*

SCHERF, U., ROSS, D.T., WALTHAM, M., SMITH, L.H., LEE, J.K., TANABE, L., KOHN, K.W., REINHOLD, W.C., MYERS, T.G., ANDREWS, D.T., SCUDIERO, D.A., EISEN, M.B., SAUSVILLE, E.A., POMMIER, Y., BOTSTEIN, D., BROWN, P.O. and WEINSTEIN, J.N. (2000): A Gene Expression Database for the Molecular Pharmacology of Cancer, *Nature Genetics, 24, 236–244.*

TORRENTE, A., KAPUSHESKY, A. and BRAZMA, A. (2005): A New Algorithm for Comparing and Visualizing Relationships Between Hierarchical and Flat Gene Expression Data Clusterings, *Bioinformatics, 21, 21, 3993–3999.*

YEOH, E.-J., ROSS, M.E., SHURTLEFF, S.A., WILLIAMS, W.K., PATEL, D., MAHFOUZ, R., BEHM, F.G., RAIMONDI, S.C., RELLING, M.V., PATEL, A., CHENG, C., CAMPANA, D., WILKINS, D., ZHOU, X., LI, J., LIU, H., PUI, C.-H., EVANS,W.E., NAEVE, C., WONG, L. and DOWNING, J.R. (2002): Classification, Subtype Discovery, and Prediction of Outcome in Pediatric Acute Lymphoblastic Leukemia by Gene Expression Profiling, *Cancer Cell, 1, 2, 133–143.*

Finding Cliques in Directed Weighted Graphs Using Complex Hermitian Adjacency Matrices

Bettina Hoser[1] and Thomas Bierhance[2]

[1] Information Services and Electronic Markets,
Universität Karlsruhe (TH), Germany; bettina.hoser@em.uni-karlsruhe.de
[2] sd&m AG, Stuttgart, Germany; thomas.bierhance@sdm.de

Abstract. The objective of this paper is to present the adaptation of the well known class of spectral graph partitioning algorithms to a new class of adjacency matrices. By the use of complex Hermitian adjacency matrices for asymmetric weighted digraphs and the subsequent application of an enhanced spectral partitioning algorithm, a better understanding of patterns within such digraphs becomes possible.

This approach was used to find cliques within online communities. To validate our approach we benchmarked against existing implementations of spectral partitioning algorithms. The major result of our research is the application of spectral partitioning in an asymmetric communication environment.

The practical implication of our work is the broadening of the use of a spectral partitioning algorithm to a new class of adjacency matrices that are able to model asymmetric weighted communication streams such as email exchange or market transactions.

1 Introduction

Analysing the structure of graphs is a common methodology to obtain insights into real world phenomena. In particular, graph partitioning is one method that is commonly used in different scientific disciplines including Very Large Scale Integration (VLSI) layouts, parallel job scheduling, image segmentation and social network analysis.

Usually, graph partitioning problems are NP-hard (Garey and Johnson (1979)), and while for small graphs ($\approx$ 100 vertices) exact solutions can be derived (Karisch et al. (2000)), the need for computationally efficient heuristics to solve such problems is evident. One class of heuristics that is widely used for partitioning graphs is called spectral partitioning. Many algorithms that use spectral partitioning have been developed to partition undirected graphs (Chan et al. (1993), Ng et al. (2002), Kannan et al. (2004), Choe and

Park (2004)). While partitioning digraphs has been of interest in recent research (see Makarychev (2006)) spectral partitioning has not been applied to digraphs as yet.

In this paper we will show that weighted digraphs can be partitioned by a spectral algorithm that uses the eigensystem of the graph's Hermitian adjacency matrix. We will show that this method can be used to reveal groups in real world social networks.

2 Spectral partitioning: Related work

Graphs can be partitioned by analysis of the eigensystem of their adjacency matrix. This method is called spectral partitioning. One of the starting points for spectral partitioning was the work realized by Fiedler (1970). He found that the second smallest eigenvalue λ_{n-1} of the graph's Laplacian matrix is a measure for a graph's algebraic connectivity. Now, the eigenvector corresponding to λ_{n-1} is the solution for a relaxed version of the minimum cut problem (Seary and Richards (1995)). Since these early works by Fiedler several algorithms have been proposed for spectral partitioning. There are mainly two different types of algorithms: those that derive k-partitions by doing recursive bipartitioning (e.g. Kannan et al. (2004) or Choe and Park (2004)) and those that directly derive k-partitions (e.g. Chan et al. (1993), Ng et al. (2002) or Alpert and Yao (1995)). Applying recursive bipartitioning can be problematic - there are examples where this method can not derive the optimal partition (Guattery and Miller (1998)) and even the first bipartition can exclude the optimal solution (Simon and Teng (1997)). The algorithms that derive k-partitions directly normally use k eigenvectors from the eigensystem (Chan et al. (1993), Ng et al. (2002)). An exception is the work by Alpert and Yao (1995) - they state that one should use as much eigenvectors as possible.

3 Clustering in Hilbert space

3.1 Hilbert space and spectral analysis of Hermitian adjacency matrices

In this section we introduce the notation and basic facts about complex numbers, Hilbert space and eigensystems of Hermitian matrices. For proofs and further reference please see Hoser and Geyer-Schulz (2005).

The complex number z can be represented in algebraic form or equivalently in exponential form $z = a + ib = |z|e^{i\phi}$ with the real part of z being denoted as $Re(z) = a$, the imaginary part as $Im(z) = b$, the absolute value as $|z| = \sqrt{a^2 + b^2}$, and the phase as $0 \leq \phi = \arccos \frac{Re(z)}{|z|} \leq \pi$, with i the imaginary unit ($i^2 = -1$). $\overline{z} = a - ib$ denotes the complex conjugate of z.

The notation will be as follows: Column vectors will be written in bold face $\mathbf{x}$, with the components $x_j, j = 1 \dots n$. Matrices will be denoted as capital letters A with a_{kl} representing the entry in the k-th row and the l-th column. Greek letters will denote eigenvalues with λ_k representing the k-th eigenvalue. The complex conjugate transpose of a vector $\mathbf{x}$ will be denoted as $\mathbf{x}^*$. The transpose of a vector $\mathbf{x}$ will be denoted as $\mathbf{x}^t$.

The outer product of two vectors $\mathbf{x}$ and $\mathbf{y}$ is defined as:

$$\mathbf{x}\mathbf{y}^* = \begin{pmatrix} x_1\overline{y_1} & \dots & x_1\overline{y_n} \\ \dots & \dots & \dots \\ x_n\overline{y_1} & \dots & x_n\overline{y_n} \end{pmatrix} \tag{1}$$

The inner product of $\mathbf{x}, \mathbf{y} \in \mathbb{C}^n$ is a defined as $\langle \mathbf{x} \mid \mathbf{y} \rangle = \mathbf{x}^*\mathbf{y} = \sum_{k=1}^{n} \overline{x_k} y_k$. The norm $\| \mathbf{x} \|$ will be defined as follows:

$$\sqrt{\langle \mathbf{x} \mid \mathbf{x} \rangle} = \| \mathbf{x} \| \tag{2}$$

Hilbert space is a complete normed inner product space with the norm defined as in Eq. 2.

A matrix H is called Hermitian, if and only if $H^* = H$ with H^* representing the conjugate complex transpose of H. This means that the matrix entries can be written as $h_{lk} = \overline{h_{kl}}$. Hermitian matrices are normal $HH^* = H^*H$.

A digraph is denoted by $G = (V, E)$ where $V = \{v_1, v_2, ..., v_n\}$ is the set of the graph's vertices and $E \subseteq V^2$ is the set of its edges. Assigning a real number to each one of the edges yields a weighted graph. The weights can be expressed by a function $c : E(G) \mapsto \mathbb{R}$. The weighted digraph can be represented by its adjacency matrix $A = (a_{ij})$ where $a_{ij} = c((v_i, v_j))$ if $(v_i, v_j) \in E(G)$ and $a_{ij} = 0$ otherwise. We consider only simple graphs. Thus $a_{ii} = 0$ for all i.

While the adjacency matrix of an undirected graph is symmetric and its eigenvalues are real, this is generally not true for digraphs. In the directed case the adjacency matrix is not necessarily symmetric and the eigenvalues can be complex. However, the standard adjacency matrix of digraph can be transformed to a Hermitian matrix. First, a complex valued adjacency matrix $A_{\mathbb{C}} \in \mathbb{C}^n$ is constructed by $A_{\mathbb{C}} = A + i \cdot A^t$. This matrix can then be rotated as in Eq.3 to obtain the Hermitian adjacency matrix H.

$$H = A_{\mathbb{C}} \cdot e^{-i\frac{\pi}{4}} \tag{3}$$

All eigenvalues of a Hermitian matrix are real and the set of all eigenvectors is orthogonal. Since all eigenvalues of a Hermitian matrix are real the interpretation of the eigenvalues does not pose any difficulty. The eigenvalues sorted by their absolute value $|\lambda_1| \geq \dots \geq |\lambda_n|$ help to identify the dominant substructures for interpretation. In addition it can be shown that $\| H \|^2 = \sum_{k=1}^{n} \lambda_k^2$ the sum of all eigenvalues defines the total variation σ^2 contained in H, thus the partial sum $\sum_{k=1}^{M} \lambda_k^2$ reveals the variation contained by all eigenspaces up to the M-th eigenspace.

For a complex Hermitian matrix with $h_{kk} = 0, \forall k$ some eigenvalues have to be negative due to the fact that $tr(H) = \sum_{k=1}^{n} h_{kk} = \sum_{k=1}^{n} \lambda_k = 0$

We identify the most central vertex in a graph by its absolute value $|x_{max,m}|$ of the eigenvector component corresponding to the largest eigenvalue $|\lambda_{max}|$. This also holds for the most central vertices in each substructure identified by the eigenvectors. These central vertices will be used later in the paper to initialize the cluster centres.

3.2 Spectral graph partitioning in Hilbert Space

The steps needed to partition a digraph consisting of n vertices into k partitions by using the Hermitian adjacency matrix' eigensystem are similar to the usual spectral partitioning for undirected graphs - mainly a dataset containing n k-dimensional vectors is built from the graph's eigensystem which is then partitioned by a clustering algorithm like the K-means algorithm. We propose the following algorithm (later on referred to as $A2$) for weighted digraphs:

1. Build the graph's complex adjacency matrix $A_{\mathbb{C}}$
2. Derive the Hermitian adjacency matrix $H = A_{\mathbb{C}} \cdot e^{i\frac{\pi}{4}}$
3. Calculate the eigenvectors $\mathbf{x_1}, \mathbf{x_2}, \ldots, \mathbf{x_k}$ corresponding to the k largest eigenvalues $\lambda_1 \geq \lambda_2 \geq \ldots \geq \lambda_k$.
4. Normalize all eigenvectors.
5. Build the matrix $R = \left(\lambda_1\mathbf{x_1}\middle|\lambda_2\mathbf{x_2}\middle|\ldots\middle|\lambda_l\mathbf{x_k}\right)$
6. Obtain $Y = (y_{ij})$ from the normalized rows of $R = (r_{ij})$: $y_{ij} = r_{ij}/||\mathbf{r_i}||$
7. Take each row $\mathbf{y_i}$ of Y as a point in $\mathbb{C}^k$.
8. Initialize the cluster centres $\mathbf{c_1}, \ldots, \mathbf{c_k}$ with $\mathbf{c_j} = \mathbf{y_i}$ such that $|x_{ij}| = max_l |x_{lj}|$.
9. Perform the cluster analysis using the K-means algorithm. Use the natural distance in Hilbert space which corresponds due to Eq. 2 to the Euclidean distance $d(\mathbf{y_i}, \mathbf{y_j}) = ||\mathbf{y_i} - \mathbf{y_j}||$.

When partitioning graphs that contain subgroups of very different size this basic algorithm will not perform very well. We found that the performance can be vastly improved by using $l > k$ eigenvectors. The cluster centres $\mathbf{c_1}, \ldots, \mathbf{c_k}$ should then be initialized with $\mathbf{c_i} = \mathbf{y_k}$ such that $\underset{j=1\ldots i}{max} \langle \mathbf{y_k}, \mathbf{c_j} \rangle$ is minimized. We will refer to this modification as $A3$.

4 Results

We compared our algorithm to the standard graph partitioning algorithm METIS (Karypis and Kumar (1998)) and the spectral graph partitioning algorithm by Ng et al. (2002). As these two can only partition undirected graphs we derived an adjacency matrix $A' = A + A^t$ prior to applying these two algorithms.

4.1 Simulation results

We performed several benchmarks using synthetic graphs and compared the results of our algorithm against the results of the algorithm by Ng et al. (NJW) and METIS. We combined star graphs and complete graphs of different size to an entire graph and added different amounts of noise. We then compared the ratio cut-size (see e.g. Chan et al. (1993)) of the resulting partitions. We also compared the number wrong assignments based on the original subgraphs. In Figure 1 the results for graphs containing three subgraphs with 8 vertices are shown. While there is almost no difference in the ratio cut-size the proposed algorithm $A2$ (as well as $A3$) performed slightly better than the others when comparing the number of wrong assignments.

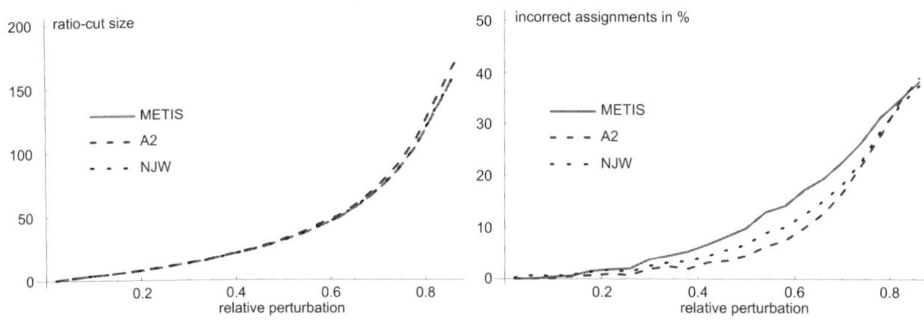

Fig. 1. Comparison with synthetic star graphs

Another interesting scenario was a set that contained graphs with three subgraphs of different size (one with 16, two with 4 vertices each). The METIS algorithm performed worse than the other two in this scenario, as it tries to return partitions of equal size. The algorithm NJW outperformed our algorithm $A2$.

This mainly happens because choosing the eigenvectors corresponding to the k largest eigenvalues is not always optimal. An example can be seen in Figure 2. It can be seen that the first and the third eigenvector both belong to the clique containing 16 vertices while the second and the fourth eigenvector belong to the cliques with 4 vertices.

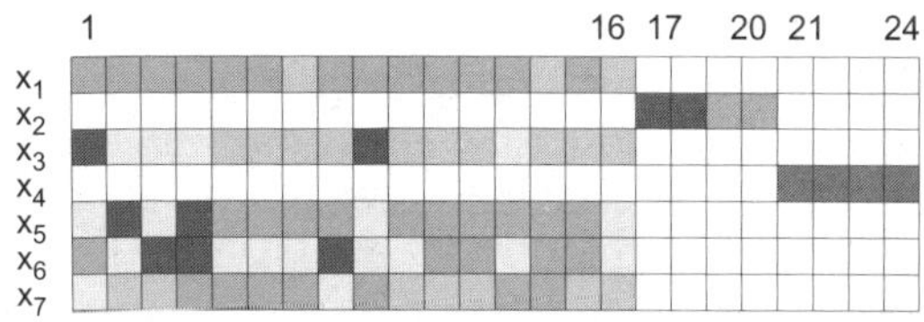

Fig. 2. Absolute values of the seven largest eigenvectors of a graph consisting of three cliques - one with 16 and two with 4 vertices each

Partitioning this graph using the first three eigenvectors will not yield the desired result. This problem is overcome by using the algorithm $A3$ that uses $l > k$ eigenvectors when constructing the matrix R. $A3$ yielded the best results in all of our simulations and outperformed METIS as well as NJW (see Figure 3).

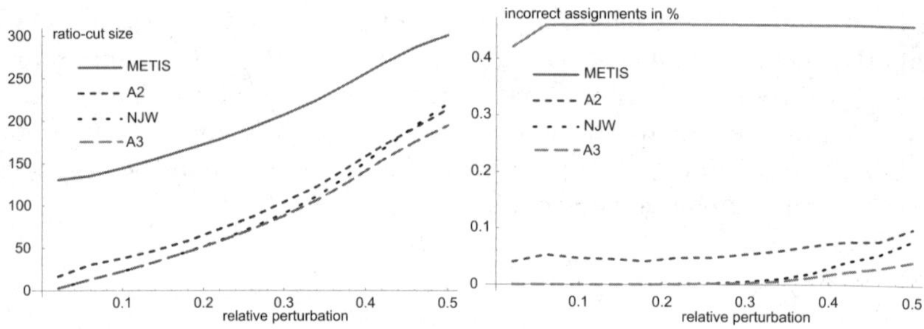

Fig. 3. Comparison with synthetic complete graphs of different size

4.2 Results with real data

As an example for a real graph we chose the social network arising from Usenet newsgroups. Each vertex represents a participant in one of these newsgroups and each vertex is connected by a directed edge if its assigned participant wrote a message in response to the message of another participant. The edges are weighted by the number of messages. The resulting graph is clearly a digraph as a participant p_1 might have replied to messages by p_2 while p_2 never replied to messages by p_1.

Our dataset consists of newsgroup messages from an open-source community (`news://news.eclipse.org`). We collected messages from three newsgroups, namely eclipse.platform.swt (SWT), eclipse.tools.jdt (JDT) and eclipse.tools.gef (GEF). The data was collected between 1st of July 2004 and 7th of February 2005. After pre-processing the data (see Bierhance (2004)) a digraph consisting of 2064 vertices was constructed as explained above. Each of the participants was assigned to one of the three newsgroups according the number of messages written to each newsgroup. Due to crosspostings six participants could not be clearly assigned to only one of the newsgroups. As can be seen in Figure 4(a) the resulting graph is highly connected - 671 participants wrote messages to more than one newsgroup.

We partitioned the graph using our algorithm $A2$, the algorithm by Ng et al. (NJW) and METIS. Each result was compared to the partitions from the original assignment (resulting from the messages by each participant to each newsgroup). The partitions derived by $A2$ matched 93.07% of the original assignment, while METIS matched 81.73% and NJW only 62.24%. The result was only slight improved by algorithm $A3$ when being compared to the result by $A2$, because the subgroups were of roughly equal size.

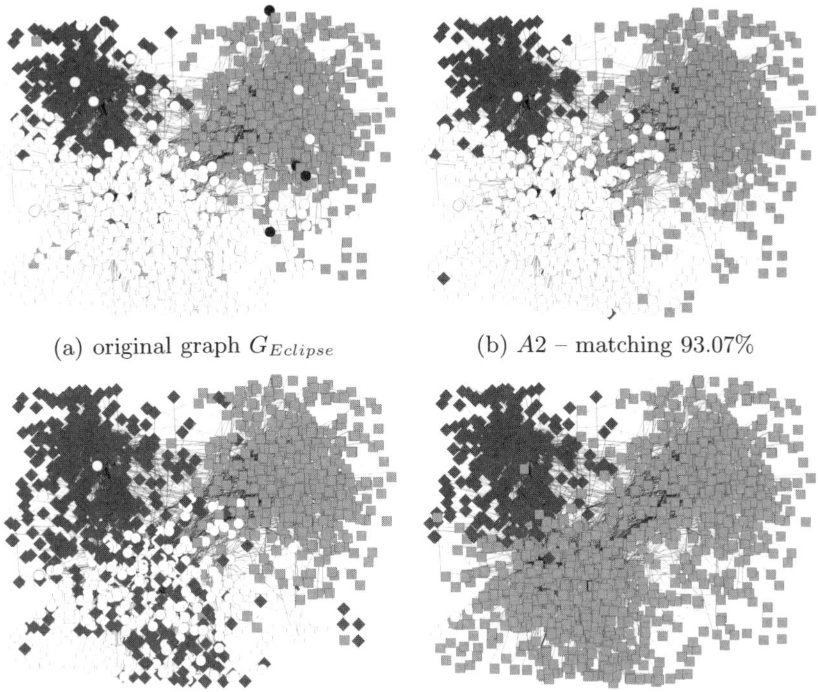

(a) original graph $G_{Eclipse}$ (b) $A2$ – matching 93.07%

(c) p-METIS – matching 81.73% (d) NJW – matching 62.24%

Fig. 4. Graph $G_{Eclipse}$ und 3-partitions by $A2$, p-METIS und NJW

5 Conclusion and further research

As was shown in 1 graph partitioning is performed successfully on undirected weighted graphs. In some instances though the information about direction of relationship is so important that it should not get lost. One example is the communication behaviour in a newsgroup or in any other network where communication can flow in both directions between any two vertices, but can be asymmetric with respect to direction. It is than necessary to discern if a cut between these two vertices can be done or not. Consider a case where the one direction between two vertices has only a medium level of communication while the reverse direction has a heavy flow of communication. How best should one decide if to cut or not? Information about direction becomes necessary. One major problem of partitioning is to define metrics that allow for measuring the distances between vertices such that an assignment to a cluster can be performed (Makarychev et al. (2006)). Hilbert space allows for a metric that includes directional information. The complex adjacency matrix described here encodes this directional behaviour. The method described yields valid results as could be shown in section 4. We hope to further improve the accuracy of this approach by taking a deeper look into the characteristics of the eigensystem both from the view on the spectrum and on the properties of the eigenvector components. We strongly believe that an approach can be found to define the optimal number of clusters, their initial centres and their final configuration from these characteristics alone.

References

ALPERT, C.J. and YAO, S. (1995): Spectral Partitioning: The More Eigenvectors, the Better. In: *Proc. of the 32nd ACM/IEEE Conf. on Design Automation*, ACM Press, 195–200.

BIERHANCE, T. (2004): *Partitionierung von gerichteten Graphen durch Eigensystemanalyse komplexwertiger hermitischer Adjazenzmatrizen.* Master's thesis. Can be obtained from Universitt Karlsruhe (TH), Information Services and Electronic Markets, Kaiserstrasse 12, 76128 Karlsruhe, Germany.

CHAN, P.K., SCHLAG, M.D.F. and ZIEN, J.Y. (1993): Spectral K-way Ratio-cut Partitioning and Clustering. In: *Proc. of the 30th Int. Conf. on Design Automation*, ACM Press, 749–754.

CHOE, T. and PARK, C. (2004): A K-way Graph Partitioning Algorithm Based on Clustering by Eigenvector. In: M. Bubak, G.D. van Albada, P.M.A. Sloot et al. (Eds.): *ICCS 2004: 4th Int. Conf., Proceedings, Part II, Volume 3037 of Lecture Notes in Computer Science*, Springer, Heidelberg, 598–601.

FIEDLER, M. (1970): Algebraic Connectivity of Graphs. *Czechoslovak Mathematical Journal, 23, 298–305.*

GAREY, M.R. and JOHNSON, D.S. (1979): *Computers and Intractability.* W. H. Freeman.

GEYER-SCHULZ, A. and HOSER, B. (2005): Eigenspectralanalysis of Hermitian Adjacency Matrices for the Analysis of Group Substructures. *Journal of Mathematical Sociology, 29, 4, 265–294.*

GUATTERY, S. and MILLER, G.L. (1998): On the Quality of Spectral Separators. *SIAM Journal on Matrix Analysis and Applications, 19, 3, 701–719.*

KANNAN, R., VEMPALA, S. and VETTA, A. (2004): On Clusterings: Good, Bad and Spectral. *Journal of the ACM, 51, 3, 497–515.*

KARISCH, S.E., RENDL, F. and CLAUSEN, J. (2000): Solving Graph Bisection Problems with Semidefinite Programming. *INFORMS Journal on Computing, 12, 3, 177–191.*

KARYPIS, G. and KUMAR, V. (1998): Multilevel K-way Partitioning Scheme for Irregular Graphs. *Journal of Parallel and Distributed Comp., 48, 1, 96–129.*

MAKARYCHEV, Y., CHARIKAR, M. and MAKARYCHEV, K. (2006): Directed Metrics and Directed Graph Partitioning Problems. In: *SODA '06: Proc. of the 17th Annual ACM-SIAM Symp. on Discrete Algorithm, 51–60.*

NG, A.Y., JORDAN, M.I. and WEISS, Y. (2002): On Spectral Clustering: Analysis and an Algorithm. In: S. Becker, T.G. Dietterich and Z. Ghahramani (Eds.): *Advances in Neural Inform. Proc. Systems Volume 14*, MIT Press, 849–856.

SEARY, A.J. and RICHARDS, W.D. (1995): Partitioning Networks by Eigenvectors. In: M. Everett and K. Rennolls (Eds.): *Proc. of the Int. Conf. on Social Networks, Volume 1: Methodology, 47–58.*

SIMON, H.D. and TENG, S. (1997): How Good is Recursive Bisection? *SIAM Journal on Scientific Computing, 18, 5, 1436–1445.*

Text Clustering with String Kernels in R

Alexandros Karatzoglou[1] and Ingo Feinerer[2]

[1] Department of Statistics and Probability Theory,
Technische Universität Wien, A-1040 Wien, Austria; alexis@ci.tuwien.ac.at

[2] Department of Statistics and Mathematics,
Wirtschaftsuniversität Wien, A-1090 Wien, Austria; h0125130@wu-wien.ac.at

Abstract. We present a package which provides a general framework, including tools and algorithms, for text mining in R using the S4 class system. Using this package and the `kernlab` R package we explore the use of kernel methods for clustering (e.g., kernel k-means and spectral clustering) on a set of text documents, using string kernels. We compare these methods to a more traditional clustering technique like k-means on a bag of word representation of the text and evaluate the viability of kernel-based methods as a text clustering technique.

1 Introduction

The application of machine learning techniques to large collections of text documents is a major research area with many applications such as document filtering and ranking. Kernel-based methods have been shown to perform rather well in this area, particularly in text classification with SVMs using either a simple "bag of words" representation (i.e., term frequencies with various normalizations) (Joachims (1999)), or more sophisticated approaches like string kernels (Lodhi et al. (2002)), or word-sequence kernels (Cancedda et al. (2003)). Despite the good performance of kernel methods in classification of text documents, little has been done in the field of clustering text documents with kernel-based methods. Kernel based clustering techniques such as kernel k-means and spectral clustering have been shown to perform well, particularly in separating non-convex clusters, while string kernels provide a novel way of mapping text into a feature space. It is thus interesting to test and compare the performance of this clustering techniques on a collection of text documents.

2 Software

R (R Development Core Team (2006)) is a natural choice for a text mining environment. Besides the basic string and character processing functions it includes an abundance of statistical analysis functions and packages and provides a Machine Learning task view with a wide range of software.

2.1 `textmin` R package

The `textmin` package provides a framework for text mining applications within R. It fully supports the new S4 class system and integrates seamlessly into the R architecture. The basic framework classes for handling text documents are:

TextDocument: Encapsulates a text document, irrelevant from its origin, in one class. Several slots are available for additional metadata, like an unique identification number or a description.

TextDocumentCollection: Represents a collection of text documents. The constructor provides import facilities for common data formats in text mining applications, like the Reuters21578 news format or the Reuters Corpus Volume 1 format.

TermDocumentMatrix: Stands for a term-document matrix with documents as rows and terms as columns. Such a term-document matrix can be easily built from a text document collection. A bunch of weighting schemes are available, like binary, term frequency or term frequency inverse document frequency. This class can be used as a fast representation for all kinds of bag-of-words text mining algorithms.

2.2 `kernlab` R package

`kernlab` is an extensible package for kernel-based machine learning methods in R. The package contains implementations of most popular kernels, and also includes a range of kernel methods for classification, regression (support vector machine, relevance vector machine), clustering (kernel k-means, spectral clustering), ranking, and principal component analysis.

3 Methods

The k-means clustering algorithm is one of the most commonly used clustering methods providing solid results but also having the drawback that it cannot separate clusters that are not linearly separable in input space. Therefore clusters found by k-means often are of poor performance as text documents with several thousands dimensions, i.e., terms, are highly non-linear.

3.1 Kernel k-means

One technique for dealing with this problem is mapping the data into a high-dimensional non-linear feature space with the use of a kernel. Kernel k-means uses a kernel function to compute the inner product of the data in the feature space. All computations are then expressed in terms of inner products thus allowing the implicit mapping of the data into this feature space. Denoting clusters by π_j and a partitioning of points as ${\pi_j}_{j=1}^k$ and if Φ is the mapping function then the k-means objective function using Euclidean distances becomes

$$\mathcal{D}({\pi_j}_{j=1}^k) = \sum_{j=1}^{k} \sum_{a \in \pi_j} \|\Phi(a) - m_j\|^2 \ , \tag{1}$$

where $m_j = \frac{1}{\|\pi_j\|} \sum_{a \in \pi_j} \Phi(a)$ and in the expansion of the square norm only inner products of the form $\langle \Phi(a), \Phi(b) \rangle$ appear which are computed by the kernel function $k(a, b)$. Unlike the normal k-means algorithm kernel k-means can separate non-convex cluster regions. Nevertheless all dimensions are considered important which leaves room for improvement, e.g., by considering only the eigenvector space.

3.2 Spectral clustering

Spectral clustering (Ng et al. (2001), Shi and Malik (2000)) works by embedding the data points of the partitioning problem into the subspace of the k largest eigenvectors of a normalized affinity matrix. The use of an affinity matrix also brings one of the advantages of kernel methods to spectral clustering, since one can define a suitable affinity for a given application. In our case we use a string kernel to define the affinities between two documents and construct the kernel matrix. The data is then embedded into the subspace of the largest eigenvectors of the normalized kernel matrix. This embedding usually leads to more straightforward clustering problems since points tend to form tight clusters in the eigenvector subspace. Using a simple clustering method like k-means on the embedded points usually leads to good performance.

3.3 String kernels

String kernels (Watkins (2000), Herbrich (2002)) are defined as a similarity measure between two sequences of characters x and x'. The generic form of string kernels is given by the equation

$$k(x, x') = \sum_{s \sqsubseteq x, s' \sqsubseteq x'} \lambda_s \delta_{s,s'} = \sum_{s \in A^*} \mathrm{num}_s(x) \mathrm{num}_s(x') \lambda_s \ , \tag{2}$$

where A^* represents the set of all non empty strings and λ_s is a weight or decay factor which can be chosen to be fixed for all substrings or can be set to a different value for each substring. This generic representation includes a large number of special cases, e.g., setting $\lambda_s \neq 0$ only for substrings that start and end with a white space character gives the "bag of words" kernel (Joachims (2002)). In this paper we will focus on the case where $\lambda_s = 0$ for all $|s| > n$ that is comparing all substrings of length less than n. This kernel will be referred to in the rest of the paper as full string kernel. We also consider the case where $\lambda_s = 0$ for all $|s| \neq n$ which we refer to as the string kernel.

4 Experiments

We will now compare the performance of the various clustering techniques on text data by running a series of experiments on the Reuters text data set.

4.1 Data

The Reuters-21578 dataset (Lewis (1997)) contains stories for the Reuters news agency. It was compiled by David Lewis in 1987, is publicly available and is currently one of the most widely used datasets for text categorization research. We used the "crude" topic with about 580 documents, the "corn" category with 280 documents and a sample of 1100 documents from the "acq" category. Our dataset consists of 1720 documents after preprocessing: We removed stop words, empty documents, punctuation and with space, converted all characters to lower case and performed stemming using the `Rstem` (Temple Lang (2005)) `Omegahat` R package.

4.2 Experimental setup

We perform clustering on the dataset using the kernel k-means and spectral clustering methods in the `kernlab` package and the k-means method in R. For the kernel k-means and spectral methods we also use the string kernels implementations provided in `kernlab`. In order to learn more about the effect of the string kernels hyper-parameters on the clustering results we run the clustering algorithms over a range of the length parameter n which controls the length of the strings compared in the two character sets and the decay factor λ. We study the effects of the parameters by keeping the value of the decay parameter λ fixed and varying the length parameter. Note that for each parameter set a new kernel matrix containing different information has to be computed. We use values from $n = 3$ to $n = 14$ for the length parameter and $\lambda = 0.2$, $\lambda = 0.5$ and $\lambda = 0.8$ for the decay factor. We also use both the string (or spectral) and the full string kernel and normalize in order to remove any bias introduced by document length. We thus use a new embedding $\hat{\phi} = \frac{\phi(s)}{\|\phi(s)\|}$ which gives rise to the kernel

$$\hat{K}(s,s') = \langle \hat{\phi}(s), \hat{\phi}(s') \rangle = \left\langle \frac{\phi(s)}{\|\phi(s)\|} \frac{\phi(s')}{\|\phi(s')\|} \right\rangle = \quad (3)$$

$$\frac{\langle \phi(s), \phi(s') \rangle}{\|\phi(s)\|\|\phi(s')\|} = \frac{K(s,s')}{\sqrt{K(s,s)K(s',s')}} . \quad (4)$$

For the classical k-means method we create a term-document matrix of the term frequencies and also an inverse term frequencies matrix.

4.3 Performance measure

We evaluate the performance of the various clustering techniques using the recall rate as the actual labels of the clustered data are known. Given a discovered cluster γ and the associated reference cluster Γ, recall R is defined as $R = \sum_{\Gamma=1}^{k} n_{\gamma\Gamma} / \sum_{\Gamma=1}^{k} N_{\Gamma}$, where $n_{\gamma\Gamma}$ is the number of documents from reference cluster Γ assigned to cluster γ, N_{Γ} is the total number of documents in cluster γ and N_{Γ} is the total number of documents in reference cluster Γ.

4.4 Results

The main goal of these experiments is to test if kernel methods along with string kernels are a viable solution for grouping text documents. From the experiments we ran it became obvious that the λ parameter influences the performance only minimally and thus we chose to look at the results in relation to the string length kernel parameter which has a more profound influence on the performance of the kernel-based clustering methods. The performance of the k-means clustering method is also very similar with both the simple document matrix and the inverse frequency document matrix. Figure 1 shows the average recall rate over 10 runs for the spectral clustering methods, and the kernel k-means method with the full string kernel compared to the reference recall rate of the inverse term document matrix clustered with a simple k-means algorithm. The plot shows that both the spectral method and kernel k-means fail to improve over the performance of the standard k-means clustering. We also note that the spectral clustering provides very stable results with almost zero variance. This can be attributed to the fact that the projection of the data into the eigenspace groups the data into tight clusters which are easy to separate with a standard clustering technique. Figure 2 displays the average recall rate of the kernel k-means and the spectral clustering method with a string kernel along with the standard k-means clustering results. It is clear that for a range of values of the string length parameter the kernel k-means functions outperforms k-means clustering and the full string kernel methods with a full string kernel. The method does not provide stable performance and the variance of the recall rate over the 10 runs is quite high. The high variance of the kernel k-means algorithm can be attributed to the fact that string kernels map the text documents into a high-dimensional feature space which increases the sensitivity of the k-means algorithm to the initial random starting points. Spectral clustering is clearly the best performing clustering method for this text document set and exhibits interesting behavior. For small lengths of substrings $(3, 4, 5)$ the performance increases monotonically and at 6 hits a threshold. For values between 6 and 10 the performance increase is much smaller and for the value of 10 the highest recall rate of 0.927 is reached. For higher values of the length parameter the performance drops sharply only to increase again for a string length value of 14. This method is very stable and exhibits minimal variance due to the fact that it projects the data into a k-dimensional eigenvector space where the data form tight clusters.

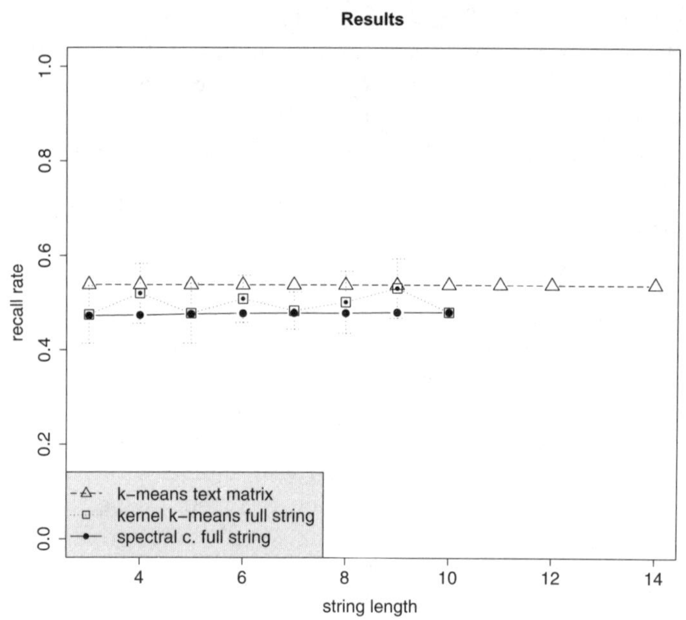

Fig. 1. Average recall rate over 10 runs for the spectral clustering, kernel k-means, with full string kernels and k-means on an inverse frequencies term matrix methods. On the y axis is the recall rate and the x axis the string length hyper-parameter of the string kernel.

4.5 Timing

We have also evaluated the methods in terms of running time. The experiments were run on a cluster of Linux machines with reference 2.6 GHz Pentium 4 CPUs. The following table provides the averaged running time for the calculation of a full kernel matrix and the averaged running time for the clustering methods. Note that the running time for the kernel-based clustering methods is the time needed to cluster data with a precomputed kernel matrix. From the results it is clear that most of the computing time is spent on the calculation of the kernel matrix.

kernel matrix calculations	≈ 2 h.
spectral clustering	≈ 20 sec.
kernel k-means	≈ 30 sec.
term matrix k-means	≈ 40 sec.

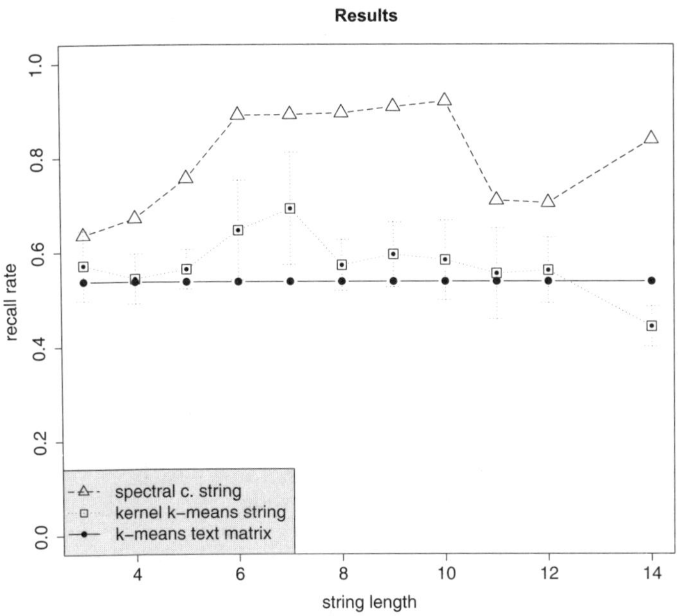

Fig. 2. Average recall rate over 10 runs for the kernel k-means and spectral clustering, with string kernels and k-means on an inverse frequency term matrix. y axis depicts the recall rate and x is the string length hyper-parameter of the string kernel.

5 Conclusions

Our results show that the spectral clustering technique combined with a string kernel outperforms all other methods and provides strong performance comparable to the classification performance of an SVM with a string kernel on a similar dataset (Lodhi et al. (2002)). This is encouraging and shows that kernel-based clustering methods can be considered as a viable text grouping method. The behavior of the kernel-based algorithms, particularly of the spectral clustering method, strongly depends on the value of the string length parameter. It is an open question if the range of good values of this parameter (6–10) on this dataset can be used on other text datasets in the same or other languages to provide good performance. It is interesting to note that a string length of 6 to 10 characters corresponds to the size of one or two words in the English language. It would be interesting to study the behavior for string lengths higher than 14. The good performance of the spectral clustering technique could be an indication that graph partitioning methods combined with string kernels could provide good results on text clustering. One drawback of the kernel based methods is the amount of time spent on the kernel matrix computation and, particularly for the spectral methods, the necessity to store a full $m \times m$ matrix with m as the number of text documents, in memory. A suffix tree based implementation of string kernels as shown by Vishwanathan and Smola (2004) combined with the Nystrom method for computing the eigenvectors of the kernel matrix as by Fowlkes et al. (2004) in using only a sample of data points could provide a solution.

References

CANCEDDA, N., GAUSSIER, E., GOUTTE, C. and RENDERS, J.M. (2003): Word-sequence Kernels. *Journal of Machine Learning Research, 3, 1059–1082.*

FOWLKES, C., BELONGIE, S., CHUNG, F. and MALIK J. (2004): Spectral Grouping Using the Nystrom Method. *Transactions on Pattern Analysis and Machine Intelligence, 26, 2, 214–225.*

HERBRICH, R. (2002): *Learning Kernel Classifiers Theory and Algorithms.* MIT Press.

JOACHIMS, T. (1999): Making Large-scale SVM Learning Practical. In: *Advances in Kernel Methods — Support Vector Learning.*

JOACHIMS, T. (2002): *Learning to Classify Text Using Support Vector Machines: Methods, Theory, and Algorithms.* The Kluwer International Series In Engineering And Computer Science. Kluwer Academic Publishers, Boston.

LEWIS, D. (1997): Reuters-21578 Text Categorization Test Collection.

LODHI, H., SAUNDERS, C., SHAWE-TAYLOR, J., CRISTIANINI, N. and WATKINS, C. (2002): Text Classification Using String Kernels. *Journal of Machine Learning Research, 2, 419–444.*

NG, A., JORDAN, M. and WEISS, Y. (2001): On Spectral Clustering: Analysis and an Algorithm. *Advances in Neural Information Processing Systems, 14.*

R DEVELOPMENT CORE TEAM (2006): *R: A Language and Environment for Statistical Computing.* R Foundation for Statistical Computing, Vienna, Austria.

SHI, J. and MALIK, J. (2000): Normalized Cuts and Image Segmentation. *Transactions on Pattern Analysis and Machine Intelligence, 22, 8, 888–905.*

TEMPLE LANG, D. (2005): *Rstem: Interface to Snowball Implementation of Porter's Word Stemming Algorithm.* R Package Version 0.2-0.

VISHWANATHAN, S. and SMOLA, A. (2004): Fast Kernels for String and Tree Matching. In: K. Tsuda, B. Schölkopf and J.P. Vert (Eds.): *Kernels and Bioinformatics.* MIT Press, Cambridge.

WATKINS, C. (2000): Dynamic Alignment Kernels. In: A.J. Smola, P.L. Bartlett, B. Schölkopf and D. Schuurmans (Eds.): *Advances in Large Margin Classifiers.* MIT Press, Cambridge, 39–50.

Automatic Classification of Functional Data with Extremal Information

Fabrizio Laurini and Andrea Cerioli

Dipartimento di Economia, Sezione di Statistica, Università di Parma, 43100 Parma, Italy; {fabrizio.laurini, andrea.cerioli}@unipr.it

Abstract. In this paper we develop statistical techniques for clustering smooth functional data based on their extremal features. Smooth functional data arise in many application fields. Our work is motivated by a problem in quality control monitoring of water supplied for human consumption, where both the *level* and the *shape* of each function are important for classification purposes.

1 Introduction

This paper is motivated by a classification problem arising in the quality monitoring of water intended for human consumption. Italian Water Utilities (WU for short) fulfill European regulations (see European Community (1998)) in order to guarantee that a number of quality requirements are satisfied by the drinking water that they provide to their customers. In particular, WU must guarantee that microbiological and chemical parameters, constituting a potential human health risk, lie below specified thresholds. For that purpose, WU are required to monitor how the quality of water intended for human consumption evolves over time, by setting up appropriate control programmes and sampling plans. When the quality standards are not attained, WU must take suitable corrective action in order to restore such standards. Cluster analysis is an important aid to these control programmes, as it can help to identify groups of sampling sites with similar microbiological and chemical trends. Detection of homogeneous clusters from the point of view of pollutant concentration and vulnerability to human health risk can considerably improve water quality management. It can result either in more efficient non-proportional sampling designs, where the available resources are allocated according to water quality at each site, or to financial savings. However, as is shown below, standard classification tools are not adequate for the purpose of detecting homogeneous trends of pollutant concentration.

To focus the discussion, we now introduce the data to be analyzed in §4; they are fully described in Cerioli et al. (2001). The data refer to NO_3 con-

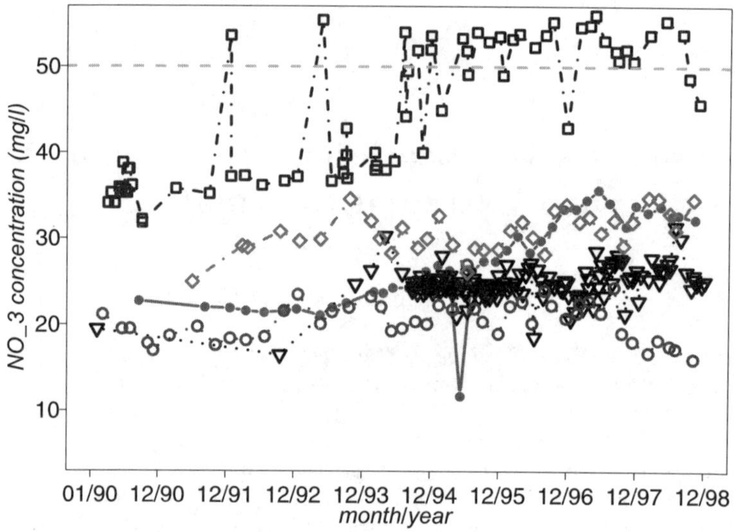

Fig. 1. Data on NO_3 monitoring from 5 wells in the period January 1990–December 1998. The dashed horizontal line is the maximum admissible concentration.

centration (in mg/l) monitored at 66 sample sites (wells) located in northern Italy, over a period of almost nine years. Water pollution from NO_3 is of great concern in northern Italy, due to the high development of farming and agricultural activities. The wells under study lie in the same administrative area (*Provincia*) but have rather heterogeneous geological features, due to varying hydro-geologic conditions, and different depths. Nevertheless, some regularities in concentration levels and trends are expected. The maximum admissible concentration (MAC) of NO_3 is 50 mg/l. If the MAC is exceeded, prompt and expensive correction actions must be taken in order to reduce the pollutant level. This implies that NO_3 must be monitored carefully and frequently at wells showing an increasing trend toward the MAC, to ensure preventive diagnosis of potential MAC exceedances. On the other hand, the overall sampling cost may be reduced by decreasing the sampling frequency at wells where the NO_3 trend is of less concern. It is then clear that managerial strategies should be oriented to the study of temporal behaviour of NO_3 concentration, bearing in mind that both the *level* and the *shape* of the pollutant trend have a prominent role.

Figure 1 provides an example of the NO_3 quality data available at five wells. Figure 1 shows many of the statistical difficulties that arise when we want to classify such data. Firstly, the actual monitoring times can be irregularly spaced, thus preventing the application of classification methods for time-series models (see, e.g., Piccolo (1990)). Secondly, the sampling rate can vary considerably from well to well and for each well over time. For instance, the increased density of triangles in the second half of the time window is due to a field campaign designed for the corresponding well, during which data were collected weekly or bi-weekly. Finally, even if the sampling rate is the same for different wells, the NO_3 data need not be collected on the same day for operational reasons. All these reasons make standard multivariate methods, including cluster analysis, unsuited for the purpose of detecting homogeneous groups of wells according to their NO_3 level and trend.

In this paper we recognize that water quality monitoring is done by discrete-time sampling of a continuous phenomenon (in our problem NO_3 concentration). Hence, we regard the concentration data available for each well as providing information about an underlying continuous function, that has to be estimated. We base our clustering approach both on metric and non-metric features of the estimated continuous functions. For this purpose, we adopt some tools from the rapidly growing field of functional data analysis (Ramsay and Silverman (2005)). These tools are sketched in §2. In §2 we propose a new dissimilarity measure that allows for the trend shape and level of each function. Application to the NO_3 monitoring problem is given in §4.

2 Reconstruction of functional data

Let y_{ij} be the j-th observed NO_3 concentration at well i. Given T_i observations available at well i and a total of n units (wells), we assume that

$$y_{ij} = \mu_i(t_{ij}) + \varepsilon_i(t_{ij}) \qquad i = 1, \ldots, n; j = 1, \ldots, T_i, \tag{1}$$

where t_{ij} is the time when y_{ij} is observed, $\mu_i(\cdot)$ is a continuous smooth function of time representing the NO_3 trend at well i, and $\varepsilon_i(t_{ij})$ is independent noise contributing to the roughness of y_{ij}. Under this model, the observed data are thus obtained by sampling the unknown function $\mu_i(t)$, where t denotes time, at fixed points t_{ij}, $j = 1, \ldots, T_i$, plus random error. Note that neither the observation times t_{ij} nor their number T_i need to be the same for all units.

The continuous smooth function $\mu_i(t)$ is the "parameter" of interest under the above functional model. It is called the true functional form of unit i and must be estimated from the observed data y_{ij}. Each function is considered as a unique entity and its estimate, $\hat{\mu}_i(t)$ say, is the basis for subsequent analysis. Ramsay and Silverman (2005) provide an overview of several techniques for analyzing such functional data. In this paper we compute $\hat{\mu}_i(t)$ through penalized cubic spline regression, by the `R` software `SemiPar` (Ruppert et al. (2003)). We choose the amount of smoothing at each well by generalized cross-validation, a popular measure in the smoothing spline literature which has nice theoretical and computational features (Ramsay and Silverman (2005, §5.4.3)). Figure 2 shows the 66 estimated smooth NO_3 concentration functions $\hat{\mu}_i(t)$.

We are interested in clustering these estimated functions according to their level and shape. Trend level is obviously important in water quality management, as it measures closeness to the MAC of NO_3. The usual distance between

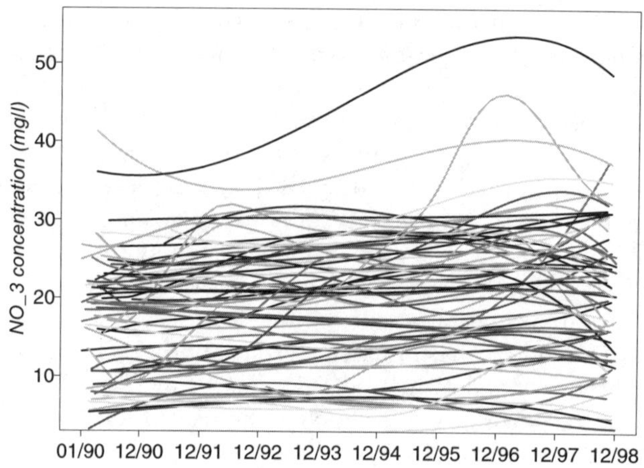

Fig. 2. Estimated functional forms of NO_3 concentration for all monitoring sites.

trend levels is the L_2 norm (Hall and Heckman (2002), Tarpey and Kinateder (2003), Ramsay and Silverman (2005, p. 21)). Consideration of trend shape and detection of upward peaks are also of considerable value for monitoring purposes, as they can highlight wells were quality is deteriorating. For this reason, metric information must be supplemented by shape information. A dissimilarity measure between functional data that is based on shape was proposed in Heckman and Zamar (2000). However, this measure was mainly intended for unimodal functions, which is not generally the case in our application. In §2, we suggest an extension of an alternative dissimilarity measure proposed in Cerioli et al. (2005). In our extension trend level is combined with shape and "true" features of $\hat{\mu}_i(t)$ are distinguished from "spurious" ones by restricting attention to the landmarks (extremes) of each function.

3 Combining trend and shape of functional data

We introduce a dissimilarity measure for functional data to be adopted as input for hierarchical classification. The proposed dissimilarity suitably takes into account both shape and level of estimated functions $\hat{\mu}_i(t)$. Thus, curves with different shape will have high dissimilarity. Similarly, curves with common shape but substantial differences in level will have large dissimilarity. A relevant feature of the proposed dissimilarity measure is that it can be computed for irregularly spaced observation times, as required in model (1).

3.1 Shape dissimilarity

The algorithm adopted for computing shape dissimilarity is similar to that of Cerioli et al. (2005). The dissimilarity is based on the actual times when a local extremal point occurs, and in our exposition we will focus on the local maxima of each curve. There is no loss of generality in this, but maxima are of major concern in water quality management. As shown by Gasser and Kneip (1995), local extrema, also called landmarks, are important tools for identifying the shape of a curve. The shape dissimilarity index is designed to collect only local maxima which have some statistical evidence of not being generated by chance alone. As discussed in Section 2, we fit a smooth function $\hat{\mu}_i(t)$ to the observed time series y_{ij}, $j = 1, \ldots, T_i$, and estimate its first derivative, say $\hat{\mu}'_i(t)$. When computing the first derivative we also consider its pointwise 95% confidence bands (Ruppert et al. (2003)). Local maxima correspond to the highest value of $\hat{\mu}_i(t)$ in the range where the confidence interval for $\hat{\mu}'_i(t)$ includes zero, after a period of significantly positive first derivative. Unlike the suggestion in Gasser and Kneip (1995) and in Cerioli et al. (2005), we allow consecutive maxima with no minimum in between due to presence of some flat zones. This is motivated by the behaviour of estimated NO_3 concentration curves, many of which are stationary for a long period and then exhibit a sudden sharp peak, after which the series tend to revert to stationarity, before another peak occurs.

Write $M_i = \{\tau_1^{(i)}, \ldots, \tau_{m_i}^{(i)}\}$ for the collection of time points where $\hat{\mu}_i(t)$ has its local maxima, and let $m_i = \#\{M_i\}$. We also adjust for boundary effects as follows. Define the lower and upper confidence bands for the derivative of $\mu_i(t)$ as $LB(\mu_i)$ and $UB(\mu_i)$, respectively. If $UB(\mu_i) < 0$ at the beginning of the observation period for unit i, then $\tau_1^{(i)} \equiv 1$; if $UB(\mu_i) > 0$ at the end of the observation period, then $\tau_{m_i}^{(i)} \equiv T_i$. If none of these conditions apply, no adjustment is needed. Moreover, we have $M_i = \emptyset$ in the case of a stationary function, i.e. if $\mu_i(t)$ does not have any local maxima.

Our shape dissimilarity measure between two smooth curves $\mu_i(t)$ and $\mu_l(t)$ is based on the comparison of the two sets of estimated local maxima, i.e. M_i and M_l, assuming that $M_i \neq \emptyset$ and $M_l \neq \emptyset$. Let $\tau_{*j}^{(l)}$ be the element of M_l which is closest to $\tau_j^{(i)}$, i.e.

$$\tau_{*j}^{(l)} = \{\tau_{j'}^{(l)} : |\tau_j^{(i)} - \tau_{j'}^{(l)}| = \min\} \qquad j = 1, \ldots, m_i$$

We first compute the average distance between maxima and scale it by the length of the window where $\hat{\mu}_i(t)$ and $\hat{\mu}_l(t)$ overlap, K_{il} say:

$$d_{il} = \frac{1}{K_{il} m_i} \sum_{j=1}^{m_i} |\tau_j^{(i)} - \tau_{*j}^{(l)}| \qquad i, l = 1, \ldots, n. \tag{2}$$

The dissimilarity $d_{il} = 0$ if $\hat{\mu}_i(t)$ and $\hat{\mu}_l(t)$ have the same shape and, clearly, $d_{ii} = 0$. Note that dissimilarity (2) is asymmetric. This is undesirable for a measure between two curves of similar importance. Hence we define our shape dissimilarity by taking the average

$$\delta_{il}^{(1)} = (d_{il} + d_{li})/2 \qquad i, l = 1, \ldots, n. \tag{3}$$

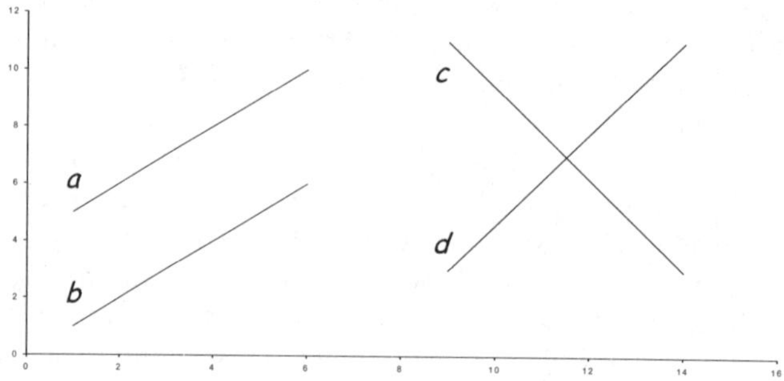

Fig. 3. Inadequacy of the L_2 norm: $\delta^{(2)}_{ab} = \delta^{(2)}_{cd}$, although shapes are very different.

3.2 Metric distance

The usual measure of distance between the continuous functions $\hat{\mu}_i(t)$ and $\hat{\mu}_l(t)$ is the L_2 norm (Ramsay and Silverman (2005, p. 21))

$$\delta^{(2)}_{il} = \left[\int \{\hat{\mu}_i(t) - \hat{\mu}_l(t)\}^2 dt \right]^{1/2} \qquad i, l = 1, \ldots, n.$$

and the integral is computed over the window where $\hat{\mu}_i(t)$ and $\hat{\mu}_l(t)$ overlap. We compute $\delta^{(2)}_{il}$ to allow for differences in the level of NO_3 contamination. However, it can easily be verified that $\delta^{(2)}_{il}$ does not take into account the shape of $\hat{\mu}_i(t)$ and $\hat{\mu}_l(t)$, being the same in both cases depicted in Figure 3.

3.3 Combined dissimilarity for shape and level

Our proposal is to combine metric and shape information about the dissimilarity between $\hat{\mu}_i(t)$ and $\hat{\mu}_l(t)$ through a Gower-type coefficient. This combined dissimilarity measure is defined as

$$\gamma_{il} = \frac{\sum_{j=1}^{2} w_j \delta^{(j)}_{il} / c_j}{\sum_{j=1}^{2} w_j} \tag{4}$$

where $w_j = 1$ if comparison is possible on dimension j, $j = 1, 2$, and $w_j = 0$ otherwise (for instance, if either $\hat{\mu}_i(t)$ or $\hat{\mu}_l(t)$ is a stationary function we have $w_1 = 0$). The scaling factor c_j is chosen to make $\delta^{(1)}_{il}$ and $\delta^{(2)}_{il}$ commensurable. For our purposes we have chosen $c_j = \max_{il}\{\delta^{(j)}_{il}\}$, so that the scaled dissimilarity $\delta^{(j)}_{il}/c_j \in [0, 1]$.

4 Application

In this section we show the performance of our combined dissimilarity measure (4) as a guide to the identification of useful clusters of wells according to

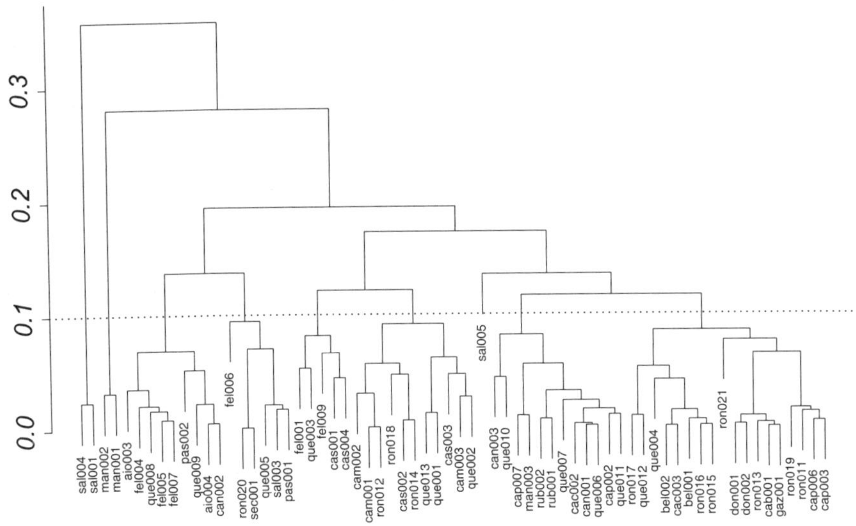

Fig. 4. Average linkage dendrogram from the combined index (4). The dotted line represents the cut at height 0.1.

NO_3 concentration trends. We use our measure as input for average-linkage clustering of the estimated functions of Figure 2. The resulting dendrogram is shown in Figure 4. The corresponding structure somehow averages those that are obtained by applying $\delta_{il}^{(1)}$ and $\delta_{il}^{(2)}$ separately (not shown).

It is possible to identify from Figure 4 several groups of relatively homogeneous wells. From inspection of the agglomeration algorithm, we choose to cut the dendrogram at dissimilarity level 0.1. This cut is represented by the horizontal dotted line in Figure 4. We correspondingly identify six homogeneous clusters and three small clusters which can be collectively thought of as "residual" clusters. Some instructive patterns for water quality management emerge from inspection of these clusters, two of which are detailed in Figure 5. This picture shows (on the same scale) estimated NO_3 trend concentrations for the wells classified in two of the main clusters. In both situations the average pollutant concentration is approximately the same, but managerial actions implied by our analysis are fairly different. The left-hand panel of Figure 5 shows functions which are relatively stable over time and decrease at the end of the observation window. The quality of water does not seem to be of great concern and the monitoring effort could thus be reduced at these sites. On the other hand, the right-hand panel of Figure 5 shows a cluster of curves exhibiting a much more oscillatory behaviour, some with more than one peak during the period and an increasing trend. Although the average of NO_3 concentration is similar, the corresponding trends are much more harm-

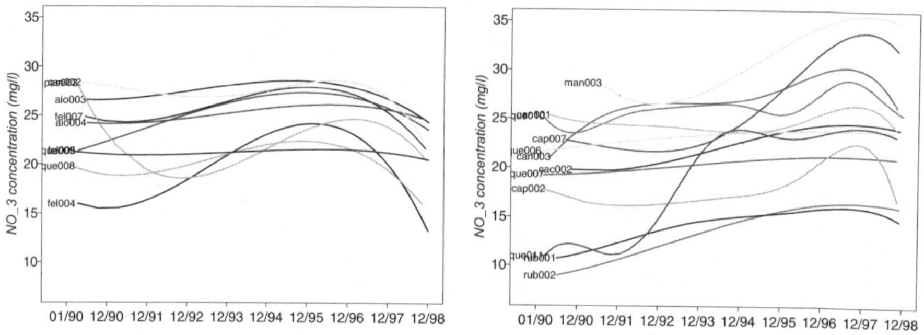

Fig. 5. Estimated NO_3 trends for two of the clusters identified from the dendrogram based on the combined dissimilarity index (4).

ful. The joint effects of an increasing trend of NO_3 concentration in the final part of the sampling window and the presence of sudden peaks deserve careful monitoring, as they can become very problematic in the future.

Similar descriptions can be obtained for the other groups not pictured here. Hence, from our analysis financial resources could be better allocated to the clusters where there is evidence that water quality is deteriorating, both in terms of NO_3 level and trend shape, at the expense of the others.

References

CERIOLI, A., GACCIOLI, M. and PEZZAROSSI, A. (2001): Metodi Statistici per la Caratterizzazione di Pozzi. *Progetto di Ricerca n. 65/P, AGAC, Reggio Emilia.*

CERIOLI, A., LAURINI, F. and CORBELLINI, A. (2005): Functional Cluster Analysis of Financial Time Series. In: M. Vichi, P. Monari, S. Mignani, and A. Montanari (Eds.): *Studies in Classification, Data Analysis, and Knowledge Organization.* Springer, Berlin, 333–342.

EUROPEAN COMMUNITY (1998): Council Directive 98/83/EC on the Quality of Water Intended for Human Consumption. *Official Journal L 330 of 05.12.98.*

GASSER, T. and KNEIP, A. (1995): Searching for Structure in Curve Samples. *Journal of the American Statistical Association, 90, 1179–1188.*

HALL, P. and HECKMAN, N.E. (2002): Estimating and Depicting the Structure of a Distribution of Random Functions. *Biometrika, 89, 145–158.*

HECKMAN, N.E. and ZAMAR, R.H. (2000): Comparing the Shapes of Regression Functions. *Biometrika, 87, 135–144.*

PICCOLO, D. (1990): A Distance Measure for Classifying ARIMA Models. *Journal of Time Series Analysis, 11, 153–164.*

RAMSAY, J.O. and SILVERMAN, B. (2005): *Functional Data Analysis.* Springer, New York (2nd Edition).

RUPPERT, D., WAND, M.P. and CARROLL, R.J. (2003): *Semiparametric Regression.* Cambridge University Press, New York.

TARPEY, T. and KINATEDER, K.K.J. (2003): Clustering Functional Data. *Journal of Classification, 20, 93–1114.*

Typicality Degrees and Fuzzy Prototypes for Clustering

Marie-Jeanne Lesot and Rudolf Kruse

Department of Knowledge Processing and Language Engineering,
Otto-von-Guericke University of Magdeburg, Universitätsplatz 2,
D-39106 Magdeburg, Germany; {lesot, kruse}@iws.cs.uni-magdeburg.de

Abstract. Fuzzy prototypes characterise data categories underlining both the common features of the category members and their discriminative features as opposed to other categories. In this paper, a clustering algorithm based on these principles is presented. It offers means to handle outliers, and a cluster repulsion effect avoiding overlapping areas between clusters. Moreover, it makes it possible to characterise the obtained clusters with prototypes, increasing the result interpretability.

1 Introduction

A prototype is an element chosen to represent a group of data, to summarise it and provide a simplified description that captures its essential characteristics. It is related to the notion of typicality, i.e. the fact that all members of a group do not have the same status: some constitute better examples and are more characteristic than others. Studies of typicality at a cognitive level (Rosch (1978)) showed that this representativeness notion depends on two complementary components, respectively called internal resemblance and external dissimilarity: a point is all the more typical as it resembles the other members of its category (internal resemblance) and differs from members of other categories (external dissimilarity).

Now these two quantities can be matched with the clustering objective, i.e. the decomposition of a data set into subgroups that are both homogeneous and distinct: homogeneity justifies the data grouping as it implies that points assigned to the same cluster do resemble one another; it requires that each point have a high internal resemblance. The fact that the clusters are distinct, or their separability, justifies the individual existence of each group as it implies that merging two clusters would lead to lose the compactness property; it requires that each point have a high external dissimilarity. Therefore, a good clustering result corresponds to a data decomposition where each point has a high typicality degree for the cluster it is assigned to. In this paper, we propose a clustering algorithm that aims at maximising the typicality degrees.

We moreover exploit the typicality framework to characterise the obtained clusters with fuzzy prototypes.

The paper is organised as follows: Section 2 recalls the formalisation of typicality degrees and Section 3 describes the proposed clustering algorithm. Section 4 presents experimental results and compares the proposed algorithm with two classic clustering algorithms, the fuzzy and possibilistic c-means.

2 Typicality degrees and fuzzy prototypes

The prototype construction method initially proposed by Rifqi (1996) relies on the typicality notion defined by Rosch (1978) presented above. It consists in computing, for each point, its internal resemblance and external dissimilarity and then its typicality degree as the aggregation of these two quantities. Prototypes are then defined as the aggregation of the most typical data.

Formally, let's denote $X = \{x_i, i = 1..n\}$ a data set, with points belonging to several categories, C a category, and x a point belonging to C. Let ρ and δ denote a resemblance and a dissimilarity measure, i.e. functions that associates point couples to values in $[0, 1]$ respectively indicating the extent to which the two points are similar or dissimilar. The internal resemblance $R(x, C)$ is then defined as the average resemblance between x and the other C members, its external dissimilarity $D(x, C)$ as its average dissimilarity to points belonging to other categories and its typicality degree $T(x, C)$ as their aggregation:

$$R(x,C) = avg(\rho(x,y), y \in C) \qquad D(x,C) = avg(\delta(x,y), y \notin C) \quad (1)$$

$$T(x,C) = \varphi(R(x,C), D(x,C)) \quad (2)$$

φ denotes an aggregation operator that expresses how the typicality degree depends on R and D: conjunctive operators e.g. allow as typical points only those with both high internal resemblance and high external dissimilarity. Variable behaviour operators, such as MICA, offer a reinforcement property: if both R and D are high, they reinforce each other to give an even higher typicality degree; if both are small, they penalise each other to give an even smaller typicality. In between, they offer a compensation property, allowing the decrease of one criterion to be compensated for by an increase of the other criterion (see Lesot et al. (2005) for a more complete discussion).

Finally significant representatives of each subgroup are deduced from typicality degrees: one can simply define the prototype as the most typical point of the group; yet this approach only defines the representative as one of the data point. A richer approach defines the prototype as a weighted mean, using as weights the typicality degrees. An even richer description defines the prototype as a fuzzy set. Indeed, a prototype can be considered as an imprecise notion: e.g. one would rather say "the typical French person is around 1.7m tall", instead of "the typical French person is 1.715m tall" (fictitious value).

Therefore the prototype should not be reduced to a single precise numerical value but take into account imprecision. This is better modeled by a fuzzy set, for instance as the fuzzy set whose core contains the points with typicality higher than 0.9, whose support contains the points with typicality higher than 0.7, and with linear interpolation in between (Lesot et al. (2005)).

3 Typicality-based clustering algorithm

In this section, the previous typicality framework is extended to unsupervised learning, to perform clustering: as indicated in the introduction, the principle is to identify a data decomposition so that each point maximises its typicality degree to the cluster it is assigned to. To that aim, as summarised in Table 1 and detailed in the following, the algorithm alternates two steps: (i) given a candidate partition of the data, compute typicality degrees with respect to this partition; (ii) given typicality degrees, update the partition so that each point becomes more typical of the cluster it is assigned to.

The expected advantages are two-fold: first the algorithm is expected to identify outliers and avoid their disturbing the results. Indeed, outliers are located far away from all data: for any cluster, they should get low internal resemblances, and thus low typicality degrees. Second, typicality degrees also offer a natural way to ensure the desired separability property and to identify regions where clusters overlap: points located in such areas are expected to get low typicality degrees, because they are similar to points assigned to other clusters, i.e. should have low external dissimilarity.

3.1 Typicality step

The first alternated step of the algorithm computes typicality degrees with respect to a candidate data partition. Contrary to the supervised learning case, typicality degrees are not computed only for the category points are assigned to, but for all of them: clusters are to be questioned, and for each point all assignments are considered. The partition is only used for the internal resemblance and external dissimilarity computation, to determine the points that should be considered as belonging to the same or to other groups.

Choice of the resemblance and dissimilarity measures

Typicality degrees rely on comparison measures to quantify resemblance and dissimilarity. Among the possible choices (Lesot (2005)), in analogy with the possibilistic c-means (Krishnapuram and Keller (1993)), we consider Cauchy functions

$$\rho(x,y) = \frac{1}{1+\left(\dfrac{d(x,y)}{\gamma_R}\right)^2} \qquad \delta(x,y) = 1 - \frac{1}{1+\left(\dfrac{d(x,y)}{\gamma_D}\right)^2} \tag{3}$$

where $d(x, y)$ denotes the Euclidean distance, and γ_R and γ_D normalisation coefficients, that correspond to reference distances: γ_R (resp. γ_D) is the distance from which the resemblance (resp. dissimilarity) between two points is smaller (resp. higher) than 0.5. They should be chosen independently one from the other: resemblance is only applied to compare points assigned to the same cluster, whereas dissimilarity only involves points of different clusters. Thus they are expected to apply on different distance scales: within-cluster distances are expected to be smaller than inter-cluster distances.

We define γ_D as a function of the data diameter, so that dissimilarity equals 0.9 for points whose distance equals half the data diameter. As regards resemblance, we define a different reference distance for each cluster, as the cluster radius, to allow handling clusters with different sizes.

Choice of the aggregation operator

Typicality degrees are then defined as the aggregation of internal resemblance and external dissimilarity. Clustering imposes constraints that reduce the possibilities as compared to the supervised case: in the latter, one may be interested in discriminative prototypes that underline the features specific of each group and are influenced by extreme group members.

In the unsupervised case, such an approach would give too much influence to outliers, that may disturb the clustering results; internal resemblance and external dissimilarity must play comparable roles. In our experiments, we consider a two-step procedure: first, we consider the minimum aggregation operator; in a second step, we apply a more flexible aggregator, in the form of the MICA operator.

3.2 Assignment step

The second alternated step of the algorithm consists in, given typicality degrees, updating the data partition so that each point becomes more typical of its cluster. A natural choice to achieve this aim consists in assigning the point to the cluster it is most typical of:

$$x \in C_r \Longleftrightarrow r = \arg\max_s T(x, C_s) \tag{4}$$

Two specific cases are handled separately: first, when the maximal typicality degree is small (smaller than 0.1 in our experiments), it does not seem justified to assign the point as it is not typical of the candidate cluster. Actually, such points are typical of no cluster, and correspond to outliers. Thus, they are assigned to a fictitious cluster instead of a regular cluster. As a consequence, they are not taken into account for the computation of internal resemblance and external dissimilarity of the other points in the following typicality step: this avoids their distorting the computation of typicality degrees, through very low resemblance values.

Table 1. Proposed typicality-based clustering algorithm

Notations: $X = \{x_i, i = 1..n\}$ the data set, c the desired number of clusters, ρ and δ a resemblance and a dissimilarity measure respectively, φ an aggregation operator
Initialisation: Apply a few steps of fuzzy c-means and assign points according to their maximal membership degrees
Loop: while assignment evolves, alternate

1. Typicality step: for each point $x \in X$ and each cluster C_r, $r = 1..c$
   a) Compute the internal resemblance $R(x, C_r) = avg(\rho(x, y), y \in C_r)$
   b) Compute the external dissimilarity $D(x, C_r) = avg(\delta(x, y), y \notin C_r)$
   c) Compute the typicality degree $T(x, C_r) = \varphi(R(x, C_r), D(x, C_r))$
2. Assignment step: for each point $x \in X$
   a) if x is typical for no cluster, i.e. $\max_r T(x, C_r) < 0.1$, assign x to a fictitious cluster, C_0
   b) else if x typicality is not clear, i.e. $T_1(x) - T_2(x) < 0.02$, where $T_i(x)$ is the i-th biggest value of $T(x, C_r)$, $r = 1..c$, assign x to the fictitious cluster C_0
   c) else assign x according to the maximal typicality degree, i.e. to C_r where $r = \arg\max_s T(x, C_s)$.

Another special case occurs when the maximal typicality degree is not clearly defined, i.e. the second biggest typicality value is very close to the biggest one (difference lower than 0.02 in our tests). In such cases, the typicality degrees are usually small, although not small enough to be handled in the previous case (they are usually around 0.3). Such points, that are generally located in cluster overlapping areas, are also assigned to the fictitious cluster.

3.3 Overall algorithm

The proposed typicality-based algorithm takes as parameter the desired number of clusters. It then computes an initial partition of the data, through a few steps of the fuzzy c-means algorithm for instance, and deduces initial values for the cluster radii. It then applies the loop indicated in Table 1, that alternatively computes typicality degrees according to a partition and updates the partition according to typicality degrees, until convergence of the partition. During this step, the aggregation operator φ is chosen to be the minimum. The loop is then performed a second time, after updating the γ_R values and changing φ to the MICA operator. In the performed tests, each convergence required only a very small number of iterations, less than 10.

At the end of the process, fuzzy prototypes can be derived from the final typicality degrees to characterise the obtained clusters.

4 Experimental results

This section illustrates the results obtained on the two-dimensional data set represented on Fig. 1a, made of three Gaussian clusters and a small outlying group. This simple example is exploited to compare in details the proposed typicality-based method to two classic clustering algorithms, the fuzzy c-means (FCM; Dunn (1973)) and the possibilistic c-means (PCM; Krishnapuram and Keller (1993)), underlining their relationships and respective properties, and interpreting the weight distribution they rely on.

Typicality-based clustering algorithm

Figure 1b and 1c represent the level lines of the typicality degree distribution and the associated fuzzy prototypes obtained when 3 clusters are searched for. Each symbol depicts a different cluster, stars represent the fictitious cluster. Figure 1b shows the expected clusters are identified, as well as the outliers that are assigned to the fictitious cluster. The latter also contains a point located in an overlapping area between clusters. This shows the method indeed takes into account both internal resemblance and external dissimilarity.

The effect of these two components can also be seen in the typicality distribution: on the one hand, the distributions are approximately centred around the group centre, due to the internal resemblance constraint. On the other hand, the distribution of the upper cluster for instance is more spread on the x-axis than on the y-axis: the overlap with the two other clusters leads to reduced typicality degrees, due to the external dissimilarity constraint.

The associated fuzzy prototypes (see Fig. 1c) provide relevant summaries: they have small support, and characterise the clusters. Indeed, they are concentrated on the central part of the clusters, but also underline their distinctive features, i.e. their particularities as compared to the other clusters: for the rightmost cluster e.g., the prototype is more spread in the bottom right region, indicating such values are specific for this cluster, which constitutes a relevant characterisation of the cluster as opposed to the two others.

Fuzzy c-means

Figure 1d represents the fuzzy sets built by FCM. As compared to Figure 1b and 1c, outliers have a bigger influence and tend to attract all three clusters in the upper left direction. This is due to the definition of the membership degrees on which FCM rely: they indicate the extent to which each point belongs to the clusters, or more precisely the extent to which it is shared between the clusters. The quantities involved in their definition are relative distances that compare the distance to a cluster centre to the distance to other centres. Now this relative definition implies the influence of a point does not decrease with its absolute distance to the centres. Outliers are considered as equally shared between all clusters, their membership degrees equal the reciprocal of the number of clusters and they influence the cluster centre positions.

Another difference between FCM and the typicality-based algorithm concerns the obtained fuzzy sets: the FCM ones are much more spread and less specific than the typicality or the prototype distributions, they cover the whole

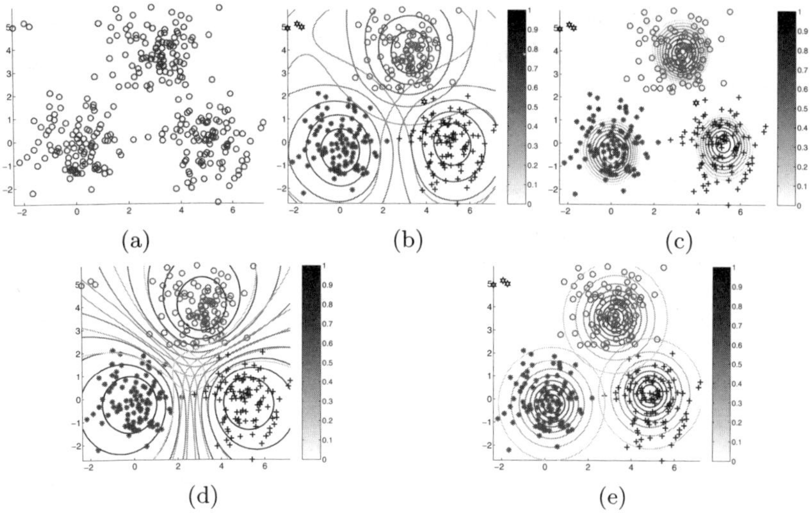

Fig. 1. (a) Considered data set, made of 3 Gaussian clusters and a small outlying group in the upper left corner, (b-e) Level lines of (b) typicality degrees, (c) fuzzy prototypes, (d) FCM membership degrees, (e) PCM possibilistic coefficients.

input space. Indeed, FCM do not aim at characterising the clusters, but at describing them as a whole, representing all data points; on the contrary, prototypes constitute representative summaries.

Possibilistic c-means

PCM relax the constraint that causes the relative definition of the membership degrees in FCM, so as to be more robust. The coefficients they rely on measure the absolute resemblance between data points and cluster centres, and not a relative resemblance. Outliers are thus associated to small coefficients, as illustrated on Fig. 1d: they are not assigned to any of the three regular clusters and do not attract them.

It can also be observed that the distributions are spherical for all clusters; indeed the possibilistic weights are decreasing functions of the distance to the cluster centres: PCM coefficients can be interpreted as measuring an internal resemblance: their semantic is that of exceptionality coefficients measuring the deviation with respect to the cluster centre.

Partially due to this fact, PCM sometimes fail to detect the expected clusters and identify several times the same clusters, whereas natural subgroups in the data are overlooked. To avoid this effect, Timm and Kruse (2002) introduce in the PCM cost function a cluster repulsion term so as to force the clusters apart. This term then influences the cluster centre definition. The proposed typicality-based approach can be seen as another solution to this problem: the external dissimilarity component leads to a cluster repulsion effect. The latter is incorporated in the coefficient definition and not only in the cluster centre expressions, which eases their interpretation and enriches the coefficient semantics.

5 Conclusion

In this paper, we presented an extension of the typicality framework to unsupervised learning, so as to identify clusters in a data set. The proposed algorithm offers means to handle outliers, as well as a cluster repulsion effect: it takes into account in its principle itself the double aim of identifying both compact and separable clusters. Moreover, due to the typicality framework, it can characterise the obtained clusters with fuzzy prototypes, increasing the result interpretability. As regards parameters, it only requires, as any partitioning clustering algorithm, the desired number of clusters.

One limitation of this algorithm is the fact that it relies on the Euclidean distance, which only makes it possible to detect spherical clusters. A considered extension concerns its adaptation to non-spherical clusters, in the form of a Gustafson-Kessel variant of the method, as well as an adaptation to non-vectorial data, such as structured data: the algorithm does not require the computation of data averages that depend on the data nature, but only on data comparisons, that can also be defined for non-vectorial data.

Acknowledgments

This research was supported by a Lavoisier grant from the French Ministère des Affaires Etrangères.

References

DAVE, R. (1991): Characterization and Detection of Noise in Clustering. *Pattern Recognition Letters, 12, 657–664.*

DUNN, J.C. (1973): A Fuzzy Relative of the Isodata Process and Its Use in Detecting Compact Well-separated Clusters. *Journal of Cybernetics, 3, 32–57.*

KRISHNAPURAM, R. and KELLER, J. (1993): A Possibilistic Approach to Clustering. *IEEE Transactions on Fuzzy Systems, 1, 98–110.*

LESOT, M.-J. (2005): Similarity, Typicality and Fuzzy Prototypes for Numerical Data. *Proc. of the 6th European Congress on Systems Science.*

LESOT, M.-J., MOUILLET, L. and BOUCHON-MEUNIER, B. (2006): Fuzzy Prototypes Based on Typicality Degrees. *Proc. of the 8th Fuzzy Days'04, 125–138.* Springer.

RIFQI, M. (1996): Constructing Prototypes from Large Databases. *Proc. of IPMU'96.*

ROSCH, E. (1978): Principles of Categorization. In: E. Rosch and B. Lloyd (Eds.): *Cognition and Categorization.* Lawrence Erlbaum Associates, 27-48.

TIMM, H. and KRUSE, R. (2002): A Modification to Improve Possibilistic Fuzzy Cluster Analysis. *Proc. of Fuzz-IEEE'02.*

On Validation of Hierarchical Clustering

Hans-Joachim Mucha

Weierstraß-Institut für Angewandte Analysis und Stochastik,
D-10117 Berlin; mucha@wias-berlin.de

Abstract. An automatic validation of hierarchical clustering based on resampling techniques is recommended that can be considered as a three level assessment of stability. The first and most general level is decision making about the appropriate number of clusters. The decision is based on measures of correspondence between partitions such as the adjusted Rand index. Second, the stability of each individual cluster is assessed based on measures of similarity between sets such as the Jaccard coefficient. In the third and most detailed level of validation, the reliability of the cluster membership of each individual observation can be assessed. The built-in validation is demonstrated on the wine data set from the UCI repository where both the number of clusters and the class membership are known beforehand.

1 Introduction

Hierarchical clustering is in some sense more general than partitioning methods because the resultant hierarchy is a sequence of nested partitions. Thus, cluster validation based on comparing partitions is a more complex task here. But then, hierarchical clustering results in a unique solution that has to be validated. In opposition, partitioning methods result in many different, locally optimal solutions depending on the (usually random) initialization of clusters.

The proposed automatic validation of hierarchical clustering was already applied successfully in different fields of research (Mucha (2004), Mucha and Haimerl (2005), Mucha (2006)). It is based on resampling techniques and therefore, it is a quite general validation tool. It consists of a three level assessment of stability. The first and most general level is decision making about the appropriate number of clusters. This most important decision is based on measures of correspondence between partitions like Rand, adjusted Rand, and Fowlkes and Mallows (Rand (1971), Fowlkes and Mallows (1983), Hubert and Arabie (1985)). Second, the stability of each individual cluster is assessed based on measures of similarity between sets, e.g. the asymmetric measure of cluster agreement or the symmetric Jaccard measure. Special properties of clusters like compactness and isolation (see, for instance Jain and Dubes (1988)) are not considered here. Generally as a preliminary remark, it should be mentioned that it makes sense to investigate the common quite different specific stability of clusters of the same clustering on the same data. Often one can observe that the clusters have a quite different stability. Some of them are very stable. Thus, they can be reproduced and confirmed to a high degree, for instance, by bootstrap simulations. They are both homogeneous inside and well separated from each other. Moreover, sometimes they are located far away from the main body of the data like outliers. On the other side, hidden and tight neighboring clusters are more difficult to detect and they cannot be reproduced to a high degree. Concerning hierarchical clustering, often a cluster remains unchanged during many steps of agglomeration (see, for instance, the cluster with 56 observations located at the top of Figure 3), but its value of stability can alter rapidly because of the altering clusters in its neighborhood. In the third and most detailed level of validation, the reliability of the cluster membership of each individual observation can be assessed.

2 The toy data "Wine"

The wine recognition data[1] are the results of a chemical analysis of wines grown in the same region in Italy but derived from three different cultivars. The analysis determined the quantities of 13 constituents (variables) found in each of the three types of wines. Altogether there are 178 wines that belong to three classes, i.e. the three types of wines. So, we know what the right answer is here. Class 1 has 59 wines, class 2 has 71 and class 3 has 48. There is a well-known hierarchical clustering technique for minimizing the within-clusters sum of squares criterion based on pairwise distances that will be applied here: Ward's method (Ward (1963)). Here we will work with ranks instead of the original values that come from scales that are not comparable one with each other. For instance, the transformation into ranks were used successfully as a preprocessing step (Mucha (1992)). Lebart et al. (1984) investigated the significance of eigenvalues of principal component analysis (PCA) of rank data by simulations.

3 How many clusters?

There are several decision criteria known for the optimal number of clusters in hierarchical clustering; see, for example, Jung (2003). Here the general approach of simulations is preferred. They are based on random resampling. For each Bootstrap sample (sampling with or without replacement), a hierarchical cluster analysis is figured out. Applying measures of correspondence

[1] `http://www.ics.uci.edu/~mlearn/MLSummary.html`

between partition like the Rand index yields values that measure the stability of the partitions into $K = 2, 3, \ldots$ clusters. Most generally, one counts the pairwise matches of observations that are "good" in the sense of similarity between partitions, and the resulting count is relativized with regard to the total number of pairwise matches. It is recommended to apply the adjusted Rand's measure instead of the Rand index because it avoids the bias towards number of clusters and has a higher sensitivity (see Hubert and Arabie (1985) and Mucha (2004) for details).

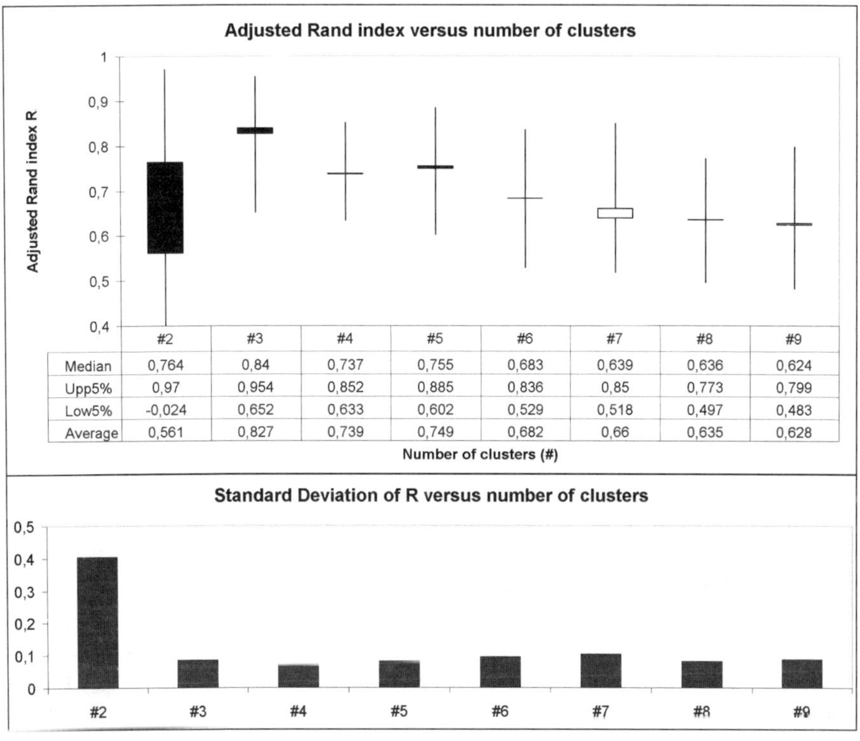

	#2	#3	#4	#5	#6	#7	#8	#9
Median	0,764	0,84	0,737	0,755	0,683	0,639	0,636	0,624
Upp5%	0,97	0,954	0,852	0,885	0,836	0,85	0,773	0,799
Low5%	-0,024	0,652	0,633	0,602	0,529	0,518	0,497	0,483
Average	0,561	0,827	0,739	0,749	0,682	0,66	0,635	0,628

Fig. 1. Graphical and numerical presentation of the summary statistics of the adjusted Rand index by taking into account 250 index values for each $K = 2,\ldots,K = 9$

The simulation results of clustering the wine data set are given in Figure 1. Here the summary statistics of 250 adjusted Rand's values R are presented for each number of clusters $K = 2, 3, \ldots, 9$. Each index value R is computed between the unique partition of clustering the set of all 178 objects and a partition of clustering the set of three-fourths randomly selected objects. The simplified boxplots at the top give a graphical representation of the lower and upper 5 percent quantile (whisker), and median, average and its difference (box). The box is filled dark when the median is greater than the average.

The applied simulation technique itself is described in the next section in more detail. Both the median and the average of the adjusted Rand values achieved their maximum at three clusters. Further on, the standard deviations of the 250 adjusted Rand values (see the bars below) decreases dramatically and the average adjusted Rand indices increases highly by going from two to three clusters. Moreover, other measures like Rand or Fowlkes and Mallows also support the decision for three clusters (see Table 1 and Section 4 below). By the way, this solution corresponds very closely to the true three classes: Seven observations are clustered into the wrong class only. Keep in mind, here we work with order information (ranks) only.

4 Cluster validation

As already mentioned above, here we do not consider special properties of clusters directly like compactness and isolation. What are stable clusters from a general statistical point of view? These clusters can be confirmed and reproduced to a high degree. To define stability with respect to the individual clusters, measures of correspondence between a cluster $\mathcal{E}$ and a cluster $\mathcal{F}$ like

$$\tau(\mathcal{E},\mathcal{F}) = \frac{|\mathcal{E}\cap\mathcal{F}|}{|\mathcal{E}\cup\mathcal{F}|}, \quad \gamma(\mathcal{E},\mathcal{F}) = \frac{|\mathcal{E}\cap\mathcal{F}|}{|\mathcal{E}|}, \quad \eta(\mathcal{E},\mathcal{F}) = \frac{|\mathcal{E}\cap\mathcal{F}|}{|\mathcal{E}|+|\mathcal{F}|} \tag{1}$$

can be defined (Hennig (2004)). ($\mathcal{E}$ and $\mathcal{F}$ are nonempty subsets of some finite set.) The autor suggests the use of the Jaccard coefficient τ in the context of a general robustness and stability theory for cluster analysis. This measure is symmetric and it attains its minimum 0 only for disjoint sets and its maximum 1 only for equal ones. As we will see later on, the symmetric Dice coefficient η behave very similar to Jaccard. The asymmetric measure γ assesses the rate of recovery of subset $\mathcal{E}$ by $\mathcal{F}$ (Mucha (2004)). It attains its minimum 0 only for disjoint sets and its maximum 1 only if $\mathcal{E}\subseteq\mathcal{F}$ holds. Obviously, it is $\tau\leq\gamma$.

Now suppose, a clustering of a set of entities $\mathcal{C} = \{1,...,i,...,I\}$ into a collection of K subsets $\{\mathcal{C}_1,...,\mathcal{C}_k,...,\mathcal{C}_K\}$ of $\mathcal{C}$ has to be investigated. Let $\mathcal{C}_k$ be one individual cluster whose stability has to be assessed. To investigate the stability, validation techniques based on random resampling are recommended. Let's consider one simulation step: Clustering of a randomly drawn sample of the set of entities $\mathcal{C}$ into a collection of K clusters $\{\mathcal{F}_1,...,\mathcal{F}_K\}$ in the same way as the whole set $\mathcal{C}$. The definition of stability of cluster $\mathcal{C}_k$ using measure γ is based on the most similar cluster

$$\gamma_k^* = \max_{\mathcal{F}_i\in\{\mathcal{F}_1,...,\mathcal{F}_K\}} \gamma(\mathcal{C}_k,\mathcal{F}_i). \tag{2}$$

By repeating resampling and clustering many times, the stability of the cluster $\mathcal{C}_k$ can be assessed, for instance, by computing the median of the corresponding values of γ_k^*. It is difficult to fix an appropriate threshold to consider a cluster as stable. The simulation itself is computationally expensive.

In this paper, subsampling (random drawing without replacement) is used consistently. Concretely, a random sample size of three-fourths of the original one is used here. To support the decision about stable regions, the clusters can often be visualized in low dimensional projections by applying methods like discriminant analysis, PCA, and multidimensional scaling.

Table 1. Assessment of a suitable number of clusters by six measures

Number of clusters	Rand	Adjusted Rand	Fowlkes and Mallows	Jaccard	Rate of Recovery	Dice
2	0.785	0.561	0.813	0.759	0.877	0.84
3	**0.922**	**0.827**	**0.886**	**0.891**	**0.942**	**0.941**
4	0.898	0.739	0.809	0.767	0.864	0.849
5	0.91	0.749	0.808	0.744	0.836	0.829
6	0.896	0.682	0.747	0.672	0.794	0.778
7	0.899	0.66	0.723	0.647	0.763	0.759
8	0.906	0.635	0.691	0.628	0.764	0.746
9	0.911	0.628	0.68	0.618	0.746	0.739

Table 2. Assessment of the stability of individual clusters by three measures

	Cluster 1			Cluster 2			Cluster 3		
Number of objects	56			74			48		
Measure	τ^*	γ^*	η^*	τ^*	γ^*	η^*	τ^*	γ^*	η^*
Median	0.912	1	0.954	0.877	0.914	0.934	0.93	0.975	0.964
Average	0.893	0.974	0.942	0.865	0.899	0.926	0.929	0.97	0.963
Standard deviation	0.071	0.039	0.041	0.073	0.081	0.044	0.049	0.033	0.027
Minimum	0.633	0.738	0.776	0.574	0.574	0.729	0.766	0.889	0.867

Table 1 shows the assessment of the stability of the partitions into $K = 2, 3, ..., 9$ clusters by the average values of the three well-known measures adjusted Rand (see also Figure 1 for details), Rand and Fowlkes and Mallows. Additionally, individual cluster validation measures τ_k^* (Jaccard), γ_k^* (rate of recovery) and η_k^* (Dice) are averaged over the corresponding clusters of a partition and give three more measures (see on this also Table 2 for the three cluster solution). The consensus decision of all six measures for three clusters is pointed out by bold format.

Figure 3 shows the graphical output of the PCA of Table 1. The plot at the left hand side summarizes the decision about the outstanding ("consensus") number of clusters. This search is done by looking for a linear combination of the six measures. Obviously, the eye-catching event along the performance axis (i.e. the first principal component) is "3" that stands for the three cluster solution. The plot at the right side compares the different measures of stability.

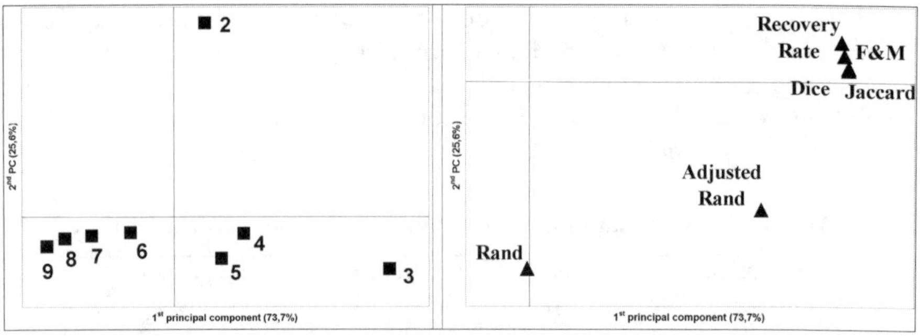

Fig. 2. PCA plots of the row points (left) and column points of Table 1

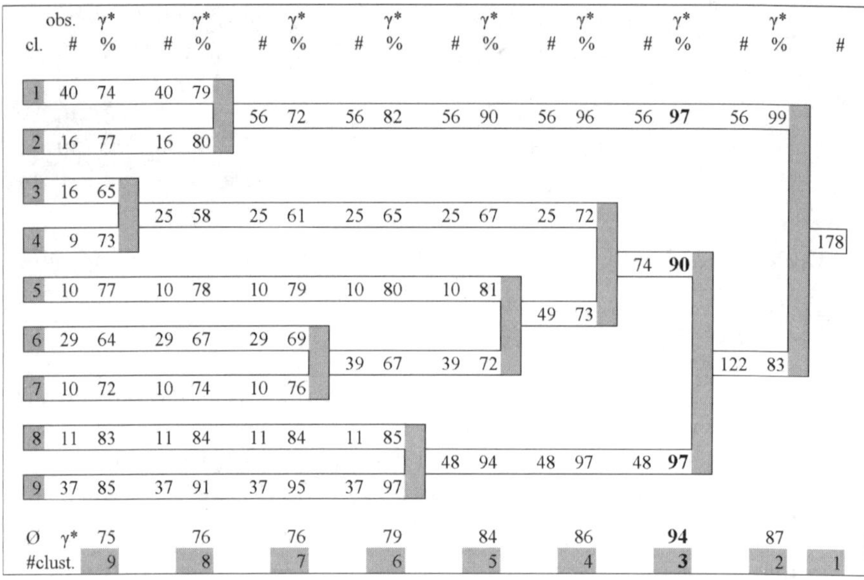

Fig. 3. Dendrogram (extract) with assessment of stability of nodes

Obviously, Dice, Jaccard, Fowlkes & Mallows and rate of recovery look quite similar for assessing the Ward's hierarchical clustering of the wine data set.

Figure 3 shows the schematic dendrogram of Ward's clustering for up to 9 clusters. Each node (cluster) of the binary tree is denoted by both the corresponding number of objects (symbol #) and the average rate of recovery (in %). The three cluster solution is emphasized in bold format. While looking for stable clusters and for an outstanding number of clusters you should keep in mind that a hierarchy is a set of nested partitions. Therefore it is recommended to walk step by step through the binary tree (dendrogram) from the right hand side (that corresponds to the root of the tree) to the left. At each step $K-1$ clusters remain unchanged and one cluster is divided only into two parts.

Usually, the higher the number of clusters K becomes during the trip through the dendrogram the smaller amount of changes of the averaged measures of stability can be expected that are given at the bottom of the Figure. Some of the clusters remain unchanged during many steps such as the cluster of 56 observations at the top of Figure 3. However, the value of stability of this cluster decreases from 99% for the partition into 2 clusters to 72% only for the partition into 7 clusters because of the altering clusters in its neighborhood.

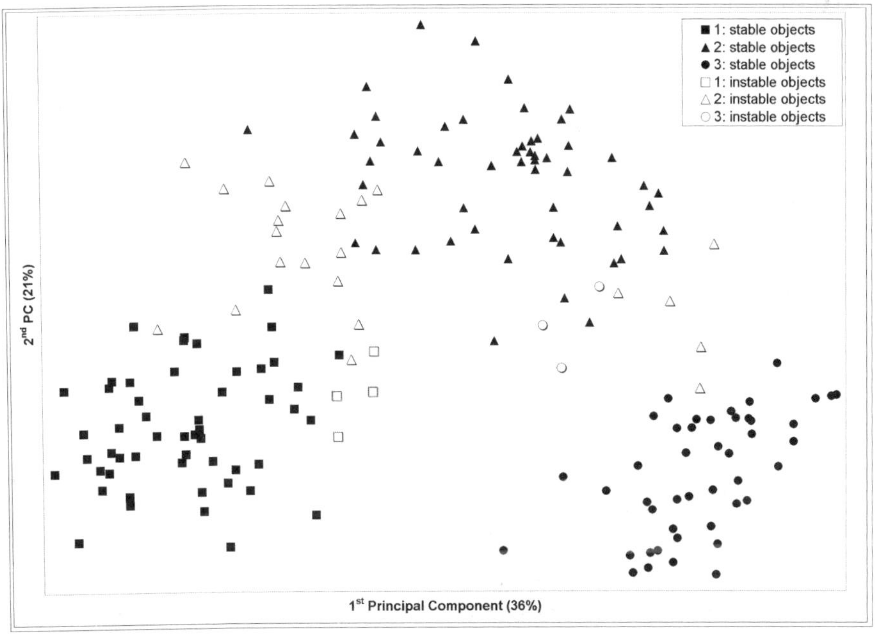

Fig. 4. PCA plot of cluster membership of stable and instable objects

During the simulations described above, the reliability of the cluster membership of each individual observation into all clusters of a hierarchy are assessed additionally. Concretely, the correct classifications of the observations into the clusters of the unique hierarchy are counted for every random subset clustering, and then the result is divided by the total number of simulations. Without loss of generality the clusters of the stable three cluster partition solution will be considered now. Figure 4 shows all observations in the first plane of the PCA based on Spearman's rank-order correlation coefficients. Stable objects with more than 90% reliability are presented in dark symbols, whereas instable ones (less than 90%) are marked by the corresponding light symbols. All seven misclassified observations (with respect to the true classes) have a reliability less than 68.1%, and thus, they are all contained in the set of instable objects.

As already seen in Table 2, cluster 2 is the most instable one with respect to the measures (1). This cluster is located in the middle of the cloud of points "in between" cluster 1 and 3 (see the PCA projection of Figure 4). The number of instable observations of cluster 2 counts to 22 out of 74 (in comparison to cluster 1 (4 of 56) and cluster 3 (3 of 48)). Obviously, clusters, that are surrounded by many neighboring clusters, are more difficult to confirm.

5 Conclusions

The stability of results of hierarchical cluster analysis can be assessed by using validation techniques based on measures of correspondence between partitions. By doing so, the number of clusters can be determined. Further on, the stability of each cluster as well as the reliability of the class membership of each observation can be assessed.

References

FOWLKES E.B. and MALLOWS, C.L. (1983): A Method for Comparing two Hierarchical Clusterings. *JASA 78, 553–569.*

HENNIG, C. (2004): A General Robustness and Stability Theory for Cluster Analysis. *Preprint, 7,*, Universität Hamburg.

HUBERT, L.J. and ARABIE, P. (1985): Comparing Partitions. *Journal of Classification, 2, 193–218.*

JAIN, A.K. and DUBES, R.C. (1988): *Algorithms for Clustering Data.* Prentice Hall, New Jersey.

JUNG, Y., PARK, H., DU, D.-Z. and DRAKE, B.L. (2003): A Decision Criterion for the Optimal Number of Clusters in Hierarchical Clustering. *Journal of Global Optimization 25, 91–111.*

LEBART, L., MORINEAU, A. and WARWICK, K.M. (1984): Multivariate Descriptive Statistical Analysis. Wiley, New York.

MUCHA, H.-J. (1992): *Clusteranalyse mit Mikrocomputern.* Akademie Verlag, Berlin.

MUCHA, H.-J. (2004): Automatic Validation of Hierarchical Clustering. In: J. Antoch (Ed.): *Proceedings in Computational Statistics, COMPSTAT 2004, 16th Symposium.* Physica-Verlag, Heidelberg, 1535–1542.

MUCHA, H.-J. (2006): Finding Meaningful and Stable Clusters Using Local Cluster Analysis. In: V. Batagelj, H.-H. Bock, A. Ferligoj and A. Ziberna (Eds.): *Data Science and Classification*, Springer, Berlin, 101–108.

MUCHA, H.-J. and HAIMERL, E. (2005): Automatic Validation of Hierarchical Cluster Analysis with Application in Dialectometry. In: C. Weihs and W. Gaul (Eds.): *Classification - The Ubiquitous Challenge*, Springer, Berlin, 513–520.

RAND, W.M. (1971): Objective Criteria for the Evaluation of Clustering Methods. *Journal of the American Statistical Association, 66, 846–850.*

WARD, J.H. (1963): Hierarchical Grouping Methods to Optimise an Objective Function. *JASA, 58, 235–244.*

Part II

Classification

Rearranging Classified Items in Hierarchies Using Categorization Uncertainty

Korinna Bade and Andreas Nürnberger

Fakultät für Informatik, Institut für Wissens- und Sprachverarbeitung, Otto-von-Guericke-Universität Magdeburg, D-39106 Magdeburg, Germany; {kbade, nuernb}@iws.cs.uni-magdeburg.de

Abstract. The classification into hierarchical structures is a problem of increasing importance, e.g. considering the growing use of ontologies or keyword hierarchies used in many web-based information systems. Therefore, it is not surprising that it is a field of ongoing research. Here, we propose an approach that utilizes hierarchy information in the classification process. In contrast to other methods, the hierarchy information is used independently of the classifier rather than integrating it directly. This enables the use of arbitrary standard classification methods. Furthermore, we discuss how hierarchical classification in general and our setting in specific can be evaluated appropriately. We present our algorithm and evaluate it on two datasets of web pages using Naïve Bayes and SVM as baseline classifiers.

1 Introduction

Classifying objects into hierarchical structures is a task needed in a growing number of applications. Current examples are the use of ontologies or keyword hierarchies in many web-based systems. This includes classifying web pages into a web directory or in a hierarchy defined by groups of users like in social web communities, e.g. the open directory project (www.dmoz.org). Despite the possibility of creating such hierarchies manually and assigning documents to it, automatic classification and hierarchy extension is of benefit for many domains as the number of documents to classify has reached huge numbers.

A common feature of most hierarchies is that data is not distributed equally across all classes. While classes in leaf nodes of the hierarchy tree usually have at least some elements assigned to it, inner tree nodes can be completely empty. Nevertheless, these nodes carry a meaning that is especially of interest when we try to classify an object, for which a specific class not yet exists. Furthermore, these inner classes are important to a user, who tries to locate information (documents or objects) in the hierarchy. The browsing behavior is guided by the class labels. The user, who is searching for information, will browse the hierarchy top-down until he finds a sub class, in which he would expect to find the information. Once he reached the most specific class describing his information need, he would start scanning the documents. If the needed information is not found, he might also browse more general classes as they also include his topic of interest. However, it is very unlikely that he would browse other specific classes in branches of the hierarchy that deal with different topics.

Flat classifiers learned over all classes in the hierarchy omit this information in the classification process. However, we argue that integrating hierarchical information into the classification process can improve the usefulness for a user. This can be done in several ways. First, the training data could be reinterpreted, being not only assigned to one class but also to parent and/or child classes in the hierarchy. Second, the hierarchy could be used directly in the classifier design. And third, the hierarchy information could be used, when interpreting the class assignments of a classifier. While following a strategy of the last type in this paper, it is also our goal to respect the user behavior in this process. This puts the following constraints on the classification. Each object that is classified in a wrong sub class, will most likely not be retrieved by the user. Each object that is classified into a class that is a generalization of the correct class, might still be retrieved by the user, depending on how much time the user is willing to spend on his search. Furthermore, a more specific class might not yet exist for a document. Here, classification in one of the most specific classes would prevent the user from retrieving the document as he would not look so deep down the hierarchy. In this case, predicting a more general class is the only way making retrieval possible. We consider these aspects throughout the paper, i.e. for the algorithm as well as for evaluation.

2 Related work

In this section, we briefly review related work for hierarchical classification. Most of it deals with integrating the hierarchy information into the classification process by adapting existing methods or building new classifiers. Cai and Hofmann (2004) try to integrate hierarchy information directly into an SVM classifier by integrating a hierarchical loss function. Different SVM classifiers for each hierarchy node are learned by Sun and Lim (2001) and Dumais and Chen (2000) to find a suitable node in the tree. McCallum et al. (1998) applied shrinkage to the estimates of a Bayes classifier. In Cesa-Bianchi et al. (2004), an incremental algorithm with performance close to SVM and also a new loss function for evaluation is presented. In Choi and Peng (2004), a greedy probabilistic hierarchical classifier is used to determine the most suitable path in the hierarchy from the root to a leaf. In a second step, another classifier is used to determine the best class along this path. The authors also suggest some criteria to create a new category. In Granitzer and Auer (2005), the performance of two Boosting algorithms, BoosTexter and CentroidBooster, is compared to the performance of support vector machines. The influence of different training sets (with and without using hierarchy information) is examined by Ceci and Malerba (2003) using Naive Bayes and Centroid learning for different numbers of extracted features. Frommholz (2001) adapts the determined weights for each category for a certain document in a post-processing step by integrating the weights of all other categories according to their proximity (the distance in the category tree/graph) to this category.

3 Performance measures for hierarchical classification

To compare results between different algorithms, it is necessary to define appropriate performance measures. For standard (flat) classification, evaluation is mostly done by precision and recall (see e.g. Hotho et al. (2005)). The combination of the two, the F-Score, is usually used to evaluate overall performance. These measures treat all classes equally. There is just one correct class and all others are wrong. However, in hierarchical classification, not all "wrong" classifications are equally "bad". As motivated with our application setting, classification in a more general class than the correct class still allows for retrieval. However, the more the concept was generalized the more unlikely it is that the user puts in the effort to retrieve this document. To integrate this notion of "not so good but still correct", we propose the following adaptation, which uses gradual membership degrees of class matches. As in the standard version, correctly classified documents are counted as a perfect match (with a value of 1) and documents that are classified in a wrong subclass are not counted at all (with a value of 0). However, documents that are classified into a more general concept are now counted with a value between 0 and 1 depending on how far they are away in the class tree. This is expressed by the hierarchical similarity between the predicted class c_p and the correct class c_c in (1), where $\mathrm{dist_H}(c_1, c_2)$ is the number of edges on the path between c_1 and c_2 in the hierarchy H and $\geq_H$ stands for 'is the same or a more general class'.

$$\mathrm{sim_H}(c_p, c_c) = \begin{cases} 1/\left(\mathrm{dist_H}(c_p, c_c) + 1\right) & \text{if } c_p \geq_H c_c \\ 0 & \text{else} \end{cases} \tag{1}$$

With the $\mathrm{sim_H}$ function, the adapted precision and recall can be determined as in (2), where $retr_c$ is the set of all documents classified (retrieved) as c and rel_c is the set of all documents that belong (are relevant) to c. The $\mathrm{sim_H}$ function could be replaced by any other function computing values between 0 and 1 to express other class relationships. If $\mathrm{sim_H}$ would be 1, only if both classes are equal, and 0 otherwise, we get standard precision and recall.

$$prec_c = \frac{\sum_{d \in retr_c} \mathrm{sim_H}(c, \mathrm{class}(d))}{|retr_c|} \qquad rec_c = \frac{\sum_{d \in rel_c} \mathrm{sim_H}(\mathrm{predClass}(d), c)}{|rel_c|} \tag{2}$$

Other researchers also proposed hierarchical performance measures, e.g. in Sun and Lim (2001). However, their focus is more on evaluating the classifier performance itself, e.g. by taking category similarities or tree distances into account. Our focus is on the usefulness of the classification from a user's point of view, which is based on user behavior in the described application scenario.

Furthermore, we introduce the idea of an *optimal node* n_o for post-processing a prediction in a class hierarchy. Let $path_c$ be the path in the hierarchy from the root node to c_c and $path_p$ the path from the root node to c_p. Then n_o of this prediction is the most specific node, for which holds $n_o \in path_c$ and $n_o \in path_p$. Note that n_o therefore is either the correct class or a parent class of it. n_o expresses the best generalization of a preliminary classifier prediction to allow for retrieval considering the described user behavior. Using this definition, we can extract three numbers to provide more information about how the data was classified in the hierarchy: the number of documents *in*, *above*, and *below* n_o.

4 An approach based on classification uncertainty

In the following, we first briefly sum up our post-processing scenario and then present an approach, which tries to solve this task. Given is a class hierarchy H together with a set of training data D, whereby not every class must have training data assigned to it. Non-leaf classes might be empty but can still be the most suitable prediction. Furthermore, we have a classifier C that computes prediction probabilities $P_C(c|d)$ for each class c for a given document d. This can either be a flat (as in our experiments) or a hierarchical classifier. Keep in mind that a flat classifier produces a probability of 0 for empty classes. In addition, the sum of the prediction probabilities for all classes does not need to sum up to 1 as the rest of the probability mass could be describing unknown classes. The goal of our approach is to find for each document a prediction that is either the correct class or a parent class of it by being as specific as possible.

The basic idea of our approach is to traverse the class hierarchy top-down. At each class, the algorithm looks for the most suitable child class of the current class and decides whether it is still reasonable to descend to it or whether it is better to return the current class as the final prediction. To be able to make this decision, we utilize two different kinds of information. The first one is based on the prediction probabilities $P_C(c|d)$ from the classifier. We integrate the hierarchical relationship between classes into these probabilities by propagating the maximum prediction probability up the hierarchy, i.e. by computing $P(c|d) = \max_{c' \leq_H c} P_C(c'|d)$. Descending the class hierarchy by always choosing the subclass with the highest value $P(c|d)$ now produces an equal prediction to choosing the class with the highest classifier prediction $P_C(c|d)$. Please note that for this purpose for each original inner class of the hierarchy a virtual leaf class is created, which is a direct sub-class of the concerning class. The classifier predictions $P_C(c|d)$ of these classes are then associated to the virtual classes. This is needed as an internal representation but will not be visible to the user, i.e. a prediction of the virtual child class or the class itself makes no difference.

Second, we use the training data to derive further class information. In Bade and Nürnberger (2005), we used class similarities. Here, we extract the probability of predicting a class correctly. We denote this as prediction accuracy of a class $P_A(c)$. By performing 10-fold crossvalidation on the training data, we build a confusion matrix M for the original classification. In a second step, we again utilize the hierarchy information to derive the prediction accuracy of each class by the following equation:

$$P_A(c) = \left(\sum_{c_1, c_2 \leq_H c} M(c_1, c_2) \right) \div \left(\sum_{c_3 \leq_H c, c_4 \in H} M(c_3, c_4) \right) \tag{3}$$

In other words, we interpret each class as the combination of it with its sub-classes. So for each class, the number of all documents is counted that belong to this class or a sub-class and that are also predicted to belong to one of these classes. This number is set in relation to all predictions of this "group" of classes.

After having determined both probabilities for each class, we can formulate a criterion to decide whether further descend into the class hierarchy should be stopped or not. The hierarchy is traversed along the highest prediction probability. If the following criteria holds, the descend is stopped:

$$\sqrt{P(c_b|d) \cdot P(c_{sb}|d)} \cdot P_A^t(\overline{c}) > P(c_b|d) \cdot P_A^t(c_b) \tag{4}$$

where $\overline{c}$ is the current class of interest and c_b and c_{sb} are the direct child classes with the highest and second highest prediction probability. t defines a threshold parameter, which can be used to tune the algorithm.

What does this equation model? At both sides of the inequality, the class accuracy and the prediction probability of a class are combined, weighted by the parameter t. The main idea is to compare the combined value of the current class $\overline{c}$ (left side of inequality) with the combined value of the best child class c_b (right side of inequality). If the value decreases down the hierarchy, the algorithm stops. However, the prediction probability in the current class is always equal to the prediction probability of the best child class due to our initialization procedure. Therefore, we decided to replace the prediction probability of the current class with the harmonic mean of the prediction probabilities of the two best child classes. The main idea is that this expresses the "purity" of the class assignment. If the two best class probabilities are almost the same, the classifier is rather undecided, which class is the best, producing an almost equal value for the prediction probability. And this is actually the kind of classification uncertainty, we want to detect and avoid by generalization. The parameter t is used to tune the strength of the influence of the class accuracy. The algorithm is summarized in Fig. 1.

```
Classify(d, C)
    For each c_i ∈ H: Compute probability estimate P_C(c_i|d) by C
    For each c_i ∈ H: Compute P(c_i|d) = max_{c' ≤_H c_i} P_C(c'|d)
    Build M by running C on training data
    For each c_i ∈ H: Compute P_A(c_i) = Σ_{c1,c2 ≤_H c_i} M(c1,c2) / Σ_{c3 ≤_H c_i, c4 ∈ H} M(c3,c4)
    c̄ = root(H)
    While c̄ has child classes:
        Determine the two child classes c_b and c_sb of c̄ with highest P(c|d)
        If sqrt(P(c_b|d) · P(c_sb|d)) · P_A^t(c̄) > P(c_b|d) · P_A^t(c_b): Break
        c̄ = c_b
    Return c̄
```

Fig. 1. Hierarchical classification based on classifier uncertainty

5 Evaluation

For evaluation, we used two different flat classifiers as reference classifiers, a standard Naïve Bayes classifier and a SVM (libSVM implementation of Chang and Lin (2001)). We evaluated our algorithm with two datasets. The first is the banksearch dataset (see also Sinka and Corne (2002)), consisting of 11000 web pages in a 2 level hierarchy (4 classes on level 1, 10 on level 2). The second is a part of the Open Directory (www.dmoz.org), consisting of 8132 web pages in a hierarchy of depth up to 4. The number of child nodes varies from 2 to 17. After parsing, we filtered out terms that occured less than 5 times, more than 10500 times, stop words, terms with less than 4 characters, and terms containing numbers. After that, we selected out of these the 100 most distinctive features per class as described in Hotho et al. (2005). Each classifier was learned with the same training data. For the banksearch data, we have chosen 300 randomly selected documents from each class, and for the Open Directory data, we have chosen 2/3 of the data in each class. The classifiers were tested with the remainder of the data.

Tables 1 and 2 summarize our results using the presented evaluation measures. As you can see with the results of the baseline classifiers, the second data is a lot more difficult to classify. The SVM classifier only has a f-score of about 60%. Reasons for this are that the hierarchy is deeper and wider, that each class has less training data, and that the data of each class is more inhomogeneous. With empirical testing, we determined an optimal value for t for each classifier and dataset by maximizing the F-Score. Due to space constraints, we only present here the results gained with this value.

For the banksearch dataset, t should be about 30 for Naïve Bayes and 6 for SVM. With this parameter selection, we get an increase in precision of about 8% and an increase in recall of about 1%, leading to an increase in the F-score of about 4% for Naïve Bayes. For SVM the improvement is smaller with an increase in precision of about 3% a slight decrease in recall of 1%, leading to an increase in the F-Score of about 1%. For the Open Directory dataset, t should be about 70 for Naïve Bayes and 4 for SVM. This leads to an increase in precision of about 30%, an increase in recall of about 10%, and an increase in the F-score of about 15% for Naïve Bayes. For SVM the improvement is again smaller with an increase in precision of about 20%, a slight decrease in recall of 1%, and an increase in the F-Score of about 5%.

The optimal node measures provides more detailed information about the effects of the post processing step. For the banksearch dataset, it can be seen that about 2/3 of the wrong predictions of the Naïve Bayes classifier are generalized to a class that allows for retrieval. For SVM, less than half of its wrong predictions could be generalized. For the Open Directory dataset, about 4/5 and more of the wrong predictions from both classifiers could be generalized. As a general problem, it can be stated that a large fraction is generalized too much.

We want to point out that our algorithm aims on generalizing predictions that are possibly wrong because the classifier is uncertain about what the correct class is. That could be e.g. documents belonging to more than one class or documents, for which a specific class not yet exists. The algorithm cannot handle problems like outliers or noise in the data, which were introduced as the dataset was generated. On the one hand that means that there does not exist a solution with all wrongly classified documents generalized without loosing a large amount of the correctly classified items. On the other hand there will always be a generalization of correctly classified items, because they also might belong to more than one class. For these documents, it might be just a coincidence that the classification algorithm and the person who created the dataset did the same class assignment. Furthermore, this means (and is shown by the experiments) that our algorithms is especially useful, if the data is difficult.

Table 1. Performance results for the banksearch data

	$prec_H$	rec_H	$fscore_H$	docs in n_o	above n_o	below n_o
Naïve Bayes	0.831	0.813	0.822	6285	0	1415
NB+pp t=30	0.917	0.820	0.866	6446	777	477
SVM	0.928	0.927	0.928	7080	0	546
SVM+pp t=6	0.955	0.917	0.936	6950	342	334

Table 2. Performance results for the open directory data

	$prec_H$	rec_H	$fscore_H$	docs in n_o	above n_o	below n_o
Naïve Bayes	0.540	0.294	0.381	1290	0	1426
NB+pp t=70	0.847	0.395	0.539	1561	846	309
SVM	0.698	0.513	0.591	1812	0	869
SVM+pp t=4	0.914	0.498	0.645	934	1445	153

6 Conclusion

In this paper, we presented an approach for improving hierarchical classification results with a post processing step rather than integrating the hierarchy information in the classifier itself. We suggested two performance measures suitable for hierarchical classification and especially our setting, an adaptation of the well-known precision and recall measure and the concept of an optimal node. With these measures and two benchmark datasets, our algorithm was evaluated, showing an improvement over the standard (flat) classification approach especially on more complex hierarchies. Since the results of this work are promising, we want to study in future work whether the generalization criterion and its parameter could be derived automatically.

References

BADE, K. and NÜRNBERGER, A. (2005): Supporting Web Search by User Specific Document Categorization: Intelligent Bookmarks. *Proc. of LIT05, 115–123.*

CAI, L. and HOFMANN, T. (2004): Hierarchical Document Categorization with Support Vector Machines. *Proceedings of 13th ACM Conference on Information and Knowledge Management, 78–87.*

CECI, M. and MALERBA, D. (2003): Hierarchical Classification of HTML Documents with WebClassII. *Proc. of 25th Europ. Conf. on Inform. Retrieval, 57–72.*

CESA-BIANCHI, N., GENTILE, C., TIRONI, A. and ZANIBONI, L. (2004): Incremental Algorithms for Hierarchical Classification. *Neural Information Processing Systems, 233–240.*

CHANG, C. and LIN, C. (2001): LIBSVM: A Library for Support Vector Machines. Software available at http://www.csie.ntu.edu.tw/~cjlin/libsvm.

CHOI, B. and PENG, X. (2004): Dynamic and Hierarchical Classification of Web Pages. *Online Information Review, 28, 2, 139–147.*

DUMAIS, S. and CHEN, H. (2000): Hierarchical Classification of Web Content. *Proceedings of the 23rd International ACM SIGIR Conference on Research and Development in Information Retrieval, 256–263.*

FROMMHOLZ, I. (2001): Categorizing Web Documents in Hierarchical Catalogues. *Proceedings of the European Colloquium on Information Retrieval Research.*

GRANITZER, M. and AUER, P. (2005): Experiments with Hierarchical Text Classification. *Proc. of 9th IASTED Intern. Conference on Artificial Intelligence.*

HOTHO, A., NÜRNBERGER, A. and PAAß, G. (2005): A Brief Survey of Text Mining. *GLDV-J. for Comp. Linguistics & Language Technology, 20, 1, 19–62.*

MCCALLUM, A., ROSENFELD, R., MITCHELL, T. and NG, A. (1998): Improving Text Classification by Shrinkage in a Hierarchy of Classes. *Proceedings of the 15th International Conference on Machine Learning (ICML98), 359–367.*

SINKA, M. and CORNE, D. (2002): A Large Benchmark Dataset for Web Document Clustering. *Soft Computing Systems: Design, Management and Applications, Volume 87 of Frontiers in Artificial Intelligence and Applications, 881–890.*

SUN, A. and LIM, E. (2001): Hierarchical Text Classification and Evaluation. *Proc. of the 2001 IEEE International Conference on Data Mining, 521–528.*

Localized Linear Discriminant Analysis

Irina Czogiel[1], Karsten Luebke[2], Marc Zentgraf[2] and Claus Weihs[2]

[1] Graduiertenkolleg Statistische Modellbildung,
Lehrstuhl für Computergestützte Statistik, Universität Dortmund,
D-44221 Dortmund, Germany; czogiel@statistik.uni-dortmund.de

[2] Lehrstuhl für Computergestützte Statistik, Universität Dortmund,
D-44221 Dortmund, Germany

Abstract. Despite its age, the Linear Discriminant Analysis performs well even in situations where the underlying premises like normally distributed data with constant covariance matrices over all classes are not met. It is, however, a global technique that does not regard the nature of an individual observation to be classified. By weighting each training observation according to its distance to the observation of interest, a global classifier can be transformed into an observation specific approach. So far, this has been done for logistic discrimination. By using LDA instead, the computation of the local classifier is much simpler. Moreover, it is ready for applications in multi-class situations.

1 Introduction

Statistical work on classification begins with the work proposed by Fisher (1936). For the dichotomous case, he suggests to reduce a multivariate classification problem to an univariate problem by linearly transforming the given observations into scalar values such that the separation of the transformed class means is maximized whilst the within class variances of the transformed observations are minimized. Although Fisher's approach is distribution-free, it does implicitly assume that the covariance structure is the same in both classes, because a pooled estimate of the common covariance matrix is used. The resulting classification rule can alternatively be derived using Bayesian argumentation although here more restrictive assumptions are made: the data within each class are assumed to be normally distributed with class-specific means and a common covariance structure. Both approaches can be extended to multi-class situations and in each case, obtaining the actual decision functions for a given data set requires the estimation of the unknown model parameters, namely the class-specific means, the class priors, and the covariance matrix. Since the estimation is carried out without taking into account the nature of the problem at hand, i.e. the classification of a specific trial point, LDA can be considered a global classifier. Hand and Vinciotti (2003) argue that an approach like this can lead to poor results if the chosen model does not exactly reflect the underlying data generating process because then, a good fit in some parts of the data space may worsen the fit in other regions. Since in classification problems accuracy is often not equally important throughout the entire data space, they suggest to improve the fit in regions where a good fit is crucial for obtaining satisfactory results – even if the fit elsewhere is degraded. For the dichotomous logistic discrimination, two approaches have been proposed to accomplish this. Hand and Vinciotti (2003) introduce a logistic discrimination model in which data points in the vicinity of the ideal decision surface are weighted more heavily than those which are far away. Another strategy is presented by Tutz and Binder (2005) who suggest to assign locally adaptive weights to each observation of the training set. By choosing the weights as decreasing in the (Euclidean) distance to the observation to be classified and maximizing the corresponding weighted (log-)likelihood, a localized version of the logistic discrimination model can be obtained. The classifier is therefore adapted to the nature of each individual trial point which turns the global technique of logistic discrimination into an observation specific approach.

In this paper, we adopt the strategy of using locally adaptive weights to the context of LDA which comprises the advantage that localizing a classification rule can be accomplished without numerical methods like Fisher scoring. In the following we call this new approach LLDA (Localized Linear Discriminant Analysis). It will be proposed in Section 2. In Section 3, the benefit of LLDA will be shown on basis of a simulated data set containing local subclasses. The application of LLDA to the real-life problem of business phase classification is described in Section 4. A summary of the main results is provided in Section 5.

2 Localized linear discriminant analysis

Let the training data consist of N observations $(\mathbf{x}_i, y_i)'$, where $\mathbf{x}_i \in \mathbb{R}^p$ is the set of explanatory variables for the ith observation and $y_i \in \{A_1, \dots, A_G\}$ denotes the corresponding class membership. The objective now is to construct a classification rule on basis of the training sample which can then be used for predicting the unknown class of a new observation. In LDA, the classification is based on the posterior class probabilities of the considered trial point $\mathbf{x}$. To calculate these, the data is assumed to be normally distributed with class-specific mean vectors $\boldsymbol{\mu}_g$ and a common covariance matrix $\boldsymbol{\Sigma}$. Let π_g denote the prior probability of A_g, choosing the class with the highest posterior probability for a given trial point $\mathbf{x}$ can then be shown to be equivalent to assigning $\mathbf{x}$ to the class with the largest value of the corresponding discriminant function

$$h_g(\mathbf{x}) = (\boldsymbol{\Sigma}^{-1}\boldsymbol{\mu}_g)'\mathbf{x} - 0.5\,\boldsymbol{\mu}_g'\,\boldsymbol{\Sigma}^{-1}\boldsymbol{\mu}_g + \ln(\pi_g).$$

In practice, the sample analogues of h_g are used:

$$\hat{h}_g(\mathbf{x}) = \left(\mathbf{S}^{-1}\bar{\mathbf{x}}_g\right)'\mathbf{x} - 0.5\,\bar{\mathbf{x}}_g'\,\mathbf{S}^{-1}\bar{\mathbf{x}}_g + \ln(p_g), \tag{1}$$

where $\bar{\mathbf{x}}_g$ denotes the mean vector and p_g denotes the proportion of the training observations belonging to A_g. The matrix $\mathbf{S}$ is the pooled estimate of the covariance matrix. A version of (1) which is adaptive to the nature of the considered trial point can be obtained by introducing weights $w_i = w(\mathbf{x}, \mathbf{x}_i)$ to the sample estimates. For the mean vector and the proportion, this can be formulated as

$$\bar{\mathbf{x}}_{g_L} = \frac{\sum_i w_i \mathbf{x}_i I_{\{\mathbf{x}_i \in A_g\}}}{\sum_i w_i I_{\{\mathbf{x}_i \in A_g\}}} \quad \text{and} \quad p_{g_L} = \frac{\sum_i w_i I_{\{\mathbf{x}_i \in A_g\}}}{\sum_i w_i}.$$

To compute an analogous variant of $\mathbf{S}$, first a weighted estimate of the covariance matrix is calculated for each class:

$$\mathbf{S}_{g_L} = \frac{1}{1 - \sum_i w_i^2 I_{\{\mathbf{x}_i \in A_g\}}} \sum_i w_i \left[(\mathbf{x}_i - \bar{\mathbf{x}}_{g_L}) I_{\{\mathbf{x}_i \in A_g\}}\right] \left[(\mathbf{x}_i - \bar{\mathbf{x}}_{g_L}) I_{\{\mathbf{x}_i \in A_g\}}\right]'.$$

These matrices are then weighted with the number of training observations of the corresponding class and aggregated to

$$\mathbf{S}_L = \frac{N}{N-G} \sum_g p_g \mathbf{S}_{g_L}.$$

As suggested by Tutz and Binder (2005), the weights are chosen to be locally adaptive in the sense that they depend on the Euclidean distance of the considered trial point $\mathbf{x}$ and the training observations $\mathbf{x}_i$. This can be accomplished by using a kernel window

$$w_i = K\left(||\mathbf{x} - \mathbf{x}_i||\right).$$

In this context, various kernels can be used and the performance of a kernel function of course depends on the nature of the problem. In this paper, we will restrict the consideration to the kernel we found most robust against varying data characteristics:

$$K(z) = \exp(-\gamma \cdot z).$$

The quantity $\gamma \in \mathbb{R}^+$ is the flexible parameter of the LLDA algorithm which should be optimized before its usage.
LLDA is based on the local estimates of the model parameters described above. Applying them in (1) yields a set of localized discriminant functions $\hat{h}_{g_L}$ which can be used to construct the classification rule

$$\hat{A}(\gamma) = \arg\max_g \hat{h}_{g_L}(\mathbf{x}). \tag{2}$$

As in classical LDA, this approach may cause numerical problems if the considered trial point $\mathbf{x}$ extremely differs from all G class-specific mean vectors since then, the posterior probabilities for all classes are approximately equal to zero. Although this case is very rare, we augmented (2) in order to account for it:

$$\hat{A}(\gamma) = \begin{cases} \arg\max\limits_g \hat{h}_{g_L}(\mathbf{x}) & , \exists\, g : \exp(-0.5(\mathbf{x}-\bar{\mathbf{x}}_g)'\mathbf{S}_L^{-1}(\mathbf{x}-\bar{\mathbf{x}}_g)) > \frac{10^{-150}}{p_{g_L}}, \\ \arg\min\limits_g ||\mathbf{x}-\bar{\mathbf{x}}_g|| & , \text{otherwise.} \end{cases}$$

If classifying $\mathbf{x}$ on basis of (2) is rather questionable because of its position in the data space, the simple classification rule 'Choose the class with the closest centroit' is applied. For programming the LLDA algorithm, we used the software package R (R Development Core Team (2006)). On the used computer, the critical value 10^{-150} reflects the square root of the machine precision of R for distinguishing a number from zero. It can be chosen computer adaptively.

3 Simulation study

In this section we use a simulated two-class discrimination problem to investigate the performance of LLDA. In our simulation study, the two classes A_1 and A_2 each consist of two subclasses, namely A_{11}, A_{12}, A_{21} and A_{22}. For the training data set, each class is chosen to contain 1000 two-dimensional observations which are equally divided into the two corresponding subclasses. The data points are generated as normally distributed with the common covariance matrix $\Sigma_{ij} = I_2\ \forall i,j$ and the following subgroup-specific means: $\mu_{11} = (1,0)'$, $\mu_{12} = (-1,0)'$, $\mu_{21} = (1.75,0)'$ and $\mu_{22} = (-1.75,0)'$. The entire data cloud building up class A_2 is then relocated by a shift defined through the vector $\mathbf{s} = (0.1, 0.3)'$ and rotated by 60°. This results in a data structure such as shown in Figure 1. The test data is generated independently from the training data in exactly the same way. It, too, consists of 1000 observations per class.

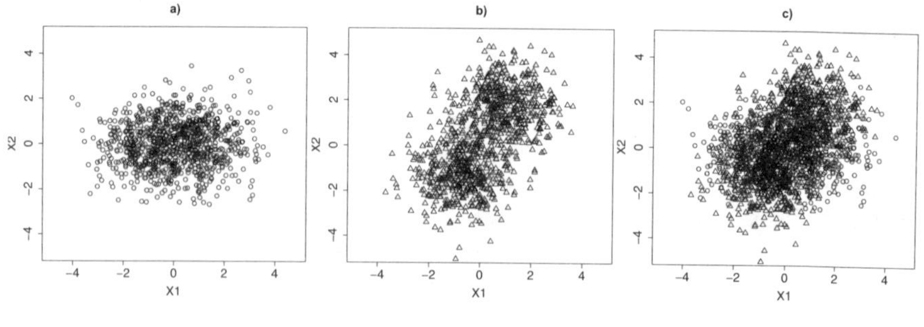

Fig. 1. Simulated Data: a) class A_1, b) class A_2 and c) combined data set

Table 1. Performances of LDA, LLDA and MDA. The error rates are averaged over 10 simulations with the standard error for the average in parenthesis.

Technique	Test Error Rate
LDA	0.4629 (0.0101)
MDA	0.2705 (0.0101)
LLDA	0.2765 (0.0089)

Due to the local structure of the simulated data, LDA is not likely to perform well. We therefore chose MDA (Mixture Discriminant Analysis) as a further competitor for LLDA since this method is particularly designed to cope with local subgroups (Hastie and Tibshirani (1996)). To evaluate the performance of LLDA, the training data is randomly divided into a learning set and a validation set containing 1333 and 667 observations respectively. The optimal value for the flexible parameter γ is then obtained by minimizing the error rate on the validation data. Having done this, the entire training data is used to create a classification rule which is then evaluated on the test data. When using LDA, no optimal parameter values have to be obtained. The classification rule here is learned on basis of the training set and used for the classification of the test observations. The same is done for MDA where the number of subclasses is set to two, i.e. the known optimum, for both classes.

Table 1 contains the results obtained by ten simulations. As expected, all classifiers perform rather badly due to the high degree of overlapping of the two classes. In particular, the global LDA classifier fails to construct a good classification rule whereas MDA and LLDA result in error rates close to the Bayes risk which for the ten simulations on average is given by 0.2747.

4 Application to business phase classification

The field of predicting economic phenomena is a diverse practical example where the data adaptive approach of LLDA is likely to outperform the classical LDA due to the fact that the economic situation develops over time. Assuming the same distribution for all observations of the same class can therefore be too restrictive. In this paper, we address the particular problem of classifying business phases. The data set we consider consists of 13 economic variables with quarterly observations describing the German business cycle from 1961/2 - 2000/4. These variables are standardized and used to classify the business cycle corresponding to a four-phase scheme (Heilemann and Muench (1996)): upswing, upper turning point, downswing and lower turning point. For such kind of time related data, the key interest often is to find a reliable classification rule for e.g. the next six quarters. In order to evaluate classifiers with respect to this request, Luebke and Weihs (2005) propose the Ex-Post-Ante error rate (EPAER).

4.1 The ex-post-ante error rate for time related data

Let the training data $\{(\mathbf{x}_t, y_t)'\}_{t=1}^{T}$ consist of T successive p-dimensional observations $\mathbf{x}_t$ with a known class membership $y_i \in \{A_1, \ldots, A_G\}$, and let *pre* denote the number of future time points for which an appropriate classification rule is required. The EPAER at time $t < T$ then has the form

$$epa(t; pre) = \frac{\sum_{i=t+1}^{t+pre} I_{\{A_i \neq \hat{A}_i^t\}}}{pre},$$

where A_i and $\hat{A}_i^t$ are the true class and the (on basis of the first t time points) estimated class for observation i respectively, i.e. the quantity $epa(t; pre)$ denotes the error rate of the classification rule which is based on the first t training observations and then applied to the next *pre* training observations. This approach therefore produces a time series $\{epa(t; pre)\}_t$ of error rates which can then be condensed to an overall accuracy measure for the classification of the next *pre* time points:

$$\hat{e}_{t_0,pre} = \sum_{t=t_0}^{T-pre} \tilde{w}(t)\, epa(t; pre), \tag{3}$$

where t_0 denotes the starting point for calculating the EPAERs. A suitable weight function for the problem at hand is

$$\tilde{w}(t; t_0, T) = \frac{t}{\sum_{t=t_0}^{T-pre} t},$$

which gives more weight to recently calculated error rates.

4.2 Results

In order to obtain a benchmark for the performance of LLDA, we first applied the classical LDA to the data set described above. The dashed line in Figure 2 shows the resulting time series of EPAERs for the prediction interval length of $pre = 6$ quarters and the starting point $t_0 = 20$. The plot reveals the interesting property that historic events which had an impact on the German business cycle can be identified by peaks of the corresponding error rates. For example the reunification of Germany (1990) changed the business cycle so that the next few phases cannot be predicted properly. Also the start of the first oil-crisis (oil price increased after 1971) and the second oil crisis (1979) cause problems for LDA. Aggregating the time series of error rates corresponding to (3) leads to an estimated overall accuracy for predicting the next six quarters of $\hat{e}_{20,6} = 0.1548$.

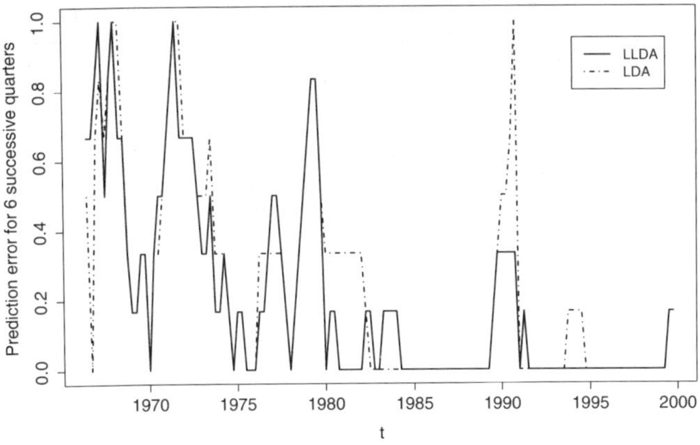

Fig. 2. Time series of ex-post-ante error rates for LDA and LLDA

As described in Section 2, when using LLDA, the performance of a classification rule is influenced by the value of γ which should therefore be optimized with respect to the chosen accuracy measure. A possible way to accomplish this is minimizing the function $\hat{e}_{t_0,pre} = \hat{e}_{t_0,pre}(\gamma)$ with respect to γ. The optimum found by doing so (setting $t_0 = 20$ and $pre = 6$) yields $\hat{e}_{20,6}(\gamma^{\text{opt}}) = 0.1166$. This result, however, is clearly due to overfitting and therefore gives an overoptimistic estimate of the accuracy. To get a more realistic impression about the benefit of optimizing γ, we applied the stepwise procedure shown in Algorithm 1. Since here in each case, the value γ_t^{opt} is obtained from data points prior to t and used for predicting the class membership of upcoming observations (which mimics a real-life situation), the resulting time series of EPAERs $\{epa(t; pre, \gamma_t^{\text{opt}})\}_t$ does not suffer from overfitting. It is shown as the solid line of Figure 2. Since its shape roughly resembles the one obtained by LDA, it, too, can be explained historically. Corresponding to

Algorithm 1 Obtaining the EPAERs based on stepwise optimal values of γ.

$t = t_0$

while t $\leq$ (T-pre) **do**

select the first t training observations $\{(\mathbf{x}_i, y_i)'\}_{i=1}^{t}$ as learning data

find the corresponding optimal value for γ:

$$\gamma_t^{\text{opt}} = \arg\min_{\gamma} \sum_{t'=[\frac{t}{5}]}^{t-pre} \tilde{w}(t'; [\tfrac{t}{5}], t) \cdot \frac{\sum_{i=t'+1}^{t'+pre} I_{\{A_i \neq \hat{A}_i^{t'}(\gamma)\}}}{pre}$$

use γ_t^{opt} to calculate the EPAER for time point t:

$$epa(t; pre; \gamma_t^{\text{opt}}) = \frac{\sum_{i=t+1}^{t+pre} I_{\{A_i \neq \hat{A}_i^{t}(\gamma_t^{\text{opt}})\}}}{pre}$$

$t \leftarrow t + 1$

end while

(3), an overall measure for the goodness of LLDA for predicting the next six quarters is given by $\hat{e}_{20,6}^{\text{opt}} = 0.1209$. Compared to the classical LDA, the observation specific approach of LLDA utilizing an optimal value of γ therefore leads to an improvement of 21.95% in terms of the Ex-Post-Ante-Error rate.

For the sake of completeness, we also applied MDA to the problem at hand. Since an optimization of the flexible MDA parameters, namely the number of local subclasses for each class, would be very time intensive, we assumed the number of subclasses to be equal to a constant s for all four classes. Optimizing s with respect to the EPAER resulted in $s^{\text{opt}} = 1$, i.e. the classical LDA approach. The MDA approach therefore does not comprise further benefit.

5 Summary

By introducing locally adaptive weights to the global LDA technique, the new observation specific classifier LLDA has been developed. Its benefit could be evaluated in two different local settings: on basis of a simulated data set containing local subclasses and in the real life application of classifying business phases. LLDA outperforms LDA in both cases as well as MDA in the business classification problem. In the simulation study, it yields similar results as MDA which is noteworthy since MDA is the method particulary developed for such data structure.

References

FISHER, R.A. (1936): The Use of Multiple Measurements in Taxonomic Problems. *Annals of Eugenics, 7, 179–188.*

HAND, D.J. and VINCIOTTI, V. (2003): Local Versus Global Models for Classification Problems: Fitting Models Where It Matters. *The American Statistician, 57(2), 124–131.*

HASTIE, T. and TIBSHIRANI, R. (1996): Discriminant Analysis by Gaussian Mixtures. *Journal of the Royal Statistical Society, Series B 58(1), 155–176.*

HEILEMANN, U. and MUENCH, J.M. (1996): West German Business Cycles 1963–1994: A Multivariate Discriminant Analysis. In: G. Poser and K. H. Oppenlander (Eds.): *Business Cycle Surveys: Forecasting Issues and Methodological Aspects Selected Papers Presented at the 22nd Ciret Conference*, CIRET-Studien 50.

LUEBKE, K. and WEIHS, C. (2005): Prediction Optimal Classification of Business Phases. *Technical Report 41/05, SFB 475, Universität Dortmund.*

R DEVELOPMENT CORE TEAM (2006): *R: A Language and Environment for Statistical Computing.* R Foundation for Statistcal Computing, Vienna.

TUTZ, G. and BINDER, H. (2005): Localized Classification. *Statistics and Computing, 15, 155–166.*

Calibrating Classifier Scores into Probabilities

Martin Gebel[1] and Claus Weihs[2]

[1] Graduiertenkolleg Statistische Modellbildung, Universität Dortmund, D-44221 Dortmund, Germany; magebel@statistik.uni-dortmund.de
[2] Lehrstuhl für Computergestützte Statistik, Universität Dortmund, D-44221 Dortmund, Germany; weihs@statistik.uni-dortmund.de

Abstract. This paper provides an overview of calibration methods for supervised classification learners. Calibration means a scaling of classifier scores into the probability space. Such a probabilistic classifier output is especially useful if the classification output is used for post–processing. The calibraters are compared by using 10–fold cross–validation according to their performance on SVM and CART outputs for four different two–class data sets.

1 Introduction

Since classifier scores are often object to post–processing, probabilistic scores are useful for further decisions. Although statistical classification methods with supervised learning already produce posterior probabilities, calibration can be necessary as methods' assumptions are not always met. Furthermore, other classifiers like the Support Vector Machine only output unnormalised scores which neither sum up to one nor lie in the interval $[0, 1]$.

This paper gives an overview of several calibration methods introduced in recent years. Widely accepted is the use of Logistic Regression (Platt (1999)) and Isotonic Regression (Zadrozny and Elkan (2002)). However, there are well–performing competitors like the use of the Beta distribution (Garczarek (2002)) or the Bayesian method (Bennett (2002)). All these calibraters are performed on classifier scores of four different two–class data sets as this is the basis of all classification methods and some calibration methods are only applicable for binary tasks. To supply reliable performance measures for generalisation 10–fold cross–validation is used in experiments.

2 Calibration of classifier scores

This section gives a short introduction to the calibration process and an overview of the currently used calibration methods. We consider a classification problem with training set $T := \{(\mathbf{x}_i, c_i), i = 1, \ldots, N\}$ where vector

$\mathbf{x}_i$ is the ith observation of random vector $\mathbf{X}$ of p feature variables and referring class $c_i \in \mathcal{C} = \{-1, +1\}$ which is the realisation of random variable C determined by a supervisor. A classifier produces an unnormalised score $s_{\text{METHOD}}(C = k|\mathbf{x})$ for every class $k \in \mathcal{C}$ and applies classification rule

$$\hat{c}(\mathbf{x}) = \arg\max_{k\in\mathcal{C}} s_{\text{METHOD}}(C = k|\mathbf{x}) . \tag{1}$$

Instead of unnormalised scores, statistical classification methods output posterior probabilities $P_{\text{METHOD}}(C = k|\mathbf{x})$ which claim to cover the confidence that an example belongs to the particular class k. In this case, the classification rule is applied to these probabilities. Probabilities are to be preferred to unnormalised scores basically for two reasons:

1. A classifier is usually just part of a decision process where decisions are associated with certain costs. If the classifier is involved in a cost–sensitive decision with costs differing between classes, it is desirable that the classifier scores reflect the assessment uncertainty.
2. Probabilistic scores simplify the comparison and combination of results from different classifiers.

Most of the statistical classification methods, like Discriminant Analysis or Logistic Regression, already output posterior probabilities, but the method's assumptions are not always met. Such failed assumptions can lead to probabilities which do not reflect the assessment uncertainty. Apart from that, Machine Learners often give no probabilistic output, for example SVM or Neural Networks. Other Machine Learners like the Tree or the Naive Bayes classifier output probabilities which tend to be too extreme (Zadrozny and Elkan (2002)).

Hence, classifier outputs need to be calibrated to score–conditional probabilities. Calibration methods can be separated into three different approaches:

- Estimate a function which maps from score $s_+ := s_{\text{method}}(+1|\mathbf{x})$ (or posterior probability) for the positive class to calibrated score–conditional probability $P_{cal}(+1|s_+)$ and determine $P_{cal}(-1|s_+)$ with the complement;
- Estimate class–conditional probabilities $P(s_+|k)$ for unnormalised scores and class priors $\pi_k := P(C = k)$ to derive $P_{cal}(k|s_+)$ with Bayes' Rule;
- Regard posterior probabilities for the assigned classes as Beta distributed random variables and optimise distributional parameters.

2.1 Calibration via mapping

Calibration via Mapping is basically the search for a function which maps an unnormalised score s_+ for the positive class to a calibrated score–conditional probability $P_{cal}(+1|s_+)$. Such mapping functions can be found by various types of regression techniques, see below. Calibrated probabilities for the negative class are estimated with the complement $P_{cal}(-1|s_+) := 1 - P_{cal}(+1|s_+)$.

Hence, such calibration methods are regularly not directly applicable for multi–class classifier outputs.

If necessary, all mapping methods can also be applied for a mapping from posterior to calibrated probabilities.

Logistic Regression method

A widely approved method for the calibration of scores is to model the log odds of calibrated probabilities

$$\log \frac{P(C=+1|s_+)}{P(C=-1|s_+)} = g(s_+) \tag{2}$$

as linear function g of scores for the positive class. (2) is transformed by using the complement $P(C=-1|s_+) = 1 - P(C=+1|s_+)$, so that the term for determining the calibrated probability becomes

$$P_{log}(C=+1|s_+) = \frac{1}{1+\exp\left[-g(s_+)\right]} \ . \tag{3}$$

The most approved choice for g is the linear function $g(s) = As + B$ with scalar parameters A and B by Platt (1999). Using this assumption leads to a calibration function with sigmoidal shape, for an example see Section 3.1.

The search for the mapping function g is an optimisation problem. The estimators $\hat{A}$ and $\hat{B}$ are found by minimising the error function

$$\mathbf{O}_{log} := -\sum_{i=1}^{N} t_i \log\left[P_{log}(+1|s_{i+})\right] + (1-t_i)\log\left[1 - P_{log}(+1|s_{i+})\right]$$

with *noisy target values*

$$t_i := \begin{cases} 1-\varepsilon_+ = \frac{N_+ + 1}{N_+ + 2} & \text{if } k = +1 \\ \varepsilon_- = \frac{1}{N_- + 2} & \text{if } k = -1 \end{cases}$$

where N_+ and N_- are the number of positive and negative observations, respectively. These modified noisy class labels are used instead of binary target values $t_i := \mathbf{I}_{[c_i=+1]} \in \{0,1\}$ with $\mathbf{I}$ as indicator function to avoid overfitting.

Isotonic Regression method

Isotonic Regression can be used to estimate a function g which describes the mapping from scores s_{i+} to score–conditional probabilities $P_{iso}(C=+1|s_{i+})$. Zadrozny and Elkan (2002) use the basic model

$$P_{iso}(C=+1|s_{i+}) = g(s_{i+}) + \varepsilon_i$$

where g is an isotonic function. Isotonic Regression is a non–parametric form of regression. The function which describes the mapping from explanatory to response variable is chosen from the class of all isotonic, i. e. non–decreasing functions. Given a training set with target values $t_i := \mathbf{I}_{[c_i=+1]} \in \{0, +1\}$ and learned scores s_{i+} an isotonic mapping function $\hat{g}$ can be found according to

$$\hat{g} := \arg\min_{\mathbf{g}} \sum_{i=1}^{N} \left[t_i - \mathbf{g}(s_{i+})\right]^2 . \tag{4}$$

The algorithm *pair–adjacent violators* (PAV) (Zadrozny and Elkan (2002)), finds the stepwise–constant non–decreasing function that best fits the training data according to (4). PAV sets the class labels sorted according to the scores for the positive class as initial functional values $\hat{g}_i$ and replaces non–isotonic sequences with their average until functional values are isotonic.

2.2 Calibration via the Bayesian theorem

While the two previously described calibraters directly estimate calibrated probabilities, the following method by Bennett (2002) contains two steps.

At first, scores s_+ for the positive class are split according to their true class into two groups, so that probabilities $P(s_+|k)$ for the score given the true class $k \in \{-1, +1\}$ can be derived. The idea of this method is to model the distribution of unnormalised scores instead of posterior probabilities. Since some of the standard classification methods only supply posterior probabilities, such probabilities are transformed to unnormalised scores by using log odds before applying the calibration procedure.

The second step is to estimate class priors π_k and determine the score–conditional probabilities by application of Bayes' Theorem

$$P_{bay}(C = k|s_+) = \frac{\pi_k \cdot P(s_+|C = k)}{\pi_- \cdot P(s_+|C = -1) + \pi_+ \cdot P(s_+|C = +1)} .$$

While class priors can be easily estimated by (smoothed) class frequencies in the training set, the crucial point in this way of calibration is choosing the distribution type of the class–conditional densities. It is not justifiable to assume a symmetric distribution e. g. the Gauss distribution for the scores, but an asymmetric one, since scores have a different distributional behaviour in the area between the modes of the distributions compared to the respective other side. The area between the modes contains scores of those observations which are difficult to classify, while the respective other halves reflect the observations for which classification is clear. This conclusion leads to the separation of scores into the three areas *obvious decision for negative class* (A), *hard to classify* (B) and *obvious decision for positive class* (C), see Figure 1.

To consider this different distributional behaviour of scores, the class–conditional densities are modelled as Asymmetric Laplace density

$$P(s_+|C = k) := f_\Lambda(s_+|\theta, \beta, \gamma) = \begin{cases} \frac{\beta\gamma}{\beta+\gamma} \exp\left[-\beta(\theta - s_+)\right] & \text{if } s_+ \leq \theta \\ \\ \frac{\beta\gamma}{\beta+\gamma} \exp\left[-\gamma(s_+ - \theta)\right] & \text{if } s_+ > \theta \end{cases}$$

with scale parameters $\beta, \gamma > 0$ and mode θ. Parameters are estimated for each class separately with Maximum Likelihood (Bennett (2002)).

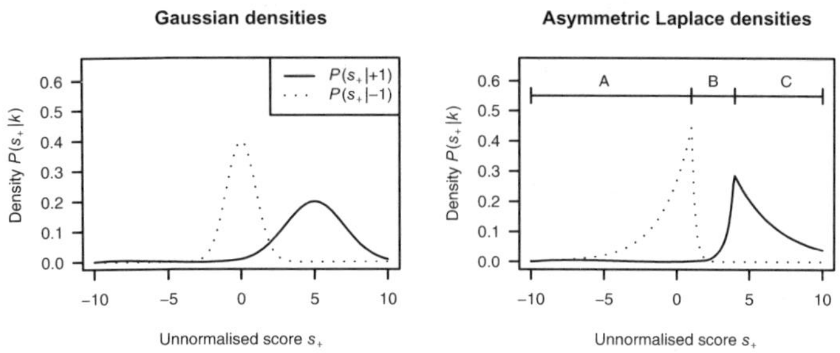

Fig. 1. Typical behaviour of class–conditional densities

2.3 Calibration via assignment values

While the Bayes method splits scores according to their true classes, the following method (Garczarek (2002)) partitions the posterior probabilities $P(k|\mathbf{x})$ by their assignment $\hat{c}(\mathbf{x}_i)$ into groups

$$T_k := \{(c_i, \mathbf{x}_i) \in T : \hat{c}(\mathbf{x}_i) = k\}$$

with potencies $N_{T_k} = |T_k|$. Unnormalised scores have to be pre–normalised before calibration, so that they meet mathematical requirements of probabilities.

Main Idea of this method is to model, in each partition separately, the *assignment values*, i. e. the posterior probabilities for the assigned classes

$$m_{ik} := \begin{cases} \max_{k \in \mathcal{C}} P(C = k|\mathbf{x}_i) \text{ if } & \hat{c}(\mathbf{x}_i) = k \\ \text{not defined} & \text{else} \end{cases} \tag{5}$$

as Beta distributed random variable $M_k \sim \mathcal{B}(p_{M_k}, N_{M_k})$ determined by an unknown expected value $p_{M_k} \in [0, 1]$ and a *dispersion* parameter $N_{M_k} \in \mathbf{N}$. Estimates are calculated with the method of moments (Garczarek (2002)).

The calibration procedure transforms the assignment values for each partition separately to Beta random variables with a new parameter pair. Since calibrated probabilities should reflect the assessment uncertainty, the local correctness rate

$$p_{T_k} := \frac{1}{N_{T_k}} \sum_{c_i \in T_k} \mathbf{I}_{[\hat{c}(\mathbf{x}_i) = c_i]}$$

is regarded as new expected value of the transformed Beta variables. With using the properties of a distribution function such transformation can be easily done so that calibrated assignment values become

$$P_{beta}(C = k|m_{ik}) := F^{-1}_{\mathcal{B}, p_{T_k}, N_{k,opt}} \left[F^{-1}_{\mathcal{B}, \hat{p}_{M_k}, \hat{N}_{M_k}}(m_{ik}) \right] \tag{6}$$

with $N_{k,opt}$ chosen as the integer $N \in \{N_{T_k}, N_{M_k}\}$ which maximises objective

$$\mathbf{O}_{beta} := \sum_{i=1}^{N} \mathbf{I}_{[\hat{c}_i = c_i]} + \mathbf{Ac} \,.$$

To avoid overfitting the number of correctly assigned observations is regularised with the performance measure accuracy **Ac**, see Section 3.2. The complement is used to estimate probabilities $P_{beta}(C \neq k|m_{ik})$ for the class which is not assigned to in the current partition.

3 Comparison of calibration methods

This section supplies a comparison of the presented calibration methods on basis of their functional behaviour and their performance.

3.1 Calibration functions

Figure 2 shows a typical behaviour of the previously presented calibration functions.

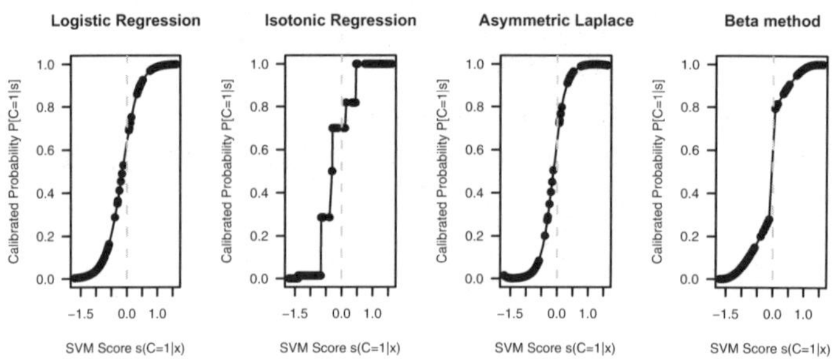

Fig. 2. Typical behaviour of calibration functions

As described in Section 2, the Logistic Regression method outputs a function with sigmoidal shape while Isotonic Regression produces a stepwise–constant function. Using the Asymmetric Laplace distribution leads to a function with three parts, each representing one area of Figure 1. The two areas with extreme probabilities refer to the observations which are easy to classify and the curve connecting those two areas stands for the "difficult to classify" observations. Finally, using assignment values for calibration leads to two independent functions, since the training set is partitioned and the functions are learned independently.

3.2 Measuring calibration performance

Performance measures can be separated into the three different concepts precision, accuracy and ability to separate (Garczarek (2002)). In the following comparison we will regard one measure for each concept.

For a calibrater it is not required to increase precision, but it is important not to decrease it. The standard measure for precision is the *correctness rate*

$$\mathbf{CR} := \frac{1}{N}\sum_{i=1}^{N} \mathbf{I}_{[\hat{c}_i = c_i]} \,. \tag{7}$$

The *accuracy* measures the effectiveness in the assignment of objects to classes by summation of squared distances between (calibrated) probabilities and an Indicator function for the true classes. The sum is standardised to achieve a measure of 1 if all observations are classified perfectly with no insecurity:

$$\mathbf{Ac} := 1 - \frac{2}{N}\sum_{i=1}^{N} \sqrt{\sum_{k \in C} \left[\mathbf{I}_{[c_i = k]} - P(C = k|s)\right]^2} \,. \tag{8}$$

Analogously, the *ability to separate* bases upon distances between (calibrated) probabilities and an Indicator function for the corresponding assigned classes:

$$\mathbf{AS} := 1 - \frac{2}{N}\sum_{i=1}^{N} \sqrt{\sum_{k \in C} \left[\mathbf{I}_{[\hat{c}_i = k]} - P(C = k|s)\right]^2} \,. \tag{9}$$

The ability to separate measures how well classes are distinguished.

3.3 Experiments

The experiments in this paper are based on the four different data sets *Pima Indians Diabetes*, *German Credit*, *Sonar* and *Ionosphere* from the UCI Machine Learning Repository. All data sets have binary class attributes.

The results for analyses with 10–fold cross–validation for calibrating SVM (radial basis kernel) and CART classifier outputs are presented in Tables 3–4.

Mapping with Logistic Regression yields good results for all data sets and classifiers. Compared to CART's posterior probabilities most of the performance measures are increased though correctness often slightly decreases. However, other calibraters are outperformed for CART on Ionosphere, Sonar and Diabetes. By using noisy target values, see Section 2.1, Logistic Regression becomes more robust compared to the other methods.

In calibrating SVM–scores the Beta method reaches best results for Sonar, Credit and Diabetes due to the pre–normalisation of scores, see Section 2.3. Lacking this advantage for CART the Beta method is beaten by Logistic Regression, but still performs well.

The Bayes method needs for its optimisations a reasonable amount of variation in scores which is naturally not given for CART probabilities and also very low for SVM on Credit (mainly consists of indicator variables) and Sonar (small sample size). Hence the Bayes method has only good results for SVM on Ionosphere and Diabetes.

Furthermore, for the previously named reasons slight overfitting occurs for SVM on Credit and Sonar which is even raised by Isotonic Regression, since this optimisation procedure lacks a model complexity term. Probabilities become too extreme, obvious from high **AS** in combination with low **CR** and **Ac**. In contrast to that, Isotonic Regression performs well, if cross–validated correctness of scores is high (Ionosphere).

Fig. 3. Performance measures for calibrating SVM–scores

	Ionosphere			Sonar			German Credit			Diabetes		
	CR	**Ac**	**AS**	**CR**	**Ac**	**AS**	**CR**	**Ac**	**AS**	**CR**	**Ac**	**AS**
$P_{log}(C\|s)$	0.943	0.838	0.889	0.778	0.365	0.446	0.737	0.403	0.690	0.772	0.412	0.631
$P_{iso}(C\|s)$	0.948	0.871	0.933	0.548	0.096	1.000	0.707	0.359	0.648	0.760	0.419	0.608
$P_{bay}(C\|s)$	0.954	0.907	0.980	0.500	0.002	0.995	0.699	0.369	0.793	0.773	0.424	0.671
$P_{beta}(C\|p)$	0.948	0.864	0.943	0.783	0.335	0.446	0.760	0.436	0.775	0.768	0.428	0.685

Fig. 4. Performance measures for calibrating CART–probabilities

	Ionosphere			Sonar			German Credit			Diabetes		
	CR	**Ac**	**AS**	**CR**	**Ac**	**AS**	**CR**	**Ac**	**AS**	**CR**	**Ac**	**AS**
$P_{\text{CART}}(C\|\mathbf{x})$	0.874	0.694	0.837	0.725	0.385	0.767	0.665	0.193	0.444	0.739	0.355	0.624
$P_{log}(C\|p)$	0.883	0.720	0.889	0.721	0.403	0.829	0.647	0.198	0.492	0.736	0.359	0.661
$P_{iso}(C\|p)$	0.877	0.737	0.884	0.721	0.343	0.664	0.656	0.186	0.440	0.727	0.345	0.600
$P_{bay}(C\|s)$	0.840	0.661	0.950	0.706	0.411	0.891	0.587	0.164	0.418	0.725	0.376	0.689
$P_{beta}(C\|p)$	0.874	0.700	0.856	0.716	0.404	0.794	0.667	0.192	0.445	0.736	0.356	0.626

4 Conclusion

Aim of this paper is to compare methods for calibration of classifier scores into probabilities that reflect the assessment uncertainty. Logistic Regression can be widely recommended based on the robust performance shown in the analyses. However, especially for calibrating SVM–scores the Beta method is a good alternative taking advance by pre–normalisation. Hence it is advisable to pre–normalise SVM–scores before applying a calibration method.

Furthermore, as in classification it is required in calibration to take overfitting into account and try to avoid it.

References

BENNETT, P. (2003): Using Asymmetric Distributions to Improve Text Classifier Probability Estimates: A Comparison of New and Standard Parametric Methods. Techn. Report CMU-CS-02-126, Carnegie Mellon, School of Computer Science.

GARCZAREK, U. (2002): Classification Rules in Standardized Partition Spaces. [http://eldorado.uni-dortmund.de:8080/FB5/ls7/forschung/2002/Garczarek]. *Dissertation*, Universität Dortmund.

PLATT, J. (1999): Probabilistic Outputs for Support Vector Machines and Comparisons to Regularized Likelihood Methods. In: A. Smola, P. Bartlett, B. Schoelkopf and D. Schuurmans (Eds.): *Advances in Large Margin Classiers.* MIT Press, Cambridge, 61–74.

ZADROZNY, B. and ELKAN, C. (2002): Transforming Classifier Scores into Accurate Multiclass Probability Estimates. In: *Proceedings of the 8th Internat. Conference on Knowledge Discovery and Data Mining.* ACM Press, Edmonton, 694–699.

Nonlinear Support Vector Machines Through Iterative Majorization and I-Splines

Patrick J.F. Groenen[1], Georgi Nalbantov[2] and J. Cor Bioch[1]

[1] Econometric Institute, Erasmus University Rotterdam, P.O. Box 1738, 3000 DR Rotterdam, The Netherlands; `groenen@few.eur.nl`, `bioch@few.eur.nl`

[2] ERIM, Erasmus University Rotterdam, P.O. Box 1738, 3000 DR Rotterdam, The Netherlands; `nalbantov@few.eur.nl`

Abstract. To minimize the primal support vector machine (SVM) problem, we propose to use iterative majorization. To allow for nonlinearity of the predictors, we use (non)monotone spline transformations. An advantage over the usual kernel approach in the dual problem is that the variables can be easily interpreted. We illustrate this with an example from the literature.

1 Introduction

In recent years, support vector machines (SVMs) have become a popular technique to predict two groups out of a set of predictor variables (Vapnik (2000)). This data analysis problem is not new and such data can also be analyzed through alternative techniques such as linear and quadratic discriminant analysis, neural networks, and logistic regression. However, SVMs seem to compare favorably in their prediction quality with respect to competing models. Also, their optimization problem is well defined and can be solved through a quadratic program. Furthermore, the classification rule derived from an SVM is relatively simple and it can be readily applied to new, unseen samples. At the downside, the interpretation in terms of the predictor variables in nonlinear SVM is not always possible. In addition, the usual formulation of an SVM is not easy to grasp. In this paper, we offer a different way of looking at SVMs that makes the interpretation much easier. First of all, we stick to the primal problem and formulate the SVM in terms of a loss function that is regularized by a penalty term. From this formulation, it can be seen that SVMs use robustified errors. Then, we propose a new majorization algorithm that minimizes the loss. Finally, we show how nonlinearity can be imposed by using I-Spline transformations.

2 The SVM loss function

In many ways, an SVM resembles regression quite closely. Let us first introduce some notation. Let $\mathbf{X}$ be the $n \times m$ matrix of predictor variables of n objects and m variables. The $n \times 1$ vector $\mathbf{y}$ contains the grouping of the objects into two classes, that is, $y_i = 1$ if object i belongs to class 1 and $y_i = -1$ if object i belongs to class -1. Obviously, the labels -1 and 1 to distinguish the classes are unimportant. Let $\mathbf{w}$ be the $m \times 1$ vector with weights used to make a linear combination of the predictor variables. Then, the predicted value q_i for object i is

$$q_i = c + \mathbf{x}_i'\mathbf{w}, \tag{1}$$

where $\mathbf{x}_i'$ is row i of $\mathbf{X}$ and c is an intercept. Consider the example in Figure 1a where for two predictor variables, each row i is represented by a point labelled '+' for the class 1 and 'o' for class -1. Every combination of w_1 and w_2 defines a direction in this scatter plot. Then, each point i can be projected onto this line. The idea of the SVM is to choose this line in such a way that the projections of the class 1 points are well separated from those of class -1 points. The line of separation is orthogonal to the line with projections and the intercept c determines where exactly it occurs. Note that if $\mathbf{w}$ has length 1, that is, $\|\mathbf{w}\| = (\mathbf{w}'\mathbf{w})^{1/2} = 1$, then Figure 1a explains fully the linear combination (1). If $\mathbf{w}$ has not length 1, then the scale values along the projection line should be multiplied by $\|\mathbf{w}\|$. The dotted lines in Figure 1a show all those points that project to the lines at $q_i = -1$ and $q_i = 1$. These dotted lines are called the margin lines in SVMs. Note that if there are more than two variables the margin lines become hyperplanes. Summarizing, the SVM has three sets of parameters that determines its solution: (1) the regression weights, normalized to have length 1, that is, $\mathbf{w}/\|\mathbf{w}\|$, (2) the length of $\mathbf{w}$, that is, $\|\mathbf{w}\|$, and (3) the intercept c. SVMs count an error as follows. Every object i from class 1 that projects such that $q_i \geq 1$ yields a zero error. However, if $q_i < 1$, then the error is linear with $1 - q_i$. Similarly, objects in class -1 with $q_i \leq -1$ do not contribute to the error, but those with $q_i > -1$ contribute linearly with $q_i + 1$. In other words, objects that project on the wrong side of their margin contribute to the error, whereas objects that project on the correct side of their margin yield zero error. Figure 1b shows the error functions for the two classes. As the length of $\mathbf{w}$ controls how close the margin lines are to each other, it can be beneficial for the number of errors to choose the largest $\|\mathbf{w}\|$ possible, so that fewer points contribute to the error. To control the $\|\mathbf{w}\|$, a penalty term that is dependent on $\|\mathbf{w}\|$ is added to the loss function. The penalty term also avoids overfitting of the data. Let G_1 and G_{-1} denote the sets of class 1 and -1 objects. Then, the SVM loss function can be written as

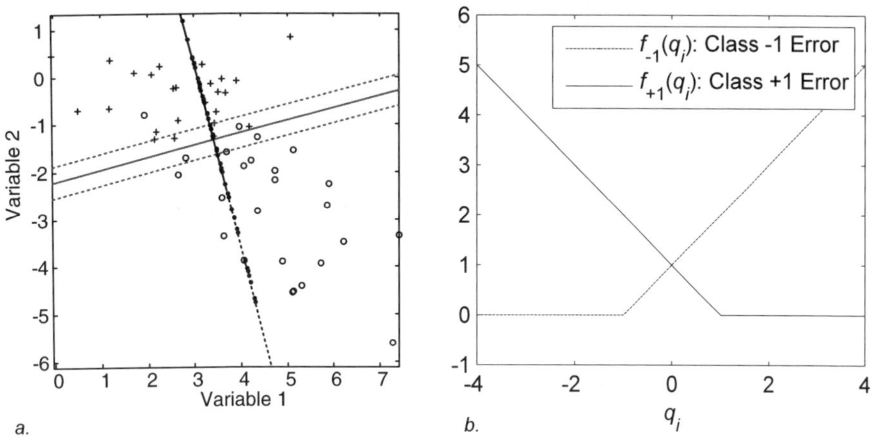

Fig. 1. Panel a projections of the observations in groups 1 (+) and −1 (o) onto the line given by w_1 and w_2. Panel b shows the error function $f_1(q_i)$ for class 1 objects (solid line) and $f_{-1}(q_i)$ for class −1 objects (dashed line).

$$
\begin{aligned}
&L_{\text{SVM}}(c, \mathbf{w}) \\
&= \textstyle\sum_{i\in G_1} \max(0, 1 - q_i) + \sum_{i\in G_{-1}} \max(0, q_i + 1) + \lambda\mathbf{w}'\mathbf{w} \\
&= \textstyle\sum_{i\in G_1} f_1(q_i) \qquad\qquad + \sum_{i\in G_{-1}} f_{-1}(q_i) \qquad\qquad + \lambda\mathbf{w}'\mathbf{w} \qquad (2)\\
&= \text{Class 1 errors} \qquad\quad + \text{Class } -1 \text{ errors} \qquad + \text{Penalty for nonzero } \mathbf{w},
\end{aligned}
$$

see, for similar expressions, Hastie et al. (2000) and Vapnik (2000). Assume that a solution has been found. All the objects i that project on the correct side of their margin, contribute with zero error to the loss. As a consequence, these objects could be removed from the analysis without changing the solution. Therefore, all the objects i that project at the wrong side of their margin and thus induce error or if an object falls exactly on the margin, then these objects determine the solution. Such objects are called support vectors as they form the fundament of the SVM solution. Note that these objects (the support vectors) are not known in advance so that the analysis needs to be carried out with all n objects present in the analysis. What can be seen from (2) is that any error is punished linearly, not quadratically. Thus, SVMs are more robust against outliers than a least-squares loss function. The idea of introducing robustness by absolute errors is not new. For more information on robust multivariate analysis, we refer to Huber (1981), Vapnik (2000), and Rousseeuw and Leroy (2003). The SVM literature usually presents the SVM loss function as follows (see Burges (1998)):

$$L_{\text{SVMClas}}(c, \mathbf{w}, \xi) = C \sum_{i \in G_1} \xi_i + C \sum_{i \in G_2} \xi_i + \frac{1}{2}\mathbf{w}'\mathbf{w}, \tag{3}$$

$$\text{subject to} \quad 1 + (c + \mathbf{w}'\mathbf{x}_i) \leq \xi_i \text{ for } i \in G_{-1} \tag{4}$$

$$1 - (c + \mathbf{w}'\mathbf{x}_i) \leq \xi_i \text{ for } i \in G_1 \tag{5}$$

$$\xi_i \geq 0, \tag{6}$$

where C is a nonnegative parameter set by the user to weight the importance of the errors represented by the so-called slack variables ξ_i. Suppose that object i in G_1 projects at the right side of its margin, that is, $q_i = c + \mathbf{w}'\mathbf{x}_i \geq 1$. As a consequence, $1 - (c + \mathbf{w}'\mathbf{x}_i) \leq 0$ so that the corresponding ξ_i can be chosen as 0. If i projects on the wrong side of its margin, then $q_i = c + \mathbf{w}'\mathbf{x}_i < 1$ so that $1 - (c + \mathbf{w}'\mathbf{x}_i) > 0$. Choosing $\xi_i = 1 - (c + \mathbf{w}'\mathbf{x}_i)$ gives the smallest ξ_i satisfying the restrictions in (4), (5), and (6). As a consequence, $\xi_i = \max(0, 1 - q_i)$ and is a measure of error. A similar derivation can be made for class -1 objects. Choosing $C = (2\lambda)^{-1}$ gives

$$\begin{aligned}
&L_{\text{SVMClas}}(c, \mathbf{w}, \xi) \\
&= (2\lambda)^{-1} \left(\sum_{i \in G_1} \xi_i + \sum_{i \in G_{-1}} \xi_i + 2\lambda \frac{1}{2}\mathbf{w}'\mathbf{w} \right) \\
&= (2\lambda)^{-1} \left(\sum_{i \in G_1} \max(0, 1 - q_i) + \sum_{i \in G_{-1}} \max(0, q_i + 1) + \lambda \mathbf{w}'\mathbf{w} \right) \\
&= (2\lambda)^{-1} L_{\text{SVM}}(c, \mathbf{w}).
\end{aligned}$$

showing that the two formulations (2) and (3) are exactly the same up to a scaling factor $(2\lambda)^{-1}$ and yield the same solution. However, the advantage of (2) is that it can be interpreted as a (robust) error function with a penalty. The quadratic penalty term is used for regularization much in the same way as in ridge regression, that is, to force the w_j to be close to zero. The penalty is particularly useful to avoid overfitting. Furthermore, it can be easily seen that $L_{\text{SVM}}(c, \mathbf{w})$ is a convex function in c and $\mathbf{w}$ because all three terms are convex in c and $\mathbf{w}$. As the function is also bounded below by zero and it is convex, the minimum of $L_{\text{SVM}}(c, \mathbf{w})$ is a global one. In fact, (3) allows the problem to be treated as a quadratic program. However, in the next section, we optimize (2) directly by the method of iterative majorization.

3 A majorizing algorithm for SVM

In the SVM literature, the dual of (3) is reexpressed as a quadratic program and is solved by special quadratic program solvers. A disadvantage of these solvers is that they may become computationally slow for large number of objects n (although fast specialized solvers exist). However, here we derive an iterative majorization (IM) algorithm. An advantage of IM algorithms is that each iteration reduces (2). As this function is convex and IM is a guaranteed descent algorithm, the IM algorithm will stop when the estimates are sufficiently close to the global minimum. Let $f(\mathbf{q})$ be the function to be minimized. Iterative majorization operates on an auxiliary function, called the majorizing function $g(\mathbf{q}, \overline{\mathbf{q}})$, that is dependent on $\mathbf{q}$ and the previous (known) estimate $\overline{\mathbf{q}}$. The majorizing function $g(\mathbf{q}, \overline{\mathbf{q}})$ has to fulfill several requirements: (1) it should touch f at the supporting point $\mathbf{y}$, that is, $f(\overline{\mathbf{q}}) = g(\overline{\mathbf{q}}, \overline{\mathbf{q}})$, (2) it should never be below f, that is, $f(\mathbf{q}) \leq g(\mathbf{q}, \overline{\mathbf{q}})$, and (3) $g(\mathbf{q}, \overline{\mathbf{q}})$ should be simple, preferably linear or quadratic in $\mathbf{q}$. Let $\mathbf{q}^*$ be such that $g(\mathbf{q}^*, \overline{\mathbf{q}}) \leq g(\overline{\mathbf{q}}, \overline{\mathbf{q}})$, for example, by finding the minimum of $g(\mathbf{q}, \overline{\mathbf{q}})$. Because the majorizing function is never below the original function, we obtain the so called sandwich inequality

$$f(\mathbf{q}^*) \leq g(\mathbf{q}^*, \overline{\mathbf{q}}) \leq g(\overline{\mathbf{q}}, \overline{\mathbf{q}}) = f(\overline{\mathbf{q}})$$

showing that the update $\mathbf{q}^*$ obtained by minimizing the majorizing function never increases f and usually decreases it. More information on iterative majorization can be found in De Leeuw (1994), Heiser (1995), Lange et al. (2000), Kiers (2002), and Hunter and Lange (2004) and an introduction in Borg and Groenen (2005). To find an algorithm, we need to find a majorizing function for (2). First, we derive a quadratic majorizing function for each individual error term. Then, we combine the results for all terms and come up with the total majorizing function that is quadratic in c and $\mathbf{w}$ so that an update can be readily derived. At the end of this section, we provide the majorization results. Consider the term $f_{-1}(q) = \max(0, q + 1)$. For notational convenience, we drop the subscript i for the moment. The solid line in Figure 2 shows $f_{-1}(q)$. Because of its shape of a hinge, this function is sometimes referred to as the hinge function. Let $\overline{q}$ be the known error q of the previous iteration. Then, a majorizing function for $f_{-1}(q)$ is given by $g_{-1}(q, \overline{q})$ at the supporting point $\overline{q} = 2$. For notational convenience, we refer in the sequel to the majorizing function as $g_{-1}(q)$ without the implicit argument $\overline{q}$. We want $g_{-1}(q)$ to be quadratic so that it is of the form $g_{-1}(q) = a_{-1}q^2 - 2b_{-1}q + c_{-1}$. To find a_{-1}, b_{-1}, and c_{-1}, we impose two supporting points, one at $\overline{q}$ and the other at $-2 - \overline{q}$. These two supporting points are located symmetrically around -1. Note that the hinge function is linear at both supporting points, albeit with different gradients. Because $g_{-1}(q)$ is quadratic, the additional requirement that $f_{-1}(q) \leq g_{-1}(q)$ is satisfied if $a_{-1} > 0$ and the derivatives at the two supporting points of $f_{-1}(q)$ and $g_{-1}(q)$ are the same. More formally, the requirements are that

$$\begin{aligned} f_{-1}(\overline{q}) &= g_{-1}(\overline{q}), \\ f'_{-1}(\overline{q}) &= g'_{-1}(\overline{q}), \\ f_{-1}(-2-\overline{q}) &= g_{-1}(-2-\overline{q}), \\ f'_{-1}(-2-\overline{q}) &= g'_{-1}(-2-\overline{q}), \\ f_{-1}(q) &\leq g_{-1}(q). \end{aligned}$$

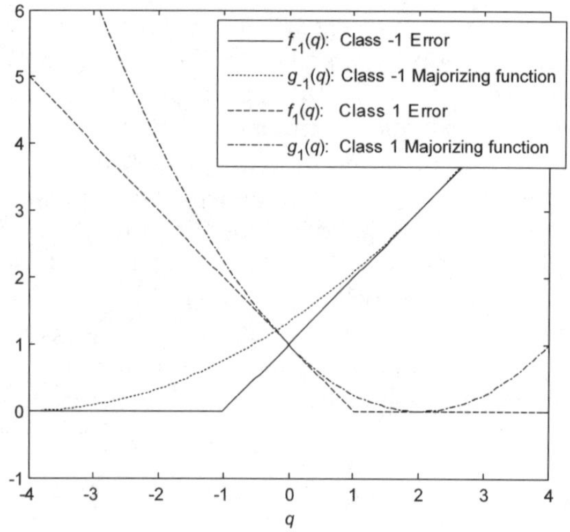

Fig. 2. The error functions of Groups -1 and 1 and their majorizing functions. The supporting point is $\overline{q} = 2$. Note that the majorizing function for Group -1 also touches at $\overline{q} = -4$ and that of Group 1 also at 0.

It can be verified that the choice of

$$a_{-1} = \tfrac{1}{4}|\overline{q}+1|^{-1}, \tag{7}$$
$$b_{-1} = -a_{-1} - \tfrac{1}{4}, \tag{8}$$
$$c_{-1} = a_{-1} + \tfrac{1}{2} + \tfrac{1}{4}|\overline{q}+1|, \tag{9}$$

satisfies all these requirements. Figure 2 shows the majorizing function $g_{-1}(q)$ with supporting points $\overline{q} = 2$ or $\overline{q} = -4$ as the dotted line. For Group 1, a similar majorizing function can be found for $f_1(q) = \max(0, 1-q)$. However, in this case, we require equal function values and first derivative at $\overline{q}$ and at $2-\overline{q}$, that is, symmetric around 1. The requirements are

$$\begin{aligned} f_1(\overline{q}) &= g_1(\overline{q}), \\ f_1'(\overline{q}) &= g_1'(\overline{q}), \\ f_1(2-\overline{q}) &= g_1(2-\overline{q}), \\ f_1'(2-\overline{q}) &= g_1'(2-\overline{q}), \\ f_1(q) &\le g_1(q). \end{aligned}$$

Choosing

$$\begin{aligned} a_1 &= \tfrac{1}{4}|1-\overline{q}|^{-1} \\ b_1 &= a_1 + \tfrac{1}{4} \\ c_1 &= a_1 + \tfrac{1}{2} + \tfrac{1}{4}|1-\overline{q}| \end{aligned}$$

satisfies these requirements. The functions $f_1(q)$ and $g_1(q)$ with supporting points $\overline{q} = 2$ or $\overline{q} = 0$ are plotted in Figure 2. Note that a_{-1} is not defined if $\overline{q} = -1$. In that case, we choose a_{-1} as a small positive constant δ that is smaller than the convergence criterion ε (introduced below). Strictly speaking, the majorization requirements are violated. However, by choosing δ small enough, the monotone convergence of the sequence of $L_{\text{SVM}}(\mathbf{w})$ will be no problem. The same holds for a_1 if $\overline{q} = 1$. Let

$$a_i = \begin{cases} \max(\delta, a_{-1i}) & \text{if } i \in G_{-1}, \\ \max(\delta, a_{1i}) & \text{if } i \in G_1, \end{cases} \tag{10}$$

$$b_i = \begin{cases} b_{-1i} & \text{if } i \in G_{-1}, \\ b_{1i} & \text{if } i \in G_1, \end{cases} \tag{11}$$

$$c_i = \begin{cases} c_{-1i} & \text{if } i \in G_{-1}, \\ c_{1i} & \text{if } i \in G_1. \end{cases} \tag{12}$$

Then, summing all the individual terms leads to the majorization inequality

$$L_{\text{SVM}}(c, \mathbf{w}) \leq \sum_{i=1}^{n} a_i q_i^2 - 2\sum_{i=1}^{n} b_i q_i + \sum_{i=1}^{n} c_i + \lambda \sum_{j=1}^{m} w_j^2. \tag{13}$$

Because $q_i = c + \mathbf{x}_i'\mathbf{w}_i$, it is useful to add an extra column of ones as the first column of $\mathbf{X}$ so that $\mathbf{X}$ becomes $n \times (m+1)$. Let $\mathbf{v}' = [c\ \mathbf{w}']$ so that $\mathbf{q} = \mathbf{X}\mathbf{v}$. Now, (2) can be majorized as

$$\begin{aligned} L_{\text{SVM}}(\mathbf{v}) &\leq \sum_{i=1}^{n} a_i (\mathbf{x}_i'\mathbf{v})^2 - 2\sum_{i=1}^{n} b_i \mathbf{x}_i'\mathbf{v} + \sum_{i=1}^{n} c_i + \lambda \sum_{j=2}^{m+1} v_j^2 \\ &= \mathbf{v}'\mathbf{X}'\mathbf{A}\mathbf{X}\mathbf{v} - 2\mathbf{v}'\mathbf{X}'\mathbf{b} + c_m + \lambda \mathbf{v}'\mathbf{K}\mathbf{v} \\ &= \mathbf{v}'(\mathbf{X}'\mathbf{A}\mathbf{X} + \lambda\mathbf{K})\mathbf{v} - 2\mathbf{v}'\mathbf{X}'\mathbf{b} + c_m, \end{aligned} \tag{14}$$

where $\mathbf{A}$ is a diagonal matrix with elements a_i on the diagonal, $\mathbf{b}$ is a vector with elements b_i, and $c_m = \sum_{i=1}^{n} c_i$, and $\mathbf{K}$ is the identity matrix except for element $k_{11} = 0$. Differentiation the last line of (14) with respect to $\mathbf{v}$ yields the system of equalities linear in $\mathbf{v}$

$$(\mathbf{X}'\mathbf{A}\mathbf{X} + \lambda\mathbf{K})\mathbf{v} = \mathbf{X}'\mathbf{b}.$$

The update $\mathbf{v}^+$ solves this set of linear equalities, for example, by Gaussian elimination, or, somewhat less efficiently, by

$$\mathbf{v}^+ = (\mathbf{X}'\mathbf{A}\mathbf{X} + \lambda\mathbf{K})^{-1}\mathbf{X}'\mathbf{b}. \tag{15}$$

Because of the substitution $\mathbf{v}' = [c\ \mathbf{w}']$, the update of the intercept is $c^+ = v_1$ and $w_j^+ = v_{j+1}^+$ for $j = 1, \ldots, m$. The update $\mathbf{v}^+$ forms the heart of the majorization algorithm for SVMs. The majorizing algorithm for minimizing the standard SVM in (2) is summarized in Figure 3. This algorithm has several advantages. First, it iteratively approaches the global minimum closer in each iteration. In contrast, quadratic programming of the dual problem need to

$t = 0$;
Set ε to a small positive value;
Set $\mathbf{w}_0$ and c_0 to a random initial values;
Compute $L_{\text{SVM}}(c_0, \mathbf{w}_0)$ according to (2);
Set $L_{-1} = L_{\text{SVM}}(c_0, \mathbf{w}_0) + 2\varepsilon$;
while $L_{t-1} - L_{\text{SVM}}(c_t, \mathbf{w}_t) > \varepsilon$ **do**
  $t = t + 1$;
  $L_{t-1} = L_{\text{SVM}}(c_{t-1}, \mathbf{w}_{t-1})$;
  Compute the diagonal matrix $\mathbf{A}$ with elements defined by (10);
  Compute the $\mathbf{b}$ with elements defined by (11);
  Find $\mathbf{v}^+$ by solving (15);
  Set $c_t^+ = v_1$ and $w_{tj}^+ = v_{j+1}^+$ for $j = 1, \ldots, m$;
end

Fig. 3. The SVM majorization algorithm

solve the dual problem completely to have the global minimum of the original primal problem. Secondly, the progress can be monitored, for example, in terms of the changes in the number of misclassified objects. Thirdly, to reduce the computational time, smart initial estimates of c and $\mathbf{w}$ can be given if they are available, for example, from a previous cross validation run. An illustration of the iterative majorization algorithm is given in Figure 4. Here, c is fixed at its optimal value and the minimization is only over $\mathbf{w}$, that is, over w_1 and w_2. Each point in the horizontal plane represents a combination of w_1 and w_2. The majorization function is indeed located above the original function and touches it at the dotted line. The w_1 and w_2 where this majorization function finds its minimum, $L_{\text{SVM}}(c, \mathbf{w})$ is lower than at the previous estimate, so $L_{\text{SVM}}(c, \mathbf{w})$ has decreased. Note that the separation line and the margins corresponding to the current estimates of w_1 and w_2 are given together with the class 1 points represented as open circles and the class -1 points as closed circles.

4 Nonlinear SVM

The SVM described so far tries to find a linear combination $\mathbf{q} = \mathbf{X}\mathbf{b}$ such that negative values are classified into class -1 and positive values into class 1. As a consequence, there is a separation hyperplane of all the points that project such that $q = 0$. Therefore, the standard SVM has a linear separation hyperplane. To allow for a nonlinear separation plane, the the classical approach is to turn to the dual problem and introduce kernels. By doing so, the relation with the primal problem $L_{\text{SVM}}(c, \mathbf{w})$ is lost and the interpretation in

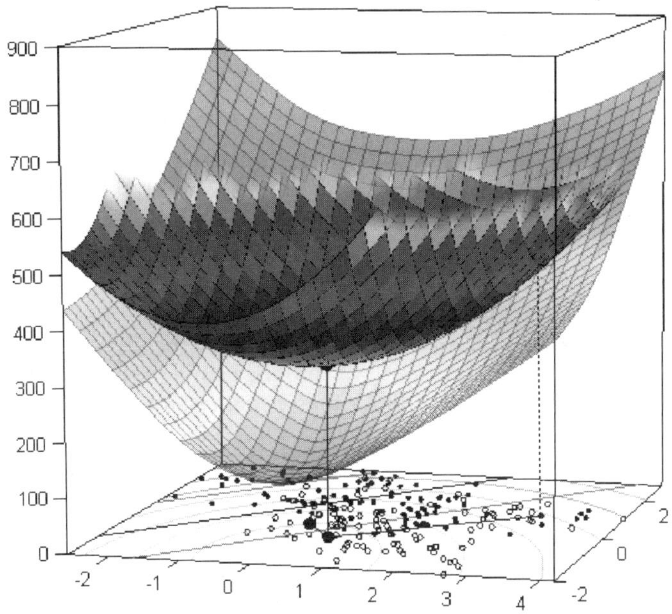

Fig. 4. Example of the iterative majorization algorithm for SVMs in action where c is fixed and w_1 and w_2 are being optimized. The majorization function touches $L_{\mathrm{SVM}}(c, \mathbf{w})$ at the previous estimates of $\mathbf{w}$ (the dotted line) and a solid line is lowered at the minimum of the majorizing function showing a decrease in $L_{\mathrm{SVM}}(c, \mathbf{w})$ as well.

terms of the original variables is not always possible anymore. To cover nonlinearity, we use the optimal scaling ideas from Gifi (1990). In particular, each predictor variable is being transformed. A powerful class of transformations is formed by spline transformation. The advantage of splines is that they yield transformations that are piecewise polynomial. In addition, the transformation is smooth. Because the resulting transformation consists of polynomials whose coefficients are known, the spline basis values can also be computed for unobserved points. Therefore, the transformed value of test points in can be easily computed. There are various sorts of spline transformations, but here we choose the I-Spline transformations (see Ramsay (1988)). An example of such a transformation is given in Figure 5f. In this case, the piecewise polynomial consists of four intervals. The boundary points between subsequent intervals are called interior knots t_k. The interior knots are chosen such that the number of observations is about equal in each interval. The degree of the polynomial d in the I-Spline transformation of Figure 5f is 2 so that each piece

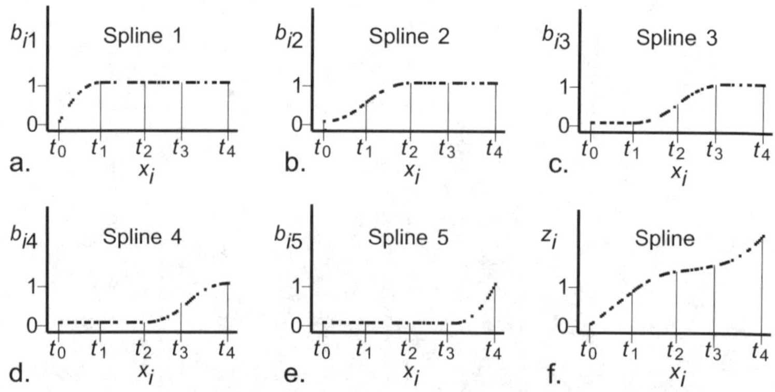

Fig. 5. Panels a to e give an example of the columns of the spline basis $\mathbf{B}$. It consists of five columns given the degree $d = 2$ and the number of interior knots $k = 3$. Panel f shows a linear sum of these bases that is the resulting I-Spline transformation $\mathbf{z}$ for a single predictor variable $\mathbf{x}$.

is quadratic in the original predictor variable $\mathbf{x}_j$. Once the number of interior knots k and the degree d are fixed, each I-Spline transformation can be expressed as $\mathbf{z}_j = \mathbf{B}_j\mathbf{w}_j$ where $\mathbf{B}_j$ is the so called spline basis of $n \times (d+k)$. For the example transformation in Figure 5f, the columns of $\mathbf{B}_j$ are visualized in Figures 5a to 5e. One of the properties of the I-Spline transformation is that if the weights $\mathbf{w}_j$ are all positive, then the transformation is monotone increasing as in our example as in Figure 5f. This property is of use to interpret the solution. To estimate the transformations in the SVM problem, we simply replace $\mathbf{X}$ by the matrix $\mathbf{B} = [\mathbf{B}_1\ \mathbf{B}_2\ \ldots\ \mathbf{B}_m]$, that is, by concatenating the spline bases $\mathbf{B}_j$, one for each original predictor variable $\mathbf{x}_j$. In our example, we have $m = 2$ variables ($\mathbf{x}_1$ and $\mathbf{x}_2$), $d = 2$, and $k = 3$, so that $\mathbf{B}_1$ and $\mathbf{B}_2$ are both matrices of size $n \times (d+k) = n \times 5$ and $\mathbf{B}$ is of size $n \times m(d+k) = n \times 10$. Then, the vector of weights $\mathbf{w}' = [\mathbf{w}_1'\ \mathbf{w}_2'\ \ldots\ \mathbf{w}_m']$ is of size $m(d+k) \times 1$ and the transformation $\mathbf{z}_j$ of a single variable $\mathbf{x}_j$ is given by $\mathbf{z}_j = \mathbf{B}_j\mathbf{w}_j$. Thus, to model the nonlinearity in the decision boundary, we extend the space of predictor variables from $\mathbf{X}$ to the space of the spline bases of all predictor variables and then search through the SVM for a linear separation in this high dimensional space. Consider the example of a mixture of 20 distributions for two groups given by Hastie et al. (2000) on two variables. The left panel of Figure 6 shows a sample 200 points with 100 in each class. It also shows the optimal Bayes decision boundary. The right panel of Figure 6 shows the results of the SVM with I-Spline transformations of the two predictor variables using $k = 5$ and $d = 2$. After cross validation, the best preforming $\lambda = .00316$ yielding a training error rate of .21. Once the SVM is estimated and $\mathbf{w}$ is known, the transformations $\mathbf{z}_j = \mathbf{B}_j\mathbf{w}_j$ are determined. Thus, each interval of the transformation is in our example with $d = 2$ a quadratic function in $\mathbf{x}_j$ for which the polynomial coefficients can

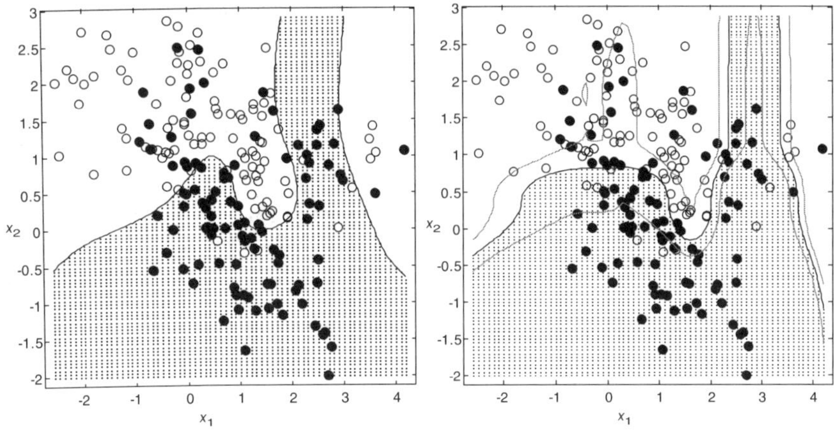

Fig. 6. The left panel shows samples from a mixture distribution of two groups (Hastie et al. (2000)) and a sample of points of these groups. The line is the optimal Bayes decision boundary and Bayes error is .21. The right panel shows the SVM solution using spline transformations of degree 2 and 5 interior knots, $\lambda = .00316$, with an training error rate of .21.

be derived. As test points, we use a grid in the space of the two predictor variables. Because the polynomial coefficients are known for each interval, we can derive the transformed (interpolated) value z_{ij} of test point i for each j and the value $q_i = c + \sum_j z_{ij}$ where c is the intercept. If q_i for the test point is positive, we classify the test point i in class 1, if $q_i < 0$ in class -1, and if $q_i = 0$ it is on the decision boundary. This classification is done for all the test points in the grid, resulting in the reconstructed boundary in the right panel of Figure 6. I-Splines have the property that for nonnegative $\mathbf{w}$ the transformation $\mathbf{z}_j$ is monotone increasing with $\mathbf{x}_j$. Let $\mathbf{w}_j^+ = \max(\mathbf{0}, \mathbf{w}_j)$ and $\mathbf{w}_j^- = \min(\mathbf{0}, \mathbf{w}_j)$ so that $\mathbf{w}_j = \mathbf{w}_j^+ + \mathbf{w}_j^-$. Then, the transformation $\mathbf{z}_j$ can be split in a monotone increasing part $\mathbf{z}_j^+ = \mathbf{B}_j\mathbf{w}_j^+$ and a monotone decreasing part $\mathbf{z}_j^- = \mathbf{B}_j\mathbf{w}_j^+$. For the mixture example, these transformations are shown in Figure 7 for each of the two predictor variables. From this figure, we see that for $\mathbf{x}_1$ the nonlinearity is caused by the steep transformations of values for $x_1 > 1$ both for the positive as for the negative part. For $\mathbf{x}_2$, the nonlinearity seems to be caused by only by the negative transformation for $x_2 < 1$.

5 Conclusions

We have discussed how SVM can be viewed as a the minimization of a robust error function with a regularization penalty. Nonlinearity was introduced by mapping the space of each predictor variable into a higher dimensional space

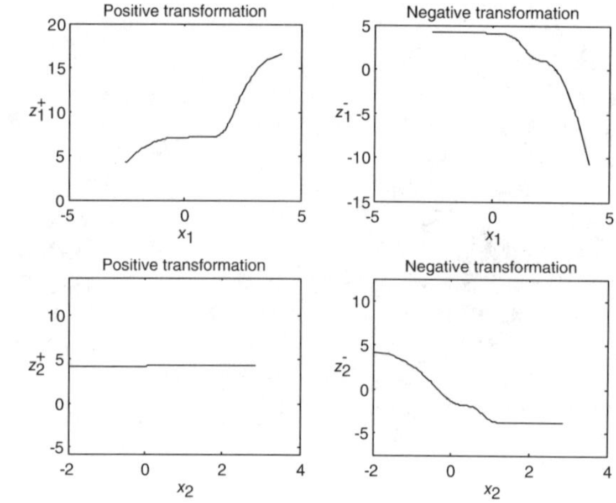

Fig. 7. Spline transformations of the two predictor variables used in the SVM solution in the right panel of Figure 6.

using I-Spline basis. The regularization is needed to avoid overfitting in the case when the number of predictor variables increases or the their respective spline bases become of high rank. The use of I-Spline transformations are useful to allow interpreting the nonlinearity in the prediction. We also provided a new majorization algorithm for the minimization of the primal SVM problem. There are several open issues and possible extensions. A disadvantage of the I-Spline transformation over the usual kernel approach is that the degree of the spline d and the number of interior knots k need to be set whereas most standard kernels just have a single parameter. We need more numerical experience to study what good ranges for these parameters are. SVMs can be extended to problems with more than two classes in several ways. If the extension has error terms of the form $f_1(q)$ or $f_{-1}(q)$ then the present majorization results can be readily applied for an algorithm. The use of splines is always possible as the columns of the spline bases replace the original predictor variables. However, to use splines makes most sense whenever the method involves taking a linear combination of the columns and when some caution against overfitting is taken. The present approach can be extended to other error functions as well. Also, there seems to be close relations with the optimal scaling approach taken in multidimensional scaling and by the work of Gifi (1990). We intend to study these issues in subsequent publications.

References

BORG, I. and GROENEN, P.J.F. (2005): *Modern Multidimensional Scaling: Theory and applications (2nd edition)*. Springer, New York.

BURGES, C.J.C. (1998): A Tutorial on Support Vector Machines for Pattern Recognition. *Knowledge Discovery and Data Mining, 2, 121–167.*

DE LEEUW, J. (1994): Block Relaxation Algorithms in Statistics. In: H.-H. Bock, W. Lenski and M. M. Richter (Eds.): *Information Systems and Data Analysis.* Springer, Berlin, 308–324.

GIFI, A. (1990): *Nonlinear Multivariate Analysis.* Wiley, Chichester.

HASTIE, T., TIBSHIRANI, R. and FRIEDMAN, J. (2000): *The Elements of Statistical Learning.* Springer, New York.

HEISER, W.J. (1995): Convergent Computation by Iterative Majorization: Theory and Applications in Multidimensional Data Analysis. In: W.J. Krzanowski (Ed.): *Recent Advances in Descriptive Multivariate Analysis.* Oxford University Press, Oxford, 157–189.

HUBER, P.J. (1981): *Robust Statistics.* Wiley, New York.

HUNTER, D.R. and LANGE, K. (2004): A Tutorial on MM Algorithms. *The American Statistician, 39, 30–37.*

KIERS, H.A.L. (2002): Setting up Alternating Least Squares and Iterative Majorization Algorithms for Solving Various Matrix Optimization Problems. *Computational Statistics and Data Analysis, 41, 157–170.*

LANGE, K., HUNTER, D.R. and YANG, I. (2000): Optimization Transfer using Surrogate Objective Functions. *Journal of Computational and Graphical Statistics, 9, 1–20.*

RAMSAY, J.O. (1988): Monotone Regression Splines in Action. *Statistical Science, 3, 4, 425–461.*

VAPNIK, V.N. (2000): *The nature of statistical learning theory.* Springer, New York.

Deriving Consensus Rankings from Benchmarking Experiments

Kurt Hornik[1] and David Meyer[2]

[1] Department of Statistics and Mathematics, Wirtschaftsuniversität Wien, A-1090 Wien, Austria; kurt.hornik@wu-wien.ac.at

[2] Department of Information Systems and Operations, Wirtschaftsuniversität Wien, A-1090 Wien, Austria; david.meyer@wu-wien.ac.at

Abstract. Whereas benchmarking experiments are very frequently used to investigate the performance of statistical or machine learning algorithms for supervised and unsupervised learning tasks, overall analyses of such experiments are typically only carried out on a heuristic basis, if at all. We suggest to determine winners, and more generally, to derive a consensus ranking of the algorithms, as the linear order on the algorithms which minimizes average symmetric distance (Kemeny-Snell distance) to the performance relations on the individual benchmark data sets. This leads to binary programming problems which can typically be solved reasonably efficiently. We apply the approach to a medium-scale benchmarking experiment to assess the performance of Support Vector Machines in regression and classification problems, and compare the obtained consensus ranking with rankings obtained by simple scoring and Bradley-Terry modeling.

1 Introduction

The past decades have featured an immense proliferation of available statistical or machine learning algorithms for supervised and unsupervised learning tasks, including decision trees, neural networks, support vector machines, and resampling methods such as bagging or boosting. With theoretical analyses of the properties of such algorithms becoming ever more challenging, detailed experiments based on suitable combinations of artificial and real-world data sets are employed to study these algorithms. In particular, performance is typically investigated using benchmarking experiments where several competing algorithms are used on a collection of data sets (e.g., from the UCI Machine Learning repository (Newman et al. (1998))).

Quite surprisingly, solid methodological frameworks for the *analysis* of the results of such benchmarking experiments are typically lacking. Often, p-values reported for assessing significant difference in the performance of algorithms are rather incorrect (e.g., necessary independence assumptions cannot be guaranteed in commonly employed experimental designs) or potentially misleading (e.g., by solely focusing on the means of performance distributions which can be considerably skewed). Hornik et al. (2005) provide a framework which allows the comparison of algorithms on *single* data sets based on classical statistical inference procedures, making it possible to test one-sided hypotheses ("Does algorithm A_i perform significantly better than algorithm A_j on data set D_b?") as well as the hypothesis of non-equivalence.

An overall analysis of the benchmarking experiment would suitably aggregate the performance "measurements" on the individual data set, resulting, e.g., in the determination of a "winner", or more generally a *consensus ranking* which orders the algorithms according to their overall performance. Clearly, conclusions drawn from such an analysis should be taken with the appropriate grain of salt: the results depend on the specific collection $\mathcal{D}$ of data sets employed and hence are primarily *conditional* on the data. They can only be "representative" across learning tasks in as much as $\mathcal{D}$ can serve this purpose. With no algorithm being able to uniformly outperform all others for all possible data sets, it is clearly impossible to use benchmark experiments to determine whether a certain algorithm is "generally" the best. Still, a chosen $\mathcal{D}$ might be reasonably representative of the needs of a group of researchers or practitioners, and there is an obvious need for a well-founded group decision based on the benchmarking results (e.g., which algorithm to deploy in a specific application).

In this paper, we indicate how consensus rankings can naturally be obtained from paired performance comparisons on the benchmark data sets. The underlying theory and computational issues are discussed in Section 2. An application to a medium-scale benchmarking experiment to assess the performance of Support Vector Machines in regression and classification problems (Meyer et al. (2003)) is given in Section 3. The obtained rankings are also compared to those provided by a simple scoring approach (Borda (1781)) and a Bradley-Terry (Bradley and Terry (1952)) model as two popular alternative approaches.

2 Consensus rankings

Consider a benchmarking experiment featuring n learning algorithms $\mathcal{A} = \{A_1, \dots, A_n\}$ and B data sets $\mathcal{D} = \{D_1, \dots, D_B\}$, and suppose that it is possible to "rank" the algorithms according to their performance on each data set D_b. Such rankings could for example be obtained based on the means or (quite surprisingly, far less popular) median performances obtained from several runs of the algorithms on suitable bootstrap samples from the data set. Note that distributions of performance measures typically exhibit considerable skewness: hence, whereas means or medians may be employed to investigate differences in location, aggregation should not be based on the "raw" values of the performance measures (but could, e.g., use the ranks or a related scoring method instead). In any case, we feel that it is both more natural and preferable to derive rankings based on the *comparisons* of performances only, in particular, basing these on a notion of one algorithm A_i performing *significantly better* than another algorithm A_j, symbolically, $A_i > A_j$. Using the experimental designs of Hothorn et al. (2005), "classical" hypothesis tests can be employed for assessing significant deviations in performance.

The collection of paired comparisons for a data set D_b induces a *relation* (more precisely, endorelation) R_b on the set of algorithms $\mathcal{A}$ which expresses either the strict preference relation as indicated above or its dual, or a "$\leq$" relation taking ties (indicating equivalent performance) into account. The collection of benchmark data sets thus induces a *profile* (ensemble) of relations $\mathcal{R} = \{R_1, \dots, R_B\}$. A consensus ranking is a suitable aggregation of the relation profile into a relation R. Hereafter, we will assume that a *linear order* is sought, i.e., that the consensus relation be an endorelation on $\mathcal{A}$ which is reflexive, asymmetric, and transitive.

There is a huge literature on consensus methods for relations, starting in the late 18th century with the approaches of Borda (1781) and Condorcet (1785) to aggregate the preferences of voters. In Borda's approach, the objects are ranked according to the so-called Borda marks, the overall numbers of "wins" in the paired comparisons. As this may result in one object being ranked above another in the consensus relation R even though it was consistenly ranked below the other in the invidual relations, Condorcet suggested to base R on a "majority" rule which ranks an object i above object j iff the number of individual wins of i over j exceeds the number of losses. This rule may result in intransitivities ("L'Effet Condorcet") even when aggregating strict preference relations, i.e., do not necessarily yield a linear order as sought. If the Condorcet solution is transitive, it agrees with the Borda solution.

The Borda and Condorcet approaches are examples of so-called *constructive* consensus methods, which simply specify a way to obtain a consensus relation. In the *axiomatic* approach (e.g., Day and McMorris (2003)), emphasis is on the investigation of existence and uniqueness of consensus relations characterized axiomatically. The *optimization* approach formalizes the natural idea of describing consensus relations as the ones which "optimally represent the profile" by providing a criterion to be optimized over a suitable set $\mathcal{C}$ of possible consensus relations. This approach goes back to Régnier (1965), who suggested to determine R by solving (a non-weighted variant of) the problem

$$\sum_{b=1}^{B} w_b d(R, R_b) \to \min_{R \in \mathcal{C}},$$

where d is a suitable dissimilarity (distance) measure. Such a relation R has also been termed the *median* (more precisely, the $\mathcal{C}$-median) of the profile (Barthélemy and Monjardet (1981)). For order relations, Kemeny and Snell (1962) have shown that there is a unique d satisfying a few natural axioms (basically, metricity and betweenness). This so-called Kemeny-Snell distance d_{KS} in fact coincides with the *symmetric difference distance* d_{Δ} between relations, i.e., the cardinality of the symmetric difference of the relations, or equivalently, the number of pairs of objects being in exactly one of the two relations. This is also the minimal path length distance d_{MPL} between the relations: in the lattice obtained by equipping the set of endorelations with its natural (pointwise incidence) order, d_{MPL} is the minimal number of moves for transforming one relation into the other along the edges of the covering graph (Hasse diagram) of the poset (Monjardet (1981)). Both characterizations suggest that d_Δ is the most natural way to measure distance between relations, and to use for the optimization-based consensus approach.

Median linear orders based on d_Δ can be computed by integer linear programming (e.g., Marcotorchino and Michaud (1982)). Write $r_{ij}(b)$ and r_{ij} for the incidences of relations R_b and R, respectively. Noting that $u = u^2$ for $u \in \{0,1\}$ and hence $|u - v| = u + v - 2uv$ for $u, v \in \{0,1\}$, we have

$$\begin{aligned}
\sum_{b=1}^{B} w_b d(R, R_b) &= \sum_b w_b \sum_{i,j} |r_{ij}(b) - r_{ij}| \\
&= \sum_b w_b \sum_{i,j} (r_{ij}(b) + r_{ij} - 2r_{ij}(b) r_{ij}) \\
&= \text{const} - \sum_{ij} \left(\sum_b (2w_b r_{ij}(b) - 1) \right) r_{ij}
\end{aligned}$$

so that, letting $c_{ij} = \sum_b (2w_b r_{ij}(b) - 1)$, the median linear order R can be obtained by solving

$$\sum_{i \neq j} c_{ij} r_{ij} \Rightarrow \max$$

with the constraints that the r_{ij} be the incidences of a linear order, i.e.,

$$\begin{array}{rll}
r_{ij} \in \{0,1\} & i \neq j & \text{(binarity)} \\
r_{ij} + r_{ji} = 1 & i \neq j & \text{(asymmetry)} \\
r_{ij} + r_{jk} - r_{ik} \leq 1 & i \neq j \neq k & \text{(transitivity)}
\end{array}$$

We note that this is a "very hard" combinatorial optimization problem (in fact, NP complete), see Wakabayashi (1998). Its space complexity is related to the number of variables and constraints which are of the orders n^2 and n^3, respectively. In fact, the asymmetry conditions imply that we can, e.g., work only with the upper diagonal part of R, i.e., $r_{ij}, i < j$, and use $r_{ij} = 1 - r_{ji}$ for $i > j$. For each triple of distinct i, j, k the 6 transitivity conditions reduce to 2 non-redundant ones for $i < j < k$. The worst case time complexity is at most of the order 2^n. Quite often, solutions can be found efficiently via Lagrangian relaxation (Marcotorchino and Michaud (1982)), i.e., by replacing the binarity constraints $r_{ij} \in \{0,1\}$ by $0 \leq r_{ij} \leq 1$, $i \neq j$, and iteratively adding "cutting planes" selectively enforcing binarity to the relaxation (Grötschel and Wakabayashi (1989)). One can also use state of the art general-purpose integer programming software, such as the open source lp_solve (Berkelaar et al. (2006)) or GLPK (Makhorin (2006)).

If the explicit asymmetry and transitivity conditions are dropped, the corresponding consensus relation can be determined immediately: obviously, the maximum is obtained by taking $r_{ij} = 1$ if $c_{ij} > 0$ and $r_{ij} = 0$ if $c_{ij} < 0$. This is exactly the Condorcet solution, as for preference relations the r_{ij} are the incidences of the wins and $\sum_b (2r_{ij}(b) - 1) > 0$ iff $\sum_b r_{ij} > B/2$, i.e., i wins over j in more than half of the comparisons in the profile. Thus, the Condorcet approach can be given an optimization ("metric") characterization as yielding the (unconstrained) median endorelation when employing symmetric difference distance.

Determining the median linear order can also be interpreted as finding the maximum likelihood paired comparison ranking (deCani (1969)). More generally, constructive consensus approaches could be based on the intrinsic or extrinsic worths (Brunk (1960)) obtained by probabilistic modeling of the paired comparison data. The Bradley-Terry model (Bradley and Terry (1952)) is the most prominent such model, representing the odds that i wins over j as α_i/α_j using worths ("abilities") α_i, or, in an equivalent logit-linear formulation, $\text{logit}(\Pr(i \text{ beats } j)) = \lambda_i - \lambda_j$ with $\lambda_i = \log(\alpha_i)$. Ordering objects according to their fitted abilities yields another simple constructive consensus approach.

3 Application: Benchmarking support vector machines

Meyer et al. (2003) report the results of a benchmark experiment of popular classification and regression methods on both real and artificial data sets. Its main purpose was to compare the performance of Support Vector Machines to other well-known methods both from the field of machine learning (such as neural networks, random forests, and bagging) and "classical" statistics (such as linear/quadratic discriminant analysis and generalized linear models). Most data sets originate from the the UCI Machine Learning repository (Blake and Merz (1998)) and are standard in benchmarking. The size and structure of the data sets cover a wide range of problems: The numbers of cases vary from 106 to 3,196, and the numbers of variables range from 2 to 166, involving a mix of dichotomous, polytomous, and metric variables. Both real and artificial data sets were employed. In total, the study involved $n_c = 17$ methods on $B_c = 21$ datasets for classification, and $n_r = 10$ methods on $B_r = 12$ datasets for regression.

All methods were repeatedly (10 times) trained and tested on all data sets, resulting in $n_c \times B_c = 357$ performance measure *distributions* for classification (misclassification rates) and 120 for regression (root mean squared errors). The error distributions were summarized by three statistics: mean, median, and interquartile range, and reported by means of 8 tables. Even using state-of-the-art visualization methods such as parallel boxplots in a trellis-layout for all data sets, it is hard to compare the performance of one method across several data sets, and to come to an overall assessment.

The method of consensus rankings provides a simple clue to further analysis: for each data set D_b, we computed two-sample t tests on the error distributions of all method pairs (A_i, A_j) to assess whether method A_i performed significantly better than A_j on data set D_b (significance level: 5%). The B relations induced by these paired comparisons were then aggregated by means of three consensus ranking methods described above (Median linear order, Borda, and the Bradley/Terry model). The resulting rankings are compared in Table 1 for classification and Table 2 for regression. Interestingly, for classification, all three methods agree at least for the top 5 methods, whereas the top rankings differ for regression. The space and time complexities for the median linear order consensus on the benchmark experiment results are summarized in Table 3. For both the classification and regression experiments, the results were immediate on a machine with a Pentium M processor with 1.6 GHz and 1 GB of memory, using the **lpSolve** interface (Buttrey (2005)) to R (R Development Core Team (2005)) for solving the integer linear programming problem. The corresponding values of the criterion function $\Phi(R) = \sum_{b=1}^{B} d(R_b, R)$ are 1,902 (median linear order), 1,916 (Borda), and 1,938 (Bradley-Terry) for the classification and 331, 355, and 333 for the regression datasets, respectively. Hence, the Borda and Bradley-Terry solutions obtained are not median linear orders, but "not too far" from these (as concerns the corresponding Φ values).

Table 1. Comparison of three consensus rankings for the classification data sets. The abbreviations are the same as in Meyer et al. (2003).

	Median	Borda	Bradley-Terry
1	svm	svm	svm
2	dbagging	dbagging	dbagging
3	randomForest	randomForest	randomForest
4	bagging	bagging	bagging
5	nnet	nnet	nnet
6	fda.mars	mart	mart
7	mart	fda.mars	fda.mars
8	multinom	multinom	multinom
9	glm	glm	glm
10	mda.mars	lda	lda
11	lda	mda.mars	mda.mars
12	rpart	knn	mda.bruto
13	lvq	rpart	fda.bruto
14	qda	lvq	knn
15	knn	mda.bruto	qda
16	mda.bruto	qda	rpart
17	fda.bruto	fda.bruto	lvq

Table 2. Comparison of three consensus rankings for the regression data sets. The abbreviations are the same as in Meyer et al. (2003).

	Median	Borda	Bradley-Terry
1	randomForest	nnet	randomForest
2	ppr	randomForest	nnet
3	nnet	ppr	ppr
4	svm	svm	svm
5	bruto	mart	bruto
6	mart	bagging	mart
7	mars	lm	mars
8	bagging	rpart	bagging
9	rpart	bruto	lm
10	lm	mars	rpart

Table 3. Space complexity in terms of number of variables and constraints and worst case time complexity for computing the median consensus from an ensemble of n learners.

n	#variables	#constraints	time complexity
10	45	330	1,204
17	136	1,632	131,072

4 Outlook

Median linear orders are only fully interpretable provided that they uniquely solve the corresponding optimization problem. This suggests employing solvers which yield *all* solutions of the underlying binary program (e.g., Branch and Bound methods), as well as considering other types of consensus relations (e.g., preorders allowing for ties, or equivalence relations giving classes of algorithms which perform "equally well"). We are currently exploring these issues, along with the development of an R package which offers computational infrastructure for relations, and methods for computing consensus rankings.

References

BARTHÉLEMY, J.-P. and MONJARDET, B. (1981): The Median Procedure in Cluster Analysis and Social Choice Theory. *Mathematical Social Sciences, 1, 235–267.*

BERKELAAR, M., EIKLAND, K. and NOTEBAERT, P. (2006): `lp_solve`. Version 5.5.0.7.

BLAKE, C.L. and MERZ, C.J. (1998): UCI Repository of Machine Learning Databases.

BORDA, J.C. (1781): Mémoire sur les Élections au Scrutin. Histoire de l'Académie Royale des Sciences.

BRADLEY, R.A. and TERRY, M.E. (1952): Rank Analysis of Incomplete Block Designs I: The Method of Paired Comparisons. *Biometrika, 39, 324–245.*

BRUNK, H.D. (1960): Mathematical Models for Ranking from Paired Comparison. *Journal of the American Statistical Association, 55, 291, 503–520.*

BUTTREY, S.E. (2005): Calling the `lp_solve` Linear Program Software from R, S-PLUS and Excel. *Journal of Statistical Software*, 14, 4.

CONDORCET, M.J.A. (1785): Essai sur l'Application de l'Analyse à la Probabilité des dÉcisions Rendues à la Pluralité des Voix. Paris.

DAY, W.H.E. and MCMORRIS, F.R. (2003): *Axiomatic Choice Theory in Group Choice and Bioconsensus.* SIAM, Philadelphia.

DECANI, J.S. (1969): Maximum Likelihood Paired Comparison Ranking by Linear Programming. *Biometrika, 56, 3, 537–545.*

GRÖTSCHEL, M. and WAKABAYASHI, Y. (1989): A Cutting Plane Algorithm for a Clustering Problem. *Mathematical Programming, 45, 59–96.*

HOTHORN, T., LEISCH, F., ZEILEIS, A. and HORNIK, K. (2005): The Design and Analysis of Benchmark Experiments. *Journal of Computational and Graphical Statistics, 14, 3, 675–699.*

KEMENY, J.G. and SNELL, J.L. (1962): *Mathematical Models in the Social Sciences*, Chapter Preference Rankings: An Axiomatic Approach. MIT Press, Cambridge.

MAKHORIN, A. (2006): *GNU Linear Programming Kit (GLPK).* Version 4.9.

MARCOTORCHINO, F. and MICHAUD, P. (1982): Agregation de Similarites en Classification Automatique. *Revue de Statistique Appliquée, XXX, 21–44.*

MEYER, D., LEISCH, F. and HORNIK, K. (2003): The Support Vector Machine under Test. *Neurocomputing, 55, 169–186.*

MONJARDET, B. (1981): Metrics on Partially Ordered Set: A Survey. *Discrete Mathematics, 35, 173–184.*

NEWMAN, D.J., HETTICH, S., BLAKE, C.L. and MERZ, C.J. (1998): UCI Repository of Machine Learning Databases.

R DEVELOPMENT CORE TEAM (2005): *R: A Language and Environment for Statistical Computing.* R Foundation for Statistical Computing, Vienna, Austria. ISBN 3-900051-07-0.

RÉGNIER, S. (1965): Sur Quelques Aspects Mathématiques des Problèmes de Classification Automatique. *ICC Bulletin, 175–191.*

WAKABAYASHI, Y. (1998): The Complexity of Computing Medians of Relations. *Resenhas, 3, 3, 323–349.*

Classification of Contradiction Patterns

Heiko Müller, Ulf Leser and Johann-Christoph Freytag

Humboldt-Universität zu Berlin, Unter den Linden 6, D-10099 Berlin, Germany;
{hmueller, leser, freytag}@informatik.hu-berlin.de

Abstract. Solving conflicts between overlapping databases requires an understanding of the reasons that lead to the inconsistencies. Provided that conflicts do not occur randomly but follow certain regularities, patterns in the form of "IF *condition* THEN *conflict*" provide a valuable means to facilitate their understanding. In previous work, we adopt existing association rule mining algorithms to identify such patterns. Within this paper we discuss extensions to our initial approach aimed at identifying possible update operations that caused the conflicts between the databases. This is done by restricting the items used for pattern mining. We further propose a classification of patterns based on mappings between the contradicting values to represent special cases of conflict generating updates.

1 Conflicts in overlapping databases

Many databases exist with overlaps in their sets of represented real-world entities. There are different reasons for these overlaps, like:

- *replication of data sources* at different sites to improve the performance of web-services and the availability of the data,
- *independent production of data* representing a common set of entities or individuals by different groups or institutions, and
- *data integration* where data is copied from sources, possibly transformed and manipulated for data cleansing, and stored in an integrated data warehouse.

Whenever overlapping data is administered at different sites, there is a high probability of the occurrence of differences. Many of these inconsistencies are systematic, caused by the usage of different controlled vocabularies, different measurement units, different data modifications for data cleansing, or by consistent bias in experimental data analysis. When producing a consistent view of the data knowledge about such systematic deviations can be used to assess the individual quality of database copies for conflict resolution.

Assuming that conflicts do not occur randomly but follow specific (but unknown) regularities, patterns of the form "IF *condition* THEN *conflict*" provide a valuable means to facilitate the identification and understanding of systematic deviations. In Müller et al. (2004) we proposed the adaptation of existing data mining algorithms to find such contradiction patterns. Evaluated by a domain expert, these patterns can be utilized to assess the correctness of conflicting values and therefore for conflict resolution. Within this paper we present a modified approach for mining contradictory data aimed at enhancing the expressiveness of the identified patterns. This approach is based on the assumption that conflicts result from modification of databases that initially where equal (see Figure 1). Conflicting values are introduced by applying different sequences of update operations, representing for example different data cleansing activities, to a common ancestor database. Given a pair of contradicting databases, each resulting from a different update sequence, we reproduce conflict generation to assist a domain expert in conflict resolution. We present an algorithm for identifying update operations that describe in retrospective the emergence of conflicts between the databases. These conflict generators are a special class of contradiction patterns. We further classify the patterns based on the mapping between contradicting values that they define. Each such class corresponds to a special type of conflict generating update operation. The classification further enhances the ability for pruning irrelevant patterns in the algorithm.

The reminder of this paper is structured as follows: In Section 2 we define conflict generators for databases pairs. Section 3 presents an algorithm for finding such conflict generators. We discuss related work in Section 4 and conclude in Section 5.

2 Reproducing conflict generation

Databases r_1 and r_2 from Figure 1 contain fictitious results of different research groups investigating a common set of individual owls. Identification of tuples representing the same individual is accomplished by the unique object identifier ID. The problem of assigning these object identifiers is not considered within this paper, i.e., we assume a preceding duplicate detection step (see for example Hernandez and Stolfo (1995)). Note that we are only interested in finding update operations that introduce conflicts between the overlapping parts of databases. Therefore, we also assume that all databases have equal sets of object identifiers.

Conflicting values are highlighted in Figure 1 and conflicts are systematic. The conflicts in attribute SPECIES are caused by the different usage of English and Latin vocabularies to denote species names, conflicts in attribute COLOR are due to a finer grained color description for male and female snowy owls (*Nyctea Scandica*) in database r_2, and the conflicts within attribute SIZE are caused by rounding or truncation errors for different species in database r_1.

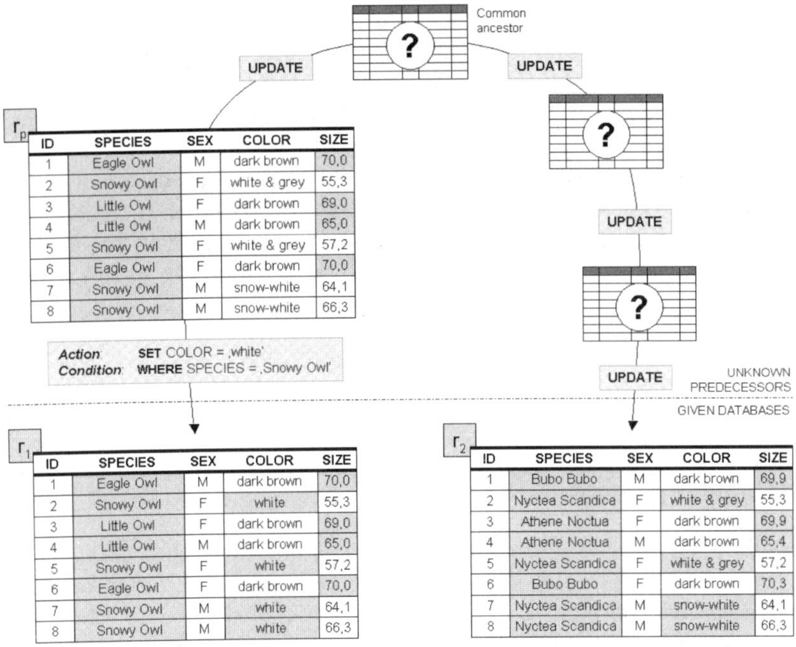

ID	SPECIES	SEX	COLOR	SIZE
1	Eagle Owl	M	dark brown	70,0
2	Snowy Owl	F	white & grey	55,3
3	Little Owl	F	dark brown	69,0
4	Little Owl	M	dark brown	65,0
5	Snowy Owl	F	white & grey	57,2
6	Eagle Owl	F	dark brown	70,0
7	Snowy Owl	M	snow-white	64,1
8	Snowy Owl	M	snow-white	66,3

ID	SPECIES	SEX	COLOR	SIZE
1	Eagle Owl	M	dark brown	70,0
2	Snowy Owl	F	white	55,3
3	Little Owl	F	dark brown	69,0
4	Little Owl	M	dark brown	65,0
5	Snowy Owl	F	white	57,2
6	Eagle Owl	F	dark brown	70,0
7	Snowy Owl	M	white	64,1
8	Snowy Owl	M	white	66,3

ID	SPECIES	SEX	COLOR	SIZE
1	Bubo Bubo	M	dark brown	69,9
2	Nyctea Scandica	F	white & grey	55,3
3	Athene Noctua	F	dark brown	69,9
4	Athene Noctua	M	dark brown	65,4
5	Nyctea Scandica	F	white & grey	57,2
6	Bubo Bubo	F	dark brown	70,3
7	Nyctea Scandica	M	snow-white	64,1
8	Nyctea Scandica	M	snow-white	66,3

Fig. 1. A model for conflict emergence in overlapping databases

Reproducing conflict generation requires the identification of possible predecessors of the given databases. We consider exactly one predecessor for each of the databases r_1 and r_2 and each non-key attribute A. A predecessor is used to describe conflict generation within attribute A by identifying update operations that modify the predecessor resulting in conflicts between r_1 and r_2. Figure 1 shows the predecessor r_p for database r_1 and attribute COLOR. Also shown is an update operation that describes the generation of conflicts by replacing the finer grained original color specifications in r_p with the more generic term '*white*' in r_1. Currently, we only consider update operations that modify the values within one attribute. We start by defining the considered predecessors and then define a representation for conflict generators.

2.1 Preceding databases

The databases within this paper are relational databases consisting of a single relation r, following the same schema $R(A_1, \ldots, A_n)$. The domain of each attribute $A \in R$ is denoted by $dom(A)$. Database tuples are denoted by t and attribute values of a tuple are denoted by $t[A]$. There exists a primary key $PK \in R$ for object identification. The primary key attribute is excluded from any modification.

Given a pair of databases r_1 and r_2, the set of potential predecessors is infinite. We restrict this set by allowing the values in the common ancestor to be modified at most once by any conflict generator. This restriction enables the definition of exactly one predecessor for each of the databases and each non-key attribute.

In the reminder of this paper we consider only conflicts within a fixed attribute $B \in R/\{PK\}$. Let r_p be the predecessor for database r_1. Database r_p equals r_1 in all attributes that are different from B. These values will not be affected by an update operation modifying attribute B. The values for r_p in attribute B are equal to the corresponding values in the contradicting database r_2. These are the values that are changed to generate conflicts between r_1 and r_2:

$$r_p = \{t \mid t \in dom(A_1) \times \cdots \times dom(A_n) \wedge \\ \exists\, t_1 \in r_1,\ t_2 \in r_2\ : \forall\, A \in R\ :\ t[A] = \begin{cases} t_1[A], & \text{if } A \neq B \\ t_2[A], & \text{else} \end{cases} \}$$

2.2 Conflict generators

A conflict generator is a (condition, action)-pair, where the condition defines the tuples that are modified and the action describes the modification itself. Conditions are represented by closed patterns as defined in the following. The action is reflected by the mapping of values between predecessor and resulting database in the modified attribute. For example, the action of the update operation shown in Figure 1 results in a mapping of values *white & grey* and *snow-white* to value *white*. We classify conflict generators based on the properties of the mapping they define.

Tuples are represented using terms $\tau : (A,\ x)$ that are (attribute, value)-pairs, with $A \in R$ and $x \in dom(A)$. Let $terms(t)$ denote the set of terms for a tuple t. For each attribute $A \in R$ there exists a term $(A, t[A]) \in terms(t)$. A pattern ρ is a set of terms, i.e., $\rho \subseteq \bigcup_{t \in r} terms(t)$. A tuple t satisfies a pattern ρ if $\rho \subseteq terms(t)$. The empty pattern is satisfied by any pattern. The set of tuples from r that satisfy ρ is denoted by $\rho(r)$. We call $|\rho(r)|$ the support of the pattern.

A pattern ρ is called a *closed pattern* if there does not exists a superset $\rho' \subset \rho$ with $\rho'(r) = \rho(r)$. We focus solely on closed patterns as conditions for conflict generators. The set of closed patterns is smaller in size than the set of patterns. Still, closed patterns are lossless in the sense that they uniquely determine the set of all patterns and their set of satisfied tuples (Zaki (2002)).

The Boolean function $conflict_B(t)$ indicates for each tuple $t_p \in r_p$ whether contradicting values exist for attribute B in the corresponding tuples $t_1 \in r_1$ and $t_2 \in r_2$ with $t_1[PK] = t_2[PK] = t_p[PK]$:

$$conflict_B(t) = \begin{cases} true, & \text{if } t_1[B] \neq t_2[B] \\ false, & \text{else} \end{cases}$$

We utilize the interestingness measures *conflict potential* and *conflict relevance* for contradiction patterns to enforce a close relationship between the tuples affected by a conflict generator and the occurrence of actual conflicts as defined in Müller et al. (2004). The *conflict potential* of a pattern ρ is the probability that a tuple $t \in r_p$ satisfying ρ has a conflict in attribute B, i.e.,

$$pot(\rho) = \frac{|\{t \mid t \in \rho(r_p) \wedge \mathit{conflict}_B(t)\}|}{|\rho(r_p)|}$$

The *conflict relevance* of a pattern ρ is the probability that a tuple $t \in r_p$ with conflict in attribute B also satisfies ρ, i.e.,

$$rel(\rho) = \frac{|\{t \mid t \in \rho(r_p) \wedge \mathit{conflict}_B(t)\}|}{|\{t \mid t \in r_p \wedge \mathit{conflict}_B(t)\}|}$$

A pattern ρ is called a *conflict generator* for attribute B if it has conflict potential and conflict relevance above given thresholds min_{pot} and min_{rel}.

A pair of tuples from r_p and r_1 with identical primary key define a mapping that reflects the modification of values in attribute B as it occurred in the transition from the predecessor to the actual database. We denote this mapping by $t[M]$ for each $t \in r_p$. It follows:

$$t[M] = (x,\, y) \;\Leftrightarrow\; \exists\, t_1 \in r_1 \;:\; t[PK] = t_1[PK] \;\wedge\; t[B] = x \;\wedge\; t_1[B] = y$$

Each conflict generator ρ defines a mapping of values for the tuples that satisfy it, denoted by $map(\rho(r_p)) = \bigcup_{t\in\rho(r_p)} t[M]$. We call a conflict generator *functional* if $map(\rho(r_p))$ defines a function where each x relates exactly to one y. A functional conflict generator is called *injective* if different x values are always mapped to different y value. We call a functional conflict generator *constant* if all x values are mapped to the same y value. This results in four classes of conflict generators, denoted by F for functional, I for injective, C for constant, and $I\&C$ for injective and constant, with $F \supseteq I$, $F \supseteq C$, and $I\&C = I \cap C$.

Regarding the description of conflicts, the action of a functional conflict generator is represented by a function $f(B)$, e.g., the rounding of values. An injective conflict generator for example represents the translation of values between different vocabularies and a constant conflict generator may represent a generalization as in the example shown in Figure 1.

3 Mining functional conflict generators

Mining conflict generators is accomplishable using closed pattern mining algorithms like CHARM (Zaki (2002)) or CARPENTER (Pan et al. (2003)). If we are interested in finding functional conflict generators term enumeration approaches like CHARM have the drawback that there is no ability for pruning based on a given mapping class during pattern mining. Therefore, we use tuple enumeration in our algorithm for REtrospective functional COnflict GeNeratIon (RECOGNIze) that is outlined in Figure 2.

Mining conflict generators using tuple enumeration is based on the following property: Each set of tuples $s \subseteq r$ defines a pattern, denoted by ρ_s, that is the set of terms common to all tuples in s, i.e., $\rho_s = \bigcap_{t \in s} terms(t)$. If ρ_s is not empty it represents a closed pattern as we cannot add any term to ρ_s without changing the set of tuples that satisfy the pattern. Different tuple sets may define the same closed pattern. Therefore, algorithms like CARPENTER efficiently enumerate those tuple sets that define the complete set of closed patterns satisfying a given support threshold.

RECOGNIze takes as parameters database r_p, conflict potential and relevance thresholds, and the requested class of the returned conflict generators (*classMapping*), i.e., F, I, C, or $I\&C$. The algorithm adopts the CARPENTER algorithm and extends the pruning in subroutine *minePattern* to avoid enumeration of tuple sets that do not represent conflict generators of the requested class. Let s_B denote the subset of r_p containing those tuples that have a conflict in attribute B. The algorithm enumerates all subsets $s_b \subseteq s_B$ that (i) have sufficient size to satisfy the relevance threshold, i.e., $|s_b| \geq min_{rel} * |s_B|$, (ii) whose resulting pattern ρ_{s_b} satisfies the conflict potential threshold, and (iii) where $map(\rho_{s_b})$ represents a valid mapping based on the specified mapping class. The enumeration is done using subroutine *minePattern*. The pa-

RECOGNIze(r_p, min_{pot}, min_{rel}, $classMapping$)

1. Initialize $CG := \{\}$ and $s_B := \{t \mid t \in r_p \wedge \textit{conflict}_B(t)\}$
2. Minimal support of conflict generators $tupsup := min_{rel} * |s_B|$
3. Call *minePattern*($\{\}$, s_B)
4. return CG

minePattern(s_b, s'_B)

1. Determine candidate tuples for extension of s_b
   $s_u := \{t | t \in s'_B \wedge t[PK] > max(s_b) \wedge terms(t) \cap \rho_{s_b} \neq \{\} \wedge compatible(t, s_b)\}$
   if $|s_u| + |s_b| < tupsup$ then *return*
2. Determine additional tuples that satisfy ρ_{s_b}
   $s_y := \{t \mid t \in s_u \wedge terms(t) \supseteq \rho_{s_b}\}$
   if $\neg\, validMapping(s_y)$ then *return*
3. if $\rho_{s_b} \in CG$ then *return*
4. Add ρ_{s_b} to the result if all constraints are satisfied
   if $|s_b| + |s_y| \geq tupsup \wedge pot(\rho_{s_b}) \geq min_{pot} \wedge validMapping(\rho_{s_b}(r_p))$ then
   $GC := CG \cup \rho_{s_b}$
5. Enumerate for the remaining candidates the tuple sets
   for each $t \in (s_u - s_y)$ do
     minePattern($s_b \cup \{t\}, (s_u - s_y) - \{t\}$)

Fig. 2. The RECOGNIze Algorithm

rameters of *minePattern* are the current tuple set s_b, and the set of tuples s'_B that are considered as possible extensions for s_b (other variables are global).

We assume that the elements in s_B are sorted in ascending order of their primary key. Tuple sets are enumerated in depth first order based on the primary key to avoid the enumeration of duplicate tuple sets. In the first step of subroutine *minePattern* the candidate tuples from s'_B for extending s_b are determined. These are the tuples that contain at least one of the terms in ρ_{s_b}. They also have to have a primary key that is greater than the maximum primary key (returned by max) of tuples in s_b to ensure depth first enumeration. In addition to CARPENTER we request that the mapping $t[M]$ is compatible with the mapping defined by ρ_{s_b}regarding the requested class (details below). If the sum of candidates and tuples in s_b is below the minimal tuple support we return, because this tuple set does not define a relevant conflict generator.

The second step determines the subset of candidates that satisfy ρ_{s_b}. It follows that $s_b \cup s_y$ defines the set of tuples from s_B that satisfy ρ_{s_b}. There is an additional ability for pruning, if the tuples in s_y do not define a mapping that is valid for the desired class. The tuples in s_y are not considered as further extensions of s_b as this would only generate identical closed patterns. Still, in Step 3 we have to check whether ρ_{s_b} is already contained in CG. Otherwise we add ρ_{s_b} to CG if all three constraints as listed above are satisfied. We then extend the current tuple set using the remaining candidates in $s_u - s_y$ and call *minePattern* recursively in order to build the complete tuple set enumeration tree.

The subroutines *compatible* and *validMapping* check whether a mapping *map*(s) is valid regarding *classMapping*. This check is trivial for either C or $I\&C$ where we test for equality of the y values (and the x values in case of $I\&C$) of the elements in *map*(s). The case of *classMapping* being F or I is described in the following: For each element in *map*(r_p), referred to as *mapping term* in the following, we maintain a list of incompatible mapping terms, denoted by *incomp*. A pair of mapping terms $(x_1, y_1), (x_2, y_2)$ is considered incompatible for conflict generators of class F if $x_1 = x_2$ and $y_1 \neq y_2$. The mapping terms are incompatible for class I if they are incompatible for class F or $x_1 \neq x_2$ and $y_1 = y_2$. In subroutine *compatible* we request *incomp*($t[M]$) to be disjoint with $\bigcup_{t \in s_p}$ *incomp*($t[M]$), i.e., the mapping term is not incompatible with any of the mapping terms currently in s_b. For $validMapping(s)$ to be true we request that $\bigcap_{t \in s}$ *incomp*($t[M]$) $= \{\}$, i.e., there exists not incompatibilities between the mapping terms of the tuples in s.

For the example in Figure 1 there are three functional conflict generators for a relevance threshold of 50% representing conflicts for femal snowy owls, male snowy owls, and snowy owls in general. They all have conflict relevance of 100%. The first two conflict generators belong to class $I\&C$ with the other belonging to C. Experiments with data sets of protein structures as used in Müller et al. (2004) show that an average of 63% of the patterns for each attribute represent functional conflict generators, with C and $I\&C$ being the most frequently subclasses.

4 Related work

Fan et al. (2001) present a method for finding patterns in contradictory data to support conflict solution. Their focus is the identification of rules that describe the conversion of contradi ting values. They do not request these rules to be associated with a descriptive condition as in our approach. We do not consider the identification of complex data conversion rules. However, the mappings defined by conflict generators could be used as input for the methods described in Fan et al. (2001). There is also a large body of work on statistical data editing, i.e., the automatic correction of conflicting values, based on the Fellegi-Holt-Model (Fellegi and Holt (1976)). These approaches rely on edits (rules) for conflict detection and determine the minimal number of changes necessary for conflict elimination. In contrast, we use object identifiers for conflict identification and currently do not consider automatic value modification.

5 Conclusion

The classification of conflict generators in this paper allows to restrict contradiction pattern mining to those patterns that represent certain situations of conflict generation, e.g., the usage of different vocabularies. The presented algorithm has the advantage of being able to prune patterns not belonging to the requested class at an earlier stage of pattern enumeration. As future work we consider, based on the idea of statistical data edits, to determine the minimal set of update operations that have to be undone in order to derive the common ancestor of a given pair of databases.

References

FAN, W., LU, H., MADNICK, S.E. and CHEUNG, D. (2001): Discovering and Reconciling Value Conflicts for Numerical Data Integration. *Information Systems, 26, 635-656.*

FELLEGI, P. and HOLT, D. (1976): A Systematic Approach to Automatic Edit and Imputation. *Journal of the American Statistical Association, 71, 17-35.*

HERNANDEZ, M.A. and STOLFO, S.J. (1995): The Merge/Purge Problem for Large Databases. *Proc. Int. Conf. Management of Data (SIGMOD).* San Jose, California.

MÜLLER, H., LESER, U. and FREYTAG, J.-C. (2004): Mining for Patterns in Contradictory Data. *Proc. SIGMOD Int. Workshop on Information Quality for Information Systems (IQIS'04).* Paris, France.

PAN, F., CONG, G., TUNG, A.K.H., YANG, J. and ZAKI, M.J. (2003): CARPENTER: Finding Closed Patterns in Long Biological Datasets. *Proc. Int. Conf. on Knowledge Discovery and Data Mining (SIGKDD).* Washington DC.

ZAKI, M.J. (2002): CHARM: An Efficient Algorithm for Closed Itemset Mining. *Proc. of the Second SIAM Int. Conf. on Data Mining.* Arlington, VA.

Selecting SVM Kernels and Input Variable Subsets in Credit Scoring Models

Klaus B. Schebesch[1] and Ralf Stecking[2]

[1] Faculty of Economics, University "Vasile Goldiş", Arad, Romania;
kbsbase@gmx.de

[2] Department of Economics, University of Bremen, D-28359 Bremen, Germany;
stecking@uni-bremen.de

Abstract. We explore simultaneous variable subset selection and kernel selection within SVM classification models. First we apply results from SVM classification models with different kernel functions to a fixed subset of credit client variables provided by a German bank. Free variable subset selection for the bank data is discussed next. A simple stochastic search procedure for variable subset selection is also presented.

1 Introduction

Classification models related to credit scoring may use a large number of potentially useful input features. However, these credit client data may or may not be effectively recorded from the past or some of the features may not be available for every client. We extend our previous work (Stecking and Schebesch (2003), Stecking and Schebesch (2006)) by introducing input variable selection using Support Vector Machines (SVM) to evaluate subsets of inputs and the comparative classification performance of different SVM-kernels on such subsets. We start out with a data set $\{x_i, y_i\}_{i=1,...,N}$, with $x_i \in \mathbb{R}^m$ the m characteristics of the ith past credit client and $y_i \in \{-1, 1\}$ the class assignment which is the observed defaulting behavior of this client. As a new credit client with personal features $x \in \mathbb{R}^m$ arrives, the SVM forecasts her defaulting behavior by the rule $y = \text{SIGN}\left(\sum_{i=1}^{N} y_i \alpha_i k(x, x_i) + b\right)$. The kernel $k(.,.)$ is chosen beforehand and $\{\alpha_i \geq 0\}_{i=1,...,N}$ as well as $b \in \mathbb{R}$ are the solutions of the SVM. For details of the SVM we refer to (Shawe-Taylor and Cristianini (2004)) and for its extensive use on credit scoring data we refer to our past work (Stecking and Schebesch (2003), Stecking and Schebesch (2006)). Here we only recall the following facts: $\alpha_i > 0$ are support vectors (SV), which in a certain sense determine the optimal separation of the two credit client classes. This separation is always to be considered relative to the effect of an externally supplied SVM-parameter, termed **capacity** $C > 0$. In view of possible overtraining by the **hard** version of the SVM, using C, one can deliberately set a level of misclassification. In this **soft** version of the SVM, an important distinction refers to **bounded** and **unbounded** SV ($\alpha_i = C$ and $0 < \alpha_i < C$, respectively). If a SVM solution contains many bounded SV, then this means that the model cannot separate many of the data points (credit clients) with high confidence into one of the defaulting classes.

The outline of this paper is as follows: First we limit our work to search for the best kernel function for an externally given **fixed** variable subset. Then **free** subset selection for a fixed number of variables, using six different kernels, is presented. Finally we use less than m randomly selected credit client features in conjunction with two different kernels $k(.,)$. In a first step, large SVM model populations resulting from a simple variant of random variable subset selection is evaluated and thereupon a simple but more informed rule of input selection is proposed.

2 Kernel selection for fixed variable subsets

The basic data set for our past credit scoring models is an equally distributed sample of 658 clients for a building and loan credit with a total number of 40 input variables, which contains 49.1% defaulting and 50.9% non defaulting credit clients (Stecking and Schebesch (2003). In the sequel a variable subset of the original credit scoring data set is used, consisting of 15 out of 40 variables. These (and only these) 15 variables are available for the whole credit portfolio of about 140 thousand clients. The goal is to establish a credit scoring function that gives good predictive results on the full data set as well as on the subset. Then, the credit scoring function, which selects the credit client, computes a real valued output for each of the 140 thousand clients of the credit portfolio and subsequently these ordered metric values can be used to build up a rating class system.

SVM with six different kernel functions are used for classifying good and bad credit clients. Detailed information about kernels, hyperparameters and tuning can be found in (Stecking and Schebesch (2006)). In Table 1 model performance for the full data set (40 inputs) and the subset (15 inputs) is compared using (i) Kernel Alignment, (ii) Vector Norm, (iii) number of (bounded) support vectors, (iv) training error and (v) leave one out (L-1-o) error. Subset selection methods include the first difference of the norm of the weight vectors $\Delta\|w\|^2 = \frac{\|w\|_f^2 - \|w\|_s^2}{\|w\|_f^2}$ between full set and subset (Rakotomamonji (2003)). Small differences then point to minimal information loss when replacing full set with subset for model building. Kernel alignment between K_f (kernel of the full data set) and K_s (kernel of the subset) is computed as $A(K_f, K_s) = \frac{\langle K_f, K_s\rangle}{\sqrt{\langle K_f, K_f\rangle\langle K_s, K_s\rangle}}$ with $\langle K_f, K_s\rangle = \mathrm{tr}(K_f^T K_s)$ as *Frobenius inner product*. Kernel alignment $A(K_1, K_2)$ in general can be interpreted as a Pearson correlation coefficient between two random variables $K_1(u, v)$ and

Table 1. Evaluation and comparison of six SVM with different kernel functions trained on the full data set with 40 input variables and on the subset selection with 15 input variables.

SVM-Kernel	No. of Inputs	No. of BSVs	Kernel Alignm.	Vector Norm	Training Error	L-1-o Error
Linear	40	316			22.64	27.20
	15	405	0.6657	0.00	30.24	36.63
Sigmoid	40	544			24.84	27.05
	15	633	0.3513	0.11	34.35	36.63
Polynomial	40	392			19.60	26.29
$(d = 2)$	15	548	0.9934	0.11	27.81	30.24
Polynomial	40	211			8.81	26.44
$(d = 3)$	15	491	0.9933	0.17	28.72	31.76
RBF	40	252			10.94	25.08
	15	424	0.9270	0.41	28.88	33.59
Coulomb	40	368			7.90	24.92
	15	448	0.9767	0.31	23.86	33.28

$K_2(u, v)$ (Shawe-Taylor and Cristianini (2004)). High kernel alignment **and** small differences in the vector norm can be found only for the two polynomial kernels. The leave one out error as an unbiased estimator of the true generalization error on the one hand shows a high increase of around 8 to 9 percentage points for linear, sigmoid, RBF and Coulomb kernel SVMs, whereas there only is a moderate increase of about 4 to 5 percentage points for the two polynomial kernel SVMs when changing from the full data set to the subset. The number of bounded support vectors as well as the training error do not give any obvious information useful to assess the quality of the subset selection.

3 Variable subset selection by ranking criteria

After treating the problem of kernel evaluation for a *fixed variable subset* now the more general case of *free variable subset selection* is considered. A selection mechanism usually includes *variable ranking* as an initial step. According to (Guyon and Elisseff (2003)) variable ranking may or may not be connected with a predictor. Without a predicting variable *missing value analysis* (MVA), tests for *multicollinearity* and *covariance* etc. can be performed. Well known ranking criteria when a predicting variable is available contain e.g. F-value and correlation coefficient. Simple ranking criteria are known to be computationally efficient and statistically robust (Hastie et al. (2001)). MVA, tests of multicollinearity and a variable selection procedure based on F-values was performed in an early step of data preprocessing, reducing the initial credit data set of about 100 to finally 40 variables used for credit scoring (Stecking and Schebesch (2003).

In the following a subset of 15 out of these 40 variables will be selected and several ranking criteria, that are based on *sensitivity analysis*, will be evaluated. Sensitivity analysis is known to be an important tool for *neural*

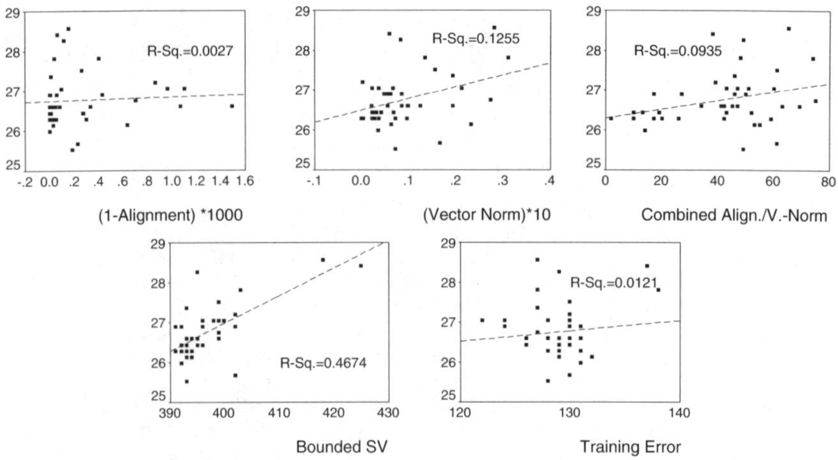

Fig. 1. Outcomes of five different sensitivity measures plotted against the leave-one-out errors for 40 input variables after elimination. A bivariate regression function (dashed line) is estimated for descriptive purpose only.

networks to control for e.g. model complexity. The changes in a derivative of the functional representation of the network (*e.g. the error function*) after elimination (or setting to a constant, e.g. mean value or zero) of a single element is estimated, and the one with the *least effect* is the first candidate for removal. "Elements" in neural networks can be units (input variables, input ensembles, layers) or weights. In a SVM setting one may concentrate on the elimination of single input variables, that cause changes in: (i) the training error of the SVM, (ii) the number of (bounded) support vectors, (iii) kernel alignment, (iv) vector norm, (v) a combination of (iii) and (iv) and finally (vi) the leave-one-out error.

In Figure 1 measures (i) to (v) (for every of the 40 input variables) are plotted against the leave-one-out error (vi) as an estimation of the "true" sensitivity of the removed variable. There is a stronger connection with the number of bounded support vectors ($R^2 = 0.4674$) and, to a lower extent, with the vector norm ($R^2 = 0.1255$). Alignment (also in combination) and training error do not show any notable dependence with the leave-one-out error.

How do the selected subsets succeed? Table 2 shows the leave one out classification errors for variable subsets selected with different criteria and trained with different SVM kernels. First it can be seen, that for a given selection criterion the classification results are quite homogenous, independent of the used kernel (rows in Table 2). Out of the essential selection criteria (alignment, vector norm, combined, bounded SVs and training error) the bounded SV criterion is most successful, followed by the vector norm criterion. The three rows *below the horizontal line* represent benchmark results: the leave-one-out

Table 2. Leave-one-out classification errors for selected subsets with each 15 variables and 658 cases. Five ranking criteria are evaluated and compared to two benchmark subsets (L-1-o and fixed subset) and to the full data set with 40 variables. Each subset is trained with six different SVM kernels.

Ranking	**Kernels**					
Criteria	**Linear**	**Sigmoid**	**Poly02**	**Poly03**	**RBF**	**Coulomb**
Alignment	36.02%	39.67%	36.02%	37.08%	35.11%	35.11%
VectorNorm	30.85%	33.13%	28.88%	30.85%	29.94%	31.00%
Combi A/V	31.16%	32.98%	32.37%	32.37%	30.40%	31.31%
No. of BSVs	27.96%	29.33%	28.72%	28.88%	26.60%	26.14%
Training	29.94%	40.12%	30.70%	30.70%	29.48%	31.16%
L-1-o	26.90%	27.66%	26.29%	25.99%	25.68%	26.44%
Fixed Subset	36.63%	36.63%	30.24%	31.76%	33.59%	33.28%
Full Set	27.20%	27.05%	26.29%	26.44%	25.08%	24.92%

criterion is computationally very expensive and supposed to give best results for subset selection based on variable ranking. The *fixed subset* consisting of 15 variables and the *full set* of 40 variables were already described in Section 2.

The bounded SV criterion performs almost as good as the leave-one-out *(L-1-o)* criterion (and even better for the Coulomb kernel), the errors are below the ones for the fixed subset and there is only a moderate distance to the classification results of the full set.

4 Variable subset selection by simple random search

As is widely acknowledged in the machine learning literature (e.g., Guyon and Eliseff (2003)), variable subsets can also have a collective influence on the performance of a classification model. This may result in an influence in addition to the isolated influence of each single variable, which is resulting from the simultaneous presence of certain variables in such a subset. Owing to the many combinations of possible subsets for large m, the collective effects of variables are difficult to analyze without extensive experimentation. In order to explore such effects by random search, for simplicity, no a priori information about single variable relevance is assumed. Recall that the basic data set for our past credit scoring models is an equally distributed sample of N=658 clients with a total number of m=40 nonredundant input features. In the models of this section we use $m < 40$ and we use only two nonlinear kernels, namely the RBF kernel $k_1(u,v) = \exp\left(-s||x_i - x_j||^2\right)$ and the Coulomb kernel $k_2(u,v) = \left(1 + \dfrac{||x_i - x_j||^2}{c}\right)^{-d}$, with $x_i, x_j, i, j \in 1, ..., N$ the input data. Kernel parameters $s, c, d > 0$ and capacity parameter $C > 0$ are also varied

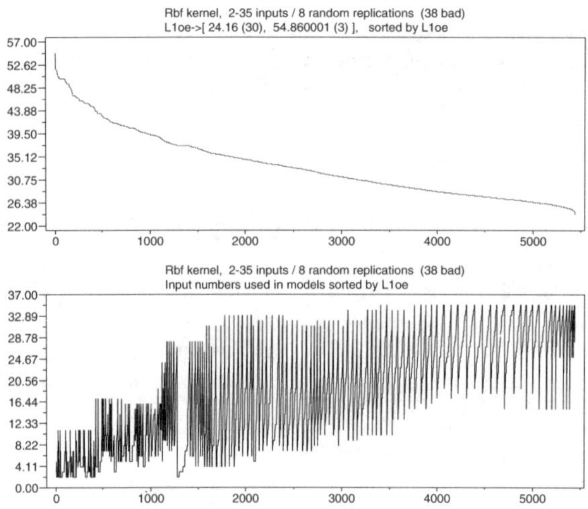

Fig. 2. SVM model population with RBF kernels for the 2-35 window. Both plots sort the models by leave-one-out error. Each entry on the abscissae represents a single model of the population, which passed the time filter (see main text). A total of 38 did not pass (inset on top of plots). The upper plot shows the sorted error while the lower plot shows the input dimension of each consectutive model (connected by lines to emphasize varaiation).

within the experiments to follow, in order to adjust the SVM. In spite of k_1 and k_2 being similar in their local behavior, preliminary modeling tells that they can produce quite different sets of support vectors $\{\alpha_i > 0\}$ (and hence different SVM-models) when varying the input dimension m. Next, subsets of $m < 40$ input features are randomly selected out of all 40 inputs in such a way that m may vary randomly within "windows" of 11-20, 15-25, 2-35 different input features respectively. For the three windows 10, 15 and 8 input samples are drawn for every m. Each sample is evaluated twenty times for different SVM parameters, which in turn are sampled around the parameter values of the most successful full ($m = 40$) kernel models. In addition, a hard time restriction (1.5 seconds) is imposed on the training of each SVM model. Hence, in the case of e.g. the 2-35 input window, a population of $34 \times 8 \times 20 = 5440$ models are set up for every kernel, and (upon passing the time restriction) each model is subjected to a leave-one-out error validation procedure, which is a good estimator of the true out of sample performance. Figure 2 depicts the resulting model population for the 2-35 window using RBF kernels. As may be expected, leave-one-out misclassification error drops with increasing input dimensions (lower plot). However, somewhat to a surprise, the variation of the input dimensions varies strongly, even for good models (small errors). Coulomb kernel models and all model populations for the other windows produce similar results (not shown). The best models from each population can reach misclassification errors close to those of the models on the full input set

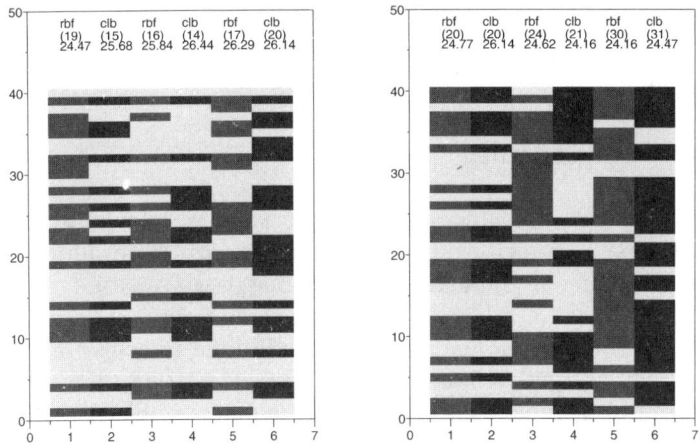

Fig. 3. Variables from the full set (of 40) effectively used by the models built by the average contribution rule (left) and the same for the best models of each population (right).

(around 24%-25%). Further combining inputs of these elite models lead as yet to no improvement. But using information from bad and good models alike can lead to models which surpass the respective elite models: In the runs for each window and for each kernel, every single input is evaluated. If the input is used in a model, one minus the leave-one-out error of the model is added to a contribution vector. These sums are then divided by the number of times the single input was effectively used in all runs. Finally, the contribution of inputs which exceed the average contribution of all inputs are forwarded as a new input feature combination. In Figure 3 the plots shows the inputs of the models selected by this rule models and by elite models. The rule models (lhs) use less input features (19,15,16,14,17,20 as opposed to 20,20,24,21,30,31 used by the elite models). The input number windows used for the six model populations (columns in both plots) are: 11-20 for populations 1 and 2, 15-25 for populations 3 and 4, 2-35 for populations 5 and 6. In the case of populations 1 and 2 the rule selected models are superior to the respective elite models in attaining a slightly lower error by using fewer inputs. This result can be reproduced for other seeds and parameter settings. However, the critical choice is the size of the input number window. For larger maximal m and for larger overall windows the selection rule seems to be biased towards input number reduction and a rule model may also jump out of the population window (14 inputs, 4th lhs model). From columns 5-6 (lhs) and 3-6 (rhs) we note that different input subsets produce models with very similar errors for our data. However, identical input subsets can also produce different errors for two kernels (columns 1-2 rhs). This fast automated process of training SVM model populations on input variable subsets yields a selection of robust models and variables. Interestingly, our selection results do not improve when trying to preselect the base population as well, e.g. when using only "elite" models from the subpopulations for each given number of inputs respectively.

5 Concluding results

Fixed variable subset evaluation leads to a kernel selection problem. It was found out, that SVM with polynomial kernels outperform all other kernels, when used for different data subsets. Changing data availability is common in the credit industry, e.g. when switching from credit applicant scoring to credit client rating. Ranking methods are very well suited to assess the importance of single credit scoring variables. Among several criteria tested, the change of the numbers of bounded support vectors (numbers of SV indicate the overall separating ability of the SVM) induced by variable removal proved to be most informative. The subsequent variable selection leads to small-sized SVM classification functions, which are competitive with the full input set models although using much less information. Furthermore, for our credit scoring data, one finds by simple stochastic search quite small variable subsets which successfully model a credit client separation with leave-one-out errors comparable to the full variable models. The model populations contain many good (competitive) models with a rather small number of inputs, which strongly enhances their utility in incremental data fusion, where credit client characteristics may be provided upon availability from different sources.

References

GUYON, I. and ELISEFF, A. (2003): An Introduction to Variable and Feature Selection. *Journal of Machine Learning Research, 3, 1157-1182.*

HASTIE, T., TIBSHIRANI, R. and FRIEDMAN, J. (2001): *The Elements of Statistical Learning.* Springer, New York.

RAKOTOMAMONJI, A. (2003): Variable Selection Using SVM-based Criteria. *Journal of Machine Learning Research, 3, 1357-1370.*

SHAWE-TAYLOR, J. and CRISTIANINI, N. (2004): *Kernel Methods for Pattern Analysis.* Cambridge University Press, Cambridge.

STECKING, R. and SCHEBESCH, K.B. (2003): Support Vector Machines for Credit Scoring: Comparing to and Combining with some Traditional Classification Methods. In: M. Schader, W. Gaul and M. Vichi (Eds.): *Between Data Science and Applied Data Analysis.* Springer, Berlin, 604-612.

STECKING, R. and SCHEBESCH, K.B. (2006): Comparing and Selecting SVM-Kernels for Credit Scoring. In: M. Spiliopoulou, R. Kruse, C. Borgelt, A. Nürnberger and W. Gaul (Eds.): *From Data and Information Analysis to Knowledge Engineering.* Springer, Berlin, 542-549.

Part III

Data and Time Series Analysis

Simultaneous Selection of Variables and Smoothing Parameters in Geoadditive Regression Models

Christiane Belitz[1] and Stefan Lang[2]

[1] Department of Statistics, University of Munich, 80539 Munich, Germany; christiane.belitz@stat.uni-muenchen.de

[2] Institute of Statistics, University of Innsbruck, 6020 Innsbruck, Austria; stefan.lang@uibk.ac.at

Abstract. In recent years a great deal of research has been devoted to developing complex regression models that allow to deal simultaneously with nonlinear covariate effects and spatial heterogeneity. Such models are referred to as geoadditive models. These models may routinely be estimated using standard statistical software packages. Much less effort, however, has been spent to model and variable selection in the context of complex regression models. In this article we develop an empirical Bayes approach for simultaneous selection of variables and the degree of smoothness in geoadditive models. Our approach allows to decide whether a particular continuous covariate enters the model linearly or nonlinearly or is removed from the model and whether a spatial effect should be added to the model or not.

1 Introduction

In recent years a considerable number of complex model classes for regression analysis have been developed. Their capabilities regarding realistic data analysis go far beyond the traditional linear or generalized linear model. A prominent example are geoadditive models (Fahrmeir and Lang (2001), Kammann and Wand (2003)) that allow for simultaneous modeling and estimation of nonlinear covariate effects, time trends, and spatial heterogeneity. Provided that the variables entering the model are known, estimation of geoadditive models may be carried out routinely using standard statistical software packages, particularly the R package mgcv (Wood (2004)) and BayesX (Brezger et al. (2005)). In this article we present an algorithm and software for *simultaneous selection of relevant variables* and the *degree of smoothness* in geoadditive models. Our algorithm is able to

- decide whether a particular covariate enters the model,
- decide whether a continuous covariate enters the model linearly or nonlinearly,

- decide whether a (smooth) spatial effect enters the model,
- select the degree of smoothness of a nonlinear or spatial effect.

The article is organized as follows: The next section gives a brief introduction to geoadditive models from a Bayesian point of view. Section 3 presents the algorithm for variable and smoothing parameter selection. Finally, section 4 illustrates the method with an application to car insurance data.

2 Geoadditive models

Suppose that observations $(y_i, \mathbf{x}_i, s_i, \mathbf{v}_i)$, $i = 1, \ldots, n$ are given, where y_i is a continuous response variable, $\mathbf{x}_i = (x_{i1}, \ldots, x_{ip})'$ is a vector of continuous covariates, s_i is a spatial index, and $\mathbf{v}_i = (v_{i1}, \ldots, v_{iq})'$ is a vector of further covariates. The values of the spatial index represent the regions or districts in connected geographical maps. Given covariates and unknown parameters, we assume that the responses y_i, $i = 1, \ldots, n$, are independent and Gaussian with geoadditive predictor

$$\eta_i = f_1(x_{i1}) + \cdots + f_p(x_{ip}) + f_{spat}(s_i) + \mathbf{v}_i'\gamma \tag{1}$$

and a common variance σ^2 across subjects. Here, $f_1, \ldots, f_p$ are unknown smooth functions of the continuous covariates, and f_{spat} is a (smooth) spatial effect that captures spatial heterogeneity due to unobserved spatially correlated covariates. The linear combination $\mathbf{v}_i'\gamma$ corresponds to the usual parametric part of the predictor. In the following, we briefly describe modeling the unknown functions $f_1, \ldots, f_p, f_{spat}$ from a Bayesian point of view. For modeling the unknown functions f_j, we follow Lang and Brezger (2004), who present a Bayesian version of P-splines introduced in a frequentist setting by Eilers and Marx (1996). The approach assumes that the unknown functions may be approximated by a polynomial spline of degree l with knots being equally spaced over the domain of x_j. The spline can be written in terms of a linear combination of K_j B-spline basis functions. Denoting the k-th basis function by B_{jk}, we obtain

$$f_j(x_j) = \sum_{k=1}^{K_j} \beta_{jk} B_{jk}(x_j).$$

To overcome the difficulties involved with regression splines, a relatively large number of knots (usually between 20 to 40) is chosen to ensure enough flexibility, and a roughness penalty on adjacent regression coefficients is defined to guarantee a sufficient amount of smoothness and to avoid overfitting. We use first or second order random walk priors defined by

$$\beta_{jk} = \beta_{j,k-1} + u_{jk}, \quad \text{or} \quad \beta_{jk} = 2\beta_{j,k-1} - \beta_{j,k-2} + u_{jk},$$

with Gaussian errors $u_{jk} \sim N(0, \tau_j^2)$ and diffuse priors $\beta_{j1} \propto const$, or β_{j1} and $\beta_{j2} \propto const$, for initial values. The amount of smoothness is controlled by the variance parameter τ_j^2. The priors for the vectors of regression coefficients $\boldsymbol{\beta}_j$ may be cast into the form of an improper normal distribution, i.e.

$$\boldsymbol{\beta}_j \mid \tau_j^2 \propto \frac{1}{(\tau_j^2)^{rk(\mathbf{K}_j)/2}} \exp\left(-\frac{1}{2\tau_j^2}\boldsymbol{\beta}_j'\mathbf{K}_j\boldsymbol{\beta}_j\right), \tag{2}$$

where $\mathbf{K}_j$ is a *penalty matrix*. The spatial effect f_{spat} may in principal be modeled by any two dimensional smoother. For data observed on a regular or irregular lattice a common approach is based on Markov random field (MRF) priors, see e.g. Besag et al. (1991). Let $s \in \{1, \ldots, S\}$ denote the pixels of a lattice or the regions of a geographical map. Then, the most simple Markov random field prior for $f_{spat}(s) = \beta_{spat,s}$ is defined by

$$\beta_{spat,s} \mid \beta_{spat,u}, u \neq s \sim N\left(\sum_{u\in\partial_s} \frac{1}{N_s}\beta_{spat,u}, \frac{\tau_{spat}^2}{N_s}\right), \tag{3}$$

where N_s is the number of regions (pixels) being adjacent to region s, and ∂_s denotes the regions which are neighbors of region s. Hence, prior (3) can be interpreted as a two dimensional extension of a first order random walk. As with P-splines, the prior for $\boldsymbol{\beta}_{spat}$ is of the form (3) with appropriate penalty matrix $\mathbf{K}_{spat}$. Alternatively, the structured spatial effect f_{spat} could be modeled by two dimensional surface estimators, see e.g. Lang and Brezger (2004) for a two dimensional version of P-splines. By defining the $n \times K_j$ design matrices $\mathbf{X}_j$ where the element in row i and column k is given by $\mathbf{X}_j(i,k) = B_{jk}(x_{ij})$, and the $n \times S$ incidence matrix $\mathbf{X}_{spat}$ we can rewrite the predictor (1) in matrix notation as

$$\eta = \mathbf{X}_1\boldsymbol{\beta}_1 + \cdots + \mathbf{X}_p\boldsymbol{\beta}_p + \mathbf{X}_{spat}\boldsymbol{\beta}_{spat} + \mathbf{V}\boldsymbol{\gamma}. \tag{4}$$

The matrix $\mathbf{V}$ is the usual design matrix for linear effects. Given the variance parameters τ_j^2, the regression parameters $\boldsymbol{\beta}_j$, respectively the functions f_j, and $\boldsymbol{\gamma}$ are estimated by maximizing the posterior mode. This is equivalent to minimizing the penalized least squares criterion

$$\begin{aligned} PLS(\boldsymbol{\beta}_1, \ldots, \boldsymbol{\beta}_p, \boldsymbol{\beta}_{spat}, \boldsymbol{\gamma}) = & \sum_{i=1}^{n}(y_i - f_1(x_{i1}) - \cdots - f_{spat}(s_i) - \boldsymbol{\gamma}'\mathbf{v}_i)^2 \\ & + \lambda_1\boldsymbol{\beta}_1'\mathbf{K}_1\boldsymbol{\beta}_1 + \cdots + \lambda_{spat}\boldsymbol{\beta}_{spat}'\mathbf{K}_{spat}\boldsymbol{\beta}_{spat} \end{aligned}$$

where $\lambda_j = \sigma^2/\tau_j^2, j = 1, \ldots, p, spat$ are smoothing parameters. The penalized least squares criterion are minimized via backfitting (Hastie and Tibshirani (1990)) using the univariate smoothers

$$\mathbf{S}_j(\mathbf{y}, \lambda_j) = \mathbf{X}(\mathbf{X}_j'\mathbf{X}_j + \lambda_j\mathbf{K}_j)^{-1}\mathbf{X}_j'\mathbf{y}, \qquad j = 1, \ldots, p, spat.$$

In the next section we present a fast algorithm for simultaneous selection of relevant terms and the amount of smoothness λ_j for the selected terms.

3 Selection of variables and smoothing parameters

Our variable selection procedure described below aims at minimizing a goodness of fit criterion. The following options are available:

- **Test- and validation sample**
  Provided that enough data are available the best strategy is to divide the data into a test- and validation sample. The test data set is used to estimate the parameters of the models. The fit of different models is assessed via the validation data set. In the case of a continuous response, typically the mean squared prediction error is minimized.
- *AIC*, *AIC*$_c$, *BIC*
  In such cases where data are sparse an estimate for the prediction error is based on goodness of fit criteria that penalize complex models. A general form for a wide range of criteria is given by $C = n\log(\hat{\sigma}^2) + penalty$ where the penalty depends on the degrees of freedom (df) of the model. The most widely used criteria are the AIC ($penalty = 2\,df$) and the BIC ($penalty = \log(n)\,df$). A bias corrected version AIC_c of AIC is widely used with regression models. Here, the penalty is given by $penalty = 2\,df + \frac{2\cdot df(df+1)}{n-df-1}$. Experience from extensive simulations suggests to use AIC_c for the models considered in this article. The degrees of freedom are given by the trace of the model hat matrix. However, since the computation of the trace is difficult and time consuming, we prefer to approximate the degrees of freedom by the sum of the degrees of freedom of the univariate smoothers S_j (Hastie and Tibshirani (1990)). Provided that correlations among covariates are not too strong the approximation works well in practice.
- **Cross validation**
  Cross validation mimics the division into test- and validation samples. The data are split into K parts (typically $K = 5$ or $K = 10$) of roughly equal size. One part is used to estimate prediction errors and the other $K-1$ parts are used to fit the models. This is done for every part and the estimated prediction errors are combined. K-fold cross validation is more time consuming than using AIC or BIC but the approximation of the degrees of freedom is not required.

Now our approach for simultaneous selection of variables and smoothing parameters works as follows:

1. **Initialisation**
   Define for every possible nonlinear term $j = 1, \dots, p, spat$ a discrete number M_j of decreasing smoothing parameters $\lambda_{j1} > \cdots > \lambda_{jM_j}$. For P-splines we define $\lambda_{j1} = \infty$ to make sure that a linear fit is included in the choice set.
2. **Start model**
   Choose a start model with current predictor

$$\hat{\boldsymbol{\eta}} = \hat{\mathbf{f}}_1 + \cdots + \hat{\mathbf{f}}_p + \hat{\mathbf{f}}_{spat}.$$

where $\hat{\mathbf{f}}_j$ is the vector of function evaluations at the observations. For example, set $\hat{\mathbf{f}}_j \equiv \mathbf{0}$, $j = 1, \ldots, p, spat$ which corresponds to the empty model. Choose a goodness of fit criteria C.

3. **Iteration**
   - For $j = 1, \ldots, p, spat$:

     For $m = 0, \ldots, M_j$:
     Compute the fits

     $$\hat{\mathbf{f}}_{jm} := \begin{cases} \mathbf{0} & m = 0 \\ \mathbf{S}_j(\mathbf{y} - \hat{\boldsymbol{\eta}}_{[j]}, \lambda_{jm}) & m = 1, \ldots, M_j \end{cases}$$

     and the corresponding predictors $\hat{\boldsymbol{\eta}}_{jm} := \hat{\boldsymbol{\eta}}_{[j]} + \hat{\mathbf{f}}_{jm}$. Here, $\hat{\boldsymbol{\eta}}_{[j]}$ is the current predictor with the j-th fit $\hat{\mathbf{f}}_j$ removed.

     Compute the updated estimate

     $$\hat{\mathbf{f}}_j = \text{argmin } C(\hat{\mathbf{f}}_{jm}),$$

     i.e. among the fits $\hat{\mathbf{f}}_{jm}$, choose the one that minimizes the goodness of fit criteria C.
   - The linear effects part $\mathbf{v}'\boldsymbol{\gamma}$ typically consists of the intercept γ_0 and dummy variables for the categorical covariates. For the moment suppose that $\mathbf{v}$ contains dummies representing only one categorical variable. Then we compare the fits $\hat{\gamma}_0 = \overline{y - \eta_{[\mathbf{v}]}}, \gamma_1 = 0, \ldots, \gamma_q = 0$ (covariate removed from the model) and $\hat{\boldsymbol{\gamma}} = (\mathbf{V}'\mathbf{V})^{-1}\mathbf{V}'(\mathbf{y} - \hat{\boldsymbol{\eta}}_{[\mathbf{v}]})$ where $\hat{\boldsymbol{\eta}}_{[\mathbf{v}]}$ is the current predictor with the linear effects removed and $\overline{y - \eta_{[\mathbf{v}]}}$ is the mean of the elements of the partial residual vector $\mathbf{y} - \hat{\boldsymbol{\eta}}_{[\mathbf{v}]}$. If more than one categorical covariate is available the procedure is repeated for every variable.
4. **Termination**
   The iteration cycle in 3. is repeated until the model, regression and smoothing parameters do not change anymore.

We conclude this section with a few remarks on the algorithm:

1. **Avoid backfitting**
   When updating the function estimate $\hat{\mathbf{f}}_j$ the other terms in the model are not re-estimated as in a backfitting procedure. Avoiding backfitting dramatically reduces computing time. Note, that the algorithm automatically collapses to backfitting as soon as the variables included in the model do not change anymore.
2. **Comparison to S-plus step.gam and mgcv of R**
   To our knowledge the only competitors of our approach are the step.gam

function of S-plus and the mgcv package of R. Simulations show (available on request), that both approaches are slower (step.gam is much slower). Our implementation is also able to deal with much larger data sets. The performance of mgcv and our algorithm is comparable, although mgcv tends to select too complex models. The S-plus step.gam approach clearly produces inferior results.

3. **Connection to boosting**
Taking two modifications our algorithm turns into a boosting approach. For a description of boosting see e.g. Tutz and Binder (2005) who discuss boosting for rigde regression which is technically similar to the methods in this paper. Boosting is obtained if we compute in every iteration only one fit for every nonlinear term. The fits must be based on very large λ_j's in order to guarantee that the resulting smoothers are "weak learners". As a second modification the smoothers are applied to the current residuals $\mathbf{y} - \hat{\boldsymbol{\eta}}$ rather than to $\mathbf{y} - \hat{\boldsymbol{\eta}}_{[j]}$.

4 Application: Car insurance data

We illustrate our approach with the analysis of a Belgian car insurance data set containing $n = 18203$ cases. The variable of primary interest is the claim size y in case of an insured event. Potential covariates explaining claim size are the age of the policyholder (*ageph*), the age of the car (*agec*), the level occupied in the 23-level Belgian bonus-malus scale (*bm*), the horsepower of the car (*hp*), the district in Belgium where the policyholder lives (*dist*) and the categorical covariates *fuel* (1 = gasoline, 0 = diesel), *cov* (1 = third party liability (TPL) only, 2 = limited material damage or theft in addition to TPL, 3 = comprehensive coverage in addition to TPL), *fleet* (1 = the vehicle belongs to a fleet, 0 = the vehicle does not belong to a fleet), *sex* (1 = male, 0 = female) and *use* (1 = professional, 0 = private use). The analysis of claim sizes is an important part of the rate-making process. The data has been analyzed before of by Denuit and Lang (2004) using geoadditive models and Markov chain Monte Carlo inference techniques. Model choice and variable selection has been done in a quite time-consuming procedure by comparing a small number of competing models via the deviance information criterion. The full model containing all possible effects is given by

$$\ln(y) = f_1(\text{agec}) + f_2(\text{bm}) + f_3(\text{ageph}) + f_4(\text{hp}) + f_{spat}(\text{dist}) + \mathbf{v}'\boldsymbol{\gamma} + \varepsilon$$

where $\mathbf{v}$ contains the categorical covariates in dummy coding. Using AIC_c as the goodness of fit criterion, the algorithm described in the previous section selects the model

$$\begin{aligned}\ln(y) = {} & \gamma_0 + f_1(\text{agec}) + f_2(\text{bm}) + f_3(\text{ageph}) + f_{spat}(\text{dist}) + \\ & \gamma_1\,\text{fleet} + \gamma_2\,\text{cov2} + \gamma_3\,\text{cov3} + \varepsilon\end{aligned}$$

where *hp* and some of the categorical covariates are not selected. Note that the dummy variables corresponding to a particular categorical covariate are included or removed jointly, i.e. it is not possible that only some of the dummies appear in the model. The selected model is identical to the model used in Denuit and Lang (2004) with slightly different smoothing parameters. The selection process, however, took only 1 minute. In case the data are divided into test and validation samples (60% versus 40 %) a slightly different model is selected with hp included as a linear effect. Posterior mode estimates including 80% and 95% pointwise credible intervals for the selected nonlinear terms are displayed in Figure 1 a)-c). Credible intervals are computed conditional on the selected model, i.e. the selection process is not considered (as is usually the case). The credible intervals are computed by running a Markov chain Monte Carlo sampler for the selected model with fixed smoothing parameters. Posterior mode estimates for the spatial effect are given in Figure 2. Results for linear effects are omitted.

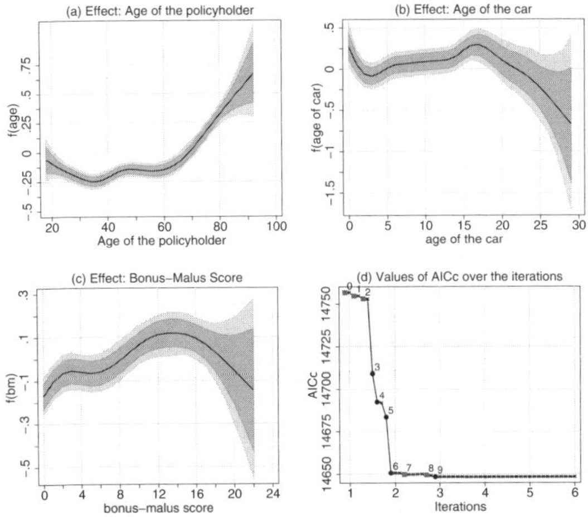

Fig. 1. Selected nonlinear effects of the continuous covariates *ageph*, *agec* and *bm* (figures a-c). Figure d) plots the trace of AIC_c.

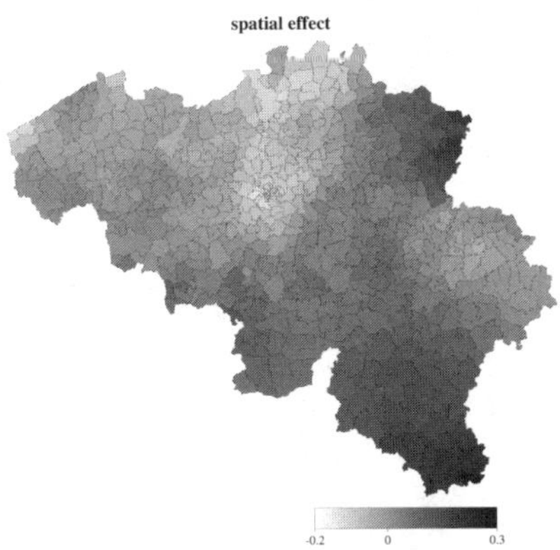

Fig. 2. Posterior mode of the spatial effect.

Figure 1 d) plots the trace of the AIC_c criterion. The algorithm terminates after 5 iterations. Within each iteration variables may be added or removed (indicated by a grey square), the smoothing parameter may be changed (black circle), or the fit of a function may be improved (cross). The numbers 1-9 indicate important changes in the model (variable included or removed, smoothing parameter changed). After the second iteration the algorithm has already collapsed to backfitting. From model change 6 on the improvement of AIC_c is almost negligible. When reporting results this information should be delivered because it gives valuable insight on competing models with comparable fit.

5 Conclusion

The article presents a fast approach for simultaneous selection of variables and smoothing parameters in geoadditive models. Software is included in the software package BayesX (Brezger et al. (2005)). Of course, an important limitation of all automated procedures for model choice should be kept in mind: Other models than the selected may have almost identical goodness of fit (see the AIC_c trace in Figure 1 d). Hence the selected model should be not treated as the only conceivable solution to the problem but more as a good starting point for further investigation.

References

BESAG, J., YORK, J. and MOLLIE, A. (1991): Bayesian Image Restoration with Two Applications in Spatial Statistics. *Annals of the Inst. of Statistical Mathematics, 43, 1-59.*

BREZGER, A., KNEIB, T. and LANG, S. (2005): BayesX Manuals. Available at http://www.stat.uni-muenchen.de/~bayesx/bayesx.html.

DENUIT, M. and LANG, S. (2004): Nonlife Ratemaking with Bayesian GAM's. *Insurance: Mathematics and Economics, 35, 627-647.*

EILERS, P.H.C. and MARX, B.D. (1996): Flexible Smoothing Using B-splines and Penalties. *Statistical Science, 11, 89-121.*

FAHRMEIR, L. and LANG, S. (2001): Bayesian Inference for Generalized Additive Mixed Models Based on Markov Random Field Priors. *Journal of the Royal Statistical Society C, 50, 201-220.*

HASTIE, T. and TIBSHIRANI, R. (1990): *Generalized Additive Models.* Chapman & Hall, London.

KAMMANN, E.E. and WAND, M.P. (2003): Geoadditive Models. *Journal of the Royal Statistical Society C, 52, 1-18.*

LANG, S. and BREZGER, A. (2004): Bayesian P-splines. *Journal of Computational and Graphical Statistics, 13, 183-212.*

TUTZ, G. and BINDER, H. (2005): Boosting Ridge Regression. *SFB 386 Discussion Paper 418.*

WOOD, S. (2004): The mgcv Package. Available at http://www.r-project.org/.

Modelling and Analysing Interval Data

Paula Brito

Faculdade de Economia/LIACC-NIAAD, Universidade do Porto,
Rua Dr. Roberto Frias, 4200-464 Porto, Portugal; mpbrito@fep.up.pt

Abstract. In this paper we discuss some issues which arise when applying classical data analysis techniques to interval data, focusing on the notions of dispersion, association and linear combinations of interval variables. We present some methods that have been proposed for analysing this kind of data, namely for clustering, discriminant analysis, linear regression and interval time series analysis.

1 Introduction

In classical data analysis, data is represented by a $n \times p$ matrix where n individuals (in rows) take exactly one value for each variable (in columns). However, this model is too restrictive to represent data with more complex information. Symbolic data analysis has extended the classical tabular model by allowing multiple, possibly weighted, values for each variable. New variable types have then been introduced - interval, categorical multi-valued and modal variables - which allow taking into account variability and/or uncertainty which is often inherent to the data (see Bock and Diday (2000)).

In this paper we focus on the analysis of interval data, that is, where individuals are described by variables whose values are intervals of $\Re$. We discuss some issues which arise when trying to apply classical data analysis techniques to interval data, and present some methods which have been proposed for analyzing this kind of data.

De Carvalho et al. (2006) have proposed a partitioning clustering method following the dynamic clustering approach and using an L_2 distance. The result of any clustering method depends heavily on the scales used for the variables, natural clustering structures can sometimes only be detected after an appropriate rescaling of variables. In this context, the standardization problem has been addressed, and three standardization techniques for interval-type variables have been proposed. Numerous methods have now been proposed for clustering interval data. Bock (2002) has proposed several clustering algorithms for symbolic data described by interval variables, based on a clustering criterion and thereby generalizing similar approaches in classical data analysis. Chavent and Lechevallier (2002) proposed a dynamic clustering algorithm for interval data where the class representatives are defined by an optimality criterion based on a modified Hausdorff distance. Souza and De Carvalho (2004) have proposed partitioning clustering methods for interval data based on city-block distances, also considering adaptive distances. Various new techniques are described in the forthcoming monograph (Diday and Noirhomme (2006)).

Based on one of the dispersion measures proposed for the standardization process, it has been possible to obtain a covariance measure and define a sample correlation coefficient r for interval data. Following this line, a linear regression model for interval data, which has r^2 as determination coefficient, has been derived. Various approaches for linear regression on interval variables have been investigated by Billard and Diday (2003) and Neto et al. (2004).

In a recent work, Duarte Silva and Brito (2006) discuss the problem of linear combination of interval variables, and compare different approaches for linear discriminant analysis of interval data. A first approach is based on the measures proposed by Bertrand and Goupil (2000) and Billard and Diday (2003), assuming an uniform distribution in each interval. Another approach consists in considering all the vertices of the hypercube representing each of the n individuals in the p-dimensional space, and then perform a classical discriminant analysis of the resulting $n \times 2^p$ by p matrix. This follows previous work by Chouakria et al. (2000) for Principal Component Analysis. A third approach is to represent each variable by the midpoints and ranges of its interval values, perform two separate classical discriminant analysis on these values and combine the results in some appropriate way, or else analyze midpoints and ranges conjointly. This follows similar work on Regression Analysis by Neto et al. (2004), and Lauro and Palumbo (2005) on Principal Component Analysis.

Perspectives of future work include modeling interval time-series data, a problem which is addressed by Teles and Brito (2005) using ARMA models.

The remaining of the paper is organized as follows: In Section 2 we define precisely interval data. Section 3 presents the dynamical clustering method and the three standardization techniques proposed in (De Carvalho et al. (2006)). In Section 4 we derive a linear regression model from one of the proposed dispersion measures. Section 5 discusses alternative definitions for the concepts of dispersion, association and linear combinations of interval variables, following (Duarte Silva and Brito (2006)). In Section 6, the three alternative discriminant analysis approaches investigated by Duarte Silva and Brito (2006) are presented. Section 7 introduces recent work on modelisation of interval time-series data. Finally, Section 8 concludes the paper, raising the main questions that remain open to future research.

2 Interval data

In classical data analysis, data is represented by a $n \times p$ matrix where n individuals (in rows) take exactly one value for each variable (in columns). Symbolic data analysis has extended the classical tabular model by allowing multiple, possibly weighted, values for each variable (see Bock and Diday (2000)).

In this paper we focus on the analysis of interval data. Given a set of individuals $\Omega = \{\omega_1, \ldots, \omega_n\}$, an interval variable is defined by an application $Y : \Omega \rightarrow T$ such that $\omega_i \rightarrow Y(\omega_i) = [l_i, u_i]$, where T is the set of intervals of an underlying set $O \subseteq \Re$. Let I be an $n \times p$ matrix representing the values of p interval variables $Y_1, \ldots, Y_p$ on Ω. Each $\omega_i \in \Omega$, is represented by a p-uple of intervals, $I_i = (I_{i1}, \ldots, I_{ip}), i = 1, \ldots, n$, with $I_{ij} = [l_{ij}, u_{ij}] = Y_j(\omega_i), j = 1, \ldots, p$.

Interval data may occur in many different situations. We may have 'native' interval data, describing ranges of variable values, for instance, daily stock prices or monthly temperature ranges; imprecise data, coming from repeated measures or confidence interval estimation; symbolic data, as descriptions of biological species or technical specifications. Interval data may also arise from the aggregation of huge data bases, when real values are generalized by intervals.

In this context, mention should also be made to Interval Calculus (Moore (1966)), a discipline that has derived rules for dealing with interval values. Given two intervals I_1 and I_2, any arithmetical operation 'op' between them is defined by I_1 op $I_2 = \{x \text{ op } y, x \in I_1, y \in I_2\}$. That is, the result of a given operation between the intervals I_1 and I_2 is an interval comprehending all possible outcome of the operation between values of I_1 and values of I_2.

3 Dynamical clustering

De Carvalho et al. (2006) have proposed a partitioning clustering method following the dynamic clustering approach and using an L_2 distance.

Dynamic clustering (Diday and Simon (1976)), generally known as (generalized) 'k-means clustering', is a clustering method that determines a partition $P = (P_1, ..., P_k)$ of a given data set $\Omega = \{\omega_1, ..., \omega_n\}$ of objects into a fixed number k of clusters $P_1, ..., P_k$ and a set $L = (\ell_1, ..., \ell_k)$ of cluster prototypes by optimizing a criterion $W(P, L)$ that evaluates the fit between the clusters and the cluster prototypes. Starting from an initial system of class representatives, or from an initial partition, the method applies iteratively an *assignment function*, which allows determining a new partition, followed by a cluster *representation function*, which defines optimum prototypes, until convergence is attained.

Let $\mathcal{P}_k$ be the family of all partitions $P = (P_1, \ldots, P_k)$ of Ω into the given number k of non-empty clusters. Also, let $\mathcal{L}$ be the set of 'admissible' class representatives or prototypes (dependent on the given data type) and denote by $\mathcal{L}_k = (\mathcal{L})^k$ the set of all systems of k prototypes $L = (\ell_1, \ldots, \ell_k)$ (one for each cluster). In our case, the 'representation space' $\mathcal{L}$ is the set $\mathcal{I}^p$ of all finite intervals from $\Re$. Consider a function $D(P_h, \ell_h)$ that measures how well the prototype ℓ_h represents the objects in class P_h (a low value indicates a good fit between ℓ_h and P_h). The clustering problem consists in finding a pair $(P^*, L^*) \in \mathcal{P}_k \times \mathcal{L}_k$ that minimizes a criterion

$$W(P,L) = \sum_{h=1}^{k} D(P_h, \ell_h) \text{ i.e. } W(P^*, L^*) = \min_{P,L} W(P,L) \tag{1}$$

In the proposed method, we use the L_2 distance between interval-vectors, $\varphi(I_i, \ell) := \sum_{j=1}^{p} \left[|l_{ij} - l_{\ell j}|^2 + |u_{ij} - u_{\ell j}|^2\right]$. Then $D(Q, \ell)$ is typically obtained as the sum over all objects ω_i from a subset (class) $Q \subset \Omega$: $D(Q, \ell) := \sum_{\omega_i \in Q} \varphi(I_i, \ell)$ for $\ell \in \mathcal{L}, Q \subset \Omega$ such that the criterion to minimize is given by

$$W(P,L) = \sum_{h=1}^{k} \sum_{\omega_i \in P_h} \sum_{j=1}^{p} \left[(l_{ij} - l_j^{(h)})^2 + (u_{ij} - u_j^{(h)})^2\right] \rightarrow \min_{P,L} \tag{2}$$

where $\ell_h = ([l_1^{(h)}, u_1^{(h)}], ..., [l_p^{(h)}, u_p^{(h)}]) \subset \Re^p$ is the prototype for the class P_h.

The dynamical clustering algorithm is then defined in terms of the following *assignment function* f and the *representation function* g that correspond to partial minimization of $W(P,L)$:

The assignment function $f : \mathcal{L}_k \longrightarrow \mathcal{P}_k$ is the function that assigns to each k-tuple $L = (\ell_1, \ldots, \ell_k)$ of class prototypes a k-partition $f(L) = P = (P_1, ..., P_k)$ of objects with classes defined by the minimum-dissimilarity rule:

$$P_h := \{\omega_i \in \Omega : D(\{\omega_i\}, \ell_h) \leq D(\{\omega_i\}, \ell_m) \text{ for } 1 \leq m \leq k\} \qquad h = 1, \ldots, k \tag{3}$$

where in the case of ties an object ω_i is assigned to the class with the smallest index.

The representation function $g : \mathcal{P}_k \longrightarrow \mathcal{L}_k$ is the function that associates to each partition in k clusters $P = (P_1, ..., P_k)$ a vector of protoptypes $g(P) = g(P_1, ..., P_k) := (\ell_1, \ldots, \ell_k)$ with p-dimensional intervals: $\ell_h = ([\bar{l}_1^{(h)}, \bar{u}_1^{(h)}], \ldots, [\bar{l}_p^{(h)}, \bar{u}_p^{(h)}])$ for $h = 1, \ldots, k$, whose lower (upper) boundaries are given by the corresponding average of lower (upper) boundaries of the data intervals $I_{ij} = [l_{ij}, u_{ij}]$ for ω_i in class P_h:

$$\bar{l}_j^{(h)} = \frac{1}{|P_h|} \sum_{\omega_i \in P_h} l_{ij} \;\; ; \;\; \bar{u}_j^{(h)} = \frac{1}{|P_h|} \sum_{\omega_i \in P_h} u_{ij} \;\; , \; j = 1, \ldots, p \; , h = 1, \ldots, k \tag{4}$$

Applying iteratively the assignment function followed by the representation function in turn decreases steadily the values $W(P,L)$ of the clustering criterion (1) until a local minimum is attained that depends on the data and on the initial configuration. The method hence minimizes $W(P, L(P)) = \sum_{h=1}^{k} \sum_{\omega_i \in P_h} \sum_{j=1}^{p} \left[(l_{ij} - \bar{l}_j^{(h)})^2 + (u_{ij} - \bar{u}_j^{(h)})^2 \right]$ with respect to partition P.

3.1 Standardization

It is clear that dissimilarity values and clustering results are strongly affected by the scales of variables. Typically, some standardization must be performed prior to the clustering process in order to attain an 'objective' or 'scale-invariant' result. Three alternative standardization methods for the case of interval data have been proposed in (De Carvalho et al. (2006)), and are described here below. The main principle is that variables $Y_j, j = 1, ..., p$ are standardized separately, each one in a linear way, with the same transformation for both the lower and the upper bound of all n component intervals $I_{ij} := [l_{ij}, u_{ij}]$, $i = 1, ..., n$.

Standardization 1: Using the dispersion of the interval centers

The first method considers the mean and the dispersion of the interval centers $(l_{ij} + u_{ij})/2$ (midpoints) and standardizes such that the resulting transformed midpoints have zero mean and dispersion 1 in each dimension. The mean value and the dispersion of all interval midpoints are given by

$$m_j := \frac{1}{n} \sum_{i=1}^{n} \frac{l_{ij} + u_{ij}}{2} \quad \text{and} \quad s_j^2 := \frac{1}{n} \sum_{i=1}^{n} \left(\frac{l_{ij} + u_{ij}}{2} - m_j \right)^2 \quad \text{respectively.} \tag{5}$$

With this notation, the data interval I_{ij} is transformed into the interval $I'_{ij} = [l'_{ij}, u'_{ij}]$ with boundaries $l'_{ij} := \frac{l_{ij} - m_j}{s_j}$ and $u'_{ij} := \frac{u_{ij} - m_j}{s_j}$ $i = 1, ..., n$ where automatically $l'_{ij} \leq u'_{ij}$ for all i, j. The new intervals I'_{ij} are standardized with $m'_j := \frac{1}{n} \sum_{i=1}^{n} \frac{l'_{ij} + u'_{ij}}{2} = 0$ and $s'^2_j := \frac{1}{n} \sum_{i=1}^{n} \left(\frac{l'_{ij} + u'_{ij}}{2} - m'_j \right)^2 = 1$.

Standardization 2: Using the dispersion of the interval boundaries

The second standardization method transforms, for each variable Y_j, the n intervals I_{ij} such that the mean and the *joint dispersion* of the rescaled interval boundaries are 0 and 1, respectively. The joint dispersion of a variable Y_j is defined by

$$\tilde{s}_j^2 = \frac{1}{n} \sum_{i=1}^{n} \frac{(l_{ij} - m_j)^2 + (u_{ij} - m_j)^2}{2} \tag{6}$$

Then, for $i = 1, ..., n$, the intervals $I_{ij} = [l_{ij}, u_{ij}]$ are transformed into $I'_{ij} = [l'_{ij}, u'_{ij}]$ with $l'_{ij} = \frac{l_{ij} - m_j}{\tilde{s}_j}$ and $u'_{ij} = \frac{u_{ij} - m_j}{\tilde{s}_j}$ with $l'_{ij} \le u'_{ij}$ for all i, j. Similarly as before, the new intervals $I'_{ij} = [l'_{ij}, u'_{ij}]$ are standardized with $m'_j = 0$ and $\tilde{s}'^2_j = \frac{1}{2n} \sum_{i=1}^{n} \left[(l'_{ij} - m'_j)^2 + (u'_{ij} - m'_j)^2 \right] = 1, \quad j = 1, ..., p.$

Standardization 3: Using the global range

The third standardization method transforms, for a given variable, the intervals $I_{ij} = [l_{ij}, u_{ij}]$ $(i = 1, ..., n)$ such that the range of the n rescaled intervals is the unit interval $[0, 1]$. Let $Min_j = Min\{l_{1j}, ..., l_{nj}\}$ and $Max_j = Max\{u_{1j}, ..., u_{nj}\}$ be the extremal lower and upper boundary values. With this notation, the interval $I_{ij} = [l_{ij}, u_{ij}]$ is transformed into the interval $I'_{ij} = [l'_{ij}, u'_{ij}]$ with boundaries $l'_{ij} = \frac{l_{ij} - Min_j}{Max_j - Min_j}$ and $u'_{ij} = \frac{u_{ij} - Min_j}{Max_j - Min_j}$ with $l'_{ij} \le u'_{ij}$ as before. It results that $Min\{l'_{1j}, ..., l'_{nj}\} = 0$ and $Max\{u'_{1j}, ..., u'_{nj}\} = 1$.

Simulation studies (De Carvalho et al. (2006)) showed that standardization greatly improves the quality of the clustering results, in terms of recovery of an imposed structure. Standardization 2 performed slightly better for ill-separated clusters where intervals have large ranges.

4 A linear regression model

Consider the dispersion measure used in the second standardization procedure, $\tilde{s}^2_j$ (see (6)), which evaluates dispersion of the interval-valued variable Y_j by the joint dispersion of the interval boundaries of its values. If we use the same principle to evaluate co-dispersion, we obtain a 'co-variance' measure by

$$\tilde{s}_{jj'} = \frac{1}{n} \sum_{i=1}^{n} \frac{(l_{ij} - m_j)(l_{ij'} - m_{j'}) + (u_{ij} - m_j)(u_{ij'} - m_{j'})}{2} \qquad (7)$$

from which we may then derive a 'correlation' measure, $r_{jj'} = \frac{\tilde{s}_{jj'}}{\tilde{s}_j \tilde{s}_{j'}}$. It is easy to show that $-1 \le r_{jj'} \le 1$.

Now, an interesting problem is to derive the linear regression between Y_j and $Y_{j'}$ for which $r^2_{jj'} = \frac{\tilde{s}^2_{jj'}}{\tilde{s}^2_j \tilde{s}^2_{j'}}$ is the determination coefficient. It may be shown that the corresponding linear regression model consists in applying the same transformation to both the lower and the upper bounds of the independent variable values, i.e.

$$\widehat{Y_{j'}(\omega_i)} = \widehat{I_{ij'}} = \alpha + \beta I_{ij} = \left[\widehat{l_{ij'}}, \widehat{u_{ij'}} \right] = [\alpha + \beta l_{ij}, \alpha + \beta u_{ij}] \qquad (8)$$

where $\beta = \dfrac{\tilde{s}_{jj'}}{\tilde{s}_j^2}$ and $\alpha = m_{j'} - \beta m_j$; furthermore α and β minimize $E = \sum_{i=1}^{n} \left[(l_{ij'} - \alpha - \beta l_{ij})^2 + (u_{ij'} - \alpha - \beta u_{ij})^2 \right]$. The regression model (8) has been independently obtained by Neto et al (2004), by direct minimization of the criterion E.

The obtained result is not surprising. It should be noticed that $\tilde{s}_j^2$ as defined in equation (6) evaluates dispersion of an interval-valued variable by the dispersion of the lower and the upper bounds of the observed intervals around the global midpoint m_j. This is in fact equivalent to considering a $2 \times n$ sample consisting of all interval boundaries, $\{l_{ij}, u_{ij}, i = 1, \ldots, n\}$; then m_j is its empirical mean and $\tilde{s}_j^2$ is its empirical variance, as computed in real-valued data analysis. Therefore, there is only one linear regression model to fit the $2 \times n$ data points, and which is obtained, as in classical analysis, by equation (8).

An important question is whether this model performs well in fitting interval data. Monte-Carlo experiences have been performed, simulating interval data with different degrees of linearity, different degrees of variability (measured by interval ranges) and quality of adjustment. The performance of the method, on the basis of MSE on lower and upper bounds, and r^2 of lower and upper bounds, was similar to that of the method applied in (Billard and Diday (2003)). Other approaches for linear regression on interval variables have been investigated by Neto et al. (2004).

5 Dispersion, association and linear combinations of interval variables

Duarte Silva and Brito (2006) have addressed the problem of linear combination of interval variables, investigating how alternative possibilities behave as concerns usual properties.

Let, as before, I be an $n \times p$ matrix representing the values of p interval variables $Y_1, \ldots, Y_p$ on a set $\Omega = \{\omega_i, i = 1, \ldots, n\}$ where each $\omega_i \in \Omega$, is represented by a p-uple of intervals, $I_i = (I_{i1}, \ldots, I_{ip}), i = 1, \ldots, n$, with $I_{ij} = [l_{ij}, u_{ij}] = Y_j(\omega_i), j = 1, \ldots, p$. Let '$\bigotimes$' denote the operator that defines linear combinations of interval variables, i.e., $Z = I \bigotimes \beta$ are r linear combinations of the Y's based on $p \times r$ real coefficients $\beta_{j\ell}, j = 1, \ldots, p; \ell = 1, \ldots, m$, stacked in a matrix β. Furthermore, let S_I be a covariance matrix of measures of dispersion (s_j^2) and association ($s_{jj'}$) for interval data. It is desired that the following basic properties are satisfied, for any $p \times r$ real matrix β:

(P1) $I_i \bigotimes \beta_\ell = \sum_{j=1}^{p} \beta_{j\ell} \times I_{ij}$ where β_ℓ denotes the ℓ-th column of matrix β.

(P2) $S_Z = S_{I \otimes \beta} = \beta^t S_I \beta$ that is, the covariance between interval variables should be a symmetric bilinear operator so that the usual formulas for variance and covariance of linear combinations still apply.

A natural definition of linear combination of interval variables, which corresponds to the regression model derived in Section 4, is Definition A: $I_i \bigotimes_A \beta_\ell = z_{i\ell A} = [\underline{z}_{i\ell A}, \overline{z}_{i\ell A}], i = 1, \ldots, n$, with

$$\underline{z}_{i\ell A} = \sum_{j=1}^{p} \beta_{j\ell}\, l_{ij} \quad ; \quad \overline{z}_{i\ell A} = \sum_{j=1}^{p} \beta_{j\ell}\, u_{ij} \tag{9}$$

Unfortunately, this quite straightforward definition does not satisfy property **(P1)** if at least one element of β_ℓ is negative. A definition of linear combination of interval variables that respects **(P1)** is given by:

Definition B: $I_i \bigotimes_B \beta_\ell = z_{i\ell B} = [\underline{z}_{i\ell B}, \overline{z}_{i\ell B}], i = 1, \ldots, n$, with

$$\underline{z}_{i\ell B} = \sum_{\beta_{j\ell}>0} \beta_{j\ell}\, l_{ij} + \sum_{\beta_{j\ell}<0} \beta_{j\ell}\, u_{ij} \quad ; \quad \overline{z}_{i\ell B} = \sum_{\beta_{j\ell}>0} \beta_{j\ell}\, u_{ij} + \sum_{\beta_{j\ell}<0} \beta_{j\ell}\, l_{ij} \tag{10}$$

Definition B is the definition obtained by applying the rules of Interval Calculus (Moore (1966)), since the resulting intervals include all possible values that are scalar linear combinations of the values within the intervals I_{ij}. However, Definition B ignores any connection that may exist between corresponding interval bounds in the original data. The existence (or lack of it) of such connection and the relevance of property **(P1)** depends on how a set of interval data ought to be interpreted. Definition A is appropriate when lower (respectively upper) bounds of different variables tend to occur simultaneously: we speak then of *Positive "Inner Correlation"*. However, as mentioned before, Definition A does not satisfy **(P1)**. Definition B satisfies **(P1)** and is appropriate in the absence of inner correlation.

An important result obtained in (Duarte Silva and Brito (2006)) shows that when we consider dispersion s_j^2 and association $s_{jj'}$ measures which depend on l_{ij} and u_{ij} symmetrically, then both Definition A and Definition B satisfy **(P2)**. It follows that variances of linear combinations are then given by quadratic forms, and ratios are maximized by a traditional eigenanalysis.

6 Discriminant analysis

In Duarte Silva and Brito (2006) three different approaches for linear discriminant analysis of interval data are compared.

6.1 Distributional approach

In a first approach it is assumed that each interval variable represents the possible values of an underlying real-valued variable. Bertrand and Goupil (2000) assume an equidistribution hypothesis, which consists in considering each observation as equally likely and that the values of the underlying variable are uniformly distributed. The empirical distribution function of an interval variable is then defined as the uniform mixture of n uniform distributions; the corresponding mean and variance may then be obtained by simple integration. Following the same reasoning, Billard and Diday (2003) have derived the joint density function of two interval variables and obtained the empirical covariance.

Assume now that the n observations are partitioned into k groups, $C_1, \ldots, C_k$. The empirical density function of variable Y_j in group C_α is defined as

$$f_{\alpha j}(\xi) = \frac{1}{n_\alpha} \sum_{\omega_i \in C_\alpha} \frac{\mathbf{1}_{I_{ij}}(\xi)}{u_{ij} - l_{ij}} \tag{11}$$

and the joint density function of two interval variables $Y_j, Y_{j'}$ in group C_α as

$$f_{\alpha jj'}(\xi_1, \xi_2) = \frac{1}{n_\alpha} \sum_{\omega_i \in C_\alpha} \frac{\mathbf{1}_{I_{ij} \times I_{ij'}}(\xi_1, \xi_2)}{(u_{ij} - l_{ij})(u_{ij'} - l_{ij'})} \tag{12}$$

It follows that the global empirical density functions are mixtures of the corresponding group specific functions, that is, they are mixtures of k mixtures of uniform laws, $f_j(\xi) = \sum_{\alpha=1}^{k} \frac{n_\alpha}{n} f_{\alpha j}(\xi)$; $f_{jj'}(\xi_1, \xi_2) = \sum_{\alpha=1}^{k} \frac{n_\alpha}{n} f_{\alpha jj'}(\xi_1, \xi_2)$.

It can easily be shown that these densities satisfy the usual properties. The global empirical mean of a variable Y_j is given by $m_j = \frac{1}{n} \sum_{i=1}^{n} \frac{l_{ij} + u_{ij}}{2} = \frac{1}{2}(\bar{l_j} + \bar{u_j})$; the empirical variance is

$$s_j^2 = \int_{-\infty}^{+\infty} (\xi - m_j)^2 f_j(\xi) d\xi = \frac{1}{3n} \sum_{i=1}^{n} (l_{ij}^2 + l_{ij} u_{ij} + u_{ij}^2) - \frac{1}{4n^2} \left[\sum_{i=1}^{n} (l_{ij} + u_{ij}) \right]^2 \tag{13}$$

and the empirical covariance between two interval variables Y_j, $Y_{j'}$ is:

$$s_{jj'} = cov(Y_j, Y_{j'}) = \int_{-\infty}^{+\infty} \int_{-\infty}^{+\infty} (\xi_1 - m_j)(\xi_2 - m_{j'}) f_{jj'}(\xi_1, \xi_2) d\xi_1 d\xi_2 =$$

$$\frac{1}{4n} \sum_{i=1}^{n} [(l_{ij} + u_{ij})(l_{ij'} + u_{ij'})] - \frac{1}{4n^2} \left[\sum_{i=1}^{n} (l_{ij} + u_{ij}) \right] \left[\sum_{i=1}^{n} (l_{ij'} + u_{ij'}) \right] \tag{14}$$

These measures of dispersion (13) and association (14) clearly treat lower and upper bounds symmetrically. From these measures a decomposition in within-groups w_{jj} and $w_{jj'}, (j \neq j')$ and between-groups $b_{jj'}$ components may be obtained. Discriminant linear functions are then given by the eigenvectors of $W^{-1}B$.

6.2 Vertices approach

In a second approach, following the work of Chouakria et al. (2000) for Principal Component Analysis, each individual is represented by the corresponding hypercube vertices. That is, the original matrix is expanded into a $(n \times 2^p) \times p$ matrix M, where each row i of I gives rise to 2^p rows of M, corresponding to all possible combinations of the limits of intervals $[l_{ij}, u_{ij}], j = 1, \ldots, p$. A classical analysis of the vertices matrix is then performed, and we obtain a factorial representation of points, one for each of the 2^p vertices. From this, we may recover a representation in the form of intervals, proceeding as Chouakria et al. (2000) in PCA: Let Q_i be the set of row indices q in matrix M which refer to the vertices of the hypercube corresponding to ω_i. For $q \in Q_i$ let $\zeta_{q\ell}$ be the value of the ℓ-th real-valued discriminant function for the vertex with row index q. The value of the ℓ-th interval discriminant variate z for ω_i is then defined by $\underline{z}_{i\ell} = \text{ Min } \{\zeta_{q\ell}, q \in Q_i\}$; $\overline{z}_{i\ell} = \text{ Max } \{\zeta_{q\ell}, q \in Q_i\}$.

6.3 Midpoints and ranges approach

A third approach consists in representing each interval I_{ij} by its midpoint c_{ij} and range r_{ij}, following similar work on Regression Analysis by Neto et al. (2004) and on Principal Component Analysis by Lauro and Palumbo (2005). Two classical analysis may then be performed: separately for $C = [c_{ij}]$ and $R = [r_{ij}]$ or jointly for the matrix $[C|R]$.

6.4 Classification rules

Classification rules are derived from discriminant space representations. These representations may assume the form of single points or intervals; classification rules will hence be based on point distances or distances between intervals, accordingly.

In the distributional approach, each observation is allocated to the group with nearest centroid in the discriminant space, according to a simple Euclidean distance. Applying Definitions A and B, linear combinations of the interval variables may be obtained, that produce interval-valued discriminant variates. In this case, classification rules may be derived by using distances between interval vectors. This is also the case in the vertices approach, where discriminant variates are interval-valued. Different interval distances δ may be used, as the Haudorff distance δ: if $z_{i\ell} = [\underline{z}_{i\ell}, \overline{z}_{i\ell}]$ and $z_{i'\ell} = [\underline{z}_{i'\ell}, \overline{z}_{i'\ell}]$, then $\delta(z_{i\ell}, z_{i'\ell}) = \text{Max } \{|\underline{z}_{i\ell} - \underline{z}_{i'\ell}|, |\overline{z}_{i\ell} - \overline{z}_{i'\ell}|\}$.

For the midpoints and ranges approach, only point distances are used to define classification rules. When two separate analysis are performed for midpoints and ranges, the discriminant variates are generally correlated, for this reason, the Mahalanobis distance should be used; when a single discriminant analysis is performed combining midpoints and ranges, the simple Euclidean distance is adequate.

Experimental results have shown (Duarte Silva and Brito (2006)) that when classes are separated by midpoints only, methods which take ranges explicitly into account have worst performance; when classes differ both in terms of midpoints and ranges, methods which take ranges explicitly into account have best performance; methods based on interval approaches capture range information in a limited extent, and therefore constitute a good compromise.

7 Interval time-series

When interval-valued data is collected as an ordered sequence through time (or any other dimension) we are in presence of an **interval time series**. Let X_{L_T} and X_{U_T} be respectively the lower and upper limits of the observed intervals with $X_{L_T} \leq X_{U_T}$. Then $[X_{L_1}, X_{U_1}], \ldots, [X_{L_N}, X_{U_N}]$ is an interval time series. Teles and Brito (2005) have addressed the problem of modelling interval time-series data, using ARMA models. It is assumed that the processes that generate the lower and the upper limits of the intervals have the same ARMA parameters and different means, with the mean of the upper limit greater than the mean of the lower limit, and that the white noise sequences of the two processes are independent with different variances. Two approaches have been considered. The first approach uses all the observations from the upper and the lower interval limits, doubling the sample size used in estimation; in a simpler approach, it is shown that the single time series of the difference of the interval limits follows another ARMA process with the same orders and ARMA parameters and standard time series methodology is applied.

Experimental results have shown that the proposed procedures perform very well both in terms of estimation accuracy and in terms of forecasting performance. As concerns estimation accuracy, the method for both interval limits shows higher accuracy and clearly outperforms the other method, being more efficient. Both methods show the same forecasting performance as single time series models, that is, in spite of the better estimation accuracy of the method for both interval limits, the forecasting performance is the same.

8 Conclusions and perspectives

The extension of classical methodologies to the analysis of interval data raises new problems: How to evaluate dispersion? How to define linear combinations? Which properties remain valid?

The concept of dispersion is a central one, and the way to evaluate dispersion of interval data is not straightforward as in the case of real-valued data. Different alternatives are possible, and the choice of one of these often determines the type of model to be used subsequently.

Representations in lower dimensional spaces may assume different forms: intervals emphasize variability inherent to each observation, point representations can display distinct contributions to the separation between groups.

The important issue in analysing interval data remains however the need for models. Without statistical modelling, no estimation or hypothesis testing is possible. This is now the main challenge facing researchers who wish to go beyond the classical framework of real-valued data.

References

BERTRAND, P. and GOUPIL, F. (2000): Descriptive Statistics for Symbolic Data. In: H.-H. Bock and E. Diday (Eds.): *Analysis of Symbolic Data.* Springer, Heidelberg, 106-124.

BILLARD, L. and DIDAY, E. (2003): From the Statistics of Data to the Statistics of Knowledge: Symbolic Data Analysis. *Journal of the American Statistical Association, 98, 462, 470–487.*

BOCK, H.-H. (2002): Clustering Algorithms and Kohonen Maps for Symbolic Data. *Journal of the Japanese Society of Computational Statistics, 15, 1–13.*

BOCK, H.-H. and DIDAY, E. (2000): *Analysis of Symbolic Data.* Springer, Heidelberg.

CHAVENT, M. and LECHEVALLIER, Y. (2002): Dynamical Clustering Algorithm of Interval Data: Optimization of an Adequacy Criterion Based on Hausdorff Distance. In: A. Sokolowski and H.-H. Bock (Eds.): *Classification, Clustering and Data Analysis.* Springer, Heidelberg, 53-59.

CHOUAKRIA, A., CAZES, P. and DIDAY, E. (2000): Symbolic Principal Component Analysis. In: H.-H. Bock and E. Diday (Eds.): *Analysis of Symbolic Data.* Springer, Heidelberg, 200-212.

DE CARVALHO, F.A.T., BRITO, P. and BOCK, H.-H. (2006): Dynamic Clustering for Interval Data Based on L_2 Distance. *Computational Statistics, 21, 2, 231–250.*

DIDAY, E. and NOIRHOMME, M. (2006): *Symbolic Data Analysis and the SODAS Software.* Wiley, to appear.

DIDAY, E. and SIMON, J.J. (1976): Clustering Analysis. In: K.S. Fu (Ed.): *Digital Pattern Recognition.* Springer, Heidelberg, 47-94.

DUARTE SILVA, A.P. and BRITO, P. (2006): Linear Discriminant Analysis for Interval Data. *Computational Statistics, 21, 2, 289–308.*

LAURO, C. and PALUMBO, F. (2005): Principal Component Analysis for Non-Precise Data. In: M. Vichi *et al* (Eds.): *New Developments in Classification and Data Analysis.* Springer, 173-184.

MOORE, R.E. (1966): *Interval Analysis.* Prentice Hall, New Jersey.

NETO, E.A.L., DE CARVALHO, F.A.T. and TENORIO, C. (2004): Univariate and Multivariate Linear Regression Methods to Predict Interval-Valued Features. In: *AI2004:Advances in Artificial Intelligence.* Lecture Notes on Artificial Intelligence, Springer, 526-537.

SOUZA, R.M.C.R. and DE CARVALHO, F.A.T. (2004): Clustering of Interval Data Based on City-Block Distances. *Pattern Recognition Letters, 25, 3, 353–365.*

TELES, P. and BRITO, P. (2005): Modeling Interval Time Series Data. In: *Proc. of the 3rd World Conference on Computational Statistics and Data Analysis.* Limassol, Cyprus.

Testing for Genuine Multimodality in Finite Mixture Models: Application to Linear Regression Models

Bettina Grün[1] and Friedrich Leisch[2]

[1] Institut für Statistik und Wahrscheinlichkeitstheorie, Technische Universität Wien, Wiedner Hauptstraße 8-10/1071, A-1040 Wien, Austria; `Bettina.Gruen@ci.tuwien.ac.at`

[2] Institut für Statistik, Ludwig-Maximilians-Universität München, Ludwigstraße 33, D-80539 München, Germany; `Friedrich.Leisch@stat.uni-muenchen.de`

Abstract. Identifiability problems can be encountered when fitting finite mixture models and their presence should be investigated by model diagnostics. In this paper we propose diagnostic tools to check for identifiability problems based on the fact that they induce multiple (global) modes in the distribution of the parameterizations of the maximum likelihood models depending on the data generating process. The parametric bootstrap is used to approximate this distribution. In order to investigate the presence of multiple (global) modes the congruence between the results of information-based methods based on asymptotic theory and those derived using the models fitted to the bootstrap samples with initalization in the solution as well as random initialization is assessed. The methods are illustrated using a finite mixture of Gaussian regression models on data from a study on spread of viral infection.

1 Introduction

Finite mixture models are a flexible method for modelling unobserved heterogeneity or approximating general distribution functions. Due to the tremendous increase in computing power in the last decades the fitting of complex models has become possible. However, these models need not be generically identifiable due to the distribution function of the components, e.g. for the binomial distribution (Teicher (1963)), or the covariate matrix involved in a regression setting (Hennig (2000)). We refer to identifiability problems as *generic* if they arise in addition to the problems caused by label switching and overfitting. Furthermore, empirical identifiability problems are encountered if the given dataset is not informative enough to distinguish between different mixture distributions. The differentiation between generic and empirical identifiability problem might be difficult or even depends on the point of view, as mixtures of binomial distributions can become identifiable by increasing the repetition parameter or mixtures of regression models by adding new covariate points.

Due to the complexity of mixture models the empirical likelihood of a given dataset is in general multimodal. Therefore, in a frequentist setting the EM algorithm (Dempster et al. (1977)) is usually repeated several times with different initializations to avoid local optima. It is hoped that the global optimum is determined by selecting the best solution of several runs. Model diagnostics are then performed on the parameters of the fitted model.

In general the empirical likelihood is used for investigating identifiability problems. However, this has the disadvantage that it entirely relies on the given dataset. As in general the aim of model fitting is to generalize from the given dataset to the underlying data generating process (DGP), model diagnostics should investigate the characteristics of the fitted models and their corresponding parameterizations depending on the DGP. If the DGP suffers from identifiability problems the distribution of these parameterizations is also multimodal. As often in practice only a single dataset is available the distribution of fitted models has to be approximated which can be done using the parametric bootstrap (BS). Using this approximation identifiability problems can be investigated.

The use of the parametric BS has been previously proposed for model diagnostics of finite mixture models including the estimation of standard errors. Basford et al. (1997) compared the estimation of standard errors of the fitted parameters of normal mixtures for information-based methods using asymptotic theory and for the parametric BS using the EM algorithm with initialization in solution. In Grün and Leisch (2004) the use of the empirical BS for detecting identifiability problems was investigated. It was shown on an artificial example that the BS procedure detected both solutions corresponding to the parameterizations of the true underlying DGP where these were known. This paper extends the previous work by formalizing the concepts of genuine multimodality and by presenting procedures which can be used without knowing the true solutions.

In Section 2 the term *genuine multimodality* is defined and methods to check for the presence of genuine multimodality are outlined. A case study where the proposed methods are applied on mixtures of linear regression models is given in Section 3 using data from a study on spread of viral infection taken from Turner (2000).

2 Genuine multimodality

In the following we consider finite mixture models of the form

$$H(\mathbf{m}y|\mathbf{m}x, \Theta) = \sum_{s=1}^{S} \pi_s F(\mathbf{m}y|\mathbf{m}x, \mathbf{m}\theta_s)$$

where H is the mixture distribution, $\mathbf{m}y$ the vector of responses, $\mathbf{m}x$ is an optional vector of covariates, S the number of components, F the component distribution function, $\mathbf{m}\theta_s$ the component specific parameters of distribution F and π_s the prior class probabilities. The component specific parameters are denoted by $\vartheta_s = (\pi_s, \mathbf{m}\theta_s)$, and $\Theta = (\vartheta_s)_{s=1,\dots,S}$ is the vector of all parameters. It is assumed that $\Theta \in \Omega$, where Ω is the set of admissible parameter vectors with

- $0 < \pi_s \leq 1, \forall s = 1, \dots, S$
- $\sum_{s=1}^{S} \pi_s = 1$ and
- $\mathbf{m}\theta_s \neq \mathbf{m}\theta_t \; \forall s \neq t$ with $s, t \in \{1, \dots, S\}$.

Because the dataset $\mathcal{X}_N$ is a random variable which depends on the DGP, the optimal model $a(\mathcal{X}_N)$ which maximizes the likelihood is also a random variable. In general a direct analysis of the distribution of $a(\mathcal{X}_N)$ is not possible, because the estimation procedure characterizes the fitted model by returning a corresponding parameterization $\Theta \in \Omega$ and therefore the model diagnostics are based on the fitted parameterizations and in fact the distribution of Θ denoted by $\mathcal{O}$ is analyzed.

The presence of genuine competing parameterizations (i.e. in addition to those induced by label switching of the components) is referred to as *genuine multimodality*:

Definition 1. *The distribution $\mathcal{O}$ of the parameters $\Theta \in \Omega$ is called* genuinely multimodal *if it holds for the set of modes $\mathcal{M}$ of $\mathcal{O}$ that*

$$\exists \Theta_1, \Theta_2 \in \mathcal{M} : \Theta_1 \neq \nu(\Theta_2) \qquad \forall \nu \in \mathit{Perm}(S)$$

A mode of a probability distribution is defined as a local maximum in the associated probability density function (Minnotte (1997)). The distribution $\mathcal{O}$ is called *genuinely unimodal* if the opposite holds for the set of modes $\mathcal{M}$.

Given a single dataset $\mathcal{X}_N$ the distribution $\mathcal{O}$ can be approximated by applying the parametric BS to the fitted model $\hat{a}(\mathcal{X}_N)$ (Hothorn et al. (2005)). In contrast to the estimation of standard errors where the EM algorithm is initialized in the solution, the EM algorithm has to be randomly initialized for investigating genuine multimodality. Initialization of the EM algorithm in the solution in general guarantees that the EM algorithm converges to a mode close to the original parameterization and label switching does in general not occur (see McLachlan and Peel (2000, p. 70)). Random initialization can be made by assigning observations to the components, i.e. to start with the M-step given a-posteriori probabilities, or to specify an initial parameter value and start with an E-step. Details can be found in McLachlan and Peel (2000, p. 54). However, if random initialization is used label switching has to be eliminated before component specific inference can be made for the BS estimates. In a frequentist setting label switching can be eliminated by relabelling the components of the BS samples in order to minimize the Kullback-Leibler divergence between the distributions of the a-posteriori probabilities of the BS model and the fitted maximum likelihood model. Minimization of this distance measure was proposed by Stephens (2000) in a Bayesian context.

In the following we propose a method for checking for genuine multimodality where solving the label switching problem is not necessary. Given the estimates of the standard errors using information-based methods 95% confidence bands of the component specific parameters ϑ_s can be determined using Bonferroni correction. As noted previously standard errors derived using information-based methods might not be reliable estimates and should be taken with care for finite mixture models, especially if the sample size is small.

For each component specific parameter estimate of the BS samples ϑ_s^b it can be determined if it lies completely within one of the confidence bands. Under the assumption that two confidence bands do not overlap for all parameters this assignment is unique. The percentage of parameters assigned is expected to be 95%. However, strong deviation is either an indication for genuine multimodality, for convergence problems of the EM algorithm or for a bad approximation of the standard errors by the information-based methods. If the third cause is applicable can be investigated by comparing the results for random initialization with initialization in solution.

The results can be visualized using parallel coordinate plots of the ϑ_s^b together with the confidence bands. To indicate the assignment to different confidence bands different colorings can be used. If identifiability problems are present the ϑ_s^b which are assigned to no component should not be spurious results where for example the EM algorithm did not converge to the global optimum but cluster around the parameters corresponding to the second mode.

The BS samples using random initialization and initialization in solution should follow the same distribution if there is no genuine multimodality present and if the EM algorithm did not get trapped in a local optimum for the random initializations. This equality of distributions can be tested using multivariate two-sample tests, as e.g. the Cramer test given in Baringhaus and Franz (2004). The Cramer test uses the difference of the sum of all the Euclidean interpoint distances between the random variables from the two different samples and one-half of the two corresponding sums of distances of the variables with the same sample as test statistics. The distribution under the null hypothesis is approximated using a normal Monte Carlo BS. This test can be applied in an automatic way and if the null hypothesis is not rejected this increases the confidence in the fitted model. However, if the null hypothesis of equality of distribution is rejected further investigations are necessary to determine if genuine multimodality is present or if for instance the EM algorithm failed to converge to the global optimum.

3 Case study: Linear regression of viral infection data

In the following a case study is given where the proposed methods are applied to a dataset taken from Turner (2000). A mixture of two linear regression models was fitted to data from a study of the spread of viral infection among potato plants by aphids. A possible explanation for the presence of two components was that some of the batches of aphids consisted of insects that had passed their "maiden" phase, i.e. they tended to settle on the first acceptable food host encountered which led to low or zero levels of transmission of the virus. 51 experiments were conducted where the number of plants in the chamber was fixed to 81 with 69 healthy and 12 infected. The independent variable is the number of aphids which range between 1 and 320. The dependent variable is the number of (additionally) infected plants recorded after 24 hours.

In order to investigate genuine multimodality this example is modified such that identifiability problems occur due to the violation of the coverage condition given in Hennig (2000) and due to the fact that the prior class probabilities are approximately the same which allows the classes to be differently combined between the two covariate points. The available values for the independent variable are restricted to 100 and 300 with 25 observations for each value and a sample is drawn given the model fitted to the original dataset and this covariate matrix.

Finite mixture models are fitted to the datasets with the EM algorithm for 10 different random initializations and the best solution is reported. Random initialization means that an a-posteriori probability of 0.9 is assigned with equal probability either to the first or the second component randomly for each observation. The estimated parameters are given in Table 1. The results are slightly different to those in Turner (2000) due to the use of different software. The standard errors estimated using information-based methods as outlined in Louis (1982) are given in round brackets. In addition standard errors are also determined using the parametric BS with initialization in solution. As noted by Basford et al. (1997) the number of BS replications B does not seem to be critical past 50 or 100 for estimating standard errors and 100 BS replications are used in the following. A direct comparison of the estimated standard errors does not indicate a strong difference between the two methods.

Table 1. Estimated parameters with standard errors estimated with information-based methods in round and with the parametric BS in square brackets.

	Original data		Modified data	
	Comp.1	Comp.2	Comp.1	Comp.2
Intercept	0.87 (0.38) [0.55]	3.46 (1.09) [1.08]	1.33 (0.62) [0.60]	1.53 (1.62) [1.49]
# aphids	0.00 (0.00) [0.00]	0.06 (0.01) [0.01]	0.00 (0.00) [0.00]	0.06 (0.01) [0.01]
σ^2	1.33 (0.44) [0.59]	10.17 (3.19) [3.45]	1.49 (0.46) [0.43]	7.89 (2.65) [2.64]
π	0.50 (0.08) [0.08]	0.50 (0.08) [0.08]	0.49 (0.08) [0.08]	0.51 (0.08) [0.08]

In Figure 1 and 2 parallel coordinate plots are given of the component specific parameter estimates of the BS samples ϑ_s^b. In addition 95% confidence bands of the parameter estimates using information-based methods are given in grey and the fitted parameters are indicated by white lines. The BS vectors are plotted in different panels depending on the component assignment. It can be clearly seen that there are more BS samples assigned to no component for random initialization than for initialization in solution. In Figure 2 the BS samples assigned to no component seem to cluster around two different parameter values indicating the identifiability problem.

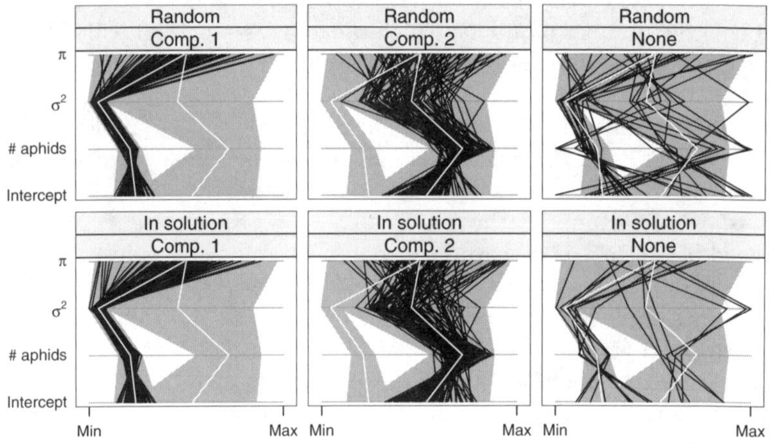

Fig. 1. Parallel coordinate plot of the parameter estimates for the parametric BS samples of the original data with assignment to the confidence bands.

The number of component specific BS estimates assigned to the components based on the confidence bands are given in Table 2. For the initialization in solution the assigned number of samples is almost equal to the expected number (i.e. 95), which also indicates a high correspondence between the information-based and the BS estimates of the standard errors. By contrast for random initialization the number is slightly smaller than expected for the original dataset which indicates that there might have been problems with the convergence of the EM algorithm. For the modified data less than 80 samples are assigned to each component for random initialization and this discrepancy is due to the identifiability problems.

The BS estimates with start in solution and random initialization are tested for equivalence using the Cramer test with 1000 BS replicates after appropriate relabelling in order to minimize the Kullback-Leibler divergence. The null hypothesis of equivalence cannot be rejected for the original dataset at a significance level of 0.05 (observed statistic = 2.96, p-value = 0.23). By contrast this test indicates a difference in distribution for the modified dataset (observed statistic = 24.56, p-value $<$ 0.001). This suggests that the model

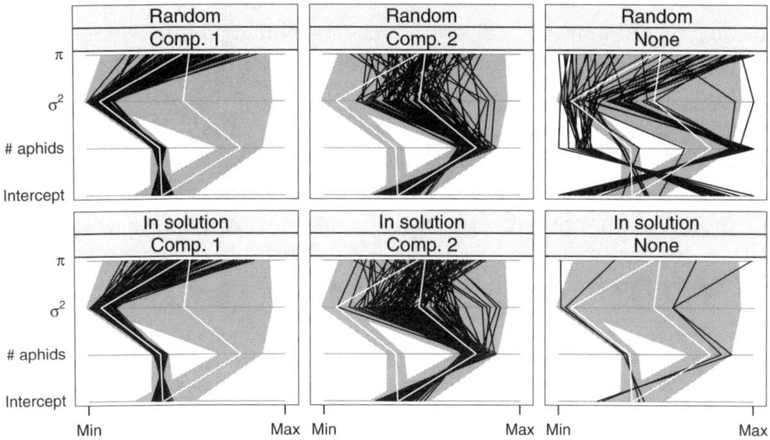

Fig. 2. Parallel coordinate plot of the parameter estimates for the parametric BS samples of the modified data with assignment to the confidence bands.

Table 2. Number of component specific parameter estimates of the parametric BS samples assigned to one of the components or to no component. The relative proportion is in round brackets.

	Original data		Modified data	
	Random	In solution	Random	In solution
Comp.1	84 (0.84)	91 (0.91)	76 (0.76)	98 (0.98)
Comp.2	87 (0.87)	94 (0.94)	78 (0.78)	98 (0.98)
None	29 (0.14)	15 (0.08)	46 (0.23)	4 (0.02)

fitted to the original data does not suffer from genuine multimodality while this cannot be excluded for the modified dataset.

If the null hypothesis is not rejected this test can be seen as an exploratory tool to increase the confidence that there is only a single genuine mode present. The rejection however only indicates a lack of correspondence in distribution between the BS results. The presence of multiple modes can be verified by inspecting the parallel coordinate plot. In this example the second detected mode however does not comply to the possible explanation where it is assumed that for one class the infection is in general higher and this mode could be therefore omitted using expert knowledge.

4 Conclusions and future work

Different methods for investigating genuine multimodality are proposed. These methods can be used for model diagnostics to investigate identifiability problems. A unique parameterization is for example of interest if the parameters are interpreted. The methods rely on the application of computational-wise demanding resampling techniques which are getting increasingly popular as more computing power becomes available.

Methods which can be automatically applied in order to ensure that the fitted model does not suffer from genuine multimodality are complemented with visualization techniques which can be used to further investigate the results if problems are indicated for the fitted model, as there are different reasons possible for rejecting the null hypothesis of genuine unimodality. In the future the proposed methods should be validated on other datasets to check if the proposed procedure gives reliable results.

Acknowledgement. This research was supported by the Austrian Academy of Sciences (ÖAW) through a DOC-FFORTE scholarship for Bettina Grün and the Austrian Science Foundation (FWF) under grant P17382.

References

BARINGHAUS, L. and FRANZ, C. (2004): On a New Multivariate Two-sample Test. *Journal of Multivariate Analysis, 88, 190–206.*

BASFORD, K.E., GREENWAY, D.R., MCLACHLAN, G.J. and PEEL, D. (1997): Standard Errors of Fitted Means Under Normal Mixture Model. *Computational Statistics, 12, 1–17.*

DEMPSTER, A.P., LAIRD, N.M. and RUBIN, D.B. (1977): Maximum Likelihood from Incomplete Data Via the EM Algorithm. *Journal of the Royal Statistical Society B, 39, 1–38.*

GRÜN, B. and LEISCH, F. (2004): Bootstrapping Finite Mixture Models. In: J. Antoch (Ed.): *Compstat 2004 — Proceedings in Computational Statistics.* Physica, Heidelberg, 1115–1122.

HENNIG, C. (2000): Identifiability of Models for Clusterwise Linear Regression. *Journal of Classification, 17, 2, 273–296.*

HOTHORN, T., LEISCH, F., ZEILEIS, A. and HORNIK, K. (2005): The Design and Analysis of Benchmark Experiments. *Journal of Computational and Graphical Statistics, 14, 3, 1–25.*

LOUIS, T.A. (1982): Finding the Observed Information Matrix When Using the EM Algorithm. *Journal of the Royal Statistical Society. Series B (Methodological), 44, 2, 226–233.*

MCLACHLAN, G.J. and PEEL, D. (2000): *Finite Mixture Models.* Wiley.

MINNOTTE, M. C. (1997): Nonparameteric Testing of the Existence of Modes. *The Annals of Statistics, 25, 4, 1646–1660.*

STEPHENS, M. (2000): Dealing with Label Switching in Mixture Models. *Journal of the Royal Statistical Society B, 62, 4, 795–809.*

TEICHER, H. (1963): Identifiability of Finite Mixtures. *The Annals of Mathematical Statistics, 34, 1265–1269.*

TURNER, T.R. (2000): Estimating the Propagation Rate of a Viral Infection of Potato Plants Via Mixtures of Regressions. *Journal of the Royal Statistical Society C, 49, 3, 371–384.*

Happy Birthday to You, Mr. Wilcoxon!*

Invariance, Semiparametric Efficiency, and Ranks

Marc Hallin**

Département de Mathématique, Institut de Statistique and ECARES,
Université libre de Bruxelles Campus de la Plaine CP210,
B-1050 Bruxelles, Belgium; mhallin@ulb.ac.be

1 Introduction

Everybody who went through a statistics course, even at introductory level, has been exposed at least to some elementary aspects of rank-based methods, and has heard about Wilcoxon's *signed rank* and *rank sum* tests.

Although early ideas of distribution-free tests can be traced back as far as John Arbuthnoth (1667-1735), Frank Wilcoxon's pathbreaking 1945 four page paper (Wilcoxon (1945)), where these two tests are described for the first time, certainly can be considered as the starting point of the modern theory of rank-based inference.

The Wilcoxon tests, and most of the subsequent theory of rank-based inference, were developed as a reaction against the pervasive presence of Gaussian assumptions in classical statistical theory. Rank tests are simple and easy to compute. Above all, they are distribution-free, hence exact and applicable under unspecified (typically, non Gaussian) densities. Moreover, they are flexible enough to adapt to a wide range of inference problems: the fifties and the sixties have witnessed an explosive but somewhat aphazard development of rank-based solutions to a variety of problems. This development was structured and systematized in the seventies, mainly on the basis of Jaroslav Hájek's fundamental contribution, leading to what can be considered as the "classical theory" of rank-based inference. This classical theory essentially addresses all testing (and estimation) problems arising in the context of general linear models with independent observations, thus covering location, scale, and regression problems, as well as analysis of variance and covariance, and

---

* The initial version of this conference was delivered in Leuven as the inaugural lecture of a Francqui Chair on October 20, 2005—thus plainly justifying the title.

** Research supported by an I.A.P. contract of the Belgian Federal Government and an Action de Recherche Concertée of the Communauté française de Belgique. Special thanks are due to Davy Paindaveine for his careful reading of the manuscript and helpful comments.

most linear experimental planning models—see the monograph by Puri and Sen (1985) for a systematic and fairly exhaustive account.

The progress since then may have been less spectacular, and the opinion is not uncommon that rank-based inference is a more or less complete—hence limited and somewhat old-fashioned—theory, the development of which has stopped in the early eighties. The objective of this nontechnical presentation is to dispel this wrong perception by showing that, quite on the contrary, ranks and their generalizations quite naturally find their ultimate expression in the modern theories of asymptotic statistical experiments and semiparametric inference. More precisely, rank-based methods, under very general assumptions, allow for semiparametrically efficient, yet distribution-free inference (testing and estimation), in a very large variety of models involving unspecified densities—much beyond the classical linear models with independent observations.

Frank Wilcoxon himself would be most surprised to see how, in slightly more than sixty years, his two tests, which he modestly considered as quick and easy tricks, to be used when everything else fails, not only have survived the many revolutions of contemporary statistics, but have turned into a timely and still growing area of modern inference, reconciling the irreconcilable objectives of efficiency and robustness.

After sixty years of unremitting service, not the slightest prospect of early retirement, thus: happy birthday to you, Mr. Wilcoxon!

2 Ranks: From distribution-freeness to group invariance

2.1 Ranks and rank tests

Let us first introduce some basic concepts and notation. Denoting by $\mathbf{X}^{(n)} := (X_1,\ X_2,\ \ldots,\ X_n)$ an n-tuple of observations, the *order statistic* is obtained by ordering the X_i's from smallest to largest: $\mathbf{X}_{(\ )} := (X_{\min} := X_{(1)}, X_{(2)}, \ldots X_{(n)} =: X_{\max})$, with $X_{(1)} \leq X_{(2)}, \leq \ldots \leq X_{(n)}$. The vector of ranks then is defined as $\mathbf{R}^{(n)} := (R_1,\ R_2,\ \ldots,\ R_n)$, with R_i such that $X_{(R_i)} = X_i$ or, equivalently, $R_i := \#\{j \mid X_j \leq X_i\}$.

This vector $\mathbf{R}^{(n)}$, provided that no ties occur (which happens with probability one as soon as $\mathbf{X}^{(n)}$ has a density), clearly is a (random) permutation of $(1,\ 2,\ \ldots,\ n)$. Assuming furthermore that the X_i's are i.i.d., with some unspecified density f over $\mathbb{R}$, the distribution of $\mathbf{R}^{(n)}$ is uniform over the $n!$ permutations of $(1, ..., n)$. An important advantage of $\mathbf{R}^{(n)}$-measurable statistics over the more general $\mathbf{X}^{(n)}$-measurable ones is thus their *distribution-freeness*: since their distributions do not depend on f, they allow for exact inference, robust to misspecification of f (hence to violations of Gaussian assumptions).

The price to be paid for this advantage is the corresponding loss of information. The observation $\mathbf{X}^{(n)}$ and the couple $(\mathbf{X}_{(\ )}, \mathbf{R}^{(n)})$ contain the same information: restricting to rank-based inference thus means throwing away the information contained in the order statistic $\mathbf{X}_{(\ )}$. Natural questions are: how crucial (in terms of efficiency) is that loss of information? what is the real cost of this conflict between robustness and efficiency? The surprising answer (an answer Wilcoxon definitely would never have dreamed of) is: in case the density f is unknown, the loss of information is nil (asymptotically), and robustness can be obtained at no cost!

Fig. 1. Frank Wilcoxon (1892-1965)

The Wilcoxon rank sum test addresses the same two-sample location problem as the classical two-sample Student test, the only difference being that the latter requires Gaussian densities. Under the null hypothesis $\mathcal{H}_0$, $X_1, \ldots, X_m, X_{m+1}, \ldots, X_n$ are i.i.d., with unspecified (nonvanishing) density f, whereas, under the alternative $\mathcal{H}_1$, i.i.d.-ness is the property of $X_1, \ldots, X_m, X_{m+1} - \theta, \ldots, X_n - \theta$, for some $\theta > 0$.

The Wilcoxon test statistic can be written as $S_W^{(n)} := \sum_{i=m+1}^{n} R_i$; unlike the Student test, the wilcoxon test does not require any assumption on the density f, and thus resists light as well as heavy tails. Wilcoxon himself in 1945 hardly realized the consequences and importance of his discovery: he mainly considered his test as a robust, "quick and easy" solution for the location shift problem—nothing powerful, though—to be used when everything else fails.

2.2 Hodges-Lehmann and Chernoff-Savage

Eleven years after Wilcoxon's seminal paper, a totally unexpected result was published by Hodges and Lehmann (1956). This result, which came as a shock to the statistical community, is the following:

$$\inf_f \text{ARE}_f \text{ (Wilcoxon / Student)} = .864 \ .$$

Recall that the ARE (asymptotic Relative Efficiency) of a sequence ($\phi_1^{(n)}$, say) of statistical procedures with respect to another one ($\phi_2^{(n)}$, say) is the limit $\mathrm{ARE}_f(\phi_1^{(n)} \;/\; \phi_2^{(n)})$, when it exists, as $n \to \infty$, of the ratio $n_2(n)/n$ of the number $n_2(n)$ of observations it takes for $\phi_2^{(n_2(n))}$ to achieve the same performance as $\phi_1^{(n)}$.

Fig. 2. Joseph L. Hodges (1922-2000) and Erich L. Lehmann (1917- —)

In the worst case, the Wilcoxon test thus only requires 13.6% more observations than the Student one in order to reach comparable power! On the other hand,

$$\sup_f \mathrm{ARE}_f \text{ (Wilcoxon / Student)} = \infty,$$

and the benefits of unrestricted validity are invaluable ...

Fig. 3. Bartel L. van der Waerden (1903-1996)

Since the Normal distribution is playing such a central role in classical statistics, the idea of considering, for the same location problem, a statistic of the form $S_{vdW}^{(n)} := \sum_{i=m+1}^{n} \Phi^{-1}\left(\frac{R_i}{n+1}\right)$ (or an equivalent *exact score* form), where Φ^{-1} denotes the standard normal quantile function, was proposed by several authors, among which Fisher, Terry, Yates, Fraser, van der Waerden, ... For simplicity, we all call them *van der Waerden statistics.*

Van der Waerden statistics are still distribution-free (since a function of ranks). In case however the actual underlying density is normal, $S_{vdW}^{(n)}$ is asymptotically equivalent to the Student statistic. Hence, at the normal, $S_{vdW}^{(n)}$ yields the same asymptotic performance as Student, which in that case is optimal.

Chernoff and Savage in 1958 (Chernoff and Savage (1958)) however established the following much stronger result, which perhaps is even more surprising than Hodges and Lehmann's:

$$\inf_f \text{ARE}_f \text{ (van der Waerden / Student)} = 1.00 ,$$

an infimum which is attained at Gaussian f only!

It follows that van der Waerden tests are always strictly better (asymptotically) than the Student one, except at the normal, where they are equally good. One thus is always better off using van der Waerden which moreover, contrary to Student, is uniformly valid. This actually should put Student and much of everyday Gaussian practice out of business!

Fig. 4. Hermann Chernoff (1923- —) and I. Richard Savage (1926-2004)

These surprising facts cannot be a mere coincidence, and raise some obvious question: what is it that makes ranks that efficient? are ranks the only statistical objects enjoying such attractive distribution-freeness/efficiency properties? Answers however are not straightforward. As we shall see, they are intimately related with the maximal invariance property of ranks with respect to certain generating groups, and the connection of such invariance with tangent space projections and semiparametric efficiency. Such answers certainly were not at hand in 1958, and only emerged quite recently (Hallin and Werker (2003)).

2.3 Group invariance

Assume that $\mathbf{X}^{(n)} = (X_1, X_2, \ldots, X_n)$ are i.i.d., with unspecified density f in the class $\mathcal{F}$ of all nonvanishing densities over $\mathbb{R}$ ($\mathbf{X}^{(n)}$ is thus *independent white noise*). Denote by $\mathrm{P}_f^{(n)}$ the joint distribution of $\mathbf{X}^{(n)}$ and let $\mathcal{P}^{(n)} := \{\mathrm{P}_f^{(n)} \mid f \in \mathcal{F}\}$.

Next consider the group of transformations (acting on $\mathbb{R}^n$)

$$\mathcal{G}, \circ := \{g_h \mid h \text{ monotone} \uparrow, \text{ continuous}, \ h(\pm\infty) = \pm\infty\}, \circ$$

mapping $(x_1, \ldots, x_n) \in \mathbb{R}^n$ onto $g_h(x_1, \ldots, x_n) := (h(x_1), \ldots, h(x_n)) \in \mathbb{R}^n$. Then, $\mathcal{G}, \circ$ is a generating group for $\mathcal{P}^{(n)}$, in the sense that for all $\mathrm{P}_{f_1}^{(n)}, \mathrm{P}_{f_2}^{(n)}$ in $\mathcal{P}^{(n)}$, there exists $g_h \in \mathcal{G}$ such that $(X_1, \ldots, X_n) \sim \mathrm{P}_{f_1}^{(n)}$ iff $g_h(X_1, \ldots, X_n) \sim \mathrm{P}_{f_2}^{(n)}$. The vector of ranks $\mathbf{R}^{(n)}$ is maximal invariant for $\mathcal{G}, \circ$, that is, $T(x_1, \ldots, x_n) = T(g_h(x_1, \ldots, x_n))$ for all $g_h \in \mathcal{G}$ iff T is $\mathbf{R}^{(n)}$-measurable.

This invariance property of ranks suggests the definition of other "ranks", associated with other generating groups. Here are a few examples:

(i) $\mathbf{X}^{(n)} := (X_1, X_2, \ldots, X_n)$ i.i.d., with unspecified density f in the class $\mathcal{F}_+$ of all nonvanishing symmetric (w. r. t. 0) densities over $\mathbb{R}$ (*independent symmetric white noise*). Let $\mathcal{P}^{(n)} = \{\mathrm{P}_f^{(n)} \mid f \in \mathcal{F}_+\}$: this family is generated by the subgroup $\mathcal{G}_+, \circ$ of $\mathcal{G}, \circ$, where $\mathcal{G}_+ := \{g_h \in \mathcal{G} \mid h(-x) = h(x)\}$. The signs and the ranks of absolute values are maximal invariant ("signed ranks");

(ii) $\mathbf{X}^{(n)} := (X_1, X_2, \ldots, X_n)$ i.i.d., with unspecified nonvanishing median-centered density f in the class $\mathcal{F}_0$ of all nonvanishing zero-median densities over $\mathbb{R}$ (*independent median-centered white noise*). Let $\mathcal{P}^{(n)} = \{\mathrm{P}_f^{(n)} \mid f \in \mathcal{F}_0\}$: this family is generated by the subgroup $\mathcal{G}_0, \circ$ of $\mathcal{G}, \circ$, where $\mathcal{G}_0 := \{g_h \in \mathcal{G} \mid h(0) = 0\}$. The signs and the ranks are maximal invariant (see Hallin et al. (2006));

(iii) $\mathbf{X}^{(n)} := (X_1, X_2, \ldots, X_n)$ independent, with unspecified nonvanishing median-centered densities $f_1, \ldots, f_n$ in the class $\mathcal{F}_0$ of all nonvanishing zero-median densities over $\mathbb{R}$ (*independent, heterogeneous median-centered white noise*). Let $\mathcal{P}^{(n)} = \{\mathrm{P}_f^{(n)} \mid f \in \mathcal{F}_0\}$; the signs are maximal invariant for the appropriate generating group (see Dufour et al. (1998));

(iv) $\mathbf{X}^{(n)} := (\mathbf{X}_1, \mathbf{X}_2, \ldots, \mathbf{X}_n)$ i.i.d., with elliptical density

$$\sigma^{-k}(\det \mathbf{V})^{-1/2} f_1(\sigma^{-1}\sqrt{(\mathbf{x}-\boldsymbol{\mu})'\mathbf{V}^{-1}(\mathbf{x}-\boldsymbol{\mu})}).$$

over $\mathbb{R}^k$ (*independent elliptical white noise* with location $\boldsymbol{\mu}$, shape $\mathbf{V}$, scale σ, and standardized radial density f_1). Write $\mathbf{X} \sim \mathrm{P}^{(n)}_{\boldsymbol{\theta};f_1}$, $\boldsymbol{\theta} = (\boldsymbol{\mu}, \sigma, \mathbf{V})$ and $\mathcal{P}^{(n)} = \{\mathrm{P}^{(n)}_{\boldsymbol{\theta};f_1} \mid f_1 \in \mathcal{F}^+\}$, where $\mathcal{F}^+$ is the class of all standardized nonvanishing densities over $\mathbb{R}^+$: the unit vectors $\mathbf{U}_i := \mathbf{V}^{-1/2}(\mathbf{X}_i - \boldsymbol{\mu})/[(\mathbf{X}_i - \boldsymbol{\mu})'\mathbf{V}^{-1}(\mathbf{X}_i - \boldsymbol{\mu})]^{1/2}$ and the ranks R_i of the "distances" $[(\mathbf{X}_i - \boldsymbol{\mu})'\mathbf{V}^{-1}(\mathbf{X}_i - \boldsymbol{\mu})]^{1/2}$ are maximal invariant (multivariate signs $\mathbf{U}_i$ and ranks R_i) for the generating group of continuous order-preserving radial transformations: see Hallin and Paindaveine (2002, 2006) for details.

It is easy to show that maximal invariants (hence, invariants) are distribution-free. As we shall see, they also have a strong connection to (semi-parametric) efficiency. This however requires some further preparation.

3 Efficiency: From parametric to semiparametric

3.1 Parametric optimality

In the sequel, we consider semiparametric models, namely, models under which the distribution of some $\mathcal{X}^n$-valued observation $\mathbf{X}^{(n)} := (X_1, X_2, \ldots, X_n)$ belongs to a family of the form $\mathcal{P}^{(n)} = \{\mathrm{P}^{(n)}_{\boldsymbol{\theta};f} \mid \boldsymbol{\theta} \in \boldsymbol{\Theta},\ f \in \mathcal{F}\}$ where $\boldsymbol{\theta} \in \boldsymbol{\Theta} \subseteq \mathbb{R}^m$ is some m-dimensional parameter of interest, and $f \in \mathcal{F}$ is a nonparametric (infinite-dimensional) nuisance. We moreover assume that $\mathcal{P}^{(n)}$ is such that all its fixed-f parametric subfamilies $\mathcal{P}^{(n)}_f := \{\mathrm{P}^{(n)}_{\boldsymbol{\theta};f} \mid \boldsymbol{\theta} \in \boldsymbol{\Theta}\}$ are LAN (see below), whereas the fixed-$\boldsymbol{\theta}$ nonparametric subfamilies $\mathcal{P}^{(n)}_{\boldsymbol{\theta}} := \{\mathrm{P}^{(n)}_{\boldsymbol{\theta};f} \mid f \in \mathcal{F}\}$ are generated by some group $\mathcal{G}^{(n)}_{\boldsymbol{\theta}}, \circ$ acting on the observation space $\mathcal{X}^n$, with maximal invariant $\mathbf{R}^{(n)}(\boldsymbol{\theta})$.

Fig. 5. Lucien Le Cam (1924-2000)

The concept of local asymptotic normality (LAN, w.r.t. $\boldsymbol{\theta}$, at given f) is due to Lucien Le Cam, and is now widely adopted as the standard structure for traditional central-limit type asymptotics. The (sub)family $\mathcal{P}^{(n)}_f$ (more precisely, the sequence of families indexed by $n \in \mathbb{N}$) is said to be LAN if, for all $\boldsymbol{\theta} \in \boldsymbol{\Theta}$, there exists a random vector $\boldsymbol{\Delta}^{(n)}_{\boldsymbol{\theta};f}$ (the *central sequence*) and a (deterministic) positive definite matrix $\boldsymbol{\Gamma}_{\boldsymbol{\theta};f}$ (the *information matrix*) such that, under $\mathrm{P}^{(n)}_{\boldsymbol{\theta};f}$, as $n \to \infty$,

(i) $\Lambda^{(n)}_{\boldsymbol{\theta}+n^{-1/2}\boldsymbol{\tau}/\boldsymbol{\theta};f} := \log\left(\dfrac{\mathrm{dP}^{(n)}_{\boldsymbol{\theta}+n^{-1/2}\boldsymbol{\tau};f}}{\mathrm{dP}^{(n)}_{\boldsymbol{\theta};f}}\right) = \boldsymbol{\tau}'\boldsymbol{\Delta}^{(n)}_{\boldsymbol{\theta};f} - \frac{1}{2}\boldsymbol{\tau}'\boldsymbol{\Gamma}_{\boldsymbol{\theta};f}\boldsymbol{\tau} + o_{\mathrm{P}}(1)$, and

(ii) $\boldsymbol{\Delta}^{(n)}_{\boldsymbol{\theta};f} \xrightarrow{\mathcal{L}} \mathcal{N}\left(\mathbf{0}, \boldsymbol{\Gamma}_{\boldsymbol{\theta};f}\right)$.

Skipping technical details, LAN implies that

(a) under $\mathrm{P}^{(n)}_{\boldsymbol{\theta}+n^{-1/2}\boldsymbol{\tau};f}$, $\boldsymbol{\tau} \in \mathbb{R}^m$, the central sequence $\boldsymbol{\Delta}^{(n)}_{\boldsymbol{\theta};f}$ is asymptotically $\mathcal{N}\left(\boldsymbol{\Gamma}_{\boldsymbol{\theta};f}\boldsymbol{\tau}, \boldsymbol{\Gamma}_{\boldsymbol{\theta};f}\right)$ as $n \to \infty$;
(b) parametric efficiency (local, at $\boldsymbol{\theta}$, and asymptotic) in the initial (fixed-f) model has the same characteristics as parametric efficiency (exact) in the Gaussian shift model $\boldsymbol{\Delta} \sim \mathcal{N}\left(\boldsymbol{\Gamma}_{\boldsymbol{\theta};f}\boldsymbol{\tau}, \boldsymbol{\Gamma}_{\boldsymbol{\theta};f}\right)$, $\boldsymbol{\tau} \in \mathbb{R}^m$, that is, for instance,
   - optimal α-level tests of $\mathcal{H}^{(n)}_0 : \boldsymbol{\theta} = \boldsymbol{\theta}_0$ achieve at $\mathrm{P}^{(n)}_{\boldsymbol{\theta}_0+n^{-1/2}\boldsymbol{\tau};f}$ asymptotic power $1 - F_{m;\boldsymbol{\tau}'\boldsymbol{\Gamma}^{-1}_{\boldsymbol{\theta}_0;f}\boldsymbol{\tau}}(\chi^2_{m;1-\alpha})$, where $F_{m;\lambda}$ stands for the noncentral chi-square distribution function with m degrees of freedom and noncentrality parameter λ, or
   - optimal estimates $\hat{\boldsymbol{\theta}}^{(n)}$ are such that $n^{1/2}(\hat{\boldsymbol{\theta}}^{(n)} - \boldsymbol{\theta}) \xrightarrow{\mathcal{L}} \boldsymbol{\Gamma}^{-1}_{\boldsymbol{\theta};f}\boldsymbol{\Delta} \sim \mathcal{N}(\mathbf{0}, \boldsymbol{\Gamma}^{-1}_{\boldsymbol{\theta};f})$.

Moreover, optimality is achieved by treating the central sequence $\boldsymbol{\Delta}^{(n)}_{\boldsymbol{\theta};f}$ exactly as one would the observation $\boldsymbol{\Delta}$ in the limit Gaussian shift model, that is, for instance,

- by basing tests for $\boldsymbol{\theta} = \boldsymbol{\theta}_0$ on the asymptotic χ^2_m null distribution of statistics of the form $Q_f := (\boldsymbol{\Delta}^{(n)}_{\boldsymbol{\theta}_0;f})'\boldsymbol{\Gamma}^{-1}_{\boldsymbol{\theta}_0;f}\boldsymbol{\Delta}^{(n)}_{\boldsymbol{\theta}_0;f}$, or
- by constructing optimal estimators $\hat{\boldsymbol{\theta}}^{(n)}$ (of the *one-step form*) such that $n^{1/2}(\hat{\boldsymbol{\theta}}^{(n)} - \boldsymbol{\theta}) = \boldsymbol{\Gamma}^{-1}_{\boldsymbol{\theta};f}\boldsymbol{\Delta}^{(n)}_{\boldsymbol{\theta};f} + o_{\mathrm{P}}(1) \xrightarrow{\mathcal{L}} \mathcal{N}(\mathbf{0}, \boldsymbol{\Gamma}^{-1}_{\boldsymbol{\theta};f})$.

Summing up, parametric efficiency (at given f and $\boldsymbol{\theta}$) is entirely characterized by the Gaussian shift model $\boldsymbol{\Delta} \sim \mathcal{N}\left(\boldsymbol{\Gamma}_{\boldsymbol{\theta};f}\boldsymbol{\tau}, \boldsymbol{\Gamma}_{\boldsymbol{\theta};f}\right)$, $\boldsymbol{\tau} \in \mathbb{R}^m$, hence by the information matrix $\boldsymbol{\Gamma}_{\boldsymbol{\theta};f}$.

3.2 Parametric efficiency in the presence of nuisance

In order to understand what is meant with semiparametric efficiency, let us first consider the concept of parametric efficiency in the presence of a parametric nuisance. In the LAN family just described, assume that the parameter breaks into $\boldsymbol{\theta}' = (\boldsymbol{\theta}'_1, \boldsymbol{\theta}'_2)$, and that inference is to be made about $\boldsymbol{\theta}_1 \in \mathbb{R}^{m_1}$, while $\boldsymbol{\theta}_2 \in \mathbb{R}^{m_2}$ is a nuisance. The central sequence $\boldsymbol{\Delta}^{(n)}_{\boldsymbol{\theta};f}$ similarly decomposes into $\begin{pmatrix} \boldsymbol{\Delta}^{(n)}_{\boldsymbol{\theta};f;1} \\ \boldsymbol{\Delta}^{(n)}_{\boldsymbol{\theta};f;2} \end{pmatrix}$, and the information matrix into $\boldsymbol{\Gamma}_{\boldsymbol{\theta};f} = \begin{pmatrix} \boldsymbol{\Gamma}_{\boldsymbol{\theta};f;11} & \boldsymbol{\Gamma}_{\boldsymbol{\theta};f;12} \\ \boldsymbol{\Gamma}_{\boldsymbol{\theta};f;12} & \boldsymbol{\Gamma}_{\boldsymbol{\theta};f;22} \end{pmatrix}$.

Inspired by exact optimality in the limit Gaussian shift, it is easy to understand that locally asymptotically optimal (efficient) inference on $\boldsymbol{\theta}_1$ should be based on the residual $\boldsymbol{\Delta}^{(n)}_{\boldsymbol{\theta};f;1} - \boldsymbol{\Gamma}_{\boldsymbol{\theta};f;12}\boldsymbol{\Gamma}^{-1}_{\boldsymbol{\theta};f;22}\boldsymbol{\Delta}^{(n)}_{\boldsymbol{\theta};f;2}$ of the regression of $\boldsymbol{\Delta}^{(n)}_{\boldsymbol{\theta};f;1}$ on $\boldsymbol{\Delta}^{(n)}_{\boldsymbol{\theta};f;2}$ in the covariance $\boldsymbol{\Gamma}_{\boldsymbol{\theta};f}$, that is, the $\boldsymbol{\Gamma}_{\boldsymbol{\theta};f}$-projection of the $\boldsymbol{\theta}_1$-central sequence parallel to the space of the $\boldsymbol{\theta}_2$-central sequence. Indeed, a local perturbation $n^{-1/2}\boldsymbol{\tau}_2$ of $\boldsymbol{\theta}_2$ induces (see (a) in Section 3.1) on the asymptotic distribution of $\boldsymbol{\Delta}^{(n)}_{\boldsymbol{\theta};f}$ a shift $\begin{pmatrix}\boldsymbol{\Gamma}_{\boldsymbol{\theta};f;12}\\ \boldsymbol{\Gamma}_{\boldsymbol{\theta};f;22}\end{pmatrix}\boldsymbol{\tau}_2$. The resulting shift for the residual $\boldsymbol{\Delta}^{(n)}_{\boldsymbol{\theta};f;1} - \boldsymbol{\Gamma}_{\boldsymbol{\theta};f;12}\boldsymbol{\Gamma}^{-1}_{\boldsymbol{\theta};f;22}\boldsymbol{\Delta}^{(n)}_{\boldsymbol{\theta};f;2}$ is thus $\boldsymbol{\Gamma}_{\boldsymbol{\theta};f;12}\boldsymbol{\tau}_2 - \boldsymbol{\Gamma}_{\boldsymbol{\theta};f;12}\boldsymbol{\Gamma}^{-1}_{\boldsymbol{\theta};f;22}\boldsymbol{\Gamma}_{\boldsymbol{\theta};f;22}\boldsymbol{\tau}_2 = \mathbf{0}$: this residual therefore is insensitive to local perturbations of $\boldsymbol{\theta}_2$. On the other hand, the asymptotic covariance of the same the residual $\boldsymbol{\Delta}^{(n)}_{\boldsymbol{\theta};f;1} - \boldsymbol{\Gamma}_{\boldsymbol{\theta};f;12}\boldsymbol{\Gamma}^{-1}_{\boldsymbol{\theta};f;22}\boldsymbol{\Delta}^{(n)}_{\boldsymbol{\theta};f;2}$ is $\boldsymbol{\Gamma}_{\boldsymbol{\theta};f;11} - \boldsymbol{\Gamma}_{\boldsymbol{\theta};f;12}\boldsymbol{\Gamma}^{-1}_{\boldsymbol{\theta};f;22}\boldsymbol{\Gamma}'_{\boldsymbol{\theta};f;12}$, whereas a perturbation $n^{-1/2}\boldsymbol{\tau}_1$ of $\boldsymbol{\theta}_1$ induces a shift $\left(\boldsymbol{\Gamma}_{\boldsymbol{\theta};f;11} - \boldsymbol{\Gamma}_{\boldsymbol{\theta};f;12}\boldsymbol{\Gamma}^{-1}_{\boldsymbol{\theta};f;22}\boldsymbol{\Gamma}'_{\boldsymbol{\theta};f;12}\right)\boldsymbol{\tau}_1$. Asymptotically efficient (at given f and $\boldsymbol{\theta}$) inference on $\boldsymbol{\theta}_1$ when $\boldsymbol{\theta}_2$ is a nuisance is characterized by the Gaussian shift model

$$\boldsymbol{\Delta} \sim \mathcal{N}\left(\left(\boldsymbol{\Gamma}_{\boldsymbol{\theta};f;11} - \boldsymbol{\Gamma}_{\boldsymbol{\theta};f;12}\boldsymbol{\Gamma}^{-1}_{\boldsymbol{\theta};f;22}\boldsymbol{\Gamma}'_{\boldsymbol{\theta};f;12}\right)\boldsymbol{\tau}, \boldsymbol{\Gamma}_{\boldsymbol{\theta};f;11} - \boldsymbol{\Gamma}_{\boldsymbol{\theta};f;12}\boldsymbol{\Gamma}^{-1}_{\boldsymbol{\theta};f;22}\boldsymbol{\Gamma}'_{\boldsymbol{\theta};f;12}\right),$$

$\boldsymbol{\tau} \in \mathbb{R}^{m_1}$ hence by the information matrix $\boldsymbol{\Gamma}_{\boldsymbol{\theta};f;11} - \boldsymbol{\Gamma}_{\boldsymbol{\theta};f;12}\boldsymbol{\Gamma}^{-1}_{\boldsymbol{\theta};f;22}\boldsymbol{\Gamma}'_{\boldsymbol{\theta};f;12}$.

3.3 Semiparametric efficiency

In the previous two sections, the density f was supposed to be correctly specified. In a semiparametric context, of course, this density f is the nuisance, playing the role of $\boldsymbol{\theta}_2$! Except for the technical details related to the infinite-dimensional nature of f (the classical reference is the monograph by Bickel et al. (1993)), this nuisance intuitively is treated in the same way as the parametric nuisance $\boldsymbol{\theta}_2$ in Section 3.2. Instead of being projected along the space of shifts induced by local variations of $\boldsymbol{\theta}_2$, however, $\boldsymbol{\Delta}^{(n)}_{\boldsymbol{\theta};f}$ is projected along the space generated by the shifts induced by variations of densities in the *vicinity* of f: the so-called *tangent space.* This projected *semiparametrically efficient central sequence* $\boldsymbol{\Delta}^{(n)*}_{\boldsymbol{\theta};f}$, with (asymptotic) covariance $\boldsymbol{\Gamma}^*_{\boldsymbol{\theta};f} \leq \boldsymbol{\Gamma}_{\boldsymbol{\theta};f}$—the *semiparametrically efficient information matrix* in turn defines a Gaussian shift model $\boldsymbol{\Delta}^* \sim \mathcal{N}\left(\boldsymbol{\Gamma}^*_{\boldsymbol{\theta};f}\boldsymbol{\tau}, \boldsymbol{\Gamma}^*_{\boldsymbol{\theta};f}\right)$, $\boldsymbol{\tau} \in \mathbb{R}^m$ which characterizes the best performance that can be expected (at f and $\boldsymbol{\theta}$) when f is unspecified.

In some models, the semiparametric information matrix $\boldsymbol{\Gamma}^*_{\boldsymbol{\theta};f}$ coincides with the parametric one $\boldsymbol{\Gamma}_{\boldsymbol{\theta};f}$: the model is *adaptive at* f, meaning that parametric and semiparametric performances are asymptotically the same at f (possibly, at all f). In general, however, $\boldsymbol{\Gamma}^*_{\boldsymbol{\theta};f} < \boldsymbol{\Gamma}_{\boldsymbol{\theta};f}$: the cost of not knowing the *true density*, at f, is strictly positive.

Although the definitions of the semiparametrically efficient (at given f) central sequence and information matrix are intuitively satisfactory, their practical value at first sight is less obvious. While $\boldsymbol{\Gamma}^*_{\boldsymbol{\theta};f} \leq \boldsymbol{\Gamma}_{\boldsymbol{\theta};f}$ provides the optimality bounds that in principle can be achieved at f, $\boldsymbol{\Delta}^{(n)*}_{\boldsymbol{\theta};f}$ heavily depend on f, and cannot be computed from the observations: $\boldsymbol{\Delta}^{(n)*}_{\boldsymbol{\theta};f}$ thus cannot be used for achieving the bound. This problem can be solved in two ways (recall that the central sequence at f—hence also the semiparametrically efficient one—only are defined up to $o_{\mathrm{P}^{(n)}_{\boldsymbol{\theta};f}}(1)$ terms).

(i) for all f in some class $\mathcal{F}$ of densities, an estimate $\hat{f}^{(n)}$ can be constructed in such a way that $\boldsymbol{\Delta}^{(n)*}_{\boldsymbol{\theta};\hat{f}^{(n)}} - \boldsymbol{\Delta}^{(n)*}_{\boldsymbol{\theta};f}$ under $P^{(n)}_{\boldsymbol{\theta};f}$ is $o_{\mathrm{P}}(1)$ as $n \to \infty$. Then, $\boldsymbol{\Delta}^{(n)}_{\boldsymbol{\theta}} := \boldsymbol{\Delta}^{(n)*}_{\boldsymbol{\theta};\hat{f}^{(n)}}$, which is a measurable function of the observations, is asymptotically equivalent to the actual efficient central sequence for any $f \in \mathcal{F}$; together with $\boldsymbol{\Gamma}^{*}_{\boldsymbol{\theta}^{(n)}} := \boldsymbol{\Gamma}^{*}_{\boldsymbol{\theta};\hat{f}^{(n)}}$, it allows for uniformly (over $\mathcal{F}$) semiparametrically efficient inference. The convergence of the distribution of $\boldsymbol{\Delta}^{(n)*}_{\boldsymbol{\theta}}$ to a $\mathcal{N}\left(\mathbf{0}, \boldsymbol{\Gamma}^{*}_{\boldsymbol{\theta};f}\right)$ one, however, may be quite slow, and unpleasant technicalities such as *sample splitting* are often required.

(ii) if, for some selected f, a distribution-free statistic $\underset{\sim}{\boldsymbol{\Delta}}^{(n)}_{\boldsymbol{\theta};f}$ can be constructed such that $\underset{\sim}{\boldsymbol{\Delta}}^{(n)}_{\boldsymbol{\theta};f} - \boldsymbol{\Delta}^{(n)*}_{\boldsymbol{\theta};f}$ under $P^{(n)}_{\boldsymbol{\theta};f}$ is $o_{\mathrm{P}}(1)$ as $n \to \infty$, then this $\underset{\sim}{\boldsymbol{\Delta}}^{(n)}_{\boldsymbol{\theta};f}$ is a version of the semiparametrically efficient central sequence at f enjoying the remarkable property of being distribution-free, hence asymptotically $\mathcal{N}\left(\mathbf{0}, \boldsymbol{\Gamma}^{*}_{\boldsymbol{\theta};f}\right)$ irrespective of the actual underlying density, thus allowing for reaching semiparametric optimality at the selected f based on exact (even under density $g \neq f$) inference. As we shall see in the next section, this is precisely what rank-based inference can provide.

4 Ranks: From tangent space to Hájek projection

A fundamental statistical principle is the Invariance Principle, stipulating that "when a statistical problem is invariant under the action of some group of transformations, one should restrict to invariant statistical procedures", that is, to statistical procedures based on invariant statistics. It has been assumed in Section 3.1 that the fixed-$\boldsymbol{\theta}$ subfamilies $\mathcal{P}^{(n)}_{\boldsymbol{\theta}}$ of $\mathcal{P}^{(n)}$ are invariant w.r.t. the groups $\mathcal{G}_{\boldsymbol{\theta}}, \circ$, with maximal invariant $\mathbf{R}^{(n)}(\boldsymbol{\theta})$ (typically, the ranks of some $\boldsymbol{\theta}$-residuals). The set of invariant statistics thus coincides with the set of $\mathbf{R}^{(n)}(\boldsymbol{\theta})$-measurable statistics (typically, the rank statistics). Since optimal (at $\boldsymbol{\theta}$ and f) inference can be based on the central sequence $\boldsymbol{\Delta}^{(n)}_{\boldsymbol{\theta};f}$, a natural idea consists in considering the invariant statistic which is closest to the central sequence by projecting $\boldsymbol{\Delta}^{(n)}_{\boldsymbol{\theta};f}$ onto the σ-field generated by $\mathbf{R}^{(n)}(\boldsymbol{\theta})$, yielding

$$\underset{\sim}{\boldsymbol{\Delta}}^{(n)}_{\boldsymbol{\theta};f} := \mathrm{E}_f\left[\boldsymbol{\Delta}^{(n)}_{\boldsymbol{\theta};f} \mid \mathbf{R}^{(n)}(\boldsymbol{\theta})\right]$$

Being $\mathbf{R}^{(n)}(\boldsymbol{\theta})$-measurable, $\underset{\sim}{\boldsymbol{\Delta}}^{(n)}_{\boldsymbol{\theta};f}$ is an invariant, hence distribution-free statistic (in the fixed-$\boldsymbol{\theta}$ submodel). The projection mapping $\boldsymbol{\Delta}^{(n)}_{\boldsymbol{\theta};f}$ onto $\underset{\sim}{\boldsymbol{\Delta}}^{(n)}_{\boldsymbol{\theta};f}$ is, in a sense, the opposite of a classical "Hájek projection"; in the sequel, as a tribute to Jaroslav Hájek, we also call it a *Hájek projection.*

The relation between the seemingly completely unrelated Hájek and tangent space projections was established by (Hallin and Werker (2003)). Under very general conditions, indeed, they show that, under $\mathrm{P}^{(n)}_{\boldsymbol{\theta};f}$, $\underset{\sim}{\boldsymbol{\Delta}}{}^{(n)}_{\boldsymbol{\theta};f} = \boldsymbol{\Delta}^{(n)*}_{\boldsymbol{\theta};f} + o_{\mathrm{P}}(1)$ as $n \to \infty$: $\underset{\sim}{\boldsymbol{\Delta}}{}^{(n)}_{\boldsymbol{\theta};f}$ is thus an invariant (rank-based) distribution-free version of the semiparametrically efficient (at $\boldsymbol{\theta}$ and f) central sequence. As explained in Section 3.2, it thus allows for distribution-free semiparametrically efficient (at $\boldsymbol{\theta}$ and f) inference on $\boldsymbol{\theta}$.

Fig. 6. Jaroslav Hájek (1926-1974)

Remark that $\underset{\sim}{\boldsymbol{\Delta}}{}^{(n)}_{\boldsymbol{\theta};f}$ is obtained as the projection of the "regular" central sequence, not the semiparametrically efficient one: Hájek projections thus are doing the same job as tangent space projections, without requiring the (often nontrivial) computation of the latter, and with the (invaluable) additional advantages of distribution-freeness. The projection $\mathrm{E}_f[\boldsymbol{\Delta}^{(n)}_{\boldsymbol{\theta};f} \mid \mathbf{R}^{(n)}(\boldsymbol{\theta})]$ is the "exact score version" of $\underset{\sim}{\boldsymbol{\Delta}}{}^{(n)}_{\boldsymbol{\theta};f}$; simpler "approximate score" versions also exist, but their form depends on the specific central sequence under study.

Uniformly semiparametrically efficient inference is also possible, by considering $\underset{\sim}{\boldsymbol{\Delta}}{}^{(n)}_{\boldsymbol{\theta};\hat{f}^{(n)}}$, where $\hat{f}^{(n)}$ is an appropriate density estimator, with the important advantage of avoiding the unpleasant technicalities, such as sample-splitting, associated with the "classical semiparametric procedures", based on $\boldsymbol{\Delta}^{(n)}_{\boldsymbol{\theta};\hat{f}^{(n)}}$. But then, $\underset{\sim}{\boldsymbol{\Delta}}{}^{(n)}_{\boldsymbol{\theta};\hat{f}^{(n)}}$ also splits the sample, into two mutually independent parts: the invariant and distribution-free part on one hand (the ranks), the "order statistic" (involved in $\hat{f}^{(n)}$ on the other, with the ranks containing the "f-free" information about the parameter $\boldsymbol{\theta}$, whereas the "order statistic" contains information on the nuisance f only.

5 Conclusion

Rank-based methods (more generally, the "maximal invariant" ones) are quite flexible, and apply in a very broad class of statistical models, much beyond the traditional context of linear models with independent observations. They are powerful—achieving semiparametric efficiency at selected density, which is the best that can be hoped for in presence of unspecified densities. In the same time, they are simpler and more robust (distribution-freeness) than "classical" semiparametric procedures. Often, they make Gaussian or pseudo-Gaussian methods non-admissible (the Chernoff-Savage phenomenon: see Hallin (1994) for time series models, Hallin and Paindaveine (2002) for elliptical location, and Paindaveine (2006) for elliptical shape).

Within their sixty years of existence, Wilcoxon's "quick and easy" tricks have grown into a full body of efficient and modern methods, reconciling the apparently antagonistic objectives of efficiency and robustness (distribution-freeness, meaning 100% resistance against misspecified densities).

Happy birthday to you, Mr Wilcoxon!

References

BICKEL, P.J., KLAASSEN, C.A.J., RITOV, Y. and WELLNER, J.A. (1993): *Efficient and Adaptive Statistical Inference for Semiparametric Models.* Johns Hopkins University Press, Baltimore.

CHERNOFF, H. and SAVAGE, I.R. (1958): Asymptotic Normality and Efficiency of Certain Nonparametric Tests. *Annals of Mathem. Statististics, 29, 972–994.*

DUFOUR, J.M., HALLIN, M. and MIZERA, I. (1998): Generalized Run Tests for Heteroscedastic Time Series. *Journal of Nonparametric Statistics, 9, 39–86.*

HALLIN, M. (1994): On the Pitman Nonadmissibility of Correlogram-based Time Series Methods. *Journal of Time Series Analysis, 16, 607–612.*

HALLIN, M. and PAINDAVEINE, D. (2002): Optimal Tests for Multivariate Location Based on Interdirections and Pseudo-Mahalanobis Ranks. *Annals of Statistics, 30, 1103–1133.*

HALLIN, M. and PAINDAVEINE, D. (2006): Semiparametrically Efficient Rank-based Inference for Shape: Optimal Rank-based Tests for Sphericity. *Annals of Statistics, 34, forthcoming.*

HALLIN, M., VERMANDELE, C. and WERKER, B.J.M (2006): Linear Serial and Nonserial Sign-and-rank Statistics: Asymptotic Representation and Asymptotic Normality. *Annals of Statistics, 34, forthcoming.*

HALLIN, M. and WERKER, B.J.M. (2003): Semiparametric Efficiency, Distribution-freeness, and Invariance. *Bernoulli, 9, 137–165.*

HODGES, J.L. and LEHMANN, E.L. (1956): The Efficiency of Some Nonparametric Competitors of the *t*-test. *Annals of Mathematical Statististics, 27, 324–335.*

PAINDAVEINE, D. (2006): A Chernoff-Savage Result for Shape. On the Non-admissibility of Pseudo-Gaussian Methods. *Journal of Multivariate Analysis, forthcoming.*

PURI, M.L. and SEN, P.K. (1985): *Nonparametric Methods in General Linear Models.* John Wiley, New York.

WILCOXON, F. (1945): Individual Comparisons by Ranking Methods. *Biometrics Bulletin, 1, 80–83.*

Equivalent Number of Degrees of Freedom for Neural Networks

Salvatore Ingrassia[1] and Isabella Morlini[2]

[1] Dipartimento di Economia e Metodi Quantitativi, Università di Catania, Corso Italia 55, 95128 Catania, Italy; s.ingrassia@unict.it

[2] Dip. di Scienze Sociali Cognitive e Quantitative, Università di Modena e Reggio E., Via Allegri 9, 42100 Reggio Emilia, Italy; morlini.isabella@unimore.it

Abstract. The notion of equivalent number of degrees of freedom (e.d.f.) to be used in neural network modeling from small datasets has been introduced in Ingrassia and Morlini (2005). It is much smaller than the total number of parameters and it does not depend on the number of input variables. We generalize our previous results and discuss the use of the e.d.f. in the general framework of multivariate nonparametric model selection. Through numerical simulations, we also investigate the behavior of model selection criteria like AIC, GCV and BIC/SBC, when the e.d.f. is used instead of the total number of the adaptive parameters in the model.

1 Introduction

This article presents the results of some empirical studies comparing different model selection criteria, like AIC, GCV and BIC (see, among others, Kadane and Lazar (2004), for nonlinear projection models, based on the *equivalent number of degrees of freedoms* (e.d.f) introduced in Ingrassia and Morlini (2005). Given a response variable Y and predictor variables $\mathbf{x} \in \mathcal{X} \subseteq \mathbb{R}^m$, throughout this paper we assume that the input-output relation can be written as $Y = \phi(\mathbf{x}) + \varepsilon$, where Y assumes values in $\mathcal{Y} \subseteq \mathbb{R}$ and ε is a random variable with zero mean and finite variance. We then assume that the unknown functional dependency $\phi(\mathbf{x}) = \mathbb{E}[Y|\mathbf{x}]$ is of the form:

$$f_p(\mathbf{x}) = \sum_{i=1}^{p} c_i \tau(\mathbf{a}_i'\mathbf{x} + b_i) + c_{p+1} \tag{1}$$

where $\mathbf{a}_1, \ldots, \mathbf{a}_p \in \mathbb{R}^m$, $b_1, \ldots, b_p, c_{p+1}, c_1, \ldots, c_p \in \mathbb{R}$ and τ is a sigmoidal function. In the following, without loss of generality, we will assume $c_{p+1} = 0$. Indeed, the expression (1) may be written in the form: $f_p(\mathbf{x}) = \sum_{i=1}^{p+1} c_i \tau(\mathbf{a}_i'\mathbf{x} + b_i)$ where the constant term c_{p+1} has been included in the summation and $\tau(\mathbf{a}_{p+1}'\mathbf{x}_i + b_{p+1}) \equiv 1$. Therefore, results presented in this article may be extended to the case $c_{p+1} \neq 0$ by simply replacing p with $p+1$. We denote by $\mathbf{A}$ the $p \times m$ matrix having rows $\mathbf{a}'_1, \ldots, \mathbf{a}'_p$, and we set $\mathbf{b} = (b_1, \ldots, b_p)$ and $\mathbf{c} = (c_1, \ldots, c_p)$. The function $f_p(\mathbf{x})$ is realized by a multilayer perceptron (MLP) having m inputs, p neurons in the hidden layer and one neuron in the output. Such quantities are called *weights* and they will be denoted by $\mathbf{w}$, so that $\mathbf{w} \in \mathbb{R}^{p(m+2)}$. It is well known that most functions, including any continuous function with a bounded support, can be approximated by models of the form (1).

2 Preliminaries and basic results

Let $\mathcal{F}$ be the set of all functions of kind (1) for a fixed p with $1 \leq p \leq N$. The problem is to find the function $f^{(0)} = f(\mathbf{w}^{(0)})$ in the set $\mathcal{F}$ which minimizes the *generalization error*:

$$\mathcal{E}(f) = \int [y - f(\mathbf{x})]^2 \, p(\mathbf{x}, y) \, d\mathbf{x} \, dy \ , \tag{2}$$

where the integral is over $\mathcal{X} \times \mathcal{Y}$. In practice, the distribution $p(\mathbf{x}, y)$ is unknown, but we have a sample $\mathcal{L} = \{(\mathbf{x}_1, y_1), \ldots, (\mathbf{x}_N, y_N)\}$, called learning set, of N i.i.d. realizations of $(\mathbf{X}, Y)$ so that we compute the *empirical error*:

$$\widehat{\mathcal{E}}(f, \mathcal{L}) = \sum_{(\mathbf{x}_n, y_n) \in \mathcal{L}} (y_n - f(\mathbf{x}_n))^2 \tag{3}$$

and estimate the least squares parameters by minimizing (3). A theoretical problem concerns the *unidentifiability* of the parameters, see Hwang and Ding (1997). That is, there exist different functions of the form (1) with a different number of parameters that can approximate exactly the same relationship function $f(\mathbf{x})$. Results due to Bartlett (1998) show that this is due to the dependency of the generalization performance of an MLP on the size of the weights rather than on the size of the model (i.e. on the number of adaptive parameters). Here an important role is played by the quantity $\|\mathbf{c}\|_1 = \sum_{i=1}^{p} |c_i|$, that is by the sum of the values of the absolute weights between the hidden layer and the output. This is justified as follows. Let $\mathcal{X}_1$ and $\mathcal{X}_2$ be two populations in $\mathbb{R}^m$ and set $\mathcal{X} = \mathcal{X}_1 \cup \mathcal{X}_2$; for each $\mathbf{x} \in \mathcal{X}$ and $y \in \{-1, +1\}$, let $y = +1$ if $\mathbf{x}$ comes from $\mathcal{X}_1$ and $y = -1$ if $\mathbf{x}$ comes from $\mathcal{X}_2$. Moreover let $f : \mathcal{X} \to \mathbb{R}$ be a discriminant function of type (1) such that $\mathbf{x}$ is assigned to $\mathcal{X}_1$ if $f(\mathbf{x}) > 0$ and to $\mathcal{X}_2$ if $f(\mathbf{x}) < 0$; in other words the function f classifies correctly the point $\mathbf{x}$ if and only if $y \cdot f(\mathbf{x}) > 0$; more generally, the function f classifies correctly the point $\mathbf{x}$ with margin $\gamma > 0$ if and only if $y \cdot f(\mathbf{x}) \geq \gamma$. For a given learning set $\mathcal{L} = \{(\mathbf{x}_1, y_1), \ldots, (\mathbf{x}_N, y_N)\}$, where $y_n = 1$ if $\mathbf{x}_n$ comes from $\mathcal{X}_1$ and $y_n = -1$ if $\mathbf{x}_n$ comes from $\mathcal{X}_2$, with $n = 1, \ldots, N$, let us consider *misclassification error with margin* γ $\widehat{\mathcal{E}}_\gamma(f, \mathcal{L}) = \#\{n : y_n f(\mathbf{x}_n) < \gamma\}/N$, where $\#\{\cdot\}$ denotes the number of elements in the set $\{\cdot\}$, which is the proportion of the number of cases which are not correctly classified with margin

γ by f. For a given constant $C \geq 1$ consider only those $\mathbf{c}$ for which $\|\mathbf{c}\|_1 \leq C$, then we have the following result:

Theorem 1 (Bartlett (1998)) Let P be a probability distribution on $\mathcal{X} \times \{-1,+1\}$, $0 < \gamma \leq 1$ and $0 < \eta \leq 1/2$. Let $\mathcal{F}$ be the set of functions $f(\mathbf{x})$ of kind (1) such that $\sum_i |c_i| \leq C$, with $C \geq 1$. If the learning set $\mathcal{L}$ is a sample of size N and has $\{-1,+1\}$-valued targets, then with probability at least $1-\eta$, for each $f \in \mathcal{F}$:

$$\mathcal{E}(f) \leq \widehat{\mathcal{E}}_\gamma(f,\mathcal{L}) + \varepsilon(\gamma, N, \eta)$$

where for a universal constant α, the quantity

$$\varepsilon(\gamma, N, \eta) = \sqrt{\frac{\alpha}{N}\left(\frac{C^2 m}{\gamma^2}\ln\left(\frac{C}{\gamma}\right)\ln^2 N - \ln\eta\right)}\,.$$

is called *confidence interval.* □

Thus the error is bounded by the sum of the empirical error with margin γ and by a quantity depending on $\|\mathbf{c}\|_1$ through C but not on the number of weights. Two other important results for our scope are given below.

Theorem 2 (Ingrassia (1999)) Let $\mathbf{x}_1, \ldots, \mathbf{x}_p$ be p distinct points in $(-r,r)^m$ with $\mathbf{x}_i \neq \mathbf{0}$ $(i = 1,\ldots,p)$ and $\mathbf{A} = (a_{ij}) \in [-u,u]^{mp}$ be a $p \times m$ matrix, with $u = 1/m$. Let τ be a sigmoidal analytic function on $(-r,r)$, with $r > 0$. Then the points $\tau(\mathbf{A}\mathbf{x}_1), \ldots, \tau(\mathbf{A}\mathbf{x}_p) \in \mathbb{R}^p$ are linearly independent for almost all matrices $\mathbf{A} = (a_{ij}) \in [-u,u]^{mp}$. □

This result proves that, given $N > m$ points $\mathbf{x}_1, \ldots, \mathbf{x}_N \in \mathbb{R}^m$, the transformed points $\tau(\mathbf{A}\mathbf{x}_1), \ldots, \tau(\mathbf{A}\mathbf{x}_N)$ generate an *over-space* of dimension $p > m$ if the matrix $\mathbf{A}$ satisfies suitable conditions. In particular, the largest over-space is attained when $p = N$, that is when the hidden layer has as many units as the number of points in the learning set. This result has been generalized as follows.

Theorem 3 (Ingrassia and Morlini (2005)) Let $\mathcal{L}$ be a given learning set and $f = \sum_{i=1}^{p} c_i \tau(\mathbf{a}_i'\mathbf{x})$. If $p = N$, then the error $\widehat{\mathcal{E}}(f,\mathcal{L})$ is zero for almost all matrices $\mathbf{A} \in [-1/m, 1/m]^{mp}$. □

3 Equivalent number of degrees of freedom

For a given $p \times m$ matrix $\mathbf{A}$, let $\mathbf{T}$ be the $N \times p$ matrix having rows $\tau(\mathbf{A}\mathbf{x}_1)', \ldots, \tau(\mathbf{A}\mathbf{x}_N)'$, with $p \leq N$. According to Theorems 2 and 3 the matrix $\mathbf{T}$ has rank p (and then it is non-singular) for almost all matrices $\mathbf{A} \in [-1/m, 1/m]^{mp}$. The empirical error $\widehat{\mathcal{E}}_\gamma(f,\mathcal{L})$ can be written as:

$$\widehat{\mathcal{E}}_\gamma(f,\mathcal{L}) = \sum_{(\mathbf{x}_n,y_n)\in\mathcal{L}} (y_n - f(\mathbf{x}_n))^2 = \sum_{(\mathbf{x}_n,y_n)\in\mathcal{L}} (y_n - \mathbf{c}'\tau(\mathbf{A}\mathbf{x}_n))$$
$$= (\mathbf{y}-\mathbf{T}\mathbf{c})'(\mathbf{y}-\mathbf{T}\mathbf{c}) = \mathbf{y}'\mathbf{y} - 2\mathbf{c}'\mathbf{T}'\mathbf{y} + \mathbf{c}'\mathbf{T}'\mathbf{T}\mathbf{c}$$

and for any fixed matrix $\mathbf{A}$, the error $\widehat{\mathcal{E}}_\gamma(f,\mathcal{L})$ attains its minimum when $\mathbf{c} = (\mathbf{T}'\mathbf{T})^{-1}\mathbf{T}'\mathbf{y}$. Thus the matrix $\mathbf{H} = \mathbf{T}(\mathbf{T}'\mathbf{T})^{-1}\mathbf{T}'$ is a projection matrix since $\hat{\mathbf{y}} = \mathbf{H}\mathbf{y}$ and $\mathbf{H}$ is symmetric, positive semidefinite, idempotent and it results:

$$\text{rank}(\mathbf{H}) = \text{trace}(\mathbf{H}) = \text{trace}\{\mathbf{T}(\mathbf{T}'\mathbf{T})^{-1}\mathbf{T}'\} = \text{trace}\{(\mathbf{T}'\mathbf{T})^{-1}\mathbf{T}'\mathbf{T}\} = p$$

so that $\hat{\mathbf{y}}$ lies in the space $\mathbb{R}^p$ and thus to the model $f(\mathbf{x}) = \sum_{i=1}^p c_i\tau(\mathbf{a}_i'\mathbf{x})$ should be assigned p *equivalent number of degrees of freedom* (e.d.f). When the error is given by the following weight decay cost function:

$$\widehat{\mathcal{E}}^*(f;\mathcal{L}) = \widehat{\mathcal{E}}(f;\mathcal{L}) + \lambda\sum w_i^2$$
$$= \mathbf{y}'\mathbf{y} - 2\mathbf{c}'\mathbf{T}'\mathbf{y} + \mathbf{c}'\mathbf{T}'\mathbf{T}\mathbf{c} + \lambda\text{tr}(\mathbf{A}\mathbf{A}') + \lambda\mathbf{c}'\mathbf{c}$$

the equivalent degrees of freedom are:

$$k = \text{tr}(\mathbf{H}_\lambda) = \text{tr}\{\mathbf{T}(\mathbf{T}'\mathbf{T} + \lambda\mathbf{I}_p)^{-1}\mathbf{T}'\} = p - \sum_{i=1}^p \frac{\lambda}{l_i + \lambda}$$

which shows that p is decreased by the quantity $\lambda\text{tr}\{(\mathbf{T}'\mathbf{T} + \lambda\mathbf{I}_p)^{-1}\}$. Since $\mathbf{T}'\mathbf{T}$ is positive semidefinite, the p eigenvalues of $\mathbf{T}'\mathbf{T}$, say $l_1,\ldots,l_p$, are non-negative. Thus $(\mathbf{T}'\mathbf{T}+\lambda\mathbf{I}_p)$ has eigenvalues $(l_1+\lambda),\ldots,(l_p+\lambda)$ and then the eigenvalues of $(\mathbf{T}'\mathbf{T} + \lambda\mathbf{I}_p)^{-1}$ are $(l_1+\lambda)^{-1},\ldots,(l_p+\lambda)^{-1}$.

4 Model selection criteria

In the general framework of model selection, we suppose there are $f_{p_1},\ldots,f_{p_K}$ models of the form (1). Since the estimation in statistical models may be thought of as the choice of a single value of the parameter chosen to represent the distribution (according to some criterion), model selection may be thought of in this framework as the estimation applied to the model f_{p_h}, with $h = 1,\ldots,K$. The only special issue is that the set of models is discrete and has a finite range. There may be occasions when one model clearly dominates the others and the choice is unobjectionable, and other occasions when there are several competing models that are supported in some sense by the data. Due to the *unidentifiability* of the parameters, there may be no particular reasons for choosing a single best model over the others according to some criterion. On the contrary, it makes more sense to "deselect" models that are obviously poor, maintaining a subset for further considerations regarding, for example, the computational costs. The following indexes are generally used for model selection since they be carried out easily and yield results that can be interpreted by most users; they are also general enough to handle with a wide variety of problems:

$$\text{AIC} := \log(\widehat{\mathcal{E}}(f)) + \frac{2k}{N} \qquad \text{BIC} := \log(\widehat{\mathcal{E}}(f)) + \frac{k\log(N)}{N}$$

$$\text{GCV} := \widehat{\mathcal{E}}(f)\left(1 - \frac{k}{N}\right)^{-2}$$

where k denotes the number of degrees of freedom of the model f. The AIC and BIC present different forms in literature, here we follow Raftery (1995). Some of these criteria obey the likelihood principle, that is they have some frequentist asymptotic justification; some others correspond to a Bayesian decision problem. It is not the goal of this paper to face the outgoing discussion about their relative importance or to bring coherence to the two different perspectives of asymptotic and Bayesian-theoretic justification. In this work, via Monte Carlo simulations, we first aim at describing the different behavior of these indexes; then, we wish to determine whether such values and the model choice are affected by how the degrees of freedoms are computed and by how the empirical error minimization is performed. In Ingrassia and Morlini (2005) a Monte Carlo study has been drawn with small data sets. For these data, BIC has been shown to select models with a smaller $k = p$ than those selected by the other criteria, in agreements with previous results (see e.g. Katz (1981), Koehler and Murphree (1988), Kadane and Lazar (2004)). A comparison with the criteria computed using the e.d.f. and $k = W$, where W is the number of all parameters in the model, has also be drawn and this shows that, when $k = W$, some indexes may assume negative values becoming useless. Values across simulations also reveal a higher variability and the presence of anomalous peaks. Another analysis concerning simulated data has shown the ability of the UEV to estimate σ^2 when $k = p$. In this work we present further results, carried out in Matlab, based on large datasets: the *Abalone* and the *Boston Housing* (www.ics.uci.edu/~mlearn/).

5 Numerical studies

The *Abalone Data* consists of 4177 instances with 7 input variables and one discrete output variable and the *Boston Housing* data consists of 506 instances concerning 13 input variables and one continuous target variable. Observations are split into a training set of dimension 3133 for the *Abalone Data* and 400 for the *Boston Housing* and a validation set of dimension 1044 for the first data set and and 106 for the second one. In order to avoid overfitting, we estimate the parameters both by minimizing the sum-of-squares error function with the stopped training strategy and by minimizing the weight decay cost function. To interpret the following numerical results, it is worth noting

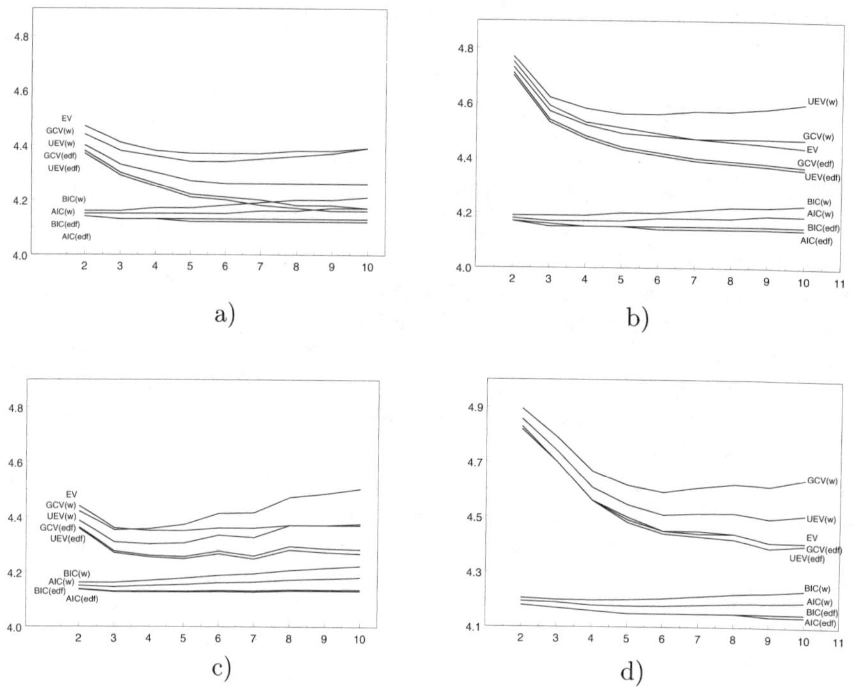

Fig. 1. Mean values of model selection criteria for the Abalone data set obtained with weight decay and a) λ =0.005, b) λ =0.05, c) λ chosen by cross validation and d) stopped training.

that when the weight decay function is used, the error on the validation set (EV) may be considered as an estimate of the generalization error since the observations are independent from those used for estimating the parameters. On the contrary, the error on the validation set is indirectly used for estimating the parameters if the stopped training stragegy is applied and cannot be considered as a generalization error estimate. For the Abalone data, the mean values obtained by repeating the estimates 100 times, with different splits of the data in the training and validation sets, are reported in Fig. 1; moreover main results referred to the Boston Housing data are reported in Table 1. The first conclusion we draw, especially evident from Table 1, is that, for different values of λ (ranging from 0.005 to 0.01) model selection criteria computed using the e.d.f., that is with $k = p$ and $k = p - \sum_{i=1}^{p} \lambda/(l_i + \lambda)$ are nearly identical and lead to the same model choice. Since $k = p - \sum_{i=1}^{p} \lambda/(l_i + \lambda)$ is not readily available in software packages, the choice $k = p$ is shown to provide a concise, simple and reliable approximation of this value. The second conclusion we draw is that BIC selects smaller models, with respect to those selected by the other criteria, when $k = W$. Indeed, it leads to the choice of the same model selected by the other indexes, when k =e.d.f. If the true

Table 1. Comparison among mean values of model selection criteria obtained with the Boston Housing data, with $k = p - \sum_{i=1}^{p} \lambda/(l_i + \lambda)$, $k = p$, $k = W$ and with $\lambda = 0.005$ and $\lambda = 0.01$. Bold values refer to the model selection.

$\lambda = 0.005$

		$k = p$			$k = p - \sum_{i=1}^{p} \frac{\lambda}{l_i+\lambda}$			$k = W$		
p	EV	AIC	BIC	GCV	AIC	BIC	GCV	AIC	BIC	GCV
2	17.89	8.56	8.59	13.08	8.55	8.59	13.06	8.76	**9.18**	16.08
3	17.92	8.42	8.46	11.29	8.40	8.45	11.26	**8.68**	9.23	14.96
4	**17.59**	**8.32**	**8.37**	**10.31**	**8.31**	**8.36**	**10.25**	**8.65**	9.35	**14.77**
5	18.36	8.32	8.38	**10.29**	8.31	8.36	**10.21**	8.71	9.55	16.00
6	19.20	8.39	8.46	11.02	8.37	8.43	10.90	8.85	9.82	18.66
7	20.10	8.32	8.40	10.31	8.30	8.36	10.17	8.84	9.96	19.10
8	20.68	8.38	8.47	10.96	8.36	8.43	10.78	8.97	10.23	22.31

$\lambda = 0.01$

		$k = p$			$k = p - \sum_{i=1}^{p} \frac{\lambda}{l_i+\lambda}$			$k = W$		
p	EV	AIC	BIC	GCV	AIC	BIC	GCV	AIC	BIC	GCV
2	17.90	8.54	8.57	12.84	8.54	8.57	12.83	8.74	**9.16**	15.79
3	17.27	8.45	8.49	11.64	8.44	8.48	11.60	8.71	9.26	15.42
4	**17.00**	**8.30**	**8.35**	**10.07**	**8.29**	**8.34**	**10.02**	**8.63**	9.32	**14.43**
5	17.95	8.31	8.37	10.17	8.29	8.35	10.09	8.70	9.54	15.80
6	19.20	8.39	8.46	11.02	8.37	8.43	10.90	8.85	9.82	18.66
7	20.10	8.32	8.40	10.31	8.30	8.36	10.17	8.84	9.96	19.10
8	20.68	8.38	8.47	10.96	8.36	8.43	10.78	8.97	10.23	22.31

underlying model is chosen to be as the one with the smallest validation error, using k =e.d.f. instead of $k = W$, leads to choices with are never considerably different and sometimes are considerably better (for example, when λ is small and BIC is used). Another conclusion we draw from Table 1 and Fig. 1 is that the GCV is always larger than the other criteria and have a smaller spread with the validation error, which is a reliable estimate of the generalization error when the weight decay approach is used. Moreover, GCV has a less smoother pattern with respect to the dimension p of the model and a scree test based on the plot of their values against p may be used to choose the optimal dimension p of the model. If the graph drops sharply, followed by a straight line with a much smaller slope, we may choose p equal to the value before the straight line begins. Fig. 1 a), b) and c) clearly indicate to choose p=3 while Fig. 1 d) suggest p=6. In the scree plots obtained from Table 1 (not reported for economy of space) there is clearly a discernible bend in slope at $p = 4$ for λ=0.005 and 0.01. In another case, with $\lambda = 0.05$ the bend in slope is at $p = 5$. In both data sets, when k =e.d.f., these criteria are nearly identical and lead to stable estimates of the generalization error and stable model choices, for different p. By comparing the results obtained with different values of λ, it is apparent that increasing the value of λ does increase the numbers of possible better models over the others and, in general, leads to less parsimonious models. In this case model choice should be based on the scree plot instead of on the basis of the absolute minimum value. The e.d.f. are still shown to work well, even if they are based on the achievement of the absolute minimum of the error function (3) which has a wider spread between the minimum of weight decay cost function, as long as λ increases.

6 Concluding remarks

Based on this computational study, we can draw conclusions about the comparisons of different degrees of freedoms given to nonlinear projection models of the form (1) and about the reliability of the model selection criteria routinely implemented by software developers. In particular, our study has shown that BIC tends to select more parsimonious models than GCV and AIC when $k = W$. The GCV criterion gives a larger value of the generalization error, which is in agreement with the empirical error computed on new independent patterns. The choice $k = p$ gives a good approximation of the trace of the projection matrix for projection models of the form (1); it leads to values of selection criteria nearly identical to those obtained with the trace. Using $k = p$ instead of $k = W$ leads to model choices which are never worst and sometimes are better (for example, when BIC is used). Using a scree test plot to select a single best model is increasingly important as long as the value of λ increases. Further simulation studies on the e.d.f. are in progress and the obtained results will be summarized in a future work.

References

BARTLETT, P.L. (1998): The Sample Complexity of Pattern Classification With Neural Networks: The Size of the Weights Is More Important Than the Size of the Network. *IEEE Transaction on Information Theory, 44, 525–536.*

HWANG, J.T.G. and DING, A.A. (1997): GPrediction Intervals for Artificial Neural Networks. *Journal of the American Statistical Association, 92, 438, 748–757.*

INGRASSIA, S. (1999): Geometrical Aspects of Discrimination by Multilayer Perceptrons. *Journal of Multivariate Analysis, 68, 226–234.*

INGRASSIA, S. and MORLINI, I. (2005): Neural Network Modeling for Small Datasets. *Technometrics, 47, 297–311.*

KADANE, J.P. and LAZAR, N.A. (2004): Methods and Criteria for Model Selection. *Journal of the American Statistical Association, 99, 279–290.*

KATZ, R.W. (1981): On Some Criteria for Estimating the Order of a Markov Chain. *Technometrics, 23, 243–249.*

KOEHLER, A.B. and MURPHREE, E.S. (1988): A Comparison of the Aikake and Schwarz Criteria for Selecting Model Order. *Applied Statistics, 37, 187–195.*

RAFTERY, A.E. (1995): Bayesian Model Selection in Social Research. *Sociological Methodology, 25, 111–163.*

Model Choice for Panel Spatial Models: Crime Modeling in Japan

Kazuhiko Kakamu[1], Wolfgang Polasek[2] and Hajime Wago[3]

[1] Stumpergasse 56, 1060, Vianna, Austria, Department of Economics & Finance, Institute for Advanced Studies; kakamu@ihs.ac.at
Machikaneyama 1-7, Toyonaka, Osaka, 460-0043, Japan, Graduate School of Economics, Osaka University; cg097kk@srv.econ.osaka-u.ac.jp
[2] Stumpergasse 56, 1060, Vianna, Austria, Department of Economics & Finance, Institute for Advanced Studies; polasek@ihs.ac.at
[3] Furoh, Chikusa, Nagoya, 464-8601, Japan, Graduate School of Economics, Nagoya University; wago@ism.ac.jp

Abstract. This paper considers the spatial patterns of crime incidents in Japan from a Bayesian point of view. We analyzed and compared different models by marginal likelihoods. From our posterior analysis, we found that the spatial pattern is different across crimes but panel SEM is selected in most of the types of crimes.

1 Introduction

Since the seminal work by Becker (1968), a large empirical literature has developed around the estimation and testing of economic models of crime incidents. In addition, spatial interaction for crime incidents were analyzed by Anselin (1988).

Kakamu et al. (2006) examined spatial interactions of crime incidents in Japan using panel data and found significant spatial interactions in 12 types of crimes. Also the importance of heteroscedasticity across prefectures were analyzed using the heteroscedasticity approach of Geweke (1993) using the panel spatial autoregressive model (SAR). According to Anselin (1988), there are other kinds of spatial models: spatial error model (SEM) and spatial Durbin model (SDM). If the model is misspecified, it may lead to different results. Therefore it is also important to determine the spatial pattern of crime incidents across prefectures in Japan.

This paper examines three different types of spatial models; panel SAR, panel SEM and panel SDM, based on the same data set with Kakamu et al. (2006) by Markov chain Monte Carlo (MCMC) methods and compares the results by marginal likelihoods. From our applications, we find that the spatial pattern is different across crimes but panel SEM is selected in most of the types of crimes.

This paper is organized as follows. In the next section, we introduce three different types of panel spatial model to examine the spatial interaction of crime incidents. Section 3 discusses our computational strategy to apply the MCMC methods and the calculation of marginal likelihoods. Section 4 presents the empirical results based on 18 types of criminal records in Japan from 1991 to 2001. Section 5 summarizes the results with concluding remarks.

2 Panel spatial models with heteroscedasticity

2.1 Panel spatial autoregressive model

Let y_{it} and x_{it} for $i = 1, \cdots, N$, $t = 1, \cdots, T$ denote dependent and independent variables, where x_{it} is a $1 \times k$ vector for the ith unit and tth period, respectively. Also let w_{ij} denote the spatial weight on jth unit with respect to ith unit. Then, the panel spatial autoregressive model with heteroscedasticity conditioned on parameters α_i, β, ρ, σ^2 and v_{it} is written as follows;

$$y_{it} = \alpha_i + x_{it}\beta + \sum_{j=1}^{N} \rho w_{ij} y_{jt} + \varepsilon_{it}, \quad \varepsilon_{it} \sim \mathcal{N}(0, \sigma^2 v_{it}). \tag{1}$$

Let $\alpha = (\alpha_1, \cdots, \alpha_N)'$ and $\theta = (\alpha', \beta')'$, then the likelihood function of the model (1) is written as follows:

$$L(Y|\theta, \rho, \sigma^2, V, Z, W)$$
$$= (2\pi\sigma^2)^{-NT/2} |I_N - \rho W|^T |V|^{-1/2} \exp\left(- \frac{e'V^{-1}e}{2\sigma^2} \right), \tag{2}$$

where $Y = (Y_1', \cdots, Y_T')'$, $Y_t = (y_{1t}, \cdots, y_{Nt})'$, $Z = (Z_1', \cdots, Z_T')'$, $Z_t = (I_N, X_t)$, I_N is $N \times N$ unit matrix, $X_t = (x_{1t}', \cdots, x_{Nt}')'$, $V = diag(V_1 \cdots, V_T)$, $V_t = diag(v_{1t}, \cdots, v_{Nt})$, W denotes weight matrix (see Anselin (1988)) and $e = Y - \rho(I_T \otimes W)Y - Z\theta$.

2.2 Panel spatial error model

The panel spatial error model with heteroscedasticity conditioned on parameters α_i, β, ρ, σ^2 and v_{it} is written as follows;

$$y_{it} = \alpha_i + x_{it}\beta + \sum_{j=1}^{N} \rho w_{ij} u_{jt} + \varepsilon_{it}, \quad \varepsilon_{it} \sim \mathcal{N}(0, \sigma^2 v_{it}), \tag{3}$$

where $u_{jt} = y_{jt} - \alpha_j - x_{jt}\beta$. Then the likelihood function of the model (3) is written as follows:

$$L(Y|\theta, \rho, \sigma^2, V, Z, W)$$
$$= (2\pi\sigma^2)^{-NT/2} |I_N - \rho W|^T |V|^{-1/2} \exp\left(- \frac{e'V^{-1}e}{2\sigma^2} \right), \tag{4}$$

where $e = Y - \rho(I_T \otimes W)Y - Z\theta + \rho(I_T \otimes W)Z\theta$.

2.3 The panel spatial Durbin model (SDM)

The panel spatial Durbin model with heteroscedasticity conditioned on parameters α_i, β_1, β_2 ρ, σ^2 and v_{it} is written as follows;

$$y_{it} = \alpha_i + x_{it}\beta_1 + \sum_{j=1}^{N} w_{ij}x_{jt}\beta_2 + \sum_{j=1}^{N} \rho w_{ij}y_{jt} + \varepsilon_{it}, \quad \varepsilon_{it} \sim \mathcal{N}(0, \sigma^2 v_{it}). \tag{5}$$

Let $\theta = (\alpha', \beta_1', \beta_2')'$, then the likelihood function of the model (5) is written as follows:

$$\begin{aligned} &L(Y|\theta, \rho, \sigma^2, V, Z, W) \\ &\quad = (2\pi\sigma^2)^{-NT/2}|I_N - \rho W|^T |V|^{-1/2} \exp\left(-\frac{e'V^{-1}e}{2\sigma^2}\right), \end{aligned} \tag{6}$$

where $e = Y - \rho(I_T \otimes W)Y - Z\theta$, $Z = (Z_1', \cdots, Z_T')'$ and $Z_t = (I_N, X_t, WX_t)$.

3 Posterior analysis

Since we use the same approach as Kakamu et al. (2006), in this section, we will introduce the MCMC strategy focusing on panel SAR briefly to save the space. However, it is easy to implement MCMC method for the other models only by changing a little.

Since we adopt a Bayesian approach, we complete the model by specifying the prior distribution over the parameters. Following Geweke (1993), a hierarchical prior $p(v_{it}^{-1})$ is assigned for all the variance parameters for $i = 1, \cdots, N$, $t = 1, \cdots, T$ and we assume $p(\theta, \rho, \sigma^2, V) = p(\theta)p(\rho)p(\sigma^2)\prod_{t=1}^{T}\prod_{i=1}^{N} p(v_{it}^{-1})$. Finally, we give the following prior distributions:

$$\begin{aligned} p(\theta) \sim \mathcal{N}(\theta_*, \Sigma_*), \quad p(\sigma^2) \sim \mathcal{G}^{-1}(\nu_*/2, \lambda_*/2), \quad p(\rho) \sim \mathcal{U}(\lambda_{min}^{-1}, \lambda_{max}^{-1}), \\ p(v_{it}^{-1}) \sim \chi^2(q_*)/q_*, \ \text{for } i = 1, \cdots, N, \ t = 1, \cdots, T, \end{aligned}$$

where $\mathcal{G}^{-1}(a, b)$ denotes an inverse gamma distribution with scale and shape parameters a and b. λ_{min} and λ_{max} denote the minimum and maximum eigenvalues of W, respectively. As is shown in Sun et al. (1999), it is well known that $\lambda_{min}^{-1} < 0$ and $\lambda_{max}^{-1} > 0$ and ρ must lie in the interval. Therefore, we restrict the prior space as $\rho \in (\lambda_{min}^{-1}, \lambda_{max}^{-1})$.

3.1 Posterior simulation

Full conditional distributions of θ, σ^2 and v_{it} for $i = 1, \cdots, N$, $t = 1, \cdots, T$ are as follows:

$$\theta|\rho, \sigma^2, V, Y, Z, W \sim \mathcal{N}(\theta_{**}, \Sigma_{**}),$$
$$\sigma^2|\theta, \rho, V, Y, Z, W \sim \mathcal{G}^{-1}(\nu_{**}/2, \lambda_{**}/2),$$
$$v_{it}^{-1}|\theta, \rho, \sigma^2, V_{-it}, Y, Z, W \sim \chi^2(q_* + 1)/(\sigma^{-2}e_{it}^2 + q_*),$$

where $\theta_{**} = \Sigma_{**}(\sigma^{-2}Z'V^{-1}(I_{NT} - I_T \otimes \rho W)Y + \Sigma_*^{-1}\theta_*)$, $\Sigma_{**} = (\sigma^{-2}Z'V^{-1}Z + \Sigma_*^{-1})^{-1}$, $\nu_{**} = NT + \nu_*$ and $\lambda_{**} = e'V^{-1}e + \lambda_*$ and V_{-it} denotes V without itth element (see Gelfand and Smith (1991)).

The full conditional distribution of ρ is written as

$$p(\rho|\theta, \sigma^2, V, Y, Z, W) \propto |I_N - \rho W|^T \exp\left(- \frac{e'V^{-1}e}{2\sigma^2} \right),$$

which cannot be sampled by standard methods (e.g. LeSage (2000)). Therefore, we adopt the random walk Metropolis algorithm with tuning parameter c (see Tierny (1994)). This tuning parameter c is used to produce an acceptance rate between 40% and 60% as is suggested in Holloway et al. (2002).

3.2 Marginal likelihood

For model M_k, let $L(Y|\theta_k, M_k)$ and $p(\theta_k|M_k)$ be the likelihood and prior for the model, respectively. Then, the marginal likelihood of the model is defined as

$$m(Y) = \int L(Y|\theta_k, M_k)p(\theta_k|M_k)d\theta_k.$$

Since the marginal likelihood can be written as

$$m(Y) = \frac{L(Y|\theta_k, M_k)p(\theta_k|M_k)}{p(\theta_k|Y, M_k)},$$

Chib (1996) suggests to estimate the marginal likelihood from the expression

$$\log m(Y) = \log L(Y|\theta_k^*, M_k) + \log p(\theta_k^*|M_k) - \log p(\theta_k^*|Y, M_k),$$

where θ_k^* is a particular high density point (typically the posterior mean or the ML estimate). He also provides a computationally efficient method to estimate the posterior ordinate $p(\theta_k^*|Y, M_k)$ in the context of Gibbs sampling and Chib and Jeliazkov (2001) provides the method in the context of Metropolis-Hasting sampling. In panel SAR, for example, we set $\theta_k = (\theta, \rho, \sigma^2, V)$ and estimate the posterior ordinate $p(\theta_k^*|Y, M_k)$ via the decomposition

$$\begin{aligned} &p(\theta_k^*|Y, M_k) \\ &\quad = p(\theta^*|\rho^*, \sigma^{*2}, V^*, Y, Z, W)p(\rho^*|\theta^*, \sigma^{*2}, V^*, Y, Z, W) \\ &\quad\quad p(\sigma^{*2}|\theta^*, \rho^*, V^*, Y, Z, W) \prod_{i=1}^{N} \prod_{t=1}^{T} p(v_{it}|\theta^*, \rho^*, \sigma^{*2}, V_{-it}^*, Y, Z, W). \end{aligned}$$

$p(\theta^*|\rho^*, \sigma^{*2}, V^*, Y, Z, W)$, $p(\sigma^{*2}|\theta^*, \rho^*, V^*, Y, Z, W)$ and $p(v_{it}|\theta^*, \rho^*, \sigma^{*2}, V_{-it}^*, Y, Z, W)$ are calculated from Gibbs output (see Chib (1996)) and $p(\rho^*|\theta^*, \sigma^{*2}, V^*, Y, Z, W)$ is calculated from Metropolis-Hasting output (see Chib and Jeliazkov (2001)).

4 Empirical results

First we would like to explain the data set used in this paper. We use the 18 types of criminal records across 47 prefectures from the period 1991 to 2001 (see Kakamu et al. (2006)). The types of crimes and the details are listed in Kakamu et al. (2006). As regressor variables, the Gross Regional Product (*GRP*), the unemployment rate (*UNEMP*), the registered foreigners (*FOREIGN*), the number of policemen (*POLICE*), the number of hotels (*HOTEL*) and the arrest rate (*ARREST*) are used and *POLICE* and *ARREST* are 1 year lagged variables. Also the same transformations and lagged variables are used. Finally, as a weight matrix, we use the matrix proposed by Kakamu et al. (2006), which considers the connection of economic activities.[1]

We assume the following hyper-parameters for the prior distributions: [2]

$$\theta_* = 0, \quad \Sigma_* = 100 \cdot I_{N+k}, \quad \nu_* = 0.01, \quad \lambda_* = 0.01, \quad q_* = 5.$$

Next, we ran the MCMC algorithm using 3000 iterations and discarding the first 1000 iterations.

Table 1 shows the results of the model choice procedure using the log marginal likelihoods. From the table, we find that the largest log marginal likelihoods are different for all type of crimes. Note that, only in *OBSCENITY* and *OTHERS*, the panel SAR is selected. The panel SDM is selected in *INJURY*, *THREAT2*, *THEFT*, *FRAUD* and *FORGERY*. The panel SEM is selected in the other 11 types of crimes.

Table 2 shows the coefficient estimates of the models, which are chosen by the marginal likelihoods. We have marked the significance of estimated parameters, by indicating if zero is included in the 95% (posterior) credible interval or not. First of all, we notice that the spatial correlation coefficients ρ are significant in 12 types of crimes. In addition, all the spatial correlations in the 12 types of crimes are positive, that is, high crime rates in the neighboring prefectures show a spill-over into other prefectures. It implies that the spatial interaction plays an important role in 2/3 of crime incidents.

Next, we will discuss the socio-economic variables briefly. In general, higher levels of *UNEMP GRP*, *POLICE* and *FOREIGN* lead to more crimes in Japan. In 5 types of crimes, the arrest rate helps to drive down the crime rate, but for *FRAUD* and *GRAFT*, the association is positive. Summarizing, we found the following tendencies: In general, negative job market and better

[1] All except one (Okinawa) Japanese prefectures are situated on the four major islands, Hokkaido, Honshu, Shikoku and Kyushu. But these four islands are connected by train and roads, despite the fact that islands are separate geographical entities. But for example, the most northern island Hokkaido is connected by the Seikan railway tunnel to Honshu. And Honshu is connected by the Awaji and Seto Bridge to Shikoku, and the southern island of Kyushu is also connected by the Kanmon Tunnel and Bridge to Honshu. Therefore, Okinawa is the only prefecture which is independent of all other prefectures.

[2] In case of the SDM, the dimension of θ_* is $2k$ and $\Sigma_* = 100 \cdot I_{N+2k}$.

Table 1. Log marginal likelihoods

	panel SAR	panel SEM	panel SDM	Selected model
MURDER	151.536	154.422	141.054	panel SEM
ROBBERY	-532.887	-531.799	-537.312	panel SEM
ARSON	-351.960	-350.250	-361.339	panel SEM
RAPE	-226.113	-225.111	-231.978	panel SEM
ASSEMBLY	942.754	943.910	924.838	panel SEM
VIOLENCE	-958.271	-956.205	-961.972	panel SEM
INJURY	-1209.674	-1217.413	-1208.621	panel SDM
THREAT1	-149.777	-148.341	-158.162	panel SEM
THREAT2	-1117.211	-1121.161	-1116.467	panel SDM
THEFT	-4250.236	-4250.346	-4246.388	panel SDM
FRAUD	-1759.962	-1763.479	-1754.681	panel SDM
EMBEZZLE	-275.736	-274.012	-283.939	panel SEM
FORGERY	-1167.172	-1164.270	-1156.179	panel SDM
GRAFT	413.354	414.936	399.545	panel SEM
TRUST	859.323	859.932	840.386	panel SEM
GAMBLING	-151.278	-149.189	-165.185	panel SEM
OBSCENITY	-836.269	-836.331	-838.359	panel SAR
OTHERS	-2731.053	-2733.374	-2731.504	panel SAR

macroeconomic conditions lead to more crimes in Japan. High crime rates are also associated with a (lagged) large police force. A higher foreigner rate is connected with a higher crime rate. In 6 types of crimes, the arrest rate helps to drive down the crime rate, but for *FRAUD* and *GRAFT*, the association is positive. Also it is important to mention that each type of crime has different features since the related socio-economic variables are different in each types of crimes.

Recall that the panel SDM has more regressors, because this model includes spatially weighted independent variables. Therefore, we can get much more information with respect to *INJURY*, *THREAT2*, *THEFT*, *FRAUD* and *FORGERY*. For *FORGERY*, negative neighbor effects can be found for *UNEMP* and *POLICE* variables and for *FRAUD*, positive effects can be found in *ARREST* variables. It implies that the higher *UNEMP* and *POLICE* in neighboring prefectures lead to higher crime rates in other prefecture and higher *ARREST* in neighboring prefectures leads to higher crime rate.

5 Conclusions

We analysed and compared different models of crime incidents in Japan from a Bayesian point of view. Using marginal likelihoods, we compared the following spatial panel models: the panel SAR, the panel SEM, and the panel SDM. The

Table 2. Empirical results: Posterior means and standard deviations (in parentheses)

	MURDER	*ROBBERY*	*ARSON*	*RAPE*	*ASSEMBLY*	*VIOLENCE*
Selected model	panel SEM	panel SEM	panel SEM	panel SEM	panel SEM	panel SEM
UNEMP	-0.180	10.497*	1.167	4.428*	-0.161	1.104
	(0.938)	(2.615)	(2.097)	(2.095)	(0.156)	(6.293)
GRP	-0.189	-3.667*	-1.163	-0.174	0.190*	3.222
	(0.550)	(1.595)	(1.429)	(0.907)	(0.078)	(2.330)
POLICE	2.265*	17.400*	6.323*	1.493	-0.002	24.593*
	(0.467)	(1.599)	(1.402)	(1.484)	(0.050)	(3.091)
FOREIGN	6.901*	5.999	12.780*	6.929	-1.114*	19.368*
	(2.454)	(7.040)	(4.948)	(4.821)	(0.460)	(8.879)
HOTEL	4.504	14.124	1.601	3.870	0.813	0.344
	(8.092)	(9.613)	(9.491)	(9.753)	(1.945)	(10.109)
ARREST	-0.134	-0.048	-0.062	-0.195	0.004	-2.954*
	(0.090)	(0.116)	(0.131)	(0.130)	(0.014)	(0.692)
ρ	0.052	0.060*	0.002	0.084*	0.068	0.109*
	(0.050)	(0.014)	(0.039)	(0.037)	(0.035)	(0.022)
σ^2	0.021	0.176	0.141	0.092	0.001	1.392
	(0.002)	(0.019)	(0.014)	(0.008)	(0.000)	(0.157)
R^2	0.992	0.989	0.957	0.974	0.886	0.988
	INJURY	*THREAT1*	*THREAT2*	*THEFT*	*FRAUD*	*EMBEZZLE*
Selected model	panel SDM	panel SEM	panel SDM	panel SDM	panel SDM	panel SEM
UNEMP	12.574	4.881*	8.529	0.808	-8.307	-6.841*
	(8.393)	(1.697)	(8.099)	(10.001)	(9.484)	(2.178)
GRP	25.527*	0.819	7.764*	14.099	31.021*	-4.139*
	(2.532)	(1.148)	(3.331)	(10.176)	(3.365)	(1.650)
POLICE	8.566	3.231*	31.298*	4.199	15.997*	4.602*
	(5.198)	(0.652)	(7.077)	(10.221)	(8.085)	(0.916)
FOREIGN	19.199*	10.490*	13.341	3.661	24.926*	24.877*
	(9.225)	(4.039)	(9.260)	(10.459)	(9.552)	(6.737)
HOTEL	1.535	10.999	3.627	0.081	0.051	-6.268
	(10.038)	(9.418)	(10.143)	(10.175)	(10.273)	(9.513)
ARREST	-10.503*	-0.022	-1.744*	3.770	5.085*	-0.096
	(2.143)	(0.079)	(0.815)	(10.406)	(1.855)	(0.123)
W-*UNEMP*	8.864		3.804	0.303	-7.669	
	(8.654)		(8.354)	(10.167)	(9.426)	
W-*GRP*	0.701		3.297	3.151	3.927	
	(3.146)		(2.878)	(10.049)	(4.224)	
W-*POLICE*	8.373		9.194	0.342	-6.319	
	(5.398)		(5.300)	(10.181)	(7.789)	
W-*FOREIGN*	-1.224		-3.461	0.841	1.028	
	(9.286)		(9.647)	(9.623)	(9.667)	
W-*HOTEL*	0.169		0.203	-0.064	-0.026	
	(10.076)		(10.158)	(9.829)	(9.852)	
W-*ARREST*	4.278		-1.004	6.123	7.003*	
	(2.498)		(0.882)	(9.919)	(2.340)	
ρ	0.195*	0.178*	0.125*	0.363*	0.076*	0.116*
	(0.030)	(0.030)	(0.034)	(0.025)	(0.030)	(0.031)
σ^2	3.933	0.057	2.915	899050.254	35.776	0.112
	(0.410)	(0.005)	(0.313)	(87950.155)	(3.325)	(0.011)
R^2	0.992	0.960	0.900	0.458	0.969	0.970
	FORGERY	*GRAFT*	*TRUST*	*GAMBLING*	*OBSCENITY*	*OTHERS*
Selected model	panel SDM	panel SEM	panel SEM	panel SEM	panel SAR	panel SAR
UNEMP	-19.237*	-1.066*	-0.005	-7.545*	17.100*	7.065
	(7.943)	(0.542)	(0.217)	(1.658)	5.426	9.856
GRP	7.410*	-0.305	-0.496*	0.245	4.650*	143.180*
	(2.613)	(0.236)	(0.165)	(0.829)	1.946	5.990
POLICE	32.250*	0.578*	0.306*	0.758	17.314*	75.175*
	(5.179)	(0.186)	(0.117)	(0.769)	3.120	8.895
FOREIGN	27.584*	0.851	1.826*	1.061	2.823	31.424*
	(9.231)	(1.022)	(0.585)	(3.883)	8.376	9.829
HOTEL	-4.332	1.785	6.232*	-20.825*	5.741	0.976
	(9.822)	(6.364)	(3.083)	(9.503)	10.100	9.903
ARREST	0.672	0.056*	-0.006	-0.006	-1.809*	-20.518*
	(0.391)	(0.025)	(0.009)	(0.083)	0.434	3.128
W-*UNEMP*	-18.147*					
	(8.336)					
W-*GRP*	-1.716					
	(2.731)					
W-*POLICE*	-11.376*					
	(4.397)					
W-*FOREIGN*	0.846					
	(9.261)					
W-*HOTEL*	-1.133					
	(10.007)					
W-*ARREST*	0.097					
	(0.710)					
ρ	0.053	0.036	0.076	0.157*	0.096*	0.117*
	(0.027)	(0.041)	(0.045)	(0.029)	(0.033)	(0.010)
σ^2	2.861	0.007	0.001	0.055	1.063	661.210
	(0.272)	(0.001)	(0.000)	(0.006)	(0.104)	(90.295)
R^2	0.989	0.549	0.792	0.915	0.981	0.959

Note: * means that the 95% credible interval does not include zero.

results show that the spatial patterns are different across all types of crimes. For the 18 types of crime, more panel SEM were chosen for the crime data in Japan. Also the socio-economic variables differ with crime types and the profiles are similar to Kakamu et al. (2006).[3] However, for the crime types *INJURY*, *THEFT*, *FRAUD*, *FORGERY* and *OTHERS* the panel SDM is selected, which leads to a richer profile since spatial lags are added in the model. Overall, we found that for the *FORGERY* model negative neighbor effects were found for the *UNEMP* and *POLICE* variables, and for the crime type *FRAUD*, positive effects can be found for the *ARREST* variable.

Our modeling approach showed that most of the crime types can be explained by models a high R^2. Only the types *THEFT* and *GRAFT* are crime types with lower R^2. These crimes seem to follow a different socio-economic pattern. Also the spatial effects are different: while *THEFT* has a high spatial correlation coefficient, *GRAFT* is spatially uncorrelated.

References

ANSELIN, L. (1988): *Spatial Econometrics: Methods and Models.* Dordrecht, Kluwer.

BECKER, G.S. (1968): Crime and Punishment: An Economic Approach. *Journal of Political Economy, 76, 169–217.*

CHIB, S. (1995): Marginal Likelihood from the Gibbs Output. *Journal of the American Statistical Association, 90, 1313–1321.*

CHIB, S. and JELIAZKOV, I. (2001): Marginal Likelihood from the Metropolis-Hastings Output. *Journal of the American Statistical Association, 96, 270–281.*

GELFAND, A.E. and SMITH, A.F.M. (1990): Sampling-based Approaches to Calculating Marginal Densities. *Journal of the American Statistical Association, 85, 398–409.*

GEWEKE, J. (1993): Bayesian Treatment of the Independent Student-t Linear Model. *Journal of Applied Econometrics, 8, 19–40.*

HOLLOWAY, G., SHANKAR, B. and RAHMAN, S. (2002): Bayesian Spatial Probit Estimation: A Primer and an Application to HYV Rice Adoption. *Agricultural Econnomics, 27, 384–402.*

KAKAMU, K., POLASEK, W. and WAGO, H. (2006): Spatial Interaction of Crime Incidents in Japan. *mimeo.*

LESAGE, J.P. (2000): Bayesian Estimation of Limited Dependent Variable Spatial Autoregressive Models. *Geographical Analysis, 32, 19–35.*

SUN, D., TSUTAKAWA, R.K. and SPECKMAN, P.L. (1999): Posterior Distribution of Hierarchical Models using CAR(1) Distributions. *Biometrika, 86, 341–350.*

TIERNEY, L. (1994): Markov Chains for Exploring Posterior Distributions (with discussion). *Annals of Statistics, 22, 1701–1762.*

[3] We omit the detail discussion about parameters and results to save the space. See Kakamu et al. (2006) for the detail discussion.

A Boosting Approach to Generalized Monotonic Regression

Florian Leitenstorfer and Gerhard Tutz

Department of Statistics, University of Munich, Akademiestrae 1, 80799 Munich, Germany; {leiten, tutz}@stat.uni-muenchen.de

Abstract. We propose a novel approach for generalized additive regression problems, where one or more smooth components are assumed to have monotonic influence on the dependent variable. The response is allowed to follow a simple exponential family. Smooth estimates are obtained by expansion of the unknown functions into B-spline basis functions, where the degree of smoothness is regularized by penalization. Monotonicity of estimates is achieved by restricting estimated coefficients to form an increasing sequence. Estimation is done by applying recently developed componentwise boosting methods for regression purposes. The performance of the new approach is demonstrated on numerical examples.

1 Introduction

In many statistical problems where generalized smooth regression methods are used, a monotonic relationship between one or more explanatory variables and the response variable has to be assumed. For instance, in studies where the influence of air pollution on mortality or illness is investigated, one expects an increase in respiratory mortality with increasing pollutant concentration. When standard smoothing techniques are applied the fitted curves may lead to unconvincing results. In the following, it is proposed to incorporate the knowledge about monotonic relationships in the estimation by using monotonic regression methods. Starting from the Pool Adjacent Violators Algorithm (PAVA) (see e.g. Robertson et al. (1988)) which produces a step function, a variety of methods has been developed to smooth the PAVA results. Alternative approaches, which will be pursued in the following, are based on the expansion of a monotonic function into a sum of basis functions, i.e. $f = \sum_j \alpha_j B_j$. To assure monotonicity of the estimate, adequate constraints have to be put on the coefficients α_j. In the examples considered here one has multiple covariates, and only for some of the covariates a monotonic effect on the conditional mean of the response has to be assumed. Furthermore, the response variables are considered as binomial or Poisson distributed. Flexible modeling tools are needed, where monotonicity restrictions can easily be incorporated into a generalized additive model (GAM) framework. Recently, boosting approaches became increasingly important in nonparametric regression, see e.g. Bühlmann and Yu (2003). In the present paper we suggest boosting based on B-spline basis functions. A special update scheme for the basis coefficients is proposed which shows good performance. It should be noted that the proposed method avoids the use of algorithms which handle inequality constraints. Procedures of this type typically are computationally burdensome and often yield unstable estimates. In Section 2 the concept of monotonic likelihood boosting based on B-splines is introduced, and an extension to multiple covariate settings is given. In Section 3, the performance of our approach is evaluated in a simulation study. In Section 4 we consider a real world data example which evaluates the association between mortality and the concentration SO_2. Note that throughout the paper, we take monotonic to mean nondecreasing.

2 Boosting B-splines in generalized monotonic regression

2.1 Monotonicity constraints for B-splines

First, we consider a generalized smooth monotonic regression problem with dependent variable y that can be non-Gaussian, and a single covariate x. It is assumed that $y_i|x_i$ has a distribution from a simple exponential family $f(y_i|x_i) = \exp\{[y_i\theta_i - b(\theta_i)]/\phi + c(y_i, \phi)\}$ where θ_i is the canonical parameter and ϕ denotes the dispersion parameter. The link between $\mu_i = E(y_i|x_i)$ and the explanatory variable x_i is determined by $\mu_i = h(\eta_i)$, where h is a given response function which is strictly monotone (the inverse of the link function $g = h^{-1}$), and the predictor $\eta_i = \eta(x_i)$ is a function of x. We assume that $\eta(x) = f(x)$ is a smooth function that satisfies the monotonicity condition $f(x) \geq f(z)$ if $x > z$. Obviously, monotonicity in η transforms into monotonicity in the means. Flexible smoothing methods based on B-splines are a common tool in statistics. Such approaches are based on an expansion of f into B-spline basis functions, where a sequence of knots $\{t_j\}$ is placed equidistantly within the range $[x_{\min}, x_{\max}]$. With $\tilde{m}$ denoting the number of interior knots, one obtains the linear term $\eta(x) = \alpha_0 + \sum_{j=1}^{m} \alpha_j B_j(x, q)$, where q denotes the degree of the B-splines and $m = \tilde{m} - 1 + q$ (the number of basis functions). Monotonicity can be assured in the following way. Suppose we have B-splines of degree $q \geq 1$. Let h be the distance between the equally spaced knots. Then the derivative $\eta'(x) = \partial\eta(x)/\partial x$ can be written as $\eta'(x) = \sum_j \alpha_j B_j'(x, q) = \frac{1}{h}\sum_j(\alpha_{j+1} - \alpha_j)B_j(x, q-1)$, for a proof see de Boor (1978). Since $B_j(x, q-1) \geq 0$, it follows from

$$\alpha_{j+1} \geq \alpha_j, \tag{1}$$

that $\eta'(x) \geq 0$ holds. This property of B-splines can be exploited in a monotonic regression setting.

2.2 An outline of the algorithm

Boosting has originally been introduced for classification problems. More recently, the approach has been extended to regression modeling with a continuous dependent variable (e.g. Bühlmann and Yu (2003), Bühlmann (2004)). The basic idea is to fit a function iteratively by fitting in each stage a "weak" learner to the current residual. In componentwise boosting as proposed by Bühlmann and Yu (2003), only the contribution of one variable is updated in one step. In contrast to these approaches we propose to update a specific simplification of the predictor which makes it easy to control the monotonicity restriction. For simplicity, in the following, the degree q of the B-splines is suppressed. In matrix notation, the data are given by $\mathbf{y} = (y_1, \ldots, y_n)'$, $\mathbf{x} = (x_1, \ldots, x_n)'$. Based on the expansion into basis functions, the data set may be collected in matrix form $(\mathbf{y}, \mathbf{B})$, where $\mathbf{B} = (B_1(\mathbf{x}), \ldots, B_m(\mathbf{x}))$, $B_j(\mathbf{x}) = (B_j(x_1), \ldots, B_j(x_n))'$. The residual model that is fitted by weak learners in one iteration step uses a grouping of B-splines. One considers for $r = 1, \ldots, m-1$, the simplified model that has the predictor

$$\eta(x_i) = \alpha_{0(r)} + \alpha_{1(r)} \left(\sum_{j=1}^{r} B_j(x_i) \right) + \alpha_{2(r)} \left(\sum_{j=r+1}^{m} B_j(x_i) \right). \tag{2}$$

When fitting model (2) the monotonicity constraint is easily checked by comparing the estimates $\hat{\alpha}_{1(r)}$ and $\hat{\alpha}_{2(r)}$, since monotonicity follows from $\hat{\alpha}_{2(r)} \geq \hat{\alpha}_{1(r)}$. Given an estimate from previous fitting, $\hat{\eta}_{\text{old}}(x_i) = \hat{\alpha}_{0,\text{old}} + \sum_{j=1}^{m} \hat{\alpha}_{j,\text{old}} B_j(x_i)$, refitting is performed by

$$\begin{aligned}\hat{\eta}_{\text{new}}(x_i) &= \hat{\eta}_{\text{old}}(x_i) + \hat{\alpha}_{0(r)} + \hat{\alpha}_{1(r)} \left(\sum_{j=1}^{r} B_j(x_i) \right) + \hat{\alpha}_{2(r)} \left(\sum_{j=r+1}^{m} B_j(x_i) \right) \\ &= \hat{\alpha}_{0,\text{old}} + \hat{\alpha}_{0(r)} + \sum_{j=1}^{r} (\hat{\alpha}_{j,\text{old}} + \hat{\alpha}_{1(r)}) B_j(x_i) + \sum_{j=r+1}^{m} (\hat{\alpha}_{j,\text{old}} + \hat{\alpha}_{2(r)}) B_j(x_i).\end{aligned}$$

It is obvious that $\hat{\eta}_{\text{new}}$ is monotonic if estimates fulfill $\hat{\alpha}_{2(r)} \geq \hat{\alpha}_{1(r)}$, provided that the previous estimate $\hat{\eta}_{\text{old}}$ was monotonic. The grouping of basis functions into $B_1, \ldots, B_r$ and $B_{r+1}, \ldots, B_m$, the effect of which is adapted by the amount $\alpha_{1(r)}$ in the first and $\alpha_{2(r)}$ in the second group, allows to control monotonicity in a simple way. The possible groupings $(r = 1, \ldots, m-1)$ are evaluated and in analogy to componentwise boosting the best refit is selected. Before giving the algorithm, which is based on likelihood-based boosting strategies, the fit of model (2) is embedded into the framework of penalized likelihood estimation. Moreover, the model is generalized to a model that contains parametric effects in addition to the smooth monotonic effects. Thus the intercept term is replaced by $\mathbf{z}_i'\boldsymbol{\alpha}_0$ where $\mathbf{z}_i$ is a vector of covariates and $\boldsymbol{\alpha}_0$ is an unknown parameter vector (possibly specifying only the intercept). It is assumed that $\mathbf{z}_i$ always contains an intercept. In order to avoid identifiability issues which arise for B-splines in connection with an intercept, the update step is split up into two parts. In the first part the smooth component is updated and in the second the parametric term. In the first part one fits by penalized likelihood. Therefore one considers

$$\mathbf{R}_{(r)} = \begin{pmatrix} \mathbf{1}_r & \mathbf{0}_r \\ \mathbf{0}_{m-r} & \mathbf{1}_{m-r} \end{pmatrix},$$

with $\mathbf{0}_r$, $\mathbf{1}_r$ denoting the vectors of length r containing 0s and 1s only. Then the linear predictor may be represented in matrix form by $\eta(x) = \mathbf{B}_{(r)}\boldsymbol{\alpha}_{(r)}$, where $\mathbf{B}_{(r)} = \mathbf{B}\mathbf{R}_{(r)}$ and $\boldsymbol{\alpha}_{(r)} = (\alpha_{1(r)}, \alpha_{2(r)})'$. It is proposed that in each boosting step, the model is estimated by one-step Fisher scoring based on generalized ridge regression, which maximizes the penalized log-likelihood $l_p(\boldsymbol{\alpha}_{(r)}) = \sum_{i=1}^{n} l_i(\boldsymbol{\alpha}_{(r)}) - \frac{\lambda}{2}\boldsymbol{\alpha}'_{(r)}\boldsymbol{\Lambda}\boldsymbol{\alpha}_{(r)}$, where $l_i(\boldsymbol{\alpha}_{(r)}) = l_i(h(\mathbf{B}_{(r)}\boldsymbol{\alpha}_{(r)}))$ is the usual log-likelihood contribution of the ith observation, $\boldsymbol{\Lambda} = \text{diag}\{1, 1\}$ and $\lambda > 0$ represents the ridge parameter. Derivation yields the corresponding penalized score function $s_p(\boldsymbol{\alpha}_{(r)}) = \mathbf{B}'_{(r)}\mathbf{W}(\boldsymbol{\eta})\mathbf{D}(\boldsymbol{\eta})^{-1}(\mathbf{y} - h(\boldsymbol{\eta})) - \lambda\boldsymbol{\Lambda}\boldsymbol{\alpha}_{(r)}$, with $\mathbf{W}(\boldsymbol{\eta}) = \mathbf{D}^2(\boldsymbol{\eta})\boldsymbol{\Sigma}(\boldsymbol{\eta})^{-1}$, $\mathbf{D}(\boldsymbol{\eta}) = \text{diag}\{\partial h(\eta_1)/\partial\eta, \ldots, \partial h(\eta_n)/\partial\eta\}$, $\boldsymbol{\Sigma}(\boldsymbol{\eta}) = \text{diag}\{\sigma_1^2, \ldots, \sigma_n^2\}$, $\sigma_i^2 = \text{var}(y_i)$, all of them evaluated at the current value of η. The monotonicity constraint from (1) is incorporated by taking into account only estimates which fulfill $\hat{\alpha}_{2(r)} \geq \hat{\alpha}_{1(r)}$. It is easily seen that the update scheme given below yields the desired nondecreasing sequences of estimates $\hat{\alpha}_1, \ldots, \hat{\alpha}_m$ in each boosting iteration. The update of the parametric term $\mathbf{z}'_i\boldsymbol{\alpha}_0$ is performed in the same way but without penalization and with the design matrix determined by $\mathbf{Z} = (\mathbf{1}, \mathbf{z}_1, \ldots, \mathbf{z}_u)$.

Generalized Monotonic Likelihood Boosting for B-splines (GMBBoost)

Step 1 (Initialization) Set $\hat{\boldsymbol{\alpha}}_0^{(0)} = (g(\bar{y}), 0, \ldots, 0)'$, $\hat{\boldsymbol{\alpha}}^{(0)} = (0, \ldots, 0)'$, $\hat{\boldsymbol{\eta}}^{(0)} = (g(\bar{y}), \ldots, g(\bar{y}))'$ and $\hat{\boldsymbol{\mu}}^{(0)} = (\bar{y}, \ldots, \bar{y})'$. *Step 2 (Iteration)* For $l = 1, 2, \ldots$

1. *Fitting step, monotone component*
   For $r = 1, \ldots, m-1$, compute the modified ridge estimate based on one-step Fisher scoring,

$$\hat{\boldsymbol{\alpha}}_{(r)} = (\mathbf{B}'_{(r)}\mathbf{W}_l\mathbf{B}_{(r)} + \lambda\boldsymbol{\Lambda})^{-1}\mathbf{B}'_{(r)}\mathbf{W}_l\mathbf{D}_l^{-1}(\mathbf{y} - \hat{\boldsymbol{\mu}}^{(l-1)}), \tag{3}$$

   where $\hat{\boldsymbol{\alpha}}_{(r)} = (\hat{\alpha}_{1(r)}, \hat{\alpha}_{2(r)})'$, $\mathbf{W}_l = \mathbf{W}(\hat{\boldsymbol{\eta}}^{(l-1)})$, $\mathbf{D}_l = \mathbf{D}(\hat{\boldsymbol{\eta}}^{(l-1)})$, and $\hat{\boldsymbol{\mu}}^{(l-1)} = h(\hat{\boldsymbol{\eta}}^{(l-1)})$. Let $A = \{r : \hat{\alpha}_{1(r)} \leq \hat{\alpha}_{2(r)}\}$ denote the candidates that fulfill the monotonicity constraint. If $A = \emptyset$, stop. Otherwise continue with step 2.

2. *Selection step and update, monotone component*
Compute the potential update of the linear predictor, $\tilde{\boldsymbol{\eta}}_{(r),\text{new}} = \hat{\boldsymbol{\eta}}^{(l-1)} + \mathbf{B}_{(r)}\hat{\boldsymbol{\alpha}}_{(r)}$, $r \in \{1, \ldots, m-1\}$. Choose $r_l \in A$ such that the deviance is minimized, i.e. $r_l = \arg\min_{r \in A} \text{Dev}(\tilde{\boldsymbol{\eta}}_{(r),\text{new}})$, where $\text{Dev}(\tilde{\boldsymbol{\eta}}_{(r),\text{new}}) = 2\sum_{i=1}^{n}[l_i(y_i) - l_i(h(\tilde{\eta}_{i,(r),\text{new}}))]$. Set

$$\hat{\alpha}_j^{(l)} = \begin{cases} \hat{\alpha}_j^{(l-1)} + \hat{\alpha}_{1(r_l)}, \ 1 \le j \le r_l, \\ \hat{\alpha}_j^{(l-1)} + \hat{\alpha}_{2(r_l)}, \ j > r_l, \end{cases} \tag{4}$$

$$\tilde{\boldsymbol{\eta}}^{(l-1)} = \hat{\boldsymbol{\eta}}^{(l-1)} + \mathbf{B}_{(r_l)}\hat{\boldsymbol{\alpha}}_{(r_l)} \quad \text{and} \quad \tilde{\boldsymbol{\mu}}^{(l-1)} = h(\tilde{\boldsymbol{\eta}}^{(l-1)}).$$

3. *Fitting step and update, parametric term*
Based on one step Fisher scoring one obtains
$\hat{\boldsymbol{\alpha}}_0 = (\mathbf{Z}'\tilde{\mathbf{W}}_l\mathbf{Z})^{-1}\mathbf{Z}'\tilde{\mathbf{W}}_l\tilde{\mathbf{D}}_l^{-1}(\mathbf{y} - \tilde{\boldsymbol{\mu}}^{(l-1)})$, where $\tilde{\mathbf{W}}_l = \mathbf{W}(\tilde{\boldsymbol{\eta}}^{(l-1)})$ and $\tilde{\mathbf{D}}_l = \mathbf{D}(\tilde{\boldsymbol{\eta}}^{(l-1)})$. Set

$$\hat{\boldsymbol{\alpha}}_0^{(l)} = \hat{\boldsymbol{\alpha}}_0^{(l-1)} + \hat{\boldsymbol{\alpha}}_0, \quad \hat{\boldsymbol{\eta}}_0^{(l)} = \tilde{\boldsymbol{\eta}}_0^{(l-1)} + \mathbf{Z}\hat{\boldsymbol{\alpha}}_0^{(l)} \quad \text{and} \quad \hat{\boldsymbol{\mu}}^{(l)} = h(\hat{\boldsymbol{\eta}}^{(l)}).$$

When using boosting techniques, the number of iterations l plays the role of a smoothing parameter. Therefore, in order to prevent overfitting, a stopping criterion is necessary. A quite common measure of the complexity of a smooth regression fit is the hat-matrix. Consequently, Bühlmann and Yu (2003) and Bühlmann (2006) developed a hat-matrix for L_2-boosting with continuous dependent variable. In the case of likelihood boosting, for more general exponential type distributions, the hat-matrix has to be approximated. For integrated splines, Tutz and Leitenstorfer (2006) give an approximation based on first order Taylor expansions, which shows satisfying properties and can be straightforwardly applied to the present case. With $\mathbf{M}_0 = \frac{1}{n}\mathbf{1}_n\mathbf{1}_n'$, $\mathbf{M}_l = \boldsymbol{\Sigma}_l^{1/2}\mathbf{W}_l^{1/2}\mathbf{B}_{(r_l)}(\mathbf{B}_{(r_l)}'\mathbf{W}_l\mathbf{B}_{(r_l)} + \lambda\boldsymbol{\Lambda})^{-1}\mathbf{B}_{(r_l)}'\mathbf{W}_l^{1/2}\boldsymbol{\Sigma}_l^{-1/2}$, where $\mathbf{W}_l = \mathbf{W}(\hat{\boldsymbol{\eta}}^{(l-1)})$ and $\boldsymbol{\Sigma}_l - \boldsymbol{\Sigma}(\hat{\boldsymbol{\eta}}^{(l-1)})$, and $\tilde{\mathbf{M}}_l = \tilde{\boldsymbol{\Sigma}}_l^{1/2}\tilde{\mathbf{W}}_l^{1/2}\mathbf{Z}(\mathbf{Z}'\tilde{\mathbf{W}}_l\mathbf{Z})^{-1}\mathbf{Z}'\tilde{\mathbf{W}}_l^{1/2}$ $\tilde{\boldsymbol{\Sigma}}_l^{-1/2}$, where $\tilde{\mathbf{W}}_l = \mathbf{W}(\tilde{\boldsymbol{\eta}}^{(l-1)})$ and $\tilde{\boldsymbol{\Sigma}}_l = \boldsymbol{\Sigma}(\tilde{\boldsymbol{\eta}}^{(l-1)})$, $l = 1, 2, \ldots$, the approximate hat-matrix is given by

$$\mathbf{H}_l = \mathbf{I} - \left[\prod_{j=1}^{l}(\mathbf{I} - \tilde{\mathbf{M}}_{l-j+1})(\mathbf{I} - \mathbf{M}_{l-j+1})\right](\mathbf{I} - \mathbf{M}_0), \tag{5}$$

with $\hat{\boldsymbol{\mu}}^{(l)} \approx \mathbf{H}_l\mathbf{y}$. By considering $\text{tr}(\mathbf{H}_l)$ as the degrees of freedom of the smoother, we use as potential stopping criterion the Akaike information criterion, $\text{AIC}(l) = \text{Dev}_l + 2\text{tr}(\mathbf{H}_l)$, where $\text{Dev}_l = 2\sum_{i=1}^{n}[l_i(y_i) - l_i(h(\hat{\eta}_i^{(l)}))]$ denotes the deviance of the model in the lth boosting step. The optimal number of boosting iterations is defined by $l_{\text{opt}}^{\text{AIC}} = \arg\min_l \text{AIC}(l)$.

2.3 Extension to generalized additive models

In biometrical or ecological problems, one is usually interested in the effect of several smooth predictor variables, some of them might have monotonic influence on y. In the following we demonstrate that the concept given above can easily be extended to a GAM setting. Let

$$\eta = \mathbf{z}'\boldsymbol{\alpha}_0 + \sum_{s=1}^{p} f_s(x_s), \tag{6}$$

where for some of the p unknown smooth functions (say $f_1, \ldots, f_v$, $v \leq p$) monotonicity constraints are assumed to hold. Using the matrix notation from above, we have a design matrix $(\mathbf{Z}, \mathbf{X})$, with the matrix of linear terms $\mathbf{Z}$ and the matrix of smooth components $\mathbf{X} = (\mathbf{x}_1, \ldots, \mathbf{x}_p)$, where $\mathbf{x}_s = (x_{1s}, \ldots, x_{ns})'$. Componentwise expansion of $\mathbf{X}$ into B-spline basis functions leads to the matrix $(\mathbf{B}^{(1)}, \ldots, \mathbf{B}^{(p)})$, where $\mathbf{B}^{(s)}$ refers to the sth predictor. It is essential to distinguish between components that are under monotonicity restrictions and those that are not. For the former, grouping of basis functions is done within each component in the same way as described in (2). For the unconstrained components, we use penalized regression splines (P-splines, cf. Eilers and Marx (1996)) as weak learners for the chosen component. Thereby, the second-order differences of the B-spline coefficients are penalized. For simplicity, it is assumed that the same number of basis functions m is used for all f_s. The vector of basis coefficients for all smooth terms in the model is then given by $\boldsymbol{\alpha} = (\alpha_{11}, \ldots, \alpha_{1m}, \ldots, \alpha_{p1}, \ldots, \alpha_{pm})'$. Thus, *step 2 (iteration)* of the algorithm described above is extended as follows: *Step 2 (Iteration):* For $l = 1, 2, \ldots$

1. *Fitting step, smooth components*
   For $s = 1, \ldots, p$,
   - If $s \in \{1, \ldots, v\}$ (monotonicity restricted), compute the estimates from (3) componentwise for the possible groupings $r = 1, \ldots, m-1$, with $\mathbf{B}^{(s)}_{(r)} = \mathbf{B}^{(s)}\mathbf{R}_{(r)}$. The set of indices for components s and split points r that satisfy the monotonicity constraint is given by

     $$A_1 = \{(s, r) \in \{1, \ldots, v\} \times \{1, \ldots, (m-1)\} : \hat{\alpha}^{(s)}_{1(r)} \leq \hat{\alpha}^{(s)}_{2(r)}\}.$$

   - If $s \in \{v+1, \ldots, p\}$ (no constraints), compute the one step Fisher scoring estimate of the P-spline,

     $$\hat{\boldsymbol{\alpha}}^{(s)} = (\mathbf{B}^{(s)\prime}\mathbf{W}_l\mathbf{B}^{(s)} + \lambda_P \boldsymbol{\Delta}_2'\boldsymbol{\Delta}_2)^{-1}\mathbf{B}^{(s)\prime}\mathbf{W}_l\mathbf{D}_l^{-1}(\mathbf{y} - \hat{\boldsymbol{\mu}}^{(l-1)}), \tag{7}$$

     where $\boldsymbol{\Delta}_2$ denotes the matrix representation of the second order differences. We set $r = 0$ and extend the selection set by $A_2 = \{(s, 0), s \in \{v+1, \ldots, p\}\}$, yielding $A = A_1 \cup A_2$.

2. *Selection step and update, smooth components*
Compute the potential update of the linear predictor, which only for the monotonic coefficients $s \leq v$ depends on the split point r. Otherwise, r is set to 0, indicating that $\tilde{\boldsymbol{\eta}}^{(s)}_{(0),\text{new}}$ is not affected by r. Choose $(s_l, r_l) \in A$ such that the deviance is minimized, i.e.
$(s_l, r_l) = \arg\min_{(s,r)\in A} \text{Dev}(\tilde{\boldsymbol{\eta}}^{(s)}_{(r),\text{new}})$. In each iteration only the basis coefficients belonging to s_l are refitted. If the selected s_l is in $\{1, \ldots, v\}$, then update $\hat{\alpha}^{(l)}_{s_l,j}$ by the refitting scheme (4). If $s_l > v$, then update $\hat{\alpha}^{(l)}_{s_l,j} = \hat{\alpha}^{(l-1)}_{s_l,j} + \hat{\alpha}^{(s_l)}_j$, with $\hat{\boldsymbol{\alpha}}^{(s_l)}$ from (7).
3. *Fitting step and update, parametric terms.* See above.

By using $\mathbf{B}^{(s)}_{(r)}$ from above, along with $\mathbf{B}^{(s)}$ and the penalty matrices for the P-spline estimates, the hat-matrix approximation from (5) and the corresponding AIC stopping criterion can be extended to the additive setting.

3 Simulation results

The performance of the proposed method is evaluated in some simulation studies. In a first setting, a unidimensional Poisson regression model is considered, with response y_i generated from $\text{P}(\exp(\eta_i))$, where $\eta_i = \eta(x_i)$ is specified by a monotonic function and $x_i \sim U[0, 5]$. We investigate a step function, $\eta(x) = 2cI(x > 2.5)$, and a plateau function, $\eta(x) = c(2/\{1 + \exp[-10(x - 1)]\} + 2/\{1 + \exp[-5(x - 4)]\} - 1)$. The strength of the signal is controlled by the constant c. For GMBBoost, $m = 22$ B-spline basis function of degree $q = 3$ are used. The ridge parameter has been chosen by $\lambda = 300$. GMBBoost is compared to unconstrained penalized regression splines as implemented in the R package `mgcv`, where the penalization parameter is determined by the UBRE criterion (see Wood (2000)). For comparability a cubic regression spline basis with a dimension of `k=22` is used. We also consider a monotonicity-constrained version of this approach, based on quadratic programming. This envolves embedding a monotone smoother in an iteratively reweighted least squares loop. For details, see Wood (1994). Ordinary PAVA is also included. A criterion for the performance of the fitting methods is the averaged Kullback-Leibler distance, which is given by $\text{AKL} = \frac{1}{n}\sum_{i=1}^{n} \text{KL}[\hat{\mu}_i, \mu_i]$. The means of AKL over $S = 250$ simulated data sets are given in Table 1 for selected sample sizes and noise levels. For the step function example, it is seen that GMBBoost is a strong competitor that clearly outperforms the unconstrained and constrained MGCV fits. PAVA does better only in the case of a stronger signal and $n = 100$. For the plateau function, monotonicity restricted MGCV and GMBBoost perform very similar, whereas PAVA does considerably worse. We investigate also a setting in higher dimensions, where only some of the components are under monotonicity constraints. A Poisson model with a linear predictor $\eta(\mathbf{x}_i) = c(\alpha_0 + \sum_{j=1}^{p} f_j(x_{ij}))$ is considered, where $p = 5$ and the last

Table 1. Poisson regression, one-dimensional setting, averaged KL error over $S = 250$ simulated data sets. The number of instances where no fit could be obtained is given in parentheses.

		Step function				Plateau function			
		MGCV	mon MGCV	PAVA	GMB (AIC)	MGCV	mon MGCV	PAVA	GMB (AIC)
$c = 0.5$	$n = 50$	0.080	0.057 [16]	0.072	**0.046**	0.068	**0.051** [3]	0.077	0.055
	$n = 100$	0.047	0.035 [12]	0.040	**0.024**	0.036	**0.029** [2]	0.046	**0.029**
$c = 1$	$n = 50$	0.156	0.114 [30]	**0.082**	0.083	0.093	0.075 [0]	0.117	**0.074**
	$n = 100$	0.099	0.079 [35]	**0.045**	0.062	0.049	0.040 [0]	0.069	**0.039**

three functions are assumed to be monotonic. See Figure 1 for the shape of the functions ($\alpha_0 = 0$). The $\mathbf{x}_i$ are drawn from a $\mathcal{N}_5(\mathbf{0}, \boldsymbol{\Sigma})$-distribution with $\boldsymbol{\Sigma} = \rho\mathbf{1}\mathbf{1}' + (1-\rho)\mathbf{I}$ ($\rho = 0.4$). We observed that a ridge parameter of $\lambda = 300$

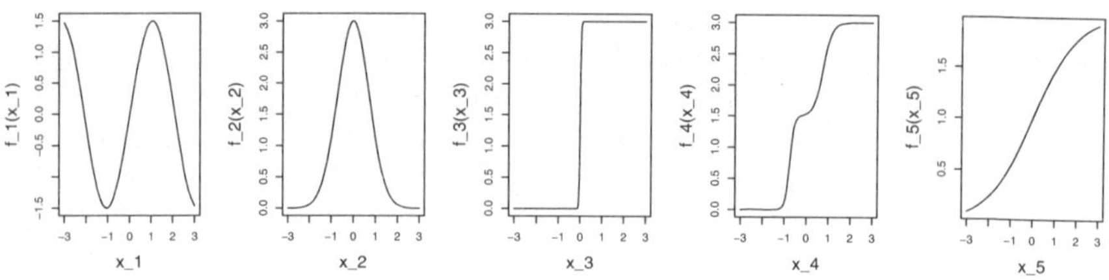

Fig. 1. Functions $f_s(.)$, $s = 1, \ldots, 5$, used for the simulation in higher dimensions. The last three function are monotonic, the first two are not.

for the grouped B-splines is large enough to yield an appropriate fit. However, we allow for more flexibility in the choice of the P-spline penalty parameter by considering $\lambda_P \in \{1000, 3000, 5000\}$. The choice of basis functions and knots is the same as in the example above. We again compare GMBBoost to MGCV which uses the same settings as before. In Figure 2, boxplots for the logarithm of the AKL are given for the various settings. It is seen that GMBBoost outperforms MGCV in most considered settings, with a distinct dominance in the lower noise case $c = 0.2$ (left panel).

4 Application

In the following we consider an example taken from Singer et al. (2002). In this study conducted from 1994 to 1997 in São Paulo, Brazil, the association between mortality of children and the concentration of the air pollutant SO_2 is investigated. Daily observations of pollutant concentration, weather variables

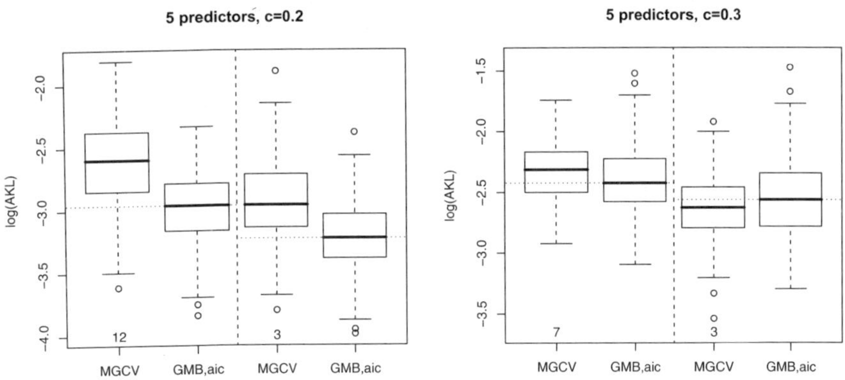

Fig. 2. Boxplots of log(AKL) for different fitting methods for the model with five predictors with different noise levels. In each panel, the results for $n = 200$ (left) and $n = 300$ (right) are given, along with the number of instances where no MGCV fit could be obtained.

and the number of daily respiratory deaths of children under five (response variable) were recorded (sample size $n = 1351$). A standard approach for data of this type is to fit a generalized additive Poisson model (core model) to control for trend, seasonality and weather variables. We consider the following the core model:

$$\eta_{\text{core}} = \log[E(\text{respiratory deaths})] = \alpha_0 + f_1(\text{time}) + f_2(\text{temp}) + \alpha_{01} \cdot \text{humidity} + \alpha_{02} \cdot \text{Monday} + \cdots + \alpha_{07} \cdot \text{Saturday} + \alpha_{08} \cdot \text{non-respiratory deaths}.$$

To investigate the effect of SO_2, we take this pollutant into the model, yielding the linear predictor $\eta = \eta_{\text{core}} + f_3(SO_2)$. Since one assumes that ambient air pollution is a good proxy for air pollution exposure, it is sensible to assume the function f_3 to be monotonic increasing. Figure 3 shows the estimated curves $\hat{f}_1$, $\hat{f}_2$ and $\hat{f}_3$ for GMBBoost ($\hat{f}_3$ under monotonicity constraint) as well as for non-restricted MGCV. From the right panel, it is seen that the MGCV fit for SO_2 is pulled down by some outliers, yielding the questionable result of decreasing mortality with increasing pollutant concentration. GMBBoost yields a curve that remains constant for high SO_2 concentrations.

5 Conclusion

A procedure is proposed that allows to use the information on monotonicity for one or more components within a generalized additive model. By using monotonicity, the procedure yields stable estimates which in contrast to GAM

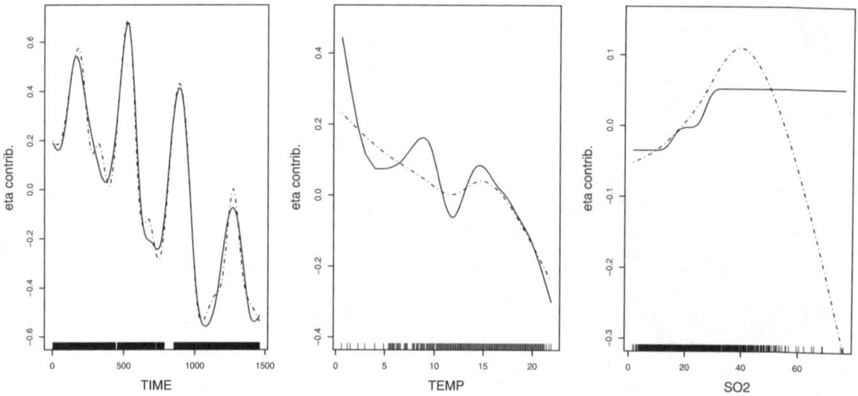

Fig. 3. Core model+$f_3(\mathrm{SO}_2)$, estimated curves for the smooth components, GMBBoost with monotonic fitting of $f_2(\mathrm{SO}_2)$ (solid) and MGCV (dash-dotted).

fitting avoid overfitting and questionable estimates. It should also be noted that the problem of choosing smoothing parameters—which in case of higher dimensional covariates is hard to tackle—is avoided by boosting techniques. The only crucial tuning parameter is the number of boosting iterations, which is chosen by the AIC criterion. Moreover, the procedure is a strong competitor to alternative approaches. An implementation in R of the approach outlined in this paper is available at `http://www.statistik.lmu.de/institut/lehrstuhl/semsto/Software/software.htm`.

References

BÜHLMANN, P. (2006): Boosting for High-dimensional Linear Models. *Annals of Statistics, 34, 559–583.*

BÜHLMANN, P. and YU, B. (2003): Boosting with the L_2-loss: Regression and Classification. *Journal of the American Statistical Association, 98, 324–339.*

DE BOOR, C. (1978): *A Practical Guide to Splines.* Springer, New York.

EILERS, P.H.C. and MARX, B.D. (1996): Flexible Smoothing with B-splines and Penalties. *Statistical Science, 11, 89–121.*

ROBERTSON, T., WRIGHT, F.T. and DYKSTRA, R.L. (1988): *Order-Restricted Statistical Inference.* Wiley, New York.

SINGER, J.M., ANDRE, C.D.S., LIMA, P.L. and CONÇEIÃO, G.M.S. (2002): Associaton between Atmospheric Pollution and Mortality in São Paulo, Brazil: Regression Models and Analysis Strategy. In Y. Dodge (Ed.): *Statistical Data Analysis Based on the L_1 Norm and Related Methods.* Birkhuser, Berlin, 439-450.

TUTZ, G. and LEITENSTORFER, F. (2006): Generalized Smooth Monotonic Regression in Additive Modeling. *Journal of Computational and Graphical Statistics* (in print).

WOOD, S.N. (1994): Monotonic Smoothing Splines Fitted by Cross Validation. *SIAM Journal on Scientific Computing, 15, 1126–1133.*

WOOD, S.N. (2000): Modelling and Smoothing Parameter Estimation with Multiple Quadratic Penalties. *Journal of the Royal Statistical Society B, 62, 413–428.*

From Eigenspots to Fisherspots – Latent Spaces in the Nonlinear Detection of Spot Patterns in a Highly Varying Background

Bjoern H. Menze, B. Michael Kelm and Fred A. Hamprecht

Interdisciplinary Center for Scientific Computing (IWR), University of Heidelberg, INF 368, 69120 Heidelberg, Germany; bjoern.menze@iwr.uni-heidelberg.de

Abstract. We present a scheme for the development of a spot detection procedure which is based on the learning of latent linear features from a training data set. Adapting ideas from face recognition to this low level feature extraction task, we suggest to learn a collection of filters from representative data that span a subspace which allows for a reliable distinction of a spot vs. the heterogeneous background; and to use a non-linear classifier for the actual decision. Comparing different subspace projections, in particular principal component analysis, partial least squares, and linear discriminant analysis, in conjunction with subsequent classification by random forests on a data set from archaeological remote sensing, we observe a superior performance of the subspace approaches, both compared with a standard template matching and a direct classification of local image patches.

1 Introduction – spot detection

In the hot and dry plains of ancient Mesopotamia and other parts of the Near East, but also in an arc stretching from the Balkans to India, small artificial mounds indicate the sites of early human settlements, some of them – as the biblic Jericho – being the remains of the first urban metropoles.

These so called "*tells*" are the result of millennia of settlement activity. Their base layers often reach as far back as 6000BC and a mud-based construction technique, prevalent to these regions, allowed some of them to raise up to significant heights during the millennia, forming characteristic landmarks. Though a large number of these mounds are well studied, the best current listings of them are neither comprehensive nor accurate. – However, in the digital elevation model of the Space Shuttle radar topography mission (SRTM), tells can be identified as small contrasting spots within the elevation pattern of the natural variation of the land surface (Sherratt (2004)).

As agricultural landuse and the growth of modern settlements impose an immanent threat to this cultural heritage and a study of the distribution of these former settlements is of high archaeological interest, we seek for a robust machine based processing of the SRTM data which allows for a fast, objective and precise guidance to tell sites in order to document them in wide regions of Turkey, Syria, Iraq and Iran.

Spot or point detection is a standard task in low level image processing. While elementary template matching is optimal for detecting point-like patterns in uncorrelated noise, other approaches exist in applications as diverse as preprocessing of microarray and gel electrophoresis image data, the detection of cars in thermal bands of satellite imagery, or peak detection in 2D mass spectrometric data (Rahnenfuehrer and Bozinov (2005), Boetticher et al. (2005)), to name a random selection. – Most of the spot detection approaches can be categorized into two classes: Parametric models are used to characterize the object, e.g. gaussian functions to model the spots, splines to fit and correct for the background. Alternatively, the detection is based on a phenomenological and nonparametric description of characteristic features, e.g. when searching for local extremes by morphological operations (watershed transformation), or evaluating the gradient images by structure tensors.

Unfortunately, a simple matched filter fails in the detection of tell-like mounds in the digital elevation model due to a high number of false positive hits. Also, the lack of positional a priori information, the variation of the spot pattern (diameter and height of the tell), and the highly variable "background", given by the natural topographic variation (ridges, walls, natural mounds), prohibit the application of spot detection algorithms as the ones mentioned above. –

Adapting ideas from face recognition, notably the concepts of *"Eigen"*- and *"Fisherfaces"* (see Belhumeur et al. (1996) and references therein), we learn adaptive templates (Section 2) from our data (Subsection 3.1), extending the idea of a (single) template matching to a multi-dimensional subspace approach for spot detection. Combined with a nonlinear classifier - random forests - we quantitatively compare (Subsection 3.2) and discuss (Section 4) different methods intermediate between *Eigen*- and *Fisherspots* for our task.

2 Subspace filters – latent spaces

The optimal filter for the detection of a signal with known shape in additive white Gaussian noise is the *matched filter* (MF) (Moon and Stirling (2000)). Convolving an image with the MF can be regarded as correlating the image with a template of the signal to be detected. From a learning perspective, and extending the idea of a signal detection to a binary classification task between (tell) pattern vs. (non-tell) background, this approach corresponds to regarding the image as a collection of (local and independent) patches. All pixels in a patch are explanatory variables with an associated label, ie.

pattern or background. In this feature space, the matched filter defines a one-dimensional linear subspace which is used to discriminate these two classes. From this point of view, the MF is very much related to linear regression methods, which motivates the approach taken in this paper and the naming *subspace filter*.

Real situations do not necessarily fulfill the ideal conditions under which the MF is proven to be optimal. Instead of seeking an optimal one-dimensional subspace and thus presuming linear separability in the feature space, we propose to perform a less restrictive dimensionality reduction, i.e. the projection onto a subspace of higher dimension followed by a nonlinear decision rule.

A common basic approach to the construction of a subspace which captures the most important variations in high dimensional data is *principal component analysis* (PCA). Its ranking criterion for the kth direction β_k is derived from the empirical covariance of the features :

$$\beta_{PCA_1,k} = \underset{\substack{||\beta||=1 \\ corr(\beta_j,\beta_k)=0, j<k}}{\arg\max} \quad var(X_1\beta) \tag{1}$$

with $corr(\beta_k, \beta_j)$ denoting the correlation between β_k and β_j; and where X_1 only holds the examples with the sought pattern. This projection compresses variation and information of the correlated spatial signal, but it neglects knowledge about the background signal X_0 and the binary character of the detection problem. In order to incorporate knowledge about X_0, PCA can be extended to derive the directions β_{PCA} from the variance of the full training data set X. This represents the prior belief that the variance of the training sample is due to interclass variations which are represented by the major eigendirections in the sample space.

The two-class information can be used explicitly as done in canonical correlation analysis (CCA). For univariate Y this is equivalent to ordinary least squares (OLS) regression (Borga (1997)) which, for the two-class problem, yields the same directions as linear discriminant analysis (LDA) (Hastie et al. (2001)). All these problems determine the optimal direction β based on the correlation between the class label Y and the projected feature scores $X\beta$. They choose directions with high discriminative power:

$$\beta_{LDA,k} = \underset{\substack{||\beta||=1 \\ corr(\beta_j,\beta_k)=0, j<k}}{\arg\max} \quad corr^2(X\beta, Y) \tag{2}$$

again with orthogonal directions β_k for linearly nonseparable problems. – OLS and LDA are known to have bad generalization performance in the presence of collinear features, i.e. they are vulnerable to overfitting (e.g. see OLS projections in Fig. 2).

Introducing a bias, forcing subspace projections to more "realistic" directions with higher data support, can help to overcome this problem. Regularization is obtained by combining the two strategies mentioned above and optimizing for covariance or equivalently for the product of variance and squared correlation (Frank and Friedman (1993)):

$$\beta_{PLS,k} = \underset{\substack{||\beta||=1 \\ corr(\beta_j,\beta_k)=0, j<k}}{\arg\max} \; cov^2(X\beta, Y) \tag{3}$$

$$= \underset{\substack{||\beta||=1 \\ corr(\beta_j,\beta_k)=0, j<k}}{\arg\max} \; corr^2(X\beta, Y)\, var(X\beta) \tag{4}$$

This forces the directions of the subspaces to have a natural "backing" in the data variation: the solution is pulled away from the OLS solution of maximal correlation towards directions of maximal variance in sample space as obtained by PCA.

Two related methods allow to vary the influence of the variance continuously. Ridge regression/penalized discriminant analysis (RR/PDA) extends the concept of OLS/LDA (Frank and Friedman (1993)):

$$\beta_{RR,k}(\gamma) = \underset{\substack{||\beta||=1 \\ corr(\beta_j^T \beta_k)=0, j<k}}{\arg\max} \; corr^2(X\beta, Y)\, \frac{var(X\beta)}{var(X\beta)+\gamma} \tag{5}$$

A generalization of PLS is continuum regression (CR) (Bjorkstrom (1999)):

$$\beta_{CR,k}(\gamma) = \underset{\substack{||\beta||=1 \\ corr(\beta_j,\beta_k)=0, j<k}}{\arg\max} \; corr^2(X\beta, Y)\, var(X\beta)^\gamma \tag{6}$$

Both approaches come at the cost of a hyperparameter γ to be tuned in addition to the optimal subspace dimension λ. Because of this and since PLS provides means to regularize LDA they will not be studied in the following.

3 Methods

3.1 Data

Tell sites. Average tells reach a height of 10-50m and have a diameter of 50-500m. In the SRTM elevation data set their patterns appear as small bright spots of one to five pixels diameter and with approximate radial symmetry (cf. Fig. 1). In the SRTM model of a North Syrian plain, the Khabur basin (Menze et al. (2006)), positions of 184 known tell sites could be identified. In addition, 50 000 locations were randomly sampled (with uniform distribution) from the same geographic region as representatives of the background class X_0. An independent test data set, comprising positions of another 133 sites, was available from an archaeological survey in the same area (Menze et al. (2006)).

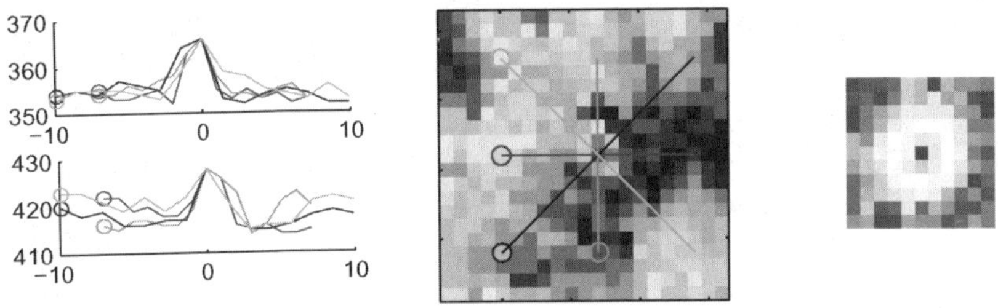

Fig. 1. Point patterns in the digital terrain model. *Left*: Profiles of different mounds, total elevation and distance in pixels (á 90m) indicated. *Center*: "Top view", profile sections indicated. *Right*: Relevant features in the classification of spots and background. The size of the image patch and filter mask are determined by the random forest Gini importance (ranked, gray/white – low/high importance). The central pixel is constant zero for all samples, see text. All figures at the same scale.

Features. Elevation data from circular regions of 1km diameter, centered around the training sites, was used as input for the classifier design (compare geometry of resulting filters: Fig. 2). To remove the absolute elevation, the feature vector contained height differences relative to the center of the image patch. The spatial extensions of the patch and therefore the optimal scale of the detection problem were assessed from the random forest Gini importance ($P = 80$, Fig. 1). Rotational symmetry was assumed for the tell pattern. Accordingly, tell patterns rotated by 90, 180 and 270 degrees were also included in the training set, increasing the number of data points within X_1 to $N_1 = 736$.

3.2 Benchmark

The performance of a number of filters were compared quantitatively: PCA on the event class (PCA_1), PCA, MF, LDA and PLS on both classes (see Table 1). The subspace scores of these filters were used for learning of the following multivariate decision rule.

Random forest (Breiman (2001)) was chosen as decision rule on the various filter responses and was also applied to the original data without intermediate dimension reduction. Random forest models the posterior probability of a class by an ensemble of trees on bootstrapped data sets. In contrast to traditional bagging, only a limited number of features is randomly chosen in the search for the optimal split at each node. Its advantage is the ease and speed of training, while its performance is comparable to other state of the art classifiers, such as support vector machines.

In the error estimation, a tenfold cross-validation over a predefined spatial grid of 60 non-overlapping boxes ($15^2 km^2$ each, covering the Khabur basin) was chosen due to the spatial correlation of the data. Before applying them to the holdout data, filter and classifier were optimized via a fivefold inner cross-validation loop, also over the spatial grid. Within this step, the subspace

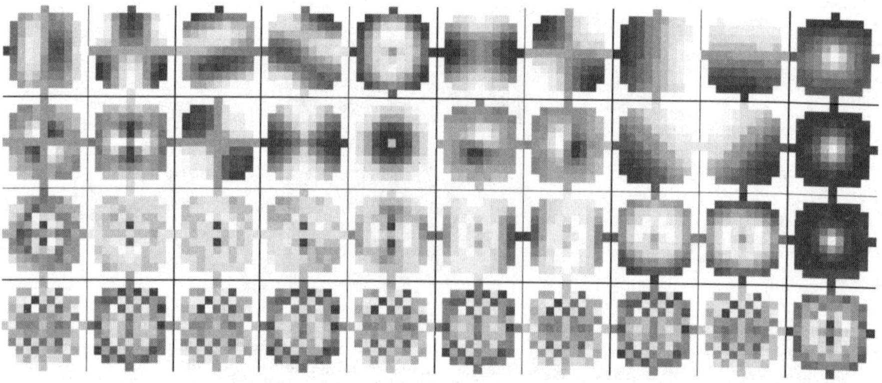

Fig. 2. First ten subspace filters for LDA, PLS, PCA_1, PCA (top to bottom, first filter left).

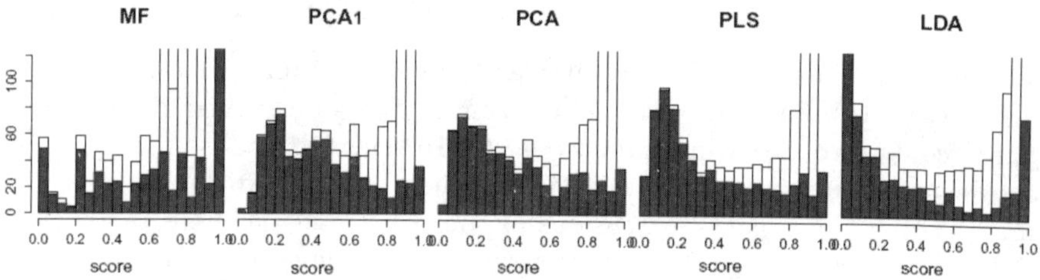

Fig. 3. Distributions of the event signals (gray) and the background (white) from the test data (score: posterior probability). Histograms are truncated, the total number of counts is 50736.

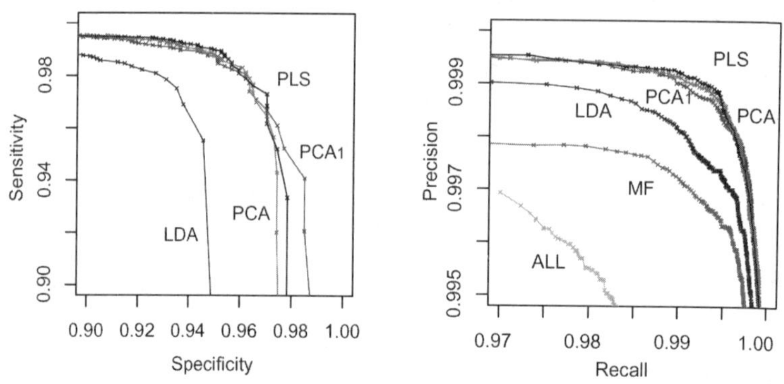

Fig. 4. Classification performance: receiver-operator-characteristic (left), precision-recall-curve (right). "ALL" denotes the direct application of the classifier without subspace filter.

dimensionality was increased from $\lambda = 1, ..., 10$, while the classifier settings were kept unchanged (300 trees, one randomly chosen variable at the nodes).

the error quantification, the area under curve of the receiver operator characteristic (ROC AUC) was used to provide an integrated measure of sensitivity and specificity. In the final evaluation also precision and recall (= sensitivity) were considered, since these measures focus on the event class.

Table 1. Classification accuracy for different thresholds. False negatives (FN) in % of the target class, false positives (FP) in ‰ of the background class (compare to Fig.4).

Thresh	PCA	PCA_1	PLS	LDA	MF	all	Thresh	PCA	PCA_1	PLS	LDA	MF	all
FN .9	7.2	8.2	**7.1**	13.0	36.5	20.7	FP .9	6.7	7.4	**6.1**	8.8	13.5	7.6
FN .95	5.0	4.9	**4.6**	11.0	33.7	16.7	FP .95	13.0	14.0	**12.0**	**12.0**	17.0	12.8
FN .99	2.6	**1.5**	2.6	6.9	25.3	13.6	FP .99	57.0	59.0	48.0	**22.0**	23.4	31.5

4 Results and discussion

Both PCA and PLS result in filter sets whose first component are similar to a *matched filter* (Fig. 2), hence their higher components indeed can be seen as higher dimensional extension to a MF. The performance of the one dimensional MF (Table 1) is exceeded by any multidimensional filter approach, while the direct application of the non-linear classifier to unfiltered data leads to a classification performance surpassed by *any* subspace approach. During resampling, the optimal dimensionality of these filters was between 5 and 7.

The application of *linear discriminant analysis* results in a distinct separation of the data and a nearly binary distribution of the scores (Fig. 3). Falsely classified signals also appear at the tails of the distribution, thus leading to the weak performance of LDA under the ROC and the precision-recall curve. The oscillating checker-board patterns in the filter set (Fig. 2) indicate an overfitting on the highly collinear image data, explaining the comparably bad generalization behavior (Table 1).

Principal component analysis performs very well in both variants (PCA, PCA_1). The distribution of the scores (Fig. 3) shows a higher variance than both PLS and LDA. The orthogonal loadings of PCA_1 are adapted to variants of the central point pattern, while loadings of PCA explain the overall variation (Fig. 2) in the data set. Classification in the PCA subspace controls false positives better than in the PCA_1 subspace (Table 1), while the latter allows the highest specificity/recall (Fig. 4) of all methods at the cost of a somewhat lower overall precision.

The shape of the *partial least squares* feature distribution is in-between the distribution of LDA (max. correlation) and PCA (max. variation), reflecting the intermediate character of PLS. On the present data, PLS is optimal under the precision/recall curve (Fig. 4) and in the control of false positive events, although the differences between PLS and PCA remain faint. –

In our data set, PCA filters obtained from both classes perform nearly as well as PCA filters learned only from the spot class (PCA_1). Based on our experience with similar problems, we argue that this a special feature of the present data set, while in general a good performance of the (two-class) PCA crucially depends on the appropriate choice of the background samples. Accordingly, we recommend to apply PCA_1 if a highly precise representation of the (spot-) pattern is sought and to consider PLS if the use of both classes and an explicit incorporation of background prototypes is desired in the definition of the subspace filters.

While the complementary concepts of Eigen- and Fisherfaces (PCA, LDA) are the most frequently applied in face recognition, we can observe an advantage of the regularized subspace filters (PCA, PLS) on our local image patches, setting the presented low level feature extraction in proximity to chemometrical data analysis rather than classical image processing. We note that the definition of the relevant scale in our detection problem – the extensions of the local image patches – by the multivariate random forest importance is novel.

Applying the PLS filter on the digital elevation model of the geographical region with the available archaeological ground truth (Menze et al. (2006)), it is possible to detect all (regular) settlement mounds higher than 5-6m (85/133) with 327 false positives in a tile of 600*1200 pixels. This allows us to use the presented spot detector in a screening of wide regions of the Near East and for a joint, machine based evaluation with other remote sensing modalities.

5 Conclusions

Extending the idea of a matched filtering (to be followed by a threshold operation) to the training of higher dimensional latent space filters combined with a subsequent nonlinear classifier proves to be a viable concept in the presented spot detection. If a (binary) training data set is available, this approach can be the appropriate choice for a detection of spot patterns in a highly varying background, supporting or replacing traditional parametric spot detectors.

References

BELHUMEUR, P.N. et al. (1996): Eigenfaces vs. Fisherfaces: Recognition Using Class Specific Linear Projections. *Proc ECCV.*

BJORKSTROM, A. (1999): A Generalized View on Continuum Regression. *Scandinavian Journal Statistics, 26, 17-30.*

BOETTICHER, G.D. et al. (2005): A SVM for Protein Spot Detection in 2-dimensional Gel Electrophoresis. *Journal of Computer Science, 1, 355-362.*

BORGA, M. (1997): A Unified Approach to PCA, PLS, MLR and CCA. Technical Report, University of Linkoping, Sweden.

BREIMAN, L. (2001): Random Forests. *Machine Learning, 45, 532.*

FRANK, I.E. and FRIEDMAN, J.H. (1993): A Statistical View of Some Chemometric Regression Tools. *Technometrics, 35, 109-148.*

HASTIE, T. et al. (2001): *The Elements of Statistical Learning.* Springer, New York.

MENZE, B.H., UR, J.A. and SHERRATT, A.G. (2006): Detection of Ancient Settlement Mounds. *Photogrammetric Engineering & Remote Sensing, 72, 321327.*

MOON, T.K. and STIRLING, W.C. (2000): *Mathematical Methods and Algorithms for Signal Processing.* Prentice Hall, New York.

RAHNENFUEHRER, J. and BOZINOV, D. (2004): Hybrid Clustering for Microarray Image Analysis. *BMC Bioinformatics, 5, online.*

SHERRATT, A. (2004): Spotting Tells from Space: *Antiquity, 77, online.*

Identifying and Exploiting Ultrametricity

Fionn Murtagh

Department of Computer Science, Royal Holloway, University of London,
Egham TW20 0EX, England; fmurtagh@acm.org

Abstract. We begin with pervasive ultrametricity due to high dimensionality and/or spatial sparsity. How extent or degree of ultrametricity can be quantified leads us to the discussion of varied practical cases when ultrametricity can be partially or locally present in data. We show how the ultrametricity can be assessed in text or document collections, and in time series signals. In our presentation we also discussed applications to chemical information retrieval and to astrophysics, in particular observational cosmology.

1 Introduction

The topology or inherent shape and form of an object is important. In data analysis, the inherent form and structure of data clouds are important. Quite a few models of data form and structure are used in data analysis. One of them is a hierarchically embedded set of clusters, – a hierarchy. It is traditional (since at least the 1960s) to impose such a form on data, and if useful to assess the goodness of fit. Rather than fitting a hierarchical structure to data, our recent work has taken a different orientation: we seek to find (partial or global) inherent hierarchical structure in data. As we will describe in this article, there are interesting findings that result from this, and some very interesting perspectives are opened up for data analysis.

A formal definition of hierarchical structure is provided by ultrametric topology (in turn, related closely to p-adic number theory). We will return to this in section 2 below. First, though, we will summarize some of our findings.

Ultrametricity is a pervasive property of observational data. It arises as a limit case when data dimensionality or sparsity grows. More strictly such a limit case is a regular lattice structure and ultrametricity is one possible representation for it. Notwithstanding alternative representations, ultrametricity offers computational efficiency (related to tree depth/height being logarithmic in number of terminal nodes), linkage with dynamical or related functional properties (phylogenetic interpretation), and processing tools based on well known p-adic or ultrametric theory (examples: deriving a partition, or applying an ultrametric wavelet transform).

Local ultrametricity is also of importance. Practical data sets (derived from, or observed in, databases and data spaces) present some but not exclusively ultrametric characteristics. This can be used for forensic data exploration (fingerprinting data sets, as we discuss below in section 5). Or, it can be used to expedite search and discovery in information spaces. Indeed we would like to go a lot further, and gain new insights into data (and observed phenomena and events) through ultrametric or p-adic representations. We see this as a program of work for the near future.

2 Quantifying degree of ultrametricity

Summarizing a full description in Murtagh (2004) we explored two measures quantifying how ultrametric a data set is, – Lerman's and a new approach based on triangle invariance (respectively, the second and third approaches described in this section).

The triangular inequality holds for a metric space: $d(x, z) \leq d(x, y) + d(y, z)$ for any triplet of points x, y, z. In addition the properties of symmetry and positive definiteness are respected. The "strong triangular inequality" or ultrametric inequality is: $d(x, z) \leq \max \{d(x, y), d(y, z)\}$ for any triplet x, y, z. An ultrametric space implies respect for a range of stringent properties (Lerman (1981)). For example, the triangle formed by any triplet is necessarily isosceles, with the two large sides equal; or is equilateral.

Firstly, Rammal et al. (1986) used discrepancy between each pairwise distance and the corresponding subdominant ultrametric. Now, the subdominant ultrametric is also known as the ultrametric distance resulting from the single linkage agglomerative hierarchical clustering method. Closely related graph structures include the minimal spanning tree, and graph (connected) components. While the subdominant provides a good fit to the given distance (or indeed dissimilarity), it suffers from the "friends of friends" or chaining effect.

Secondly, Lerman (1981) developed a measure of ultrametricity, termed H-classifiability, using ranks of all pairwise given distances (or dissimilarities). The isosceles (with small base) or equilateral requirements of the ultrametric inequality impose constraints on the ranks. The interval between median and maximum rank of every set of triplets must be empty for ultrametricity. We have used extensively Lerman's measure of degree of ultrametricity in a data set. Taking ranks provides scale invariance. But the limitation of Lerman's approach, we find, is that it is not reasonable to study ranks of real-valued distances defined on a large set of points.

Thirdly, our own measure of extent of ultrametricity (Murtagh (2004)) can be described algorithmically. We assume a Euclidean metric. (In view of the use of scalar product in the definition of an angle, we assume a Hilbert space for our data.) We examine triplets of points (exhaustively if possible, or otherwise through sampling), and determine the three angles formed by the associated triangle. We select the smallest angle formed by the triplet points. Then we check if the other two remaining angles are approximately equal. If they are equal then our triangle is isosceles with small base, or equilateral (when all triangles are equal). The approximation to equality is given by 2 degrees (0.0349 radians). Our motivation for the approximate ("fuzzy") equality is that it makes our approach robust and independent of measurement precision.

Studies are discussed in Murtagh (2004) showing how numbers of points in our clouds of data points are irrelevant; but what counts is the ambient spatial dimensionality. Among cases looked at are statistically uniformly (hence "unclustered", or without structure in a certain sense) distributed points, and statistically uniformly distributed hypercube vertices (so the latter are random 0/1 valued vectors). Using our ultrametricity measure, there is a clear tendency to ultrametricity as the spatial dimensionality (hence spatial sparseness) increases (Murtagh (2004)).

3 Ultrametricity and dimensionality

3.1 Distance properties in very sparse spaces

Murtagh (2004), and earlier work by Rammal et al. (1985, 1986), has demonstrated the pervasiveness of ultrametricity, by focusing on the fact that sparse high-dimensional data tend to be ultrametric. One reason for this is as follows.

As dimensionality grows, so too do distances (or indeed dissimilarities, if they do not satisfy the triangular inequality). The least change possible for dissimilarities to become distances has been formulated in terms of the smallest additive constant needed, to be added to all dissimilarities (Torgerson (1958), Cailliez and Pagès (1976), Cailliez (1983), Neuwirth and Reisinger (1982)). Adding a sufficiently large constant to all dissimilarities transforms them into a set of distances. Through addition of a larger constant, it follows that distances become approximately equal, thus verifying a trivial case of the ultrametric or "strong triangular" inequality. Adding to dissimilarities or distances may be a direct consequence of increased dimensionality.

For a close fit or good approximation, the situation is not as simple for taking dissimilarities, or distances, into ultrametric distances. A best fit solution is given by De Soete (1986) (and software is available in R, Hornik (2005)). If we want a close fit to the given dissimilarities then a good choice would avail either of the maximal inferior, or subdominant, ultrametric; or the minimal superior ultrametric. Stepwise algorithms for these are commonly known as, respectively, single linkage hierarchical clustering; and complete link hierarchical clustering (see Benzécri (1979), Lerman (1981), Murtagh (1985) and other texts on hierarchical clustering).

3.2 Very high dimensions are naturally ultrametric

Bellman's (1961) "curse of dimensionality" relates to exponential growth of hypervolume, and hence complexity, as a function of dimensionality. Problems become tougher as dimensionality increases. In particular problems related to proximity search in high-dimensional spaces tend to become intractable.

In a way, a "trivial limit" (Treves (1997)) case is reached as dimensionality increases. This makes high dimensional proximity search very different, and given an appropriate data structure – such as a binary hierarchical clustering tree – we can find nearest neighbors in worst case $O(1)$ or constant computational time (Murtagh (2004)). The proof is simple: the tree data structure affords a constant number of edge traversals.

The fact that limit properties are "trivial" makes them no less interesting to study. Let us refer to such "trivial" properties as (structural or geometrical) regularity properties (e.g. all points lie on a regular lattice). First of all, the symmetries of regular structures in our data may be of importance. Secondly, "islands" or clusters in our data, where each "island" is of regular structure, may be exploitable. Thirdly, the mention of exploitability points to the application areas targeted: in this article, we focus on search and matching and show some ways in which ultrametric regularity can be exploited in practice. Fourthly, and finally, regularity by no means implies complete coverage (e.g., existence of all pairwise linkages) so that interesting or revealing structure will be present in real data sets.

Thus we see that in very high dimensions, and/or in very (spatially) sparse data clouds, there is no longer a "curse of dimensionality".

4 Approximating local ultrametricity

Now we look at data where some triangles are consistent with ultrametric properties, while others are not.

It has long been known (Chávez et al. (2001), van Rijsbergen (1979)) that forms of data structuring, and more particularly data clustering, can be used to expedite search problems in high dimensions. Some of the work of Chávez and Navarro and their colleagues provides an explanation as to why and how clustering can be exploited for high dimensional proximity search.

In large data sets, i.e. large n or number of observations, a clever way to expedite proximity searching (in particular nearest neighbor finding) in metric spaces is as follows. The metric property implies that the triangular inequality holds. We have a given point and we are looking for its nearest neighbor. We use a third point, called a pivot point. Such a pivot point is carefully selected at the start of the processing, and all necessary distances to it are stored. Through the triangular inequality, we then form a bound on the best potential nearest neighbor distance. Thereby we limit the region within which the search is carried out. See Chávez et al. (2000, 2001, 2003), Bustos et al. (2003)). As pointed out in Murtagh (2004), the bounding rule, or rejection rule, that ensues, is forcing retained triangles to be isosceles. This is interesting because it can be viewed as finding locally ultrametric relationships.

In Chávez et al. (2000, 2001) the ambient spatial dimension is termed the "representational dimension", or embedding dimension, m. (This is dimensionality, m: we have for example $x \in \mathbb{R}^m$.) Search is subject to the curse of dimensionality when addressed in all generality in $\mathbb{R}^m$. However there is often a smaller "intrinsic dimensionality", or average local dimensionality (e.g. when the data are clustered, or lie on a surface of dimension $< m$). This can be exploited to provide fast proximity searching opportunities. However it is difficult in general to define the intrinsic dimensionality.

These authors (Chávez et al. (2000, 2001)) define intrinsic dimensionality of a metric space as: $\rho = \frac{\mu^2}{2\sigma^2}$ where μ and σ^2 are, respectively, the mean and variance of the distances.

So, firstly, the intrinsic dimensionality grows with the mean distance. We have observed that ultrametricity increases with average distance both by simulations in Murtagh (2004), and also through the argument of a simple additive transformation (in section 3.1 above). Secondly, the intrinsic dimensionality grows with inverse variance. Small variance of distances implies equilateral triangles between point triplets, and therefore implies ultrametricity.

We see therefore that the intrinsic dimensionality of Chávez et al. (2000, 2003) affords another definition of ultrametricity. We have already observed how their fast, pivot-based proximity rule can be interpreted as local enforcement of the ultrametric inequality. We conclude from these observations that local or global ultrametricity (i.e., high values of Chávez and Navarro's ρ, or high local contributions to ρ) permit fast proximity search.

5 Increasing ultrametricity through data recoding

5.1 Ultrametricity of time series

In Murtagh (2005a) we use the following coding to show that chaotic time series are less ultrametric than, say, financial, biomedical or meteorological time series; random generated (uniformly distributed) time series data are remarkably similar in their ultrametric properties; and ultrametricity can be used to distinguish various types of biomedical (EEG) signals. Our methodology is empirical: we took 44 time series, and investigated different user parameters.

A time series can be easily embedded in a space of dimensionality m, by taking successive intervals of length m, or a delay embedding of order m. Thus we define points

$$\mathbf{x}_r = (x_{r-m+1}, x_{r-m+2}, \ldots, x_{r-1}, x_r)^t \in \mathbb{R}^m$$

where t denotes vector transpose.

Given any $\mathbf{x}_r = (x_{r-m+1}, x_{r-m+2}, \ldots, x_{r-1}, x_r)^t \in \mathbb{R}^m$, let us consider the set of s such contiguous intervals determined from the time series of overall size n. For convenience we will take $s = \lfloor n/m \rfloor$ where $\lfloor . \rfloor$ is integer truncation. The contiguous intervals could be overlapping but for exhaustive or near-exhaustive coverage it is acceptable that they be non-overlapping. In our work, the intervals were non-overlapping. The quantification of the ultrametricity of the overall time series is provided by the aggregate over s time intervals of the ultrametricity of each $\mathbf{x}_r$, $1 \leq r \leq s$.

We seek to directly quantify the extent of ultrametricity in time series data. In Rammal et al. (1986) and Murtagh (2004) it was shown how increase in ambient spatial dimensionality leads to greater ultrametricity. However it is not satisfactory from a practical point of view to simply increase the embedding dimensionality m insofar as short memory relationships are of greater practical relevance (especially for prediction). The greatest possible value of m is the total length of the time series, n. Instead we will look for an ultrametricity measurement approach for given and limited sized dimensionality m. Our experimental results for real and for random data sets are for "window" lengths $m = 5, 10, \ldots, 105, 110$.

We seek local ultrametricity, i.e. hierarchical structure, by studying the following: Euclidean distance squared, $d_{jj'} = (x_{rj} - x_{rj'})^2$ for all $1 \leq j, j' \leq m$ in each time window, $\mathbf{x}_r$.

We enforce sparseness (Rammal et al. (1985), Rammal et al. (1986), Murtagh (2004)) on our given distance values, $\{d_{jj'}\}$. We do this by linearly approximating each value $d_{jj'}$, in the range $\max_{jj'} d_{jj'} - \min_{jj'} d_{jj'}$, by an integer in $1, 2, \ldots p$. Note that the range is chosen with reference to the currently considered time series window, $1 \leq j, j' \leq m$. Note too that the value of p must be specified. In our work we set $p = 2$. Thus far, the recoded value, $d'_{jj'}$ is not necessarily a distance. With the extra requirement that $d'_{jj'} \longrightarrow 0$ whenever $j = j'$ it can be shown that $d'_{jj'}$ is a metric (Murtagh (2005a)).

To summarize, in our coding, a small pairwise transition is mapped onto a value of 1; and a large pairwise transition is mapped onto a value of 2. A pairwise transition is defined not just for data values that are successive in time but for any pair of data values in the window considered.

This coding can be considered as (i) taking a local region, defined by the sliding window, and (ii) coding pairwise "change" = 2, versus "no change" = 1, relationships. Then, based on these new distances, we use the ultrametric triangle properties to assess conformity to ultrametricity. The average overall ultrametricity in the time series, quantified in this way, allows us to fingerprint our time series.

5.2 Ultrametricity of text

In Murtagh (2006a), words appearing in a text (in principle all, but in practice a set of the few hundred most frequent) are used to fingerprint the text. Rare words in a text corpus may be appropriate for querying the corpus for relevant texts, but such words are of little help for inter-text characterization and comparison. We also use entire words, with no stemming or other preprocessing. A full justification for such an approach to textual data analysis can be found in Murtagh (2005b).

So our methodology for studying a set of texts is to characterize each text with numbers of terms appearing in the text, for a set of terms. The χ^2 distance is an appropriate weighted Euclidean distance for use with such data (Benzécri (1979), Murtagh (2005b)). Consider texts i and i' crossed by words j. Let k_{ij} be the number of occurrences of word j in text i. Then, omitting a constant, the χ^2 distance between texts i and i' is given by $\sum_j 1/k_j(k_{ij}/k_i - k_{i'j}/k_{i'})^2$. The weighting term is $1/k_j$. The weighted Euclidean distance is between the *profile* of text i, viz. k_{ij}/k_i for all j, and the analogous *profile* of text i'. (Our discussion is to within a constant because we actually work on *frequencies* defined from the numbers of occurrences.)

Correspondence analysis allows us to project the space of documents (we could equally well explore the terms in the *same* projected space) into a Euclidean space. It maps the all-pairs χ^2 distance into the corresponding Euclidean distance. In the resulting factor space, we use our triangle-based approach for quantifying how ultrametric the data are.

We did this for a large number of texts (novels – Jane Austen, James Joyce, technical reports – airline accident reports, fairy tales – Brothers Grimm, dream reports, Aristotle's *Categories*, etc.), finding consistent degree of ultrametricity results over texts of the same sort.

Some very intriguing ultrametricity characterizations were found in our work. For example, we found that the technical vocabulary of air accidents did not differ greatly in terms of inherent ultrametricity compared to the Brothers Grimm fairy tales. Secondly we found that novelist Austen's works were distinguishable from the Grimm fairy tales. Thirdly we found dream reports to be have higher ultrametricity level than the other text collections.

5.3 Data recoding in the correspondence analysis tradition

If the χ^2 distance (see above, section 5.2) is used on data tables with constant marginal sums then it becomes a weighted Euclidean distance. This is important for us, because it means that we can directly influence the analysis by equi-weighting, say, the table rows in the following way: we double the row vector values by including an absence (0 value) whenever there is a presence (1 value) and vice versa. Or for a table of percentages, we take both the original value x and $100 - x$. In the correspondence analysis tradition (Benzécri (1979), Murtagh (2005b)) this is known as *doubling* (*dédoublement*).

More generally, booleanizing, or making qualitative data in this way, for a varying (value-dependent) number of target value categories (or modalities) leads to the form of coding known as *complete disjunctive form*.

Such coding increases the embedding dimension, and data sparseness, and thus may encourage degree of ultrametricity. That it can do more we will now show.

The iris data has been very widely used as a toy data set since Fisher used it in 1936 (taking from a 1935 article by Anderson) to exemplify discriminant analysis. It consists of 150 iris flowers, each characterized by 4 petal and sepal, width and breadth, measurements. On the one hand, therefore, we have the 150 irises in $\mathbb{R}^4$. Next, each variable value was recoded to be a rank (all ranks of a given variable considered) and the rank was boolean-coded (viz., for the top rank variable value, $1000\ldots$, for the second rank variable value, $0100\ldots$, etc.). Following removal of zero total columns, the second data set defined the 150 irises in $\mathbb{R}^{123}$. Actually, this definition of the 150 irises is in fact in $\{0,1\}^{123}$.

Our triangle-based measure of the degree of ultrametricity in a data set (here the set of irises), with $0 =$ no ultrametricity, and $1 =$ every triangle an ultrametric-respecting one, gave the following: for irises in $\mathbb{R}^4$, 0.017; and for irises in $\{0,1\}^{123}$: 0.948.

This provides a nice illustration of how recoding can dramatically change the picture provided by one's data. Furthermore it provides justification for data recoding if the ultrametricity can be instrumentalized by us in some way (e.g. to facilitate fast proximity search).

6 Conclusions

It has been our aim in this work to link observed data with an ultrametric topology for such data. The traditional approach in data analysis, of course, is to impose structure on the data. This is done, for example, by using some agglomerative hierarchical clustering algorithm. We can always do this (modulo distance or other ties in the data). Then we can assess the degree of fit of such a (tree or other) structure to our data.

For our purposes, here, this is unsatisfactory.

Firstly, our aim was to show that ultrametricity can be naturally present in our data, globally or locally. We did not want any "measuring tool" such as an agglomerative hierarchical clustering algorithm to overly influence this finding. (Unfortunately Rammal et al. (1986) suffers from precisely this unhelpful influence of the "measuring tool" of the subdominant ultrametric.)

Secondly, let us assume that we did use hierarchical clustering, and then based our discussion around the goodness of fit. This again is a traditional approach used in data analysis, and in statistical data modeling. But such a discussion would have been unnecessary and futile. For, after all, if we have ultrametric properties in our data then many of the widely used hierarchical clustering algorithms will give precisely the same outcome, and furthermore the fit is by definition exact.

In linking data with an ultrametric view of it we have, in this article, proceeded a little in the direction of exploiting this achievement. While some applications, like discrimination between time series signals, or texts, have been covered here, other areas have just been opened up, e.g. search and discovery in massive biochemical databases; hierarchical structures in cosmology (Murtagh (2006b)); and automated ontology creation for semantic web applications. In the distance there looms the challenge of analysis of networks of enormous size (internet, or biological). There is a great deal of work to be accomplished.

References

BELLMAN, R. (1961): *Adaptive Control Processes: A Guided Tour.* Princeton University Press, Princeton.

BENZÉCRI, J.P. (1979): *L'Analyse des Données, Tome I Taxinomie, Tome II Correspondances.* 2nd ed., Dunod, Paris.

BUSTOS, D., NAVARRO, G. and CHÁVEZ, E. (2003): Pivot Selection Techniques for Proximity Searching in Metric Spaces. *Pattern Recognition Letters, 24, 2357–2366.*

CAILLIEZ, F. and PAGÈS, J.P. (1976): *Introduction à l'Analyse de Données.* SMASH (Société de Mathématiques Appliquées et de Sciences Humaines), Paris.

CAILLIEZ, F. (1983): The Analytical Solution of the Additive Constant Problem. *Psychometrika, 48, 305–308.*

CHÁVEZ, E. and NAVARRO, G. (2000): Measuring the Dimensionality of General Metric Spaces. Technical Report TR/DCC-00-1, Department of Computer Science, University of Chile.

CHÁVEZ, E., NAVARRO, G., BAEZA-YATES, R. and MARROQUÍN, J.L. (2001): Proximity Searching in Metric Spaces. *ACM Computing Surveys, 33, 273–321.*

CHÁVEZ, E. and NAVARRO, G. (2003): Probabilistic Proximity Search: Fighting the Curse of Dimensionality in Metric Spaces. *Information Processing Letters, 85, 39–56.*

DE SOETE, G. (1986): A Least Squares Algorithm for Fitting an Ultrametric Tree to a Dissimilarity Matrix. *Pattern Recognition Letters, 2, 133–137.*

FISHER, R.A. (1936): The Use of Multiple Measurements in Taxonomic Problems. *The Annals of Eugenics, 7, 179–188.*

HORNIK, K. (2005): A CLUE for CLUster Ensembles. *Journal of Statistical Software, 14, 12.*

LERMAN, I.C. (1981): *Classification et Analyse Ordinale des Données.* Dunod, Paris.

MURTAGH, F. (1985): *Multidimensional Clustering Algorithms.* Physica, Würzburg.

MURTAGH, F. (2004): On Ultrametricity, Data Coding, and Computation. *Journal of Classification, 21, 167–184.*

MURTAGH, F. (2005a): Identifying the Ultrametricity of Time Series. *European Physical Journal B, 43, 573–579.*

MURTAGH, F. (2005b): *Correspondence Analysis and Data Coding with R and Java.* Chapman & Hall/CRC, Florida.

MURTAGH, F. (2006a): A Note on Local Ultrametricity in Text, *Literary and Linguistic Computing*, submitted.

MURTAGH, F. (2006b): From Data to the Physics using Ultrametrics: New Results in High Dimensional Data Analysis. In: A.Yu. Khrennikov, Z. Rakić and I.V. Volovich (Eds.): *p-Adic Mathematical Physics*, American Institute of Physics Conf. Proc. Vol. 826, 151–161.

NEUWIRTH, E. and REISINGER, L. (1982): Dissimilarity and Distance Coefficients in Automation-Supported Thesauri. *Information Systems, 7, 47–52.*

RAMMAL, R., ANGLES D'AURIAC, J.C. and DOUCOT, B. (1985): On the Degree of Ultrametricity. *Le Journal de Physique – Lettres, 46, L-945–L-952.*

RAMMAL, R., TOULOUSE, G. and VIRASORO, M.A. (1986): Ultrametricity for Physicists. *Reviews of Modern Physics, 58, 765–788.*

TORGERSON, W.S. (1958): *Theory and Methods of Scaling*, Wiley, New York.

TREVES, A. (1997): On the Perceptual Structure of Face Space. *BioSystems, 40, 189–196.*

VAN RIJSBERGEN, C.J. (1979): *Information Retrieval*, 2nd ed., Butterworths.

Factor Analysis for Extraction of Structural Components and Prediction in Time Series

Carsten Schneider and Gerhard Arminger

Fachbereich B - Wirtschaftsstatistik,
Bergische Universität Wuppertal, 42097 Wuppertal, Germany;
{schneider, arminger}@wwst09.wiwi.uni-wuppertal.de

Abstract. In this paper, factor analysis is used for the dimensional reduction of complex time series. If the structure within data is too complex to use e.g. ARIMA-models, factor analysis can be used for simplification without relevant loss of explained variation. The result are data with simple structure that can be forecasted by a standard prediction model. To give an example for this approach we predict the electricity demand per quarter of an hour of industrial customers in Germany. The data have a rather complex structure with 96 observations per day and possibly different cyclical variations during the day regarding different weekdays.

1 Introduction and problem description

In 1998 the German electricity market was deregulated by a new energy-market law (EnWG (1998)). As a result, all customers, industrial as well as private consumers, can freely choose their energy provider. This opening of the market even to distributors without their own electricity network or without any power plants results in a serious difficulty: the whole physical infrastructure remains in the possession of the original energy distributors with their former regional monopolies. All providers of power without an own power grid must have the possibility to use foreign networks and have to pay a certain amount for this use (Burmeister (2006)). A new European Guideline (2003/54/EG) has been transposed to national law by a modified german energy-market law in 2005, but both of these modifications do not result in changes to the technical requirements to predictions of electricity demand.

As electricity is not storable over time, it must be assured that the demand for and the supply of power are equal at any time (Laumanns (2005)). To guarantee the performance of the electricity networks, the regional providers need to know who will send how much power through their net at any point in time and will receive a fixed amount of money for every unit of power that is sent through their network (Fritz and König (2000)). Therefore, electricity providers have to predict the demand of their customers, not only for all customers but for every regional network. For simplification, the electricity demand in this region is measured every quarter of an hour and it is assumed to be constant during the following 15 minutes. The data consists therefore of a multivariate time series for each day and 96 observation points per day which have to be predicted.

We will introduce a prediction method that uses factor analysis for the recognition of the structural daily pattern in the electricity demand of any regional group of customers. Using factor analysis decreases the dimension of the multivariate time series, that is the number of daily data points, and makes the prediction more stable. The identified factors are predicted using SAR-models and finally the procedure of calculating factors is reversed, as valid forecasts have to be made quarter-hourly for the following day. We compare the forecasts of this procedure to those of standard time series methods.

2 Factor analysis for the representation of the daily structure

As basic data for the following factor analysis we assume a homogeneous daily pattern of electricity demand. Therefore all weekdays can be regarded separately or can be classified in a preprocessing step to day types d (Schneider et al. (2005)). For the followings steps structural homogeneity is assumed, which means that the factor analysis and the prediction have to be computed separately for every day type. Therefore the index $d = 1, ..., D$ with D as maximal number of regarded weekdays will be disregarded for simplicity in the following.

For each day type d, there exists an intradaily cyclical pattern of electricity demand y_{ij} where i denotes the observation day ($i = 1, ..., n$) and j denotes the measurement time in quarters of an hour, $j = 1, ..., m$ with $m = 96$ (Harvey and Koopman (1993)). In the first step, the demands are standardized by using mean $\bar{y}_j$ and standard deviation s_j as follows:

$$z_{ij} = \frac{y_{ij} - \bar{y}_j}{s_j} \tag{1}$$

This standardization is done for all days within a day type d. In the second step, the correlation matrix of the y_{ij} is computed using the standardized values:

$$r_{jk} = \frac{1}{n-1} \sum_{i=1}^{n} z_{ij} z_{ik}, \quad j, k = 1, ..., m, \ j \neq k \tag{2}$$

The result is summarized in the correlation matrix $\mathbf{R} = \frac{1}{n-1}\mathbf{Z}\mathbf{Z}'$. This is the 96×96 correlation matrix of the intradaily consumption demands for each day type $d = 1, ..., D$.

The basic idea of factor analysis is to reduce the m observed variables (demands at times $j = 1, ..., 96$) to as few as possible latent (hypothetical) variables that reproduce the stochastic dependencies and therefore the correlation matrix $\mathbf{R}$ of the standardized observations (Arminger (1979)). The factor analytic model is described by the linear combination of $Q \leq m$ factor scores p_{iq} and factor loadings a_{jq}:

$$z_{ij} = \sum_{q=1}^{Q} a_{jq} p_{iq}, \quad i = 1, ..., n, \ j = 1, ..., m \text{ and } q = 1, ..., Q \tag{3}$$

with

p_{iq} : factor score for factor q concerning object i

a_{jq} : factor loading of variable j for factor q

For identification, it is assumed that the factor scores themselves are standardized, that is with mean value $\bar{p}_q = 0$ and standard deviation $s_q = 1, \ q = 1, ..., Q$. Consequently the covariance (and correlation) matrix of the factors is the $Q \times Q$ identity matrix $\mathbf{I}_Q$. The properties of the factor model can more easily be seen using the matrix formulation

$$\mathbf{Z} = \mathbf{P}\mathbf{A}' \tag{4}$$

where $\mathbf{Z}$ is the $n \times m$ matrix of standardized observations, $\mathbf{P}$ is the $n \times Q$ matrix of factor scores p_{iq} and $\mathbf{A}$ is the $m \times Q$ matrix of factor loadings a_{qj}. The special adaption of the factor model for this kind of time series is the interpretation of the electricity demand per quarter hour as variable j and of the day as object i. As a result of the standardization and non-correlation of the factors, the correlation matrix $\mathbf{R}$ of observations may be written as

$$\mathbf{R} = \frac{1}{n-1}\mathbf{Z}\mathbf{Z}' = \frac{1}{n-1}\mathbf{A}\mathbf{P}'\mathbf{P}\mathbf{A}' = \mathbf{A}\mathbf{I}_Q\mathbf{A}' = \mathbf{A}\mathbf{A}' \tag{5}$$

Therefore, $\mathbf{R}$ is only a function of the factor loading matrix $\mathbf{A}$. This is the so called "*Fundamental Concept of the Factor Analysis*" (Thurstone (1941), Kim and Mueller (1994)) which is only applicable to explanatory factor analysis. Note that the factor loadings are only identified up to an orthogonal transformation $\mathbf{T} \sim Q \times Q$ with $\mathbf{T}\mathbf{T}' = \mathbf{I}_Q$ because the transformation of $\mathbf{A}$ into $\mathbf{A}^\star = \mathbf{A}\mathbf{T}$ yields the same result ($\mathbf{A}^\star\mathbf{A}^{\star\prime} = \mathbf{A}\mathbf{T}\mathbf{T}'\mathbf{A}' = \mathbf{A}\mathbf{A}'$).

From equation 5 one can deduce immediately that the variance of each standardized observation variable may be written as the sum of squared factor loadings, that is

$$1 = \sigma_j^2 = \sum_{q=1}^{Q} a_{jq}^2 \tag{6}$$

If Q is set to m, the decomposition of $\mathbf{R}$ into Q factors in $\mathbf{R} = \mathbf{A}\mathbf{A}'$ is called principal components analysis (Jolliffe (1986)). In this case, the variance is completely explained by the Q factors (components). However, very often only a few components, for instance Q_1, suffice to explain a great proportion of the existing variance. If these components are collected as the first Q_1 factors, then the partial sum

$$h^2_{j,Q_1} = \sum_{q=1}^{Q_1} a^2_{jq} \leq 1 \tag{7}$$

is called "*Communality*" (variance explained by the first Q_1 factors). The remaining $Q - Q_1$ factors are interpreted as specific factors comparable to the function of an error term in regression analysis.

In the case of modelling the intradaily cyclical variation, one aims to reduce the complexity of 96 data points down to only as few as possible common factors that determine the main features of the observed variation. To calculate the number of common factors one starts with $h^2_{j,Q_1} = 1$ and computes a principle components analysis with $Q = m$ components as first solution. The algorithms for this principle components analysis are found in Jolliffe (1986).

The next aspect is determining the right number of extracted factors for further use in a model of electricity demand. This question is equivalent to finding the optimal balance between the maximization of that part of the variance in the data that can be explained by the extracted factors [Q_1 increasing] versus the simplification of the used model [Q_1 decreasing]. As seen above, the communality in equation (7) describes the part of the variance of the variable of interest explained by Q_1 factors. On the other hand the part of the explained variance by the one special factor q concerning all variables is given by the squared eigenvalue:

$$\lambda^2_q = \sum_{j=1}^{m} a^2_{jq} \quad \forall \;\; q = 1, ..., Q \tag{8}$$

Following the KAISER-criterion only factors q with eigenvalues $\lambda_q > 1$ are chosen (Webb (1999, p. 254)). This means that the factor q explains a greater part of the variance in the data than a single variable does. All other factors with lower explanatory power and a corresponding eigenvalue $\lambda_q < 1$ are not extracted for the final model. After this identification of relevant factors the matrix $\mathbf{A}$ of factor loadings is reduced from Q to Q_1 factors. Additionally, the factors are now sorted by relevance which means that the loadings of the factor q with the greatest eigenvalue build the first column of the new matrix $\mathbf{A}^\star \sim m \times Q_1$ (Schneider (2002)).

The factor extraction for modelling the daily demand pattern has to be calculated for every classified day type and will be based on a quarter of an hour average electricity demand over all days belonging to this day type. In this case the factor values p_{iq} for every day of type d in the prediction period of length h, $h = 1, ..., H$ have to be calculated by solving

$$z_{ij} = \sum_{q=1}^{Q_1} a^\star_{jq} p_{iq} + \eta_{ij} \tag{9}$$

In this model the non-explained part of the variance is interpreted as error term η_{ij}. If for example Monday and Tuesday belong to the same day type d, this will result in identical factor values for all Mondays and Tuesday within the prediction period for all possible forecast horizons H.

3 Forecasting using SAR-models

Our final aim is the prediction of the original time series. So far, the structure of this time series is simplified by calculating factor values and factor loadings. The few factors that represent the variation during the day instead of 96 time points now have to be forecasted. Seasonal Autoregressive Models (Ghysels and Osborn (2001)) are used to predict the factor values, the seasonal component here refers to the weekly structure and does not consider any intradaily behavior of the data.

For the prediction step based on the factor analysis the factor values for every day type d have to be predicted. It is assumed that the prediction horizon h is not larger than the seasonal length s. The following seasonal autoregressive model $SAR(1,1)_7$ with factor-specific parameters $\rho_{1,q}$ and $\rho_{2,q}$ serves as forecasting model for the factor scores $p_{i,q}$ with a seasonal length of $s = 7$ as weekly seasonality may be regarded apart from the identification of day types.

$$p_{i+h,q} = \rho_{1,q} p_{i,q} + \rho_{2,q} p_{i+h-s,q} + \varepsilon_{i+h,q} \tag{10}$$

The specification of the order of the SAR-model is motivated by the possible correlations and the contents of the observed data. The estimation of parameters is calculated using the OLS-estimator and is computed separately for every extracted factor $q = 1, ..., Q_1$. The results are Q_1 predictions $\hat{p}_{i+h,q}$ for every day $i + h$. If any additional information like for instance holidays or extended machinery is available as external regressor for several times, this information can be included by using an SARX-model instead.

Forecasts of electricity demand have to be calculated for every quarter hour of the following day. Therefore the predicted factor values have to be recomputed to 96 separated prediction values per day. For this reversion the predicted model according to equation (9) is written as

$$\hat{z}_{i+h,j} = \sum_{q=1}^{Q_1} a^{\star}_{j,q} \hat{p}_{i+h,q} \tag{11}$$

These predictions have to be corrected by reversing the standardization which has been done for computing the factor analysis. Therefore the predicted electricity demand

$$\hat{y}_{i+h,j} = \hat{z}_{i+h,j} s_j + \bar{y}_j \tag{12}$$

is calculated for every quarter hour $j = 1, ..., 96$ during the predicted days $i + 1, ..., i + h$.

4 Results

Before the calculation of dimensional reduction and prediction can be done, the observed data is analyzed in a preprocessing step concerning the homogeneity of the daily electricity demand. As first result it is assumed that for all weekdays (Monday to Friday) the demand has the same daily pattern. Therefore these five days are merged as one day type. In this example the weekend is not regarded furthermore. The data then consists of the electricity demand during 250 days, each day having 96 time points. If the data would be regarded as one longitudinal time series it had 24000 time points. The structure would be a double seasonal behavior of weekly seasonality regarding the weekdays and intradaily seasonality regarding the time points during the day.

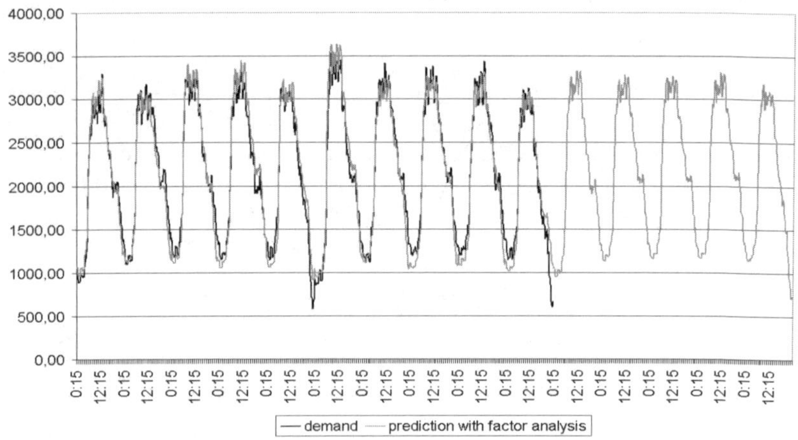

Fig. 1. Comparison of demand and prediction using factor analysis and SAR-model

The dimensional reduction using the Kaiser criterion for the factor analysis results in 8 factors explaining 90.2% of the variance in the data for the weekdays. This leads to a dimensional reduction of 91.7% compared to 96 time points that had to by regarded otherwise. The loss of information therefore is less than 10% which should be negligible. These 8 factors are used for prediction and the resulting forecasts are shown in Figure 1. Note that the weekends are missing, the data for Friday are followed immediately by the data for Monday.

As measure for the prediction fit the mean absolute percentage error (MAPE) is chosen

$$MAPE = \frac{1}{m \cdot h} \sum_{i=t+1}^{t+h} \sum_{j=1}^{m} \frac{|y_{i,j} - \hat{y}_{i,j}|}{y_{i,j}} \tag{13}$$

The 5 step-ahead prediction using factor analysis and the introduced SAR-model results in forecasts for the weekdays of the next week. These forecasts can be compared to the results of practically used simple forecasting models like some kind of exponential smoothing or a simple seasonal AR-model. The result of computing the forecasts for one week (5 workdays) and calculating the MAPE is given in Table 1.

Table 1. Comparison of forecasting procedures

model	MAPE
factor + $SAR(1,1)_5$	7.47 %
$SAR(1,1)_{96}$	20.87 %
exponential smoothing	10.40 %

Exponential smoothing as it is used by electricity providers due to its simplicity separates the time series into 96 completely unlinked series and calculates forecasts neglecting any correlations between originally neighboring observations

$$\hat{y}_{i+1,j} = \alpha_j y_{i,j} + (1-\alpha_j)\hat{y}_{i,j} \ , \ j = 1, ..., m \ . \tag{14}$$

In this framework the forecasting horizon stays 5 but the procedure has to be computed $m = 96$ times.

An alternative for practical use is the simple AR-model. This model regards the electricity demand as one time series y_τ with 15 minutes distance between two neighboring observations. Therefore 480 steps refer to forecasts for the weekdays of a whole week. For modelling the seasonal behavior during the day a seasonal AR-model with $s = 96$ has to be computed

$$\hat{y}_{\tau+1} = \hat{\phi}_1 y_\tau + \hat{\phi}_2 y_{\tau-96} \ . \tag{15}$$

The second dimension of seasonality would be a lag of length 480, referring to the same quarter hour one week ago, but unfortunately this length is hardly computable using standard procedures. A standard program treats the weekly season as daily season of length 5 in this case which leads to computational problems. An additional disadvantage of this model is that different kinds of correlations between neighboring observations, e.g. basic load during the night or arising demand in the morning, can hardly be modelled in this simple SAR-framework.

5 Conclusion

The error rates using the proposed dimensional reduction instead of the other practically implemented forecasting procedures decrease. It takes into account the correlations between neighboring observations and holds the forecasting horizon for the time series procedure as short as possible. Furthermore, the simple SAR-model used to predict the factor scores can easily be supplemented by additional information if available. This information such as holidays or new machinery can be added as exogenous variable.

If an interpretable structure of the data is needed additionally, for example to identify breaks or to separate shift times from basic demand times, the factor analysis can be complemented by a factor rotation. In this case, the dimensional reduction and the identified daily pattern stay identical but the factor loadings become a structure that can be interpreted easily. If for example oblimin rotation (Arminger (1979, p. 100)) is used for our data, the basic load during the night is represented by one factor and the arising demand in the morning is represented by a second factor. The representation in this context can be seen by relative high values for the factor loadings of the representing factors for several interrelated quarter hours and low loadings for all other times.

References

ARMINGER, G. (1979): *Faktorenanalyse.* Teubner, Stuttgart.

BURMEISTER, T. (2006): Netznutzung und Bilanzkreissystem. In: K.-P. Horstmann and M. Cieslarczyk (Eds.): *Energiehandel.* Heymanns, Köln.

EnWG (1998): *Energiewirtschaftsgesetz vom 24.04.1998.* Bundesgesetzblatt, Berlin.

FRITZ, W. and KÖNIG, S. (2000): Der liberalisierte Strommarkt - eine Einführung. In: M. Kahmann and S. König (Eds.): *Wettbewerb im liberalisierten Strommarkt.* Springer, Berlin.

GHYSELS, E. and OSBORN, D.R. (2001): *The Econometric Analysis of Seasonal Time Series.* Cambridge University Press, Cambridge.

HARVEY, A.C. and KOOPMAN, S.J. (1993): Forecasting Hourly Electricity Demand Using Time-Varying Splines. *Journal of the American Statistical Association, 88, 425, 1228–1236.*

JOLLIFFE, I.T. (1986): *Principle Component Analysis.* Springer, New York.

KIM, J.-O. and MUELLER, C.W. (1994): Introduction to Factor Analysis. In: M.S. Lewis-Beck (Ed.): *Factor Analysis and Related Techniques.* Sage, London.

LAUMANNS, U. (2005): Technische Grundlagen der Energiepolitik. In: D. Reiche (Ed.) *Grundlagen der Energiepolitik.* Lang, Frankfurt/Main.

SCHNEIDER, C. (2002): *Kostenoptimale Prognose von Lasten in der Energiewirtschaft.* Eul, Lohmar.

SCHNEIDER, C., ARMINGER, G. and SCHWARZ, A. (2005): Using Analysis of Variance and Factor Analysis for the Reduction of High Dimensional Variation in Time Series of Energy Consumption. *Allgemeines Statistisches Archiv, 89, 4, 403–418.*

THURSTONE, L. (1941): *Factorial Studies of Intelligence.* University of Chicago Press, Chicago/Ill.

WEBB, A. (1999): *Statistical Pattern Recognition.* Arnold, London.

Classification of the U.S. Business Cycle by Dynamic Linear Discriminant Analysis

Roland Schuhr

Institut für Empirische Wirtschaftsforschung, Universität Leipzig,
04109 Leipzig, Germany; schuhr@wifa.uni-leipzig.de

Abstract. Linear discriminant analysis (LDA) was well established by Meyer and Weinberg and by Heilemann and Münch as a technique for the analysis of business cycles. The technique, however, ignores the chronological order of the underlying time series data. This paper presents a dynamic version (DLDA) of linear discriminant analysis and a dynamic measure of separation as additional instruments for business cycle analysis.[1]

1 Introduction

The economic literature of the 20th century includes several works about the decomposition of the business cycle into different successive phases, see e.g. Tichy (1994). In 1975 Meyer and Weinberg proposed a 4-phase decomposition scheme for the United States business cycle, including the phases 'recession', 'recovery', 'demand pull' and 'stagflation'. The authors presented classification results - i.e. the assignment of months to cycle phases - for the post World War II period. For classification as well as for selection of the classifying variables they used linear discriminant analysis. Recently Heilemann and Münch (2002, 2005) updated the classification results up to the year 2000.

The decomposition of the business cycle into phases and the multivariate representation of the cycle by a set of classifying variables offer a deeper understanding of economic fluctuations and may help to determine 'stylized facts' regarding the subject. The empirical findings by Heilemann and Münch (2005) suggest, however, that the significance of the phases and the classification power of the variables change in course of time. Unfortunately classical linear discriminant techniques are constructed to analyze cross-sectional data and are only of limited value in analyzing the development of a time-dependent phenomenon. To overcome limitations to obtaining a more detailed empirical analysis of changes in the structure of the U.S. business cycle, this paper

[1] The author wishes to thank Ullrich Heilemann, Universität Leipzig, for his comments and his support and the anonymous referees for their comments.

presents a dynamic version of linear discriminant analysis and a dynamic measure of the separation power of the classifying variables.

2 Dynamic linear discriminant analysis (DLDA)

There are two different approaches to justifying linear discriminant techniques. Fisher (1936) employed the least squares approach in order to construct linear functions which project multivariate observations from different classes onto a low-dimensional discriminant space such that the between-class variance is maximized relative to the within-class variance. The resulting classification procedure is distribution-free, but it implies a common covariance structure for all classes, because a pooled estimate of the within-class covariance matrix is used. A second approach based on a Bayes procedure was proposed by Wald (1944). An observation point with unknown class membership is assigned to the class with maximum a posteriori class probability. Wald applied the Bayes theorem under the assumption of normally distributed observations within each class. The multivariate normal distributions have class-specific mean vectors and a common covariance matrix. In the case of uniform a priori class probabilities, Wald's procedure is statistically equivalent to Fisher's distribution-free procedure (see e.g. Wasserman (2004, p. 356) for a proof concerning the special case of discrimination into one of two classes).

Fisher and Wald examined the problem of classification with cross-sectional data. If we use classical linear discriminant analysis (LDA) for business phase classification, the techniques will ignore the chronological order of the underlying time series data. Therefore, Wald's approach will be modified here to a dynamic linear discriminant analysis procedure (DLDA). DLDA differs from other extensions of LDA to time series data (see e.g. Shumway (1982)) that assign a stationary time series to one of several classes of stationary time series rather than a single time period to one of several phases of an evolutionary economic process.

The specific problem of classification can be outlined as follows. Let

$$\mathbf{x}_t' = (X_{1t}, X_{2t}, \ldots, X_{pt}) \qquad (t = 1, 2, 3, \ldots)$$

denote a multiple time series. The vector $\mathbf{x}_t$ is a vector of classifying variables $X_{1t}, \ldots, X_{pt}$ representing the multivariate structure of the business cycle in time period t (e.g. a month). Based on an observed value of $\mathbf{x}_t$ it is to decide whether the period t of the observation corresponds to phase 1, 2, 3 or 4 of the cycle (1 = recovery, 2 = demand-pull, 3 = stagflation, 4 = recession).

DLDA solves the classification problem under the assumption of $N(\boldsymbol{\mu}_{kt}, \boldsymbol{\Sigma}_t)$ normally distributed vectors $\mathbf{x}_t$. The mean vectors $\boldsymbol{\mu}_{kt}$ $(k = 1, \ldots, 4)$ vary from phase to phase of the business cycle. The covariance matrices $\boldsymbol{\Sigma}_t$ are phase-invariant. In contrast to classical LDA, means and covariances change through time. According to the Bayesian decision rule, a time period t will be assigned to the cycle phase k with maximum a posteriori phase probability

$$P(k|\mathbf{x}_t) = \frac{\pi_{kt} \cdot p_{kt}(\mathbf{x}_t)}{\sum_{i=1}^{4} \pi_{it} \cdot p_{it}(\mathbf{x}_t)} \qquad (k = 1, \ldots, 4), \tag{1}$$

where $p_{kt}(\bullet)$ is the density of the multivariate normal distribution with parameters $\boldsymbol{\mu}_{kt}$ and $\boldsymbol{\Sigma}_t$. π_{kt} denotes the a priori probability of phase k in time period t. Equivalently, the linear discriminant functions can be evaluated

$$\delta_{kt} = -\frac{1}{2}\boldsymbol{\mu}'_{kt}\boldsymbol{\Sigma}_t^{-1}\boldsymbol{\mu}_{kt} + \boldsymbol{\mu}'_{kt}\boldsymbol{\Sigma}_t^{-1}\mathbf{x}_t + \ln \pi_{kt} \qquad (k = 1, \ldots, 4). \tag{2}$$

Period t will be assigned to phase k with maximum value δ_{kt}.

The distribution parameters $\boldsymbol{\mu}_{kt}\,(k = 1, \ldots, 4)$ and $\boldsymbol{\Sigma}_t$ are estimated recursively by exponential smoothing techniques. For any period t the observed values of $(\mathbf{x}_\tau, y_\tau)$ $(\tau = 1, 2, \ldots, t-1)$ are used as the information set. The variable y_τ with value set $\{1, 2, 3, 4\}$ indicates the a priori classification of a former period $\tau < t$ with respect to the cycle phase. Therefore, the estimation results in period t are conditioned to past observations of the time series, and the classification of period t becomes an ex ante classification. Let

$$\hat{\boldsymbol{\mu}}_{kt} := \mathbf{m}_{k,t-1} \quad , \quad \hat{\boldsymbol{\Sigma}}_t := \mathbf{S}_{t-1} \tag{3}$$

denote the estimators of $\boldsymbol{\mu}_{kt}\,(k = 1, \ldots, 4)$ and $\boldsymbol{\Sigma}_t$. The smoothed statistics $\mathbf{m}_{k,t-1}\,(k = 1, \ldots, 4)$ and $\mathbf{S}_{t-1}$ follow the recursion equations

$$\mathbf{m}_{k\tau} = \begin{cases} \alpha \cdot \mathbf{x}_\tau + (1-\alpha) \cdot \mathbf{m}_{k,\tau-1} & \text{, if } y_\tau = k \\ \mathbf{m}_{k,\tau-1} & \text{, otherwise} \end{cases} \qquad (\tau = 1, \ldots, t-1) \tag{4}$$

and

$$\mathbf{S}_\tau = \beta \cdot \mathbf{z}_\tau \mathbf{z}'_\tau + (1-\beta) \cdot \mathbf{S}_{\tau-1} \qquad (\tau = 1, \ldots, t-1) \tag{5}$$

with $\mathbf{z}_\tau = \mathbf{x}_\tau - \mathbf{m}_{k\tau}$ if $y_\tau = k$. The equations are initiated by some starting values $\mathbf{m}_{k0}$ $(k = 1 \ldots 4)$ and $\mathbf{S}_0$. The smoothed statistics are weighted moving averages of past observations. The weights decrease exponentially with the age of the data. The smoothing parameters α and β control the speed of the decrease. These constants must be fixed carefully in the ranges $0 < \alpha < 1$ and $0 < \beta < 1$, respectively. Large values cause the smoothed statistics to react quickly to changes of the mean and of the covariance structure of the time series - but also to random fluctuations. The smaller the values, the slower the response. A reasonable procedure is to carry out DLDA on the data set for a fine grid of different parameter values and choose the values that minimize the number of classification errors. Typically, the procedure selects parameter values significantly less than 0.1.

DLDA should not only classify time periods, but also inform one about the separation (classification) power of the classifying variables. In context of the analysis of variance, the Lawley-Hotelling trace criterion is a well-known multivariate measure of separation (see Ahrens and Läuter (1974, p. 108)).

The measure can easily be adapted for the purpose of business cycle analysis based on time series data. Using the smoothed statistics (4) and (5) we obtain

$$T^2(X_{1t},\ldots,X_{pt}) = \frac{1}{t-4} \cdot \sum_{i=1}^{4} n_{kt} \cdot (\mathbf{m}_{kt} - \mathbf{m}_{\bullet t})' \, \mathbf{S}_t^{-1} (\mathbf{m}_{kt} - \mathbf{m}_{\bullet t}), \tag{6}$$

where $\mathbf{m}_{\bullet t} = \frac{1}{t} \cdot \sum_{k=1}^{4} n_{kt} \cdot \mathbf{m}_{kt}$ is the weighted average of the estimated phase specific mean vectors. Furthermore, n_{kt} denotes the number of periods from set $\{1, 2, ..., t\}$ with cycle phase k. The statistic (6) measures the distances of the estimated phase-specific means $\mathbf{m}_{kt}$ to the 'overall' mean $\mathbf{m}_{\bullet t}$ of the classifying variables $X_{1t}, \ldots, X_{pt}$. It always holds that $T^2 \geq 0$. If $T^2 = 0$ the variables completely fail to separate the phases of the cycle in period t. The larger T^2, the larger the separation power. The statistic allows the comparison of the separation power of the variable set in successive time periods.

The multivariate measure of separation can be defined for any non-empty subset of the p-dimensional variable set too. For a single variable X_{it}, the resulting univariate measure of separation is

$$T^2(X_{it}) = \frac{1}{(t-4) \cdot s_{ii,t}} \cdot \sum_{i=1}^{4} n_{kt} \cdot (m_{i,kt} - m_{i,\bullet t})^2, \tag{7}$$

where $s_{ii,t}$ is the (i,i)-th element of $\mathbf{S}_t$ and $m_{i,kt}$ is the i-th element of $\mathbf{m}_{kt}$. Similar to the usual F-statistic of univariate analysis of variance, the ratio (7) measures the sum of squares between the classes relative to the sum of squares within the classes in a single dimension. The statistic allows comparison of the separation power of different scalar variables in one time period or comparison of the separation power of one variable in different time periods.

3 Empirical DLDA results for the U.S. business cycle

Meyer and Weinberg (1975) used an iterative procedure for classifying time periods according to their 4-phase business cycle scheme. For each month in the sample period from February 1947 to September 1973 they chose an initial phase assignment: They started with 2-phase classifications by the National Bureau of Economic Research (NBER) and separated their new phases demand-pull and stagflation from the two phases upswing and recession by economic deliberations. Then they used these initial assignments as a priori classifications for LDA. Boundary months of cycle phases were iteratively re-assigned according to the classifications by LDA until the number of misclassifications at the boundaries was minimized. The LDA was based on 20 variables. Meyer and Weinberg (1975, p. 176) selected the variables through a general survey of literature and chose variables 'that had figured prominently in the development of formal econometric models of the U.S. economy or had been singled out as particularly sensitive cyclical indicators...'. The variable

Table 1. Classification of U.S. business cycles, 1948-5 to 2000-12

	Cycle	Starting month of ...			
		Recovery	Demand-Pull	Stagflation	Recession
1	1948-05 to 1949-10	...	...	1948-05	1948-12
2	1949-11 to 1954-07	1949-11	1950-07	1951-01	1953-11
3	1954-08 to 1958-04	1954-08	1955-03	-	1957-09
4	1958-05 to 1961-01	1958-05	-	-	1960-06
5	1961-02 to 1970-11	1961-02	1965-05	1967-12	1970-01
6	1970-12 to 1975-03	1970-12	1973-01	-	1974-10
7	1980-10 to 1982-12	1975-04	1978-07	-	1979-07
8	1975-04 to 1980-09	1980-10	1981-04	-	1981-10
9	1983-01 to 1991-12	1983-01	1984-04	1987-11	1990-11
10	1991-12 to [2000-12]	1991-12	...	...	...

Source: Heilemann and Münch (2005)

set includes: unemployment rate, nominal and real gross national product (GNP), government surplus or deficit, gross government expenditures, net exports, output per man-hour, prime rate, money supply M1 and M2, average yields on corporate bonds, N.Y. Stock Exchange composite price index, index of unit labor cost, compensation per man-hour, GNP price deflator and five additional indices for consumer and wholesale prices. The observations of variables with an underlying trend were transformed into change rates (per cent changes against previous year or month) to moderate distorting trend effects on LDA. Heilemann and Münch (2002, 2005) used the same procedure - now based on 19 instead of 20 variables - to update the classifications up to December 2000. All classification results are summarized in Table 1.

Here, DLDA was applied to monthly data from December 1970 to December 2000 in order to analyze changes in the multivariate structure of the U.S. business cycle. The sample period corresponds to the cycles 6 to 10 of the post World War II era. The last cycle is incomplete. The computations are based on the same data as the computations by Heilemann and Münch (2002, 2005). Following the authors all a priori phase probabilities were set to 0.25. The classification results in Table 1 were used as a priori classifications of the months for the training of DLDA. Data from May 1948 to November 1970 were used to estimate the initial smoothed statistics. The smoothing parameters were fixed at $\alpha = 0.05$ and $\beta = 0.08$. These choices minimized the classification error rate by DLDA.

For the sample period, Heilemann and Münch (2005) reported classification results by classical linear discriminant analysis. The 'in-the-sample-error-rate' is 14.4%. Not all too surprising, the classification performance of DLDA is superior to LDA: The error rate of the ex-ante classifications is 5.8% ('out-of-the-sample-error-rate', see also Table 2). The DLDA results allow two conclusions. First, the set of classifying variables proposed by Meyer and Weinberg

Table 2. Ex-ante classifications of the periods 1970-12 to 2000-12 by DLDA

		Predicted phase			
Actual phase	No.	Recovery	Demand-Pull	Stagflation	Recession
Recovery	194	187	1	1	5
Demand-Pull	82	8	74	0	0
Stagflation	36	0	2	34	0
Recession	49	0	3	1	45

has still a reasonably high classification power in the younger history of the U.S. business cycle. Second, to guarantee a low ex-ante classification error rate the smoothing parameters α and β had to be fixed at relatively large values that enable DLDA to react fast to changes of the means and the covariances of the classifying variables. This indicates substantial changes in the multivariate structure of the business cycle during the sample period. The second conclusion is supported by the sequentially computed values of the multivariate measure of separation (6) for the complete variable set (see Figure 1). Substantial decreases of the separation power in November 1987, November 1990 – December 1991 and September 1998 – April 1999 indicate at least three sudden changes in the multivariate structure. The decreases coincide with the stock price crisis in October 1987, the first Gulf War and the 1998 currency crises in South East Asia, South America and Russia.

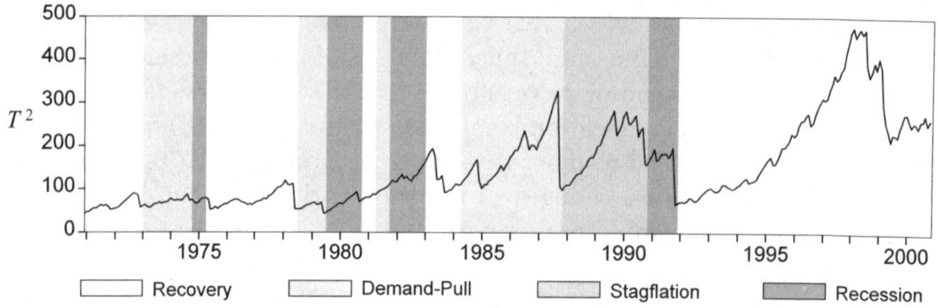

Fig. 1. Multivariate measure of separation for the periods 1970-12 to 2000-12

The analysis of the classification powers of single variables gives a deeper insight into the structural changes. Because of the large amount of results, only some selected results based on the univariate measure of separation (7) are presented here (Figure 2). For computing the values of (7), the smoothing parameters α and β were set equal 0.02 instead of $\alpha = 0.05$ and $\beta = 0.08$. This new specification lead to a stronger smoothing of fluctuations of the measure from period to period and ease the interpretation of Figure 2.

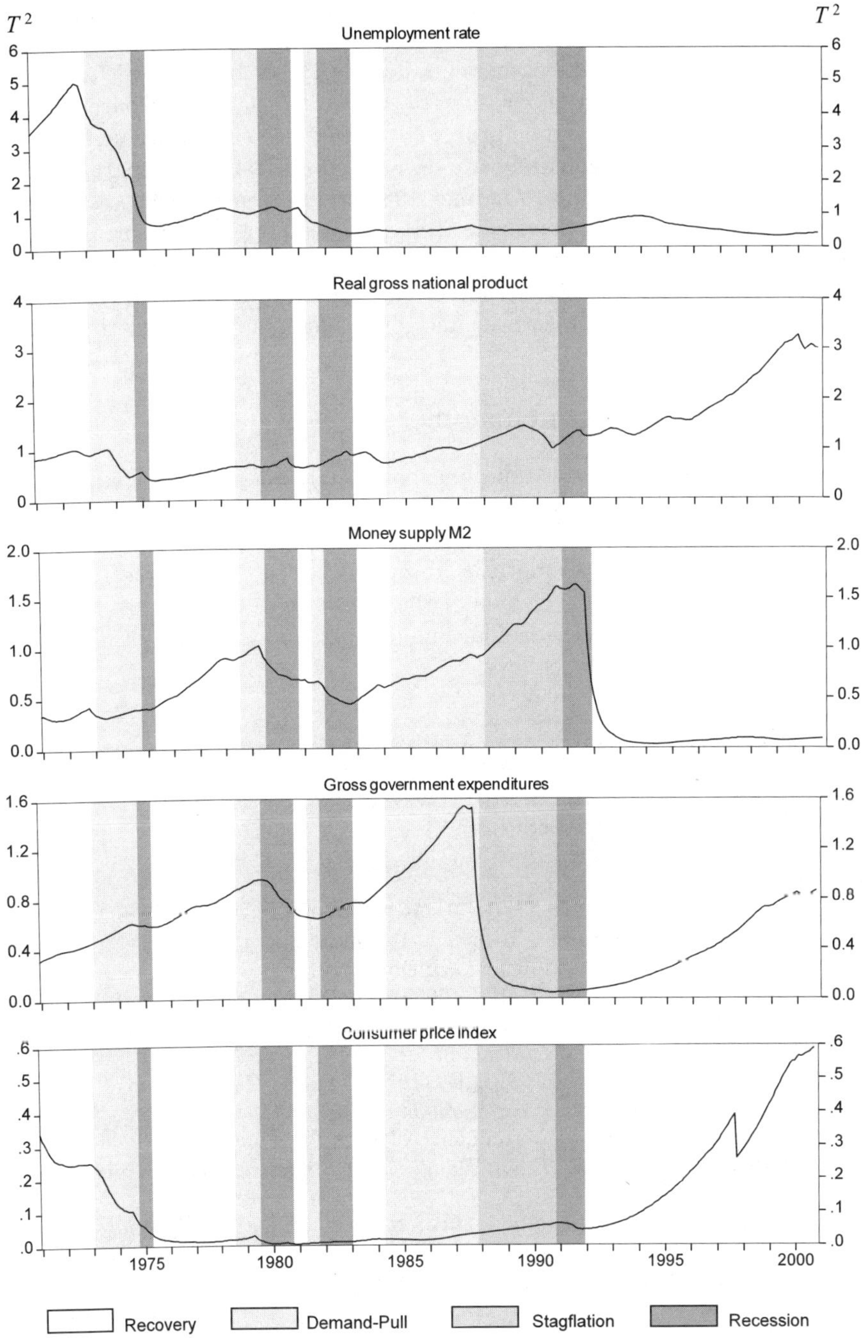

Fig. 2. Univariate measure of separation for selected classifying variables and the periods 1970-12 to 2000-12

According to (7) the unemployment rate was the most powerful classifying variable at the begin of cycle 6. In the course of time the variable lost separation power. It was surpassed in the following cycles by the real GNP, which became the most important classifier of the variable set. The separation power of the gross government expenditures and the money supply M2 increased over a time period of approximately two decades. Both lost their separation power rapidly in the periods after October 1987 and November 1991, respectively. After a long period of decreasing separation power the importance of the indices for consumer, wholesale and industrial prices as classifying variables increased again in the last decade of the 20th century. The development of the statistic (7) for the U.S. consumer price index is typical for this group of variables.

4 Summary and conclusions

This paper proposes a dynamic linear discriminant procedure as an instrument for the exploration of business cycle evolution. The recursive estimation of means and covariances of multiple time series enables DLDA to adapt and to detect changes in the multivariate cycle structure. Dynamic measures of separation describe the evolution of the separation power of classifying variables. The potentials of the new technique are demonstrated by the statistical analysis of the U.S. business cycle.

References

AHRENS, H. and LÄUTER, J. (1974): *Mehrdimensionale Varianzanalyse.* Akademie-Verlag, Ost-Berlin.

FISHER, R.A. (1936): The Use of Multiple Measurements in Taxomonic Problems. *Annals of Eugenics, 7, 179-188.*

HEILEMANN, U. and MÜNCH, H.J. (2002): Classifying U.S. Business Cycles 1948 to 1997 - Meyer/Weinberg revisited. Technical Report 29/2002, SFB 475, Department of Statistics, University of Dortmund.

HEILEMANN, U. and MÜNCH, H.J. (2005): The Clinton Era and the U.S. Business Cycle: What Did Change? Technical Report 12/2005, SFB 475, Department of Statistics, University of Dortmund.

MEYER, J.R. and WEINBERG, D.H. (1975): On the Classification of Economic Fluctuations. *Explorations in Economic Research, 2, 167-202.*

SHUMWAY, R.H. (1982): Discriminant Analysis for Time Series. In: P.R. Krishnaiah and L.N. Kanal (Eds.): *Handbook of Statistics, Vol. 2.* North-Holland, Amsterdam, 1-46.

TICHY, G.J. (1994): *Konjunktur - Stilisierte Fakten, Theorie, Prognose.* Springer, Berlin.

WALD, A. (1944): On a Statistical Problem Arising in the Classification of an Individual into One of Two Groups. *Annals of Mathematical Statistics, 15, 145-162.*

WASSERMAN, L.A. (2004): *All of Statistics - A Concise Course in Statistical Inference.* Springer, New York.

Examination of Several Results of Different Cluster Analyses with a Separate View to Balancing the Economic and Ecological Performance Potential of Towns and Cities

Nguyen Xuan Thinh[1], Martin Behnisch[2] and Alfred Ultsch[3]

[1] Leibniz Institute of Ecological and Regional Development (IOER); Ng.thinh@ioer.de
[2] IFIB, University of Karlsruhe (TH); Martin.Behnisch@email.de
[3] Department of Mathematics & Computer Science, University of Marburg; ultsch@informatik.uni-marburg.de

Abstract. The objective of this paper is to compare cluster analyses conducted by using different methods for 116 administratively autonomous municipalities (kreisfreie Staedte) in Germany. The cluster analyses aim to provide answers to the question as to the impact of land-use structures on the performance potential of towns and cities. Drawing on the database established, 11 attribute variables for the analysis were selected that significantly characterise a city's land-use structures and go a long way towards moulding its economic and ecological performance. We show that no cluster structure exists in the data set. Therefore we investigate the data set by using Gaussian Mixture-Models estimating by Expectation-Maximization (EM) algorithm. This indicates that three or two variables suffice to classify the cities. The next step in our exploratory research is to conduct and to compare the results of different classification algorithms for these three and two variables. The classification based on EM algorithm allows us to identify 8 classes. We discuss them and compare this result with results of some cluster analyses with a separate view to balancing the economic and ecological performance potential of towns and cities.

1 Introduction

In the focus of research about cities and their similarities there are several trials to reduce the multitude of cities to a few types of cities. Comparisons of cities and typological grouping processes are methodical instruments to develop statistical scales and criteria about urban phenomenons. These analyses enable a closer look to the ranking of cities. Classification is used as a method to build models about the reality with a special intention to a selected point of view. A team of the Leibniz Institute of Ecological and Regional Development (IOER) produced a dataset embracing indicators and information for 116 administratively autonomous municipalities (Regional Cities), the entirety of such entities in Germany with the exception of Eisenach, and investigated effects of urban land-use structures on the degree of sealing and the price of land (Arlt et al. (2001)). Cluster analyses for the cities and subsequent analysis of the structural characteristics of clusters can help extrapolate benchmark values and recommendations for sustainable urban development (Behnisch (2005), Thinh et al. (2002)). In this paper different approaches are presented in detail to classify the 116 objects. It is fundamental that the classes represent issues of the urban phenomena and the classes should be well defined and separated. Several possibilities are discussed to reduce the number of variables and properties of different cluster results are shown. A comparative visualisation of classification results supports the interpretation.

2 Inspection of data and methods

Drawing on the database established, 11 attribute variables were selected that significantly characterise a city's land-use structures and go a long way towards moulding its economic and ecological performance (see Table 1).

Table 1. The set of variables

No	Name of the variable	Abbreviation
(1)	settlement and transport landtake as a percentage of the urban area,	Svproz
(2)	population density in inhabitants per square kilometre of city area,	Bdichte
(3)	settlement density in inhabitants per ha of settlement areas and transport land,	Sdichte
(4)	recreation area provision measured in square metres per inhabitant,	Ejeew
(5)	open space provision measured in square metres per inhabitant,	Fjeew
(6)	degree of sealing in urban nucleus as a percentage,	Vgstadt
(7)	eco-value of urban nucleus (non-dimensional),	Oewstadt
(8)	gross value-added measured in EUR per square metre of urban space,	Boges
(9)	gross value-added in EUR per m^2 of settlement areas and transport land,	Bosv
(10)	land price in the form of purchase values for developed land (EUR/m^2), and	Bopreis
(11)	unemployment rate as percentage.	Alquote

The inspection of data includes the visualisation in form of histograms, Q-Q-Plots, Box-Plots and Scatter-Plots. The authors decided to use transformation measurements such as ladder of power (Hartung (2005)) to take into account the different distributions of variables and the restrictions of multivariate statistics. In conclusion to the logical interpretation of the data and the calculated correlation matrix the authors decide to exclude variable 2 (population density in inhabitants per square kilometre of city area) and variable 5 (open space provision measured in square metres per inhabitant). Some hierarchical algorithms (WARD, SINGLE-LINKAGE, AVERAGE-LINKAGE) and one partitioning algorithms (K-MEANS) show that clustering of the cities is not possible. A seven cluster solution by WARD-algorithm is displayed in Figure 1. For distance based cluster algorithms it is hard to detect correct boundaries for the clusters. The MDS has characterized one cloud of objects and just only some special objects beside.

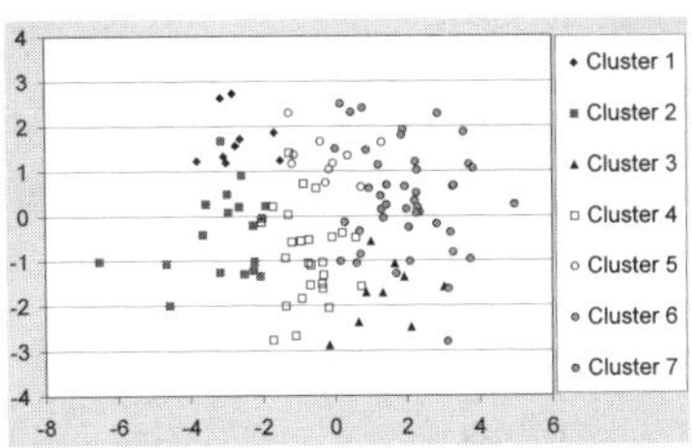

Fig. 1. MDS of the data (116 objects, 9 variables)

For these reasons the authors investigated the data set by using Gaussian Models estimating by Expectation-Maximization (EM)-algorithm. Gaussian mixtures (Lauritzen (1996)) are combinations of a finite number of Gaussian distributions. They are used to model complex multidimensional distributions. A mixture of Gaussians can be written as a weighted sum of Gaussian densities. By varying the number of Gaussians K, the weights ω_k, and the parameters μ_k and covariance matrix Σ_k of each Gaussian density function, Gaussian mixtures can be used to describe any complex probability density function. Recall the d-dimensional Gaussian probability density function (PDF):

$$g_{(\mu,\Sigma)}(x) = \frac{1}{\sqrt{2\pi}^d \sqrt{\det(\sum)}} e^{-\frac{1}{2}(x-\mu)^T \sum^{-1}(x-\mu)} \qquad (1)$$

A weighted mixture of K Gaussians can be written as

$$gm(x) = \sum_{k=1}^{K} \omega_k \cdot g_{(\mu_k,\Sigma_k)}(x), \qquad (2)$$

where the weighted are all positive and sum to one.

The EM-algorithm is an ideal candidate for solving parameter estimation problems for the Gaussian Mixture Models (GMM) or other neural networks (Bilmes (1997)). When there is a need to learn parameters of the Gaussian mixture, the EM algorithm starts with initial values for all parameters and they are re-estimated iteratively. The aim is to optimize the likehood that the given data points are generated by a mixture of Gaussians (Redner and Walker (1984)). The numbers next to the Gaussians give the relative importance (amplitude) of each component. It is crucial to start with 'good' initial parameters as the algorithm only finds a local, and not a global optimum. The procedure of optimization includes several calculations that vary from 2 up to a value of 5 Gaussian Estimations. The GMM are verified by Pareto Density Estimation (PDE) and probability density functions (PDF). Density based clustering algorithms have drawn much attention in the last years within the context of data mining (Ultsch (1999), Xu et al. (1998)).

3 Results

The task is to find a classification which is especially affected by economic and ecological performance potentials. Figure 2 shows the whole variable set with their Gaussian distributions.

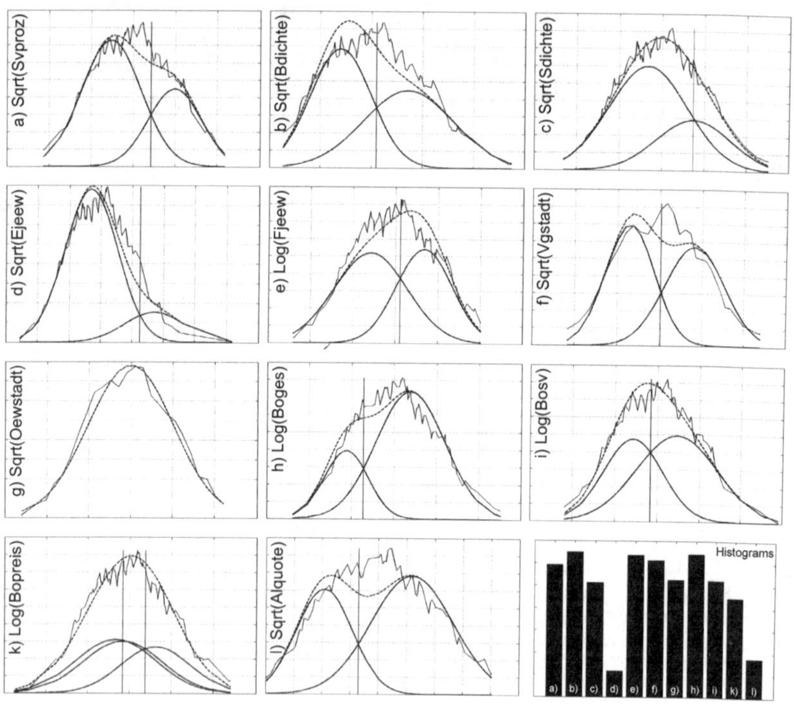

Fig. 2. GMM and sum of correlation-coefficient for each of the 11 variables, PDE = Black(fine-line), Gaussian Mixture = Dashed, Single Gaussian (mode)=Black

For each variable it is possible to devide the 116 objects into several groups by defining different levels (1 to 3). Based on three criteria: (1) multi-modal yes or no, (2) symmetry of the Gaussian Mixtures, and (3) how good the Gaussian Mixtures approximate the curve of the probability density function (dashed curve), the authors can ascertain that the following seven variables Svproz (1), Bdichte (2), Fjeew (5), Vgstadt (6), Bosv (9), Bopreis (10) and Alquote (11) can contribute to the classification of the cities. The first four variables correlate with each other very high. The degree of sealing is an important indicator for the ecological performance of a city therefore the authors prefer to use the variable Vgstadt (6). There are high correlations between Bosv (9) and Bopreis (10) and between Bopreis (10) and Alquote (11). Also the variables Bosv (9) and Alquote (11) correlate, but not so high. The authors can use just Bopreis (10) or just Bosv (9) or Bosv (9) and Alquote (11). All conceivable variables can be devided into different groups characterized by the Gaussian Mixture Models and values of Baysian Decision boundary. Thereby the data of unemployment rate can be divided into 2 groups (low/high, marginal value: 13 %) and the degree of sealing cuts also into 2 groups (low/high, marginal value: 14.5 %). The land price is represented by 3 groups (low/middle/high, marginal value: 80 EUR/m^2, 160 EUR/m^2) and the gross value-added implements 2 groups (low/high, marginal value: 82 EUR/m^2.). The classification presented here is based on three variables (Alquote, Vgstadt, Bosv) that form a general basis for a new grouping process. The aggregation of the grouping process (cross-classified table) leads to a structure of 8 classes. Table 2 contains a description of each class.

Table 2. Characteristics of the 8 class solution

Class	Label	ID [-]	Objects [-]	Vgstadt [%]	Bosv [EUR/m^2]	Alquote [%]
1	Cities with high ecological performance	111	13	low	low	low
2	Cities with weak economic performance and high unemployment	112	26	low	low	high
3	Sustainable city class	121	16	low	high	low
4	Special city class I	122	4	low	high	high
5	Special city class II	211	4	high	low	low
6	Cities with weak ecological and economic performance and high unemployment	212	18	high	low	high
7	Regions with strong economic development	221	20	high	high	low
8	Strong urbanized regions	222	15	high	high	high

Figure 3 shows the spatial distribution of the 116 German Regional Cities. It verifies the description of the labelling process, e.g., low developed cities (class 2) are mainly located in the Eastern Part of Germany and the class of regions with strong economic development (class 7) include exclusively cities

Fig. 3. Spatial distribution of 8 city classes

in the West Germany. Cities of the class 3 are characterised by high ecological and economic performance and low unemployment rate, and therefore a good ecological, economic and social development. Such cities can be named as sustainable cities.

Another approach for the classification of Regional cities as presented here displays the results of different classification algorithms by just two variables. The authors carried out cluster analysis with the two variables degree of sealing (Vgstadt) and gross value-added in EUR per m^2 of settlement areas and transport land (Bosv). The different partitions are shown in Figure 4 both in deterministic and fuzzy restriction.

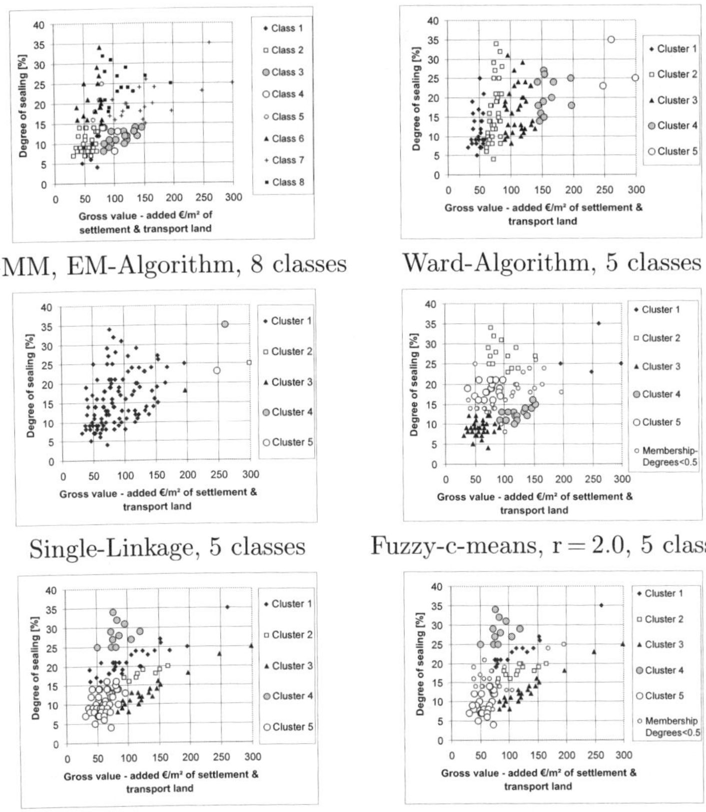

GMM, EM-Algorithm, 8 classes — Ward-Algorithm, 5 classes

Single-Linkage, 5 classes — Fuzzy-c-means, r = 2.0, 5 classes

Gath&Geva, r = 2.0, 5 classes — Gustafson Kessel, r = 2.0, 5 classes

Fig. 4. Comparison of different classification results

4 Conclusion

In this paper the authors present an approach for Regional City Classification by using different clustering techniques and comparing several results. The central issue of the grouping processes are the ecological and economical performance potentials. First the authors examine the pool of data and important variables were extracted. Typical clustering approaches lead to fail in finding logical structures in consideration of 9 variables. An EM-algorithm has been derived for the case when the input data is a mixture density of several data classes, with each data class dependent on a different set of parameters. The application of graphical Gaussian models was used to find meaningful pairwise interactions among sets of cities. Graphical Gaussian modeling has the advantage of being able to model conditional distributions of continuous variables. As such, the authors believe that this method complements the typical clustering approaches used to analyze cities or urban areas. By using two variables several classification processes are calculated and compared to get more information about the advantages and disadvantages of algorithms in order to the central problem. The results must be in a symbolic representation, they are useable for urban planners, regional policy and knowledge acquisition systems.

Acknowledgements

We thank Rico Vogel for his preparation of the LaTeX-document for our text.

References

ARLT, G., GÖSSEL, J., HEBER, B., HENNERSDORF, J., LEHMANN, I. and THINH, N.X. (2001): *Auswirkungen städtischer Nutzungsstrukturen auf Bodenversiegelung und Bodenpreis.* IÖR-Schriften 34, Dresden, 1 CD-ROM.

BEHNISCH, M. (2005): Bestandsorientiertes Klassifikatormodell - Ein Informations- und Analysewerkzeug zur Untersuchung von Gebäuden und Stadt. In: Wittmann J. and Thinh N. X. (Hrsg.): *Simulation in Umwelt- und Geowissenschaften - Workshop Dresden 2005*, Shaker, Aachen.

BILMES, J. (1997): *A Gentle Tutorial on the EM Algorithm and its Application to Parameter Estimation for Gaussian Mixture and Hidden Markov Models.* Technical Report, University of Berkeley, ICSI-TR-97-021. http://crow.ee.washington.edu/people/bulyko/papers/em.pdf.

HARTUNG, J. (2005): *Statistik - Lehr- und Handbuch der angewandten Statistik.* Oldenbourg, München.

LAURITZEN, S. (1996): *Graphical Models.* Oxford University Press.

REDNER, R. and WALKER, H. (1984): Mixture Densities, Maximum Likehood and the EM Algorithm. *SIAM Review, 26, 2.*

THINH, N.X., ARLT, G., HEBER, B., HENNERSDORF, J. and LEHMANN, I. (2002): Evaluation of Urban Land-use Structures with a View to Sustainable Development. *Environmental Impact Assessment Review, 22, 5, 475–492.*

ULTSCH, A. (1999): *Data Mining and Knowledge Discovery with Emergent Self-Organizing Feature Maps for Multi-variate Times Series.* Kohonen Maps, 33–46.

XU, X., ESTER, M., KRIEGEL, H. and SANDER, J. (1998): A Distribution-Based Clustering Algorithm for Mining in Large Spatial Databases. *Proc. 14th Int. Conf. on Data Engineering (ICDE'98)*, Orlando, 324–331.

Part IV

Visualization and Scaling Methods

VOS: A New Method for Visualizing Similarities Between Objects

Nees Jan van Eck and Ludo Waltman

Econometric Institute, Faculty of Economics, Erasmus University Rotterdam, P.O. Box 1738, 3000 DR Rotterdam, The Netherlands;
{nvaneck, lwaltman}@few.eur.nl

Abstract. We present a new method for visualizing similarities between objects. The method is called VOS, which is an abbreviation for *visualization of similarities*. The aim of VOS is to provide a low-dimensional visualization in which objects are located in such a way that the distance between any pair of objects reflects their similarity as accurately as possible. Because the standard approach to visualizing similarities between objects is to apply multidimensional scaling, we pay special attention to the relationship between VOS and multidimensional scaling.

1 Introduction

In this paper, a new method for visualizing similarities between objects is presented. The method is called VOS, which is an abbreviation for *visualization of similarities*. The aim of VOS is to provide a low-dimensional visualization in which objects are located in such a way that the distance between any pair of objects reflects their similarity as accurately as possible. Objects that have a high similarity should be located close to each other, whereas objects that have a low similarity should be located far from each other. Because the standard approach to visualizing similarities between objects is to apply multidimensional scaling (MDS) (Borg and Groenen (2005)), the relationship between VOS and MDS is given special attention in this paper.

The paper is organized as follows. In Section 2, a description of VOS is provided. In Section 3, VOS and MDS are applied to a simple example data set. The results that are obtained demonstrate an interesting property of VOS. In Section 4, the relationship between VOS and MDS is analyzed theoretically. Finally, some conclusions are drawn in Section 6.

2 Description of VOS

In this section, we provide a description of VOS. Let there be n objects, denoted by $1, \ldots, n$. Let there also be an $n \times n$ similarity matrix $\mathbf{S} = (s_{ij})$

satisfying $s_{ij} \geq 0$, $s_{ii} = 0$, and $s_{ij} = s_{ji}$ for all $i, j \in \{1, \ldots, n\}$. Element s_{ij} of $\mathbf{S}$ denotes the similarity between the objects i and j. It is assumed that the similarities in $\mathbf{S}$ can be regarded as measurements on a ratio scale. VOS aims to provide a low-dimensional Euclidean space in which the objects $1, \ldots, n$ are located in such a way that the distance between any pair of objects i and j reflects their similarity s_{ij} as accurately as possible. Objects that have a high similarity should be located close to each other, whereas objects that have a low similarity should be located far from each other. The $n \times m$ matrix $\mathbf{X}$, where m denotes the number of dimensions of the Euclidean space, contains the coordinates of the objects $1, \ldots, n$. The vector $\mathbf{x}_i = (x_{i1}, \ldots, x_{im}) \in \mathbb{R}^m$ denotes the ith row of $\mathbf{X}$ and contains the coordinates of object i. The idea of VOS is to minimize a weighted sum of the squared distances between all pairs of objects. The higher the similarity between two objects, the higher the weight of their squared distance in the summation. To avoid solutions in which all objects are located at the same coordinates, the constraint is imposed that the sum of all distances must equal some positive constant. In mathematical notation, the objective function to be minimized in VOS is given by

$$E(\mathbf{X}; \mathbf{S}) = \sum_{i<j} s_{ij} \|\mathbf{x}_i - \mathbf{x}_j\|^2, \tag{1}$$

where $\|\cdot\|$ denotes the Euclidean norm. Minimization of the objective function is performed subject to the following constraint

$$\sum_{i<j} \|\mathbf{x}_i - \mathbf{x}_j\| = 1. \tag{2}$$

Note that the distances $\|\mathbf{x}_i - \mathbf{x}_j\|$ in the constraint are not squared.

To provide further motivation for the above constrained optimization problem, we note that when visualizing similarities it seems natural to expect that each object i is located close to what we call its ideal coordinates, which are given by

$$c_i(\mathbf{X}, \mathbf{S}) = \frac{\sum_j s_{ij} \mathbf{x}_j}{\sum_j s_{ij}}. \tag{3}$$

In other words, each object i may be expected to be located close to a weighted average of the coordinates of all other objects, where the coordinates of objects more similar to object i are given higher weight in the calculation of the weighted average. Locating each object i exactly at its ideal coordinates $c_i(\mathbf{X}, \mathbf{S})$ is only possible by locating all objects at the same coordinates, which clearly does not result in a useful solution. Rather than locating each object exactly at its ideal coordinates, VOS seems to have the tendency to locate objects close to their ideal coordinates. This can be seen as follows. Suppose that the coordinates of all objects except some object i are fixed, and forget for the moment about the constraint in (2). Minimization of the objective function in (1) then reduces to minimization of

$$E_i(\mathbf{x}_i; \mathbf{X}, \mathbf{S}) = \sum_j s_{ij} \|\mathbf{x}_i - \mathbf{x}_j\|^2. \tag{4}$$

Minimization of (4) can be performed analytically and results in the solution $\mathbf{x}_i = c_i(\mathbf{X}, \mathbf{S})$. In other words, if the coordinates of all objects except some object i are fixed and if the constraint in (2) is not taken into consideration, then VOS will locate object i exactly at its ideal coordinates. Of course, objects do not have fixed coordinates, and solutions depend not only on the objective function but also on the constraint. For these reasons, VOS generally does not locate objects exactly at their ideal coordinates. However, the situation with fixed coordinates and without the constraint at least seems to indicate that VOS has the tendency to locate objects close to their ideal coordinates.

Finally, we mention some approaches that are closely related to VOS. The idea of visualizing similarities by locating objects close to their ideal coordinates can also be found in our earlier research (Van den Berg et al. (2004), Van Eck et al. (2005)). In this research, instead of the constraint in (2) some penalty function is used to avoid solutions in which all objects are located at the same coordinates. Davidson et al. (1998) take an approach that visualizes similarities between objects by solving a constrained optimization problem. The objective function in their approach is exactly the same as in VOS, but the constraints are different. The constraints of Davidson et al. have the advantage that they allow the optimization problem to be solved as an eigenvalue problem. In our experience, however, the constraints of Davidson et al. result in less satisfactory visualizations than the constraint that is used in VOS.

3 Application to a simple example data set

In this section, we consider a simple example data set of similarities between objects. The data set is also studied by Kendall (1971) and Mardia et al. (1979), who find that a so-called horseshoe effect occurs when the similarities in the data set are visualized using multidimensional scaling (MDS). Kendall and Mardia et al. apply MDS without weights. We refer to MDS without weights as standard MDS throughout this paper. In this section, we first reproduce the result obtained by Kendall and Mardia et al. by applying standard MDS to the data set. We then apply VOS to the data set and demonstrate that VOS does not produce a horseshoe effect.

The data set consists of a 51×51 similarity matrix $\mathbf{S} = (s_{ij})$ given by

$$s_{ij} = \begin{cases} 8 & \text{if } 1 \leq |i-j| \leq 3 \\ 7 & \text{if } 4 \leq |i-j| \leq 6 \\ \cdots\cdots\cdots & \\ 1 & \text{if } 22 \leq |i-j| \leq 24 \\ 0 & \text{otherwise.} \end{cases} \tag{5}$$

Visualization of the similarities in **S** using standard MDS results in the solution shown in Figure 1. This solution was obtained using the PROXSCAL program available in SPSS. The similarities were treated as ordinal data. The option offered by PROXSCAL to untie tied observations was not used. As can be seen in Figure 1, standard MDS provides a solution in which the objects lie on a curve in the form of a horseshoe. (Treating the similarities as interval data rather than ordinal data results in a very similar solution.) The objects lie in the expected order, that is, object 1 is followed by object 2, object 2 is followed by object 3, and so on. However, due to the horseshoe form, there is a problem with the distances between the objects. This problem is sometimes referred to as the horseshoe effect (Mardia et al. (1979)). Consider, for example, the objects 1 and 51, which are the objects lying at the ends of the horseshoe. Object 1 lies closer to object 51 than to many other objects, like object 40. Based on the solution from standard MDS, one would therefore expect object 1 to be more similar to object 51 than to object 40. However, this expectation is incorrect, since both the similarity between the objects 1 and 51 and the similarity between the objects 1 and 40 equal 0. Moreover, both object 1 and object 40 have a positive similarity with the objects 16 to 25, whereas there are no objects with which both object 1 and object 51 have a positive similarity. Therefore, if indirect similarities via third objects are taken into account, then object 1 is more similar to object 40 than to object 51. This is exactly opposite to the impression given by the solution from standard MDS.

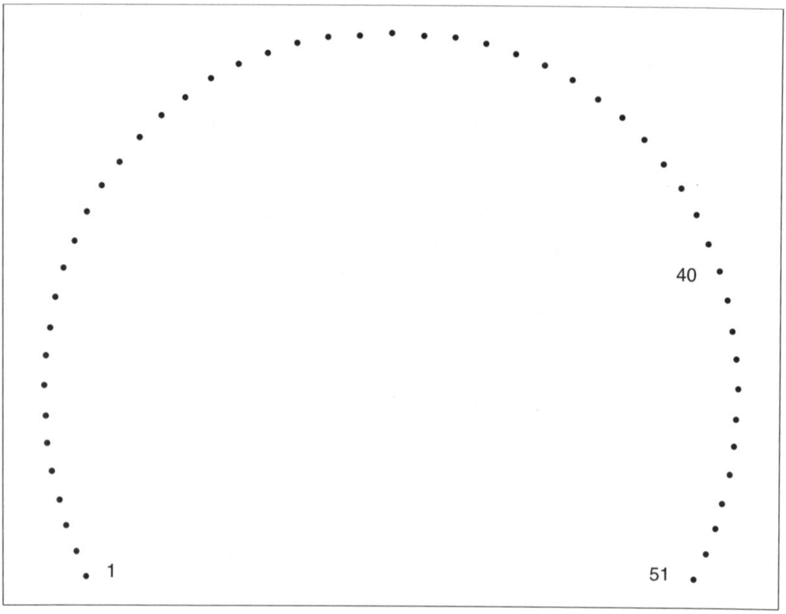

Fig. 1. Visualization of the similarities s_{ij} in (5) obtained using standard MDS.

We now consider the result of visualizing the similarities s_{ij} in (5) using VOS. The solution provided by VOS is shown in Figure 2. In this solution, the objects lie almost on a straight line. They also lie in the expected order, with object 1 followed by object 2 and so on. Interestingly, in contrast to the solution from standard MDS, the solution from VOS does not suffer from the horseshoe effect. In fact, if indirect similarities via third objects are taken into account, then the distances between the objects in the solution from VOS very accurately reflect the similarities between the objects. For example, the objects 1 and 51 lie further from each other than the objects 1 and 40. This is exactly what one would expect based on the objects' indirect similarities. Both the objects 1 and 51 and the objects 1 and 40 have a similarity of 0, but the objects 1 and 51 do not have third objects with which they both have a positive similarity, whereas the objects 1 and 40 do have such objects, namely the objects 16 to 25. Object 1 is therefore more similar to object 40 than to object 51, and this is exactly what is reflected by the distances in the solution from VOS.

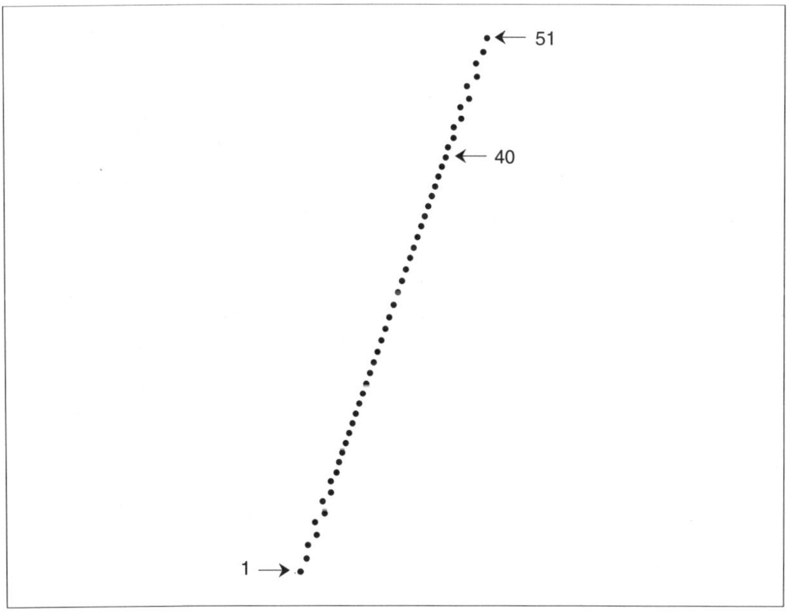

Fig. 2. Visualization of the similarities s_{ij} in (5) obtained using VOS.

The results presented in this section indicate that VOS and standard MDS may provide very different solutions. In applications in which indirect similarities via third objects may contain relevant information, VOS probably provides better solutions than standard MDS. An example of an application where the use of VOS may be more appropriate than the use of standard MDS is the visualization of associations between concepts based on co-occurrence data (e.g. Van den Berg et al. (2004), Van Eck et al. (2005), Van Eck et al. (2006)). Typically, many pairs of concepts do not co-occur at all, and these pairs of concepts then have a similarity of 0. Standard MDS aims to provide a visualization in which for each pair of concepts with a similarity of 0 the distance between the concepts is the same. VOS seems to pay more attention to indirect similarities via third concepts and may therefore locate concepts with a high indirect similarity closer to each other than concepts with a low indirect similarity. Because of this property, we expect VOS to provide more insightful visualizations of concept associations than standard MDS.

4 Relationship with multidimensional scaling

VOS and standard MDS may provide very different solutions, as we have shown in Section 3. In this section, we provide a theoretical analysis of the relationship between VOS and MDS. More specifically, we show that under certain conditions VOS is equivalent to Sammon mapping (Sammon (1969)), which is a special variant of MDS. The mathematical notation in this section is the same as in Section 2. In addition, $\mathbf{D} = (d_{ij})$ is used to denote an $n \times n$ dissimilarity matrix satisfying $d_{ij} > 0$ and $d_{ij} = d_{ji}$ for all $i, j \in \{1, \ldots, n\}$. Element d_{ij} of $\mathbf{D}$ denotes the dissimilarity between the objects i and j. Like standard MDS, Sammon mapping aims to provide a low-dimensional space in which the objects $1, \ldots, n$ are located in such a way that the distance between any pair of objects i and j reflects their dissimilarity d_{ij} as accurately as possible. Objects that have a high dissimilarity should be located far from each other, whereas objects that have a low dissimilarity should be located close to each other. If similarities rather than dissimilarities are available, the similarities have to be transformed into dissimilarities before Sammon mapping can be applied. We note that Sammon mapping and VOS have a very similar purpose, the difference being that Sammon mapping uses dissimilarities whereas VOS uses similarities. In Sammon mapping, the following objective function is minimized

$$\sigma(\mathbf{X}; \mathbf{D}) = \sum_{i<j} \frac{(d_{ij} - \|\mathbf{x}_i - \mathbf{x}_j\|)^2}{d_{ij}}. \tag{6}$$

Sammon mapping differs from standard MDS because of the division by d_{ij} in the summation in (6).

The following theorem states the equivalence, under certain conditions, of VOS and Sammon mapping.

Theorem 1. *Let $s_{ij} > 0$ for all i and j ($i \neq j$), and let similarities be transformed into dissimilarities using $d_{ij} = s_{ij}^{-1}$ ($i \neq j$). VOS and Sammon mapping are then equivalent in the sense that VOS solutions and Sammon mapping solutions differ only by a multiplicative constant.*

Due to space limitations, we do not prove the theorem here. A proof of the theorem is available in our working paper (Van Eck and Waltman (2006)).

We note that Sammon mapping in the way it is discussed in this section is equivalent to weighted MDS where to each pair of objects i and j a weight is given that equals d_{ij}^{-1}. It therefore follows from Theorem 1 that there also exists an equivalence, under certain conditions, between VOS and weighted MDS.

5 Conclusions

In this paper, we have presented VOS, which is a new method for visualizing similarities between objects. VOS aims to provide a low-dimensional visualization in which objects are located in such a way that the distance between any pair of objects reflects their similarity as accurately as possible. As we have discussed in this paper, VOS has the following three properties. First, VOS seems to have the tendency to locate objects close to what we have called their ideal coordinates. The ideal coordinates of an object i are defined as a weighted average of the coordinates of all other objects, where the coordinates of objects more similar to object i are given higher weight in the calculation of the weighted average. Second, VOS seems to pay more attention to indirect similarities via third objects than standard MDS. For example, if two objects i and j have a similarity of 0, the distance between the objects in a visualization obtained using VOS seems to depend on the number of third objects with which the objects i and j both have a positive similarity. The higher the indirect similarity via third objects, the closer the objects i and j are located to each other. Third, although VOS and standard MDS may provide very different visualizations, VOS is, under certain conditions, equivalent to a special variant of MDS called Sammon mapping. Furthermore, if weights are used in MDS and these weights are chosen in the appropriate way, then there also exists an equivalence, under certain conditions, between VOS and MDS.

Finally, we would like to refer the reader who is interested in a practical application of VOS to Van Eck et al. (2006a, 2006b). Van Eck et al. apply VOS to the visualization of associations between concepts based on co-occurrence data.

Acknowledgement

We would like to thank Jan van den Berg and Patrick Groenen. Our discussions with them have contributed significantly to the ideas presented in this paper. We are also grateful to Joost van Rosmalen for his valuable comments on an earlier draft of the paper.

References

BORG, I. and GROENEN, P.J.F. (2005): *Modern Multidimensional Scaling.* Springer, Berlin.

DAVIDSON, G.S., HENDRICKSON, B., JOHNSON, D.K., MEYERS, C.E. and WYLIE, B.N. (1998): Knowledge Mining with VxInsight: Discovery through Interaction. *Journal of Intelligent Information Systems, 11, 259–285.*

KENDALL, D.G. (1971): Seriation from Abundance Matrices. In: F.R. Hodson, D.G. Kendall and P. Tautu (Eds.): *Mathematics in the Archaeological and Historical Sciences.* Edinburgh University Press, 215–252.

MARDIA, K.V., KENT, J.T. and BIBBY, J.M. (1979): *Multivariate Analysis.* Academic Press.

SAMMON, J.W. (1969): A Nonlinear Mapping for Data Structure Analysis. *IEEE Transactions on Computers, C- 18, 5, 401–409.*

VAN DEN BERG, J., VAN ECK, N.J., WALTMAN, L. and KAYMAK, U. (2004): A VICORE Architecture for Intelligent Knowledge Management. In: *Proceedings of the KDNet Symposium on Knowledge-Based Services for the Public Sector, 63–74.*

VAN ECK, N.J. and WALTMAN, L. (2006): VOS: A New Method for Visualizing Similarities Between Objects. Technical Report ERS-2006-020-LIS, Erasmus University Rotterdam, Erasmus Research Institute of Management.

VAN ECK, N.J., WALTMAN, L. and VAN DEN BERG, J. (2005): A Novel Algorithm for Visualizing Concept Associations. In: *Proceedings of the 16th International Workshop on Database and Expert Systems Applications, 405–409.*

VAN ECK, N.J., WALTMAN, L., VAN DEN BERG, J. and KAYMAK, U. (2006a): Visualizing the WCCI 2006 Knowledge Domain. In: *Proceedings of the 2006 IEEE International Conference on Fuzzy Systems, 7862–7869.*

VAN ECK, N.J., WALTMAN, L., VAN DEN BERG, J. and KAYMAK, U. (2006b): Visualizing the Computational Intelligence Field. *IEEE Computational Intelligence Magazine.* Accepted for publication.

Multidimensional Scaling of Asymmetric Proximities with a Dominance Point

Akinori Okada[1] and Tadashi Imaizumi[2]

[1] Department of Industrial Relations, School of Social Relations, Rikkyo (St. Paul's) University, 3-34-1 Nishi Ikebukuro, Toshima-ku Tokyo, 171-8501 Japan; okada@rikkyo.ac.jp

[2] School of Management and Information Sciences, Tama University, 4-4-1 Hijirigaoka, Tama city, Tokyo, 206-0022 Japan; imaizumi@tama.ac.jp

Abstract. The purpose of the present study is to introduce a model and the associated nonmetric algorithm of multidimensional scaling for analyzing one-mode two-way (object × object) asymmetric proximities. In the model each object is represented as a point in a multidimensional Euclidean space, and a point, called the dominance point, is also embedded in the same multidimensional Euclidean space. The dominance point governs the asymmetry in the proximity relationships among objects, and represents the whole one-mode two-way asymmetric proximities dealt with in the analysis. An application to car switching data is presented.

1 Introduction

As researchers becomes aware of the importance of the asymmetry in proximity relationships, several models and algorithms of multidimensional scaling, which can deal with asymmetry, have been introduced in the last three decades (Borg and Groenen (2005, pp. 495-518), Zielman and Heiser (1996)). While some of the multidimensional scaling can deal with two-mode three-way asymmetric proximities (Chino et al. (1996), DeSarbo et al. (1992), Okada and Imaizumi (1997), Zielman (1991), Zielman and Heiser (1993)), most of them deal with one-mode two-way asymmetric proximities, which represent asymmetric proximity relationships among a set of objects (object × object).

Okada and Imaizumi (1987) has introduced a model and the associated nonmetric algorithm to analyze one-mode two-way asymmetric proximities (Carroll and Arabie (1980)). A mode is defined as a particular class of entities (e.g., objects, individuals, variables, subjects, time points), and an N way means the data are comprised of the Cartesian product (where some of the modes can be repeated) of N modes (Carroll and Arabie (1980), p. 610). In the present application, 12 car categories constitute a mode. The data represent the similarity among a mode or 12 car categories, which means the data have one mode. And the data is comprised of the Cartesian products of two modes or the repetition of the mode (12 car categories), i.e. two-way diadic data of "car category × car category" relationships. Thus the present data is one-mode two-way proximities. One mode two-way asymmetric proximities can be represented as a two-way array or a proximity matrix, where the (j, k) element of the matrix shows the proximity from objects j to k.

In the model of Okada and Imaizumi (1987), each object is represented as a point and a hypersphere in a multidimensional Euclidean space. The radius of the hypersphere represents the asymmetry of the proximity relationships among objects. Okada and Imaizumi (1997) extended the model and the algorithm to deal with two-mode three-way asymmetric proximities (object × object × individual or source). There are two modes; a set of objects and a set of individuals or sources. The mode of objects is repeated twice, and the mode of individuals or sources is not repeated to form the Cartesion product of the data. Two-mode three-way asymmetric proximities are regarded as a set of one-mode two-way asymmetric proximities or a set of proximity matrices given by a set of individuals or sources (hereafter the term source is used), where each proximity matrix (one-mode two-way asymmetric proximities) comes from a source. Several models based on the extended model for two-mode three-way asymmetric proximities have been introduced since then. All of these models inherit the characteristic that each object is represented as a point and a hyperellipsoid in a multidimensional Euclidean space.

Okada and Imaizumi (2003a, 2003b, 2005) introduced a new sort of models to deal with two-mode three-way asymmetric proximities. This is a joint space model where objects and sources are represented in the same multidimensional Euclidean space. In the model each object is represented as a point and a hypersphere, but the radius of the hypersphere of an object for a source varies according to the distance between the point representing the object and the point representing the source.

Another class of the joint space model for two-mode three-way asymmetric proximity was introduced, where each object is represented as a point in a multidimensional Euclidean space, and does not have its own hypersphere (or all radii of the hyperspheres are assumed to be unity). Each source is also represented as a point in the same multidimensional Euclidean space. The asymmetry of an object for a source is determined only by the distance between the point representing the object and that representing the source (Okada and Imaizumi (2004)). While the model was applied successfully to some kind of data, the relationships between objects and sources (or the asymmetry of the proximity relationships among objects) are not easy to understand, because the effect of the distance between the object and the source on the proximity relationships among objects for the source are rather complex.

The purpose of the present study is to develop a joint space model and an associated nonmetric algorithm of one-mode two-way asymmetric multidimensional scaling. In the model, each object is represented only as a point in a multidimensional Euclidean space and does not have its hypersphere. In the same multidimensional Euclidean space a point called the dominance point, which corresponds to the whole one-mode two-way asymmetric proximities to be dealt with in the analysis or the proximity matrix, is embedded, so that asymmetric proximity relationships among objects are more easily be interpreted than in Okada and Imaizumi (2004).

2 The model

The model consists of (a) a joint configuration consists of points representing objects and a point representing a proximity matrix to be dealt with, and (b) the asymmetry weight. As mentioned earlier, the point representing the proximity matrix is called the dominance point. The dominance point characterizes the property of the asymmetry in the proximity relationships among objects of the proximity matrix. The dominance point represents a hypothetical object whose similarity to the other objects is always smaller than that from the other objects to the hypothetical object. As will be mentioned later, in the case of the brand switching the dominance point represents the hypothetical brand having the strongest competitive power. This is the reason why the point is called the dominance point.

The joint configuration represents the relationships among objects, and those between objects and the dominance point. The asymmetry weight represents the salience of the asymmetry in proximity relationships among objects. Figure 1 shows a two-dimensional joint configuration. Object j is represented as a point (x_{j1}, x_{j2}), where x_{js} is the coordinate of object j along dimension s of the joint configuration. The dominance point is represented as a point (y_1, y_2), where y_s is the coordinate of the dominance point along dimension s of the joint configuration. d_{jk} is the Euclidean distance between two-points representing objects j and k

$$d_{jk} = \sqrt{\sum_{s=1}^{p} (x_{js} - x_{ks})^2}, \tag{1}$$

where p is the dimensionality of the joint space. And d_j is the distance between the point representing object j and the dominance point

$$d_j = \sqrt{\sum_{s=1}^{p} (y_s - x_{js})^2}. \tag{2}$$

Let s_{jk} be the observed proximity from objects j to k. It is assumed that s_{jk} is monotonically decreasing (when s_{jk} depicts similarity) or increasing (when s_{jk} depicts dissimilarity) related to m_{jk} which is defined as

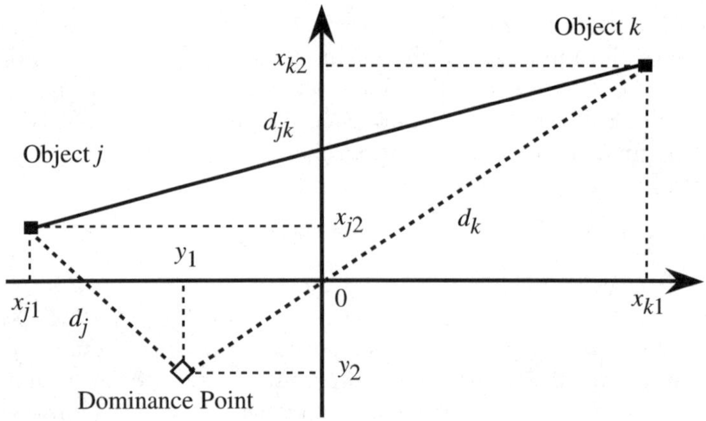

Fig. 1. The common joint configuration

$$m_{jk} = d_{jk} - u(d_j - d_k), \tag{3}$$

where u ($u > 0$) is the asymmetry weight which represents the salience of the asymmetry in proximity relationships among objects, or represents the magnitude of the effect of the distance between the point representing the object and the dominance point upon the proximities among objects. When the dominance point is closer to the point representing objects j than to k, d_j is smaller than d_k. Then m_{jk} is larger than m_{kj}, cf. Okada and Imaizumi (1987), (1997). The definition of m_{jk} is similar to that of Okada and Imaizumi (1987, 1997).

3 The algorithm

An associated algorithm to derive the joint configuration of objects and the dominance point and the asymmetry weight u from observed one-mode two-way asymmetric proximities was developed. Let n be the number of objects. A nonmetric iterative algorithm to derive the joint configuration ($x_{js}; j = 1, \cdots, n; s = 1, \cdots, p : y_s; s = 1, \cdots, p$) and the asymmetry weight u from observed proximities s_{jk} ($j, k\,[j \neq k] = 1, \cdots, n$) was extended from the one for Okada and Imaizumi (1987) which had been developed based on Kruskal's nonmetric algorithm (Kruskal (1964)). The badness-of-fit measure called stress

$$S = \sqrt{\sum_{\substack{j=1}}^{n} \sum_{\substack{k=1 \\ j \neq k}}^{n} (m_{jk} - \hat{m}_{jk})^2 \Big/ \sum_{\substack{j=1}}^{n} \sum_{\substack{k=1 \\ j \neq k}}^{n} (m_{jk} - \bar{m})^2} \tag{4}$$

is defined, where $\hat{m}_{jk}$ is the monotone transformed s_{jk}, and $\bar{m}$ is the mean of m_{jk}

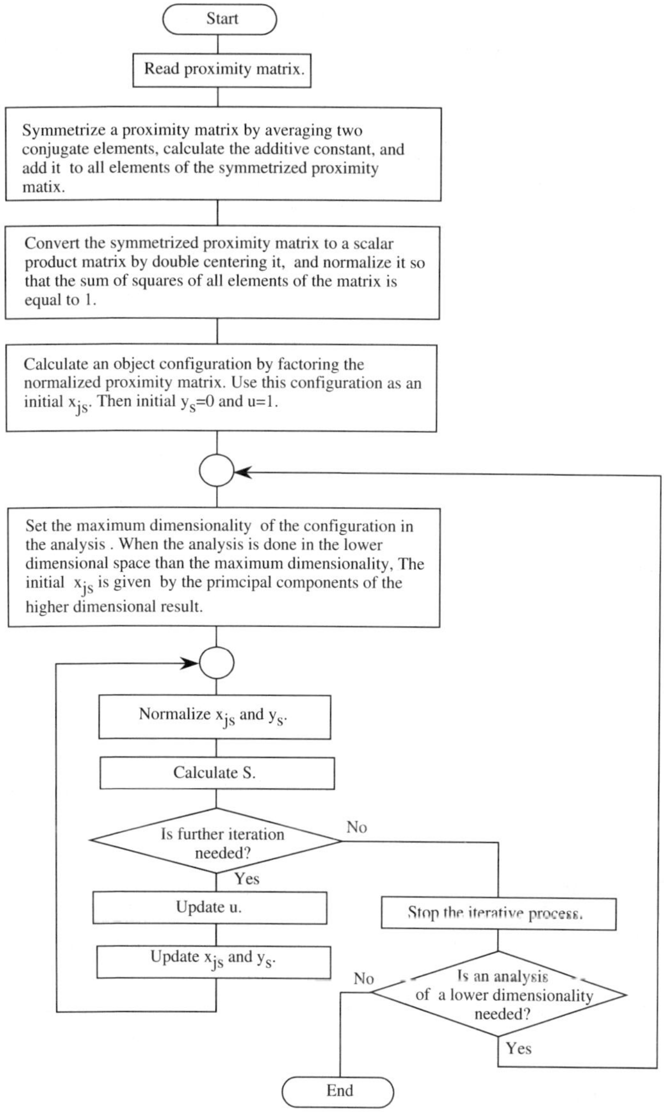

Fig. 2. Flow of the algorithm

$$\bar{m} = \sum_{\substack{j=1 \\ j \neq k}}^{n} \sum_{k=1}^{n} m_{jk}/(n(n-1)). \tag{5}$$

The joint configuration and the asymmetry weight u which minimize the stress are sought for a given dimensionality iteratively. The iteration stops when either (a), (b), or (c) is satisfied; (a) the number of the iterations reaches the maximum number of the iterations given before the analysis (100-200 iterations), (b) the stress is less than or equal to 0.1E-05, or (c) the stress of the present iteration does not decrease or increase less than or equal to 0.1E-07 compared with the stress of the previous iteration. In each iteration, the present x_{js} and y_s are normalized so that the origin is at the centroid of n points representing objects and the dominance point;

$$\sum_{j=1}^{n}\sum_{s=1}^{p} x_{js} + \sum_{s=1}^{p} y_s = 0, \tag{6}$$

and that the sum of squared coordinates of objects and of the dominance point along p dimensions is equal to $n + 1$;

$$\sum_{j=1}^{n}\sum_{s=1}^{p} x_{js}^2 + \sum_{s=1}^{p} y_s^2 = n + 1. \tag{7}$$

The joint configuration; the coordinate of the object x_{js} and of the dominance point y_s, and the asymmetry weight u are improved by the steepest descent method,

$$x_{js}^{(l+1)} = x_{js}^{(l)} - \alpha^{(l)} G_x^{(l)} \tag{8}$$

$$y_s^{(l+1)} = y_2^{(l)} - \alpha^{(l)} l G_y^{(l)} \tag{9}$$

$$u^{(l+1)} = u^{(l)} - \beta^{(l)} G_u^{(l)} \tag{10}$$

where, G_x, G_y and G_u is a gradient of S with respect to x_{js}, y_s, and u, respectively, (l) is the iteration number, and the step-size parameters $\alpha^{(l)}$ and $\beta^{(l)}$ are calculated by the linear search method.

The orientation of dimensions of the joint configuration of the present model is determined uniquely only up to the orthogonal rotation, because the orthogonal rotation does not change the distance and therefore does not affect the badness-of-fit measure S defined by Equation (4). The flow chart of the present algorithm is shown in Figure 2.

4 An application

The present asymmetric multidimensional scaling was applied to analyze car switching data among car categories (Harshman et al. (1982)). The original car switching data consist of frequencies of the car trade-in and purchase data among 16 car categories. In the present analysis, 12 car categories, obtained by eliminating four specialty car categories, were dealt with. The reasons for the elimination is described in elsewhere (Okada (1988)). The 12 car categories are; (a) Subcompact domestic (SUBD), (b) Subcompact captive imports (SUBC), (c) Subcompact imports (SUBI), (d) Low price compact (COML), (e) Medium price compact (COMM), (f) Import compact (COMI),

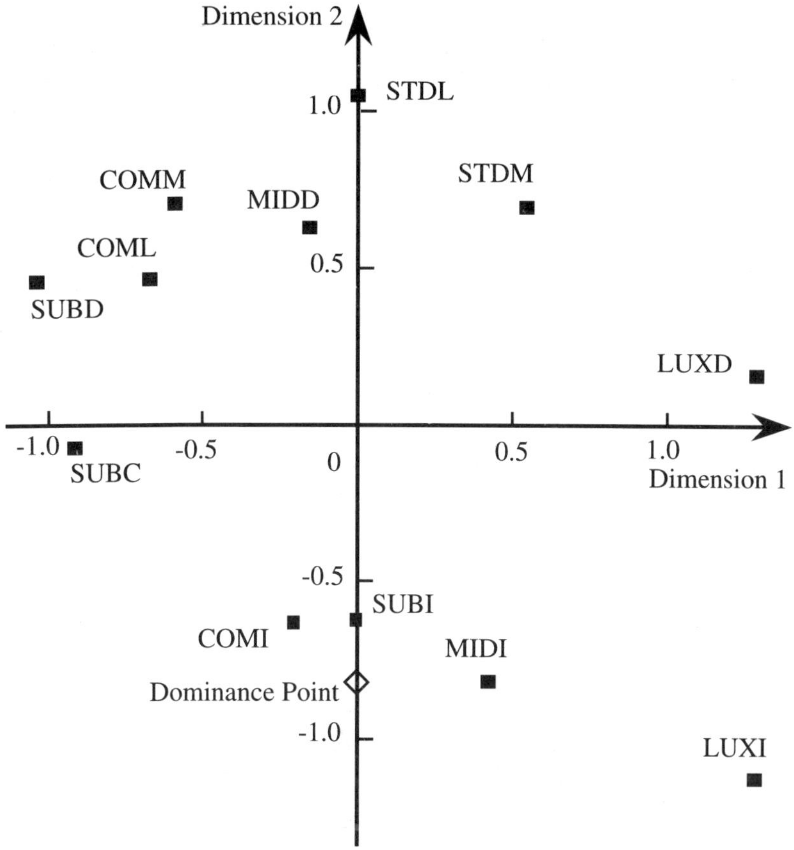

Fig. 3. The two-dimensional joint configuration of 12 car categories and the dominance point

(g) Midsize domestic (MIDD), (h) Midsize imports (MIDI), (i) Low price standard (STDL), (j) Medium price standard (STDM), (k) Luxury domestic (LUXD), and (l) Luxury import (LUXI). Abbreviations in parentheses introduced by Harshman et al. (1982) will be used to represent these car categories hereafter. The data were rescaled to remove the differences in the share of the 12 car categories (Harshman et al. (1982); Okada and Imaizumi (1997)). The rescaled car switching data were regarded to be similarities from the traded-in car categories to the newly purchased car categories. Then the rescaled car switching data are one-mode two-way asymmetric similarities among 12 car categories.

The analysis of the similarities by using the present model and the algorithm was done in five- through one-dimensional spaces. The smallest stress obtained by analyzing the one-mode two-way asymmetric similarities among 12 car categories for five- through one-dimensional spaces were 0.441, 0.453,

0.453, 0.481, and 0.614 respectively. These figures suggest adopting the two-dimensional result as the solution. While the three- and the two-dimensional results have the same stress values, it is appropriate to choose the more parsimonius two-dimensional result as the solution among the results having the same stress value. Figure 3 shows the obtained two-dimensional solution. In Figure 3, each car category is represented as a point, and the dominance point is also represented as a point in the two-dimensional Euclidean space. The obtained asymmetry weight has value u=0.126. The configuration shown in Figure 3 was derived by orthogonally rotating the originally obtained configuration so that two dimensions can easily be interpreted, because the originally obtained configuration is determined uniquely up to the orthogonal rotation.

The vertical dimension seems to represent the difference between domestic and imported car categories, because domestic car categories are located in the upper area of the configuration, and imported car categories are in the lower area of the configuration. The subcompact captive imports car category (SUBC) is located between the domestic and the imported car categories. The horizontal dimension seems to correspond to the size or the price of the car categories, because less expensive or smaller car categories are located in the left area of the configuration, and expensive and larger car categories are in the right area of the configuration. These two dimensions are compatible with those derived by earlier studies (Okada (1988), Okada and Imaizumi (1987), Zielman and Heiser (1993).

The dominance point is located near to SUBI and COMI. Car categories located nearer to the dominance point have the stronger competitive power in the car switching than those located farther from the dominance point have. This tells that the car switching from larger domestic car categories to smaller imported car categories is larger than that from smaller imported car categories to larger domestic car categories. The asymmetry weight u of 0.126 tells that the second term of the left side of Equation (3) which represents the asymmetry between m_{jk} and m_{kj} is defined by the difference between two distances d_j and d_k times 0.126. The two distances d_j and d_k representing the asymmetry affects 0.126 times smaller than d_{jk} does.

5 Discussion

The present one-mode two-way asymmetric multidimensional scaling is based on a joint space model, where both objects and the dominance point are represented in a same multidimensional Euclidean space (Coombs (1964)). Although the joint space model has been employed by some of the earlier models for two-mode three-way asymmetric multidimensional scaling, the present model employs the joint space model for one-mode two-way asymmetric multidimensional scaling.

While in the present model each object is represented only as a point in a multidimensional Euclidean space, the present model has close relationships

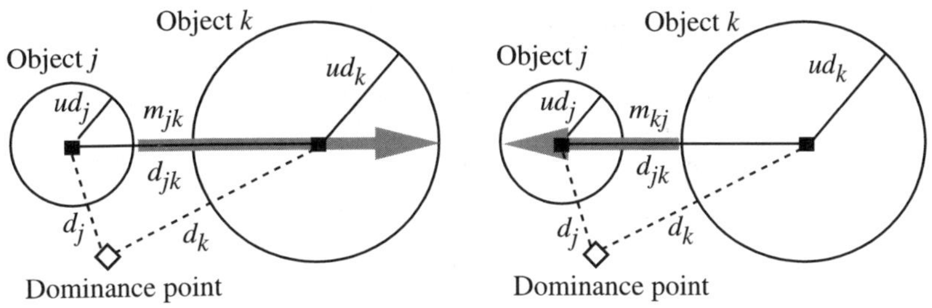

Fig. 4. Each panel shows two objects j and k in a two-dimensional Euclidean space. The panel shows interpreting the present model by considering ud_j as the radius r_j of the circle representing object j, and ud_k as the radius r_k of the circle representing object k. The gray bold arrow in the left panel shows m_{jk}, and the gray bold arrow in the right panel shows m_{kj}.

with Okada and Imaizumi (1987) where each object is represented as a point and a hypersphere. And the present model has inherited several aspects from Okada and Imaizumi (1987). Equation (3) can be rewritten as

$$m_{jk} = d_{jk} - ud_j + ud_k, \tag{11}$$

ud_j corresponds to r_j, and ud_k corresponds to r_k of the model of Okada and Imaizumi (1987). Equation (11) can also be rewritten as

$$m_{jk} = d_{jk} - ud_j r_j + ud_k r_k, \tag{12}$$

where r_j and r_k are assumed to be unity. The model represented by Equation (12) assumes that all objects have a radius of unity, and the unit radius is uniformly stretched or shrunk by ud_j and ud_k; the distance between the point representing object j and the dominance point multiplied with the asymmetry weight u and the distance between the point representing object k and the dominance point multiplied with the asymmetry weight u (cf. Okada and Imaizumi (2004)).

In the present model, ud_j (or the radius of an object j in Figure 4) is proportional to the distance between the point representing object j and the dominance point. In Figure 4 the radius of an object is almost the half of the distance between the point representing the object and the dominance point (the asymmetry weight u is almost equal to 0.5); r_j is almost equal to $0.5 \times d_j$. The term ud_j has the same meaning r_j has (the larger ud_j suggests the larger outward tendency from object j to the other objects, and the smaller inward tendency into object j from the other objects). Because the distance between the point representing an object and the dominance point is proportional to the magnitude of the asymmetry of the object or the radius, it is easier to understand the asymmetric proximity relationships among objects than in the case of earlier models for two-mode three-way asymmetric multidimensional scaling (e.g. Okada and Imaizumi (2004)).

The model of Okada and Imaizumi (2004) employs the joint space model where each object is represented as a point in a multidimensional Euclidean space and each source is represented as a point in the same multidimensional Euclidean space as well. In the model each object does not have its own hypersphere, and the term representing the asymmetry of object j is defined by

$$1 - \exp(-d_{ij}^2),$$

while in the present model the term is defined by ud_j as shown by Equations (3) or (11). In the model of Okada and Imaizumi (2004), the symmetric relationships are represented by the Euclidean distance defined by Equation (1), but the asymmetric relationships are represented by the distance including an exponential function as shown above. This means that two terms or two distances corresponding to symmetric and asymmetric relationships have different characteristics, suggesting the difficulty in interpreting the derived configuration especially comparing the symmetric and asymmetric relationships. On the other hand, in the present model both of the two terms or the two distances are the Euclidean distance. This means that the two terms or the two distances corresponding to symmetric and asymmetric relationships have the same characteristics, suggesting the interpretation of the derived configuration is easier than that derived by Okada and Imaizumi (2004).

The dominance point of the present model governs the asymmetry of the proximity relationships among objects of the data dealt with in the analysis. The asymmetry in proximity relationships between objects j and k becomes larger as the difference of the two distances increases; one is d_j, the distance between the point representing object j and the dominance point, and the other is d_k, the distance between the point representing object k and the dominance point. When the dominance point is on the perpendicular bisector of the line connecting two points representing objects j and k, two distances d_j and d_k are equal. Then m_{jk} is equal to m_{kj}, and there is no asymmetry in the relationship between objects j and k. When object j is located at the exactly same position of the dominance point, the distance between the point representing object j and the dominance point is zero. Then ud_j is equal to zero. The dominance point representing a hypothetical object whose asymmetry term (the second term of the right side of Equation (11)) or the radius is zero, suggesting the hypothetical object having the strongest competitive power among the objects (cf. Okada and Imaizumi (2005)). This tells that the more an object is preferred, the nearer the dominance point is located to the object, suggesting the object has stronger competitive power. In this case the dominance point has the largest preference like the ideal point.

The present model can be extended so that two-mode three-way asymmetric proximities (object $\times$ object $\times$ source) can be dealt with. Several extensions seem to be possible. It is assumed that s_{jki} (similarity from objects j to k for source i) is monotonically decreasing or increasing related to m_{jki}. One interesting extension is representing each source or each proximity matrix as a dominance point and introducing u_i, and define m_{jki} as

$$m_{jki} = d_{jk} - u_i(d_{ij} - d_{ik}), \tag{13}$$

where u_i is the asymmetry weight for source i, and d_{ij} is the distance between the point representing object j and the dominance point representing source i;

$$d_{ij} = \sqrt{\sum_{s=1}^{p} (y_{is} - x_{js})^2}. \tag{14}$$

In the obtained joint configuration, objects and sources are represented as points in a same multidimensional Euclidean space, and each source does not need to have its own configuration of objects (cf. Carroll and Chang (1970)). In the joint configuration the relationships among sources, as well as the relationships between objects and sources, and those among sources are described. It seems easy to understand the differences among sources or individual differences in relation to asymmetric relationships among objects.

Acknowledgments

The authors would like to express their gratitude to the anonymous referee for the variable review which was very helpful to improve the earlier version of the present paper.

References

BORG, I. and GROENEN, P.J.F. (2005): *Modern Multidimensional Scaling.* Springer, New York.

CARROLL, J.D. and ARABIE, P. (1980): Multidimensional Scaling. In: M.R. Rosenzweig and L.W. Porter (Eds.): *Annual Review of Psychology, 31.* Annual Reviews, Palo Alto, 607-649.

CARROLL, J.D. and CHANG, J.J. (1970): Analysis of Individual Differences in Multidimensional Scaling Via an N-way Generalization of 'Eckart-Young' Decomposition. *Psychometrika, 35, 283–319.*

COOMBS, C.H. (1964): *A Theory of Data.* John Wiley, New York.

CHINO, N., GROROUD, A. and YOSHINO, R. (1996): Complex Analysis of Two-Mode Three-Way Asymmetric Relational Data. *Proceedings of the Fifth Conference of the International Federation of Classification Societies.* 83-86.

DeSARBO W.S., JOHNSON, M.D., MANRAI, A.K., MANRAI, L.A. and EDWARD, E.A. (1992): TSCALE: A New Multidimensional Scaling Procedure Based on Tversky's Contrast Model. *Psychometrika, 57, 43–69.*

HARSHMAN, R.A., GREEN, P.E., WIND, Y. and LUNDY, M.E. (1982): A Model for the Analysis of Asymmetric Data in Marketing Research. *Marketing Science, 1, 205–242.*

KRUSKAL, J.B. (1964): Nonmetric Multidimensional Scaling: A Numerical Method. *Psychometrika, 29, 115–129.*

OKADA, A. (1988): Asymmetric Multidimensional Scaling of Car Switching Data. In: W. Gaul and M. Schader (Eds.): *Data, Expert Knowledge and Decisions.* Springer, Heidelberg, 279-290.

OKADA, A. and IMAIZUMI, T. (1987): Nonmetric Multidimensional Scaling of Asymmetric Proximities. *Behaviormetrika, 21, 81–96.*

OKADA, A. and IMAIZUMI, T. (1997): Asymmetric Multidimensional Scaling of Two-mode Three-way Proximities. *Journal of Classification, 14, 95–224.*

OKADA, A. and IMAIZUMI, T. (2003a): Joint Space Model for Multidimensional Scaling of Asymmetric Proximities. *Abstracts of the 27th Annual Conference of the German Classification Society.* 134.

OKADA, A. and IMAIZUMI, T. (2003b): Asymmetric Multidimensional Scaling Based on Joint Space Model. *Proceedings of the 13th International Meeting and the 68th Annual American Meeting of the Psychometric Society.*

OKADA, A. and IMAIZUMI, T. (2004): A Joint Space Model of Asymmetric Multidimensional Scaling. *Proceedings of the International Meeting and the 69th Annual American Meeting of the Psychometric Society.*

OKADA, A. and IMAIZUMI, T. (2005): Joint Space Model for Multidimensional Scaling of Two-Mode Three-Way Asymmetric Proximities. In: D. Baier and K.-D. Wernecke (Eds.): *Innovation in Classification, Data Science, and Information Systems.* Springer, Berlin Heidelberg, 371-378.

ZIELMAN, B. (1991): Three-Way Scaling of Asymmetric Proximities. *Research Report RR91-01.* Department of Data Theory, University of Leiden.

ZIELMAN, B. and HEISER, W.J. (1993): Analysis of Asymmetry by a Slide-Vector. *Psychometrika, 58,101–114.*

ZIELMAN, B. and HEISER, W.J. (1996): Models for Asymmetric Proximities. *British Journal of Mathematical and Statistical Psychology, 49, 127–146.*

Single Cluster Visualization to Optimize Air Traffic Management

Frank Rehm[1], Frank Klawonn[2] and Rudolf Kruse[3]

[1] German Aerospace Center, Lilienthalplatz 7, 38108 Braunschweig, Germany; frank.rehm@dlr.de
[2] University of Applied Sciences Braunschweig/Wolfenbüttel, Salzdahlumer Strasse 46/48, 38302 Wolfenbüttel, Germany; f.klawonn@fh-wolfenbuettel.de
[3] Otto-von-Guericke-University of Magdeburg, Universitätsplatz 2, 39106 Magdeburg, Germany; kruse@iws.cs.uni-magdeburg.de

Abstract. In this paper we present an application of single cluster visualization (SCV) a technique to visualize single clusters of high-dimensional data. This method maps a single cluster to the plane trying to preserve the relative distances of feature vectors to the corresponding prototype vector. Thus, fuzzy clustering results representing relative distances in the form of a partition matrix as well as hard clustering partitions can be visualized with this technique. The resulting two-dimensional scatter plot illustrates the compactness of a certain cluster and the need of additional prototypes as well. In this work, we will demonstrate the visualization method on a practical application.

1 Introduction

Evaluation of clustering partitions turned out to be challenging. Common prototype-based clustering algorithms minimize an objective function (Bezdek (1981)). These methods always fit the clusters to the data, even if the cluster structure is not adequate for the problem. Evaluating a partition by the value of the objective function is thusly not meaningful. To cope with this problem many validity measures are developed to analyze clustering partitions (Davies and Bouldin (1979), Dunn (1974), Höppner et al. (1999), Rubens (1992), Windham (1981)).

Unfortunately, many of the measures condense the whole issue to a single value which is associated with certain loss of information. Aggravating the situation, different measures provide contradictory results for the same partition. In the recent years visual techniques are developed to enlighten encoded information in the partition matrix, a matrix that holds membership degrees of feature vectors to prototype vectors, or to visualize clusters on low-dimensional mappings (Abonyi and Babuska (2004), Hathaway and Bezdek (2003), Huband et al. (2005), Klawonn et al. (2003)).

In this paper we apply single cluster visualization (SCV), a recently proposed method (Rehm et al. (2006)), on a practical example. SCV is used to visualize partitions of weather data that will be needed to predict flight durations at Frankfurt Airport. The rest of the paper is organized as follows. In section 2, we will briefly describe fuzzy clustering as a common representative for prototype-based clustering. Section 3 recalls the visualization technique. Results on the practical data will be given in section 4. Finally we conclude with section 5.

2 Clustering

Clustering techniques aim at finding a suitable partition for a given data set. Prototype-based clustering methods, like k-means (for crisp clustering) or fuzzy c-means (for fuzzy clustering), represent clusters by means of centre (or prototype) vectors. A partition matrix U describes a partitioning extensively holding for every feature vector $\mathbf{x}_j$ a single membership degree u_{ij} to a certain prototype vector $\mathbf{v}_i$. For crisp clustering techniques a partition matrix can be easily computed after the partitioning process. Fuzzy clustering algorithms provide membership degrees directly as one part of the clustering result. In this section, we briefly describe fuzzy c-means as a common representative for fuzzy clustering.

Fuzzy c-means aims at minimizing an objective function J that describes the sum of weighted distances d_{ij} between c prototypes vectors $\mathbf{v}_i$ and n feature vectors $\mathbf{x}_j$ of the data set $\mathbf{X} = \{\mathbf{x}_1, \mathbf{x}_2, \ldots, \mathbf{x}_n\}$ in the feature space $\mathbb{R}^p$

$$J = \sum_{i=1}^{c} \sum_{j=1}^{n} u_{ij}^m d_{ij}. \tag{1}$$

By means of the fuzzifier $m \in (1, \infty]$ one can control how much the clusters overlap. In order to avoid the trivial solution assigning no data to any cluster by setting all u_{ij} to zero and avoiding empty clusters, the following constraints are required:

$$u_{ij} \in [0,1] \qquad 1 \leq i \leq c, \;\; 1 \leq j \leq n \tag{2}$$

$$\sum_{i=1}^{c} u_{ij} = 1 \qquad 1 \leq j \leq n \tag{3}$$

$$0 < \sum_{j=1}^{n} u_{ij} < n \qquad 1 \leq i \leq c. \tag{4}$$

The Euclidean norm

$$d_{ij} = d^2(\mathbf{v}_i, \mathbf{x}_j) = (\mathbf{x}_j - \mathbf{v}_i)^T(\mathbf{x}_j - \mathbf{v}_i)$$

is used for fuzzy c-means as distance measure. Other distance measures can be applied resulting in clustering techniques which can adopt different cluster shapes (Gath and Geva (1989), Gustafson and Kessel (1979)). The minimization of the functional (1) represents a nonlinear optimization problem that is usually solved by means of Lagrange multipliers, applying an alternating optimization scheme (Bezdek (1980)). This optimization scheme considers alternatingly one of the parameter sets, either the membership degrees

$$u_{ij} = \frac{1}{\sum_{k=1}^{c}\left(\frac{d_{ij}}{d_{kj}}\right)^{\frac{1}{m-1}}} \tag{5}$$

or the prototype parameters

$$\mathbf{v}_i = \frac{\sum_{j=1}^{n}(u_{ij})^m \mathbf{x}_j}{\sum_{j=1}^{n}(u_{ij})^m} \tag{6}$$

as fixed, while the other parameter set is optimized according to equations (5) and (6), respectively, until the algorithm finally converges.

3 Single cluster visualization

We apply in this paper a recently presented method to visualize clustering results of high-dimensional data on the plane (Rehm et al. (2006)). The main concepts of Single Cluster Visualization (SCV) are the visualization of the data set from the perspective of a certain cluster while preserving the fuzzy membership degrees when mapping the data onto the plane. Thus, the challenge is to determine representative distances of feature vectors to respective cluster prototypes that preserve the fuzzy membership degrees approximately.

As already mentioned, membership degrees describe the feature vector's gradual membership to a certain cluster and can be easily determined for any prototype based clustering technique using equation (5). Preserving membership degrees instead of preserving original distances allows a very efficient computation of meaningful transformations.

In general, membership degrees cannot be preserved exactly when dimensionality reduction is carried out. A helpful step to preserve membership degrees approximately is the adaptation of the noise clustering idea (Davé (1991), Davé and Krishnapuram (1997)). Noise clustering is based on the introduction of an additional cluster – the noise cluster – that is supposed to contain all noisy or outlying vectors. Noise is defined over a fixed distance, the so-called noise distance, and denoted by δ. The prototype $\mathbf{v}_c$ of such a noise cluster is rather virtual and thusly has no parameters. The clustering scheme differs only in that point from fuzzy c-means, that membership degrees to the noise cluster will be obtained considering $d_{cj} = \delta^2$ in equation (5). Note, the noise clustering aspect is only used here as a dodge to reduce the number of variables and not for outlier detection. A second step required for our visualisation purposes is a relaxation in that point that mainly the membership degrees to the two mostly competing prototypes will be regarded. All other prototypes will be regarded as one composed noise cluster.

When mapping a clustering result onto the plane two coordinates are needed for each data point. To achieve this, the usual computation of membership degrees according to equation (5) is considered which provides a simple connection between membership degrees and distances:

$$\frac{u_{ij}}{u_{\ell j}} = \frac{\frac{1}{\sum_{k=1}^{c}\left(\frac{d_{ij}}{d_{kj}}\right)^{\frac{1}{m-1}}}}{\frac{1}{\sum_{k=1}^{c}\left(\frac{d_{\ell j}}{d_{kj}}\right)^{\frac{1}{m-1}}}} = \left(\frac{d_{\ell j}}{d_{ij}}\right)^{\frac{1}{m-1}}. \tag{7}$$

Cluster i is defined to be the cluster that is to be visualized. Note, the complete data set will be plotted by SCV, however, from the perspective of cluster i. With cluster ℓ we regard a second cluster, which is a virtual cluster, since it contains all feature vectors with the highest membership degree apart from u_{ij}. Membership degrees to this cluster will be denoted by $u_{\ell j}$. In this sense, this second cluster indicates for each data object, how much cluster i has to compete for this data object with another cluster. Data objects that are assigned with a high membership degree to cluster i will be placed close to projected cluster centre of cluster i, whereas data objects assigned to a high degree to any other cluster will be placed near the centre of the virtual (combined) cluster ℓ. Moreover, a noise cluster is defined covering all other clusters aside from i and ℓ. Since original distances are not preserved but representative distances by means of membership degrees, cluster i and cluster ℓ can be initially placed on arbitrary positions on the plane. To place prototype $\mathbf{v}_i$ at $(0,0)$ and prototype $\mathbf{v}_\ell$ at $(1,0)$ as proposed in (Rehm et al. (2006)) suggests to define the noise distance by $\delta = 1$. Keeping in mind that only two clusters plus the noise cluster are considered, we have $u_{noisej} = 1 - u_{ij} - u_{\ell j}$. According to equation (7) this leads to

$$\frac{u_{ij}}{u_{noisej}} = \left(\frac{1}{\hat{d}_{ij}}\right)^{\frac{1}{m-1}}. \tag{8}$$

The distance between cluster i and feature vector $\mathbf{x}_j$ on the plane is denoted by $\hat{d}_{ij}$ to emphasize the fact that it is not dealt with original distances any more but with representative distances with respect to the according membership degrees. Solving equation (8) for $\hat{d}_{ij}$ one obtains

$$\hat{d}_{ij} = \left(\frac{u_{noisej}}{u_{ij}}\right)^{m-1}. \tag{9}$$

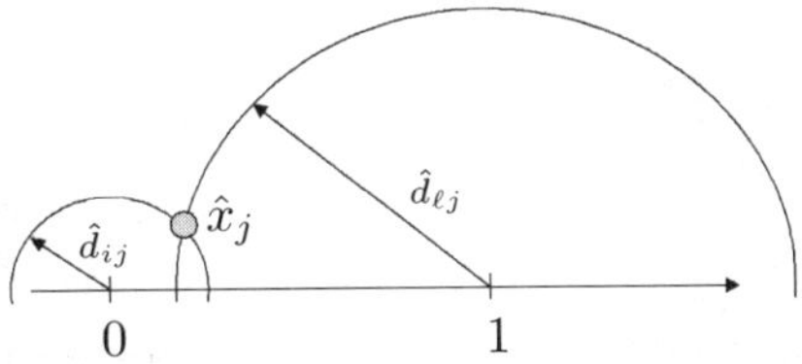

Fig. 1. Placement of $\hat{x}_j$ in the plane

Analogously, one obtains for the second cluster ℓ

$$\hat{d}_{\ell j} = \left(\frac{u_{noisej}}{u_{\ell j}} \right)^{m-1} . \tag{10}$$

Figure 1 illustrates this approach. With equation (9) one can compute the representative distance of each feature vector x_j to the cluster i, so that it is possible to draw a circle around $(0, 0)$, the position of cluster i, as one hint for the feature vector's position in the plane. With the distance to the other cluster $(1, 0)$ that we get from equation (10), one could draw another circle around the cluster centre of cluster ℓ. The intersection point of these two circles would be the position of the new feature vector $\hat{x}_j$ in the plane.

Demonstrative examples are given in (Rehm et al. (2006)). In the next section, we discuss the results of an application of SCV on a problem that arises when predicting aircraft delay as a function of weather.

4 Results

In this section, we discuss the results of SCV applying it to a weather data set that is extensively described in (Rehm et al. (2005)). The data describes some weather factors captured by various sensors present at Frankfurt Airport. An ensemble of the different weather factors at one point in time, such as atmospheric pressure, temperature, wind speed, precipitation, height of cloud layers, etc., forms a weather report. Such a report is usually released every thirty minutes. In case of rapidly changing weather the frequency is increased.

In addition to the weather data flight durations of arriving aircraft are available. More precisely, we consider flight durations in the Terminal Management Area (TMA) - a controlled airspace over the airport - and classify them into short, medium and long flights. This traffic data as well as the weather data is available for one year. Earlier studies analyzed the same data set to predict flight durations of arriving aircraft (Rehm (2004), Rehm et al. (2005)). A variety of methods were applied and some weather factors affecting flight duration could be discovered. However, these predictions were always associated with a wide variance. The visualizations shown on figure 2 reveal the reason for that.

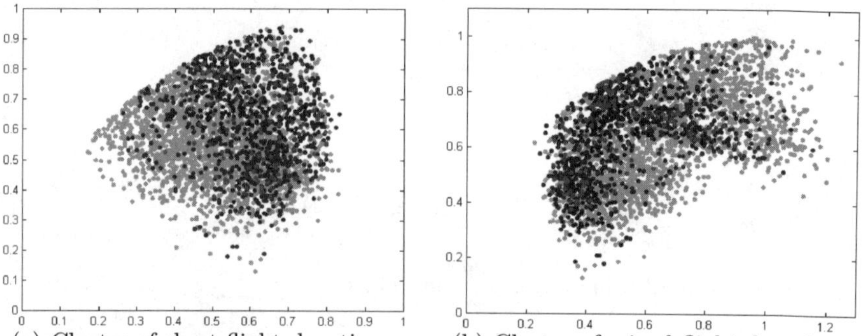

(a) Cluster of short flight durations (b) Cluster of mixed flight durations

Fig. 2. Visualization of weather clusters

Figure 2(a) shows one cluster of a three-cluster-partition. Note, that the axis labels are intentionally omitted here, since the features of the new 2-dimensional feature space are combined attributes of the original space. All feature vectors left from 0.5 on the x-axis have their highest membership degree to the cluster we try to visualize here. On the first sight it is visible that no compact cluster could be found. The changeover from cluster i to cluster ℓ is quite fluent. According to the visualization no clear border between cluster i and another cluster can be drawn. A second cluster, depicted in figure 2(b), represents flights of all three categories. Estimating flight durations based on the cluster's average flight duration produces in comparable cases a considerable variance and poor predictions accordingly. Borders between cluster i and cluster ℓ cannot be decided.

These visualizations reveal that flight durations can be partly classified using weather data as figure 2(a) evinces. However, the whole data set should not be analyzed only by partitioning methods. Some regions in the feature space seem to be more complicated and flight duration categories cannot be separated linearly. Recent studies applying support vector machines could improve prediction quality and underline our assumptions (Lesot et al. (2006)).

5 Conclusions

We presented in this paper the application of SCV – an efficient technique to map clustering results of high-dimensional data onto the plane. We showed results on a complex weather data set that is used to predict aircraft delay. Earlier studies have already shown that delay prediction using this weather data is fairly complicated. The results in this work reveal reasons for that and give hints how to overcome the problem.

References

ABONYI, J. and BABUSKA, R. (2004): FUZZSAM - Visualization of Fuzzy Clustering Results by modified Sammon Mapping. *Proceedings of the IEEE International Conference on Fuzzy Systems, 365–370.*

BEZDEK, J.C. (1980): A Convergence Theorem for the Fuzzy ISODATA Clustering Algorithms, *IEEE Transactions on Pattern Analysis and Machine Intelligence, 2, 1–8.*

BEZDEK, J.C. (1981): *Pattern Recognition with Fuzzy Objective Function Algorithms.* Plenum Press, New York.

DAVE, R.N. (1991): Characterization and Detection of Noise in Clustering. *Pattern Recognition Letters, 12, 657–664.*

DAVE, R.N. and KRISHNAPURAM, R. (1997): Robust Clustering Methods: A Unified View. *IEEE Transactions on Fuzzy Systems, 5, 270–293.*

DAVIES, D.L. and BOULDIN, W. (1979): A Cluster Separation Measure. *IEEE Transactions on Pattern Analysis and Machine Intelligence, 1, 224–227.*

DUNN, J.C. (1974): Well Separated Clusters and Optimal Fuzzy Partitions. *Journal of Cybernetics, 4, 95–104.*

GATH, I. and GEVA, A.B. (1989): Unsupervised Optimal Fuzzy Clustering. *IEEE Transactions on Pattern Analysis and Machine Intelligence, 11, 773–781.*

GUSTAFSON, D.E. and KESSEL, W.C. (1979): Fuzzy Clustering with a Fuzzy Covariance Matrix. *Proceedings of the IEEE Conference on Decision and Control,* San Diego, 761–766.

HATHAWAY, R.J. and BEZDEK, J.C. (2003): Visual Cluster Validity for Prototype Generator Clustering Models. *Pattern Recognition Letters, 24, 9–10.*

HÖPPNER, F., KLAWONN, F., KRUSE, R. and RUNKLER, T. (1999): *Fuzzy Cluster Analysis.* John Wiley & Sons, Chichester.

HUBAND, J.M., BEZDEK, J.C. and HATHAWAY, R.J. (2005): bigVAT: Visual Assessment of Cluster Tendency for Large Data Sets. *Pattern Recognition Letters, 38, 1875–1886.*

KLAWONN, F., CHEKHTMAN, V. and JANZ, E. (2003): Visual Inspection of Fuzzy Clustering Results. In: Benitez, J., Cordon, O., Hoffmann, F. and Roy, R. (Eds.): *Advances in Soft Computing: Engineering Design and Manufacturing.* Springer, London, 65–76.

LESOT, M.J., REHM, F., KLAWONN, F. and KRUSE, R. (2006): Prediction of Aircraft Flight Duration. *Proceedings of the 11th IFAC Symposium on Control in Transportation Systems, Delft, 107–112.*

REHM, F. (2004): Prediction of Aircraft Delays as a Function of Weather. *2nd WakeNet2-Europe Workshop on Capacity Gains as Function of Weather and Weather Prediction Capabilities*, Langen.

REHM, F., KLAWONN, F. and KRUSE, R. (2005): Learning Methods for Air Traffic Management. In: Godo, L. (Ed.): *ECSQARU 2005, LNAI, 3571.* Springer, Berlin, 992–1001.

REHM, F., KLAWONN, F. and KRUSE, R. (2006): Visualization of Single Clusters. *LNCS, 4029*, Springer, Berlin, 663–671.

RUBENS, M. (1992): Fuzzy Clustering Algorithms and their Cluster Validity. *European Journal of Operational Research, 10, 294–301.*

WINDHAM, M.P. (1981): Cluster Validity for Fuzzy Clustering Algorithms. *Fuzzy Sets and Systems, 5, 177–185.*

Rescaling Proximity Matrix Using Entropy Analyzed by INDSCAL

Satoru Yokoyama[1] and Akinori Okada[2]

[1] School of Science for Open and Environmental Systems, Graduate School of Science and Technology, Keio University, 3-14-1 Hiyoshi, Kohoku-ku Yokohama, 223-8522 Japan; satoru_y@ae.keio.ac.jp

[2] Department of Industrial Relations, School of Social Relations, Rikkyo (St. Paul's) University, 3-34-1 Nishi Ikebukuro, Toshima-ku Tokyo, 171-8501 Japan; okada@rikkyo.ac.jp

Abstract. Yokoyama and Okada (2005, in press) suggested a new method that rescales a brand-switching data matrix using entropy. These studies applied the rescaling method to car-switching data, and the configuration derived by Kruskal's multidimensional scaling was interpreted as the circumplex. In the present paper, we apply that rescaling method to intergenerational occupational mobility data for four years, and analyzed the resluts by Krsukal's multidimensional scaling. As a result, the configurations are also interpreted as the circumplex. Furthermore, we also find that the result of the analysis of these rescaled data by INDSCAL is interpreted as the circumplex.

1 Introduction

Recently, multidimensional scaling (MDS) models have been applied to various kinds of data. In particular, some studies have analyzed brand-switching data by MDS. The result of the analysis is strongly affected by the factors extraneous to the data, as Harshman et al. (1982) pointed out.

Furthermore, Harshman et al. (1982) also proposed a treatment for brand-switching data (e.g., car-switching data). The large differences in the overall sums of the rows and columns of a brand-switching data matrix are due primarily to extraneous factors, such as market share. These overall differences reflect factors of interchangeability and the relative attractiveness of objects to different kinds of individuals. Hence, that method removes all extraneous size differences in order to reveal the structure of the data more clearly.

Yokoyama and Okada (2005, in press) suggested a new method that rescales a brand-switching data matrix using entropy, and the resulting configuration of the analysis by Kruskal's MDS (Kruskal (1964)) can be interpreted as a circumplex structure. The present method corresponds to transforming a proximity matrix into the variance of each row and column using entropy. The present paper applies the present method to two-mode three-way proximity data, analyzes the data by INDSCAL (Carroll and Chang (1970)), and investigate the characteristics of the rescaling method. Using the present rescaling method, it is thought that we can determine the uncertainty of a brand-switching. The configurations appear to be interpreted by circumplex with two or three dimensions (cf. Yokoyama and Okada (in press)). In addition, we discuss the possibility of other interpretations and make comparisons with other rescaling methods.

2 Rescaling method using entropy

Yokoyama and Okada (2005, in press) suggested a rescaling method using the uncertainty of a brand-switching matrix. In this method, brand-switching of each row and each column is considered to be a probability before we calculate the uncertainty from these probabilities using entropy, and then rescale the matrix, whose elements represent the uncertainty of the corresponding row and column, by combining the entropy from the row and from the column. Hence, we calculate the rescaling matrix by uncertainty.

The present rescaling method is represented by the following algorithm. Let $\mathbf{X} = [x_{jk}]$ be the raw brand-switching data matrix, and $\mathbf{P}^r = [p^r_{jk}]$ and $\mathbf{P}^c = [p^c_{jk}]$ be the row and column probability of each brand-switch,

$$\mathbf{P}^r = \begin{pmatrix} \frac{x_{11}}{\sum_{k=1}^n x_{1k}} & \cdots & \frac{x_{1n}}{\sum_{k=1}^n x_{1k}} \\ \vdots & \ddots & \vdots \\ \frac{x_{n1}}{\sum_{k=1}^n x_{nk}} & \cdots & \frac{x_{nn}}{\sum_{k=1}^n x_{nk}} \end{pmatrix}, \quad \mathbf{P}^c = \begin{pmatrix} \frac{x_{11}}{\sum_{j=1}^n x_{j1}} & \cdots & \frac{x_{1n}}{\sum_{j=1}^n x_{jn}} \\ \vdots & \ddots & \vdots \\ \frac{x_{n1}}{\sum_{j=1}^n x_{j1}} & \cdots & \frac{x_{nn}}{\sum_{j=1}^n x_{jn}} \end{pmatrix}. \tag{1}$$

Then $\mathbf{H}^r$ and $\mathbf{H}^c$ define the entropy measures calculated from $\mathbf{P}_r$ and $\mathbf{P}_c$ respectively,

$$\mathbf{H}^r = \begin{pmatrix} -\sum_{l=1}^n p^r_{1l} \log_2 p^r_{1l} \\ \vdots \\ -\sum_{i=l}^n p^r_{nl} \log_2 p^r_{nl} \end{pmatrix},$$

$$\mathbf{H}^c = \left(-\sum_{l=1}^n p^c_{l1} \log_2 p^c_{l1} \ \cdots \ -\sum_{l=1}^n p^c_{ln} \log_2 p^c_{ln} \right). \tag{2}$$

Finally, the rescaled matrix is berived by combing $\mathbf{H}^r$ and $\mathbf{H}^c$ as follows:

$$\mathbf{H} = \mathbf{H}^r \mathbf{1}^t + \mathbf{1}\mathbf{H}^c. \tag{3}$$

If a value of the (j, k) element of $\mathbf{H}$ is large, it is thought to have two meanings. Either, it might be uncertain to brand-switch row j and/or column k or, object j is not similar to object k. Therefore, the relation between the "value of elements of $\mathbf{H}$" and the "proximities of corresponding objects" can be summarized as in Table 1. In the original (raw) data or in Harshman et al. (1982)'s rescaling method, a larger (or smaller) value of the (j, k) element shows that objects j and k are similar (or dissimilar). By contrast in the present rescaling, each element expresses the uncertainty of the corresponding row and column, and the uncertainty shows that corresponding two objects are dissimilar. Therefore a larger (or smaller) value of the (j, k) element shows that objects j and k are dissimilar (or similar). In addition, each element of the rescaled matrix is in the interval $[0, 2\log_2 n]$.

Table 1. Relations of value of elements and proximity

	Value of elements	Proximity
Original (Raw) data, Harshman et al. (1982)'s rescaling method	Large	Similarity
	Small	Dissimilarity
Proposed rescaling method method	Large	Dissmilarity
	Small	Similarity

3 Appling Kruskal's MDS

3.1 Car-switching data

In this section, we introduce the result (applying the present rescaling method) to one-mode two-way data, i.e., car-switching data (Harshman et al. (1982) p. 221, Table 4). The car-switching data consist of the car trade-in frequency of U.S. buyers in 1979, and cars are classified in 16 segments. We show the result briefly, while Yokoyama and Okada (in press) discusses it in more detail. The data were rescaled using the proposed method, and analyzed by Kruskal's MDS.

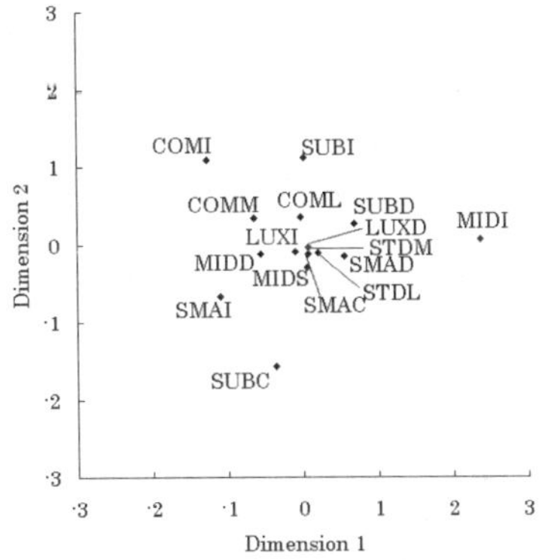

Fig. 1. Configuration of the car-switching data

As a result, the stress values in five- through unidimensional space were 0.565, 0.576, 0.595, 0.629, and 0.711, respectively. The two-dimensional result seems appropriate as the solution. Fig. 1 shows the configuration of the analysis. In Fig. 1, import categories (SUBC, SUBI, SMAI, COMI, MIDI), domestic categories (SUBD, SMAD, COML, COMM, MIDD) and luxury and specialty categories (SMAC, MIDS, STDL, STDM, LUXD, LUXI) of car segments are placed from the periphery to the center (origin). In addition, Medium Price Compact (COMM) and Import Compact (COMI) are the same direction from the center, while Low Price Compact and Subcompact Imports are also in the same direction, although it differs from the first. These two directions are "Compact Categories". Therefore, Fig. 1 is interpreted as the distance and the direction from the origin. In other words, this configuration appears to be interpreted by circumplex (Degerman (1970, 1972)).

In the analysis using Harshman et al. (1982)'s rescaling method, the axes of the solution can be interpreted as either a size or price dimension and an imports/domestic dimension (in Okada and Imaizumi (1987), Yokoyama and Okada (in press)). It is thought that the present result can extend the managerial implications slightly.

3.2 Intergenerational occupational mobility data

The present rescaling method is applied to the intergenerational occupational mobility data for four years (1955, 1965, 1975, and 1985) for Japan (Seiyama et al. (1990, pp. 46–47, Table 2.12))[1]. These data have eight occupational categories (objects), Professional is intellectual occupations; Non-manual large means office workers employed by the large companies; Non-manual small means office workers employed by the small companies; Non-manual self means self-employed office workers; Manual large means physical labors employed by the large companies; Manual small means physical labors employed by the small companies; Manual self means self-employed physical labors; Farmer means agricultural occupations.

First, in the present paper, these four data matrices are rescaled using the present method, and then analyzed by Kruskal's MDS. The analyses with the maximum dimension were ten through five dimensions, and the two-dimensional results were chosen as the solution because of the simplicity in the interpretation.

Fig. 2 shows the two-dimensional configuration of four analyses. The upper-left panel is the result for 1955, upper-right that for 1965, lower-left that for 1975 and lower-right that for 1985. The resulting minimized stress values, S, were 0.636, 0.519, 0.758, and 0.791 respectively.

In 1955, Professional and Farm were very close to each other, and the other six occupations were located around these two occupations. This suggests that

[1] In the present paper, each category is named as in Okada and Imaizumi (1997, p. 214).

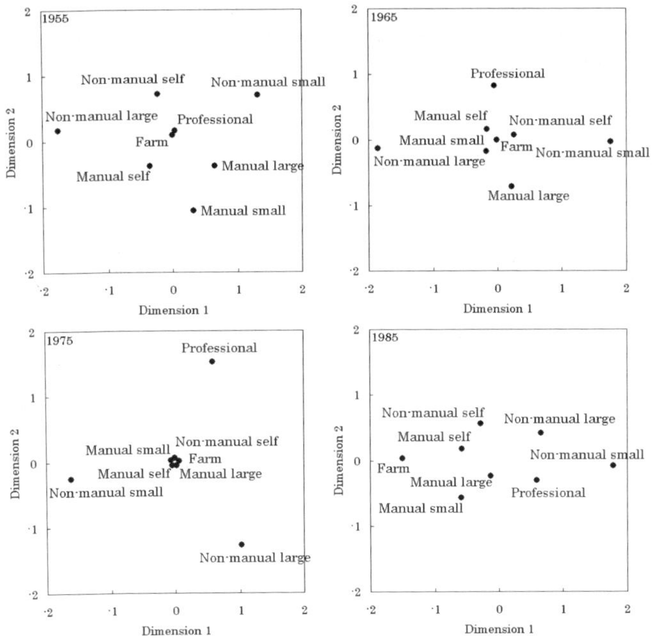

Fig. 2. Configurations of the intergenerational occupational data analyzed by Kruskal's MDS: the upper-left is the result for 1955, upper-right that for 1965, lower-left that for 1975 and lower-right that for 1985

both occupations were close because sons inherited from their fathers. The vertical dimension represents the difference between non-manual and manual occupations.

In 1965, Farm, Non-manual self, Manual small, and Manual self appear in the center, and the others are in the periphery because these four occupational categories have relatively small uncertainties.

In 1975, the configuration differs from the other configurations, Professional, Non-manual small, and Non-manual large appear in the periphery. These occupational mobilities appear uncertain, because in this year many fathers had manual occupations, while their sons increasingly chose non-manual occupations.

In 1985, the horizontal dimension appears represent the difference between manual and non-manual. The uncertainties of Farm and Non-manual small are relatively large because of the occupation shifts to the right side from the left side of this configuration in 1980. Therefore, these three non-manual occupations and Farm appear on opposite sides of the horizontal dimension.

In these analyses, the stress value might be large. However, when the same data were analyzed using Kruskal's MDS after using the rescaling method suggested by Harshma et al. (1982), the respective stress values were 0.745, 0.782,

0.791, and 0.777. Therefore, we choose the result for which the configration could be interpreted. In fact, these four configurations seem to be interpretable as the circumplex because some occupations appear in the center and the others in the periphery. In addition, it is possible to interpret all configurations using the direction from the center.

The analyses of the car-switching data and intergenerational occupational mobility data using the proposed rescaling method tells that the proximity of the objects that appear in the periphery is relatively large, while the proximity of the objects that appear in the center is small. In other words, for objects appearing from the center to outward, the rescaled proximity is increasing. Therefore, it seems possible to interpret it as the circumplex.

4 Applying INDSCAL

In this section, the four matrices of the intergenerational occupational mobility data that were analyzed in Sec. 3.2 are regarded as two-mode three-way data. Four matrices of rescaled data are analyzed by INDSCAL. The analysis with rational initial configurations and values are done using maximum dimensionalities of eight through five. The one variance accounted for (VAF) is obtained in eight-dimensional space, two VAF are obtained in seven-dimensional space, and so on until four VAF in five- through unidimensional spaces. The largest VAF in each dimension is chosen as the maximized VAF in that dimension. The resulting maximized VAF in five through unidimensions were 0.849, 0.716, 0.575, 0.424, and 0.227.

Considering these VAF and interpretation, the three dimensional result is chosen as the solution. Table 2 shows the coordinates of the subject space, and Fig. 3 is the configuration of the group stimulus space.

In Table 2, while the weights obtained are not exactly on an diagonal line or a line through the origin, four points seem to be close to the diagonal. Therefore, the orientation of the dimensions might have indeterminacy. Nevertheless, as shown below, the configurations can be interpreted as the circumplex. The uniqueness of the orientation of the dimensions is discussed in the next section.

In Fig. 3, manual self and Non-manual self appear near the origin, Professional, Manual small, and Farm appear more peripherally, and Manual large, Non-manual large, and Non-manual small appear in the most peripheral area. The objects in these last two groups have large uncertainties compare with other objects. In Fig 3, the three axes can be interpreted in the following way: dimension 1 distinguishes Non-manual large and Non-manual small from the other six occupations, dimension 2 distinguishes Non-manual large and Non-manual small from the other six, and dimension 3 distinguishes Manual large from the other seven. Here, it is more important that the configurations can be interpreted as the circumplex, similar to the result of the one-mode two-way analysis.

Table 2. Coordinates of subject space

Year	dim 1	dim 2	dim 3
1955	0.484	0.448	0.384
1965	0.507	0.456	0.378
1975	0.455	0.434	0.391
1985	0.459	0.437	0.398

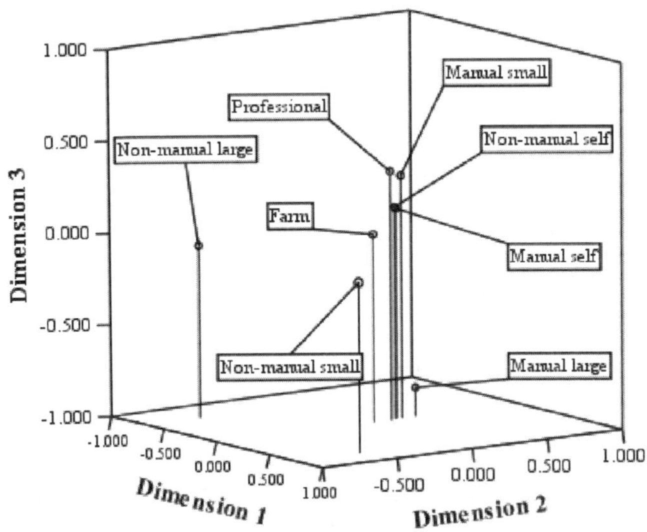

Fig. 3. Group stimulus space of the occupational data

5 Discussion

In the present paper, we presented a rescaling method which uses entropy and an application to two-mode three-way proximity data, as well as an application to one-mode two-way data. The analysis of the intergenerational occupational mobility data by INDSCAL suggested the group stimulus space was interpreted using the circumplex structure as in the analysis of the one-mode two-way data by Kruskal's MDS.

As shown in the subject space in Table 2, four points lie close to the diagonal line. The indeterminacy of the orientation of the dimensions is considered here. As noted earlier, the VAF in three-dimensional space was 0.575. The dimensions of the resulting three-dimensional group stimulus space were rotated 15 degrees in any two dimensions, clock and counter-clockwise. Then, the subject space that maximizes the VAF was derived for the rotated group stimulus spaces. For all cases, the VAF was nearly 0.575 and the orientation of the dimensions appeared to have indeterminacy. Therefore, the coordinates of the subject space do not seem to have an important meaning and the group stimulus space is to define the direct sum of the four configurations of Fig. 2 (see Carroll and Chang (1970, p. 306)). Therefore, comparing Figs. 2 and 3, when the group stimulus space is projected onto dimension 3, the projected space is similar to the configuration for 1975, and when it is projected onto dimension 2, the projected space is similar to that for 1965's. In addition, it seems possible to obtain the configuration for the other years by projecting the group stimulus space onto any arbitrary direction. Therefore, an advantage of this analysis is that it is able to interpret the group stimulus space including the results of Kruskal's MDS. However, we think that a detailed discussion of the relation between the subject space and the orientation of the dimensions of the group stimulus space is needed.

The present paper shows that the proximity data rescaled using the proposed method results in a configuration that can be interpreted as the circumplex after analyses by Kruskal's MDS and INDSCAL, where the more uncertain objects appear in more peripheral areas.

References

CARROLL, J.D. and CHANG, J.J. (1970): Analysis of Individual Differences in Multidimensional Scaling via an N-way Generalization of 'Eckart-Young' Decomposition. *Psychometrika, 35, 283–319.*

DEGERMAN, R.L. (1970): Multidimensional Analysis of Complex Structure: Mixtures of Class and Quantitative Variation. *Psychometrika, 35, 475–491.*

DEGERMAN, R.L. (1972): The Geometric Representation of Some Simple Structures. In: R.N. Shepard, A.K. Romney and S.B. Nerlove (Eds.): *Multidimensional Scaling: Theory and Applications in the Behavioral Sciences Vol. 1 Theory.* Seminar Press, New York, 105-155.

HARSHMAN, R.A., GREEN, P.E., WIND, Y. and LUNDY, M.E. (1982): A Model for the Analysis of Asymmetric Data in Marketing Research. *Marketing Science, 1, 205–242.*

KRUSKAL, J.B. (1964): Nonmetric Multidimensional Scaling: A Numerical Method. *Psychometrika, 29, 115–129.*

OKADA, A. and IMAIZUMI, T. (1987): Nonmetric Multidimensional Scaling of Asymmetric Proximities. *Behaviormetrika, 21, 81–96.*

OKADA, A. and IMAIZUMI, T. (1997): Asymmetric Multidimensional Scaling of Two-mode Three-way Proximities. *Journal of Classification, 14, 195–224.*

SEIYAMA, K., NAOI, A., SATO, Y., TSUZUKI, K. and KOJIMA, H. (1990): Stratification Structure of Contemporary Japan and Its Trend. In: A. Naoi and K. Seiyama (Eds.): *Social Stratification in Contemporary Japan, Vol.1. Structure and Process of Social Stratification.* Tokyo University Press, Tokyo, 15-50.

YOKOYAMA, S. and OKADA, A. (2005): Rescaling Proximity Matrix Using Entropy [Summary]. *Proceedings of the 29th Annual Conference of the German Classification Society, 254.*

YOKOYAMA, S. and OKADA, A.: Rescaling a Proximity Matrix Using Entropy in Brand-switching Data. *The Japanese Journal of Behaviormetrics, forthcoming.*

Part V

Information Retrieval, Data and Web Mining

Canonical Forms for Frequent Graph Mining

Christian Borgelt

European Center for Soft Computing, c/ Gonzalo Gutiérrez Quirós s/n,
33600 Mieres, Spain; christian.borgelt@softcomputing.es

Abstract. A core problem of approaches to frequent graph mining, which are based on growing subgraphs into a set of graphs, is how to avoid redundant search. A powerful technique for this is a canonical description of a graph, which uniquely identifies it, and a corresponding test. I introduce a family of canonical forms based on systematic ways to construct spanning trees. I show that the canonical form used in gSpan (Yan and Han (2002)) is a member of this family, and that MoSS/MoFa (Borgelt and Berthold (2002), Borgelt et al. (2005)) is implicitly based on a different member, which I make explicit and exploit in the same way.

1 Introduction

Recent years saw an intense and still growing interest in the problem how to find common subgraphs in a database of (attributed) graphs, that is, subgraphs that appear with a user-specified minimum frequency. For this task—which has applications in, for example, biochemistry, web mining, and program flow analysis—several algorithms have been proposed. Some of them rely on principles from inductive logic programming and describe graph structures by logical expressions (Finn et al. (1998)). However, the vast majority transfers techniques developed originally for frequent item set mining.[1] Examples include MolFea (Kramer et al. (2001)), FSG (Kuramochi and Karypis (2001)), MoSS/MoFa (Borgelt and Berthold (2002)), gSpan (Yan and Han (2002)), CloseGraph (Yan and Han (2003)), FFSM (Huan et al. (2003)), and Gaston (Nijssen and Kok (2004)). A related approach is used in Subdue (Cook and Holder (2000)). The basic idea of these approaches is to grow subgraphs into the graphs of the database, adding an edge and maybe a node in each step, counting the number of graphs containing each grown subgraph, and eliminating infrequent subgraphs.

[1] See, for example, Goethals and Zaki (2003, 2004) for details and references on frequent item set mining.

While in frequent item set mining it is trivial to ensure that the same item set is checked only once, in frequent subgraph mining it is a core problem how to avoid redundant search. The reason is that the same subgraph can be grown in several different ways, adding the same nodes and edges in different orders. Although multiple tests of the same subgraph do not invalidate the result, they can be devastating for the execution time of the algorithm.

One of the most promising ways to avoid redundant search is to define a canonical description of a (sub)graph. Together with a specific way of growing the subgraphs, such a canonical description can be used to check whether a given subgraph has been considered in the search before and thus need not be extended. This approach underlies the gSpan algorithm (Yan and Han (2002)) and its extension CloseGraph (Yan and Han (2003)). In this paper I generalize the canonical form of gSpan, thus arriving at a family of canonical descriptions. I also show that a competing algorithm called MoSS/MoFa (Borgelt and Berthold (2002), Borgelt et al. (2005)) is implicitly based on a canonical description from this family, which is different from the gSpan one. By making this canonical form explicit, it can be exploited in MoSS/MoFa in the same way as in gSpan, leading to a significant improvement of the MoSS/MoFa algorithm.

2 Finding frequent subgraphs

Generally, a graph database is searched for frequent subgraphs as follows: Given an initial node (for which all possibilities have to be tried), a subgraph is grown by adding an edge and, if necessary, a node in each step. In this stepwise extension process one usually restricts the search to connected subgraphs (which suffices for most applications). In its most basic form the search considers all possible extensions of the current subgraph. (It will be shown later how the set of extensions can be reduced by exploiting a canonical description.)

Note that, as a consequence of the above, the search produces a numbering of the nodes in each subgraph: the index of a node simply reflects the order in which it was added. In the same way it produces an order of the edges—again the order in which they were added. Even more: the search builds a spanning tree of the subgraph, which is enhanced by additional edges (closing cycles).

3 Canonical forms of attributed graphs

In this section I describe the family of canonical descriptions that is introduced in this paper, using the special cases employed in gSpan and MoSS/MoFa as examples and pointing out alternatives. How these canonical forms define an extension strategy and thus a search order is discussed in the next section.

3.1 General idea

The core idea underlying a canonical form is to construct a code word that uniquely identifies a graph up to isomorphism and symmetry (i.e. automorphism). The characters of this code word describe the edges of the graph. If the graph is attributed or directed, they also comprise information about edge attributes and/or directions as well as attributes of the incident nodes.

While it is straightforward to capture the latter information about an edge (i.e. attributes and edge direction), how to describe the connection structure is less obvious. For this, the nodes of the graph must be numbered (or more generally: endowed with unique labels), because we need a way to specify the source and the destination node of an edge. Unfortunately, different ways of numbering the nodes of a graph yield different code words, because they lead to different specifications of an edge (simply because the indices of the source and the destination node differ). In addition, the edges can be listed in different orders. How these two problems can be treated is described in the following: the different possible solutions give rise to different canonical forms.

However, given a (systematic) way of numbering the nodes of a (sub)graph and a sorting criterion for the edges, a canonical description is generally derived as follows: each numbering of the nodes yields a code word, which is the concatenation of the sorted edge descriptions (details are given in Section 3.4). The resulting list of code words is sorted lexicographically. The lexicographically smallest code word is the canonical description. (Note that the graph can be reconstructed from this code word.)

3.2 Constructing spanning trees

From the review of the search process in Section 2 it is clear that we can confine ourselves to numberings of the nodes of a (sub)graph that result from spanning trees, because no other numberings will ever occur in the search. Even more: specific systematic ways of constructing a spanning tree suffice. The reason is that the basic search algorithm produces *all* spanning trees of a (frequent) (sub)graph, though usually in different branches of the search tree.[2] Since the extensions of a (sub)graph need to be checked only once, we may choose to form them only in the branch of the search tree, in which the spanning tree of the (sub)graph has been built in the chosen way. This can also be used to ensure that the canonical description has the *prefix property*, meaning that each prefix of a canonical code word is a canonical code word itself. Since in the search we extend only graphs in canonical form, the prefix property is needed to make sure that all (sub)graphs can be reached.

The best-known systematic methods for constructing a spanning tree of a graph are, of course, depth-first and breadth-first search. Thus it is not surprising that gSpan uses the former to define its canonical form (Yan and

[2] They occur in the same search tree node only if the graph exhibits some symmetry, i.e., if there exists an automorphism that is not the identity.

Han (2002)). However, the latter (i.e., breadth-first search) may just as well be chosen as a basis for a canonical form. And indeed: as will turn out later, the (heuristically introduced) local extension order of the MoSS/MoFa algorithm (Borgelt and Berthold (2002), Borgelt et al. (2005)) can be justified from a canonical form that is based on a breadth-first search tree. Thus MoSS/MoFa can be seen as implicitly based on this canonical form.

Other methods include a spanning tree construction that first visits all neighbors of a node (like breadth-first search), but then chooses the next node to extend in a depth-first manner (This may be seen as a variant of depth-first search.). However, in the following I confine myself to (standard) depth-first and breadth-first search trees to keep things simple. Nevertheless, it should be kept in mind that there are various other possibilities one may explore.

It should be noted that there is no restriction on the order in which the neighbors of a node are visited in the search. Hence there is generally a large number of different spanning trees, even if the root node is fixed. As a consequence choosing a method for constructing a spanning tree is not sufficient to avoid redundant search. Since usually several spanning trees of a (sub)graph can be constructed in the chosen way, there are several search tree branches that qualify for an extension of the (sub)graph. Although this freedom will be reduced below by exploiting edge and node attributes, it cannot be eliminated completely, since there are no local (i.e. node-specific) criteria that allow for an unambiguous decision in all cases (Borgelt and Berthold (2002)) gives an example). Therefore we actually need to construct and compare code words to avoid all redundancy.

3.3 Edge sorting criteria

Once we have a numbering of the nodes, we can set up edge descriptions and sort them. In principle, the edge descriptions can be sorted using any precedence order of the edge's properties (i.e. attribute of the edge and attributes and indices of the source and destination node). However, we can exploit the purpose for which the canonical form is intended to find appropriate sorting criteria. Recall that we construct different spanning trees of the same (sub)graph in different branches of the search tree. Each of these gives rise to a numbering of the nodes and thus a code word. In addition, recall that the canonical form is intended for confining the extensions of a (sub)graph to one branch of the search tree. Hence we need a way of checking whether the code word resulting from the node numbering in a search tree node is minimal or not: if it is, we descend into the search tree branch, otherwise we prune it.

In order to carry out this test, we could construct all other possible code words for a (sub)graph and compare them to the one resulting from the node numbering in the current search tree node. However, such a straightforward approach is much too costly. Fortunately, it can be made much more efficient, since the code words are compared lexicographically. Hence we may not need to know the full code words in order to decide which of them is lexicographically smaller—a prefix may suffice. This immediately gives rise to the idea to check all code words in parallel that share the same prefix. However, whether this is (easily) possible or not, depends on how we sort the edges.

Fortunately, for both canonical forms, depth-first and breadth-first search, there is a sorting criterion that yields such an order. The core idea is to define the order of the edges in such a way that they are sorted into the order in which they are added to the (sub)graph in the search. This has three advantages: in the first place, it ensures the prefix property. Secondly, we need no sorting to obtain the code word that results from the node numbering in the current search tree node. Since it is easiest to implement the search by always appending the added edge to a list of contained edges, this edge list already yields the code word. Thirdly, we can carry out the search for alternative code words in basically the same way as the whole search for frequent subgraphs. Doing so makes it possible to compare the prefixes of the code words after the addition of every single edge, thus making the test of a code word maximally efficient. Details about the comparisons are given below, after the exact form of the code words for the two canonical forms are defined.

3.4 Code words

In my definition of a code word I deviate slightly from the definition of gSpan (Yan and Han (2002)), where code words are simple lists of edge descriptions, each of which comprises all information about the edge and the incident nodes. The reason is that it is not necessary to compare the attribute of the source node, except for the first edge that is added. In other words, we may precede the sequence of edge descriptions by a character that specifies the attribute of the root of the spanning tree, while at the same time we cancel the attribute of the source node from the following edge descriptions. Then the general forms of code words (as regular expressions with non-terminal symbols) are:

- Depth-First Search: $a\,(i_d\,\underline{i_s}\,b\,a)^m$
- Breadth-First Search: $a\,(i_s\,b\,a\,i_d)^m$

Here a is a node attribute and b an edge attribute. i_s is the index of the source and i_d the index of the destination node of an edge. (Source and destination of an edge are defined by the relation $i_s < i_d$, that is, the incident node with the smaller index is the source.) Parentheses are for grouping characters; each parenthesized sub-word stands for one edge. The exponent m means that there are m repetitions of the group of characters to which it is attached.

The describing properties of an edge are compared in the order in which they appear in the parenthesized expressions. All characters are compared ascendingly, with the exception of the underlined i_s in the depth-first search form, which is compared descendingly. Note that the parenthesized expressions (that is, the edge descriptions) are sorted and are concatenated afterwards to form the code word. It is easy to see that in this way the code word describes how the edges have been added in the search process.

It should be noted that one may let spanning tree edges take absolute precedence over other edges. That is, the code word may start with the spanning tree edges, and only after all of them the other edges (which lead to cycles) are listed. Here, however, I make no such distinction of edge types. The order of the edge descriptions in the code words is defined by the stated edge properties alone and thus spanning tree edges may be intermingled with edges closing cycles. The reason is that I want the edges to be in exactly the order in which they have been added to the (sub)graph in the search tree. However, one may also choose to find spanning trees first before closing cycles. As the ideas underlying the Gaston algorithm (Nijssen and Kok (2004)) suggest, there may be good reasons for adopting such a strategy, as it may speed up the search.

3.5 Checking for canonical form

After the code words are defined, the test whether a code word is a canonical description of a (sub)graph can be stated formally. The pseudocode below describes the procedure. w is the code word to be tested, $G = (V, E)$ is the corresponding (sub)graph. Each node $v \in V$ has an attribute $v.a$, which I assume to be coded as an integer, and an index field $v.i$, which is filled by the algorithm. Likewise each edge $e \in E$ has an attribute $e.a$, again assumed to be coded as an integer, and a marker $e.i$, which is used to record whether it was visited. Since apart from node and edge attributes a code word contains only indices of nodes, it can thus be represented as an array of integers.

```
function isCanonical (w: int array, G: graph) : boolean;
var v : node;                      (* to traverse the nodes of the graph *)
    e : edge;                      (* to traverse the edges of the graph *)
    x : node array;                (* to collect the numbered nodes *)
begin
  forall v ∈ G.V do                (* traverse all nodes and *)
    v.i := −1;                     (* clear their indices *)
  forall e ∈ G.E do                (* traverse all edges and *)
    e.i := −1;                     (* clear their markers *)
  forall v ∈ G.V do begin          (* traverse the potential root nodes *)
    if v.a < w[0] then return false;   (* abort on smaller root nodes *)
    if v.a > w[0] then continue;       (* skip larger root nodes *)
    v.i := 1; x[0] := v;           (* number and record the root node *)
    if not rec(w, 1, x, 1, 0)      (* check the code word recursively *)
    then return false;             (* abort if a smaller word is found *)
    v.i := −1;                     (* clear the node index again *)
  end
  return true;                     (* the code word is canonical *)
end
```

This function is the same, regardless of whether a depth-first or a breadth-first search canonical form is used. The difference lies only in the implementation of the function "rec", mainly in the order in which edge properties are compared. Here I confine myself to the implementation for breadth-first search. However, for depth-first search the function can be implemented in a very similar way.

The basic idea is to add one edge in each level of the recursion. The description of this edge is generated and if it already allows to decide whether the generated code word is larger or smaller (prefix test!), the recursion is terminated. Only if the edge description coincides with the one found in the code word to check, the edge and the node at the other end (if necessary) are marked/numbered and the function is called recursively. Note that the loop over the edges incident to the node $x[i]$ in the pseudocode below assumes that the edges are considered in sorted order, that is, the edges with the smallest attribute are tested first, and among edges with the same attribute, they are considered in increasing order of the attribute of the destination node.

```
function rec (w: int array, k : int, x: node array, n: int, i: int) : boolean;
var d : node;                          (* node at the other end of an edge *)
    j : int;                           (* index of destination node *)
    u : boolean;                       (* flag for unnumbered destination *)
    r : boolean;                       (* buffer for a recursion result *)
begin
  if k ≥ length(w) return true;        (* full code word has been generated *)
  while i < w[k] do begin              (* check for an edge with a *)
    forall e incident to x[i] do       (* source node having a smaller index *)
      if e.i < 0 then return false;
    i := i + 1;                        (* go to the next extendable node *)
  end
  forall e incident to x[i] (in sorted order) do begin
    if e.i > 0 then continue;          (* skip visited incident edges *)
    if e.a < w[k + 1] then return false;      (* check the *)
    if e.a > w[k + 1] then return true;       (* edge attribute *)
    d := node incident to e other than x[i];
    if d.a < w[k + 2] then return false;      (* check destination *)
    if d.a > w[k + 2] then return true;       (* node attribute *)
    if d.i < 0 then j := n else j := d.i;
    if j < w[k + 3] then return false;        (* check destination *)
    if j = w[k + 3] then begin                (* node index *)
      e.i := 1; u := d.i < 0;          (* mark edge and number node *)
      if u then begin d.i := j; x[n] := d; n := n + 1; end
      r := rec(w, k + 4, x, n, i);            (* check recursively *)
      if u then begin d.i := −1; n := n − 1; end
      e.i := −1;                       (* unmark edge (and node) again *)
      if not r then return false;
    end                                (* evaluate the recursion result *)
```

```
  end
  return true;                    (* return that no smaller code word *)
end                               (* than w could be found *)
```

3.6 A simple example

In order to illustrate the code words defined above, Figure 1 shows a simple molecule (no chemical meaning attached; it was constructed merely for illustration purposes). This molecule is represented as an attributed graph: each node stands for an atom and each edge for a bond between atoms. The nodes carry the chemical element of the corresponding atom as an attribute, the edges are associated with bond types. To the right of this molecule are two spanning trees for this molecule, both of which are rooted at the sulfur atom. Spanning tree edges are depicted as solid lines, edges closing cycles as dashed lines. Spanning tree A was built with depth-first search, spanning tree B with breadth-first search, and thus correspond to the two considered approaches.

If we adopt the precedence order S $\prec$ N $\prec$ O $\prec$ C for chemical elements (derived from the frequency of the elements in the molecule) and the order - $\prec$ = for the bond types, we obtain the two code words shown in Figure 2. It is easy to check that these two code words are actually minimal and thus are the canonical description w.r.t. a depth-first and breadth-first search spanning tree, respectively. In this case it is particularly simple to check this, because the root of the spanning tree is fixed, as there is only one sulfur atom.

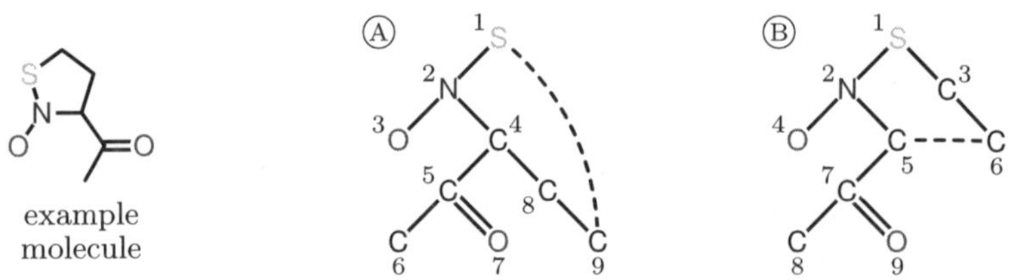

Fig. 1. An example fragment/molecule and two possible canonical forms: A – form based on a depth-first search tree, B – form based on a breadth-first search tree.

```
A:  S 21-N 32-O 42-C 54-C 65-C 75=O 84-C 98-C 91-S
B:  S 1-N2 1-C3 2-O4 2-C5 3-C6 5-C6 5-C7 7-C8 7=O9
```

Fig. 2. Code words describing the two canonical forms shown in Figure 1. Note that in form A the edges are sorted descendingly w.r.t. the second entries (i.e., i_s).

4 Restricted extensions

Up to now canonical descriptions were only used to test whether a search tree branch corresponding to a (sub)graph has to be descended into or not *after* the (sub)graph has been constructed. However, canonical forms can also be used to restrict the possible extensions directly. The idea is that for certain extensions one can see immediately that they lead to a code word that is not minimal. Hence one need not construct and test the (sub)graph, but can skip the extension right way. For the two special cases I consider here (depth-first and breadth-first search spanning trees), the allowed extensions are:

- Depth First Search: *Rightmost Extension* (Yan and Han (2002))

  Only nodes on the rightmost path of the spanning tree of the (sub)graph may be extended, and if the node is no leaf, it may be extended only by edges whose descriptions do not precede the description of the downward edge on the rightmost path. That is, the edge attribute must be no less than the attribute of the downward edge and if the edge attribute is identical, the attribute of its destination node must be no less than the attribute of the downward edge's destination node. Edges between two nodes that are already in the (sub)graph must lead from a node on the rightmost path to the rightmost leaf (that is, the deepest node on the rightmost path). In addition, the index of the source node of such an edge must precede the index of the source node of an edge already incident to the rightmost leaf.

- Breadth First Search: *Maximum Source Extension* (Borgelt and Berthold (2002))

  Only nodes having an index no less than the maximum source index of an edge already in the (sub)graph may be extended.[3] If the node is the one having the maximum source index, it may be extended only by edges whose descriptions do not precede the description of any downward edge already incident to this node. That is, the attribute of the new edge must be no less than that of any downward edge, and if it is identical, the attribute of the new edge's destination node must be no less than the attribute of any corresponding downward edge's destination node (where corresponding means that the edge attribute is the same). Edges between two nodes already in the (sub)graph must start at an extendable node and must lead "forward", that is, to a node having a larger index.

In both cases it is easy to see that an extension violating the above rules leads to a (sub)graph description that is not in canonical form (that is, a numbering of the nodes of the (sub)graph that does not lead to the lexicographically smallest code word). This is easy to see, because there is a depth-first or breadth-first search numbering, respectively, *starting at the same root node*,

[3] Note that if the (sub)graph contains no edge, there can only be one node, and then, of course, this node may be extended without restriction.

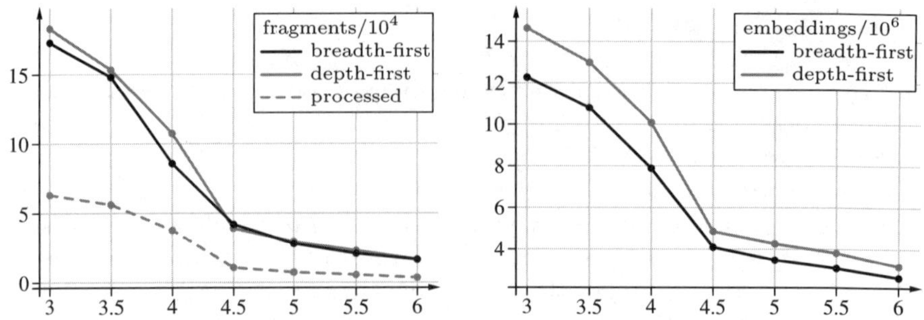

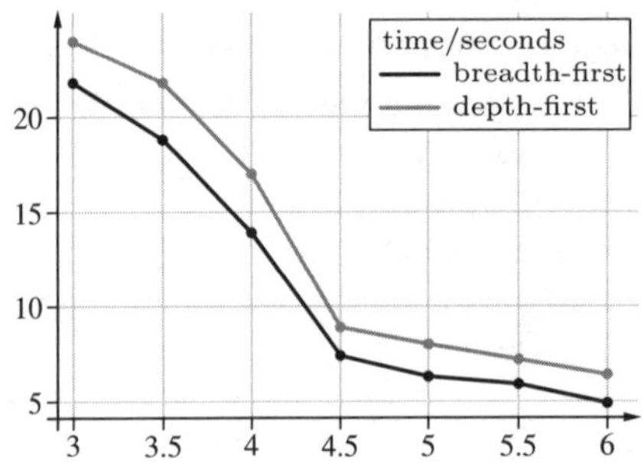

Fig. 3. Experimental results on the IC93 data. The horizontal axis shows the minimal support in percent. The curves show the number of generated and processed fragments (top left), number of generated embeddings (top right), and the execution time in seconds (bottom left) for the two canonical forms/extension strategies.

which leads to a lexicographically smaller code word (which may or may not be minimal itself—all that matters here is that it is smaller than the one derived from the current node numbering of the (sub)graph). In order to find such a smaller word, one only has to consider the extension edge. Up to this edge, the construction of the code word is identical, but when it is added, its description precedes (w.r.t. the defined precedence order) the description of the next edge in the code word of the unextended (sub)graph.

It is pleasing to see that the *maximum source extension*, which was originally introduced in the MoSS/MoFa algorithm based on heuristic arguments (Borgelt and Berthold (2002), Borgelt et al. (2005)), can thus nicely be justified and extended based on a canonical form.

As an illustration of these rules, consider again the two search trees shown in Figure 1. In the depth-first search tree A atoms 1, 2, 4, 8, and 9 are extendable (rightmost extension). On the other hand, in the breadth-first search tree B atoms 7, 8, and 9 are extendable (maximum source extension). Other restrictions are, for example, that the nitrogen atom in A may not be extended by a bond to another oxygen atom, or that atom 7 in B may not be extended by a single bond. Tree A may not be extended by an edge between two nodes already in the (sub)graph, because the edge $(1, 9)$ already has the smallest possible source (duplicate edges between nodes may be allowed, though). Tree B, however, may be extended by an edge between atoms 8 and 9.

5 Experimental results

In order to test the search tree pruning based on a breadth-first search canonical form, I extended the MoSS/MoFa implementation described in (Borgelt et al. (2005)). In order to compare the two canonical forms discussed in this paper, I also implemented a search based on rightmost extensions and a corresponding test for a depth-first search canonical form (that is, basically the gSpan algorithm (Yan and Han (2002)), with the exception of the minimal difference in the definition of the code word pointed out above). This was done in such a way that only the functions explicitly referring to the canonical form are exchanged (extension generation and comparison and canonical form check), so that on execution the two algorithms share a maximum of the program code.

Figure 3 shows the results on the 1993 subset of the INDEX CHEMICUS (IC (1993)), Figure 4 the results on a data set consisting of 17 steroids. In all diagrams the grey solid line describes the results for a depth-first canonical form, the black solid line the results for a breadth-first canonical form. The diagram in the top left shows the number of generated and actually processed fragments (note that the latter, shown as a dashed grey line, is necessarily the same for both algorithms). The diagram in the top right shows the number of generated embeddings, the diagram in the bottom left the execution times.[4]

As can be seen from these diagrams, both canonical forms work very well (compared to an approach without canonical form pruning and an explicit removal of found duplicates, I observed speed-ups by factors between about 2.5 and more than 30). On the IC93 data the breadth-first search canonical form performs slightly better, needing about 10–15% less time. As the other diagrams show, this is mainly due to the lower numbers of fragments and embeddings that are generated. On the steroids data the depth-first search canonical form performs minimally better at low support values. Again this is due to a smaller number of generated fragments, which, however, is outweighed by a larger number of generated embeddings for higher support values.

6 Conclusions

In this paper I introduced a family of canonical forms of graphs that can be exploited to make frequent graph mining efficient. This family was obtained by generalizing the canonical form introduced in the gSpan algorithm (Yan and Han (2002)). While gSpan's canonical form is defined with a depth-first search tree, my definition allows for any systematic way of obtaining a spanning tree. To show that this generalization is useful, I considered a breadth-first search spanning tree, which turned out to be the implicit canonical form

[4] Experiments were done with Sun Java 1.5.0_01 on a Pentium 4C@2.6GHz system with 1GB main memory running S.u.S.E. Linux 9.3.

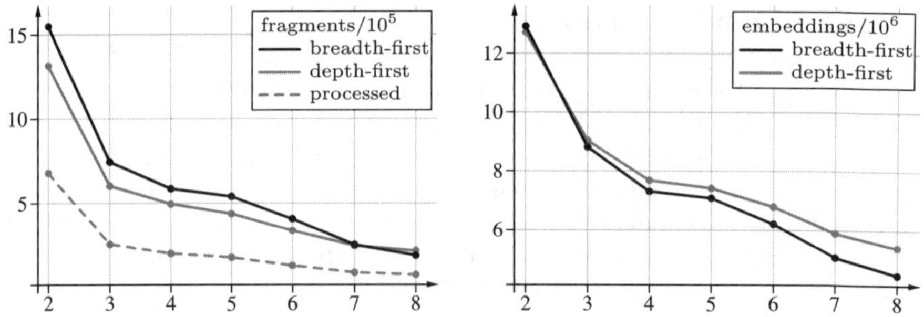

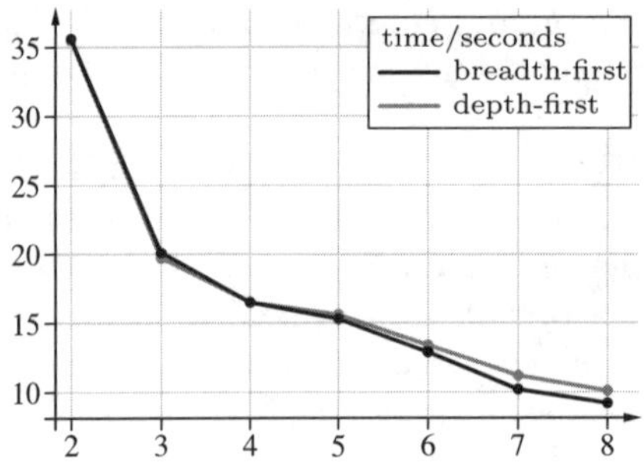

Fig. 4. Experimental results on the steroids data. The horizontal axis shows the absolute minimal support. The curves show the number of generated and processed fragments (top left), number of generated embeddings (top right), and the execution time in seconds (bottom left) for the two canonical forms/extension strategies.

underlying the MoSS/MoFa algorithm (Borgelt and Berthold (2002), Borgelt et al. (2005)). Exploiting this canonical form in MoSS/MoFa in the same way as the depth-first search canonical form is exploited in gSpan leads to a considerable speed up of this algorithm. It is pleasing to see that based on this generalized canonical form, gSpan and MoSS/MoFa can nicely be described in the same general framework, which also comprises a variety of other possibilities (an example alternative was pointed out above).

References

BORGELT, C. and BERTHOLD, M.R. (2002): Mining Molecular Fragments: Finding Relevant Substructures of Molecules. *Proc. 2nd IEEE Int. Conf. on Data Mining.* IEEE Press, Piscataway, 51–58.

BORGELT, C., MEINL, T. and BERTHOLD, M.R. (2004): Advanced Pruning Strategies to Speed Up Mining Closed Molecular Fragments. *Proc. IEEE Conf. on Systems, Man and Cybernetics, CD-ROM.* IEEE Press, Piscataway.

BORGELT, C., MEINL, T. and BERTHOLD, M.R. (2005): MoSS: A Program for Molecular Substructure Mining. *Proc. Open Source Data Mining Workshop.* ACM Press, New York, 6–15.

COOK, D.J. and HOLDER, L.B. (2000): Graph-based Data Mining. *IEEE Trans. on Intelligent Systems 15, 2, 32–41.*

FINN, P.W., MUGGLETON, S., PAGE, D. and SRINIVASAN, A. (1998): Pharmacore Discovery Using the Inductive Logic Programming System PROGOL. *Machine Learning 30, 2-3, 241–270.*

GOETHALS, B. and ZAKI, M. (2003/2004): Proc. 1st and 2nd IEEE ICDM Workshop on Frequent Itemset Mining Implementations. *CEUR Workshop Proceedings 90 and 126.* Sun SITE Central Europe and RWTH Aachen http://www.ceur-ws.org/Vol-90/, http://www.ceur-ws.org/Vol-126/.

HUAN, J., WANG, W. and PRINS, J. (2003): Efficient Mining of Frequent Subgraphs in the Presence of Isomorphism. *Proc. 3rd IEEE Int. Conf. on Data Mining.* IEEE Press, Piscataway, 549–552.

INDEX CHEMICUS — Subset from 1993. Institute of Scientific Information, Inc. (ISI). Thomson Scientific, Philadelphia.

KRAMER, S., DE RAEDT, L. and HELMA, C. (2001): Molecular Feature Mining in HIV Data. *Proc. 7th ACM SIGKDD Int. Conf. on Knowledge Discovery and Data Mining.* ACM Press, New York, 136–143.

KURAMOCHI, M. and KARYPIS, G. (2001): Frequent Subgraph Discovery. *Proc. 1st IEEE Int. Conf. on Data Mining.* IEEE Press, Piscataway, 313–320.

NIJSSEN, S. and KOK, J.N. (2004): A Quickstart in Frequent Structure Mining can Make a Difference. *Proc. 10th ACM SIGKDD Int. Conf. on Knowledge Discovery and Data Mining.* ACM Press, New York, 647–652.

WASHIO, T. and MOTODA, H. (2003): State of the Art of Graph-based Data Mining. *SIGKDD Explorations Newsletter 5, 1, 59–68.*

YAN, X. and HAN, J. (2002): gSpan: Graph-based Substructure Pattern Mining. *Proc. 2nd IEEE Int. Conf. on Data Mining.* IEEE Press, Piscataway, 721–724.

YAN, X. and HAN, J. (2003): CloseGraph: Mining Closed Frequent Graph Patterns. *Proc. 9th ACM SIGKDD Int. Conf. on Knowledge Discovery and Data Mining.* ACM Press, New York, 286–295.

Applying Clickstream Data Mining to Real-Time Web Crawler Detection and Containment Using ClickTips Platform

Anália Lourenço and Orlando Belo

Department of Informatics, School of Engineering, University of Minho, Campus de Gualtar, 4710-057 Braga, PORTUGAL; {analia, obelo}@di.uminho.pt

Abstract. Web crawler uncontrolled widespread has led to undesired situations of server overload and contents misuse. Most programs still have legitimate and useful goals, but standard detection heuristics have not evolved along with Web crawling technology and are now unable to identify most of today's programs. In this paper, we propose an integrated approach to the problem that ensures the generation of up-to-date decision models, targeting both monitoring and clickstream differentiation. The ClickTips platform sustains Web crawler detection and containment mechanisms and its data webhousing system is responsible for clickstream processing and further data mining. Web crawler detection and monitoring helps preserving Web server performance and Web site privacy and clickstream differentiated analysis provides focused report and interpretation of navigational patterns. The generation of up-to-date detection models is based on clickstream data mining and targets not only well-known Web crawlers, but also camouflaging and previously unknown programs. Experiments with different real-world Web sites are optimistic, proving that the approach is not only feasible but also adequate.

1 Introduction

Presently, Web crawlers play a crucial role within the information retrieval arena (Pant et al. (2004)). In addition to general-purpose crawlers, an ever growing number of focused crawlers selectively seek out documents relevant to pre-defined sets of subjects (Almeida et al. (2001)).

Easily, Web crawlers can be confused with regular users or even impersonate other Web crawlers or legitimate users. These programs do not have any strict regulation enforcing their identification and limiting their actions. So far, Web crawler identification and containment has relied on the compliance with standard exclusion protocol (http://www.robotstxt.org/wc/exclusion.html) suggestions and the experience of Webmasters (Tan and Kumar (2002)). However, this procedure has proved to be prone to fail and it is clearly unable to keep up with the steady generation of new kinds of crawlers. Web crawlers are becoming smarter, obfuscating their actions and purposes whenever they want to pass by unnoticed, overcoming conventional traps. Web server overload and privacy or copyrights violation are not exactly unseen within the crawling scenario and the lack of information about these programs makes clickstream differentiation quite inefficient. Heuristics based on ethical assumptions and crawler-alike session metrics cannot prevent these situations and identification catalogues cannot follow the pace of crawler evolving.

These needs and limitations drove our research towards the conception and validation of more adequate means of identification. The established goals were three-fold: to detect crawling activities at their earliest stage and to perform their containment when it seems fit; to provide up-to-date clickstream flow differentiation, making it possible to sustain focused processing and data mining; and, to apply usage mining techniques to the interpretation of crawler navigational patterns and to the creation and update of detection models. Non-trivial knowledge about how Web crawlers operate within a site, identifying their purposes and primary traversal patterns, will allow the maintenance of up-to-date detection heuristics. Regular analysis will target actual visitors, directing site restructuring and marketing campaigns into the right way. Meanwhile, crawler analysis will support the detection model update and will provide an insight about the vulnerability of the sites and highlighting the visibility of the site. In order to achieve such goals we have designed and implemented a Java platform, called ClickTips, which can be shortly defined as an integrated differentiating clickstream processing and analysis platform. Several experiments with real-world Web sites have shown that ClickTips is able to obtain highly accurate crawler detection models as well as to sustain differentiated clickstream processing and analysis. Results show that the induced models are able to detect crawler visits at their earliest stage, minimizing potential server overload, site privacy, or copyrights violation.

2 ClickTips: A differentiated clickstream processing and analysis platform

2.1 General description

Our approach follows the same assumption and basic steps of the technique proposed by Tan and Kumar (Tan and Kumar (2002)): regular and crawler Web sessions are different in essence and thus, the mining of Web session primary metrics may provide the means to differentiate the two groups. However, we introduce an integrated approach where crawler detection is actually applied to focused clickstream processing and analysis and hazard identification and containment. Tan and Kumar work was limited to the definition of a crawler detection technique based on navigation pattern mining, reporting a data mining experiment on clickstream controlled data. We, on the

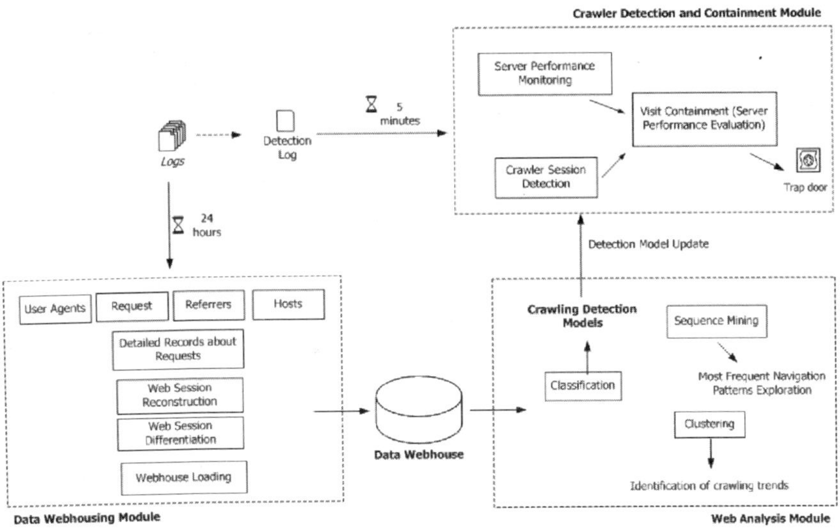

Fig. 1. Functional architecture of ClickTips

other hand, designed, implemented and tested an integrated differentiating clickstream processing and analysis platform that sustains crawler detection within a wider scope of site preservation, control and analysis.

ClickTips is a JAVA based platform devoted to differentiated clickstream processing and Web analysis. Special emphasis was set on Web crawler detection, containment and clickstream differentiation, which turned ClickTips into a novel, integrated platform. It benefited from older experience concerning general clickstream processing (Cooley et al. (1999)) and our research on crawler trends and purposes, providing a whole new level of clickstream processing and an adequate workbench for Web crawler analysis and control. Moreover, it as a generic platform, capable of embracing site specificities when it seems relevant, but not requiring such information for providing a standard (more than acceptable) level of processing and analysis. Presently, it generates overall and differentiated profile reports and recurs to an open-source data mining tool for advanced usage mining such as crawler detection and the analysis of most common navigation patterns. The platform embraces three main modules (Fig. 1): the data webhousing module, the Web analysis module and the detection and containment module.

2.2 Data webhousing module

ClickTips' data webhouse deals with standard and state-of-the-art clickstream processing issues. The general processes ensure overall data processing while the usage differentiation process allows the execution of specialised processing that deals with the specificities and needs of both crawler and regular analysis.

Standard clickstream processing involves host and user recognition and location, referrer characterization, request parsing, user agent identification and Web session reconstruction (Kimball and Merz (2000), Sweiger et al. (2002)). In Tan and Kumar (2002) the sessionizing schema encompassed both parallel and multi-agent activities, sustaining a catalog with the identity of known proxies. Instead, we chose a schema that decomposes parallel and multi-agent visits into single machine and single-agent visits, discriminating the activity of each access program. The use of multiple IP addresses (parallel activities) or different access programs (e.g. a browser or a download manager) should not interfere with crawler identification. A multi-agent session may contain a crawler sub-session that urges containment, but the other sub-sessions should not suffer any action. For example, a session from a proxy server involves many sessions, most of them from regular users, and the containment of the IP address would result on the unnecessary loss of users. Our sessioninzing schema ensures that only the specific set of user agents coming from that IP address and tagged as to contain will actually suffer access restrictions, preserving regular usage at all times.

```
Let S be the set of reconstructed Web sessions.
Let C_BD be the crawler identification database.
Let ToInspect be the sessions that need manual inspection.

For each s in S Do
        selfIdentified <- robotsFileRequested (s.requests)
        foundMatch <- checkUserAgent(C_BD, s.userAgent)
        If (foundMatch or selfIdentified)
                s.agentType <- "Crawler"
        Else
                s.agentType <- "?"
                addSession(ToInspect, s)
        End If
End For
For each t in ToInspect Do
        manualUserAgentCheck(t.userAgent)
End For
finalSessionLabelling(S, ToInspect)
```

Fig. 2. Semi-automatic session differentiation schema

Regarding usage differentiation, we had two main concerns: to implement a basic schema capable of sustaining the process while there is no decision model; to conceive and implement a model generation process that is able to induce up-to-date decision models whenever it detects meaningful crawler evolving, i.e., when navigation patterns suffer a significant change. Initially, differentiation is based on a semi-automatic labelling schema, based on known identifications and ethical assumptions (Fig. 2). As stated earlier, such a schema is prone to fail, but, when deploying ClickTips for the first time over a giving site, there is no better way of performing differentiation. Afterwards, navigation pattern mining ensures the production of up-to-date identification models that will sustain both clickstream differentiation and crawler detection. Differentiated streams are submitted to focused sessionizing, trying to identify specific parallel and multi-agent activities, and crawler sessions are kept under surveillance. Meanwhile, all acquired data are loaded into the data webhouse.

Table 1. Navigation pattern attributes

robots.txt file was requested?
Session occurred between 12am to 8am (local time)?
Reoccurrence rate of file requests.
Percentage of invalid requests.
Percentage of HTTP request methods (GET, POST, HEAD,...).
Percentage of unassigned referrer hits.
Percentage of each document class (internet, image, text, media, ...).
Total number of requests.
Total number of page requests.
Session duration in milliseconds.
Total amount of transferred bytes.
Average and standard deviation time between two page requests.
Crawler or Browser label (assigned class).

ClickTips sustains three star schemas, each one meeting particular analysis commitments. The overall, differentiated usage schema supports visit evaluation towards the generation and update of crawler detection models. The regular usage schema keeps regular users navigational patterns over time, providing the means to interpret and foresee users needs and trends. Finally, the crawler usage schema provides insights about crawler purposes, similarities and the impact of their actions over Web servers and the Web itself. Each star schema supports three levels of analysis: hit, navigation pattern and overall session characterisation.

2.3 Web analysis module

The clickstream data mining module generates conventional usage reports, provides the means to study the navigational patterns and generates decision models towards crawler detection. Reported information embraces overall hit and session statistics, request type classes distribution, HTTP method calls distribution, host domain and location information and referrer identification and location, among others. Unlike well-known usage report tools like Analog (http://www.analog.cx/) or Webalizer (http://www.mrunix.net/webalizer/), ClickTips' reports are driven by analysis and not debug needs. Furthermore, ClickTips sustains usage differentiated analysis which grants more focused and accurate reports, providing reliable information about regular usage trends and purposes as well as insights about the kind of agents that are crawling the site. Presently, the mining requirements of the platform do not justify the appliance of state-of-the-art techniques nor the implementation of mining algorithms. So far, ClickTips needs are fulfilled by the Weka open-source data mining platform (Witten and Frank (2005)), which is also written in JAVA and thus, fairly easy to integrate with.

```
Let S be the set of reconstructed, labelled Web sessions.
Let n be the maximum length of session.
Let x be the accuracy threshold.

stopMining <- false
If (!stopMining)
    For i=1, ..., n Do
        subSessions_i <- gatherPageRequestsTill(S, i)
        decisionModel_i <- C45Mining(subSessions_i)
        If (i>1)
            If (|decisionModel_i.accuracy- decisionModel_i.accuracy| < x)
                stopMining <- true
            End If
        End If
    End For
End If
```

Fig. 3. Crawler-oriented incremental data mining

Our focus was set on the deployment of an incremental clickstream data mining approach capable of sustaining up-to-date crawler detection (Fig. 3), inspired in Tan and Kumar former approach. Both approaches rely on navigation pattern mining and recur to Classification techniques in order to solve the problem. Yet, our Web analysis goals and philosophy established different procedures at both processing and dataset labelling stages and, therefore, although the mining steps are basically the same, the set of session metrics is not (Table 1). Our experiments with real-world Web sites have helped on the refinement of the approach and have proved both its adequacy and its efficiency (Table 2). Results show that detection models are able to tag known, previously unknown or undercover crawlers within the earliest stages of their visits and clickstream data mining is more than capable of keeping such models up-to-date with both crawler and site evolving.

2.4 Detection and containment module

The Web crawler detection and containment module involves a decision model, a monitor and a containment script. The first two components provide the means to control server and site crawler-related hazards. The decision model supports the detection of crawler visits while they are hapenning and at their earliest stage, while the monitor triggers containment actions whenever server performance suffers anomalous deterioration. Identifying crawling traverses is not enough for preserving server functionality and site accessibility. The decision model may not detect certain crawlers and the overall set of active sessions at a given moment may represent an overwhelming load in spite of their origin. Monitoring server performance allows the timely identification of such situations and the information delivered by the decision model helps determining the best solution.

Table 2. Accuracy of different Classification algorithms

No. Requests	ZeroR	J48	NaiveBayes	DecisionTable
1	63.56	94.50	88.97	94.54
2	82.19	96.99	93.08	96.51
3	88.23	97.87	95.34	97.29
4	89.93	98.16	96.12	97.62
5	90.12	98.31	95.95	97.97
6	90.01	98.44	96.08	97.74

At first, detection relies on available crawler identification databases and crawler ethic behaviour to perform its work. But, the appliance of incremental data mining over the profiling data webhouse provides detection models that keep up with the latest notices about crawler evolving patterns. In turn, in the two first days of deployment, the monitor gathers information about relevant server metrics such as the percentage distribution of hits, the percentage distribution of status codes and the percentage distribution of each class of requests. Such data is recorded per minute and grouped per half an hour in order to monitor server response along time. Afterwards, it evaluates these metrics, comparing values each half an hour to the corresponding values of mean and standard deviation of the last two days. This comparison enables the identification of anomalous peaks of visits that are deteriorating server performance.

Basically, in a (pre-)overload situation, crawler sessions are the first candidates for termination, staring with known malicious crawlers and crawlers with long traverses. The containment of regular sessions is considered only if there are no more crawler sessions and the overload persists. In this regard, the container maintains a black list of visitors that should be banished from the site (agents that consume too many server resources, snoop around private contents or seem to be attacking the server) as well as a list of imediate containment. Containment actions are support by trap doors that re-direct the specified requests to a dead site area.

3 Final remarks

This work remarks the relevance of Web crawler detection to both general Web usage profiling and Web crawler management, stating the necessity of identifying and controlling Web crawling activities in order to focus usage profiling, improve legitimate information retrieval tasks and prevent malicious visits. In the past, heuristic-based approaches tackled the problem by conjugating representative elements like host IP addresses and user agents with behaviour-specific metrics. Nowadays, the increasing diversity of crawlers, their constant evolving and the fact that many crawlers are deliberately delaying their visits and using fake identifications makes such detection very hard and prone-to-fail. Our proposal aims to provide more suitable detection through navigation pattern mining. Specifically, we propose an integrated differentiating clickstream processing and analysis platform, named ClickTips, which targets detection and containment, data webhouse processing and usage mining. It supports not only the control and study of crawling activities as it deploys valid and focused clickstream processing and analysis. In fact, what sets apart ClickTips from other tools is its up-to-date clickstream usage differentiation and according processing along with its site-independent, interoperable and integrated design and implementation. So far, no other tool has claimed its appliance at such complete extent.

Acknowledgements

The work of Anlia Loureno was supported by a grant from Fundao para a Cincia e Tecnologia - SFRH/BD/8242/2002.

References

ALMEIDA, V., MENASCE, D.A., RIEDI, R.H., PELIGRINELLI, F., FONSECA, R.C. and MEIRA Jr., W. (2001): Analyzing Web Robots and Their Impact on Caching. *In Proceedings of the 6th Web Caching and Content Delivery Workshop. Boston MA.*

COOLEY, R., MOBASHER, B. and SRIVASTAVA, J. (1999): Data Preparation for Mining World Wide Web Browsing Patterns. *Knowledge and Information Systems, 1, 1.*

KIMBALL, R. and MERZ, R. (2000): *The Data Webhouse Toolkit: Building the Web-Enabled Data Warehouse.* Wiley, New York.

PANT, G., SRINIVASAN, P. and MENCZER, F. (2004): Crawling the Web. In: M. Levene and A. Poulovassilis (Eds.): *Web Dynamics: Adapting to Change in Content, Size, Topology and Use.* Springer, Berlin.

SWEIGER, M., MADSEN, M.R., LANGSTON, J. and LOMBARD, H. (2002): *Clickstream Data Warehousing.* Wiley, New York.

TAN, P.N. and KUMAR, V. (2002): Discovery of Web Robot Sessions Based on their Navigational Patterns. *Data Mining and Knowledge Discovery, 6, 1, 9-35.*

WITTEN, I.H. and FRANK, E. (2005): *Data Mining: Practical Machine Learning Tools and Techniques*, 2nd Edition. Morgan Kaufmann, San Francisco.

Plagiarism Detection Without Reference Collections

Sven Meyer zu Eissen, Benno Stein and Marion Kulig

Faculty of Media, Media Systems, Bauhaus University Weimar, 99421 Weimar, Germany; {sven.meyer-zu-eissen, benno.stein}@medien.uni-weimar.de

Abstract. Current research in the field of automatic plagiarism detection for text documents focuses on the development of algorithms that compare suspicious documents against potential original documents. Although recent approaches perform well in identifying copied or even modified passages (Brin et al. (1995), Stein (2005)), they assume a closed world where a reference collection must be given (Finkel (2002)). Recall that a human reader can identify suspicious passages within a document without having a library of potential original documents in mind.

This raises the question whether plagiarized passages within a document can be detected automatically if no reference is given, e. g. if the plagiarized passages stem from a book that is not available in digital form. This paper contributes right here; it proposes a method to identify potentially plagiarized passages by analyzing a single document with respect to changes in writing style. Such passages then can be used as a starting point for an Internet search for potential sources. As well as that, such passages can be preselected for inspection by a human referee. Among others, we will present new style features that can be computed efficiently and which provide highly discriminative information: Our experiments, which base on a test corpus that will be published, show encouraging results.

1 Introduction

A recent large-scale study on 18,000 students by McCabe (2005) reveals that about 50% of the students admit to plagiarize from extraneous documents. Plagiarism in text documents happens in several forms: one-to-one copies, passages that are modified to a greater or lesser extent, or even translated passages. Figure 1 shows a taxonomy of plagiarism delicts along with possible detection methods.

1.1 Some background on plagiarism detection

The success of current approaches in plagiarism detection varies according to the underlying plagiarism delict. The approaches stated in Brin et al. (1995)

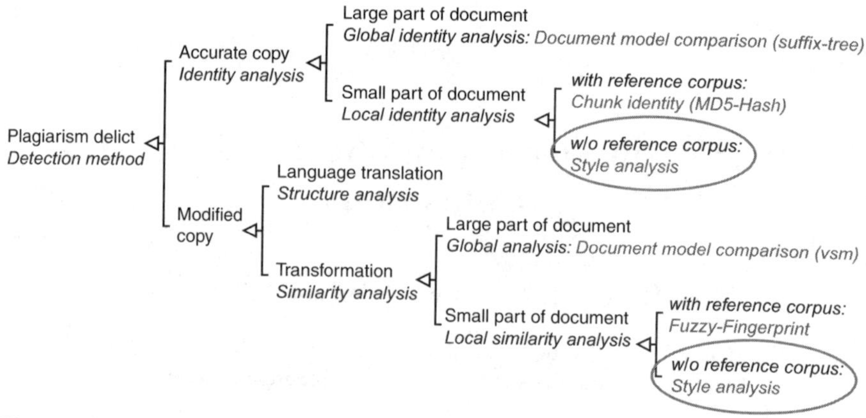

Fig. 1. A taxonomy of plagiarism delicts and analysis methods according to Stein and Meyer zu Eissen (2006). The encircled parts indicate our contributions: the detection of a plagiarism delict without having a reference corpus at hand.

and Hoad and Zobel (2003) employ cryptographic hash functions to generate digital fingerprints of so-called text chunks, which are compared against a database of original text passage fingerprints. Since cryptographic fingerprints identify a text chunk exactly, the quality of these approaches depends on offsets and sizes of chunks within both plagiarized and original texts. An approach introduced in Stein (2005) overcomes these limitations: unlike cryptographic fingerprints, the proposed method generates fingerprints that are robust against modifications to some extent.

However, the mentioned approaches have one constraint in common: they require a reference collection with original documents. Observe that human readers may identify suspicious passages within a document without having a library of reference documents in mind: changes between brilliant and baffling passages, or the change of person narrative give hints to plagiarism. Situations where such an *intrinsic plagiarism detection* can be applied are shown encircled in Figure 1.

1.2 Contributions of the paper

Basically, the power of a plagiarism approach depends on the quality of the quantified linguistic features. We introduce features which measure—simply put—the customariness of word usage, and which are able to capture a significant part of style information. To analyze the phenomenon of intrinsic plagiarism detection we have constructed a base corpus from which various application corpora can be compiled, each of which modeling plagiarism delicts of different severity. Section 3 reports on experiments that we have conducted with this corpus.

2 Quantification of writing style

Intrinsic plagiarism detection can be operationalized by dividing a document into "natural" parts, such as sentences, paragraphs, or sections, and analyzing the variance of certain style features. Note in this connection that within the experiments presented in Section 3 the size of a part is chosen rather small (40-200 words), which is ambitious from the analysis standpoint, but which corresponds to realistic situations.

2.1 Stylometric features

Each author develops an individual writing style; i. e. he or she employs consciously or subconsciously patterns to construct sentences and uses an individual vocabulary. Stylometric features quantify style aspects, and some of them have been used successfully in the past to discriminate between texts with respect to authorship (Koppel and Schler (2004), Sorensen (2005)). Most stylometric features are based on the following semiotic features:

1. Text statistics, which operate at the character level.
   Examples: number of commas, question marks, word lengths.
2. Syntactic features, which measure writing style at the sentence level.
   Examples: sentence lengths, use of function words
3. Part-of-speech features to quantify the use of word classes.
   Examples: number of adjectives or pronouns
4. Closed-class word sets to count special words.
   Examples: number of stopwords, foreign words, "difficult" words
5. Structural features, which reflect text organization.
   Examples: paragraph lengths, chapter lengths

Based on these features, formulas can be constructed that quantify the characteristic trait of an author's writing style. Almost all of the developed formulas aim at a quantification of the educational background, i. e., they quantify an author's vocabulary richness or style complexity, or a reader's grading level that is required to understand a text. Figure 2 classifies style-quantifying formulas according to their intention.

Widely employed grading measures include the Flesch Kincaid Grade Level (Flesch (1948), Kincaid et al. (1975)) and the Dale-Chall formula (Dale and Chall (1948)). The former, which is used among others by the US Government Department of Defense, combines the average number of syllables per word, ASW, with average sentence length, ASL, as follows: $FK = 0.39 \cdot ASL + 11.8 \cdot ASW - 15.59$. The resulting number shall be an estimate for the number of years a reader has to spend in school before being able to understand the text.

The Dale-Chall formula employs a closed-class word list containing 3000 familiar words usually known by 4th grade children. The formula combines the percentage of difficult words that do not appear in the list with the average sentence length and defines a monotonic function that maps onto a grading level.

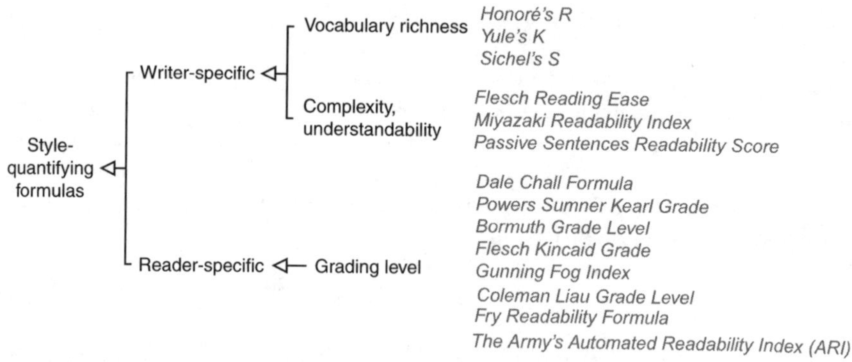

Fig. 2. A classification of the most well-known style-quantifying formulas with respect to their application range and underlying concept.

Methods to measure an author's vocabulary richness are often based on the ratio between the number of different words and the total number of words within a document; well-known examples include Yule's K (Yule (1948)) and Honore's R (Honore (1979)). However, it was reported by Tweedie and Baayen (1997) and Stamatatos et al. (2001) that these measures depend significantly on document length or passage length. As a consequence, they are not suited to compare passages of varying lengths and deliver unreliable results for short passages, which is a disqualifying criterion for plagiarism analysis.

We now introduce a new vocabulary richness statistic, the averaged word frequency class, which turned out to be the most powerful and stable concept with respect to intrinsic plagiarism detection that we have encountered so far.

2.2 Averaged word frequency class

The frequency class of a word is directly connected to Zipf's law and can be used as an indicator of a word's customariness. Let $\mathcal{C}$ be a text corpus, and let $|\mathcal{C}|$ be the number of words in $\mathcal{C}$. Moreover, let $f(w)$ denote the frequency of a word $w \in \mathcal{C}$, and let $r(w)$ denote the rank of w in a word list of $\mathcal{C}$, which is sorted by decreasing frequency.

In accordance with (University of Leipzig (1995)) we define the word frequency class $c(w)$ of a word $w \in \mathcal{C}$ as $\lfloor \log_2(f(w^*)/f(w)) \rfloor$, where w^* denotes the most frequently used word in $\mathcal{C}$. In the Sydney Morning Herald Corpus, w^* denotes the word "the", which corresponds to the word frequency class 0; the most uncommonly used words within this corpus have a word frequency class of 19. A document's averaged word frequency class tells us something about style complexity and the size of an author's vocabulary—both of which are highly individual characteristics (Meyer zu Eissen and Stein (2004)).

Note that, based on a lookup-table, the averaged word frequency class of a text passage can be computed in linear time in the number of words. Another salient property is its small variance with respect to text length, which renders it ideal for our purposes.

3 Experimental analysis

This section reports on experiments related to a plagiarism analysis without reference collections; it addresses the following questions:

1. Which vocabulary richness measure is suited best?—which leads us to the question: How stable is a measure with respect to text length?
2. To which extent is the detection of plagiarized text portions possible?

The first question can be answered by analyzing the characteristic of the vocabulary richness measures concerning single author (= non plagiarized) documents. The second question can be reformulated as a document classification task, given a reference corpus with plagiarized and non plagiarized documents.

3.1 Evaluation of vocabulary richness measures

As pointed out above, changes in vocabulary richness across paragraphs are a good indicator for plagiarism. Confer in this connection the left plot in Figure 3, which contrasts the averaged word frequency class of four different authors.

Plagiarism analysis requires a measure that works reliably at the *paragraph level.* Put another way, when analyzing a portion of text from a single-author document the ideal vocabulary richness measure should behave fairly constant—regardless of the portion's position and size. An according comparison of Honore's R, Yule's K, and the average word frequency class is shown in the right plot of Figure 3; here, the analyzed text portion varies between 10% and 100% of the entire document. Observe that the average word frequency class is stable even for small paragraphs, which qualifies the measure as a powerful instrument for intrinsic plagiarism analysis.

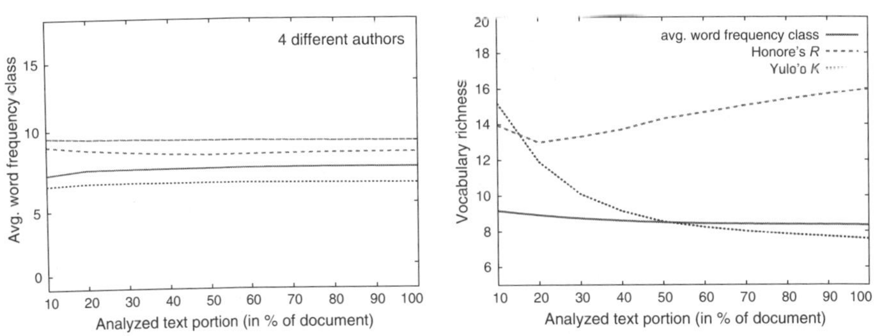

Fig. 3. Average word frequency class of four different authors (left plot). The right plot shows the development of Honore's R, Yule's K, and the average word frequency class of a single-author document for different text portions. For a better readability the values of Honore's R and Yule's K are divided by 100 and 10 respectively.

3.2 Corpus construction

Since no reference collection is available for classification experiments, we have compiled a new corpus, which will be made available for all interested researchers. Its construction is oriented at the following corpus-linguistic criteria described in Garside et al. (1997):

1. authenticity and homogeneity
2. possibility to include many types of plagiarism
3. easy processable for both human and machine
4. clear separation of text and annotations

We chose genuine computer science articles from the ACM digital library, which were "plagiarized" by hand with both copied as well as reformulated passages from other ACM computer science articles, contributing to criterion 1. To separate annotations from text and to allow both maintenance for human editors and standardized processing for machines, all documents in the corpus are represented in XML-syntax (cf. criteria 2-4). They validate against the following DTD, which declares a mixed content model and provides element types for plagiarism delict and plagiarism source among others.

```
<!ELEMENT document (#PCDATA|plagiarized)*>
<!ATTLIST document source CDATA #REQUIRED>
<!ELEMENT plagiarized (#PCDATA)>
<!ATTLIST plagiarized type (copied|mod|trans) source CDATA #REQUIRED>
```

An XML document with k plagiarized passages defines a template from which 2^k instance documents can be generated, depending on which of the k plagiarized parts are actually included. Instance documents contain no XML tags in order to ensure that they can be processed by standard algorithms. Instead, a meta information file is generated for each, specifying the exact position of plagiarized passages.

3.3 Classification experiments

For the results presented here more than 450 instance documents were generated each of which containing between 3 and 6 plagiarized passages of different lengths. During the plagiarism analysis each instance document was decomposed into 50 - 100 passages, and for each passage a paragraph-specific feature vector $\mathbf{f}_p$ was computed. The feature set includes average sentence length, 18 part-of-speech features, average stopword number, the Gunning Fog index, Flesch-Kincaid Grade Level, the Dale-Chall formula, Honore's R, Yule's K, and the averaged word frequency class.

Since we are interested in the detection of writing style variations, a document-specific feature vector, $\mathbf{f}_d$, was computed and compared to each of the $\mathbf{f}_p$. The rationale is that the relative differences between $\mathbf{f}_d$ and the feature vectors of the plagiarized passages reflect possible writing style changes.

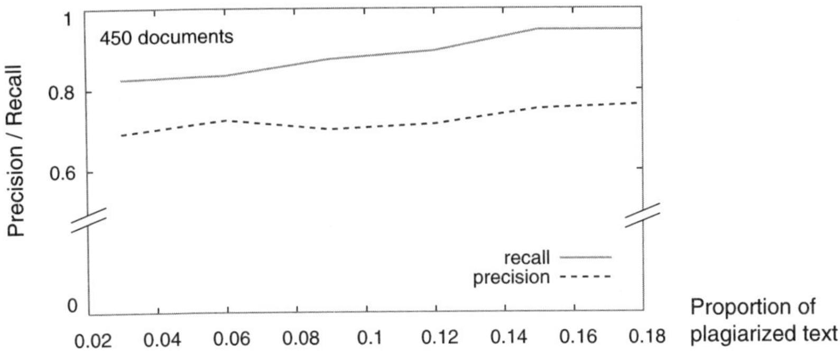

Fig. 4. Detection performance versus severity of plagiarism delict: The plot shows the averaged values for precision and recall of a series of experiments, where the sizes of the plagiarized passages are successively increased. The values are averaged using a ten-fold cross-validation.

The vector of these relative differences along with the class information (plagiarized or not) formed the input for different machine learning approaches. Figure 4 summarizes the results: We obtained good detection rates for plagiarism delicts in terms of precision and recall, irrespective of the plagiarism severity. These results were achieved using a classical discriminant analysis; however, an SVM classification showed similar results. Table 1 quantifies the discrimination power of the best features.

Table 1. Significance scores for the three best-discriminating features. Lower Lambda-values and higher F-ratios indicate a better performance.

	Wilks Lambda	F-Ratio	significant
av. word frequency class	0.723	152.6	yes
av. preposition number	0.866	61.4	yes
av. sentence length	0.880	54.0	yes

4 Summary

This paper presented an approach to detect plagiarized passages within a document if no reference collection is given against which the suspicious document can be matched. This problem, which we call "intrinsic plagiarism detection", is related to the identification of an author's writing style, for which various measures have been developed in the past. We presented new style features and showed their usability with respect to plagiarism detection: Classification experiments on a manually constructed corpus delivered promising precision and recall values, even for small plagiarized paragraphs.

Another result of our research shall be emphasized: A vocabulary richness measure qualifies for intrinsic plagiarism detection only, if is has a small variance subject to the analyzed text portion's size. Our experiments revealed that the introduced averaged word frequency class outperforms other well-known measures in this respect.

References

BRIN, S., DAVIS, J. and GARCIA-MOLINA, H. (1995): Copy Detection Mechanisms for Digital Documents. In: *Proceedings of SIGMOD '95.*

DALE, E. and CHALL, J.S. (1948): A Formula for Predicting Readability. *Educ. Res. Bull., 27.*

FLESCH, R. (1948): A New Readability Yardstick. *Journal of Applied Psychology, 32, 221–233.*

GARSIDE, R., LEECH, G. and MCENERY, A. (1997): *Corpus Annotation: Linguistic Information from Computer Text Corpora.* Longman.

HOAD, T.C. and ZOBEL, J. (2003): Methods for Identifying Versioned and Plagiarised Documents. *JASIST, 54, 3, 203–215.*

HONORE, A. (1979): Some Simple Measures of Richness of Vocabulary. *Association for Literary and Linguistic Computing Bulletin, 7, 2, 172–177.*

KINCAID, J., FISHBURNE, R.P., ROGERS, R.L. and CHISSOM, B.S. (1975): Derivation of New Readability Formulas for Navy Enlisted Personnel. *Research Branch Report 85*, US Naval Air Station.

KOPPEL, M. and SCHLER, J. (2004): Authorship Verification as a One-class Classification Problem. In *Proceedings of ICML 04*, Banff, Canada. ACM Press.

MCCABE, D. (2005): Research Report of the Center for Academic Integrity. `http://www.academicintegrity.org`.

MEYER ZU EISSEN, S. and STEIN, B. (2004): Genre Classification of Web Pages: User Study and Feasibility Analysis. In: *KI 2004, LNAI.* Springer.

SORENSEN, J. (2005): A Competitive Analysis of Automated Authorship Attribution Techniques. `http://hbar.net/thesis.pdf`.

STAMATATOS, E., FAKOTAKIS, N. and KOKKINSKIS, G. (2001): Computer-based Authorship Attribution without Lexical Measures. *Computers and the Humanities, 35, 193–214.*

STEIN, B. (2005): Fuzzy-Fingerprints for Text-Based Information Retrieval. In the *Proceedings I-KNOW 05, Graz,* J.UCS, 572–579. Know-Center.

STEIN, B. and MEYER ZU EISSEN, S. (2006): Near Similarity Search and Plagiarism Analysis. In *Proc. 29th Annual Conference of the GfKl,* Springer, Berlin.

TWEEDIE, F.J. and BAAYEN, R.H. (1997): Lexical "Constants" in Stylometry and Authorship Studies. In *Proceedings of ACH-ALLC '97.*

UNIVERSITY OF LEIPZIG (1995): Wortschatz. `http://wortschatz.uni-leipzig.de`.

YULE, G. (1944): *The Statistical Study of Literary Vocabulary.* Cambridge University Press.

Putting Successor Variety Stemming to Work

Benno Stein and Martin Potthast

Faculty of Media, Media Systems, Bauhaus University Weimar, 99421 Weimar, Germany; {benno.stein, martin.potthast}@medien.uni-weimar.de

Abstract. Stemming algorithms find canonical forms for inflected words, e. g. for declined nouns or conjugated verbs. Since such a unification of words with respect to gender, number, time, and case is a language-specific issue, stemming algorithms operationalize a set of linguistically motivated rules for the language in question. The most well-known rule-based algorithm for the English language is from Porter (1980).

The paper presents a statistical stemming approach which is based on the analysis of the distribution of word prefixes in a document collection, and which thus is widely language-independent. In particular, our approach tackles the problem of index construction for multi-lingual documents. Related work for statistical stemming either focuses on stemming quality (such as Bachin et al. (2002) or Bordag (2005)) or investigates runtime performance (Mayfield and McNamee (2003) for example), but neither provides a reasonable tradeoff between both. For selected retrieval tasks under vector-based document models we report on new results related to stemming quality and collection size dependency.

Interestingly, successor variety stemming has neither been investigated under similarity concerns for index construction nor is it applied as a technology in current retrieval applications. The results show that this disregard is not justified.

1 Introduction

Most of the words in a text document have various morphological variants. Since the variants have a similar semantics they can be considered as equivalent for the purpose of many retrieval tasks. Consider for example the words "connecting" and "connect": they are not recognized being equivalent without having them reduced to their stem. A *stem* is the portion of a word that is common to a set of inflected forms; it is not further analyzable into meaningful elements and carries the principle portion of meaning of the words in which it functions. *Stemming* is the process of reducing a word to its stem, and a *stemmer* or a stemming algorithm is a computer program that automates the task of stemming. As illustrated in Figure 1 stemming happens at an early stage in the text processing chain.

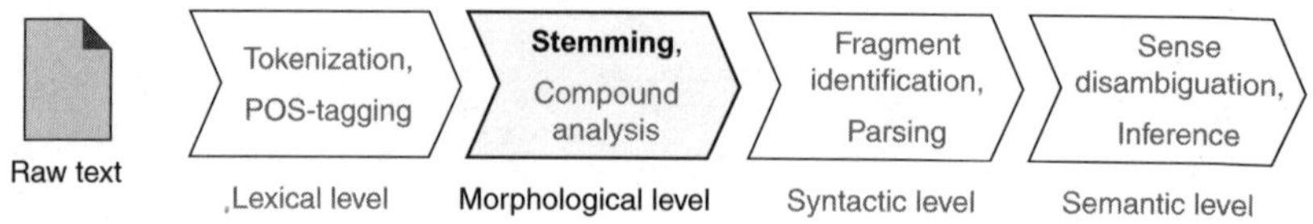

Fig. 1. The role of stemming in the text processing chain.

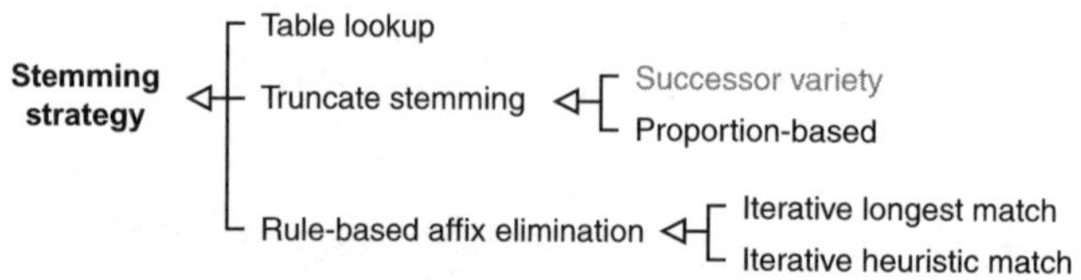

Fig. 2. Taxonomy of stemming strategies.

1.1 On stemming

Since natural languages are irregular, stemming algorithms have a heuristic component and hence are subject to generating incorrect stems. One refers to overstemming if too much of a word is removed, and to understemming if words could be conflated to the same stem but remained distinct after stemming. An example for overstemming is "relati-ve" / "relati-vistic", an example for understemming is "sensib-le" / "sensibili-ty". In the literature on the subject stemming algorithms are often judged by counting the number of produced stemming mistakes. However, we believe that it is reasonable to measure the power of a stemming approach by its impact on the performance of dedicated information retrieval tasks.

The different stemming strategies developed in the past can be classified according to the following scheme (cf. also Figure 2):

- *Table Lookup.* Stores to each stem all flections in a hash table. Problems with this approach include the handling of non-standard words and storage requirements.
- *Truncate Stemming.* Retains the first k letters of a word, where k is a suitable integer; a word with less than k letters is simply returned.
- *Rule-based Affix Elimination.* Removes the match related to a precondition of a rule and possibly repairs the resulting stem. An important variant is iterative longest match stemming, which forms the base of several well-known stemming algorithms, such as Lovins (1968), Porter (1980), Paice (1990), and Krovetz (1993). Note that these approaches are tailored to the English language; a recent development is the Snowball initiative of Porter (2001), which employs language-specific rules.

Successor variety analysis is a special form of truncate stemming that applies a variable, say, word-specific k computed from the underlying collection. Compared to rule-based affix elimination the successor variety analysis is a purely syntactic approach and hence it should be inferior to a knowledge-based stemmer. At closer inspection the situation looks differently: (*i*) Successor variety analysis is adaptive and conservative by nature since it identifies and applies collection-specific stemming rules. (*ii*) A rule-based approach is problematic if, for instance, the document language is unknown, if no language-specific stemming rules are at hand, or if a document combines passages from several languages. Successor variety analysis is unaffected by this.

1.2 Contributions

Despite its advantages successor variety analysis is not applied as a technology for index refinement in current retrieval applications. In addition, only few and rather outdated analyses have considered this approach in their evaluations. The paper in hand addresses this gap. We have developed an implementation for successor variety stemming along with new pruning heuristics, which is compared to well-known rule-based stemming implementations. In particular, the following questions are investigated:

1. How far is successor variety stemming behind rule-based stemming for the languages English and German?
2. What is the quantitative connection between the quality of successor variety stemming and corpus size?[1]

To answer these questions we have set up a large number of text categorization experiments with different clustering algorithms. Since these algorithms are susceptible to various side effects, we will also present results that rely on an objective similarity assessment statistic: the measure of expected density, $\overline{\rho}$, proposed by Stein et al. (2003).

The remainder of the paper is organized as follows. Section 2 introduces and discusses successor variety stemming, and Section 3 reports on selected classification experiments and similarity analysis.

2 Successor variety stemming

Given a set of a word's morphological variants, a potential stem can be derived heuristically, by a skillful analysis of prefix frequency and prefix length among the variants. E. g., the longest common prefix of the words "connection", "connect", "connectivity", "connecting" is "connect", which is also the stem. This principle applies to other languages like German as well: "verbinden", "Verbindung", "verbinde", "verbindend" all share the same stem "verbind". Of course there are exceptions in the German language which fall not into this scheme, such as the past principle "verbunden" in the example.[2]

[1] This question addresses also the issue of a "break-even point", above which one gets a pay-off from one or the other strategy.

[2] This idea extends naturally to the identification of compound word concepts. If sequences of n words occur significantly often then it is likely that these words

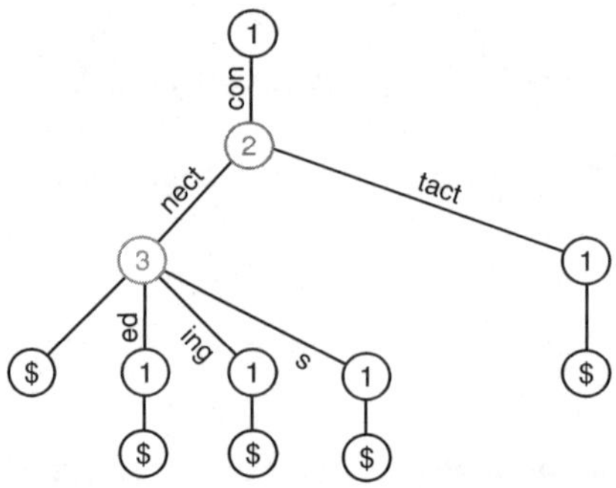

Fig. 3. A suffix tree at the character level for the words "connect", "connected", "connecting", "connects", and "contact". The inner nodes hold the frequency information, the $-sign denotes the end of string.

To identify significant prefixes we have developed an analysis method based on Weiner's (1973) suffix tree data structure, into which each word w of each document d of a collection D is inserted. The construction of a suffix tree is straightforward: A word w is inserted by testing whether some edge emanating from the root node is labeled with w's first character c_1. If so, this edge is traversed and the procedure is repeated by testing whether some edge of the successor node is labeled with c_2. If, at some depth k, a node n without a matching edge is reached a new node is created and connected to n with an edge labeled c_{k+1}. An in-depth view about data structure is given by Gusfield (1997). Frequency information is updated during suffix tree insertion. Figure 3 shows a suffix tree for morphological variants of the word "connect"; inner nodes with an outdegree of 1 are omitted. Note that suffix trees are used here because of their small memory footprint; with respect to runtime they may be outperformed by Morrisons's (1968) Patricia data structure.

It remains to be answered how *reasonable* stems can be identified using a suffix tree. Obviously, a possible stem lies on a path that starts at the root, and, inner nodes that have a high outdegree—compared to their immediate predecessors—may be good stemming candidates. Frakes (1992) proposes strategies to operationalize this observation:

- *Cutoff Method.* Take all subsequences of w as candidates that start at the root and end in a node with more successors than a given threshold. It remains unclear how the threshold has to be chosen; particularly it will depend on w's length.
- *Peak and Plateau Method.* Choose subsequences of w that end in a node that has more successors than its predecessor.

form a concept. Stemming and concept identification are essentially the same—the granularity makes the difference: the former means frequency analysis at the character level; likewise, the latter means frequency analysis at the word level.

- *Complete Word Method.* Define as stem of a word w the shortest substring w' of w, which also occurs as a word in a document.
- *Entropy Method.* Choose subsequences w' of w whose entropy is significantly higher than the entropy of its immediate predecessor.

The strategy that has been applied in this paper is an enhancement of the peak and plateau method with respect to different subsequence lengths. To accept the substring of a word up to the k-th letter as suitable stem, the successor variety values v_{k-1}, v_k, and v_{k+1} must fulfill the following inequalities:

$$k > x \cdot m \quad \text{with} \quad x \in (0;1) \tag{1}$$

$$\frac{v_k}{v_{k-1}} > y \quad \text{with} \quad y > 0 \tag{2}$$

$$\frac{v_{k+1}}{v_k} < z \cdot \frac{v_k}{v_{k-1}} \quad \text{with} \quad z > 0 \tag{3}$$

The first inequality ensures a minimum length for a stem. The second inequality is a refinement of the peak and plateau method that simply claims the constraint $y = 1$. In the form used here especially the case $0 < y < 1$, where the successor variety v_k is significantly high but not larger than v_{k-1}, can be recognized. The third inequality ensures that a word is not overstemmed given the case that a succeeding character provides a reasonable clipping position as well. Note that all computations within the analysis can be done during a single depth-first traversal of the suffix-tree, which leads to an overall time complexity comparable to Porter's algorithm.

Figure 4 shows stems obtained by our successor variety analysis, the respective stems returned by Porter's algorithm, and the optimum stems.

Successor variety		**Porter**		**Optimum**
Stem(s)	Affix(es)	Stem(s)	Affix(es)	
minist	- ers, erial, er, - ry, ries	minist ministri	- ers, er - $, es	minist
oper operat operati	 - ors, or, es, e - onal, ons, ng, on	oper	- $, ators, ational, - ator, ates, ations, - ating, ation, ate	oper operat
exten	- sion, d, ds, ding, - t, ded, sive	extend extens extent	- $, s, ing, ed - ion, ive	exten

Fig. 4. The table contrasts stems obtained by our successor variety analysis and by Porter's algorithm with optimum stems. The $-symbol denotes the empty string.

3 Quantitative analysis of stemming approaches

Only few experiments related to stemming were made in the past;[3] in particular, existing research investigates neither the quality of successor variety stemming nor its language independence.[4] In this connection the purpose of our stemming analyses is twofold. (*i*) We want to analyze the potential of successor variety stemming compared to rule-based stemming. (*ii*) Since successor variety stemming is expected to be collection-size-dependent by nature, the trade-off between stemming quality and collection size shall be revealed. The employed document model is the vector space model; stopwords are omitted and term weighting is done according to the $tf \cdot idf$-scheme.

The analyses illustrate the impact of a stemming strategy in two ways: *indirectly*, by comparing the classification performance in a categorization task, expressed as F-Measure value, and *directly*, by comparing the intrinsic similarity relations captured in a document model, expressed as expected density $\overline{\rho}$. The latter is defined as follows. Let $\mathcal{C} = \{C_1, \ldots, C_k\}$ be a categorization of D, and let $G = \langle V, E, \varphi \rangle$ be the underlying similarity graph of D. Based on the global edge density θ of G, $\overline{\rho}$ averages the class-specific density improvement within the subgraphs $G_i = \langle V_i, E_i, \varphi \rangle$ induced by the categories C_i:

$$\overline{\rho}(\mathcal{C}) = \sum_{i=1}^{k} \frac{|V_i|}{|V|} \cdot \frac{\sum_{e \in E_i} \varphi(e)}{|V_i|^{\theta}}, \quad \text{where } \theta \text{ computes from } |V|^{\theta} = \sum_{e \in E} \varphi(e)$$

If for a collection of documents the $\overline{\rho}$-value under document model M_2 is larger than the $\overline{\rho}$-value under document model M_1, then M_2 captures more of the intrinsic similarity structure of the collection (Stein et al. (2003)).

The experiments rely on RCV1, the "Reuters Corpus Volume 1", described by Rose et al. (2002), as well as on a corpus of German newsgroup postings. RCV1 contains about 800,000 documents each of which consisting of a few hundred up to several thousands words. The documents are tagged with meta information like category (also called topic), geographic region, or industry sector. The German newsgroup corpus has been compiled in our working group and comprises 26,000 postings taken from 20 different newsgroups. From these corpora the samples were formed as follows. For the analysis of the categorization tasks the sample sizes were 1000 documents, chosen from 10 categories. For the analysis of the intrinsic similarity relations the sample sizes were ranging from 200 to 1000 documents, chosen from 5 categories.

Figure 5 shows selected analysis results for the two languages English (left-hand side) and German (right-hand side). The tables comprise the effects of the different document models within the categorization task, expressed as

[3] Studies were conducted by Frakes (1984), Harman (1991), Frakes and Baeza-Yates (1992), Braschler and Ripplinger (2004), and Abdou et. al. (2005).

[4] Language independence, however, applies primarily to languages whose flections base on suffixes, prefixes, or circumfixes.

Stemming strategy	F-min (sample size 1000, 10 categories)	F-max	F-av.
without		—baseline—	
Porter	-11%	14%	6%
successor variety	-15%	14%	4%

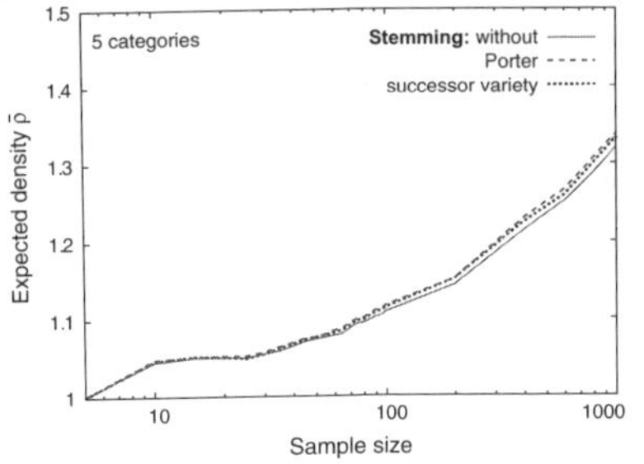

Stemming strategy	F-min (sample size 1000, 10 categories)	F-max	F-av.
without		—baseline—	
Snowball	-8%	26%	10%
successor variety	-9%	23%	6%

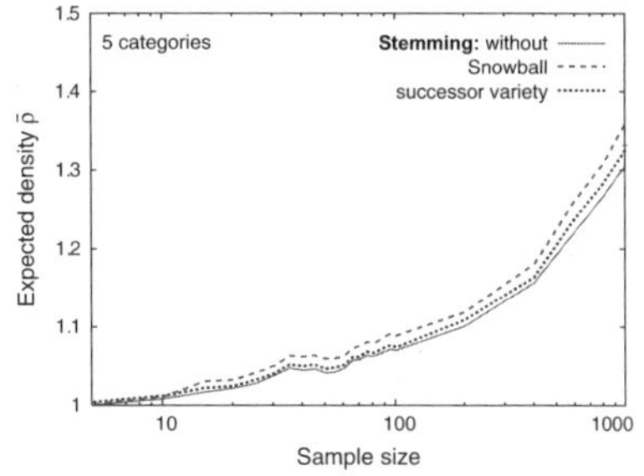

Fig. 5. The effect of no stemming, rule-based stemming, and successor variety stemming on the vector space model, given an English collection (left) and a German collection (right). The tables (top) comprise the effects within the categorization task; the graphs below show the relation between the collection size and the similarity relations captured within the document model.

improvements in the F-Measure-value for different cluster algorithms. The graphs show that—independent of the stemming strategy—the collection size is positively correlated with the expected density $\bar{\rho}$. This means that additional documents add similarity information rather than noise. Altogether, rule-based stemming performs slightly better than successor variety stemming, which shows a reasonable performance though.[5]

4 Discussion

This paper reported on a comparison between rule-based stemming and successor variety stemming by evaluating the retrieval performance of the respective vector space models. Our approach to successor variety stemming is based on a suffix tree data structure, and controlled by new pruning heuristics that skillfully analyze the successor variety of the inner tree nodes.

For the performance evaluation both an indirect method and a direct method has been applied. The former relies on the application of clustering algorithms in a text categorization task; the latter relies on the similarity graph of a document collection and quantifies improvements between the inter-class and the intra-class variance of the edge weights. The following analysis results shall be emphasized:

1. The effect of stemming with respect to the vector space model is less than commonly expected.

[5] Files with meta information have been recorded which describe our sample collections; they are available to other researchers upon request.

2. Compared to rule-based stemming, the retrieval performance of our optimized successor variety stemming is only slightly worse. Note that for the German language this performance can be further improved by applying the same strategy to identify prefixes and by adjusting the pruning heuristics to identify compound words.

Two salient advantages of successor variety stemming are its language independence and its robustness with respect to multi-lingual documents. An obvious disadvantage may be the necessary statistical mass: successor variety stemming cannot work if only few, very small document snippets are involved. This effect could directly be observed in our experiments.

References

ABDOU, S., RUCH, P. and SAVOY, J. (2005): Evaluation of Stemming, Query Expansion and Manual Indexing Approaches for the Genomic Task. In: *Proc. TREC'05.*

BACCHIN, M., FERRO, N. and MELUCCI, M. (2002): Experiments to Evaluate a Statistical Stemming Algorithm. In: *Proc. of CLEF'02.*

BORDAG, S. (2005): Unsupervised Knowledge-free Morpheme Boundary Detection. In: *Proc. of RANLP'05.*

BRASCHLER, M. and RIPPLINGER, B. (2004): How Effective is Stemming and Decompounding for German Text Retrieval? *Information Retrieval, 7, 3-4, 291–316.*

FRAKES, W.B. (1984): Term Conflation for Information Retrieval. In: *Proc. SIGIR'84.*

FRAKES, W.B. and BAEZA-YATES, R. (1992): *Information Retrieval: Data Structures and Algorithms.* Prentice-Hall, Upper Saddle River.

GUSFIELD, D. (1997): *Algorithms on Strings, Trees, and Sequences: Computer Science and Computational Biology.* Cambridge University Press.

HARMAN, D. (1991): How Effective is Suffixing? *J. of the ASIS&T., 42, 1, 7–15.*

KROVETZ, R. (1993): Viewing Morphology as an Inference Process. In: *Proc. SIGIR'93.*

LOVINS, J.B. (1968): Development of a Stemming Algorithm. *Mechanical Translation and Computation Linguistics, 11, 1, 23–31.*

MAYFIELD, J. and MCNAMEE, P. (2003): Single n-gram Stemming. In: *Proc. SIGIR'03.*

MORRISON, D.R. (1968): PATRICIA—Practical Algorithm to Retrieve Information Coded in Alphanumeric. *Journal of the ACM, 15, 4, 514–534.*

PAICE, C.D. (1990): Another Stemmer. In: *SIGIR Forum, 24, 3, 56–61.*

PORTER, M.F. (1980): An Algorithm for Suffix Stripping. *Program, 14, 3, 130–137.*

PORTER, M. (2001): Snowball. `http://snowball.tartarus.org/`.

ROSE, T.G., STEVENSON, M. and WHITEHEAD, M. (2002): The Reuters Corpus Volume 1 - From Yesterday's News to Tomorrow's Language Resources. In: *Proc. of LREC'02.*

STEIN, B., MEYER ZU EISSEN, S. and WISSBROCK, F. (2003): On Cluster Validity and the Information Need of Users. In: *Proc. of AIA'03.*

WEINER, P. (1973): Linear Pattern Matching Algorithm. In: *Proc. of the 14th IEEE Symp. on Switching and Automata Theory.*

Collaborative Filtering Based on User Trends

Panagiotis Symeonidis, Alexandros Nanopoulos, Apostolos Papadopoulos and Yannis Manolopoulos

Aristotle University, Department of Informatics, Thessaloniki 54124, Greece;
{symeon, alex, apostol, manolopo}@delab.csd.auth.gr

Abstract. Recommender systems base their operation on past user ratings over a collection of items, for instance, books, CDs, etc. Collaborative Filtering (CF) is a succesful recommendation technique. User ratings are not expected to be independent, as users follow trends of similar rating behavior. In terms of Text Mining, this is analogous to the formation of higher-level concepts from plain terms. In this paper, we propose a novel CF algorithm which uses Latent Semantic Indexing (LSI) to detect rating trends and performs recommendations according to them. Our results indicate its superiority over existing CF algorithms.

1 Introduction

The "information overload" problem affects our everyday experience while searching for valuable knowledge. To overcome this problem, we often rely on suggestions from others who have more experience on a topic. In Web case, this is more manageable with the introduction of Collaborative Filtering (CF), which provides recommendations based on the suggestions of users who have similar preferences.

Latent Semantic Indexing (LSI) has been extensively used in informational retrieval, to detect the latent semantic relationships between terms and documents. Thus, higher level concepts are generated from plain terms. In CF, this is analogous to the formation of users' trends from individual preferences.

In this paper, we propose a new algorithm that is based on LSI to produce a condensed model for the user-item matrix. This model comprises a matrix that captures the main user trends removing noise and reducing its size.

Our contribution and novelty are summarized as follows: (i) based on Information Retrieval, we include the pseudo-user concept in CF (ii) We implement a novel algorithm, which tunes the number of principal components according to the data characteristics. (iii)We generalize the recommendation procedure for both user- and item-based CF methods. (iv) We propose a new top-N generation list algorithm based on SVD and the Highest Prediction Rated items.

The rest of this paper is organized as follows. Section 2 summarizes the related work, whereas Section 3 contains the analysis of the CF factors. The proposed approach is described in Section 4. Experimental results are given in Section 5. Finally, Section 6 concludes this paper.

2 Related work

In 1992, the Tapestry system (Goldberg et al. (1992)) introduced Collaborative Filtering (CF). In 1994, the GroupLens system (Resnick et al. (1994)) implemented a CF algorithm based on common users preferences. Nowadays, it is known as user-based CF algorithm. In 2001, another CF algorithm was proposed. It is based on the items' similarities for a neighborhood generation of nearest items (Sarwar et al. (2001)) and is denoted as item-based CF algorithm.

Furnas et al. (1988) proposed Latent Semantic Indexing (LSI) in Information Retrieval area to detect the latent semantic relationships between terms and documents. Berry et al.(1994) carried out a survey of the computational requirements for managing (e.g., folding-in, which is a simple technique that uses existing SVD to represent new information.) LSI-encoded databases. He claimed that the reduced-dimensions model is less noisy than the original data.

Sarwar et al. (2000) applied dimensionality reduction for only the user-based CF approach. In contrast to our work, Sarwar et al. included test users in the calculation of the model as were apriori known. For this reason, we introduce the notion of pseudo-user in order to insert a new user in the model (folding in), from which recommendations are derived.

3 Factors affecting CF

In this section, we identify the major factors that critically affect all CF algorithms. Table 1 summarizes the symbols that are used in the sequel.

Table 1. Symbols and definitions.

Symbol	Definition	Symbol	Definition
k	number of nearest neighbors	$p_{u,i}$	predicted rate for user u on item i
N	size of recommendation list	c	number of singular values
$\mathcal{I}$	domain of all items	A	original matrix
$\mathcal{U}$	domain of all users	S	Singular values of A
u, v	some users	V'	Right singular vectors of A
i, j	some items	A^*	Approximation matrix of A
$\mathcal{I}_u$	set of items rated by user u	$\mathbf{u}$	user vector
$\mathcal{U}_i$	set of users rated item i	$\mathbf{u}_{\mathbf{new}}$	inserted user vector
$r_{u,i}$	the rating of user u on item i	n	number of train users
$\overline{r}_u$	mean rating value for user u	m	number of items
$\overline{r}_i$	mean rating value for item i	U	Left singular vectors of A

Similarity measures: A basic factor for the formation of user/item neighborhood is the similarity measure. The most extensively used similarity measures are based on correlation and cosine-similarity (Sarwar et al. (2001)). Specifically, user-based CF algorithms mainly use Pearson's Correlation (Equation 1), whereas for item-based CF algorithms, the Adjusted Cosine Measure is preferred (Equation 2)(McLauglin and Herlocker (2004)).

$$\text{sim}(u,v) = \frac{\sum\limits_{\forall i \in S} (r_{u,i} - \overline{r}_u)(r_{v,i} - \overline{r}_v)}{\sqrt{\sum\limits_{\forall i \in S} (r_{u,i} - \overline{r}_u)^2} \sqrt{\sum\limits_{\forall i \in S} (r_{v,i} - \overline{r}_v)^2}}, S = \mathcal{I}_u \cap \mathcal{I}_v. \quad (1)$$

$$\text{sim}(i,j) = \frac{\sum\limits_{\forall u \in T} (r_{u,i} - \overline{r}_u)(r_{u,j} - \overline{r}_u)}{\sqrt{\sum\limits_{\forall u \in \mathcal{U}_i} (r_{u,i} - \overline{r}_u)^2} \sqrt{\sum\limits_{\forall u \in \mathcal{U}_j} (r_{u,j} - \overline{r}_u)^2}}, T = \mathcal{U}_i \cap \mathcal{U}_j. \quad (2)$$

Generation of recommendation list: The most often used technique for the generation of the top-N list, is the one that counts the frequency of each item inside the found neighborhood, and recommends the N most frequent ones. Henceforth, this technique is denoted as Most-Frequent item recommendation (MF). MF can be applied to both user-based and item-based CF algorithms.

Evaluation metrics: Mean Absolute Error (MAE) has been used in most of related works. Although, it has received criticism as well (McLauglin and Herlocker (2004)). Other extensively used metrics are *precision* (ratio of relevant and retrieved to retrieved) and *recall* (ratio of relevant and retrieved to relevant). We also consider F_1, which is another popular metric for CF algorithms and combines the two previous ones.

4 Proposed method

Our approach applies Latent Semantic indexing in CF process. To ease the discussion, we will use the running example illustrated in Figure 1 where I_{1-4} are items and U_{1-4} are users. As shown, the example data set is divided into train and test set. The null cells(no rating) are presented as zeros.

	I_1	I_2	I_3	I_4
U_1	4	1	1	4
U_2	1	4	2	0
U_3	2	1	4	5

(a)

	I_1	I_2	I_3	I_4
U_4	1	4	1	0

(b)

Fig. 1. (a) train set ($n \times m$), (b) test set.

Applying SVD on train data: Initially, we apply SVD on train data $n \times m$ matrix A that produces three matrices. These matrices obtained by SVD can give by performing multiplication the initial matrix as the following Equation 3 and Figure 2 show:

$$A_{n \times m} = U_{n \times n} \cdot S_{n \times m} \cdot V'_{m \times m} \tag{3}$$

4	1	1	4
1	4	2	0
2	1	4	5

$A_{n \times m}$

-0.61	0.28	-0.74
-0.29	-0.95	-0.12
-0.74	0.14	0.66

$U_{n \times n}$

8.87	0	0	0
0	4.01	0	0
0	0	2.51	0

$S_{n \times m}$

-0.47	-0.28	-0.47	-0.69
0.11	-0.85	-0.27	0.45
-0.71	-0.23	0.66	0.13
-0.52	0.39	-0.53	0.55

$V'_{m \times m}$

Fig. 2. Example of: $A_{n \times m}$ (initial matrix A), $U_{n \times m}$ (left singular vectors of A), $S_{n \times m}$ (singular values of A), $V'_{m \times m}$ (right singular vectors of A).

Preserving the principal components: It is possible to reduce the $n \times m$ matrix S to have only c largest singular values. Then, the reconstructed matrix is the closest rank-c approximation of the initial matrix A as it is shown in Equation 4 and Figure 3:

$$A^*_{n \times m} = U_{n \times c} \cdot S_{c \times c} \cdot V'_{c \times m} \tag{4}$$

2.69	0.57	2.22	4.25
0.78	3.93	2.21	0.04
3.17	1.38	2.92	4.78

$A^*_{n \times i}$

-0.61	0.28
-0.29	-0.95
-0.74	0.14

$U_{n \times c}$

8.87	0
0	4.01

$S_{c \times c}$

-0.47	-0.28	-0.47	-0.69
0.11	-0.85	-0.27	0.45

$V'_{c \times m}$

Fig. 3. Example of: $A^*_{n \times m}$ (approximation matrix of A), $U_{n \times c}$ (left singular vectors of A^*), $S_{c \times c}$ (singular values of A^*), $V'_{c \times m}$ (right singular vectors of A^*).

We tune the number, c, of principal components (i.e., dimensions) with the objective to reveal the major trends. The tuning of c is determined by the information percentage that is preserved compared to the original matrix. Therefore, a c-dimensional space is created and each of the c dimensions corresponds to a distinctive rating trend. We have to notice that in the running example we create a 2-dimensional space using 83% of the total information of the matrix (12,88/15,39).

Inserting a test user in the c-dimensional space: It is evident that, for user-based approach, the test data should be considered as unknown in the c-dimensional space. Thus a specialized insertion process should be used. Given the current ratings of the test user u, we enter pseudo-user vector in the c-dimensional space using the following Equation 5 (Furnas et al. (1988)). In the current example,we insert U_4 into the 2-dimensional space, as it is shown in Figure 4:

$$\mathbf{u_{new}} = \mathbf{u} \cdot V_{m \times c} \cdot S^{-1}_{c \times c} \tag{5}$$

$\mathbf{u_{new}}$

-0.23	-0.89

u

1	4	1	0

$V_{m\times c}$

-0.47	0.11
-0.28	-0.85
-0.47	-0.27
-0.69	0.45

$S^{-1}_{c\times c}$

0.11	0
0	0.25

Fig. 4. Example of: $\mathbf{u_{new}}$ (inserted new user vector), **u** (user vector), $V_{m\times c}$ (two left singular vectors of V), $S^{-1}_{c\times c}$ (two singular values of inverse S).

Generating the neighborhood of users/items: Having a reduced dimensional representation of the original space, we form the neighborhoods of users/items in that space. For the user-based approach, we find the k nearest neighbors of pseudo user vector in the c-dimensional space. The similarities between train and test users can be based on Cosine Similarity. First, we compute the matrix $U_{n\times c} \cdot S_{c\times c}$ and then we perform vector similarity. This $n \times c$ matrix is the c-dimensional representation for the n users.

For the item-based approach, we find the k nearest neighbors of item vector in the c-dimensional space. First, we compute the matrix $S_{c\times c} \cdot V_{c\times m}$ and then we perform vector similarity. This $c \times m$ matrix is the c-dimensional representation for the m items.

Generating and evaluating the recommendation list: As it is mentioned in Section 3, existing ranking criteria, such as MF, are used for the generation of the top-N list in classic CF algorithms. We propose a ranking criterion that uses the predicted values of a user for each item. Predicted values are computed by Equations 6 and 7, for the cases of user-based and item-based CF, respectively. These equations have been used in related work for the purpose of MAE calculation, whereas we use them for generation of top-N list.

$$p_{u,i} = \overline{r}_u + \frac{\sum_{v\in U} \text{sim}(u,v)(r_{v,i} - \overline{r}_v)}{\sum_{v\in U} |\text{sim}(u,v)|} \tag{6}$$

$$p_{u,i} = \overline{r}_i + \frac{\sum_{j\in \mathcal{I}} \text{sim}(i,j)(r_{u,j} - \overline{r}_j)}{\sum_{j\in \mathcal{I}} |\text{sim}(i,j)|} \tag{7}$$

Therefore, we sort (in descending order) the items according to predicted rating value, and recommend the first N of them. This ranking criterion, denoted as Highest Predicted Rated item recommendation (HPR), is influenced by the good accuracy of prediction that existing related work reports through the MAE. HPR opts for recommending the items that are more probable to receive a higher rating.

5 Experimental configuration

In the sequel, we study the performance of the described SVD dimensionality reduction techniques against existing CF algorithms. Both user-based and item-based algorithms are tested. Factors, that are treated as parameters, are the following: the neighborhood size (k, default value 10), the size of the recommendation list (N, default value 20),and the size of train set (default value 75%). The metrics we use are recall, precision, and F_1.

We perform experiments in a real data set that has been used as benchmarks in prior work. In particular, we examined MovieLens data set with 100,000 ratings assigned by 943 users on 1,682 movies. The range of ratings is between 1(bad)-5(excellent) of the numerical scale. Finally, the value of an unrated item is considered equal to zero.

5.1 Results for user-based CF algorithm

Firstly, we compare existing user-based CF algorithm that uses Pearson similarity against two representative SVD reductions(SVD50 and SVD10). These are percentages of the initial information we keep from the initial user-item matrix after applying SVD. The results for precision and MAE vs. k are displayed in Figure 5a and b, respectively.

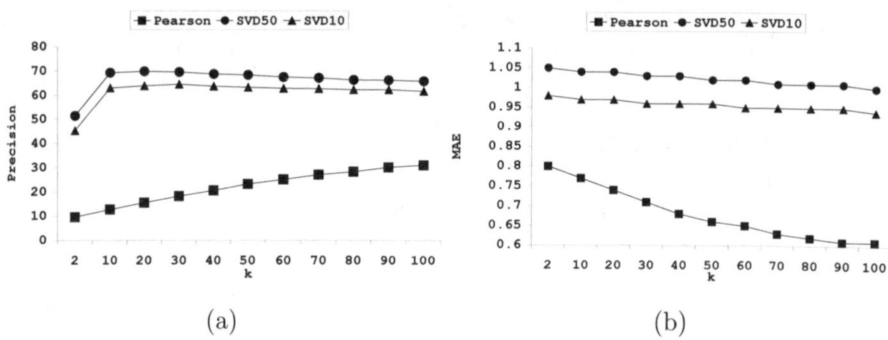

Fig. 5. Performance of user-based CF vs. k: (a) precision, (b) MAE.

As shown, the existing Pearson measure performs worst than SDV reductions. The reason is that the MovieLens data set is sparse and relatively large (high n value). The SVD reductions reveal the most essential dimensions and filter out the outliers and misleading information.

We now examine the MAE metric. Results are illustrated in Figure 5b. Pearson yields the lowest MAE values. This fact indicates that MAE is good only for the evaluation of prediction and not of recommendation, as Pearson measure did not attain the best performance in terms of precision.

Finally, we test the described criteria for the HSR top-N list generation algorithm. The results for precision and recall are given in Figure 6. As shown, the combination of the SVD similarity measure with HPR as list generation algorithm, clearly outperforms the Pearson with HPR. This is due to the fact that in the former the remaining dimensions are the determinative ones and outliers users have been rejected. Note that in the SVD50 we preserve only 157 basic dimensions instead of 708 train users for the latter.

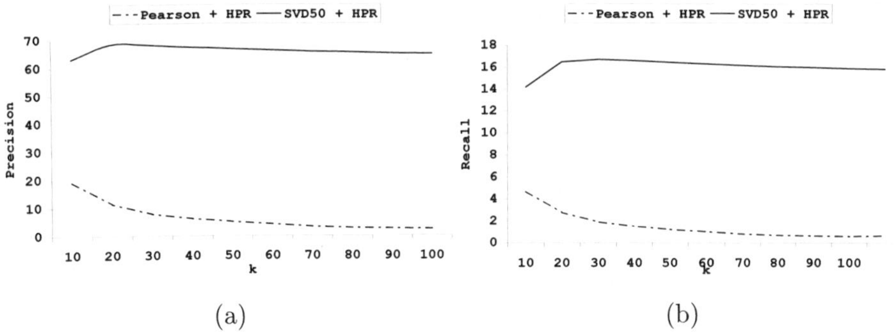

Fig. 6. Comparison HPR criteria for the generation of top-N (a) precision, (b) recall list for user-based CF vs. k.

5.2 Results for item-based CF algorithms

We perform similar measurements for the case of item-based CF. Thus, we examine the precision and recall for the existing Adjusted Cosine Measure (considers co-rated items) against SVD50 and SVD10 for the item-based case. The results are depicted in Figure 7 and are analogous to those of the user-based case. MAE and HPR results are also analogous to the user-based case.

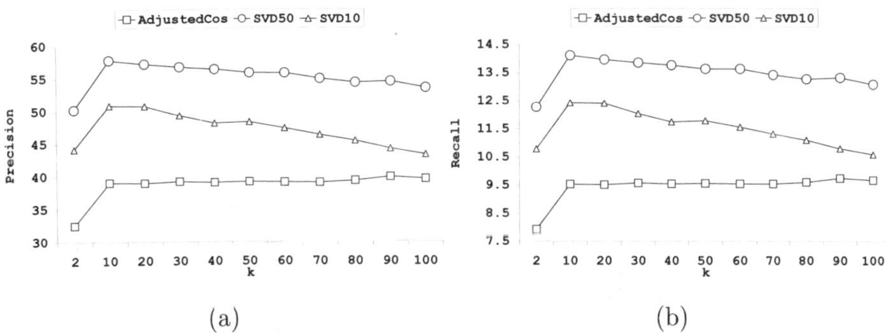

Fig. 7. Performance of item-based CF vs. k: (a) precision, (b) recall.

5.3 Comparative results

We compared user-based and item-based SVD50. For the generation of the top-N list, for both of them we use MF. The results for F_1 are displayed in Figure 8a, which demonstrate the superiority of user-based SVD50 against item-based SVD50.

Regarding the execution time, we measured the wall-clock time for the on-line parts for all test users. The results vs. k are presented in Figure 8b. Item based CF needs less time to provide recommendations than user-based CF. The reason is that a user-rate vector in user-based approach has to be inserted in the c-dimensional space. The generation of top-N list for the user-based approach further burdens the CF process.

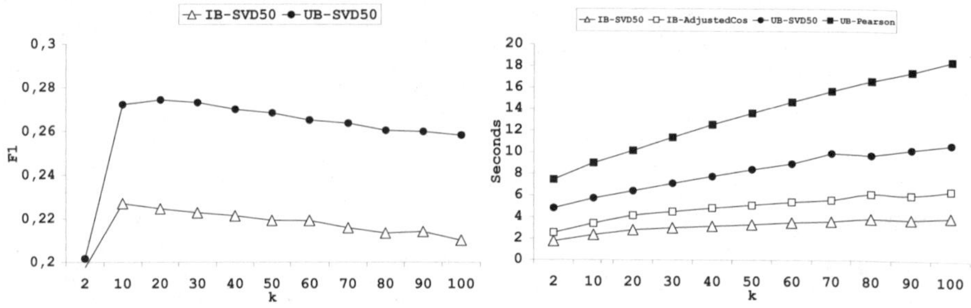

(a) (b)

Fig. 8. Comparison between item-based and user-based CF in terms of precision vs. k and in terms of execution times.

6 Conclusions

We performed experimental comparison of the proposed method against well known CF algorithms, like user-based or item-based methods (that do not consider trends), with real data sets. Our method show significant improvements over existing CF algorithms, in terms of accuracy (measured through recall/precision). In terms of execution times, due to the use of smaller matrices, execution times are dramatically reduced. In our future work we will consider the issue of an approach that would combine user and item based approaches, attaining high accuracy recommendations in the minimum responding time.

References

BERRY, M., DUMAIS, S. and OBRIEN, G. (1994): Using Linear Algebra for Intelligent Information Retrieval. *SIAM Review, 37, 4, 573-595.*

FURNAS, G., DEERWESTER, S., DUMAIS, S. et al. (1988): Information Retrieval Using a Singular Value Decomposition Model of Latent Semantic Structure. In: *Proc. ACM SIGIR Conf., 465-480.*

GOLDBERG, D., NICHOLS, D., BRIAN, M. and TERRY, D. (1992): Using Collaborative Filtering to Weave an Information Tapestry. *ACM Communications, 35, 12, 61-70.*

MCLAUGLIN, R. and HERLOCHER, J. (2004): A Collaborative Filtering Algorithm and Evaluation Metric That Accurately Model the User Experience. In: *Proc. ACM SIGIR Conf., 329-336.*

RESNICK, P., IACOVOU, N., SUCHAK, M., BERGSTROM, P. and RIEDL, J. (1994): Grouplens- An Open Architecture for Collaborative Filtering on Netnews. In: *Proc. Conf. Computer Supported Collaborative Work, 175-186.*

SARWAR, B., KARYPIS, G., KONSTAN, J. and RIEDL, J. (2000): Application of Dimensionality Reduction in Recommender System- A Case Study. In: *ACM WebKDD Workshop.*

SARWAR, B., KARYPIS, G., KONSTAN, J. and RIEDL, J. (2001): Item-based Collaborative Filtering Recommendation Algorithms. In: *Proc. WWW Conf., 285-295.*

Investigating Unstructured Texts with Latent Semantic Analysis

Fridolin Wild and Christina Stahl

Institute for Information Systems and New Media,
Vienna University of Economics and Business Administration,
Augasse 2-6, A-1090 Vienna, Austria; fridolin.wild@wu-wien.ac.at

Abstract. Latent semantic analysis (LSA) is an algorithm applied to approximate the meaning of texts, thereby exposing semantic structure to computation. LSA combines the classical vector-space model — well known in computational linguistics — with a singular value decomposition (SVD), a two-mode factor analysis. Thus, bag-of-words representations of texts can be mapped into a modified vector space that is assumed to reflect semantic structure. In this contribution the authors describe the *lsa* package for the statistical language and environment R and illustrate its proper use through examples from the areas of automated essay scoring and knowledge representation.

1 Introduction to latent semantic analysis

Derived from latent semantic indexing, LSA is intended to enable the analysis of the semantic structure of texts. The basic idea behind LSA is that the collocation of terms of a given document-term vector space reflects a higher-order — latent semantic — structure, which is obscured by word usage (e.g., by synonyms or ambiguities). By using conceptual indices that are derived statistically via a truncated singular value decomposition, this variability problem is believed to be overcome (Deerwester et al. (1990)).

In a typical LSA process, first a document-term matrix M is constructed from a given text base of n documents containing m terms. The term 'textmatrix' will be used throughout the rest of this contribution to denote this type of document-term matrices. This textmatrix M of the size $m \times n$ is then resolved by the singular value decomposition into the term-vector matrix T (constituting the left singular vectors), the document-vector matrix D (constituting the right singular vectors) being both orthonormal and the diagonal matrix S. These matrices are then reduced to a particular number of dimensions k, giving the truncated matrices T_k, S_k and D_k — the latent semantic space. Multiplying the truncated matrices T_k, S_k and D_k results in a new matrix M_k which is the least-squares best fit approximation of M with k singular values.

M_k is of the same format as M, i.e., rows represent the same terms, columns the same documents.

To keep additional documents from influencing a previously calculated semantic space or to simply re-use the structure contained in an already existing factor distribution, new documents can be folded-in after the singular value decomposition. For this purpose, the add-on documents can be added to the pre-exisiting latent semantic space by mapping them into the existing factor structure. Moreover, folding-in is computationally a lot less costly, as no singular value decomposition is needed. To fold-in, a pseudo-document vector $\hat{m}$ needs to be calculated in three steps (Berry et al. (1995)): after constructing a document vector v from the additional documents containing the term frequencies in the exact order constituted by the input textmatrix M, v can be mapped into the latent semantic space by applying (1) and (2).

$$\hat{d} = v^T T_k S_k^{-1} \tag{1}$$

$$\hat{m} = T_k S_k \hat{d} \tag{2}$$

Thereby, T_k and S_k are the truncated matrices from the previously calculated latent semantic space. The resulting vector $\hat{d}$ of Equation (1) represents an additional column of D_k. The resulting pseudo-document vector $\hat{m}$ from Equation (2) is identical to an additional column in the textmatrix representation of the latent semantic space.

2 Influencing parameters

Several classes of adjustment parameters can be functionally differentiated in the latent semantic analysis process. Every class introduces new parameter settings that drive the effectiveness of the algorithm. The following classes have been identified so far by Wild et al. (2005): textbase compilation and selection, preprocessing methods, weighting schemes, choice of dimensionality, and similarity measurement techniques (see Figure 1).

Different texts create a different factor distribution. Moreover, texts may be splitted into components such as sentences, paragraphs, chapters, bags-of-words of a fixed size, or even into context bags around certain keywords. The document collection available may be filtered according to specific criteria such as novelty or sampled into a random sample, so that only a subset of the existing documents will actually be used in the latent semantic analysis. The textbase compilation and selection options form one class of parameters.

Document preprocessing comprises several operations performed on the input texts such as lexical analysis, stop-word filtering, reduction to word stems, filtering of keywords above or below certain frequency thresholds, and the use of controlled vocabularies (Baeza-Yates (1999)).

Weighting schemes have been shown to significantly influence the effectiveness of LSA (Wild et al. (2005)). Weighting schemes in general can be

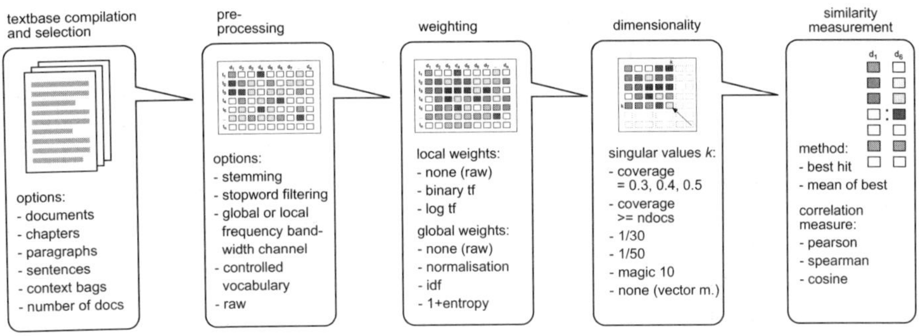

Fig. 1. Parameter classes influencing the algorithm effectiveness.

differentiated into local (lw) and global (gw) weighting schemes, which may be combined as follows:

$$\check{m} = lw(m) \cdot gw(m) \tag{3}$$

Local schemes only take into account term frequencies within a particular document, whereas global weighting schemes relate term frequencies to the frequency distribution in the whole document collection. Weighting schemes are needed to change the impact of relative and absolute term frequencies to, e.g., emphasize medium-frequency terms as they are assumed to be most representative for the documents described. Especially when dealing with narrative text, high-frequency terms are often semantically meaningless functional terms (e.g., 'the', 'it') whereas low-frequency terms can in general be considered to be distractors—generated, for example, through the use of metaphors. See Section 3 for an overview on common weighting mechanisms.

The choice on the ideal number of dimensions is responsible for the effect that distinguishes LSA from the pure vector-space model: if all dimensions are used, the original matrix will be reconstructed and an unmodified vector-space model is the basis for further processing. If less dimensions than available non-zero singular values are used, the original vector space is approximated. Thereby, relevant structure information inherent in the original matrix is captured, reducing noise and variability in word usage. Several methods to determine the optimal number of singular values to be used have been proposed. Wild et al. (2005) report a new method for calculating the number via a share between 30% and 50% of the cumulated singular values to show best results.

How the similarity of document or term vectors is measured forms another class of influencing parameters. Both, the similarity measure chosen and the similarity measurement method affects the outcomes. Various correlation measures have been applied in LSA. Among others, these comprise the simple crossproduct, the Pearson correlation (and the nearly identical cosine measure), and Spearman's Rho. The measurement method can, for example, simply be a vector to vector comparison or the average correlation of a vector with a particular vector set.

3 The LSA package for R

In order to facilitate the use of LSA, a package for the statistical language and environment R has been implemented by Wild (2005). The package is open-source and available via CRAN, the Comprehensive R Archive Network.

A higher-level abstraction is introduced to ease the application of LSA. Five core methods perform the direct LSA steps. With **textmatrix**(), a document base can be read in from a specified directory. The documents are converted to a textmatrix (i.e., document-term matrix, see above) object, which holds terms in rows and documents in columns, so that each cell contains the frequency of a particular term in a particular document. Alternatively, pseudo documents can be created with **query**() from a given text string. The output in this case is also a textmatrix, albeit it has only one column (the query).

By calling **lsa**() on a textmatrix, a latent semantic space is constructed, using the singular value decomposition as specified in Section 1. The three truncated matrices from the SVD are returned as a list object. A latent semantic space can be converted back to a textmatrix object with **as.textmatrix**(). The returned textmatrix has the same terms and documents, however, with modified frequencies, that now reflect inherent semantic relations not explicit in the original input textmatrix.

Additionally, the package contains several tuning options for the core routines and various support methods which help setting the influencing parameters. Some examples are given below, for additional options see Wild (2005).

Considering text preprocessing, **textmatrix**() offers several argument options. Two stop-word lists are provided with the package, one for German language texts (370 terms) and one for English (424 terms), which can be used to filter terms. Additionally, a controlled vocabulary can be specified, sort order will be sustained. Support for Porter's Snowball stemmer is provided through interaction with the Rstem package (Lang (2004)). Furthermore, a lower boundary for word lengths and minimum document frequencies can be specified via an optional switch.

Methods for term weighting include the local weightings (lw) raw, log, binary, and the global weightings (gw) normalisation, two versions of the inverse document frequency (idf), and entropy in both the original Shannon as well as in a slightly modified, more popular version (Wild (2005)).

Various methods for finding a useful number of dimensions are offered in the package. A fixed number of values can be directly assigned as an argument in the core routine. The same applies for the common practise to use a fixed fraction of singular values, e.g., 1/50th or 1/30th. Several support methods are offered to automatically identify a reasonable number of dimension: a percentage of the cumulated values (e.g., 50%); equalling the number of documents with a share of the cumulated values; dropping all values below 1.0 (the so called 'Kaiser Criterion'); and finally the pure vector model with all available values (Wild (2005)).

4 Demonstrations

In the following section, two examples will be given on how LSA can be applied in practise. The first case illustrates how LSA may be used to automatically score free-text essays in an educational assessment setting. Typically, if conducted by teachers, essays written by students are marked through careful reading and evaluation along specific criteria, among others their content. 'Essay' thereby refers to "a test item which requires a response composed by the examinee, usually in the form of one or more sentences, of a nature that no single response or pattern of responses can be listed as correct" (Stalnaker (1951)).

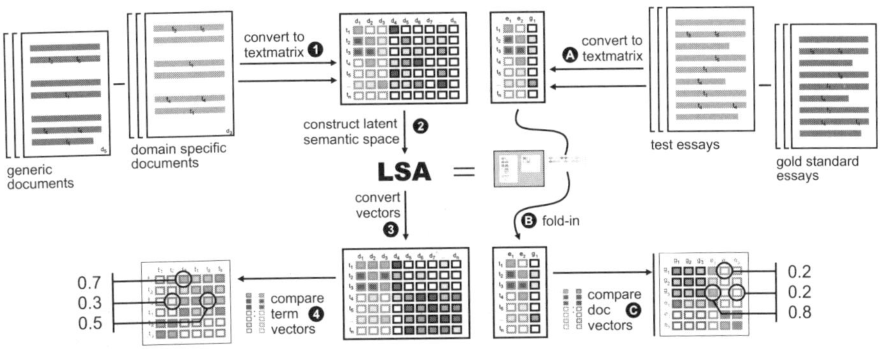

Fig. 2. LSA Process for both examples.

When emulating human understanding with LSA, first a latent semantic space needs to be trained from domain-specific and generic background documents. Generic texts thereby add a reasonably heterogeneous amount of general vocabulary whereas the domain-specific texts provide the professional vocabulary. The document collection is therefore converted into a textmatrix object (see Figure 2, Step 1). Based on this textmatrix, a latent semantic space is constructed in Step 2. Ideally, this space is an optimal configuration of factors calculated from the training documents and is able to evaluate content similarity. To avoid the essays to be tested and a collection of best-practise examples (so called 'gold-standard essays') from influencing this space, they are folded in after the SVD. In Step A they are converted into a textmatrix applying the vocabulary and term order from the textmatrix generated in Step 1. In Step B they are folded into this existing latent space (see Section 1). As a very simple scoring method, the Pearson Correlation between the test essays and the gold-standard essays can be used for scoring as indicated in Step C. A high correlation equals a high score. See Listing 1 for the R code.

Listing 1. Essay scoring with LSA.

```
library("lsa") # load package
# load training texts
trm = textmatrix("trainingtexts/")
trm = lw_bintf(trm) * gw_idf(trm) # weighting
space = lsa(trm) # create LSA space
# fold-in test and gold standard essays
tem = textmatrix("essays/", vocabulary=rownames(trm))
tem = lw_bintf(tem) * gw_idf(tem) # weighting
tem_red = fold_in(tem, space)
# score essay against gold standard
cor(tem_red[,"gold.txt"], tem_red[,"E1.txt"]) # 0.7
```

The second case illustrates, how a space changes behavior, when both, corpus size of the document collection and number of dimensions, are varied. This example can be used for experiments investigating the two driving parameters 'corpus size' and 'optimal number of dimensions'.

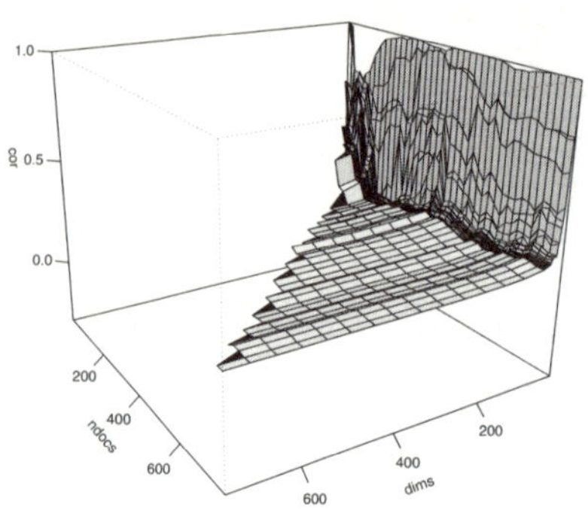

Fig. 3. Highly frequent terms 'eu' vs. 'oesterreich' (Pearson).

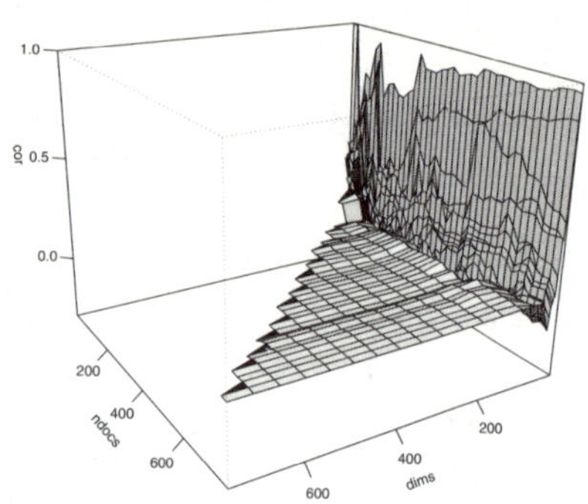

Fig. 4. Highly frequent terms 'jahr' vs. 'wien' (Pearson).

Therefore, as can be seen in Figure 2, a latent semantic space is constructed from a document collection by converting a randomised full sample of the available documents to a textmatrix in Step 1 (see Listing 2, Lines 4-5) and by applying the **lsa**() method in Step 2 (see Line 16). Step 3 (Lines 17-18) converts the space to textmatrix format and measures the similarity between two terms. By varying corpus size (Line 9 and 10-13 for sanitising) and dimensionality (Line 15), behavior changes of the space can be investigated.

Figure 4 and Figure 4 show visualisations of this behavior data: the terms of Figure 4 were considered to be highly associated and thus were expected to be very similar in their correlations. Evidence for this can be derived from the chart when comparing with Figure 4, visualising the similarities from a term pair considered to be unrelated ('jahr' = 'year', 'wien' = 'vienna'). In fact the base level of the correlations of the first, highly associated term pair is visibly higher than that of the second, unrelated term pair. Moreover, at the turning point of the cor-dims curves, the correlation levels have an even increased distance which already stabilises for a comparatively small number of documents.

Listing 2. The geometry of meaning.

```
tm = textmatrix("texts/",stopwords=stopwords_de)
# randomize document order
rndsample = sample(1:ncol(tm))
sm = tm[,rndsample]
# measure term-term similarities
s = NULL
for (i in (2:ncol(sm))) {
  # filter out unused terms
  if (any(rowSums(sm[,1:i])==0)) {
    m = sm[-(which(rowSums(sm[,1:i])==0)),1:i]
  } else { m = sm }
  # increase dims
  for (d in 2:i) {
    space = lsa(m, dims=d)
    redm = as.textmatrix(space)
    s = c(s, cor(redm["jahr",],redm["wien",]))
  }
}
```

5 Evaluating algorithm effectiveness

Evaluating the effectiveness of LSA, especially with changing parameter settings, is dependent on the application area targeted. Within an information retrieval setting, the same results may lead to a different interpretation than in an essay scoring setting. One evaluation option is to externally validate by comparing machine behavior to human behavior (see Figure 5). For the essay scoring example, the authors have evaluated machine against human scores, finding a man-machine correlation (Spearman's Rho) of up to .75, significant at a level below .001 in nine exams tested. In comparison, human-to-human interrater correlation is often reported to vary around .6 (Wild et al. (2005)). In the authors own tests, the highest human rater intercorrelation was found to be .8 (for the same exam as the man-machine correlation mentioned above), decreasing rapidly with dropping subject familiarity of the raters.

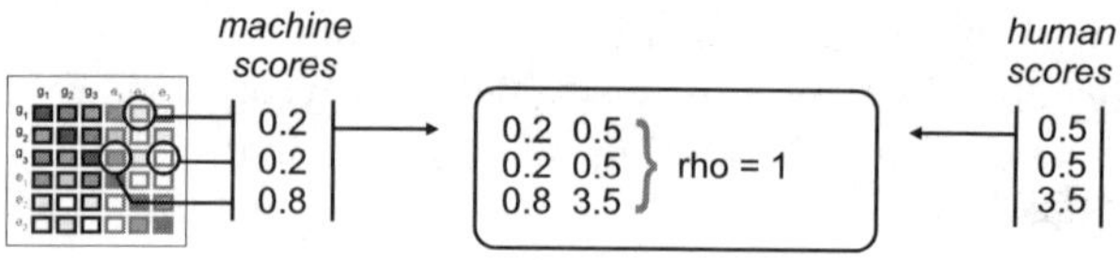

Fig. 5. Evaluating the algorithm.

6 Conclusion and identification of current challenges

An overview over latent semantic analysis and its implementation in the R package 'lsa' has been given and has been illustrated with two examples. With the help of the package, LSA can be applied using only few lines of code. As rolled out in Section 2, however, the various influencing parameteres may hinder users in calibrating LSA to achieve optimal results, sometimes even lowering performance below that of the (quicker) simple vector-space model.

In general, LSA shows greater effectiveness than the pure vector space model in settings that benefit from fuzziness (e.g., information retrieval, recommender systems). However, in settings that have to rely on more precise representation structures (e.g., essay scoring, term relationship mining), better means to predict behavior under certain parameter settings could ease the applicability and increase efficiency by reducing tuning times. For the future, this can be regarded as the main challenge: an extensive investigation of the influencing parameters, their settings, and their interdependancies to enable a more effective application.

References

BAEZA-YATES, R. and RIBEIRO-NETO, B. (1999): *Modern Information Retrieval.* ACM Press, New York.

BERRY, M., DUMAIS, S. and O'BRIEN, G. (1995): Using Linear Algebra for Intelligent Information Retrieval. *SIAM Review, 37, 573–595.*

DEERWESTER, S., DUMAIS, S., FURNAS, G., LANDAUER, T. and HARSHMAN, R. (1990): Indexing by Latent Semantic Analysis. *JASIS, 41, 391–407.*

LANG, D.T. (2004): *Rstem.* R Package Version 0.2-0.

STALNAKER, J.M. (1951): The Essay Type of Examination. In: E.F. Lindquist (Ed.): *Educational Measurement.* George Banta, Menasha, 495–530.

WILD, F., STAHL, C., STERMSEK, G. and NEUMANN, G. (2005): Parameters Driving Effectiveness of Automated Essay Scoring with LSA. In: M. Danson (Ed.): *Proceedings of the 9th CAA.* Prof. Development, Loughborough, 485–494.

WILD, F. (2005): *lsa: Latent Semantic Analysis.* R Package Version 0.57.

Part VI

Marketing, Management Science and Economics

Heterogeneity in Preferences for Odd Prices

Bernhard Baumgartner and Winfried J. Steiner

Department of Marketing, University of Regensburg, 93040 Regensburg, Germany; {bernhard.baumgartner, winfried.steiner}@wiwi.uni-r.de

Abstract. The topic of odd pricing has attracted many researchers in marketing. Most empirical applications in this field have been conducted on the aggregate level, thereby assuming homogeneity in consumer response to odd pricing. This paper provides the first empirical study to measure odd pricing effects at the individual consumer level. We use a Hierarchical Bayes mixture of normals model to estimate individual part-worths in a conjoint experiment, and demonstrate that preferences of consumers for odd and even prices can be very heterogeneous. Methodologically, our study offers new insights concerning the selection of the appropriate number of components in a continuous mixture model.

1 Introduction

It is well known that retailers have a tendency to set so called odd prices. Price endings just below a round number, especially prices ending in '9', are more frequently observed than even or round prices ending in '0' (Holdershaw and Gendall (1997)). The widespread practice of setting prices ending in '9' is based on the assumption that price thresholds exist in consumers' responses to price changes from a 9-ending price to a 0-ending price. The marketing literature has developed several theories supporting the existence of such price thresholds.

Level effects address potential biases in consumers' price perception and price processing, resulting in an underestimation of odd prices. Consumers may round down prices thereby ignoring the right most digit(s), compare prices from left to right or have limited memory capacity remembering only the more important left hand digits (e.g., Coulter (2001)). *Image effects* assume that the right most digits do have a meaning to the consumer. Two types of image effects can be distinguished: price image and quality image effects. According to the *price image effect*, a price ending in '9' signals a price discount to the consumer and therefore an opportunity to buy the product cheaper (e.g., Schindler (2003)). According to the *quality image effect*, a price ending in '0' signals high product quality (e.g., Stiving (2000)). Hence, contrary to level effects and the price image effect, the existence of a quality image effect would advocate the setting of even prices.

While the theories mentioned above provide some justification for the widespread use of odd prices, the results of empirical studies are anything else than conclusive with regard to the existence of odd pricing effects. The impact of price endings in '9' on price perception, price recall and stated or revealed preferences (purchases) was investigated in several studies. Gedenk and Sattler (1999) provide an overview of those studies and conclude that there is a lot of ambiguity about the effect of 9-ending prices. Whereas some studies indicate positive odd pricing effects supporting level and/or price image effects, an almost equal number of surveys didn't find any significant effects, and still other studies uncovered negative effects of 9-ending prices supporting the quality image hypothesis.

It is important to note that almost all empirical studies dealing with odd pricing effects have been conducted on an aggregate level. It has therefore been assumed that all respondents perceive or respond to odd prices homogeneously. We claim that consumers may respond very differently to odd pricing, and that this unobserved heterogeneity in price response has been one main reason for the inconclusive empirical results. For example, if some customers respond positively to odd prices, but other customers associate a lower quality with them, an aggregate analysis may yield insignificant results about the effects. To the best of our knowledge, only Wedel and Leeflang (1998) have considered heterogeneity in consumers' responses to odd prices in a Gabor-Granger type price study using a binary logit model embedded in a latent class segmentation approach. Their findings provide the first empirical verification that consumer segments may vary in their responsiveness to odd pricing.

In the following, we present the first empirical application to uncover heterogeneity in preferences for odd pricing at the individual consumer level using a hierarchical Bayes mixture of normals procedure. Our findings indicate a large amount of heterogeneity of respondents in their response to odd and even prices. We also study the influence of individual background variables on individual price response, and address level effects by varying the time available for respondents to choose one item from a choice set. Our study further offers new insights concerning the selection of the appropriate number of components in a continuous mixture model.

2 Conjoint study

2.1 Experimental design

We applied our conjoint study to the product category chocolate drinks. Following Gendall et al. (1998), we used the two attributes price (with five different levels) and brand name (with three different levels representing well-known national brands) for stimuli construction. The price levels were specified to represent one upper and one lower price boundary in the product category as well as three test prices, with one price ending in 99 cents and one round price ending in 00 cents among them. The test price ending in 95 cents was added to contrast the presumably strong low price image of the 99-cent ending not only with an even price but also with another just below a round figure price. The upper and lower price levels were set equidistant from the round price, thereby constituting anchor points to respondents in the product category. Table 1 shows price and brand attribute levels used for stimuli construction.

Table 1. Attribute levels used in the conjoint study

Attribute Levels	Product Category Chocolate Drinks
Brand 1	Suchard
Brand 2	Nesquick
Brand 3	Kaba
Low 'Anchor' Price	Euro 1.90
Price '95'	Euro 1.95
Price '99'	Euro 1.99
Price '00'	Euro 2.00
High 'Anchor' Price	Euro 2.10

Stimuli presentation within choice sets was based on a fractional factorial design. Every choice set contained the three brands at a specified price level, and respondents were asked to choose one of the alternatives from each choice set. The sample for our study consisted of 144 consumers.

It seems reasonable to expect that respondents who are under greater time pressure pay less attention to each price and therefore are more likely to ignore the right most digit(s). In order to address this level effect hypothesis, we divided the sample randomly into two subsamples and varied the time (10 versus 16 seconds) available to respondents to choose one of the items from each choice set.

2.2 Estimation of individual part-worth utilities

As usual in choice-based conjoint analysis, we use the Multinomial Logit Model (MNL) to estimate respondents' part-worth utilities. Accordingly, the conditional choice probability P_{ijc} that respondent i ($i = 1, \ldots, I$) will choose alternative/stimuli j ($j = 1, \ldots, J$) from choice set c ($c = 1, \ldots, C$) is:

$$P_{ijc} = \frac{\exp(V_{ij})}{\sum_{n \in c} \exp(V_{in})} \tag{1}$$

The deterministic utility V_{ij} of respondent i for alternative j is obtained from an additive part-worth function for the predictor variables, which are represented by the brand names and the price levels in our application:

$$V_{ij} = \beta_{B1,i} x_{B1,j} + \beta_{B2,i} x_{B2,j} + \beta_{P95,i} x_{P95,j} + \beta_{P99,i} x_{P99,j} + \beta_{P00,i} x_{P00,j} + \beta_{Ph,i} x_{Ph,j} \quad (2)$$

$B1$ and $B2$ denote the brands Suchard and Nesquick, and $P95$, $P99$, $P00$ and Ph denote the price levels ending in 95, 99 and 00 cents and the upper boundary price level. $x_{..,j}$ are dummy variables indicating the presence ($x_{..,j} = 1$) or absence ($x_{..,j} = 0$) of the respective attribute level for alternative j. The parameters β are the part-worth utilities to be estimated for the brand and price levels. Note that the part-worth utilities for Kaba and the price level Euro 1.90 have been scaled to zero to constitute the reference categories.

The index i of the parameters accounts for the fact that we estimate part-worth utilities β_i for each individual respondent using a hierarchical Bayes (HB) procedure. Allenby et al. (1995) and Allenby and Ginter (1995) were the first to demonstrate HB's high potential to recover individual part-worths with high accuracy in conjoint analysis. Accordingly, we assume that the individual-level parameters β_i come from a distribution with unknown population mean B and unknown variance matrix Ω (also called population-level parameters):

$$\beta_i \sim N(B, \Omega) \quad (3)$$

In addition, the parameters B and Ω are supplied themselves with prior distributions, known as hyperpriors. Frequently, a normal prior for B and an inverted Wishart prior for Ω are used. This way, convenient forms for the (conditional) posteriors on B and Ω from which taking draws is easy are obtained. For details, we refer to Train (2003) and Rossi and Allenby (2003). The first-stage prior (3) can be extended to consider that the individual part-worth utilities depend on individual-specific covariates z_i. Formally, this is accomplished by linking a matrix Γ of additional parameters (representing the effects of an individual's purchase frequency of chocolate drinks and her/his stated importance of price in our study) to the individual part-worth utilities β_i (e.g., Allenby and Ginter (1995)):

$$\beta_i = \Gamma \cdot z_i + \zeta_i, \qquad \zeta_i \sim N(0, \Omega) \quad (4)$$

In standard HB applications of the MNL, the normal distribution is commonly assumed for the first stage prior. A potential drawback of this assumption is that the normal distribution has thin tails, which means that there is only a small probability for individual parameter values to deviate from the population mean to a larger extent. In other words, heterogeneity of individuals may not be captured adequately. To overcome this problem with a straightforward model extension, mixtures of normals can be used for the first stage prior. Mixtures of normals can accommodate heavy-tailed as well as skewed distributions and provide a still higher flexibility to capture heterogeneity in consumers' preferences (Rossi et al. (2005)). This leads to:

$$\zeta_i \sim N(\mu_{ind_i}, \Omega_{ind_i}), \qquad ind_i \sim multinomial_K(pvec), \quad (5)$$

with ind_i denoting the component $(1, \ldots, K)$ individual i is from, and $pvec$ denoting a vector of mixing probabilities. For the specification of conjugate priors for $pvec$ and the component-specific population-level parameters μ_k and Ω_k $(k = 1, \ldots, K)$, we refer to Rossi et al. (2005).

Draws from the joint posterior distribution of the parameters are taken via Monte Carlo Markov Chain simulation. We applied the MCMC procedure as described by Rossi et al. (2005) using the R-Code provided by the authors on Greg Allenby's homepage (http://fisher.osu.edu/ allenby 1/). We simulated 25.000 draws from the chain as burn-in to ensure convergence and retained every 10th draw from the succeeding 10.000 draws for a total of 1000 draws from the posterior. To decide about the optimal number of components K, approximations to Bayes factors like the Newton Raftery Criterion (NRC), or a comparison of sequence plots of log-likelihood values for models with different numbers of components have been suggested (e.g., Rossi et al. (2005)). Applied to our data, however, both criteria failed to provide an unambiguous solution. Especially the NRC turned out to be very unstable for increasing K. We therefore relied on a recommendation provided by Fruewirth-Schnatter et al. (2004) to identify a unique labeling via graphical inspection of the posterior simulations of component-specific parameters. Accordingly, we compared two-dimensional scatterplots of draws of component-specific part-worths for models with different numbers of components, with the dimensions referring to the part-worths for a specific brand-price combination. The plots clearly supported a 3-component-solution.

3 Empirical results

We summarize the results of our conjoint study by using two-dimensional plots of the estimated part-worth preferences. Figure 1 displays the individual part-worth utilities for the brands *Suchard* and *Nesquick*. Remember that the reference level *Kaba* has been scaled to have a part-worth utility of zero, and that the estimated part-worths for Suchard and Nesquick must be interpretated relative to the reference category. A point above the x-axis (right to the y-axis) therefore represents a respondent who prefers Suchard to Kaba (Nesquick to Kaba). If a respondent's part-worth utility for one of the three brands turned out to be larger than her/his part-worths for both other brands in at least 90 percent of the 1000 MCMC draws, we interprete this as a "clear" preference for this brand. Clear preferences are indicated by squares (Suchard), triangles (Nesquick) and diamonds (Kaba). Figure 1 reveals strong heterogeneity in respondents' brand preferences for chocolate drinks, with 61 percent of the respondents showing a clear preference for one of the brands.

Figure 2 represents the individual preferences for the price levels of interest, the price ending in 99 (Euro 1.99), the second odd price ending in 95 (Euro 1.95) and the round price ending in 00 (Euro 2.00). By using (a) the difference between a respondent's part-worth utilities for the prices end-

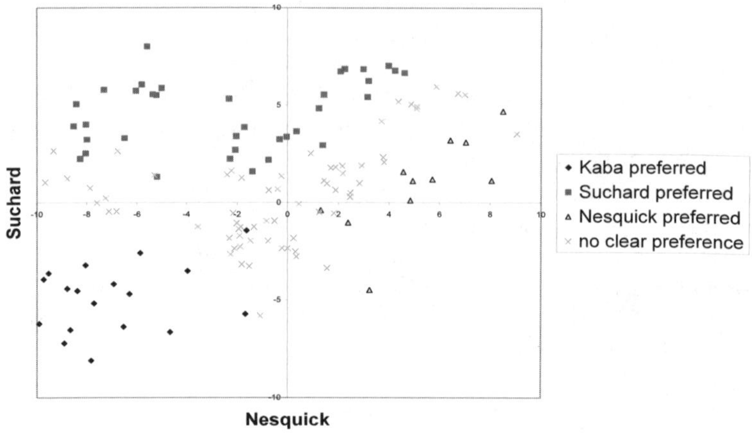

Fig. 1. Part-worth utilities for brands

ing in 99 and 95 on the x-axis (i.e., $\beta_{P99,i} - \beta_{P95,i}$) and (b) the difference between a respondent's part-worth utilities for the prices ending in 00 and 99 (i.e., $\beta_{P00,i} - \beta_{P99,i}$) on the y-axis, we can display all the relevant information on the respondents' preferences for odd and even prices again in one two-dimensional plot. In a similar way as described for the brand part-worths above, we further indicate whether a respondent has a clear preference among the three price levels. In particular, a diamond (triangle) denotes that the difference $\beta_{P00,i} - \beta_{P99,i}$ has been positive (negative) for respondent i in at least 90 percent of the 1000 draws in the Markov chain. In other words, respondent i's part-worth utility for the price ending in 00 turned out to be higher (lower) than her/his part-worth for the 99-ending price in at least 900 of the 1000 draws. Correspondingly, a square characterizes a respondent with a clear preference for the 99-ending price compared to the price ending in 95.

Respondents who are rational in the sense that they prefer lower prices to higher prices are displayed in the lower left quadrant, because here both $\beta_{P00,i} - \beta_{P99,i}$ (the difference between the part-worths concerning the round and the 99-ending price) and $\beta_{P99,i} - \beta_{P95,i}$ (the difference between the part-worths concerning the prices ending in 99 and 95) are negative. If a respondent has a linear price response function in the area between Euro 1.95 and Euro 2.00, her/his part-worths would lie along the dashed line in this sector. Only 26 percent of the respondents are in the lower left sector and show a rational behavior with regard to price.

In the upper left quadrant, respondents preferring both the prices ending in 95 and 00 over the 99-ending price are shown. In other words, here are those respondents who dislike (double) 9-ending prices, probably because they

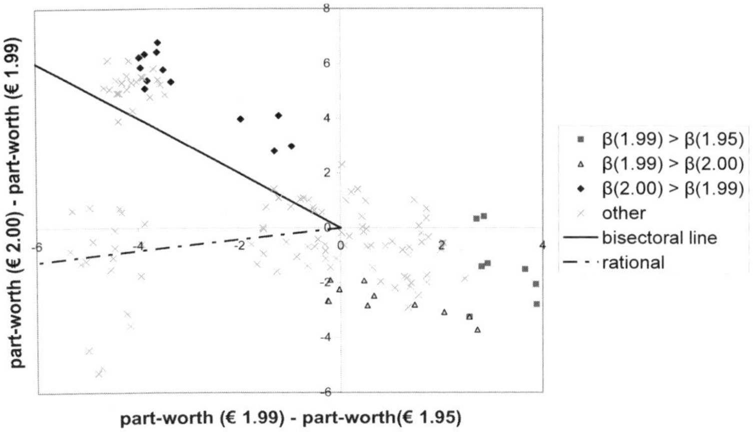

Fig. 2. Part-worth utilities for odd and even price levels

assign a low-quality image to them. In addition, most of these respondents lie above the bisectoral line implying that they prefer the round price most, just like respondents in the upper right quadrant do. 38 percent of the respondents in our category chocolate drinks are located in this area (above the bisectoral line in the upper left sector *and* upper right sector), and we can observe one rather clearly separated segment of consumers in the upper left corner. These consumers obviously use the round price as a quality indicator.

Finally, 29 percent of the respondents are located in the lower right quadrant. These respondents just prefer the 99-ending price over both the price ending in 95 and the round price. They seem to use the (double) 9-ending as a signal for a deal discount offering a good opportunity to buy a brand cheaper. Note that some respondents lie in the area below the bisectoral line in the upper left quadrant. That is why the percentages (26 percent, 38 percent and 29 percent) do not add up to 100 percent.

The results regarding the influence of background variables on preferences can be summarized as follows: (1) With an increasing consumption rate, the preference for the most renowned brand Kaba increases relative to Suchard or Nesquick. (2) The preference for higher prices decreases with an increase in the stated importance of price. (3) Respondents with less time available in a choice situation were more attracted by the odd prices (Euro 1.95 and Euro 1.99) as compared to the round price (Euro 2.00). This finding confirms the existence of level effects in the sense that people who are under greater time pressure may be more prone to ignore the right most digit(s) than people who have sufficient time for their decision.

4 Conclusions

Empirical studies are highly inconclusive about the effects of odd pricing. This may be due to the fact that almost all of these studies have been conducted on the aggregate level, thereby assuming homogeneity in consumers' responses. In this paper, using conjoint analysis and a continuous mixture of normals model for estimation of part-worths, we demonstrated that consumers may be very heterogeneous in their preferences for odd and even prices. We further addressed the problem of selecting the optimal number of components K in continuous mixture models. According to our experiences, the graphical model selection procedure suggested by Fruewirth-Schnatter et al. (2004) is highly promising, while other criteria failed to provide a solution for our data.

References

ALLENBY, G.M., ARORA, N. and GINTER J.L. (1995): Incorporating Prior Knowledge into the Analysis of Conjoint Studies. *Journal of Marketing Research, 89, 152-162.*

ALLENBY, G.M. and GINTER J.L. (1995): Using Extremes to Design Products and Segment Markets. *Journal of Marketing Research, 32, 392-403.*

COULTER, K.S. (2001): Odd-Ending Price Underestimation: An Experimental Examination of Left-to-Right Processing Effects. *Journal of Product and Brand Management, 10, 276-292.*

FRUEHWIRTH-SCHNATTER, S., TUECHLER, R. and OTTER, T. (2004): Bayesian Analysis of the Heterogeneity Model. *Journal of Business and Economic Statistics, 22, 2-15.*

GEDENK, K. and SATTLER H. (1999): The Impact of Price Thresholds on Profit Contribution - Should Retailers Set 9-ending Prices? *Journal of Retailing, 75, 1311-1330.*

GENDALL, P., FOX M.F. and WILTON, P. (1998): Estimating the Effect of Odd Pricing. *Journal of Product and Brand Management, 7, 421-432.*

HOLDERSHAW, J. and GENDALL, P. (1997): The Widespread Use of Odd Pricing in the Retail Sector. *Marketing Bulletin, 8, 53-58.*

ROSSI, P.E. and ALLENBY, G.M. (2003): Bayesian Statistics and Marketing. *Marketing Science, 22, 304-328.*

ROSSI, P.E., ALLENBY, G.M. and MCCULLOCH, R.E. (2005): *Bayesian Statistics and Marketing.* Wiley, New York.

SCHINDLER, R.M. (2003): The 99 Price Ending as a Signal of a Low-Price Appeal. *Advances in Consumer Research, 30, 270-276.*

STIVING, M. (2000): Price-Endings When Prices Signal Quality. *Management Science, 46, 1617-1629.*

TRAIN, K. (2003): *Discrete Choice Methods with Simulation.* Cambridge University Press, Cambridge.

WEDEL, M. and LEEFLANG, P.S.H. (1998): A Model for the Effects of Psychological Pricing in Gabor-Granger Price Studies. *Journal of Economic Psychology, 19, 237-260.*

Classification of Reference Models

Robert Braun and Werner Esswein

Lehrstuhl für Wirtschaftsinformatik, insb. Systementwicklung,
Technische Universität Dresden, D-01062 Dresden, Germany;
{robert.braun, werner.esswein}@tu-dresden.de

Abstract. The usefulness of classifications for reuse, especially for the selection of reference models is emphasised in the literature. Nevertheless, an empirical classification of reference models using formal cluster analysis methods is still an open issue. In this paper a cluster analysis is applied on the latest and largest freely available reference model catalogue. In the result, based on at last 9 selected variables, three different clusters of reference models could be identified (*practitioner reference models*, *scientific business process reference models* and *scientific multi-view reference models*). Important implications of the result are: a better documentation is generally needed to improve the reusability of reference models and there is a gap between scientific insights (regarding the usefulness of multi-view reference models and regarding the usefulness of concepts for reuse and customisation) and their application as well as tool support in practice.

1 Introduction

The main objective of reference models for business system analysis is to streamline the design of individual models by providing a generic solution for these models (see Becker and Delfmann (2004), Fettke and Loos (2003)). Thereby, the "... application of reference models is motivated by the 'Design by Reuse' paradigm. Reference models accelerate the modelling process by providing a repository of potentially relevant business processes and structures." (see Rosemann and van der Aalst (2006, p. 2) and Figure 1)

The usefulness of classifications for reuse, especially for the selection of reference models is emphasised in the literature (see Fettke and Loos (2003), Fettke et al. (2006)) and although some catalogues of reference models already exist (see Fettke and Loos (2003, p. 37) for an overview), an empirical classification (for this term see Bortz and Döring (2002, p. 382)) of these reference models using formal cluster analysis methods is still an open issue. The state of the art on indexing of reference models - based always on different defined

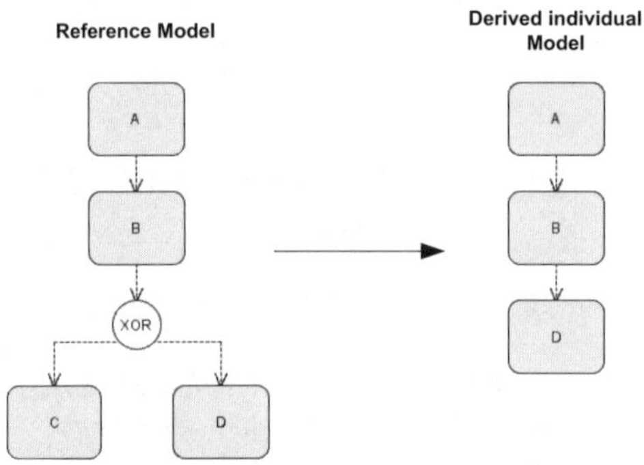

Fig. 1. Example for referencing (adapted from Rosemann and van der Aalst (2006, p. 5))

variables - can be viewed as a special kind of classification where each variable defines a specific classification class (see Fettke and Loos (2003, p. 49)).

The advantage of cluster analysis where more than one variable can be used simultaneously for clustering objects (see Backhaus et al. (2006, p. 510)) has not been applied. In this paper, however, the idea is to use multivariate data analysis for reference models by which the resulting clusters will improve the reusability concerning the selection of reference models. Beyond this idea, the study is driven by the following research questions: Which clusters of reference models exist in practice and how is the allocation of the reference models? The answers to these questions can provide useful insights into further reference modelling research. Therefore, a classification of reference models based on cluster analysis methods will be introduced and discussed in this paper.

The paper is structured as follows. Section 2 describes the data set. In section 3 the classification procedure is explained. The classification results and their implications are presented and discussed in section 3. Finally, conclusions and an outlook to further research are given in section 5.

2 Data set

The data is taken from the latest and largest freely available reference model catalogue, where 38 reference models are indexed with 20 variables (see Fettke and Loos (2004, pp. 30–39); an additional documentation of this reference model catalogue can be found here: `http://wise.wiwi.tu-dresden.de/systementwicklung/profil/publikation/artikel_buchbeitraege/downloads/gfkl2006/src/`).

The three variables *Number* (an unique consecutive number), *Name* (the name of the reference model) and *Literature* (primary and secondary literature, which are describing the reference model) were excluded as they are only used for identification of the reference models. Therefore, 17 variables remain as candidates for using them for classification.

However, the description and values of these 17 variables show that a reasonable empirical classification with these partially very complex variables is not possible (see Fettke and Loos (2004, pp. 21–24, pp. 30–39), Fettke et al. (2006, pp. 470–472, pp. 482–483)).

For instance, the variable *Responsibility for Modeling* has the values 'Autor', 'Autorin', 'Autorenteam', 'Siemens AG', 'Office of Government Commerce', ... but only 'Autor' and 'Autorenteam' occur more than once in the data set. To improve the classification results the values should be further generalised.

The values of the variable *Domain* are not standardised short descriptions of the person(s) or organisation responsible for developing the reference model. Thus reference models with slightly different descriptions of the same domain will not be identified as similar.

The variable *Modelling Grammar(s)* stores in just one value the names of all used modelling grammars which makes it impossible to identify reference models using one equal and one different grammar as similar by the equal one. The variable *Modelling Grammar(s)* contains even wrong values using the names of the diagrams ('Datenmodell', 'Objektmodell'). Furthermore, some values are on a different level of abstraction ('UML' on the one hand and 'UML Klassendiagramm' on the other hand).

The last examples for the partially very complex variables are *Evaluation/ Inter-subjective Verifiable* and *Use Case(s)/Inter-subjective Verifiable* where two information are stored in each variable with 'Evaluation' respectively 'Number of use cases' and 'Inter-subjective Verifiable'.

Because of the before-mentioned problems, the 17 original variables were modified and adequately prepared for classification. This and the remaining classification procedure is described in the next section.

3 Classification procedure

First of all, the 17 original variables were split into 73 binary-coded variables (a documentation of the modified reference model catalogue can be found here: `http://wise.wiwi.tu-dresden.de/systementwicklung/profil/publikation/artikel_buchbeitraege/downloads/gfkl2006/bin/`).

However, 73 variables are far too many for a classification of only 38 objects (reference models). Therefore, the amount of variables has been reduced to 22. The selected 22 variables are marked in the modified reference model catalogue that is referenced above.

To minimise the information loss as consequence of this reduction of variables, the cluster analysis has been run first with all 73 variables and afterwards with the 22 only. The results of the cluster analysis were similar for both cases using the same proximity and cluster algorithm. Only two objects (reference models) were placed in different groups. Therefore, the 22 binary variables are used for classification (range: $\{0 = \text{false}; 1 = \text{true}\}$):

- *General Characterisation*
  - Responsibility for Modeling - Science
  - Access - Open
  - Tool Support - Yes
  - Domain Differentiation by Institution
- *Construction*
  - Domain of the Reference Model - Software Development
  - Domain of the Reference Model - Re-Organisation
  - Modelling Grammar - Entity-Relationship-Model (ERM)
  - Modelling Grammar - Function-Tree
  - Modelling Grammar - Event-driven Process Chain (EPC)
  - Modelling Grammar - Unified Modeling Language (UML)
  - Modelling Grammar - Object-Modeling-Technique (OMT)
  - Modelling Framework - Used
  - Number of used Diagrams (Presentations) ≤ 20
  - Number of used Views > 1
  - Structure-related Size (number of classes or entity-types and relationship-types) ≤ 100
  - Function-related Size ≤ 100
  - Process-related Size (number of process steps (activities)) ≤ 100
  - Construction Method - Stated
  - Evaluation - Reference Model is evaluated
- *Application*
  - Application Method(s) - Stated
  - Reuse and Customisation - Concepts defined
  - Use Cases > 1

Note: In some cases further information was needed for the values of the variables than the original reference model catalogue contained. In these cases, the referred literature in Fettke and Loos (2004) has been studied in order to retrieve the missing information. But it was not possible to get all missing information, especially if access to a reference model is restricted. Hence, the binary-coded variables in the modified reference model catalogue got also the value '0' if the appropriate variable in the original reference model catalogue has the value 'not specified' and the missed information could not be retrieved. For that reason, some reference models have probably been misplaced.

For the cluster analysis the procedure described in Backhaus et al. (2006, p. 492) has been used. Simple Matching and the Squared Euclidian Distance have been used as proximities but the latter one has been chosen in order to use Ward's algorithm for combining the clusters of the hierarchical button up clustering (see Opitz (1980, p. 68)). Ward's algorithm has been used because it usually finds better clusters and the objects are more adequately grouped than by using any other cluster algorithm (see Bergs cited in Backhaus et al. (2006, p. 528)).

At first, the single-linkage algorithm was used for combining clusters in order to identify outliers. However, none was found. In a second step, Ward's algorithm has been used for merging clusters of all 38 reference models. After the analysis of the dendrogram, the elbow diagram and the sum of the squared error values, the decision fell upon a two cluster solution. To identify the variables responsible for the formation of the clusters, a discriminant analysis was applied. 9 out of the 22 variables showed a significant influence on the formation of the two clusters (see Table 1).

Table 1. The variables with significant influence (significance level = 5 %)

No.	Variable	Significance
1	Responsibility for Modeling - Science	.014
2	Access - Open	.000
3	Tool Support - Yes	.004
4	Number of Diagrams $\leq$ 20	.000
5	Number of Views > 1	.000
6	Structurerelated Size $\leq$ 100	.002
7	Functionrelated Size $\leq$ 100	.005
8	Construction Method - Stated	.000
9	Reuse and Customisation - Concepts defined	.011

Finally, only these 9 variables were used for a second cluster analysis. Because the decision was made to concentrate this (first) study on these variables. The Squared Euclidian Distance as proximity and Ward's algorithm for the fusion process were used again. The analysis of the dendrogram, the elbow diagram and the sum of the squared error values showed similar results as the solution with 22 variables. However, focussing on the homogeneity of the cluster solution, the preference shifted to the three cluster solution. To check the significance of the variables, a discriminant analysis was performed again. The result remains the same with all 9 variables playing a significant role in the formation of the three clusters.

4 Results and implications

4.1 Results

The resulting three clusters are depicted in Table 2. The first cluster (A) contains only reference models developed by practitioners (companies, government departments or other associations) and is the smallest cluster. The other two clusters (B and C) contain only reference models developed by scientists. While reference models in the largest cluster (B) use only one view (focussing primarily on business processes), cluster C contains reference models using multiple views.

Table 2. Reference model cluster

Cluster	Description	Objects	(%)
A	Practitioner Reference Models	8	(21.1)
B	Scientific Business Process Reference Models	16	(42.1)
C	Scientific Multi-View Reference Models	14	(36.8)

4.2 Implications

By taking a closer look at the averages of the 9 variables (see Table 3), different characteristics of the three clusters can be identified:

- *Access:* Access to reference models developed by practitioners (A) is mostly restricted in contrast to reference models developed by scientists (B and C).
- *Tool Support:* Practitioner reference models (A) and - as described above - scientific business process reference models (B) concentrate primarily on the process view. There is tool support for these reference models. However, practitioner reference models are much more tool supported than the scientific ones in cluster B. The multi-view reference models of cluster C often lack tool support. This observation leads to the assumption that multi-view modelling is not sufficiently supported by tools or the current modelling tools cannot cope with the requirements for multi-view modelling. A current survey supports this thesis (see Sarshar et al. (2006, p. 124)). Another explanation could be that scientists prefer ambitious solutions and disregard increased usability of reference models by tool support.
- *Construction Method:* In case of multi-view reference models the construction method is more often explicitly stated. This could lead to the assumption that in this case such a method has been used frequently.
- *Reuse and Customisation:* Multi-view reference models define more often concepts of reusability and customisation which can be an indicator for the more ambitious aim of the developed reference model. This can lead to the assumption that insufficient support for the definition of concepts for reuse and customisation by current modelling tools exists (see again Sarshar et al. (2006, p. 124) for support of this thesis).

Because of the restricted access to cluster A's reference models an interpretation of the other variables for this cluster is theoretically questionable (see the corresponding note in section 3).

Table 3. Averages (a) and standard deviations (s) of the 9 significant variables (see Table 1)

		Variable (No.)								
Cluster		**1**	**2**	**3**	**4**	**5**	**6**	**7**	**8**	**9**
A	a	.0000	.1250	.6250	.0000	.0000	.0000	.0000	.0000	.2500
	s	.00000	.35355	.51755	.00000	.00000	.00000	.00000	.00000	.46291
B	a	1.0000	.5000	.3125	.1250	.0000	.0625	.0000	.2500	.0000
	s	.00000	.51640	.47871	.34157	.00000	.25000	.00000	.44721	.00000
C	a	1.0000	1.0000	.0000	.6429	.8571	.4286	.2857	.7143	.4286
	s	.00000	.00000	.00000	.49725	.36314	.51355	.46881	.46881	.51355
All	a	.7895	.6053	.2632	.2895	.3158	.1842	.1053	.3684	.2105
	s	.41315	.49536	.44626	.45961	.47107	.39286	.31101	.48885	.41315

5 Conclusion and outlook

The following conclusions can be drawn from these results:

1. A classification of reference models might improve their reusability, retrieval and selection. Consequently, a better documentation of the reference models is generally needed. The documentation has to be based on specific variables which require a) significance and b) strict ranges.
2. Empirical research should be applied to identify these variables.
3. There is a gap between scientific insights (regarding the usefulness of multi-view reference models and regarding the usefulness of concepts for reuse and customisation) and their application as well as tool support in practice. This shows that multi-view modelling, concepts for reuse and customisation and their tool support are further research topics (see Fettke and Loos (2004, p. 28)).
4. A discussion about the term 'reference model' (see e. g. Fettke and Loos (2004, pp. 9–12), Thomas (2006)) can also be stimulated by insights about the usage of reference models in practice. This study has shown a tendency of a more theoretical usage of this term (reference models as statements by scientists).

Besides the results shown in this first study, further research about the empirical classification of reference models has to be done. This includes a repetition of this study with other variables (or coded in another form), other proximities and cluster algorithms as well as other cluster solutions and another or more comprehensive data set.

References

BACKHAUS, K., ERICHSON, B., PLINKE, W. and WEIBER, R. (2006): *Multivariate Analysemethoden: Eine anwendungsorientierte Einführung.* Springer, Berlin, 11th edition.

BECKER, J. and DELFMANN, P. (Eds.) (2004): *Referenzmodellierung: Grundlagen, Techniken und domänenbezogene Anwendung.* Physica, Heidelberg.

BORTZ, J. and DÖRING, N. (2002): *Forschungsmethoden und Evaluation für Human- und Sozialwissenschaftler.* Springer, Berlin.

FETTKE, P. and LOOS, P. (2003): Classification of Reference Models: A Methodology and Its Application. *Information Systems and e-Business Management, 1, 1, 35–53.*

FETTKE, P. and LOOS, P. (2004): Referenzmodellierungsforschung - Langfassung eines Aufsatzes. Working Papers of the Research Group Information Systems & Management, No. 16, Johannes Gutenberg-Universität Mainz.

FETTKE, P., LOOS, P. and ZWICKER, J. (2006): Business Process Reference Models: Survey and Classification. In: C. Bussler and A. Haller (Eds.): *Business Process Management Workshops: BPM 2005 International Workshops, BPI, BPD, ENEI, BPRM, WSCOBPM, BPS, Nancy, Revised Selected Papers*, Springer, Berlin, 469–483.

OPITZ, O. (1980): *Numerische Taxonomie.* Fischer, Stuttgart.

ROSEMANN, M. and VAN DER AALST, W.M.P. (2006): A Configurable Reference Modelling Language. *Information Systems* (In Press).

SARSHAR, K., WEBER, M. and LOOS, P. (2006): Einsatz der Informationsmodellierung bei der Einführung betrieblicher Standardsoftware: Eine empirische Untersuchung bei Energieversorgerunternehmen. *Wirtschaftsinformatik, 48, 2, 120–127.*

THOMAS, O. (2006): Das Referenzmodellverständnis in der Wirtschaftsinformatik: Historie, Literaturanalyse und Begriffsexplikation. Veröffentlichungen des Instituts für Wirtschaftsinformatik, No. 187, Universität des Saarlandes.

Adaptive Conjoint Analysis for Pricing Music Downloads

Christoph Breidert[1] and Michael Hahsler[2]

[1] PONTIS Venture Partners, Löwelstr. 12, A-1010 Vienna, Austria; christoph@breidert.net

[2] Department of Information Systems and Operations, Vienna University of Business Administration and Economics, A-1090 Vienna, Austria; hahsler@ai.wu-wien.ac.at

Abstract. Finding the right pricing for music downloads is of ample importance to the recording industry and music download service providers. For the recently introduced music downloads, reference prices are still developing and to find a revenue maximizing pricing scheme is a challenging task. The most commonly used approach is to employ linear pricing (e.g., iTunes, musicload). Lately, subscription models have emerged, offering their customers unlimited access to streaming music for a monthly fee (e.g., Napster, RealNetworks). However, other pricing strategies could also be used, such as quantity rebates starting at certain download volumes.

Research has been done in this field and Buxmann et al. (2005) have shown that price cuts can improve revenue. In this paper we apply different approaches to estimate consumer's willingness to pay (WTP) for music downloads and compare our findings with the pricing strategies currently used in the market.

To make informed decisions about pricing, knowledge about the consumer's WTP is essential. Three approaches based on adaptive conjoint analysis to estimate the WTP for bundles of music downloads are compared. Two of the approaches are based on a status-quo product (at market price and alternatively at an individually self-stated price), the third approach uses a linear model assuming a fixed utility per title. All three methods seem to be robust and deliver reasonable estimations of the respondent's WTPs. However, all but the linear model need an externally set price for the status-quo product which can introduce a bias.

1 Introduction

Download services for digital music have gained popularity over the last years, most notably Apple's successful iTunes music download store. Single track downloads have doubled and grown to 353 million in 2005, and CD sales are gradually substituted by music downloads (IFPI (2006)). For music download services, pricing schemes for individual songs and especially for bundles of

Table 1. Attributes and levels used in the conjoint analysis.

Package	Distribution channel	Sound quality	Booklet	Price
✓ 1 title	✓Record store	✓Radio (64 kbs)	✓No booklet	✓ 1 €
✓ 3 titles	✓Mail	✓CD (128+ kbs)	✓Booklet	✓ 5 €
✓ 6 titles	✓Download			✓10 €
✓12 titles				✓15 €
				✓20 €

songs are still developing. Currently, most online services employ linear pricing and the prices in Europe vary between 0.99 € and 1.49 € per title. Finding optimal prices which maximize revenue is of great interest to practitioners and to researchers. A survey in the U.S. market by Jupiter Research (2003) concludes that at \$0.99 a market reach of 74% is achieved. However, the margins for music download service providers are very small (Jupiter Research (2006)). About 3% of \$0.99 go to the service providers. The rest goes to the recording industry (47%), the credit card companies (25.3%), the collecting societies (12.1%), the artists (8.3%), and network carriers (4%).

Researchers have tried to estimate consumer's WTP to improve the pricing schemes for music downloads. For example Buxmann et al. (2005) use self-stated WTP to estimate demand curves for online music in Germany with the conclusion that special prices and rebates could improve sales. Bamert et al. (2005) conducted a conjoint study for pricing online music in the Swiss market with the result that price is the most important attribute and usage restrictions (digital rights management), offered range of titles, and method of payment are less important. Using a direct survey, the authors also found out that at a price of 0.99 Swiss Francs (0.32 €) 16 songs can be sold to the average user.

In this paper we discuss three approaches based on adaptive conjoint analysis for pricing music downloads. After we present the design of the study, we compare the results of the three approaches with the current pricing practices in the market.

2 Setup of the conjoint interview

The interview was performed as a supervised adaptive conjoint analysis (ACA, cf. Green et al. (1991), Johnson (1991)) using non-price attributes that discriminate between buying music in conventional record stores and downloading music bought online to estimate respondents' utility structures and responsiveness to price changes. The interview was carried out among students of the Vienna University of Business Administration and Economics in spring 2005. In this paper the results for a sample of 99 respondents are reported.

Table 2. Time since the participants bought their last CD.

Months	1	3	6	12	13+
Participants	10	23	11	15	40

The design of the conjoint study is shown in Table 1. The levels of the attributes "Sound quality," "Price" and "Package" have a natural ordering (e.g., more songs are better) and thus do not need to be ranked by the respondents. The order of the levels of the attributes "Distribution channel" and "Booklet" do not have such a clear ordering and thus were elicited in a ranking task. After completion of the ranking scene the respondents rated the attributes in an importance scene. Finally, the respondents were presented a series of paired comparison scenes following the ACA procedure.

All interview scenes were explained to the respondents before the start of the interview, and a supervisor was present during the interview, in case of comprehension problems.

3 Results of the interviews

After the conjoint interview some socio-demographic information about the participants was elicited. 60 of the 99 participants were female. The average age was 23.72 years with a standard deviation of 2.65 years. As their preferred audio system 47 students mentioned their CD player and 52 already preferred to use their personal computer. Most of the participants (69) had access to a broad band internet connection while 25 used a modem to dial-in. 5 did not have access to the internet at home.

The participants were also asked about their music shopping behavior. Table 2 summarizes when the participants last bought a CD in a store. The data shows that the majority (55 participants) did not buy a CD within the last 6 months. However, 24 participants stated that they use file-sharing services often to obtain music and another 47 stated that they use such services occasionally. This illustrates the importance that music distributed via internet already has reached.

We checked whether significant relationships between the variables exist and found out that female participants prefer using a CD player while the male participants prefer to play their music with a personal computer. There is also a significant relationship between using personal computers for playback and using file-sharing tools. However, the use of file-sharing tools in the sample is not gender specific.

Table 3 compares the importance of the attributes calculated from the results of the conjoint analysis. The attribute booklet has by far the lowest importance across all respondents. In the decision making process the attributes price and package (number of songs) have the highest contribution to the valuation of product offerings. Measured in terms of conjoint utilities,

Table 3. Importance of attributes.

	Minimum	Median	Maximum
Price	1.727	4.670	7.772
Package	1.378	4.592	7.235
Sound quality	0.054	2.036	5.020
Distribution	0.083	1.986	5.144
Booklet	0.070	1.163	4.760

price and package are around four times more important than the booklet. The attributes sound quality and distribution channel have a higher importance than booklet, but compared to price and package they are still relatively low.

Measuring quantitative attribute levels with conjoint analysis can sometimes result in reversals (cf. Orme (2002)). If all attributes stay the same and only the attribute package is changed, the utility should increase monotonically with the number of titles contained in the package. The same holds for the attribute price. However, in real survey data this is not always the case. Even though the attributes package and price were pre-ordered when the conjoint analysis was initialized, reversals for package were observed for 36 respondents and for 51 respondents for price. The part-worth utilities with and without reversals are shown in Figure 1. In the plots to the right it can be seen that most reversals represent minor fluctuations which can be attributed to measurement and estimation errors. Major reversals seem to be the result of non-cooperative respondents (e.g., random choices to finish the interview faster). Such respondents need to be removed from the dataset.

For the price attribute practitioners suggest to only use few price points and to apply linear interpolation heuristics in order to avoid reversals in the data (Orme (2002)). We estimated an exchange rate between utility and price by fitting a linear model to the estimated utility scores and the price levels (cf. Kohli and Mahajan (1991), Jedidi and Zhang (2003)). For eight respondents the reversals are so strong that the estimated exchange rate is negative, which means that the respondents would be willing to pay more money for less utility. Since this is not plausible, these eight respondents were removed from the dataset. In the remaining dataset the linear models fit the data well (mean R-squared value of 0.87 across the respondents).

4 Estimation of willingness-to-pay

Three different approaches were used to estimate the respondents WTPs. The first approach is the classical approach with a fixed status-quo product (cf. Kohli and Mahajan (1991)). As fixed status-quo product usually a typical product is used for which the market price is known. The price of the other product configurations in the conjoint study is calculated by translating the

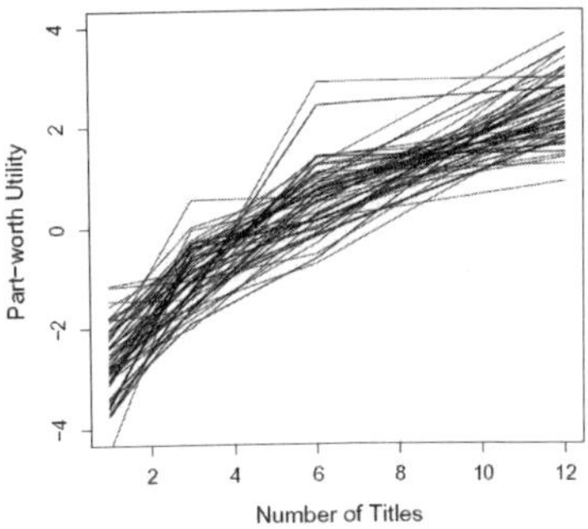

(a) Title (no reversals)

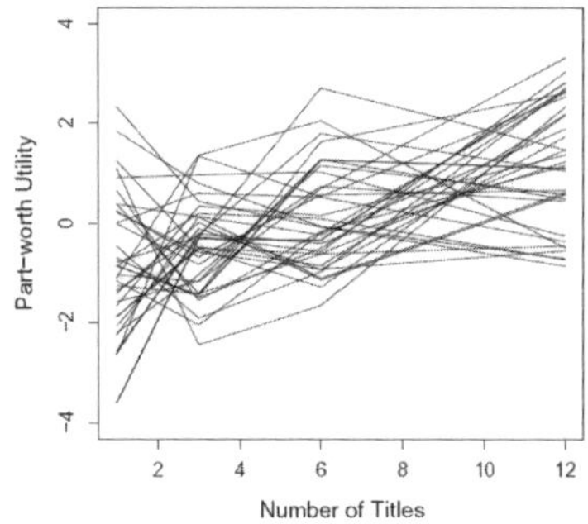

(b) Title (reversals)

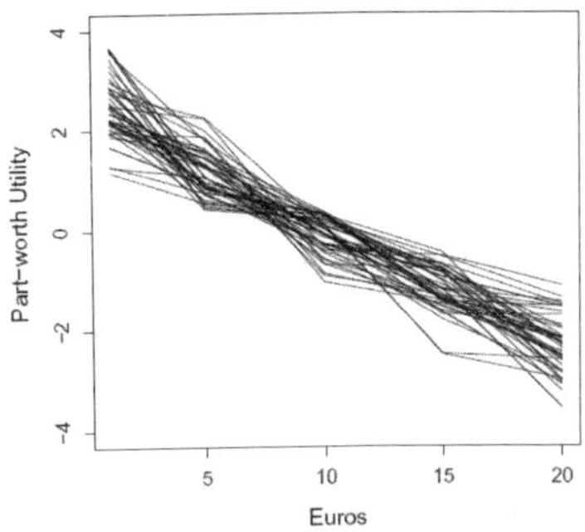

(c) Price (no reversals)

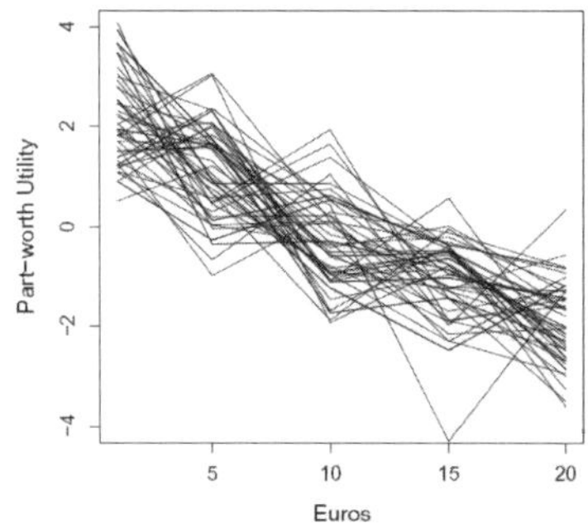

(d) Price (reversals)

Fig. 1. Part-worth utilities for title and price without and with reversals.

utility differences to the status-quo product into a price difference using the utility-price exchange rate. If the respondents are willing to pay the price for the status-quo product, they are also willing to pay the estimated prices for the other products. For our calculations a price of 0.99 € for the download of one song was used, because this is the price that is currently charged by most download services in Europe. Based on this price the WTPs for the different packages 3, 6, and 12 titles were estimated.

As a second approach an individual, self-stated WTP for the status-quo product was used for each respondent to calculate the WTPs for the other package sizes. The self-stated WTPs were elicited from the respondents at the end of the conjoint interview by directly asking them to state a price they were willing to pay for the download of one song. Three users were removed from the dataset because they stated an unreasonably high WTP of over 5 € for a single title.

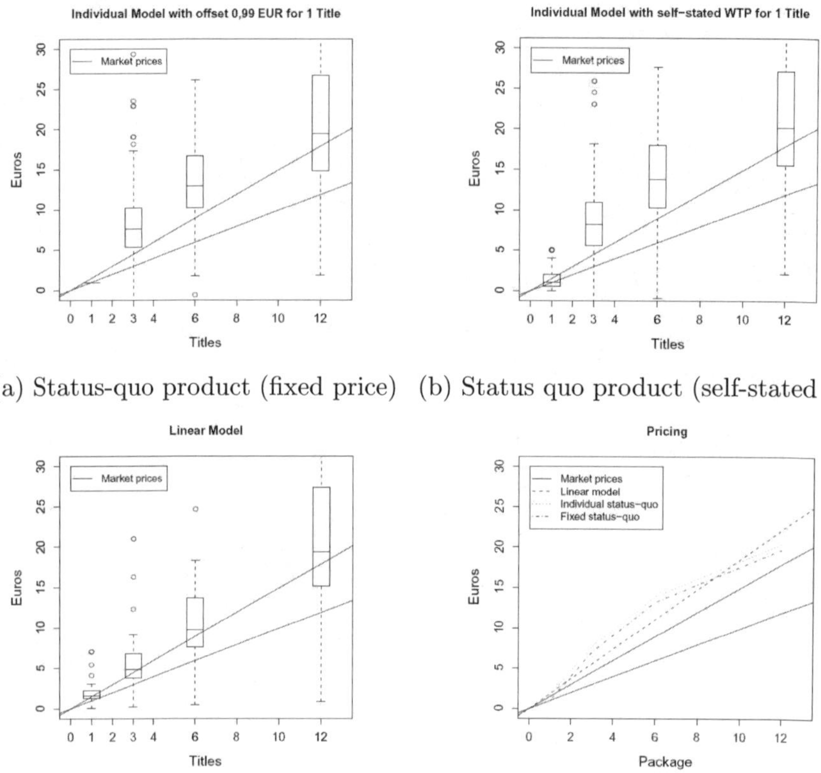

(a) Status-quo product (fixed price) (b) Status quo product (self-stated price)

(c) Linear model (d) Comparison of average WTPs

Fig. 2. Estimations of WTPs based on different estimation approaches.

For the third approach a linear model was applied to the estimated part-worth utilities to calculate the marginal utility for an additional title. In our dataset (see Figure 1) a linear model for the range between 1 and 12 titles seems to be a reasonable simplification. The models' mean R-squared value is 0.80 across the respondents. Using the marginal utility and the exchange rate between utility and price the corresponding monetary equivalent which represents an estimate of the respondent's WTP per title can be calculated.

$$WTP\ per\ title = exchange\ rate\ \frac{price}{utility} \cdot marginal\ utility\ \frac{utility}{title} \tag{1}$$

With this procedure a status-quo product is not needed. A similar approach was used for quantitative attributes by Jedidi and Zhang (2003).

The results of the different approaches are plotted in Figure 2. To give a reference, we indicate the price range of 0.99 € to 1.49 € which is currently used by the market leaders in each plot by two solid lines. Figures 2(a) and (b) show boxplots of the WTPs based on a status-quo product ((a) priced at

Table 4. Average estimated WTPs in € by method (*values not estimated).

	1 title	3 titles	6 titles	12 titles
Status-quo product (fixed price)	0.99*	6.96	13.20	19.69
Status quo product (self-stated price)	2.39*	8.36	14.60	21.09
Linear model	1.84	5.53	11.06	22.12

0.99 € and (b) using the individually self-stated WTP). Since both models are based on the same utility structure, both plots look similar. Only the off-set price for one song is different with an on average higher self-stated price. In the two plots the decreasing marginal price for additional songs is clearly visible. At 12 titles already a part of the interquartile range (represented by the box) is below the upper market price line indicating that many respondents are not willing to pay 1.49 € per song for 12 or more songs.

Figure 2(c) shows the estimates based on the linear model. In the linear model the marginal price for one song is fixed which would mean that a person willing to pay the market price for the first song, would buy all available songs. This is obviously not possible. However, for a small number of songs (≤ 12) the linear model provides a useful approximation. The big advantage of the linear model is that it allows us to estimate the price for one song without using a status-quo product.

In Figure 2(d) the average WTPs of the different approaches are compared. The average estimated WTPs are given in Table 4. The prices for one song for the methods using status-quo products are not estimated but are the market price or self-stated. Only with the linear model the WTPs for one song can be calculated for each respondent. The average of 1.84 € seems a reasonable result with a value between the market price and the self-stated prices.

The results of the three approaches can be used together with the linear pricing scheme of 0.99 € per title currently used in the market. Based on the estimations with a fixed status quo product 72 respondents would be willing to pay the market price for the download of three songs (2.97 €), 58 would be willing to buy 6 songs (5.94 €), and 18 would be willing to buy 12 songs (11.88 €). When the self-stated status quo product is used for each respondent, more songs could be sold given the linear pricing scheme used in the market. 73 respondents would accept the price for 3 songs, 66 the price for 6 songs, and 18 the price for 12 songs. With the linear model a single WTP is estimated for the download of one song. Using this estimation 75 respondents would be willing to pay 0.99 € for the download of a title.

5 Conclusion

We investigated the valuation of music downloads and purchasing a music CD at a record store with adaptive conjoint analysis for a group of students. Practitioners believe that consumers generally value a CD notably higher than the download of music (IFPI (2006)). However, this seems not to be true for the participants of our interview. Our investigation showed that the main differentiators booklet and distribution are only of little importance to the interviewed students. Our data also shows that the marginal WTPs per title decreases with larger package sizes. Therefore linear pricing strategies as found in the market seem not to be optimal to maximize profits.

We compared three approaches to estimate the willingness-to-pay from conjoint data. Two approaches use externally obtained prices for a status-quo product. Obtaining these prices can introduce a bias (e.g., not all customers buy the status-quo product at the market price). The third approach uses a linear approximation to compute a fixed utility per title which eliminates the need for an external price. However, the linear model cannot represent the fact that for a given customer the marginal utility of additional songs decreases (e.g., after the most favorite songs have been purchased).

An idea for future research is to use the linear model with the data for package sizes 1 to 6 where the linear approximation yields a good estimate for the WTPs of one song. These WTPs could then be used as the price for the status-quo product to offset the WTPs for larger package sizes. This combination would eliminate the need of an external price and at the same time reflect the decreasing marginal utility of buying additional songs.

References

BAMERT, T., MEIER-BICKEL, T.S. and RÜDT, C. (2005): Pricing Music Downloads: A Conjoint Study, *Proceedings of the 2005 European Academy of Management Annual Conference.*

BUXMANN, P., POHL, G., JOHNSCHER, P. and STRUBE, J. (2005): Strategien für den digitalen Musikmarkt – Preissetzung und Effektivität von Maßnahmen gegen Raubkopien. *Wirtschaftsinformatik, 47, 2, 118–125.*

GREEN, P.E., KRIEGER, A.M. and AGARWAL, M.K. (1991): Adaptive Conjoint Analysis: Some Caveats and Suggestions. *Journal of Marketing Research, 28, 2.*

IFPI (2006): IFPI:06 Digital Music Report. *266/http://www.ifpi.org/site-content/library/digital-music-report-2006.pdf.*

JEDIDI, K. and ZHANG, Z.J. (2002): Augmenting Conjoint Analysis to Estimate Consumer Reservation Price. *Management Science, 48, 10.*

JOHNSON, R.M. (1991): Comment on "Adaptive Conjoint Analysis: Come Caveats and Suggestions". *Journal of Marketing Research, 28, 2.*

JUPITER RESEARCH (2003): Pricing Music Downloads: Online Music's Volume-Driven Future, Concept Report, Jupiter Research, January 30, 2003.

JUPITER RESEARCH (2006): European Digital Music Value Chain, Vision Report, Jupiter Research, February 1, 2006.

KOHLI, R. and MAHAJAN, V. (1991): A Reservation-Price Model for Optimal Pricing of Multiattribute Products in Conjoint Analysis. *Journal of Marketing Research 28, 3, 347–354.*

ORME, B. (2002): Formulating Attributes and Levels in Conjoint Analysis. Technical report, Sawtooth Software Inc.

Improving the Probabilistic Modeling of Market Basket Data

Christian Buchta

Institute for Tourism and Leisure Studies, Vienna University of Economics and Business Administration, A-1090 Vienna, Austria;
christian.buchta@wu-wien.ac.at

Abstract. Current approaches to market basket simulation neglect the fact that empty transactions are typically not recorded and therefore should not occur in simulated data. This paper suggest how the simulation framework without associations can be extended to avoid empty transactions and explores the possible consequences for several measures of interestingness used in association rule filtering.

1 Introduction

Researchers in the field of association rule mining, a popular data mining technique, are interested in the properties of different measures of interestingness as these can be used to filter the multitude of rules that are typically generated. A recent approach to studying this problem is to investigate the properties of transaction data from a probabilistic point of view (Hahsler et al. (2006)). In this contribution a framework for generation of transaction data without associations is proposed and shown to produce distributions of several measures of interest that are comparable to real world data. However, this simple model does not take into account that empty transactions are typically not recorded and therefore should not occur in simulated data. Thus, two research questions arise: 1) how strong is the bias introduced by the presence of empty transactions and 2) how can we improve the current framework while maintaining its advantage to be able to specify a model with desired distributional properties.

This paper proposes a simple extension of the simulation framework without associations to distributions without empty transactions. In Section 2 an overview of approaches to generating binary deviates is given. In Section 3 we introduce the problem of excluding empty baskets from deviate generation with the independence model and in the following section we provide a proof for a condition of the existence of a generator. Finally, the properties of the proposed model and the consequences for several measures of interestingness are discussed in Section 5.

2 Approaches to binary deviate generation

Transaction or market basket data can be simulated by drawing from a multivariate distribution of variables that indicate the presence (absence) of items in a basket. Therefore models of binary variables are of interest. The usual approach to generating uncorrelated binary deviates $x_i \in \{0,1\}, i = 1,2,\dots,n$ is to specify their marginal probabilities $p_i = P(x_i = 1)$, and use these as threshold on uniform deviates $u \sim U(0,1)$:

$$x_i = \begin{cases} 1 \text{ if } u \leq p_i \\ 0 \text{ otherwise} \end{cases} . \tag{1}$$

In Emerich and Piedmont (1991) it is shown how this approach can be extended to obtain correlated binary deviates by using correlated normal deviates: the cutoff values are now the quantiles of the standard normal distribution for the given marginal probabilities. Leisch et al. (1998) use Monte Carlo simulation to obtain a correlation matrix of the normal distribution that corresponds with a given correlational structure of the binary variables. Further, a heuristic approach to the problem of finding a feasible correlation matrix of binary variables is presented in Orasch et al. (1999). However, consistency of a binary model specification does not imply that a multivariate normal distribution and thus a model for deviate generation does exists.

A different approach is to specify an autologistic model and generate the binary deviates using a Gibbs sampler, i.e. a model of the form $p(x) = C\exp(x'\Theta x)$, $x \in \{0,1\}^n$ where $C = p(0)$ is a normalizing constant which ensures that the probabilities sum to one. A promising feature of this class of models is that they are estimable from observed data using a neural network approach, i.e. a set of logistic regression models with constraints $\theta_{ij} = \theta_{ji}$. Unfortunately, the sample size is a limiting factor for the size of models identifiable. Thus, for market basket data with their usually large number of items we must settle for small, mutually independent, models that focus on the most prominent dependencies. An approach that deals with the model selection problem was suggested by Hruschka (1990). Alternatively, as in the normal distribution approach we could search for a feasible autologistic model given the first and second order marginal probabilities. However, such a model may again not exist and the need to sum over $2^n - 1$ terms further restricts any heuristic search to small models.

In Lee (1993) a linear programming approach is presented where the first and second order marginal probabilities p are given and the distribution function q has to satisfy $Aq = p$ with $a_{ij} \in \{0,1\}$ defining the marginal sums. Deviates can be generated using the inversion method $x_j : z_j \leq u < z_{j+1}$ with

$z_j = z_{j-1} + q_j$ and $j = 0, 1, \ldots, 2^n - 1$ indexing binary sequences of length n. Other than in the previous approaches, for a consistent model specification we always have a method for deviate generation. However, this approach is again applicable to small problems only.

Finally, a result of Bahadur (1961) shows that if all higher order correlations, i.e. between more than two variables, are assumed to be equal to zero only the identity matrix will be a consistent correlation matrix as the model size $n \to \infty$. Therefore, as the number of items in transaction data usually is large using the simple independence model for data generation is a reasonable simplification.

However, all the models discussed so far include deviates with all-zeros which are typically not observed in market basket data. As it is not clear whether this introduces a considerable bias into simulation studies on association rule mining, or not we need to develop models without empty baskets.

3 The basket exclusion problem

Let the random variable $X_i = 1$ indicate that item i is contained in a basket, and $X_i = 0$ otherwise. Let $0 < p_i < 1$ denote the marginal probability that $X_i = 1$, and $1 - p_i$ that $X_i = 0$. Further, let us assume that the variables are independent, $p_{ij} = p_i p_j$. The probability of observing basket $X = x$ can then be expressed as the product of marginal probabilities

$$p(x) = \prod_i p_i^{x_i} \tag{2}$$

and, more specifically, the probability of the empty basket $X_i = 0, \forall i$ is $p(0) = \prod_i (1 - p_i)$.

We observe that if we exclude the empty basket from the distribution the marginal probabilities increase

$$\frac{p_i}{1 - \prod_k (1 - p_k)} > p_i \;\; \forall i \tag{3}$$

because the probability of a basket with at least one item is less than one.

Now let us assume there exists a factor γ that accounts for the exclusion of the empty basket such that the marginal probabilities of the deviates remain the same as in the independence model:

$$\frac{\gamma}{1 - \prod_k (1 - \gamma p_k)} p_i = p_i \;\; \forall i. \tag{4}$$

That is, we use $q_i = \gamma p_i$ as the thresholds for deviate generation (compare Equation (1) from the previous section) and reject deviates with all zeros $X_i = 0, \forall i$.

We observe that $0 < \gamma < 1$ because at the lower bound we would reject any deviate whereas at the upper bound the marginal probabilities of the deviates would be too large as the probability of the all-zero deviates must satisfy $0 < p(0) < 1$.

For example assume a model with two items and marginal probabilities $p_1 = 0.5$ and $p_2 = 0.6$. As shown in the left panel of Table 1 a distribution with a zero probability of the empty basket $p(0) = 0$ exists and, as can easily be verified, it is unique. The right panel shows a corresponding model that satisfies Equation 4 and the range constraint on γ. More interestingly, as will be shown in the next section, the generator model is unique.

Table 1. A feasible model with $p(0)$=0 and $\gamma = 0.33\dot{3}$.

	$X_1 = 0$	$X_1 = 1$	Σ
$X_2 = 0$	0	0.4	0.4
$X_2 = 1$	0.5	0.1	0.6
Σ	0.5	0.5	1

	$X_1 = 0$	$X_1 = 1$	Σ
$X_2 = 0$	0.66	0.13	0.8
$X_2 = 1$	0.16	0.03	0.2
Σ	0.83	0.16	1

As another example, assume the marginal probabilities are given by $p_1 = p_2 = 0.5$. Now Table 2 shows only the distribution with a zero probability of the empty basket as a model that satisfies Equation 4 and the range constraint on γ does not exist.

Table 2. A feasible model with $p(0) = 0$ and no generator.

	$X_1 = 0$	$X_1 = 1$	Σ
$X_2 = 0$	0.0	0.5	0.5
$X_2 = 1$	0.5	0.0	0.5
Σ	0.5	0.5	1

As the examples suggest, the important question is can we formulate constraints that guarantee the existence of a model for deviate generation. More specifically, can we express them in terms of the marginal probabilities?

4 The root finding problem

Let us rewrite Equation (4) from the previous section as

$$f(x) = 1 - \prod_i (1 - xp_i) - x \tag{5}$$

We observe that $f(0) = 0$, so 0 is a root of the function, and that $f(1) < 0$. Further, if $f(x) = 0$ then the root equals the probability of a deviate with at least one item, γ in Equation (4). Thus, as we reject $1-\gamma$ percent of the deviates and retain γ, the latter quantity can be interpreted as the *efficiency* of the generator.

For example, Figure 1 shows in the left panel the marginal probabilities for a typical market basket simulation with 250 items (compare Hahsler et al. (2006)). The right panel shows Equation (5) for this data. We see that the function is concave and that it has two roots, one at $x = 0$ which is at the lower bound and another at $x = 0.952$ which is within the feasible range of γ as indicated by the vertical lines. The probability of the empty basket is therefore $1 - x = 0.041$ which equals the percentage of deviates we would reject on average.

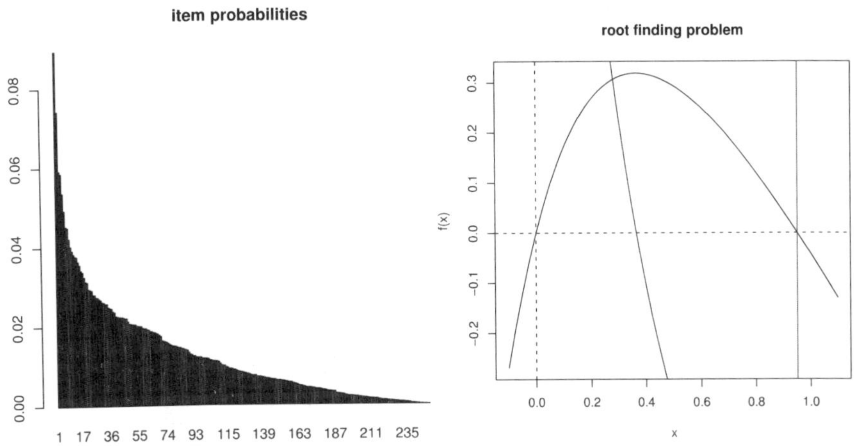

Fig. 1. A typical market basket simulation scenario.

More formally, we can derive that $f'(0) = \sum_i p_i - 1$ and thus a sufficient condition that a root on the interval $(0, 1)$ exists if and only if $\sum_i p_i > 1$. In the example above the sum of the item probabilities is 3.142. Further, we can establish that the condition is also necessary by showing that $f''(z) < 0$ on the interval $(0, 1)$, i.e. there is only one such root. This is indicated by the line for the first derivative in the left panel of Figure 1.

For a proof we note that Equation (5) expands to

$$f(x) = -z + z\sum_i p_i - z^2 \sum_{j>i} p_i p_j + z^3 \sum_{k>j>i} p_i p_j p_k - \dots \tag{6}$$

and after rearranging we obtain for the second derivative

$$f''(z) = -\sum_{j \neq i} p_i p_j \prod_{k \neq i,j} (1 - z p_k). \tag{7}$$

Thus, $f''(z) < 0$ for $0 \leq z \leq 1$ and at least one pair i, j with $p_i, p_j > 0$. $\square$

5 Model properties

The model introduces negative correlations between the item variables as the joint probabilities that two items are in the same basket are lower than expected

$$p_i p_j > \frac{\gamma^2}{1 - \prod_k (1 - \gamma p_k)} p_i p_j = \gamma p_i p_j \quad \forall i, j \tag{8}$$

and the correlations reduce to the following expression

$$\rho_{ij} = -(1-\gamma)\frac{\sqrt{p_i p_j}}{\sqrt{(1-p_i)(1-p_j)}} \quad \forall i, j \tag{9}$$

More generally, several measures of interest in association rule filtering are based on marginal 2×2 tables for two itemsets $A, B \subset I, A \cap B = \{\}$ with $\neg A = I \setminus A$ and $\neg B = I \setminus B$, respectively. For the extended independence model the probabilistic structure of these tables is shown in Table 3.

Table 3. The probabilistic structure of marginal tables.

	$\neg A$	A	Σ
$\neg B$	$\frac{(1-\gamma p_A)(1-\gamma p_B)-(1-\gamma)}{\gamma}$	$p_A(1-\gamma p_B)$	$1-p_B$
B	$(1-\gamma p_A)p_B$	$\gamma p_A p_B$	p_B
Σ	$1-p_A$	p_A	1

For instance we can easily verify that the ϕ coefficient, a measure of the correlation of binary variables $\frac{s(A,B)s(\neg A,\neg B)-s(A,\neg B)s(\neg A,B)}{\sqrt{s(A)s(B)s(\neg A)s(\ negB)}}$, with $s(\cdot)$ indicating the support of an itemset, has the same expectation as the correlation coefficient. As another example, the expectation of the lift of an association rule is $\frac{s(A,B)}{s(A)s(B)} = \gamma < 1$.

For example, Figure 2 shows the histogram of the correlations for the simulation scenario from the previous section. As expected there is a bias towards negative correlations but it is low, $i\rho_{ij} \geq -0.004$.

As a result we can state that the expectation of several measures of interest which are based on marginal 2×2 tables do not change considerably if we exclude (or include) empty transactions from a simulation with many low probability items. On the other hand a model that contains only a few items may be considerably biased unless we exclude the empty baskets. Especially, this needs to be dealt with in a model for generating correlated deviates as positive correlations get attenuated and negative correlations amplified.

6 Conclusions

In this paper we showed how the framework of simulating multivariate binary deviates without associations can be extended to avoid the generation of all-

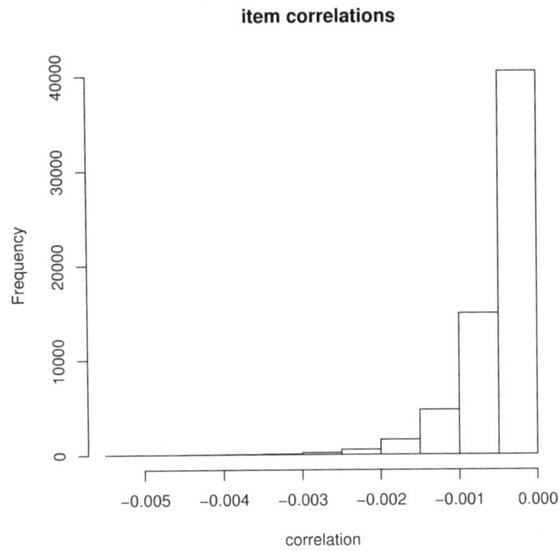

Fig. 2. Histogram of item correlations in a typical simulation scenario.

zero deviates. We indicated that the approach is reasonable and provided a proof for a sufficient and necessary condition for the existence of a model. We also discussed the consequences for several measures of interestingness that are used for filtering of association rules in market basket analysis: the correlations introduced by the model are negative and typically negligible.

However, we left the problem of modelling dependencies between itemsets for future work. The need for models with dependencies is indicated by the observation that the current simulation framework with no associations does not produce statistics of the basket size that are comparable to what we observe in typical market basket data. This might be due to higher order dependencies among the items. Under the assumption that we model only a few items the results indicate that non-exclusion of empty transactions could bias a model considerably.

From a technical as well as an application point of view it might also be interesting to consider generation frameworks where more than one transaction could be constrained to have zero probability mass. Actually, as consumers exhibit similar purchase patterns over time and/or the set of all possible transactions is typically large in comparison, zero-mass transactions are the rule and not the exception.

Finally, real world market basket analysis could benefit from measuring or estimating actual numbers of empty transactions. For example surveys in retailing on choice behavior across different outlets could provide such figures.

References

BAHADUR, R.R. (1961): A Representation of the Joint Distribution of Responses to n Dichotomous Items. In: H. Solomon (Ed.): *Studies in Item Analysis and Prediction.* Standford Mathematical Studies in the Social Sciences VI. Stanford University Press, Stanford.

EMERICH, L.J. and PIEDMONT, M.R. (1991): A Method for Generating High-dimensional Multivariate Binary Deviates. *Statistical Computing, 45, 302–304.*

HAHSLER, M., HORNIK, K. and REUTTERER, T. (2006): Implications of Probabilistic Data Modelling for Mining Association Rules. In: M. Spiliopoulou, R. Kruse, C. Borgelt, A. Nürnberger and W. Gaul (Eds.): *From Data and Information Analysis to Knowledge Engineering.* Springer, Berlin, 598–605.

HRUSCHKA, H. (1990): Bestimmung der Kaufverbundenheit mit Hilfe eines probabilistischen Messmodells. *zfbf, 418–434.*

LEE, A.J. (1997): Some Simple Methods for Generating Correlated Categorical Deviates. *Computational Statistics & Data Analysis, 25, 133–148.*

LEE, A.J. (1993): Generating Random Binary Deviates Having Fixed Marginal Distributions and Specified Degrees of Association. *The American Statistician, 47, 3, 209–215.*

LEISCH, F., WEINGESSEL, A. and HORNIK, K. (1998): On the Generation of Correlated Artificial Binary Data. *Technical Report 13, SFB Working Paper Series.*

OMAN, S.D. and ZUCKER, D.M. (2001): Modelling and Generating Correlated Binary Variables. *Biometrica, 88, 1, 287–290.*

ORASCH, M., LEISCH, F. and WEINGESSEL, A. (1998): On Specifying Correlation Matrices for Binary Data. *Technical Report 53, SFB Working Paper Series.*

Classification in Marketing Research by Means of LEM2-generated Rules

Reinhold Decker and Frank Kroll

Department of Business Administration and Economics, Bielefeld University, D-33501 Bielefeld, Germany; {rdecker, fkroll}@wiwi.uni-bielefeld.de

Abstract. The vagueness and uncertainty of data is a frequent problem in marketing research. Since rough sets have already proven their usefulness in dealing with such data in important domains like medicine and image processing, the question arises, whether they are a useful concept for marketing as well. Against this background we investigate the rough set theory-based LEM2 algorithm as a classification tool for marketing research. Its performance is demonstrated by means of synthetic as well as real-world marketing data. Our empirical results provide evidence that the LEM2 algorithm undoubtedly deserves more attention in marketing research as it is the case so far.

1 Introduction

Classification is a common problem in marketing, particularly in market segmentation (Which consumers belong to which market segment?), sales force management (Which customers should be targeted with which customer care program?), and direct marketing (Which advertising materials should be sent to which customer?). One way to solve such marketing tasks is the deduction of decision rules from appropriate survey data.

A promising approach for generating decision rules is the LEM2 algorithm (Learning from Examples Module Version 2), which is based on the rough set theory. The concepts of rough set theory have been used successfully in medicine and in bio-informatics, among other things. Therefore, it is quite surprising that – to our knowledge – there is still no publication applying LEM2 to classification tasks in marketing. The present paper aims at filling this gap. Substantiating the adequacy of LEM2 for the above mentioned tasks opens new options for knowledge discovery in marketing databases and the development of decision support systems for marketing planning.

The rough set theory was introduced by Pawlak (1982) and is a mathematical approach to deal with vagueness and uncertainty in data. The main idea of rough set theory, which recently aroused increasing interest in data mining, is to describe a given set $\mathcal{X}$ with fuzzy boundaries by two sets called the lower and the upper approximation. Rough sets enable to partition a given universe $\mathcal{U}$, i.e. the objects considered, into equivalence classes, which are also called elementary sets. The objects belonging to those elementary sets are indiscernible with regard to a given set of attributes $\mathcal{B}$ and set $\mathcal{X}$. The lower approximation $\underline{\mathcal{B}}\mathcal{X}$ contains all objects, whose elementary set is completely contained in $\mathcal{X}$ (with regard to the set of attributes $\mathcal{B}$). The upper approximation $\overline{\mathcal{B}}\mathcal{X}$, in return, contains all those objects, for which at least one element of the associated elementary set is contained in $\mathcal{X}$. Those objects are possible elements of $\mathcal{X}$. The set $\mathcal{X}$ is called rough, if the lower approximation does not equal the upper approximation, otherwise $\mathcal{X}$ is called crisp. Therefore, the main idea is to approximate a rough set $\mathcal{X}$ by two crisp sets, namely the very lower and upper approximation. For a more detailed introduction into rough set theory we refer to Pawlak (1991).

The remainder of the paper is structured as follows: First, a brief description of the LEM2 algorithm is given. Then, the LEM2 algorithm is empirically compared to alternative methods based on three different classification tasks in marketing.

2 Remarks on the LEM2 algorithm

The LEM2 algorithm was published in 1992 by Jerzy W. Grzymala-Busse as a part of the LERS (Learning from Examples based on Rough Sets) system. The basic principle of this approach needed for the marketing examples will be sketched in the following (see also Grzymala-Busse (1997) for details).

Let $\mathcal{T}$ be a set of pairs of attributes $a \in \mathcal{B}$ and corresponding attribute values v. Then set $\mathcal{X}$ depends on set $\mathcal{T}$ if and only if:

$$\emptyset \neq [\mathcal{T}] = \bigcap_{(a,v)\in\mathcal{T}} [(a,v)] \subseteq \mathcal{X}, \tag{1}$$

where $[(a,v)]$ is the equivalence class or elementary set of pair (a,v). Furthermore, set $\mathcal{T}$ is a minimal complex of $\mathcal{X}$ if and only if $\mathcal{X}$ depends on $\mathcal{T}$ and no proper subset $\mathcal{Q} \subset \mathcal{T}$ exists such that $\mathcal{X}$ depends on $\mathcal{Q}$. Because of the possible existence of various combinations of attributes a and values v, set $\mathcal{X}$ can have several minimal complexes.

In rough set theory data is represented as a decision system $\mathcal{S}_d$. A simple decision system referring to a hypothetical car evaluation is shown in Table 1. The example at hand includes two condition attributes, i.e. $\mathcal{B}$ = {seating capacity, vibration} and one decision attribute d = {quality}. A set of objects with the same decision attribute value relation (e.g. (quality, low)) is called a decision class.

Table 1. A simple decision system $\mathcal{S}_d$ for car evaluation

	Seating capacity	Vibration	Quality
c_1	5	medium	low
c_2	4	medium	low
c_3	5	medium	low
c_4	5	low	medium
c_5	2	low	medium
c_6	4	medium	high
c_7	4	low	high
c_8	2	low	high

If we consider the decision class (quality, low), for example, this results in the set $\mathcal{X} = \{c_1, c_2, c_3\}$. In addition, let $\mathcal{T}$ be a set of pairs (seating capacity, 5) and (vibration, medium). The elementary sets of these two pairs are $\{c_1, c_3, c_4\}$ and $\{c_1, c_2, c_3, c_6\}$. Then, according to Equation (1), $[\mathcal{T}] =$ [(seating capacity, 5)] $\cap$ [(vibration, medium)] $= \{c_1, c_3, c_4\} \cap \{c_1, c_2, c_3, c_6\} = \{c_1, c_3\} \subseteq \{c_1, c_2, c_3\} = \mathcal{X}$. Thus $\mathcal{X}$ depends on $\mathcal{T}$. Furthermore, $\mathcal{T} = \{$(seating capacity, 5); (vibration, medium)$\}$ is a minimal complex because there is no proper subset $\mathcal{Q} \subset \mathcal{T}$, which depends on $\mathcal{X}$.

Let $\mathbb{T}$ be a nonempty set of minimal complexes. Then $\mathbb{T}$ is a local covering of $\mathcal{X}$ if and only if the following conditions are satisfied (Grzymala-Busse (1997)):

a) Each $\mathcal{T} \in \mathbb{T}$ is a minimal complex of $\mathcal{X}$.
b) Each object in $\mathcal{X}$ is covered by at least one minimal complex, i.e. $\bigcup_{\mathcal{T} \in \mathbb{T}} [\mathcal{T}] = \mathcal{X}$ holds.
c) $\mathbb{T}$ is minimal and no proper subset of $\mathbb{T}$ exists, which satisfies conditions a) and b).

If all conditions are satisfied, then set $\mathcal{X}$ is covered by the smallest set of pairs (a, v) it depends on. In addition each minimal complex of the local covering represents a *decision rule* for set $\mathcal{X}$.

So far, we solely considered the decision class associated with the pair (quality, low). To generate rules for the whole decision system, the algorithm sketched above has to be run for each decision class and each approximation of $\mathcal{X}$. If the data set is inconsistent, then the lower approximation $\underline{\mathcal{B}}\mathcal{X}$ is used to determine certain decision rules, whereas the upper approximation $\overline{\mathcal{B}}\mathcal{X}$ is used to determine uncertain ones. However, if the data set is consistent, then $\underline{\mathcal{B}}\mathcal{X} = \overline{\mathcal{B}}\mathcal{X}$ applies and we solely generate certain decision rules by using the lower approximation $\underline{\mathcal{B}}\mathcal{X}$. For a more detailed description of the corresponding pseudo-code see Grzymala-Busse (1997).

Continuing the example given in Table 1, the lower approximation of $\mathcal{X}$ is $\underline{\mathcal{B}}\mathcal{X} = \{c_1, c_3\}$. With all conditions a) to c) satisfied, the set $\mathbb{T} = \{\{$(seating capacity, 5); (vibration, medium)$\}\}$ is a local covering of $\mathcal{X}$ and the minimal complex $\mathcal{T}$ represents a certain decision rule: (seating capacity, 5) $\wedge$ (vibration, medium) $\rightarrow$ (quality, low).

3 Synthetic data example

The data set used in the following is a synthetic one published by Zupan (1997). It is used to demonstrate the general performance of the LEM2 algorithm and comprises 1,728 car evaluations based on six condition attributes, such as estimated safety (values: low, medium, high) and the number of doors (values: 2, 3, 4, 5 and more) and one decision attribute, namely the degree of acceptance of the car (values: no, rather no, rather yes, yes). The data set is consistent since the quality of classification $\mathcal{X}$ equals 100 %. That is to say, there are no objects with identical condition attribute values but different decision attribute values. All attributes were included in the analysis.

The computations were made with the rough set analyzing tool RSES (Bazan et al. 2004). The method for resolving conflicts was standard covering, hence the decision rule with the highest support fires. To validate the generated decision rules 10-fold cross validation was applied. Thereto, the data set was randomly divided into ten similar subsets. Then, nine subsets were used to learn and one subset to validate the generated rules. This procedure was repeated until each subset was validated.

To assess the performance of the LEM2 algorithm three benchmarks were employed for comparison purposes, namely the decomposition tree approach by Nguyen (1999), the Local Transfer Function Classifier (LTF-C), and linear discriminant analysis as the standard classification method in marketing practice. LTF-C is an artificial neural network approach particularly developed to solve classification problems which has already proven its outstanding performance in a recent study by Wojnarski (2003).

Table 2 shows that LEM2 outperforms linear discriminant analysis by 22 % and LTF-C by 16.4 % and results in a hit rate of almost 100 %, i.e. nearly four times as high as random assignment. The difference between the decomposition tree and LEM2 is quite marginal.

Table 2. Performance based on the synthetic data

Method	**Hit rate**	**Random assignment**	**Rank**
LEM2	98.0 %	25.0 %	1
Decomposition tree	97.8 %		2
LTF-C	81.6 %		3
Discriminant analysis	76.0 %		4

By using the LEM2 algorithm altogether 269 decision rules could be generated from the available car data. Some of these decision rules are depicted in Table 3. The numbers in squared brackets indicate how often the respective rule was identified in the data set. For example, if the purchase price of a car is high and the estimated safety is high and the seating capacity is 4 persons and the maintenance costs are medium then the respective offer would be accepted by the respondents. This decision rule is supported by 12 respondents. On the other hand, reductions in safety can not be compensated by a lower purchase price (see last rule). Obviously, safety ranks higher than inexpensiveness.

Table 3. Selected LEM2 decision rules

(estimated safety = low) → (acceptance = no) [576]
(purchase price = high) ∧ (maintenance costs = very-high) → (acceptance = no) [108]
(seating capacity = 5 and more) ∧ (number of doors = 2) ∧ (luggage volume = small) → (acceptance = no) [48]
(purchase price = high) ∧ (estimated safety = high) ∧ (seating capacity = 4) ∧ (maintenance costs = medium) → (acceptance = yes) [12]
(purchase price = low) ∧ (estimated safety = medium) ∧ (seating capacity = 4) ∧ (maintenance costs = medium) ∧ (luggage volume = large) → (acceptance = rather no) [4]

4 Real world data examples

After having demonstrated the basic suitability of the LEM2 approach by means of synthetic marketing data, we will now consider real world marketing data. The data set at hand was generated from a customer survey conducted by a German car manufacturer. Each respondent had purchased a new car within the last three months before the survey started. Considering only completed questionnaires, the data set contains 793 objects (respondents), each of them being characterized by 45 condition attributes (namely the items referred to in the questionnaire) and one decision attribute (namely the car currently owned: model A, B, C or D).

4.1 Customer selection in direct mailing

The first example is a typical problem of direct mailing: Assuming that the car manufacturer wants to send out different information brochures on the four models (A, B, C and D) to potential new customers, a customized shipping suggests itself to minimize the marketing costs and to enhance the likelihood of use. Each recipient should get only the brochure of that model he presumptively is interested in most. Therefore, we used the following condition attributes to generate appropriate decision rules and to draw a comparison between the methods considered: pre-owned car brand, sex, employment status, and size of household. In contrast to the synthetic data used above the present data has a low quality of classification $\mathcal{X}$ of only 9.7 %, which corresponds with a high level of inconsistency.

10-fold cross validation resulted in the hit rates depicted in Table 4. Once again, the LEM2 algorithm clearly outperforms the benchmarks. Altogether, 92 decision rules could be generated, e.g., (pre-owned car brand = Porsche) $\wedge$ (employment status = employee) $\rightarrow$ (model = A).

Table 4. Performance based on the direct mailing data

Method	Hit rate	Random assignment	Rank
LEM2	61.7 %	25.0 %	1
Decomposition tree	38.2 %		4
LTF-C	49.1 %		2
Discriminant analysis	44.9 %		3

4.2 Customer advisory service

In the second example we assume that a car retailer wants to improve his recommendation policy with regard to the four models. The basis for these recommendations is a set of indicators regarded to be crucial in car purchase decision making. From the above mentioned car manufacturer survey such information is available in terms of 21 binary coded condition attributes like driving comfort, styling and handling. The question belonging to the first condition attribute, for example, reads: "Has driving comfort been a crucial car attribute when making your last car purchase decision?". The decision attribute was composed of two variables, namely model (values: model A, B, C, D) and customer satisfaction (values: satisfied, not satisfied). Therefore, there are eight decision classes with attribute values (A, satisfied), (A, not satisfied),..., (D, not satisfied).

Since the quality of classification $\mathcal{X}$ equals 95 % the data set is only slightly inconsistent. Once again the LEM2 algorithm outperforms LTF-C and linear discriminant analysis (see Table 5). But the differences are less distinct than in the example before. The difference between LEM2 and the decomposition tree is negligible. Altogether, 416 decision rules could be generated from the available data, e.g., (economy = yes) $\wedge$ (driving comfort = no) $\wedge$ (reliability = yes) $\wedge$ (brand image = yes) $\rightarrow$ (model = A, customer = satisfied).

Table 5. Performance based on the customer advisory service data

Method	Hit rate	Random assignment	Rank
LEM2	37.5 %	12.5 %	2
Decomposition tree	37.7 %		1
LTF-C	30.9 %		3
Discriminant analysis	29.6 %		4

Since the number of variables or rather attributes included in a survey usually directly correlates with the willingness of the respondents to participate in the questioning and, thus, with the survey costs in general, it is interesting to see what happens if the number of condition attributes is reduced successively. The result of a random reduction of condition attributes is shown in Figure 1.

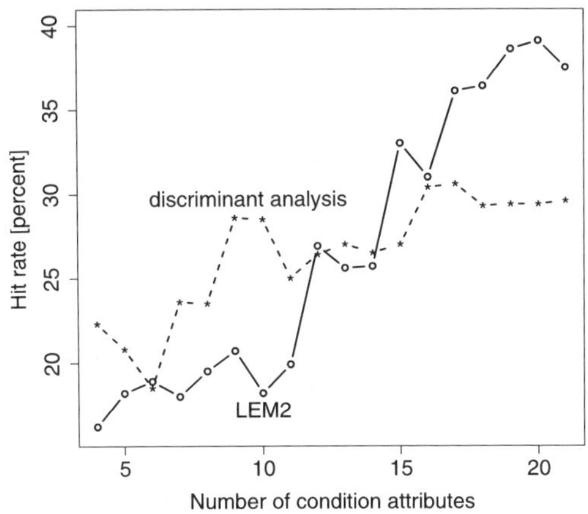

Fig. 1. Hit rates for different numbers of condition attributes

From the two curves we see that popular linear discriminant analysis outperforms the LEM2 algorithm if the number of condition attributes is small, at least in the present case of binary coded data. But with an increasing number of condition attributes LEM2 performs better and better and finally clearly outperforms the standard classification method in marketing research.

Obviously, LEM2 needs an appropriate amount of information to adequately learn the existing causalities. In the previous example this problem did not arise due to the higher explanatory power of the condition attributes considered. In so far LEM2 is not necessarily the best choice but seems to be promising if the number of variables to be included in the classification process is large. Furthermore, it should be taken into account if easy-to-interpret decision rules are required, e.g., as a foundation of customer advisory service optimization. Here, the availability of natural language like rules is often more helpful than abstract discriminant functions, for example.

5 Concluding remarks

The present study has shown that the LEM2 algorithm is a promising tool to solve common classification problems in a user-friendly way by providing shortest possible, easy-to-interpret decision rules. Furthermore, those decision rules are distinguishable into certain and uncertain ones.

It could be shown that the LEM2 algorithm outperforms both linear discriminant analysis and LTF-C in all data examples considered. Only the decomposition tree slightly outperformed the LEM2 algorithm once. The direct mailing example has shown that already small numbers of condition attributes can lead to adequate classification results. But, as obvious from the customer advisory service example, the LEM2 algorithm is not necessarily the best choice and requires an adequate amount of information if more complex causalities have to be uncovered.

Unlike other methods with similar foci like neural networks and association rules an a priori parameterization is not required. A more technical advantage of LEM2 is its low computational costs. On the other hand, if the variables to be included are continuous a discretization must precede, which – in some cases – may reduce their expressiveness. Without this step, each object would be an elementary set and the generation of meaningful rules becomes impossible.

Further research should be devoted to more extensive comparisons which also include the recent modification of the original LEM2 algorithm suggested by Grzymala-Busse (2003), which combines discretization and rule generation. Furthermore, the application of the rough set theory to new domains in marketing and business administration in general seems to be worth closer consideration. A very up-to-date challenge in this respect would the application to web-based recommender systems, where the concept of the lower and upper approximation of sets can be used to suggest items (e.g. consumer goods) in a customized way.

References

BAZAN, J.G., SZCZUKA, M.S., WOJNA, A. and WOJNARSKI, M. (2004): On the Evolution of Rough Set Exploration System. In: S. Tsumoto, R. Slowinski, J. Komorowski, and J.W. Grzymala-Busse (Eds.): *Rough Sets and Current Trends in Computing, Proceedings of the 4th International RSCTC'04 Conference.* Springer, Berlin, 592–601.

GRZYMALA-BUSSE, J.W. (1997): A New Version of the Rule Induction System LERS. *Fundamenta Informaticae, 31, 1, 27–39.*

GRZYMALA-BUSSE, J.W. (2003): MLEM2 – Discretization During Rule Induction. In: M.A. Klopotek, T.T. Wierzchon, and K. Trojanowski (Eds.): *Intelligent Information Processing and Web Mining, Proceedings of the International IISPWM'03 Conference.* Springer, Berlin, 499–508.

NGUYEN, S.H. (1999): *Data Regularity Analysis and Applications in Data Mining.* Ph.D. Thesis, Department of Mathematics, Computer Science and Mechanics, Warsaw University, Warsaw.

PAWLAK, Z. (1982): Rough Sets. *International Journal of Computer and Information Sciences, 11, 5, 341–356.*

PAWLAK, Z. (1991): *Rough Sets: Theoretical Aspects of Reasoning about Data.* Kluwer Academic Publishers, Dordrecht.

WOJNARSKI, M. (2003): LTF-C: Architecture, Training Algorithm and Applications of New Neural Classifier. *Fundamenta Informaticae, 54, 1, 89–105.*

ZUPAN, B., BOHANEC, M. and DEMSAR, J. (1997): Machine Learning by Function Decomposition, In: D.H. Fisher (Ed.): *Proceedings of the 14th International Conference on Machine Learning.* Morgan Kaufmann Publishers, Nashville, 421–429.

Pricing Energy in a Multi-Utility Market

Markus Franke[1], Andreas Kamper[2] and Anke Eßer[3]

[1] Institute for Information Systems and Management; maf@em.uni-karlsruhe.de
[2] Institute of Applied Informatics and Formal Description Methods; aka@aifb.uni-karlsruhe.de
[3] Institute for Industrial Production, Universität Karlsruhe (TH), D-76128 Karlsruhe, Germany; anke.esser@wiwi.uni-karlsruhe.de

Abstract. We present a solution to the problem of tariff design for an energy supplier (utility). The tariffs for electricity and – optionally – heat created with our pricing model are optimal in terms of the utility's profit and take into account the consumers' predicted behavior, their load curve, the utility's generation prices, and prices for trading electricity on a day-ahead market like the European Energy Exchange (EEX). Furthermore, we analyze the repercussions of different assumptions about consumer behavior on a simulated market with four competing utilities.

Consumer demand is modeled using an attraction model that reflects consumer inertia. Consumers will not always change their supplier, even if the total energy bill could be reduced by doing so: First, motivation to search for lower prices and to actually switch one's supplier is low, given the small possible savings. Second, legal constraints may demand a minimal contract duration in some countries.

The resulting nonlinear profit optimization problem of the suppliers is solved with a genetic algorithm. By varying the attraction parameters and thus representing different degrees of inertia, we observe different developments of the market.

1 Introduction

Today's energy markets are marked by a growing insecurity in the wake of their liberalization: Due to an increasing customer volatility it is becoming more and more difficult for utilities to plan their investments for the next decades. In addition, the problem of predicting their consumers' behavior and adapting their tariffs accordingly gains importance. We present an approach for creating tariffs for different groups of energy consumers depending on the suppliers' costs, the tariffs published by their competitors and – most important – a model for the consumers' behavior. This approach is then used to simulate and analyze an electricity market where four big suppliers compete for consumers.

An extension of the tariff model exists for multi-utility markets where suppliers may sell bundles of energy products, e.g. heat and electricity. For the scope of this paper, we will however limit the discussion to one product, electricity, in order to facilitate the analysis of the market.

We will start this paper with a description of the optimization model in section 2 detailing especially the components for consumer behavior, generation related aspects, market activities and taxes. The simulation approach for the market together with the results is presented in section 3 and is followed by the conclusion and an outlook in section 4.

2 Description of the model

For the scenario discussed in this paper, two groups of participants are important: Consumers and suppliers. Let us assume that the consumers can be clustered into classes that correspond to typical load curves for their respective members. Let I be the set of consumer classes. In our model, a consumer always belongs to exactly one class $i \in I$. N_i is the total number of consumers that belong to class i. The load curve of a member of class i is represented by variables x_{it} denoting the demand of one consumer of this type at a given period of time t. This demand is derived from the synthetic load curves published by Tiedemann and Fünfgeld (2001) for the VDEW. On the other side, the set H contains all suppliers. Each of them offers a tariff menu denoted by J^h; $j \in J^h$ is a single tariff. Let us further assume that for each customer group i there is exactly one tariff tailored for the members of this group that is denoted by j^i. A typical tariff might for instance be targeted at single households, large families or small businesses. The fit between customer group i and its tariff j^i is guaranteed by a self-selection criterion (Eq. 9).

In addition to this rudimentary personalization, i.e. price discrimination according to consumer groups, the model includes a time dependency of electricity prices. Today, two models are common for small consumers in Germany: A flat tariff with a base price and one price for electricity consumption that is independent of the point in time at which energy is consumed. The other one has three components: The base rate, one price for consumption during the day, and another, lower one for consumption during the night. This tariff is especially suited for customers with night storage heaters that are able to store heat generated during the night – i.e. during off-peak hours – and emit it during the day when it is usually needed.

Such time dependent tariffs aim at establishing incentives for consumers to shift load to off-peak hours. If prices are calculated correctly, both sides profit from such a change in behavior: The customer is charged less for the same amount of energy, the supplier is able to make better use of its cheaper base load plants and to reduce expensive peak load generation or purchases.

The objective of our model is to support suppliers that are willing to introduce time-dependent and personalized tariff models. In this case, given the complexity of a perfect price discrimination, we recur to the segmentation into user classes and time slots. The tariffs generated by our approach partition the year into 54 different time slots: The year consists of six seasons or two-month intervals. Inside each interval, we distinguish between working days and weekends or holidays. A working day has five slots, a weekend day has four. The set of time slots is denoted by T. If one of its elements $t \in T$ is used as a subscript in a variable (e.g. b_t), the variable refers to all periods that match the description of the time slot t. The total number of hours in slot t is denoted by $|t|$. The influence of such time dependent tariffs has already been evaluated and tested in the field, as the publications for instance by Morovic et al. (1998) and Eßer et al. (2006) show.

2.1 Modeling the consumer behavior

In the household sector of the German energy market, the visible consequences of liberalization are very limited. Since the liberalization, prices have dropped – at first that is – but only a very small share of consumers have ventured to switch their electricity supplier. According to the general belief this is mainly due to two factors: First, electric energy is so omnipresent that it is rarely perceived as a good whose consumption generates costs. For most customers, the only occasion when the cost of energy is noticed is the arrival of the annual energy bill. The second major problem is the inconvenience of switching one's supplier that comes from two sources. First, the switch from one supplier to another is only possible with long delays due to legal restrictions. Furthermore, it is a common fear that, if some problem arises during the switch, the home's power will be cut, and that, maybe, quality of service is worse than with the old, usually local supplier. Both fears are, of course, irrational, but cannot be ignored for the consumer model.

This is the situation that strongly influences the model for consumer behavior: First, there is a negligible elasticity in everyday use. As a consequence, even if the prices are temporally differentiated, it is still safe to assume that the standard load profiles published by Tiedemann and Fünfgeld (2001) can be used to estimate the real load curves. This is very important since there are virtually no real detailed consumption data: the consumption of a single household is usually metered once per year in Germany; data with a high temporal resolution usually refer only to whole areas. The monetary criterion for the consumer's choice is the total energy bill for one year as given in Eq. 5.

Second, consumer behavior is quite inert, many consumers will simply not switch their supplier, even if they can realize savings. It is for these reasons that we chose the attraction model proposed by Lilien et al. (1992). Although it is well adapted to the situation in the energy market – inert consumers that make their choice based on one or several numerical influence factors – it has not yet been used and tested for its consequences in such a setting.

For the market described here, we have considered only one attraction parameter for the analysis, the total energy bill as detailed above, since we wanted to analyze the influence of tariffs on the market in absence of other factors.

The market share that a given tariff j can achieve in consumer group i compared with other tariffs $j' \in J$ is defined as

$$s_{ij} = \frac{(p_{ij}^{\text{total}})^{-a_j}}{\sum_{j' \in J} (p_{ij'}^{\text{total}})^{-a_{j'}}} \tag{1}$$

Multiplying s_{ij} with N_i gives n_{ij}, the number of customers of type i choosing tariff j (Eq. 4). With this, the total revenue r_h can be computed according to Eq. 6 for supplier h.

Contrary to many models, demand is not given exogenously for this optimization problem. Instead, the demand function is endogenized into the model in a fashion similar to the method proposed by Geyer-Schulz (1997).

2.2 Generation

The generation-related aspects of the model are taken from a synthetic power plant mix developed by Tietze-Stöckinger et al. (2004) that is representative of a big supplier in Germany. The model provides generation costs and capacities for a given projected demand. We approximated the cost function subject to changing demand volumes by repeatedly sampling points on it with runs of the model and obtained a linear cost function by interpolating these points.

The output of all power plants controlled by supplier h in period t is summed up in the variable b_{ht}, its upper limit is denoted by $b_{ht}^{\max}$. The approximated cost function is given by $c_{ht}(b_{ht})$ – the time dependency is motivated by the fact that not all power plants are online at all times but have to be switched off for example for maintenance usually once per year. The cost function reflects both the actual generation cost as well as the costs incurred for transporting the energy to the customer using transport networks.

The fixed costs for connecting a consumer of type i to the local distribution network charged by the local utility are given as c_i^{connect}.

2.3 Market activities

In addition to their own generation capacities, suppliers have access to an energy market – for instance the EEX in Leipzig – where they can sell excess production or buy electricity they are either not able or willing to produce themselves. The price for one kWh of electricity delivered at time t is p_t^{spot}. The volume traded by supplier h at time t is denoted by m_{ht}. If m_{ht} is negative, the supplier sells energy, if it is positive, energy is bought. Thus the total contribution of the energy trade to the objective function is

$$-\sum_{t \in T} p_t^{\text{spot}} m_{ht} \,. \tag{2}$$

The prices were taken from the once-public price database of the EEX (2003) from which we extracted the prices for 2003.

2.4 Taxes

The energy consumption at the consumers' site is subject to two kinds of taxes. The first kind, given by d^{total}, applies to the total sales volume, as does e.g. value added tax (VAT). Other taxes, like the German energy tax, are only levied on the actual consumption part. This tax rate is denoted by d_i^{var} and may depend on the consumer type. In total, the taxes amount to

$$\frac{d^{\text{total}} r_h}{1-d^{\text{total}}} + \sum_{i\in I} \frac{d_i^{\text{var}} \sum_{j\in J}\sum_{t\in T} n_{ij} x_{it} p_{jt}}{1+d_i^{\text{var}}}. \tag{3}$$

2.5 The optimization problem

The optimization problem is as follows:

$$\max \text{profit} = r_h - \sum_{t\in T} p_t^{\text{spot}} m_{ht} - \sum_{t\in T} c_{ht}(b_{ht})$$
$$- \left(\frac{d^{\text{total}} r_h}{1-d^{\text{total}}} + \sum_{i\in I} \frac{d_i^{\text{var}} \sum_{j\in J}\sum_{t\in T} n_{ij} x_{it} p_{jt}}{1+d_i^{\text{var}}} \right) - \sum_{i\in I}\sum_{j\in J} n_{ij} c_i^{\text{connect}}$$

$$\text{s.t. } n_{ij} = \frac{(p_{ij}^{\text{total}})^{-a_j}}{\sum_{j'\in J}(p_{ij}^{\text{total}})^{-a_{j'}}} N_i \tag{4}$$

$$p_{ij}^{\text{total}} = \sum_{t\in T} p_{jt} x_{it} \tag{5}$$

$$r_h = \sum_{j\subset J}\sum_{i\in I} n_{ij} p_{ij}^{\text{total}} \tag{6}$$

$$\forall t \in T : \sum_{i\in I}\sum_{j\in J} n_{ij} x_{ijt} = b_{ht} + m_{ht} \tag{7}$$

$$\forall t \in T : b_{ht} \le b_{ht}^{\max} \tag{8}$$

$$\forall i \in I, j \in J^h \backslash \{j^i\} : p_{ij^i}^{\text{total}} < p_{ij}^{\text{total}} \tag{9}$$

$$\forall t \in T : b_{ht} \ge 0 \tag{10}$$

$$\forall j \in J, t \in T : p_{jt} \ge 0 \tag{11}$$

$$\forall j \in J^h : p_j^{\text{base}} \ge 0 \tag{12}$$

The components of the objective function as well as the meaning of Eqs. 4 to 6 and 9 have already been introduced in the corresponding sections. Eq. 7 guarantees that the demand of all customers is covered either by energy generation or purchases, respecting the constraint that the total energy production cannot be higher than the maximum generation capacity in the respective time slots (Eq. 8). Finally, prices and generation should be nonnegative (Eqs. 10-12). Storage power stations are not considered in the scope of this model.

3 Simulation and first results

To analyze the reverberations that this pricing model might have on an energy market we simulated a hypothetic oligopoly market with four big suppliers. Since the optimization problem is nonlinear, we used the java implementation of the gags framework for genetic algorithms by Merelo and Prieto (1996). It is of special interest which consequences the choice of the attraction model's inertia parameter a has on the development of this market. In order to isolate these effects as well as possible, we assumed that four identical suppliers exist, resulting in a symmetric market, each of them using the model with a as sole influence factor for the attraction model and the same parameter set.

These four suppliers take turns in creating and publishing new tariffs. As a primary measure for the character of the market we chose the profit that a supplier expects when optimizing the tariff menu. Five experiments were conducted for each value of a between zero and 45. An experiment consists of 500 rounds; in each round, every supplier can generate and evaluate 10,000 tariff menu generations with the genetic algorithm, and publish the best tariff.

The simulation results plotted in Fig. 1 show three distinctive zones.

When $a = 0$, there is absolutely no reaction from the customers' part – all suppliers get a market share of 25%, which allows them to increase prices infinitely or at least to the limits set by a regulator.

When a is between one and about 15 to 20, profits are stable with little variation; the market is in an equilibrium point. Obviously, this point is easily found by the genetic algorithm.

Interesting is the third zone, starting around $a = 15$ to 20, where volatility and average profits increase considerably. In order to investigate the events in

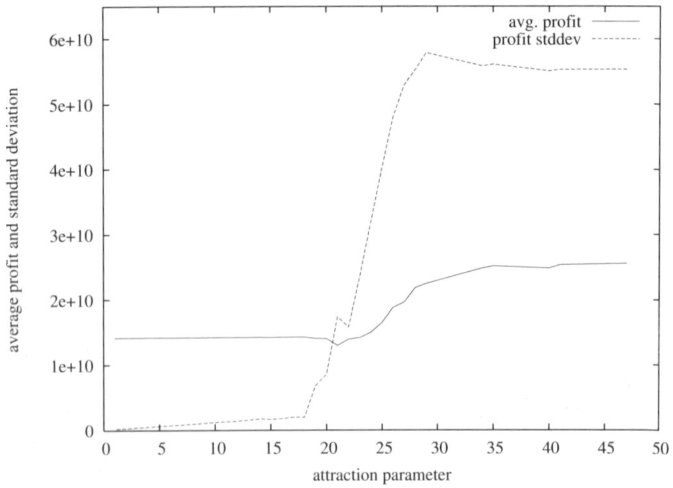

Fig. 1. Average profits and standard deviation

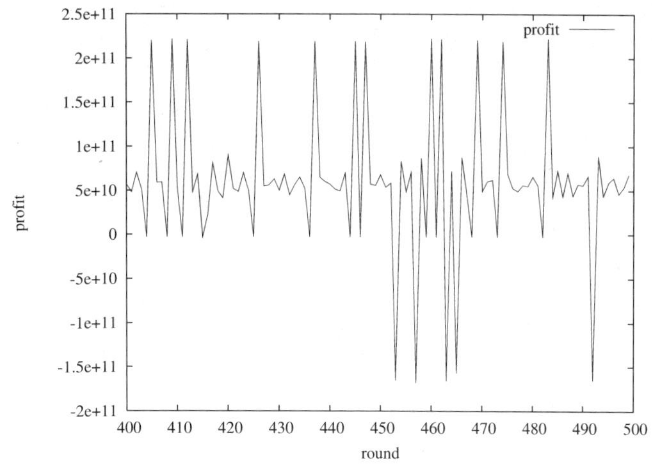

Fig. 2. Profit for supplier 0, $a = 25$

this zone a bit closer, consider Fig. 2 for the profit of the first supplier during the last 100 rounds when the attraction parameter $a = 25$ is used.

The volatility indicated by the overall standard deviation is visible in this plot: With growing consumer flexibility, it obviously becomes harder to find a sensible or even an admissible solution. This leads to the volatility observable in the figures. This is a typical effect in dynamic systems: Small deviations from optimal strategies – as they may happen with genetic algorithms – lead to huge distortions in the overall outcome.

In this specific setting, the average sum over all profits is higher if one of the suppliers publishes a tariff that does not find any customers (the rounds with zero profit in Fig. 2). This is to be expected, since this event effectively increases the attractivity of the remaining three players. These results show that the market becomes inherently instable when consumer inertia is low. Note that there is practically no trend over the rounds shown here, instead, the profit level oscillates around the same level.

The negative peaks in the profit plot are an indicator for the genetic algorithm's difficulties – these mark penalty costs for not finding a solution respecting the self selection constraint in Eq. 9. These peaks become more and more frequent as a increases.

4 Conclusion and outlook

We have presented a model for the generation of optimal tariff menus for energy suppliers that especially allows for the modeling of customer reactions. The results of the simulation show that stable markets can only be expected if consumer inertia is high. If, on the contrary, consumers react sensitively to price changes, it becomes more and more difficult to keep the market in a stable state. This, at the same time, shows the importance of reliable estimations for the consumers' demand elasticity used for the attraction model. If the inertia is underestimated by all suppliers, the tariffs may lead to the oscillations just because the suppliers anticipate such a consumer reaction.

Future work will be dedicated to a more detailed quantitative analysis of the situation and its causes. It will be interesting to experimentally replace the attraction model with some other, less volatile variants. For example, a temporal dampening factor might help alleviate the effects. Alternatively, the switching behavior could be modeled using Markov chains.

Additionally, it will be interesting to further explore the parameter space for higher values of a. We conjecture that there might be a zone in which suppliers are forced to iteratively lower prices to maintain their customer base, the classical race to the bottom that can often be found as a warning example for detrimental pricing strategies, for instance in Simon's book (1992).

Acknowledgment

We gratefully acknowledge the funding of the project "SESAM" by the Bundesministerium für Bildung und Forschung (BMBF).

References

EßER, A., FRANKE, M., KAMPER, A., MÖST, D. and RENTZ, O. (2006): *Analyse der Auswirkungen von Strompreissignalen auf das Lastverhalten ausgewählter Verbraucher.* Technical Report, Universität Karlsruhe (TH).

EUROPEAN ENERGY EXCHANGE (2003): *Marktdaten der EEX für 2003*, Leipzig.

GEYER-SCHULZ, A. (1997): Learning Strategies for Managing New and Innovative Products. In: R. Klar and O. Opitz (Eds.): *Classification and Knowledge Organization*, Springer, Heidelberg, 262–269.

LILIEN, G.L., KOTLER, P. and MOORTHY, K.S. (1992): *Marketing Models.* Prentice-Hall, Upper Saddle River.

MERELO, J.J. and PRIETO, A. (1996): Gags, a Flexible Object-oriented Library for Evolutionary Computation. In: P. Isasi and D. Borrajo (Eds.): *MALFO96, Proceedings of the First International Workshop on Machine Learning, Forecasting and Optimization*, 99–105.

MOROVIC, T., PILHAR, R. and MÖHRING-HÜSER, W. (1998): *Dynamische Stromtarife und Lastmanagement – Erfahrungen und Perspektiven.* Technical Report, Forschungsgesellschaft für umweltschonende Energieumwandlung und -nutzung mbH, Kiel.

SIMON, H. (1992): *Preismanagement: Analyse, Strategie, Umsetzung.* Gabler, Wiesbaden.

TIEDEMANN, R. and FÜNFGELD, C. (2001): *Die Repräsentativen VDEW-Lastprofile – Der Fahrplan.* Technical Report, TU Cottbus.

TIETZE-STÖCKINGER, I., FICHTNER, W. and RENTZ, O. (2004): *Kurzbeschreibung PERSEUS-MUM und Ergebnisse.* Technical Report, Universität Karlsruhe (TH).

Disproportionate Samples in Hierarchical Bayes CBC Analysis

Sebastian Fuchs and Manfred Schwaiger

Institute for Market-based Management (IMM), Munich School of Management, Ludwig-Maximilians-Universität, D-80539 Munich, Germany;
{fuchs, schwaiger}@bwl.uni-muenchen.de

Abstract. Empirical surveys frequently make use of conjoint data records, where respondents can be split up into segments of different size. A lack of knowledge how to handle such random samples when using Hierarchical Bayes-regression gave cause to a more detailed observation of the preciseness of estimation results. The study on hand comprises a survey on the effects of disproportionate random samples on the calculation of part-worths in choice-based conjoint analyses. An explorative simulation using artificial data demonstrates that disproportionate segment sizes have mostly negative effects on the goodness of part-worth estimation when applying Hierarchical Bayes-regression. These effects vary depending on the degree of disproportion. This finding could be generated due to the introduction of a quality criterion designed to compare both true and estimated part-worths, which is applied on a flexible range of sample structure. Subsequent to the simulation, recommendations will be issued how to best handle disproportionate data samples.

1 Introduction

In order to picture all different characters of a target group, surveying the total population would need to be aspired. This is not practical for financial, organizational and time reasons alone (cp. Berekoven et al. (2004, p. 51)). However, since people are similar in various matters, they can be split up into groups, e.g. by means of cluster analysis.

Given a satisfactory clustering result those segments are homogenous within but heterogeneous between segments with respect to the attributes determining the cluster structure (cp. Malhotra and Birks (2005) or Opitz (1980)). Considering preferences for example, clustering into segments provides the opportunity to better extract utilities assigned to products and services than calculating them for the entire population without segmentation. Prerequisite for this is that segment sizes of the random sample proportionally correspond to those in the parent population. As long as this is not given, the sample is said to be of a disproportionate structure (cp. Berekoven et al. (2004, p. 54)). The term of disproportion, as it is applied in the following, describes the fact that segments of a random sample are **unequally sized**, no matter if this disproportion is justified in the sense of representativeness or not. To point out this important fact again: This is a fact appearing in almost every empirical sample. We examine, how estimation-methods perform when a sample shows a more or less disproportionate structure. This is getting especially interesting when researchers try to estimate individual data (like utilities) from disproportionate samples. To be able to generate individual part-worths, advanced users prefer the Hierarchical Bayes-method in choice-based conjoint analysis. Via an iterative process part-worths are estimated using the so called "Gibbs sampler" (cp. Sawtooth (2003, p. 8)) belonging to the Markov-Chain-Monte-Carlo algorithms (cp. e.g. Chib and Greenberg (1996, p. 410) or Gelman et al. (1995, pp. 320-321)). This algorithm is able to generate quasi-individual, thus personal and non-group-related part-worths, among others even for attribute levels not evaluated by the corresponding person. Choice-based conjoint analysis is a method that shows only a fragment of all combinatorial possible stimuli to one interviewee. The respondent does not evaluate enough constellations of attribute-levels to let a researcher calculate her or his exact preferences regarding the particular characteristics of the stimuli.

To fill up information gaps regarding one single individual the HB-algorithm takes parameters from the whole sample **assuming normal distribution concerning preferences** (cp. Teichert (2001, p. 190ff.)). In this context the main question of this study arises: Does the HB-method actually fulfil its purpose and generate correct part-worths given a random sample is disproportionate in structure? In detail this question means: What happens if the assumption of normal distribution of preferences has to be abandoned regarding a sample at hand? Which samples are qualified to extract quasi-individual data, according to the fact that there are big and small segments?

2 Simulation design

2.1 Test setup

The problem on hand requires a quality criterion which is able to identify how exactly the HB-regression works in regard to different data records (using the software-tool Sawtooth CBC/HB^{TM}. Consequently, the setup of an adequate data record, serving these particular purposes, is essential. The criterion used in this study, aPWD, is the City Block distance (cp. Lehmann et al. (1998) or Opitz (1980)) between estimated and true part-worths in the sample. Hence, smaller values indicate better fit. In order to calculate aPWD we have to create artificial data based on given part-worths and compare those with the ones being calculated by the HB-regression. aPWD was preferred to popular goodness-of-fit measures because it is able to avail the exceptional fact that the exact utilities (forming the data basis) are known. As opposed to other approaches, we not only show the frequencies of non-appropriate estimations, in addition we now can measure their absolute deviance.

We define:

$$\mathrm{aPWD}_c = \frac{\sum_{k=1}^{K}\sum_{q_c=1}^{Q_c} \left|\mathrm{PW}_{\mathrm{kq_c[HB]}} - \mathrm{PW}_{\mathrm{kq_c[true]}}\right|}{Q_c \cdot K} \quad \text{with } \mathrm{c} \in [1;\mathrm{C}] \tag{1}$$

where aPWD_c is the average part-worth deviation in the segment of interviewees c, $\mathrm{PW}_{kq_c[\mathrm{HB}]}$ are the part-worths for attribute level k ($k = 1, \ldots, K$) and respondent q_c in segment c ($q_c = 1, \ldots, Q_c$) calculated by means of the HB-algorithm, $\mathrm{PW}_{kq_c[\mathrm{true}]}$ are the true part-worths used to create the artificial choice-dependent data, and C is describing the number of distinguishable segments of respondents.

From the structure of the quality criterion it can be concluded that the observation of main effects is being focused on. It evaluates deviations between part-worths of separate feature characteristics, not deviations between combinations of feature characteristics (interaction effects). For the test runs, artificial choice-dependent data in various forms is fed in the CBC/HB. The sample structure is varied in diverse manners (see subsection 2.4). This way, it can be shown how the sample structure influences the quality of estimation.

2.2 Assumptions

If one segment is very dominantly represented, the estimation process consists of particularly many similar individuals for this group of interviewees. The presumption stands to reason that the amount of information is larger here, compared to more weakly represented segments. This might result in minor estimation errors for the disproportionately large segment. The question arises whether a threshold exists, at which the disproportion spoils individual part-worth estimation. What happens, if two segments are over-represented? What if there exists only one single minority segment? Assuming that the results for disproportionate data records are relatively equal to those of proportionate, is there a compensation between large and small segment in terms of goodness of fit? In this case it needs to be clarified if large, small or medium-sized segments benefit from disproportion.

2.3 Design

In order to correctly answer the questions posed above, a test design is required, which excludes disturbing influences from the survey. For reasons of external validity, the design has to resemble typical studies employed in the practice of market research. The data result from fictitious choice tasks evaluating four stimuli with four attributes having four levels each. Every fictitious individual passes four random-choice tasks including a so called "none"-option (cp. Sawtooth (1999, p. 7)). Only first-choice and main-effects are observed. According to Johnson and Orme (1996, pp. 5-), this design displays a classic choice-data record without harmful characteristics concerning validity. Based on test runs with Sawtooth, the optimal number of interviewees is 40. Hence, 40 fictitious people are sufficient for the mutual alignment of all attribute characteristics and to setup a nearly perfect orthogonal design. The efficiency coefficient used by Sawtooth, showing the proximity towards a "perfect design" in the interval $[0;1]$, was calculated close to the maximum (≥ 0.991). Perfect design in this case means, that a maximum of combinatorial possible constellations of attribute-levels are evaluated by a minimum of respondents.

2.4 Data generation

In the experience of Orme (1998, p. 8), a data record consisting of several analysable customer segments should at least include 200 individuals per segment. Regarding four segments leads to 800 cases. Under the condition of 40 interviewees being essential for an efficient design, a subdivision of the respective segments in groups of 40 interviewees each is generated. Consequently, a segment of 200 interviewees consists of five times 40 persons. Hence, each questioning setting in the calculation is considered five times. Due to this proceeding, it is possible to admit the simulation of even extreme variations of data records which depart from a postulated structure. Despite this, the efficiency of the choice design does not suffer from these variations.

To make all segments distinguishable from one another, it is implied that all fictitious individuals of a segment do have quite similar utility structures. This means that, for instance, people of segment 1 show part-worths within the interval $[-5;-2.5]$ each, in regard to all four attributes (A to D) and attribute level 1 (A1, B1, C1, D1). For all four attribute levels 2 (A2, B2, C2, D2) they feature part-worths $\in [-2.5;0]$, for levels 3 (A3, B3, C3, D3) $\in [0;2.5]$ and concerning level 4 $\in [2.5;5]$. All part-worths stem from uniformly distributed random numbers of respectively given intervals. All intervals are of equal size, so that relative importance (the range of the part-worths, cp. Malhotra and Birks (2005)) of all characteristics is kept equal as well. Between the segments there is no overlapping of preferences to keep away from structures being alike normal distributed.

2.5 Versions of analysis

Based on so-called pseudo-OLS estimation (cp. Sawtooth (2003, p. 3)) 81 calculation cycles plus pre-tests were conducted. Simulating diverse intensities of disproportion, numerous data records were generated. Under the conditions listed in section 2.4 this happens by reducing segments in steps of 40 persons and increasing other segments the same way. Thereby the samples of 800 individuals remain comparable and the efficiency of the choice design persists.

Data sets containing one large and three small segments, two large and two small segments, four segments declining in size, and data records containing merely one minority segment. Besides processing a data record as a whole, the segments were also estimated separately. In addition, combinations of separate and common estimations were conducted. Further versions have been made changing the setting of prior variance (1.0 and 2.0). It determines the acceptance of the preconditioned variance-covariance-matrix (cp. Orme (2003), pp. 3-4). The higher the value, the more strictly the estimation sticks to individual data.

3 Simulation results

3.1 Explored tendencies

Basically, five tendencies can be observed for the process:

1. In general, the estimation deteriorates with an increasing over-dimensionality of particular segments. This is the case for tests including just one group of interviewees being disproportionately large, as well as for tests showing two or three "oversized" segments. Figures 1 and 2 chart the goodness of estimation in form of the inverse quality criterion aPWD for the whole sample each. However, this tendency is not valid for the case

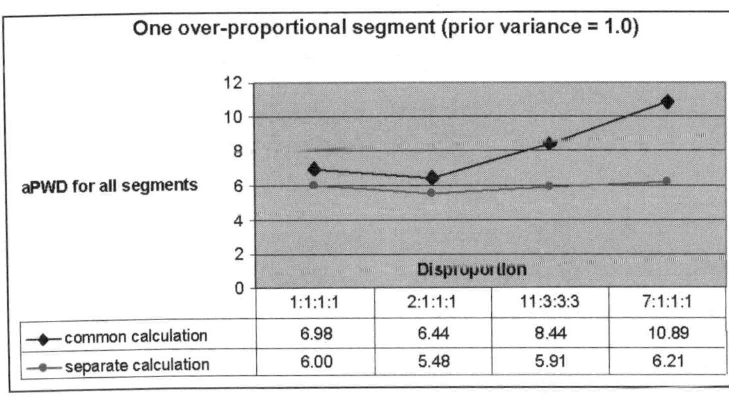

Fig. 1. aPWD for one disproportionately large segment (prior variance = 1.0)

regarding a disproportion with one disproportionately large segment in the proportion of 2 : 1 : 1 : 1. The same is valid with two disproportionately large segments within the degree of 3 : 3 : 2 : 2 and 7 : 7 : 3 : 3, while employing a prior variance of 1.0. Concerning this, a non-explainable, marginal improvement of the overall estimation occurs given a common estimation.

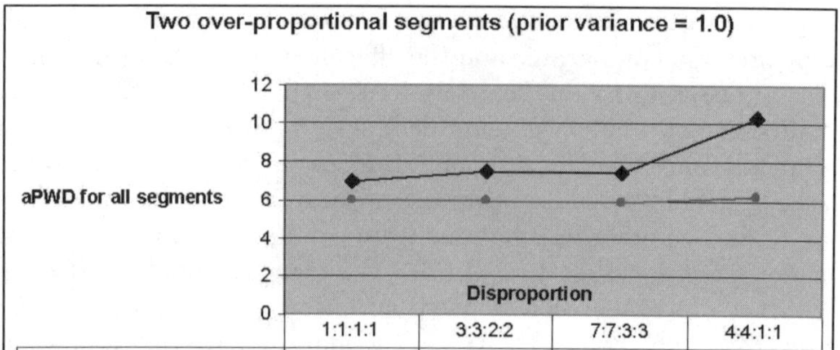

	1:1:1:1	3:3:2:2	7:7:3:3	4:4:1:1
common calculation	6.98	7.46	7.45	10.32
separate calculation	6.00	5.97	5.90	6.21

Fig. 2. aPWD for two disproportionately large segments (prior variance = 1.0)

2. In regard to samples containing two "larger" segments, a de-concentration seems to be advantageous for the quality of an estimation process. The probable reason for this is that distributing over-representation among two segments leads up to better results because it slightly better fits the HB's assumption of normal distribution.

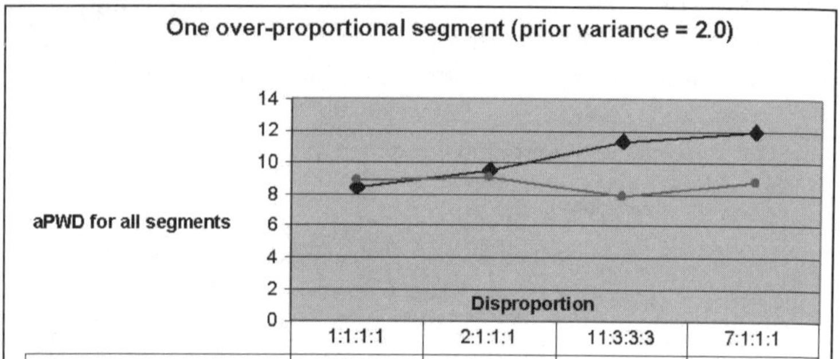

	1:1:1:1	2:1:1:1	11:3:3:3	7:1:1:1
common calculation	8.42	9.51	11.38	11.99
separate calculation	8.91	9.07	7.85	8.75

Fig. 3. aPWD for one disproportionately large segment (prior variance = 2.0)

3. Globally seen, better results are achieved if segments undergo the HB-algorithm separately. However, given the employment of higher variances, this is solely valid for disproportionately structured samples, as it can be seen in figures 3 and 4.
4. If figures 3 and 4 are compared with the results in figures 1 and 2, it becomes apparent, how the prior variance influences the quality of the estimations. The higher the variance a priori installed, the worse the estimation of the process turns out to be. Little information is gained per person on the respective behaviour towards all attribute characteristics.

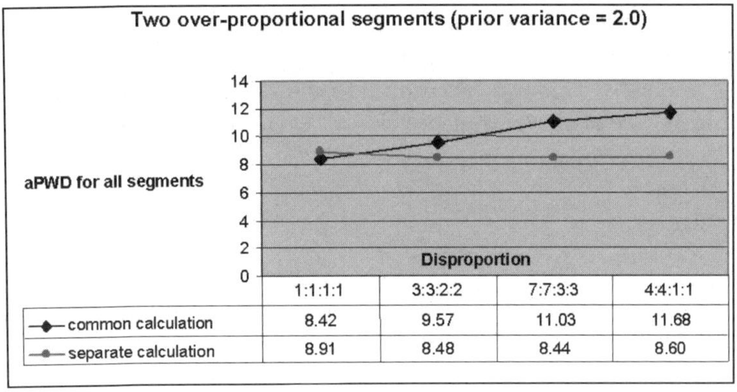

Fig. 4. aPWD for two disproportionately large segments (prior variance = 2.0)

This forces the HB-algorithm to fall back on information gained from the whole sample. Using even higher prior variances worsens outcomes without any exemption, as further tests have shown.

5. Given a common estimation, disproportionately large segments showed the worst performance at any time. It further can be observed that, with an increasing disproportion, the estimation of the minority segments got better constantly. Test runs using data records that merely contain one single minority group do achieve poor results as well, given a common estimation of all segments. This is being verified by test samples, whose segment sizes decline from one segment to the next.

3.2 Conclusion and recommendations

The main-challenge of the HB-regression in choice-based conjoint-analysis could be observed well in this simulation. It is able to generate quasi-individual data to fill information gaps. But the success in doing so depends on the adequateness of the data record. However, experiences concerning this method are needed in order to be capable to handle extreme cases, as being observed here. Hence, the HB-process is not to be seen as a panacea for the estimation of individual preferences, as it is stated by numerous sources (cp. e.g. Johnson and Orme (1996) or Orme (1998)). Based on our simulation results, some general recommendations can be made:

- Increasing disproportion leads to a deterioration in quality of the estimation. Segments can be calculated in common without probl2ems, if dissimilarity in between them is low. However, if strong disproportion is present – e.g., if one segment is twice as large as one of the other segments – a separate calculation of segments should be preferred.
- The best results concerning disproportionate samples can be obtained, if all segments on hand are being calculated in common at first. The

results of minority segments should be reused as opposed to the results of disproportionately large segments. They should be calculated separately in a second step. This way, according to the test runs on hand, results for all segments are achieved that are closest to the true part-worths underlying.

References

BEREKOVEN L., ECKERT W. and ELLENRIEDER, P. (2004): *Marktforschung. Methodische Grundlagen und Praktische Anwendung.* 10. Auflage, Gabler, Wiesbaden.

CHIB, P. and GREENBERG, E. (1996): Markov Chain Monte Carlo Simulation Methods in Econometrics. *Econometric Theory, 12, 409–431.*

GELMAN, A., CARLIN, J., STERN H. and RUBIN, D. (1995): *Bayesian Data Analysis.* Chapman & Hall, London.

JOHNSON, R. and ORME, B. (1996): *How Many Questions Should You Ask in Choice-Based Conjoint Studies?* Sawtooth Software Research Paper Series, Sawtooth Software Inc., Conference Proceedings of the ART Forum, Beaver Creek.

LEHMANN, D., GUPTA, S. and STECKEL, J. (1998): *Marketing Research.* Addison-Wesley, Reading.

MALHOTRA, N. and BIRKS, D. (2005): *Marketing Research.* Prentice Hall, Harlow.

OPITZ, O. (1980): *Numerische Taxonomie.* UTB, Stuttgart.

ORME, B. (1998): *Sample Size Issues for Conjoint Analysis Studies, Sawtooth Software Research Paper Series.* Sawtooth Software Inc., Sequim, Washington.

ORME, B. (2003): *New Advances Shed Light on HB Anomalies.* Sawtooth Software Research Paper Series, Sawtooth Software Inc., Sequim, Washington.

SAWTOOTH SOFTWARE INC. (1999): *Choice-based Conjoint (CBC).* Technical Paper, Sawtooth Software Technical Paper Series, Sequim, Washington.

SAWTOOTH SOFTWARE INC. (2003): *CBC Hierarchical Bayes Analysis.* Technical Paper (Version 3.1), Sawtooth Software Technical Paper Series, Sequim, Washington.

TEICHERT, T. (2001): *Nutzenschätzung in Conjoint-Analysen. Theoretische Fundierung und empirische Aussagekraft.* Dt. Univ.-Verl., Gabler, Wiesbaden.

Building on the Arules Infrastructure for Analyzing Transaction Data with R

Michael Hahsler[1] and Kurt Hornik[2]

[1] Department of Information Systems and Operations, Wirtschaftsuniversität, A-1090 Wien, Austria; hahsler@wu-wien.ac.at

[2] Department of Statistics and Mathematics, Wirtschaftsuniversität, A-1090 Wien, Austria; kurt.hornik@wu-wien.ac.at

Abstract. The free and extensible statistical computing environment R with its enormous number of extension packages already provides many state-of-the-art techniques for data analysis. Support for association rule mining, a popular exploratory method which can be used, among other purposes, for uncovering cross-selling opportunities in *market baskets,* has become available recently with the R extension package **arules**. After a brief introduction to transaction data and association rules, we present the formal framework implemented in **arules** and demonstrate how clustering and association rule mining can be applied together using a market basket data set from a typical retailer. This paper shows that implementing a basic infrastructure with formal classes in R provides an extensible basis which can very efficiently be employed for developing new applications (such as clustering transactions) in addition to association rule mining.

1 Introduction

An aim of analyzing transaction data is to discover interesting patterns (e.g., association rules) in large databases containing transaction data. Transaction data can originate from various sources. For example, POS systems collect large quantities of records (i.e., transactions, *market baskets*) containing products purchased during a shopping trip. Analyzing market basket data is called *Market Basket Analysis* (Russell et al. (1997), Berry and Linoff (1997)) and is used to uncover unexploited selling opportunities. Table 1 depicts a simple example for transaction data. Formally, let $I = \{i_1, i_2, \ldots, i_n\}$ be a set of n binary attributes called *items*. Let $\mathcal{D} = \{t_1, t_2, \ldots, t_m\}$ be a set of transactions called the *database*. Each transaction in $\mathcal{D}$ has a unique transaction ID and contains a subset of the items in I.

Categorical and/or metric attributes from other data sources (e.g., in survey data) can be mapped to binary attributes and thus be treated in the same way as transaction data (Piatetsky-Shapiro (1991), Hastic et al. (2001)). Here interesting relationships between values of the attributes can be discovered.

Table 1. Example of market basket data represented as (a) shopping lists and as (b) a binary purchase incidence matrix where ones indicate that an item is contained in a transaction.

transaction ID	items
1	milk, bread
2	bread, butter
3	beer
4	milk, bread, butter
5	bread, butter

(a)

trans. ID	items			
	milk	bread	butter	beer
1	1	1	0	0
2	0	1	1	0
3	0	0	0	1
4	1	1	1	0
5	0	1	1	0

(b)

Agrawal et al. (1993) stated the problem of mining association rules from transaction data (e.g., database of market baskets) as follows:

A *rule* is defined as an implication of the form $X \Rightarrow Y$ where $X, Y \subseteq I$ and $X \cap Y = \emptyset$. The sets of items (for short *itemsets*) X and Y are called *antecedent* (left-hand-side or lhs) and *consequent* (right-hand-side or rhs) of the rule. To select interesting rules from the set of all possible rules, constraints on various measures of strength and interestingness can be used. The best-known constraints are minimum thresholds on *support* and *confidence*. The support $\mathrm{supp}(X)$ of an itemset X is defined as the proportion of transactions in the data set which contain the itemset. The confidence of a rule is defined $\mathrm{conf}(X \Rightarrow Y) = \mathrm{supp}(X \cup Y)/\mathrm{supp}(X)$. *Association rules* are required to satisfy both a minimum support and a minimum confidence constraint at the same time.

An infrastructure for mining transaction data for the free statistical computing environment R (R Development Team (2005)) is provided by the extension package **arules** (Hahsler et al. (2005, 2006)). In this paper we discuss this infrastructure and indicate how it can conveniently be enhanced (by providing functionality to compute proximities between transactions) to create application frameworks with new data analysis capabilities.

In Section 2 we give a very brief overview of the infrastructure provided by **arules** and discuss calculating similarities between transactions. In Section 3 we demonstrate how the **arules** infrastructure can be used in combination with clustering algorithms (as provided by a multitude of R extension packages) to group transactions representing similar purchasing behavior and then to discover association rules for interesting transaction groups. All necessary R code is provided in the paper.

2 Building on the arules infrastructure

The **arules** infrastructure implements the formal framework presented by the S4 class structure (Chambers (1998)) in Figure 1. For transaction data the

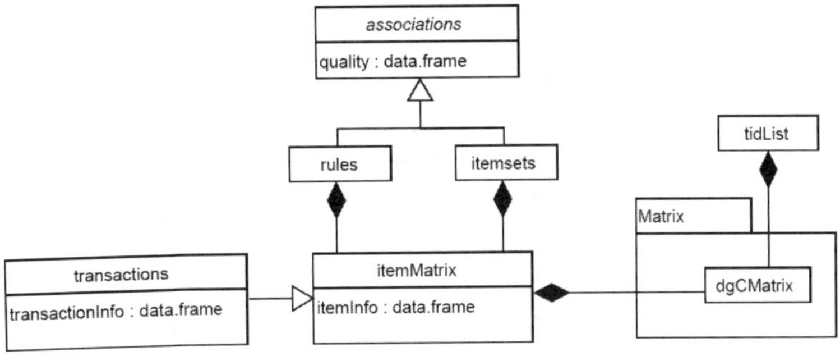

Fig. 1. Simplified UML class diagram (see Fowler (2004)) of the **arules** package.

classes transactions and tidLists (transaction ID lists, an alternative way to represent transaction data) are provided. When needed, data formats commonly used in R to represent transaction data (e.g., data frames and matrices) are automatically transformed into transactions.

For mining patterns, the popular mining algorithm implementations of Apriori and Eclat (Borgelt (2003)) are used in **arules**. Patterns are stored as itemsets and rules representing sets of itemsets or rules, respectively. Both classes directly extend a common virtual class called associations which provides a common interface. In this structure it is easy to add a new type of associations by adding a new class that extends associations.

For efficiently handling the typically very sparse transaction data, items in associations and transactions are implemented by the itemMatrix class which provides a facade for the sparse matrix implementation dgCMatrix from the R package **Matrix** (Bates and Maechler (2005)). To use sparse data (e.g. transactions or associations) for computations which need dense matrices or are implemented in packages which do not support sparse representations, transformation into dense matrices is provided. For all classes standard manipulation methods as well as special methods for analyzing the data are implemented. A full reference of the capabilities of **arules** can be found in the package documentation (Hahsler et al. (2005)).

With the availability of such a conceptual and computational infrastructure providing fundamental data structures and algorithms, powerful application frameworks can be developed by taking advantage of the vast data mining and analysis capabilities of R. Typically, this only requires providing some application-specific "glue" and customization. In what follows, we illustrate this approach for finding interesting groups of market baskets, one of the core value-adding tasks in the analysis of purchase decisions. As grouping (clustering) is based on notions of proximity (similarity or dissimilarity), the glue needed is functionality to compute proximities between transactions, which can be done by using the asymmetric Jaccard dissimilarity (Sneath (1957)) often used for binary data where only ones (corresponding to purchases in our application) carry information. Alternatively, more domain specific proximity measures like *Affinity* (Aggarwal et al. (2002)) are possible. Extensions to proximity measures for associations are straightforward.

In a recent version of **arules**, we added the necessary extensions for clustering. In the following example, we illustrate how conveniently this now allows for combining clustering and association rule mining.

3 Example: Clustering and mining retail data

We use 1 month (30 days) of real-world point-of-sale transaction data from a typical local grocery outlet for the example. The items are product categories (e.g., *popcorn*) instead of the individual brands. In the available 9835 transactions we found 169 different categories for which articles were purchased. The data set is included in package **arules** under the name `Groceries`. First, we load the package and the data set.

```
R> library("arules")
R> data("Groceries")
```

Suppose the store manager wants to promote the purchases of beef and thus is interested in associations which include the category *beef*. A direct approach would be to mine association rules on the complete data set and filter the rules which include this category. However, since the store manager knows that there are several different types of shopping behavior (e.g., small baskets at lunch time and rather large baskets on Fridays), we first cluster the transactions to find promising groups of transactions which represent similar shopping behavior. From the distance-based clustering algorithms available in R, we choose *partitioning around medoids* (PAM; Kaufman and Rousseeuw (1990)) which takes a dissimilarity matrix between the objects (transactions) and a predefined number of clusters (k) as inputs and returns cluster labels for the objects. PAM is similar to the well-known k-means algorithm, with the main differences that is uses medoids instead of centroids to represent cluster centers and that it works on arbitrary dissimilarity matrices. PAM is available in the recommended R extension package **cluster** (Maechler (2005)).

To keep the dissimilarity matrix at a manageable size, we take a sample of size 2000 from the transaction database (Note that we first set the random number generator's seed for reasons of reproducibility). For the sample, we calculate the dissimilarity matrix using the function `dissimilarity()` with the method `"Jaccard"` which implements the Jaccard dissimilarity. Both `dissimilarity()` and `sample()` are recent "glue" additions to **arules**. With this dissimilarity matrix and the number of clusters pre-set to $k = 8$ (using expert judgment by the store manager), we apply PAM to the sample.

```
R> set.seed(1234)
R> s <- sample(Groceries, 2000)
R> d <- dissimilarity(s, method = "Jaccard")
R> library("cluster")
```

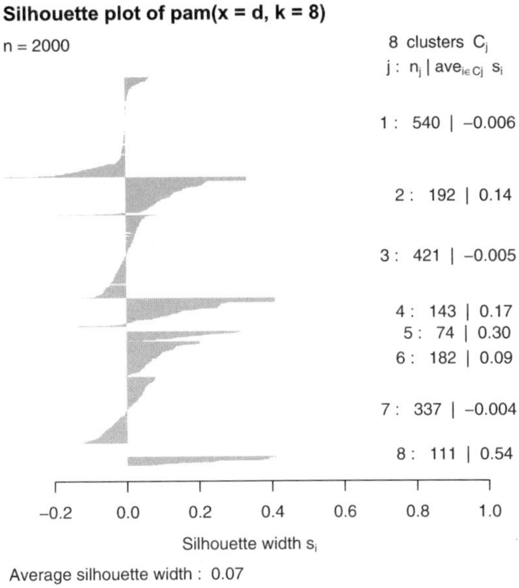

Fig. 2. Silhouette plot of the clustering.

```
R> clustering <- pam(d, k = 8)
R> plot(clustering)
```

Visualization is employed to assess the obtained clustering. The silhouette plot (Kaufman and Rousseeuw (1990)) in Figure 2 displays the silhouette widths for each object (transaction) ordered by cluster as horizontal bars. The silhouette width is a measure of how well an object belongs to its assigned cluster. Compact clusters exclusively consist of objects with high silhouette widths. In the plot we see, that cluster 8 is by far the most compact cluster. Several other clusters have objects with negative silhouette widths which indicates dispersed clusters. However, this is typical for clustering in high-dimensional space.

To predict labels for the whole data set based on the clustered sample, we use the nearest neighbor approach. Package **arules** provides now a new "glue" `predict()` method which can be used to find the labels for all transactions in the Groceries database given the cluster medoids. With labels for all transactions, we can generate a list of transaction data sets, one for each cluster.

```
R> allLabels <- predict(s[clustering$medoids], Groceries,
+     method = "Jaccard")
R> cluster <- split(Groceries, allLabels)
```

The transaction data set for each cluster can now be analyzed and used independently. We will demonstrate this by choosing two different clus-

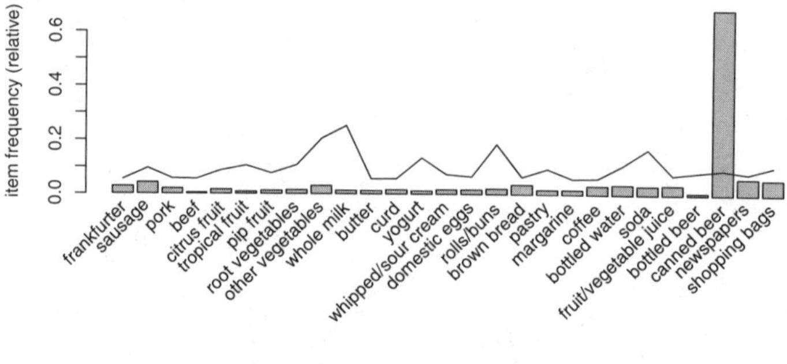

Fig. 3. Item frequencies in cluster 8.

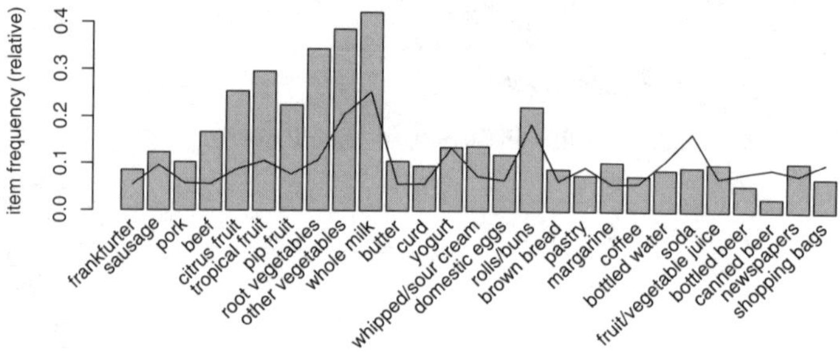

Fig. 4. Item frequencies in the cluster 3.

ters, clusters number 8 and 3. For visualization of the clusters, we use `itemFrequencyPlot()` from **arules** which produces a cluster profile where bars are used to represent the relative frequency of product categories in the cluster and a line is used for the relative frequency of the categories in the whole data set. Large differences between the data set and the cluster are interesting since they indicate strong cluster-specific behavior. For better visibility, we only show the categories with a support greater than 5%.

```
R> itemFrequencyPlot(cluster[[8]], population = s, support = 0.05)
R> itemFrequencyPlot(cluster[[3]], population = s, support = 0.05)
```

The cluster profile of the compact cluster 8 (Figure 3) shows a group of transactions which almost entirely consists of canned beer.

Cluster 3 consists of many transactions containing a large number of items. The cluster's profile in Figure 4 shows that almost all product categories are on average bought more often in the transactions in this group than in the whole data set. This cluster is interesting for association rule mining since the transactions contain many items and thus represent high sales volume.

As mentioned above, we suppose that the store manager is interested in promoting beef. Because beef is not present in cluster 8, we will concentrate on cluster 3 for mining association rules. We choose relatively small values for support and confidence and, in a second step, we filter only rules which have the product category beef in the right-hand-side.

```
R> rules <- apriori(cluster[[3]], parameter = list(support = 0.005,
+     confidence = 0.2), control = list(verbose = FALSE))
R> beefRules <- subset(rules, subset = rhs %in% "beef")
```

Now the store manager can use a wide array of methods provided by **arules** to analyze the found 181 rules. As an example, we show the 3 rules with the highest confidence values.

```
R> inspect(head(SORT(beefRules, by = "confidence"), n = 3))

  lhs                       rhs     support confidence  lift
1 {tropical fruit,
   root vegetables,
   whole milk,
   rolls/buns}          => {beef} 0.006189     0.3889 2.327
2 {tropical fruit,
   other vegetables,
   whole milk,
   rolls/buns}          => {beef} 0.006631     0.3846 2.302
3 {tropical fruit,
   root vegetables,
   other vegetables,
   rolls/buns}          => {beef} 0.005747     0.3824 2.288
```

4 Conclusion

In this contribution, we showed how the formal framework implemented in the R package **arules** can be extended for transaction clustering by simply providing methods to calculate proximities between transactions and corresponding nearest neighbor classifications. Analogously, proximities between associations can be defined and used for clustering itemsets or rules (Gupta et al. (1999)). Provided that item ontologies or transaction-level covariates are available, these can be employed for interpreting and validating obtained clusterings. In addition, stability of the "interesting" groups found can be assessed using resampling methods, as e.g. made available via the R extension package **clue** (Hornik (2005, 2006)).

References

AGGARWAL, C.C., PROCOPIUC, C.M. and YU, P.S. (2002): Finding Localized Associations in Market Basket Data. *Knowledge and Data Engineering, 14, 1, 51–62.*

AGRAWAL, R., IMIELINSKI, T. and SWAMI, A. (1993): Mining Association Rules Between Sets of Items in Large Databases. In: *Proceedings of the 1993 ACM SIGMOD International Conference on Management of Data.* ACM Press, 207–216.

BATES, D. and MAECHLER, M. (2005): *Matrix: A Matrix Package for R.* R package version 0.95-5.

BERRY, M. and LINOFF, G. (1997): *Data Mining Techniques for Marketing, Sales and Customer Support.* John Wiley & Sons.

BORGELT, C. (2003): Efficient Implementations of Apriori and Eclat. In: *FIMI'03: Proceedings of the IEEE ICDM Workshop on Frequent Itemset Mining Implementations.*

CHAMBERS, J.M. (1998): *Programming with Data.* Springer, New York.

FOWLER, M. (2004): *UML Distilled: A Brief Guide to the Standard Object Modeling Language.* Addison-Wesley Professional, third edition.

GUPTA, G.K., STREHL, A. and GHOSH, J. (1999): Distance Based Clustering of Association Rules. In: *Proceedings of the Artificial Neural Networks in Engineering Conference, 1999, St. Louis.* ASME, 9, 759–764.

HAHSLER, M., GRÜN, B. and HORNIK, K. (2005): arules – A Computational Environment for Mining Association Rules and Frequent Item Sets. *Journal of Statistical Software, 14, 15, 1–25.*

HAHSLER, M., GRÜN, B. and HORNIK, K. (2006): *arules: Mining Association Rules and Frequent Itemsets.* R package version 0.2-7.

HASTIE, T., TIBSHIRANI, R. and FRIEDMAN, J. (2001): *The Elements of Statistical Learning.* Springer, Berlin.

HORNIK, K. (2005): A CLUE for CLUster Ensembles. *Journal of Statistical Software, 14(12).*

HORNIK, K. (2006): *CLUE: CLUster Ensembles.* R package version 0.3-3.

KAUFMAN, L. and ROUSSEEUW, P. (1990): *Finding Groups in Data.* Wiley-Interscience Publication.

MAECHLER, M. (2005): *cluster: Cluster Analysis Extended Rousseeuw et al.* R package version 1.10.2.

PIATETSKY-SHAPIRO, G. (1991): Discovery, Analysis, and Presentation of Strong Rules. In: G. Piatetsky-Shapiro and W. J. Frawley (Eds.): *Knowledge Discovery in Databases*, AAAI/MIT Press, Cambridge, MA.

R DEVELOPMENT CORE TEAM (2005): *R: A Language and Environment for Statistical Computing.* R Foundation for Statistical Computing, Vienna, Austria. ISBN 3-900051-07-0.

RUSSELL, G.J., BELL, D., BODAPATI, A., BROWN, C.L., JOENGWEN, C., GAETH, G., GUPTA, S. and MANCHANDA, P. (1997): Perspectives on Multiple Category Choice. *Marketing Letters, 8, 3, 297–305.*

SNEATH, P.H. (1957): Some Thoughts on Bacterial Classification. *Journal of General Microbiology, 17, 184–200.*

Balanced Scorecard Simulator – A Tool for Stochastic Business Figures

Veit Köppen, Marina Allgeier and Hans-J. Lenz

Institute of Information Systems, Free University Berlin, D-14195 Berlin, Germany; {koeppen, lslenz}@wiwiss.fu-berlin.de, m.allgeier@yahoo.de

Abstract. Long term business success is highly dependent on how fast the business reacts on the changes in the market situation. Those who want to be successful need relevant, in-time and accurate information. *Balanced Scorecard Simulator* is a management tool that can be used efficiently in the processes of planning, decision and controlling. Based on the Balanced Scorecard concept the program combines imprecise data of business figures with forward and backward computation. It is also possible to find out whether or not the data are consistent with the BSC model. The visualization of the simulation results is done by a Kiviat diagram. The aim of the design is a software tool based on a BSC model and MCMC methods but is easy to handle.

1 Introduction

Hard competitive conditions continuously challenge enterprizes to improve the quality of information on which entrepreneurial decisions are based. For this reason management tools that support these processes play an important role. The Balanced Scorecard (BSC) is today a well-known and widely used concept from Kaplan and Norton (1996). It was successfully integrated in the management processes of many companies, because it introduces the performance from different economic and business perspectives and helps to understand and recognize the factors on which company prosperousness depends. BSC figures are regularly computed from retrospective information and are used to be compared with a corporate goal, to analyze the differences and to set new targets. The *Balanced Scorecard Simulator* is an easy-to-use management tool which supports the Markov Chain Monte Carlo (MCMC) simulation of Balanced Scorecard indicators. It also improves the forecast quality of business figures. The quality of the simulation results can be achieved through modeling stochastic indicators on the one hand and of their functional dependencies designed as an equation system on the other hand. It detects contradictions primarily in data which should fulfill a system of equations. The goal is to obtain realistic characteristic quantities which cannot be measured with hundred per cent precision (Friedag and Schmidt (2004)). Consequently, with the *Balanced Scorecard Simulator* complete and precise information about indicators which is based on non-contradictory data can be achieved and erroneous decisions avoided. The performance measurement system in the simulator is based on the Balanced Scorecard of Kaplan and Norton (1996). Some tools which are based on imprecise business figures already exist, for a possiblistic approach cf. Müller et al. (2003) and for a probabilistic approach cf. Schmid (1979) and Lenz and Rödel (1991). However the *Balanced Scorecard Simulator* is an absolutely novel approach that not only brings financial and non-financial indicators together in an equation system but also uses MCMC techniques as an appropriate instrument for the simulation.

2 Model

The MCMC simulation is based on a stochastic model of the selected BSC indicators, cf. Fig. 1.

The model is characterized by the following three features:

- 24 indicators (variables) and three system constants are assigned to the four BSC perspectives (employee and innovation, internal business process, customer and financial perspective). The indicators were selected following the procedures applied in an earlier case study (Todova and Ünsal (2005)).
- All variables of the model are connected to each other by the four basic arithmetical operations;
- 15 equations with up to three variables exist. Obviously, each equation can be uniquely solved for each existing variable ("separation principle").

3 Prior information

It is assumed that for some BSC indicators there exists prior information which refers to the respective probability distribution $\pi(x)$ of the corresponding variable x. Five classic types of distribution are considered:

- Normal distribution: $\pi(x) \sim N(\mu, \sigma^2)$
- Uniform distribution: $\pi(x) \sim U(a, b)$
- Triangular distribution: $\pi(x) \sim Tr(a, b, c)$
- Exponential distribution: $\pi(x) \sim Exp(\lambda)$
- Cauchy distribution: $\pi(x) \sim Ca(a, b)$

These distributions are implemented in the *Balanced Scorecard Simulator* because they are the most commonly used distributions. But other distributions can easily be implemented. At present, the specification of a distribution from

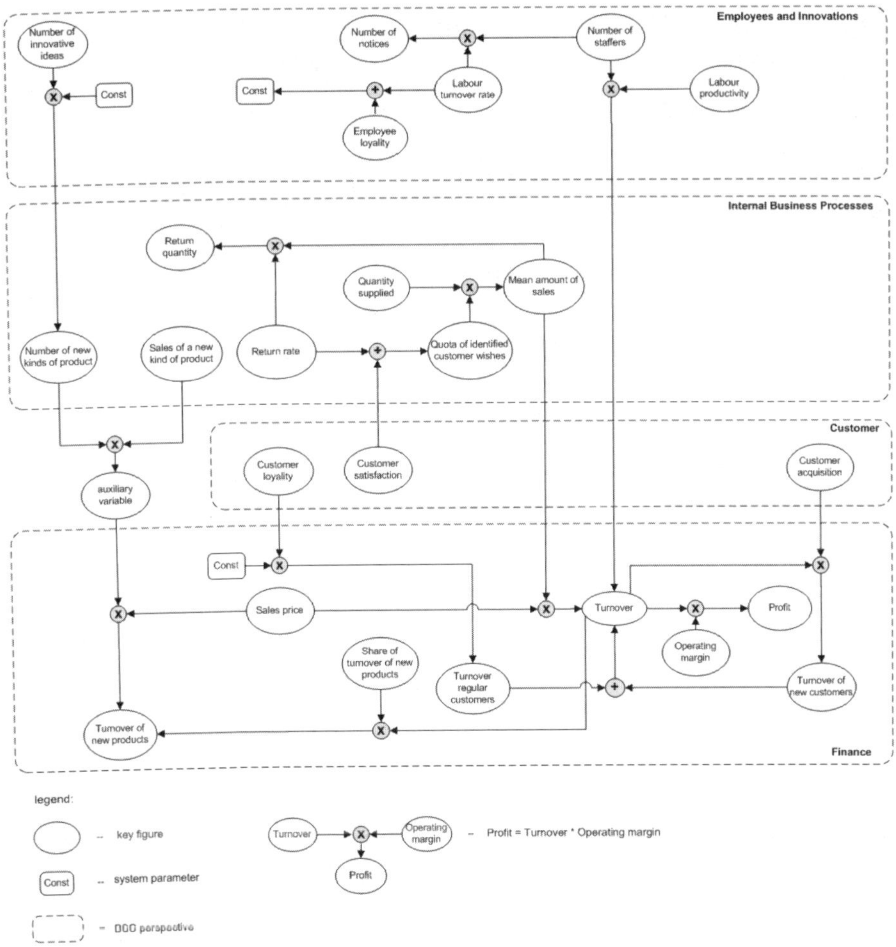

Fig. 1. Balanced scorecard model graph

data and the parameter estimation is not included in the software. All constants and target values of the variables are consequently regarded as known.

4 Simulation

The process of simulation is carried out in two steps. In the first step a sample is drawn from every known distribution of an indicator. An imputation of unknown variables is based on the whole equation system. The fusion of several samples for a variable that exists in more than one equation is carried out in the projection step. The results are tabulated as well as visualized in a Kiviat diagram.

4.1 Metropolis Hastings algorithm

The simulation of the BSC indicators is carried out using the Metropolis Hastings procedure. The Metropolis Hastings (MH) algorithm is a kind of Markov Chain Monte Carlo (MCMC) method. The MH procedure is suitable for the generation of random numbers of any arbitrary probability density function. The resulting sample is denoted by $x = \{x_1, \ldots, x_T\}$. The MH algorithm which is used in the *Balanced Scorecard Simulator* is described by the algorithm 1 (Köppen and Lenz (2005)).

Algorithm 1 MH Algorithm

Input: $\pi()$ – target function
 $q(\cdot,\cdot)$ – proposal distribution, transition kernel
 T – sample size (number of iterations)
Output: $x = \{x_1, \ldots, x_T\}$ – sample

t = 0
initialize x_t
repeat
 increment t
 generate $y \sim q(x_{t-1}, \cdot)$
 generate $u \sim U(0,1)$
 calculate $\alpha(x_{t-1}, y) \leftarrow min\left\{1, \frac{\pi(y)}{\pi(x_{t-1})}\frac{q(x_{t-1},y)}{q(y,x_{t-1})}\right\}$
 if $\alpha(x_{t-1}, y) \geq u$ **then**
  $x_t \leftarrow y$
 else
  $x_t \leftarrow x_{t-1}$
 end if
until $t = T$

The random numbers are not drawn directly from the target function $\pi()$, but from a proposal distribution $q()$ instead, where it is easier to draw samples from. The proposal distribution $q(x, y)$ is selected from a set of methods that exist for variants of the MH procedure. The Independence Chain Method used in the *Balanced Scorecard Simulator* allows to draw candidates from an independent proposal distribution $q(x, y) = q(\cdot)$. Consequently, the value x_{t-1} does not influence the candidate y drawn at time t. Therefore the acceptance probability for the candidate y in the above algorithm is modified as follows, cf. Liu (2001):

$$\alpha(x_{t-1}, y) = min\left\{1, \frac{\pi(y)}{\pi(x_{t-1})}\frac{q(x_{t-1})}{q(y)}\right\} \tag{1}$$

The proposal distribution is defined at the start of the simulation and depends on the distribution type of the target function. The density of the normal distribution $N(\mu_N, \delta_N^2)$ serves as a proposal distribution for almost all target functions. The parameters of the proposal distribution are denoted as $\mu_N = E_\pi[X]$ and $\delta_N^2 = Var_\pi[X]$. $E_\pi[X]$ represents the expected value and $Var_\pi[X]$ represents the variance. In the case of a Cauchy distribution $q(\cdot,\cdot)$ becomes a Cauchy density as a proposal function with the same location parameter but a 110% scaling parameter in comparison with the given Cauchy function.

4.2 Projection step

Multiple samples $x, \hat{x}_1, \dots, \hat{x}_k$ referring to a selected indicator are computed in a simulation run, where k depends on the number of equations where the indicator occurs. The samples they have to be merged and a projection step is necessary at the end of the simulation to obtain an estimator from these samples. First of all the product space is spanned:

$$dom\hat{X} = domX \times dom\hat{X}_1 \times \dots \times dom\hat{X}_k \qquad (2)$$

As the indicators have to fulfill each single equation, a projection on the subspace $X - \hat{X}_1 - \dots - \hat{X}_k = 0$ is performed. The projection is carried out in step 9 of algorithm 2.

Algorithm 2 Projection step

Input: $\left(X, \hat{X}_1, \dots, \hat{X}_k\right)$ for each indicator

Output: $\hat{f}_{\hat{X}}$, $E\left[\hat{X}\right]$, $Var\left[\hat{X}\right]$

1: $\underline{q_{max}} \leftarrow max\left\{\underline{q_1}, \dots, \underline{q_k}, \underline{q}\right\}$ { Determine the maximum of all lower 1% quantiles $\underline{q_i}$ of all samples $X, \hat{X}_1, \dots, \hat{X}_k$}
2: $\overline{q_{min}} \leftarrow min\left\{\overline{q_1}, \dots, \overline{q_k}, \overline{q}\right\}$ { determine the minimum of all upper 99% quantiles $\overline{q_i}$ of all samples $X, \hat{X}_1, \dots, \hat{X}_k$}
3: **for all** Variables **do**
4:     **if** $\underline{q_{max}} > \overline{q_{min}}$ **then**
5:         $\underline{I_q} \leftarrow \emptyset$, mark system as inconsistent
6:     **else**
7:         $I_q \leftarrow \left[\underline{q_{max}}, \overline{q_{min}}\right]$
8:     **end if**
9:     Calculate $f\hat{X}(x) \leftarrow c \cdot f_X(x) \cdot f_{\hat{X}_1}(x) \cdot \ldots \cdot f_{\hat{X}_k}(x) \qquad \forall x \in I_q, c \in \Re_+$
10:     Calculate $E\left[\hat{X}\right] \leftarrow \sum_{j=1}^{k} x_j \cdot \hat{f}_{\hat{X}}(x_j)$
11:     Calculate $Var\left[\hat{X}\right] \leftarrow \sum_{j=1}^{k} \left(x_j - E\left[\hat{X}\right]\right)^2 \cdot \hat{f}_{\hat{X}}(x_j)$
12: **end for**

The lower $(\underline{q})$ and upper $(\overline{q})$ quantiles and the median are determined for the graphic visualization of the results, cf. Fig. 1.

5 The software

The BSC model was implemented as the *Balanced Scorecard Simulator*. The simulator connects several applications with an Excel GUI and VBA to input prior information. The simulation is coded by the statistical programming language R which communicates with Excel via the R-Excel server. The simulation results are presented as a predetermined Excel table and as a Kiviat diagram. The Kiviat diagram maps all simulated indicators with the target values as entered by the user. Additionally, it is possible to examine the sensitivity of the indicators dependent on the given prior information, before running the simulation. An automatic storage of intermediate results allows for an analysis of inconsistencies in the BSC system. Furthermore, the performance measurement system can be extended by a further indicator if this indicator is assigned to one of the four perspectives of the given BSC.

6 Simulation example

In this section we illustrate the *Balanced Scorecard Simulator* by an example. In our example full prior information is provided for all 24 characteristics, cf. Tab. 2, columns 2 and 3. The simulation is primarily used to identify inconsistencies of the BSC data. In addition, a new indicator is established, namely "costs" related to the financial perspective, which is defined as $costs = turnover - profit$.The indicator *turnover regular customers of the previous period* is constant and set to 1900.

The CPU operating time of a test run with a number of iterations $T =$ 100.000 is approx. 5 min. The data set does not contradict the given model. The columns $\hat{\mu}$, $\hat{\sigma}$ and planned target values in Tab. 2 show the result of the simulation. An analysis of the computed expected values and standard deviations for every indicator provides evidence that the imprecision (error intervals) of the simulated quantities are reduced. The observed values are adjusted in the way that a shift is proportional to the prior variance of a variable.

The results of simulation are visualized in a Kiviat diagram, cf. Fig. 1. This chart type is suitable for the representation of multiple indicators. If a system of equations is classified as inconsistent, only the number and the names of the incorrect indicators are reported by a result notification. This makes it easier for the user to identify the causes of inconsistency in the data.

7 Summary

The adaptation and modeling respectively of key figures as random variables in a BSC enhances the information content and brings the BSC closer to reality, (cf. Müller et al. (2003)). Randomness happens due to a kind of 'natural'

Table 1. Prior information of the BSC characteristic quantities

BSC indicator	distribution	target value	$\hat{\mu}$	$\hat{\sigma}$	planned target value
Number of innovative ideas	$N(18, 2^2)$	18	18.00	1.86	20.00
Employee loyalty	$U(0.97, 0.99)$	0.99	0.99	0.0004	0.99
Number of notices	$U(1, 3)$	2	2.06	0.24	2.05
Number of staffers	$N(205, 5^2)$	205	201.14	3.76	203.70
Labour turnover rate	$N(0.01, 0.001^2)$	0.01	0.01	0.0006	0.01
Labour productivity	$N(20, 2^2)$	20	19.94	0.42	20.00
Number of new kinds of product	$U(3, 5)$	4.5	4.47	0.86	5.00
Sale set of a new kind of product	$U(45, 50)$	50	47.49	1.43	50.00
Returns quantity	$N(89, 5^2)$	90	89.75	4.40	100.00
Return rate	$N(0.19, 0.02^2)$	0.2	0.17	0.01	0.20
Quantity supplied	$N(550, 10^2)$	556	548.73	9.30	625.00
Mean amount of sales	$N(440, 10^2)$	445	439.45	9.41	500.00
Quota of identified customer wishes	$N(0.8, 0.01^2)$	0.8	0.80	0.01	0.80
Customer loyalty	$N(2, 0.01^2)$	2	2.00	0.01	2.04
Customer satisfaction	$U(0.6, 0.7)$	0.6	0.65	0.03	0.60
Customer acquisition	$N(0.05, 0.01^2)$	0.05	0.05	0.0003	0.05
Turnover of new products	$U(1990, 2010)$	2000	1999.40	5.67	2037.04
Sales price	$N(8, 2^2)$	9	8.50	0.57	8.15
Share of turnover of new products	$N(0.45, 0.001^2)$	0.5	0.45	0.00	0.50
Turnover regular customers	$N(3800, 50^2)$	3800	3795.20	44.36	3874.07
Turnover of new customers	$N(199, 10^2)$	200	198.78	6.09	200.00
Turnover	$N(4000, 25^2)$	4000	3994.38	20.54	4074.07
Operating margin	$N(0.25, 0.02^2)$	0.27	0.25	0.01	0.27
Profit	$N(1000, 25^2)$	1100	999.25	22.09	1100
Costs	unknown	unknown	3013.77	129.07	2974.07

longitudinal derivation of indicators, errors in observations and measurements of actual data of BSC indicators or evaluation tolerance in target data.

The software *Balanced Scorecard Simulator* is an equally useful management tool for planning, decision making and controlling, i.e. it allows the computation of consistent target values, it produces coherent, analytical data and supports the controller to detect incoherencies in the values of the BSC.

Planned future investigations in the field of simulation of stochastic business figures concern the following topics:

- Simulation of arbitrary performance measurement systems or an individual adjustment of the indicators and perspectives in a business-specific BSC;
- Performance improvement of the simulation method, efficient estimation of distributions and parameters and sampling from multi-dimensional distributions;
- Improvement in the debit quantity regulation in the *Balanced Scorecard Simulator*.

Further improvements of the software can be achieved by implementing multivariate density estimation of real data of a company's data warehouse. Due to the fact, that the Metropolis Hastings algorithm is used for sampling, this does not influences the simulation.

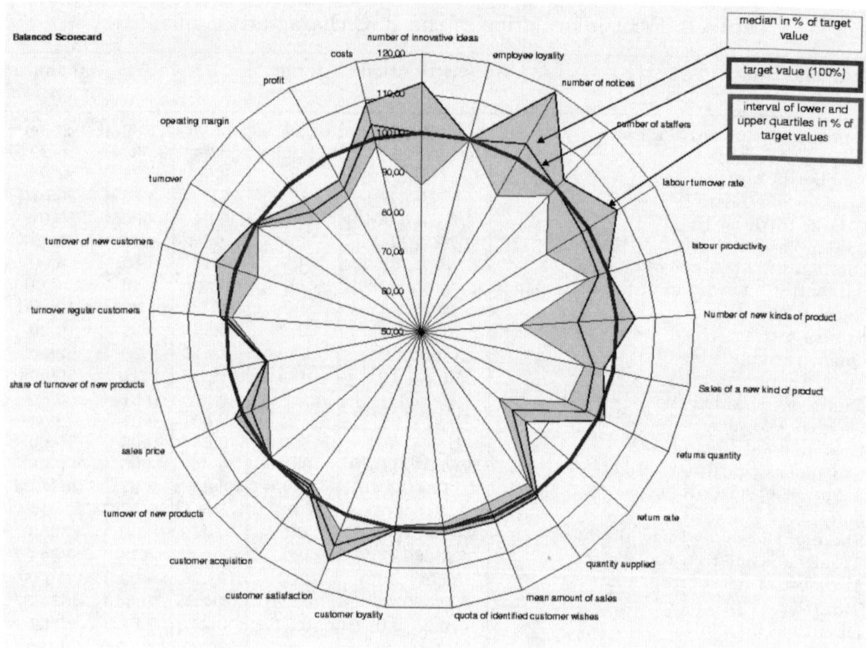

Fig. 2. Simulation results of BSC characteristic quantities in Kiviat diagram form

References

FRIEDAG, H.R. and SCHMIDT, W. (2004): *Balanced Scorecard.* 2nd edition, Haufe, Planegg.

KAPLAN, R.S. and NORTON, D.P. (1996): *The Balanced Scorecard. Translating Strategy Into Action.* Harvard Business School Press, Harvard.

KÖPPEN, V. and LENZ, H.-J. (2005): Simulation of Non-linear Stochastic Equation Systems. In: S.M. Ermakov, V.B. Melas and A.N. Pepelyshev (Eds.): *Proceeding of the Fifth Workshop on Simulation.* NII Chemistry Saint Petersburg University Publishers, St. Petersburg.

LENZ, H.-J. and RÖDEL, E. (1991): Data Quality Control. In: *Proceedings of the Annual Meeting of GÖR.* Trier.

LIU, J.S. (2001): *Monte Carlo Strategies in Scientific Computing.* Springer, Berlin.

MÜLLER, R.M., LENZ, H.-J. and RUHNKE, K. (2003): Ein fuzzybasierter Ansatz zur Durchführung analytischer Prüfungen bei der Existenz von Schätzspielräumen. *Die Wirtschaftsprüfung, 10, 56, 532–541.*

SCHMID, B. (1979): Bilanzmodelle. ETH Zürich, Zürich.

TODOVA, A. and ÜNSAL, A. (2005): *Analyse und Modellierung von Einflussgrössen in der Balanced Scorecard.* MA Thesis, Freie Universität, Berlin.

Integration of Customer Value into Revenue Management

Tobias von Martens and Andreas Hilbert

Professur für Wirtschaftsinformatik, insb. Informationssysteme im Dienstleistungsbereich, TU Dresden, D-01062 Dresden;
{martens, hilbert}@wiid.wiwi.tu-dresden.de

Abstract. This paper studies how information related to the customer value can be incorporated into the decision on the acceptance of booking requests. Information requirements are derived from the shortcomings of transaction-oriented revenue management and sources of this information are identified in the booking process. Afterwards, information related to customer value is integrated into the network approach of inventory control.

1 Problem definition and outline

Revenue-optimized utilization of limited and perishable capacity is of great importance in the services sector. Revenue management, commonly known as yield management, provides various methods for allocating capacity optimally on different customer segments and attaining this allocation with the aid of booking mechanisms.

Regardless of the increasing deployment in practice, transaction-oriented optimization methods of revenue management exhibit deficiencies and risks. A booking request is accepted if the price of the requested booking class exceeds the opportunity costs of the capacity utilization. Since the offering price is the decisive criterion, neither the value of the customer for the company nor the negative consequences resulting from the rejection of requests are paid attention to. Hence, revenue is most often optimized only in the short-term while establishing profitable relationships to valuable customers is hardly possible by applying transaction-oriented inventory control.

For several years, efforts have been made to focus on customer value in revenue management, too. Noone and Kimes (2003), for instance, segment demand within the hotel business into four groups according to two loyalty dimensions and propose segment-specific revenue management strategies. Esse (2003) classifies customers corresponding to their value and suggests to allocate contingents of capacity to those classes. Hendler and Hendler (2004) try to approximate customer value on the basis of casino expenditure forecasts and compare this value with the opportunity costs to decide whether to offer customers a room for free. However, all these approaches are either limited to a certain application area or give general suggestions without paying attention to certain optimization methods.

In contrast, this paper proposes a cross-industry approach to identify information related to customer value in the booking process and to incorporate this information into the network programming model of inventory control (see Bertsimas and Popescu (2003)). An examination of potential benefits and challenges as well as an outlook on remaining research round off the paper.

2 Information requirements

This approach aims at considering value-related information in addition to the price of the requested booking class when deciding on the acceptance of booking requests. Beside the customer value itself, information about customer behaviour is important for inventory control. While the customer value is determined by several influencing variables (see Rudolf-Sipoetz and Tomczak (2001)), customer behaviour can be described by price-, booking- and rejection-related information. Some of these aspects are shortly explained below. Figure 1 represents exemplary information related to customer value required for revenue management.

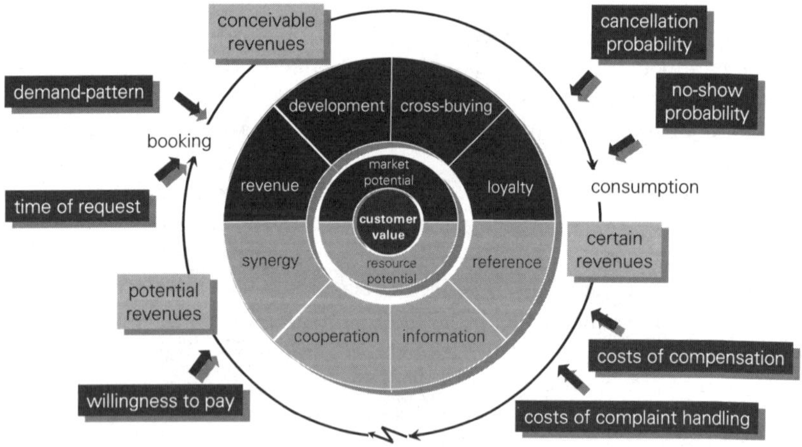

Fig. 1. Information requirements in the context of value-oriented inventory control (adapted from Rudolf-Sipoetz and Tomczak (2001))

The revenue potential is largely determined by the willingness to pay. However, the range of available booking classes has an impact on the extent to which the company is able to utilize the customer's willingness to pay at the time of booking. Buy-up and buy-down effects can arise, i.e. customers switch to higher- or lower-valued booking classes in case of unavailability of their preference (see Talluri and van Ryzin (2004)). On the other hand, two types of behaviour, yieldable demand (selection of the original preference) and priceable demand (selection of the lowest available fare), can be distinguished (see Boyd and Kallesen (2004)). Hence, information about the individual demand-pattern is required for inventory control.

Information about booking behaviour facilitates the company to plan the integration of the customer as an external factor into the service realization. Among others, the time of booking as well as the probability of cancellations and no-shows (customers do not utilize their reservation) are of importance for inventory control.

In order to use as much of the available capacity as possible, overbooking is applied, i.e. more reservations are accepted than could be served in fact (see Karaesmen and van Ryzin (2004)). Due to the fact that customers could be refused despite their reservation, information about their rejection behaviour, e.g. expected costs of compensation and complaint handling, is relevant as well.

3 Sources of information

Potential sources of information satisfying the information requirements described above are identified based on a generic booking process represented in Figure 1.

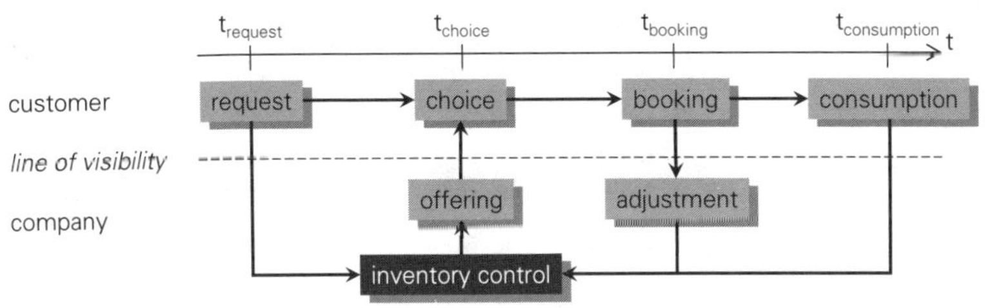

Fig. 2. Generic booking process of revenue management

Typically, customers specify the requested service initially, e.g. by selecting a connection and a date when booking a flight. Subsequently, they may choose one of the available booking classes and complete booking in order to be eligible for the service. The company, on the other hand, can control the utilization of its capacity in two ways: Firstly, by determining available booking classes based on information available at the time of request, and secondly, by adjusting inventory control after booking has been completed.

The latter may have an impact on the acceptance of future requests in the booking period.

In addition to the steps of the generic booking process, conceivable booking scenarios impact which information is available, and at which time. In this paper, the following scenarios are distinguished:

1. Customers remain completely anonymous.
2. Customers book for the first time and identify themselves at the time of booking.
3. Customers book repeatedly and identify themselves at the time of request.

Value-related information is often not directly available from the sources described above but can be predicted on the basis of indicators. Figure 2 represents exemplary scenario-dependent sources of information that can be used to generate such indicators. Afterwards, some of the indicators are explained briefly.

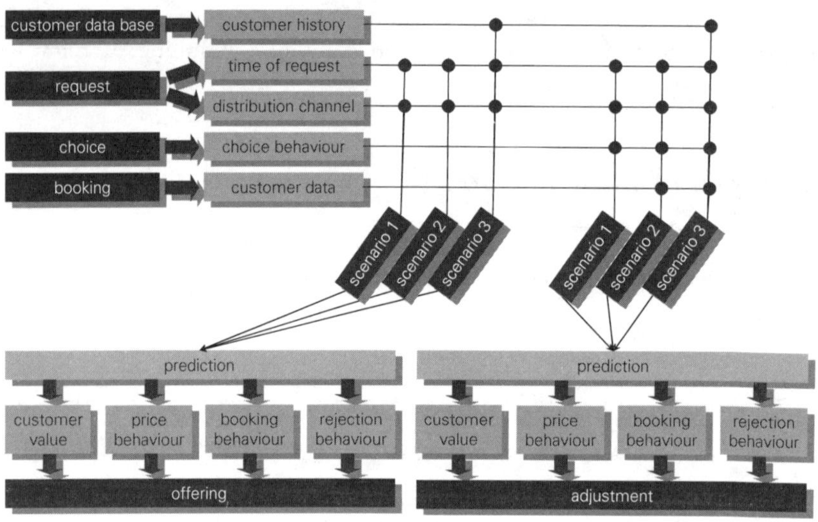

Fig. 3. Sources of information for predicting information related to customer value

The customer data base of the company can provide information regarding the customer history and allow for an analysis of the correlations between indicators and value-related information in past booking processes.

By means of the customer request as a source of information, the time of request and the chosen distribution channel are available as indicators without the need for customer identification. Information related to customer value, e.g. willingness to pay or cancellation probability, can be predicted with the aid of the respective indicator if a-priori probabilities are accordingly assigned to them (see Dacko (2004)). Referring to empirical findings concerning the time of request, Boyd and Bilegan (2003) discuss different models for the arrival of booking requests and propose to consider segment-specific Markov-processes. By doing so, customer requests can be classified to a certain segment by means of the time of request without value-related information being available.

The selection made by customers out of a set of available booking classes can indicate their willingness to pay, taking into account their demand-pattern, i. e. the potential discrepancy between the chosen booking class and the original preference. The prediction of the willingness to pay and the demand-pattern can be improved by analyzing the customer's choice behaviour given a certain set of available booking classes across several comparable booking processes, assuming constant customer behaviour (see Boyd and Kallesen (2004)).

If customers have to reveal customer data to complete their booking, their demographical characteristics can provide a basis for classifying customers to customer segments and to adjust inventory control accordingly (see Gourville and Soman (2002)).

4 Generation of value-related information

For the prediction of value-related information, two sets of attributes are regarded:

- a set M_i of indicators, e.g. indicators of customer value, time of request, distribution channel, choice behaviour, and customer data,
- a set M_c of value-related information, e.g. the customer value itself, willingness to pay, demand-pattern, costs of compensation and complaint handling as well as cancellation and no-show probabilities.

During a training step across several comparable booking processes, the demand is segmented on the basis of M_i and M_c whereas ideally only the set M_c is used to distinguish the customer segments. Subsequently, forecasting models for predicting attributes out of M_c on the basis of M_i are developed in each segment, e.g. with the aid of appropriate data mining methods (see Weatherford et al. (2003)).

In the deployment step, i.e. during a booking process, the customer is classified to a customer segment on the basis of available attributes out of M_i (see Esse (2003)). Afterwards, attributes out of M_c are predicted on the basis of the indicators using the segment-specific forecasting models.

5 Segment-specific inventory control

Out of the numerous optimization methods for inventory control, this approach exemplary extends the network programming model (see Bertsimas and Popescu (2003)) by the segment-specific consideration of parameters, e.g. demand and cancellation probabilities, and by the integration of customer value. The network approach is based on the following decision rule: a booking request regarding a product j is accepted if the revenue r_{dj} associated with the sale of the respective product in the customer segment d is at least equal to the opportunity costs OC_j of the capacity utilization:

$$r_{dj} \geq OC_j(n,t) \tag{1}$$

In order to approximate the opportunity costs, the value V of the remaining capacity n (in case of the request being rejected) and $n - A^j$ (if the request consuming A^j units of capacity is accepted) is calculated given the (expected) aggregated demand D^{t-1} in the subsequent periods. The difference between the two capacity values indicates the opportunity costs. n represents the capacity units that are remaining of the undiminished capacity N of various resources, e.g. seats on flight legs or rooms on certain nights. A describes the amount of capacity units that are required by a product, e.g. a flight path consisting of several flight legs, on a resource, e.g. a specific flight leg. In case of only one resource, A is a vector and n is a real number. In case of multiple resources, A is a matrix and n is a vector.

$$OC_j(n,t) \cong V(n, D^{t-1}) - V(n - A^j, D^{t-1}) \tag{2}$$

The product can be booked as from the time $t = T$ and is consumed at the time $t = 0$. The length of a period is set in a manner that at most one request can arrive during a period.

One possibility to approximate the value of the remaining capacity units n at the time t is to solve the following deterministic linear programming model (see Pak (2005)):

$$\begin{aligned} V(n, D^t) &= \max R \cdot Y \\ \text{s.t.} \quad A \cdot Y &\leq n \\ 0 \leq Y &\leq D^t \end{aligned} \tag{3}$$

Thereby, R denotes the revenues of selling the products j in the various customer segments d, whereas the decision variable Y assigns contingents of the respective products to the customer segments. The aggregated and segment-specific future demand is represented by corresponding expected values in D^t.

In order to consider demand distribution in addition to its expected value, it is proposed to simulate k different demand scenarios $\hat{D}_i^t$ in a Monte Carlo study and to weight the calculated opportunity costs in each case with a factor α_i:

$$OC_j(n,t) = \sum_{i=1}^{k} \alpha_i \cdot OC_j(n, \hat{D}_i^{t-1}) \tag{4}$$

6 Integration of customer value

In contrast to the traditional procedure of measuring the value contribution of a booking request only by the price of the product requested, the approach at hand incorporates a segment-specific revenue r_{dj} into the decision. This consists of a short-term component CV_{dj}^{ST} as well as a long-term component CV_{d}^{LT} which is transformed by a risk-preference or utility function $u(x)$. Both are balanced by a segment-specific weighting factor $\alpha_d = [0;1]$. The higher α_d the more emphasis is on the transaction-related component. While the short-term component equals the offering price, the long-term customer value can be approximated using one of various approaches, e.g. Gierl and Kurbel (1997).

$$r_{dj} = \alpha_d \cdot CV_{dj}^{ST} + (1 - \alpha_d) \cdot u(CV_{d}^{LT}) \tag{5}$$

7 Conclusions

From the deficiencies of transaction-oriented approaches of revenue management being predominant so far, this paper derives the motivation to incorporate information related to customer value into inventory control. This takes into account different valuations of customer segments by deciding on the acceptance of a booking request not merely on the basis of the offering price. Rather, it is possible to carry out a segment-specific weighting between a transaction-related and a value-oriented revenue contribution of the customer. The integration of value-related information into revenue management expands the revenue optimization beyond a single booking period toward the customer lifecycle looked at.

The complexity of the optimization problem has risen due to the use of segment-specific variables, e.g. value contributions and cancellation probabilities. Hence, the underlying dynamic, stochastic optimization problems are often formulated as static, deterministic problems (see Bertsimas and Popescu (2003)) as done in this paper, and heuristics, e.g. evolutionary algorithms, are applied to approximate the optimal solution adequately with acceptable efforts (see Pulugurtha and Nambisan (2003)). Moreover, data base infrastructure and tool support have to be sufficient to facilitate both recording relevant indicators in the booking process, e.g. time of request and choice behaviour, and analyzing correlations between indicators and value-related information.

It remains to be studied to what extent information related to customer value can be predicted by means of indicators in the booking process and information in the customer data base. In addition, the simplifying assumption that customers are lost to competitors in case of being rejected requires further investigation.

Up to now, scientific contributions to revenue management at most focused on determining available booking classes and an optimal offering price in the booking process respectively. The adjustment of inventory control, e.g. by re-classifying accepted requests, based on information not available until the completion of the booking has to be looked at more intensely. Nevertheless, attention has to be paid on customer perceptions regarding transparency and fairness of inventory control and strategic customer behaviour.

References

BERTSIMAS, D. and POPESCU, I. (2003): Revenue Management in a Dynamic Network Environment. *Transportation Science, 37, 257–277.*

BOYD, E.A. and BILEGAN, I.C. (2003): Revenue Management and E-Commerce. *Management Science, 49, 1363–1386.*

BOYD, E.A. and KALLESEN, R. (2004): The Science of Revenue Management when Passengers Purchase the Lowest Available Fare. *Journal of Revenue & Pricing Management, 3, 171–177.*

DACKO, S.G. (2004): Marketing Strategies for Last-Minute Travel and Tourism. *Journal of Travel & Tourism Marketing, 16, 7–20.*

ESSE, T. (2003): Securing the Value of Customer Value Management. *Journal of Revenue & Pricing Management, 2, 166–171.*

GIERL, H. and KURBEL, T. (1997): Möglichkeiten zur Ermittlung des Kundenwertes. In: J. Link, D. Braendli, C. Schleuning, and R.E. Hehl (Eds.): *Handbuch Database Marketing.* IM Marketing-Forum, Ettlingen.

GOURVILLE, J. and SOMAN, D. (2002): Pricing and the Psychology of Consumption. *Harvard Business Review, 80, 90–96.*

HENDLER, R. and HENDLER, F. (2004): Revenue Management in Fabulous Las Vegas: Combining Customer Relationship Management and Revenue Management to Maximise Profitability. *Journal of Revenue & Pricing Management, 3, 73–79.*

KARAESMEN, I. and VAN RYZIN, G. (2004): Overbooking with Substitutable Inventory Classes. *Operations Research, 52, 83–104.*

NOONE, B.M., KIMES, S.E. and RENAGHAN, L.M. (2003): Integrating Customer Relationship Management and Revenue Management: A Hotel Perspective. *Journal of Revenue & Pricing Management, 2, 7–21.*

PAK, K. (2005): *Revenue Management: New Features and Models.* Dissertation, Erasmus Universiteit, Rotterdam.

PULUGURTHA, S.S. and NAMBISAN, S.S. (2003): A Decision-Support Tool for Airline Yield Management Using Genetic Algorithms. *Computer-Aided Civil & Infrastructure Engineering, 18, 214–223.*

RUDOLF-SIPOETZ, E. and TOMCZAK, T. (2001): *Kundenwert in Forschung und Praxis.* Thesis, St. Gallen.

TALLURI, K. and VAN RYZIN, G. (2004): Revenue Management under a General Discrete Choice Model of Consumer Behavior. *Management Science, 50, 15–33.*

WEATHERFORD, L.R., GENTRY, T.W. and WILAMOWSKI, B. (2003): Neural Network Forecasting for Airlines: A Comparative Analysis. *Journal of Revenue & Pricing Management, 1, 319–331.*

Women's Occupational Mobility and Segregation in the Labour Market: Asymmetric Multidimensional Scaling

Miki Nakai

Department of Social Sciences, College of Social Sciences, Ritsumeikan University, 56-1 Toji-in Kitamachi, Kita-ku, Kyoto 603-8577 Japan;
mnakai@ss.ritsumei.ac.jp

Abstract. The aim of this paper is to examine the career dynamics among women using asymmetric multidimensional scaling. Based upon a national sample in Japan in 1995, we analyze the occupational mobility tables of 1405 women among nine occupational categories obtained at various time periods. We find that asymmetric career changes within similar occupational categories take place a lot in one's 30s or later. Women, unlike in the case of men, appear to change their occupational status in mid-career.

1 Introduction

Social mobility, or movement or opportunities for movement between different social groups has been a central concern for sociologists and economists, to comprehend the mechanisms by which classes maintain their advantages and disadvantages. Much of the attention has been focused on intergenerational social mobility, which refers to the social mobility between generations, not intragenerational mobility, which refers to the social mobility within a generation, in sociology. Furthermore, most studies of intragenerational social mobility have not included data on female mobility until recently.

However, recent accumulation of female data and increasing concern over the way in which labour market structures affect women's employment as well as men's made studies of the pattern of female career mobility an essential topic of investigation (Scott (1994), Sato (1998)). A few studies have investigated both men and women, but most include only working women and do not adequately report the status of unemployment.

Some research, on the other hand, place most attention on women's career patterns especially regarding exit and reentry into the labour market (Nakai and Akachi (2000)). Those studies revealed that the spouse's socioeconomic status is one of the key determinants of women's employment exit and reentry, suggesting that women's career patterns are very different from that of men. On the other hand, certain research has suggested that the rising importance of the service sector relaxes women's identification with the familial role, and allows women to design career scenarios and life-cycle destinies independent of any male partner in post-industrial societies (Esping-Andersen (1993)). Despite its importance for theory and policy, empirical research on women's career developments and occupational trajectories of individuals over the life course remains scarce.

Another reason why the issue of women's employment career mobility has attracted increased attention in recent years is that it may be necessary to provide insights into the structural characteristics of the career mobility of women at a time when the labour force is shrinking.

The aim of this paper is to examine the occupational mobility of women in Japan. We take into account women's occupational or career shifts among the occupational segments and investigate the occupational mobility tables measured at various time periods between the ages of 20 and 45. The analysis of occupational trajectories may provide interesting points that are helpful in further understanding the structural barriers to mobility or gender segregation and segmentation of the work force. In the present case, the data matrix is asymmetric. Therefore, using asymmetric multidimensional scaling, an exploratory method which allows to develop better understanding the structure of asymmetric relations between entries, we explore the structure of the mobility of women over the life course. Some researchers have analyzed the intergenerational occupational mobility of men using asymmetric multidimensional scaling (Okada and Imaizumi (1997), De Rooij (2002)). However, they did not deal with intragenerational nor female mobility.

2 Data

The data is from a nationally representative survey conducted in 1995 of social stratification and mobility in Japan. Of approximately 4000 men and women sampled, aged 20-69, 2653 (1248 men and 1405 women) were successfully interviewed. This data provides a complete occupational history allowing a lifetime analysis of entire occupational careers.

We utilize the occupational mobility that occurs between specific ages at intervals of five, ten and twenty years for each woman between the ages of 20 and 45, specifically between 20 and 25, 25 and 30, 30 and 35, 35 and 40, 20 and 30, 25 and 35, 30 and 40, 35 and 45, 20 and 40 and 25 and 45 years of age. The contingency tables show intragenerational mobility between nine occupational categories including 'not employed' status. The nine occupational categories are:

(1) Professional occupations **(Professional)**,
(2) Nonmanual occupations in large organizations **(Nonmanual large)**,

(3) Nonmanual occupations in small organizations **(Nonmanual small)**,
(4) Nonmanual self-employed occupations **(Nonmanual self)**,
(5) Manual occupations in large organizations **(Manual large)**,
(6) Manual occupations in small organizations **(Manual small)**,
(7) Manual self-employed occupations **(Manual self)**,
(8) Farm occupations **(Farm)**, and
(9) Not employed or homemaker **(Not employed)**.

The words in bold in parentheses are used to represent each occupational category in the following figure.

The women's occupational mobility is shown in Table 1. The entries in the table indicate the number of women whose occupational position at the time of origin is i and the destination occupation is j within a given age range (five, ten and twenty years). Some suggest that the job-shifting pattern of early years of employment is strikingly different from that of later careers (Seiyama (1988)). Also, it is obvious that career change may be related to a specific life stage. The data is appropriate to clarify this issue.

In previous studies, analysis of comparison of two points in time, for example, the first job and market position at age forty, has widely been used because its main focus was upon men's opportunities for upward mobility to top-level job positions. Some studies, on the other hand, analyzed career

Table 1. Part of intragenerational occupational mobility tables.

	Occupation at age 25								
Occupation at age 20	1	2	3	4	5	6	7	8	9
1 Professional	49	4	4	3	0	3	1	0	27
2 Nonmanual large	3	91	16	13	1	5	2	2	98
3 Nonmanual small	1	6	93	19	1	11	4	9	120
4 Nonmanual self	0	0	0	21	0	0	1	0	7
5 Manual large	0	2	3	3	27	8	1	3	24
6 Manual small	0	1	7	3	2	57	8	2	42
7 Manual self	0	0	0	1	0	0	8	0	5
8 Farm	0	0	2	1	0	2	1	47	10
9 Not employed	57	28	27	21	5	11	2	8	231
Occupation at age 25	Occupation at age 30								
1 Professional	69	1	3	1	0	0	1	0	25
2 Nonmanual large	1	67	4	3	0	3	1	0	37
3 Nonmanual small	2	0	61	3	1	5	1	1	58
4 Nonmanual self	0	0	1	74	0	0	1	0	4
5 Manual large	1	0	2	0	19	2	0	0	11
6 Manual small	0	0	2	1	0	64	4	3	20
7 Manual self	0	0	0	1	0	0	19	0	7
8 Farm	0	0	0	1	1	2	1	64	3
9 Not employed	6	4	21	8	3	22	5	0	455

mobility based on an index of annual flow percentages across occupations. However, this does not indicate the timing of the likelihood of career change. Social or occupational mobility can be ascertained through dynamic analysis of mobility patterns across the life course. The study of mid-career mobility after marriage or later, i.e. during child-rearing phase, as well as early career mobility, is important to understand career development and the possible barriers women encounter. The reason why the focus here is primarily on this age range is as follows: This age bracket is critical because in Japan, up to now, the participation of the female labour force is strikingly different from that of other countries, giving rise to the so-called "M-shaped" working pattern. The Japanese woman shows a drop in labour force participation in her late twenties and resurgence by her late thirties, producing the M-shaped curve. This means that women's career mobility inclusive of exit and reentry the labour force occurs between 20 and 45 years-of-age, and considerably different work histories emerge among individuals.

The set of ten 9×9 mobility tables between various periods of time, which is two-mode three-way proximities, describe asymmetric relations. These mobility tables were analyzed by asymmetric multidimensional scaling (Okada and Imaizumi (1997)).

3 Results

Based upon stress S (stress$= 0.484$), or the measure of badness-of-fit, obtained for a range of dimensionalities, the two-dimensional solution was selected as best representing the structure of data. The two-dimensional space for the intragenerational occupational mobility for women, which is the common object configuration, is presented in Figure 1. The solution is quite interpretable. In the common object configuration, each object (occupational category) is represented by a point and a circle in a multidimensional space. Both symmetric and asymmetric proximity relationships are represented in the same configuration space. The distance between the points represents the symmetry, and the difference between the radii represents the asymmetry.

The main dimension, Dimension 1 (horizontal) appears to separate among self-employed, farmers, employees of small organizations, employees of large organizations and professionals, with one exception: self-employed nonmanual can be found located on the opposite side of the self-employed manual of Dimension 1. This finding represents the polarity of self-employment activities. Some studies point to heterogeneity among self-employed women, suggesting that there are different forms of self-employment participation; non-professional/professional and managerial self-employment (Budig (2006), Arum and Müller (2004)). The evidence of our study supports this position.

Professional equals direction as in employees of a large organization given its custom of long-term and stable employment.

The second dimension (vertical) is clearly interpretable; it represents a nonmanual/manual dimension. There are four nonmanual occupational cate-

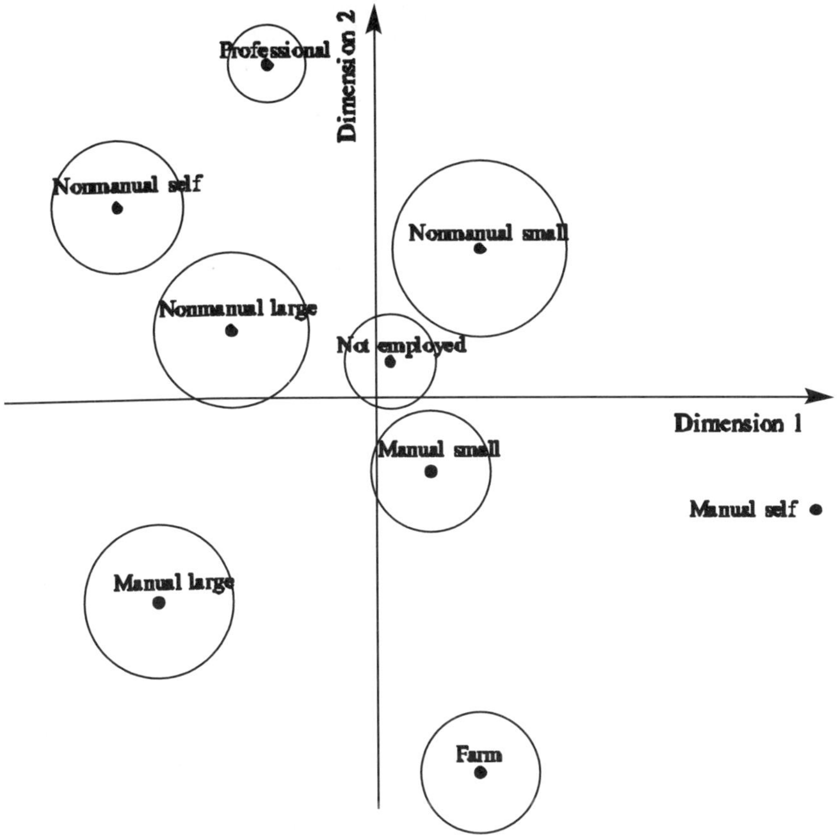

Fig. 1. The common object configuration of nine occupational categories.

gories in the upper half, and in the lower half there are four manual occupational categories.

Not employed has an intermediate location on both dimensions. Not employed, nonmanual large/small and manual small appear close together, suggesting that there are a large number of transitions among these statuses. On the other hand, self-employed, farm, professional are located at the periphery of the configuration. Therefore, specific characteristics of a job affect the likelihood of labour force withdrawal and re-entry among women. Self-employed and professional are distant from not employed, by reason that women with better market skills or property are more likely than others to stay in the workforce.

The resulting radius of nonmanual occupations in small organizations is the largest, and nonmanual in large organizations has the second largest radius. In the present analysis, a large radius indicates a lower tendency for women to stay in the same occupational categories at the later stage as that

Table 2. Symmetry weight and asymmetry weight.

Period of age	Symmetry weight	Asymmetry weight	
		Dimension 1	Dimension 2
20-25	1.000	0.364	0.343
25-30	1.000	0.276	0.685
30-35	0.930	2.332	1.208
35-40	0.995	1.269	0.206
20-30	1.002	0.202	0.364
25-35	0.997	0.648	0.312
30-40	0.908	2.466	1.673
35-45	0.974	5.780	0.278
20-40	0.994	0.900	0.347
25-45	0.998	0.637	0.276

in the initial stage during a certain period. A large radius also means that there are fewer entrants from other categories. Therefore, the above results show that the women whose occupational position in the initial stage is nonmanual in small organizations have the largest tendency to turn over during a given time period. Nonmanual workers in large organizations have somewhat same tendency as nonmanual workers in small organizations.

On the other hand, not employed, professional and self-employed manual have a smaller radius than other categories, the smallest is self-employed manual. This means that the inflow into these categories exceed the outflow. The previous study has shown, instead, that professional is an exclusive occupational category based on the analysis of male occupational trajectories (Hara (1979)). However, a certain proportion of women whose former status was not professional could enter into this field in the later stage. This suggests that expertise and special skills should facilitate women's re-entry into the labour force. The number of workers categorized as self-employed manual is relatively small, but career mobility into it does not seem to present a significant barrier. This finding is in close agreement with the result that determined the flow from nonmanual occupations of large firms to self-employment among female workers (Sato (1998)). Moreover, it has often been argued that self-employment or working at home is a possible arrangement for married women (Edwards and Hendry (2002)). The results can be understood in the light of these arguments.

Next, the question concerning symmetric and asymmetric aspects for each period of time arises. Table 2 shows the symmetry and asymmetry weights. Symmetry weights remain fairly constant throughout the career period, although those of women's middle career (in the early thirties and thirties) are a little smaller than at other time periods.

With regard to asymmetric attributes, asymmetry weights along both dimensions during the twenties are small, showing that there is relatively little asymmetric occupational mobility along both dimensions in the early career stage. At the late twenties stage, although the magnitude of asymmetry weight is not very large, asymmetry weight of Dimension 2 is larger than that of Dimension 1, suggesting that asymmetric occupational mobility along Dimension 2 is greater than that along Dimension 1. It might be inferred from the result that there are some job transitions between nonmanual and manual, but within the organizations of similar size at the late twenties stage. It is expected that knowledge flow and transfers take place among firms of similar size due to the difference of educational or other credentials that specific sizes of organizations deem necessary. It would also appear that this asymmetry represents women's entry into professional jobs, as well as withdrawal from the labour force in their late twenties.

Asymmetry weights along Dimension 1 begin to show a marked increase and asymmetry weights along Dimension 2 also modestly increase in the early thirties age range, suggesting that asymmetric career changes within both nonmanual and manual occupational categories occur frequently in people's early thirties. Turnover between nonmanual and manual is rather limited. This tendency is especially pronounced in the late thirties and early forties age range as the sizable weight indicates. Asymmetric job mobility in mid-career takes place within similar occupational categories, in which the nature of the work has a degree of commonality. This also suggests that nonmanual jobs form one of the segments and that manual jobs are segmented into another segment.

The above asymmetric career changes seem to reflect a career shift from nonmanual work in large organizations to self-employed nonmanual work. It has been suggested that women use self-employment as a strategy to balance work and family commitments, while, in contrast, men use self-employment to advance their career (Edwards and Hendry (2002)). The finding of our analysis is consistent with this argument. Unlike in the case of men who have the feature of job stability and low occupational mobility in their mid-career period, women tend to change occupational status to a much greater degree in their thirties.

4 Discussion and conclusion

The analyses reported here suggest two main conclusions. First, unlike in the case of other welfare regimes in Western Europe and men, Japanese women seem to change their working pattern, inclusive of employment/non-employment transition, a great deal according to their household responsibilities in their thirties or later. Second, it may be concluded that there is a divergence of nonmanual and manual mobility pattern in the middle career period. Among women working in relatively high-prestige occupations, mainly nonmanual jobs in the initial stage of their work career, there exist two heterogeneous groups. Some stay in the labour market working as professional or self-employed, while the other is likely to leave the labour force when entering motherhood. On the other hand, women working in low-prestige occupations such as manual labour would be less likely to exit employment. Quite a number of women may be in contingent jobs with lower position and lower pay.

The policy implications of these results are clear. The result suggests that the thirties mark a critical period for women's working life as well as for their family life. Therefore, it would be important that all companies implement an effective work-life balance program. Although employees in the large organizations are entitled to paid maternity leave, working women actually have very little access to it. In particular, for women working in small organizations, there are practically no family friendly policies available. Furthermore, the results also suggest the need for policies to offer updating education and training to facilitate women re-enter the workforce and any programs to encourage women to become entrepreneurs.

This paper has achieved a major breakthrough in gaining insight into career mobility among women in Japan, especially asymmetric turnover in mid-career so that we could advance the hypotheses for future research. Our study has one limitation. We did not differentiate atypical form of employment from full-time work. More appropriate occupational schema might be useful to highlight occupational segregation and segmentation in the labour market. Further research must be undertaken in these areas to compare and assess the change in the career mobility patterns across cohorts and of subsequent generations by utilizing the data of a follow-up study.

References

ARUM, R. and MÜLLER, W. (2004): *The Reemergence of Self-Employment.* Princeton Univ. Press.

BUDIG, M.J. (2006): Intersections on the Road to Self-Employment: Gender, Family, and Occupational Class. *Social Forces, 84, 4, 2223–2239.*

DE ROOIJ, M. (2002): Distance Association Models for the Analysis of Repeated Transition Frequancy Tables. *Statistica Neerlandica, 55, 2, 157–181.*

EDWARDS, L.N. and HENDRY, E.F. (2002): Home-based Work and Women's Labor Force Decisions. *Journal of Labor Economics, 20, 170–200.*

ESPING-ANDERSEN, G. (1993): *Changing Classes: Stratification and Mobility in Post-Industrial Societies.* Sage.

HARA, J. (1979): Analysis of Occupational Career. In: K. Tominaga (Ed.): *Social Stratification Structure in Japan.* Univ. of Tokyo Press, Tokyo.

NAKAI, M. and AKACHI, M. (2000): Labor Market and Social Participation. In: K. Seiyama (Ed.): *Gender, Market, and Family.* Univ. of Tokyo Press, Tokyo.

OKADA, A. and IMAIZUMI, T. (1997): Asymmetric Multidimensional Scaling of Two-Mode Three-Way Proximities. *Journal of Classification, 14, 195–224.*

SATO, Y. (1998): Change in Job Mobility Patterns in the Postwar Japanese Society. In: Y. Sato (Ed.): *Social Mobility and Career Analysis*, 45–64.

SCOTT, A.M. (1994): *Gender Segregation and Social Change.* Oxford Univ. Press.

SEIYAMA, K. (1988): Analysis of Occupational Mobility. In: T. Imada and K. Seiyama (Eds.): *Structure and Process of Social Stratification*, 251–305.

Multilevel Dimensions of Consumer Relationships in the Healthcare Service Market M-L IRT vs. M-L SEM Approach

Iga Rudawska[1] and Adam Sagan[2]

[1] Department of Marketing, University of Szczecin, 71-101 Szczecin, Poland; igita@sus.univ.szczecin.pl

[2] Chair of Market Analysis and Marketing Research, Cracow University of Economics, 31-510 Cracow, Poland; sagana@ae.krakow.pl

Abstract. The aim of the paper is to compare two measurement models: IRT multilevel and SEM multilevel model of patients - physicians relationships. These relationships are nested in the institutional context of healthcare units. The Likert-type scale was developed and the nature of the constructs discussed. This scale was adopted on individual (patients) and well as institutional (units) level along with between variable that describes cluster specific characteristics. CTT and IRT multilevel random intercept models are discussed.

1 Customer relationships in the healthcare services

The encounters between patients and healthcare providers is a critical component of service quality. Healthcare services have deeply interpersonal nature and are described by high degree of person-to-person interaction. The patient-healthcare provider relationship is also strongly asymmetric.

The patients' dependence is based on the physicians knowledge and experience. Medical care belongs also to credence-based services. As a consequence, trust and commitment are considered essential for the understanding the patient-healthcare provider relationship.

Berry (2000) stresses that any relationship can be divided into three levels based on financial, social and structural bonds. Financial bonds are considered the weakest form and may only lead to spurious relationships. Social bonds involve regular communication with customers and service continuity through a personal service representatives. Structural bonds offer value-adding problem solutions that are independent from individual service representatives, and which are difficult to copy for competitors. According to Wilson (1995) social bonding can be defined as the degree of mutual personal friendship and liking shared by the buyer and seller. Such a strong personal friendship tends to hold a relationship together. Numerous studies done by medical sociologists (Ben-Sira (1982)) also have revealed that the socio-emotional aspects of care are even more important in treating the patient than technical one.

In this empirical study it is assumed that patient-provider relationship can be described along a continuum, ranging from weak and formal to true relationship bonds, based on trust and commitment. Rudawska (2006) listed 13 types of relationship bonds that can reveal 3 main categories of latent variables: formal (financial, technical and organizational), social (individualization, emotional, psychological, empowerment) and structural (knowledge, communication, personal, reputation, ethical, partnership) one . Formal bonds represent rather conventional exit barriers like the monopoly in the local market or standard price incentives. They tie the patient to the service provider and maintain the relationship on the basic level. Social bonds arise when the patient and the service personnel know each other well. They have a positive lock-in effect and make the contact easy and more satisfying. The most advanced structural bonds are based on positive commitment both by the patient and service provider. It is important to note that structural bond can be based not only on real, behavioral relations, but also on symbolic and ideological ones. All of 13 variables have been tested both on patients and healthcare providers using 5-point Likert scale. Relational bonds have been used to predict continuous dependent variable that is perceived probability of healthcare unit (HCU) choice.

These categories of relational bonds can be regarded as a latent variables. The relations between the latent variables and its manifest indicators usually takes into account two main perspectives: formative and reflexive (Jarvis et al. (2003)).

According to the first view, latent variables are formed by the manifest indicators. In this case particular forms of relational bonds (i.e. formal) consist of predefined subcategories (i.e. financial, technical, organizational).

The second view stresses that latent variable is a true cause of observed intercorrelations of its underlying manifest variables (i.e. financial, formal and organizational aspects of relations are explained by the formal bond construct).

Additionally, in the latter perspective also the two views are possible. The manifest indicators of relational bonds can be viewed in the context of common factor model and classical test theory. It means that common factor is the source of variation of manifest variables (indicators) that are regarded as a parallel and equivalent measures of latent variable. This assumption suggests existing three latent variables (common factors) measured on each level by proposed indicators.

On the other hand, the global relational bond can be conceptualized as a unidimensional construct that differentiates possible responses on underlined, cumulative set of items in the context of item response theory (IRT). It means that indicators reflect also the intensity of given subject's responses on the continuum of latent variable.

2 Multilevel models of relational bonds in healthcare service market

2.1 The factor models of the relational bonds

The study of relational bonds in the healthcare service market involves the hierarchical as well as institutional nature of the discussed relationships: patients are nested in basic healthcare units (HCU) and the content and nature of relations are strongly influenced by institutional forces. It means that properties of HCU (localization, infrastructure, atmosphere etc.) may cause the meaningful variation in the means of the dependent variables across HCU. Therefore the latent variables on the cluster level may explain the covariation among the observed indicators (random slopes and/or intercepts coefficients). This aspect of the relationships is tested by multilevel structural equation models.

Muthén's (1994) six-step approach to multilevel latent variable models is adopted in the process of modeling: 1/basic analysis of intraclass correlations (ICC), pooled within-cluster covariance matrix, and estimated between covariance matrix, 2/ factor analysis of total covariance matrix ignoring clustering, 3/ factor analysis correcting for standard errors, 4/ CFA on pooled within-cluster covariance matrix, 5/ CFA on estimated between covariance matrix, 6/ full two-level CFA/SEM model . M*plus* 4.0 was used in the estimation.

Basic information

In the basic step the ICC and design effects (*deff*) has been calculated as well as pooled within covariance matrix and between covariance matrix. The (*deff*'s) and ICC's for manifest variables and ICC's for the manifest and latent variables are given respectively as follows: Financial (Fin) 3.95; 0.252, Technical (Tech) 5.64; 0.396, Organizational (Org) 4.45; 0.294, Knowledge (Know) 8.37; 0.628, Communication (Comm) 8.16; 0.611, Individualization (Ind) 7.90; 0.588, Emotional (Emot) 9.01; 0.691, Psychological (Psych) 9.68; 0.741, Empowerment (Emp) 9.35; 0.712, Personal (Pers) 9.55; 0.729, Reputation (Rep) 8.49; 0.639, Ethical (Eth) 8.95; 0.678, Partnership (Part) 7.80; 0.580, Formal (F) 0.38, Social (So) 0.74, Structural (St) 0.77.

The results show that dominant share of social and structural bond (as well as formal with a little smaller size) is due to the HCU level effects. The higher level of ICC and *deff* (usually above 2), the more cluster-level characteristics influence variation among observable variables. High intraclass correlations and large design effects suggest that high level of variance in the observed variables is attributable to membership in their clusters. The number of clusters is 119 (the number of sampled HCU from the population of HCU's in the west-northern region of Poland) and average cluster size is 12.73. Total sample size was 1515 respondents (patients).

SEM complex and ignoring clustering models

Suggested 3-factor model was used for structural equation models of relationship bonds. Two kinds of the models was built: the first based on total sample covariance data and the second - total sample correcting for standard errors (s.e.). The parameters of both models are presented in Table 1.

Table 1. SEM model of relationship bonds

Variables	Estimates	S.E. basic	S.E. complex
Formal bond	Cronbach's alpha=0.64		
Fin	1.00	0.00	0.00
Tech	1.23	0.07	0.09
Org	0.83	0.05	0.13
Social bond	Cronbach's alpha=0.90		
Ind	1.00	0.00	0.00
Emot	1.23	0.04	0.11
Psych	1.26	0.04	0.08
Emp	1.10	0.04	0.08
Structural bond	Cronbach's alpha=0.89		
Know	1.00	0.00	0.00
Comm	1.05	0.03	0.12
Pers	1.18	0.03	0.13
Rep	1.10	0.03	0.12
Part	1.08	0.03	0.12
Formal with Social	0.54	0.05	0.15
Social with Structural	0.68	0.04	0.14
Formal with Structural	0.52	0.04	0.14
Choice on Formal	0.01	0.08	0.09
Choice on Social	0.02	0.18	0.19
Choice on Structural	0.35	0.11	0.13

The chi-square of basic model (ignoring clustering) equals 509.31 with $df = 51$ ($p = 0.00$), CFI = 0.96 and RMSEA = 0.077. The chi-square of complex sample model equals 172.44 with $df = 1$ ($p = 0.00$), CFI = 0.99 and RMSEA = 0.04. The goodness of fit of complex model is better than of basic model.

The structural parameters are supplemented with two kind of s.e.: ignoring clustering (first column) and correcting for complex sampling (second column). The reliability of constructs seem to be acceptable (Cronbach's alpha for the latent variables except formal bond is above 0.7) and factor loadings suggest that parallel, reflexive items are used for the measurement of these three latent factors. Moreover, s.e. under complex model are greater than for the model ignoring clustering where s.e. are (spuriously) lower.

Correlations between constructs suggests also simplex structure, because correlations between the adjacent constructs (F-So, So-St) are higher than between distant constructs (F-St). The only significant structural path is between structural bond and the choice of HCU. This finding confirms the hypothesis that formal bond has the weakest influence on the choice of HCU provider in comparison to structural (partnership, reputation etc.).

IRT models

The concept of relational bonding can be viewed also from the IRT perspective. This is suggested by the structure of the correlations between the items. Rasch 1-parameter model based on total sample and complex sample data was estimated. Table 2 shows that the item thresholds (loadings are fixed to 1.00) are similar but s.e. of thresholds are higher for the complex sample model. The Loevinger's H for the scale is 0.53 and indicates acceptable reliability of

Table 2. Rasch model of relational bonds

Variables	Thresholds (s.e.) - basic	Thresholds (s.e.) - complex	Loevinger's H
Fin	2.08 (0.17)	2.09 (0.27)	0.37*
Org	0.89 (0.16)	0.90 (0.22)	0.34*
Tech	0.37 (0.15)	0.38 (0.27)	0.48*
Emot	1.04 (0.15)	1.05 (0.34)	0.57
Ind	0.85 (0.15)	0.86 (0.34)	0.52
Emp	0.62 (0.15)	0.63 (0.33)	0.57
Psych	0.44 (0.15)	0.45 (0.35)	0.62
Rep	0.48 (0.15)	0.49 (0.34)	0.54
Eth	0.18 (0.15)	0.18 (0.33)	0.56
Part	0.14 (0.15)	0.15 (0.33)	0.54
Comm	0.03 (0.15)	0.04 (0.31)	0.59
Pers	-0.39 (0.15)	-0.39 (0.33)	0.62
Know	-1.19 (0.15)	-1.18 (0.32)	0.60

*indicates the worst items that violate the assumption of double monotonicity

the cumulative scale and confirms also the lower reliability of the formal bond measures[1].

CFA on pooled within-cluster and estimated between covariance matrix

Confirmatory factor analytic model on pooled-within covariance matrix compared to CFA on full-covariance matrix was used to assess the level of ecological fallacy.

[1] Loevinger's H is used in the context of nonparametric Mokken scaling. The analysis of scale reliability was done using the MSPWIN5 program - Mokken Scale Analysis for Polytomous Items.

Table 3. CFA on pooled-within, between and full covariance matrices

Matrices	Within		Between		Full	
Variables	Estimates	(s.e.)	Estimates	(s.e.)	Estimates	(s.e.)
Fin	1.00	(0.00)	1.00	(0.00)	1.00	(0.00)
Tech	1.31	(0.14)	1.21	(0.14)	1.23	(0.07)
Org	0.99	(0.09)	0.81	(0.16)	0.83	(0.05)
Ind	1.00	(0.00)	1.00	(0.00)	1.00	(0.00)
Emot	2.19	(0.41)	1.32	(0.09)	1.22	(0.04)
Psych	3.03	(0.55)	1.31	(0.09)	1.26	(0.03)
Emp	1.99	(0.36)	1.07	(0.10)	1.10	(0.03)
Know	1.00	(0.00)	1.00	(0.00)	1.00	(0.00)
Comm	1.16	(0.09)	1.06	(0.08)	1.07	(0.03)
Pers	0.94	(0.08)	1.24	(0.08)	1.18	(0.03)
Rep	0.87	(0.08)	1.06	(0.09)	1.05	(0.03)
Eth	0.56	(0.07)	1.17	(0.09)	1.08	(0.03)
Part	0.79	(0.08)	1.11	(0.08)	1.09	(0.03)
Formal with Social	0.03	(0.00)	0.41	(0.07)	0.56	(0.04)
Formal with Structural	0.09	(0.01)	0.40	(0.07)	0.54	(0.03)
Social with Structural	0.03	(0.00)	0.54	(0.08)	0.68	(0.03)

Table 3 shows factor loadings, standard errors and factor correlations respectively. Loadings for pooled within matrix are noticeably lower and its s.e. higher than for full covariance matrix. Also the factor correlations for pooled within matrix are almost zero with comparison to full covariance matrix. One can suggest that these differences are due to confounding effects of cluster level variables on individual level attributes. This effects appear also in the context of estimated between covariance matrix. The factor loadings in both type of analysis (between and full) are similar as well as correlations between factors. Higher factor correlations reflected on between covariance matrix indicate an ecological correlations between relationship bonds.

2.2 ML-SEM and ML-IRT models of relationship bonds

In the two-level structural equation model, CTT (classical test theory) is the measurement model nested in the first level of the analysis. The parameters of the model are estimated simultaneously on both levels.

Figure 1 presents 3-factor SEM model with probability of HCU choice as a dependent observed variable. Chi-square value for this model is 310.08 with df 162 and p-value = 0.00. The basic model fit indicators CFI = 0.861 and RMSEA = 0.026. The goodness-of-fit shows significant Chi-square statistics, moderately good CFI and acceptable RMSEA indices. In the within part of

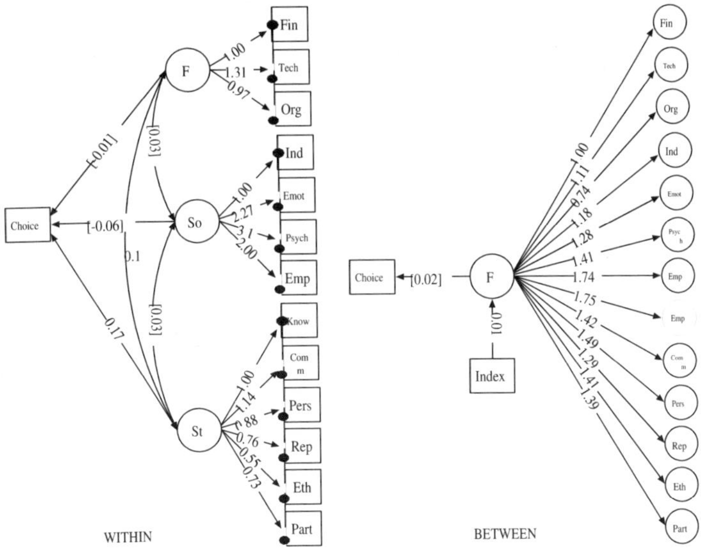

Fig. 1. ML-SEM model
(bullets represent the random intercept)

the model the relational bonds are uncorrelated with each other and only significant relation between structural bond and choice of HCU exists.

On the between part of the model, the HCU - level variation in dimensions of relational bonds is a function of single latent variable that is the general climate of HCU. We can conclude that patients' perception across the HCUs' is explained by this general factor. Between level covariate that is the index of healthcare service perception of HCU's providers (manifest exogenous variable "Index") has significant but weak influence on the general HCU climate on between level[2]. The climate of HCU's has no influence on the subjective probability of choice.

The second multilevel model was based on IRT measurement model that becomes very popular in multilevel SEM modeling (Fox 2005). The loglikelihood of the best model is 8132.16 and AIC = 16298.33. Within part of the model assumes unidimensional and cumulative nature of the items. The formal bond items was excluded from the analysis because of low reliability. The factor loadings of 2-parameter IRT indicate that the highest discriminant power on the unidimensional continuum represent the items for social bond and little lower for structural bond. The numbers in the brackets represent the items's thresholds that show the general increase in difficulty of items from structural to social. As in ML-SEM model, the variable Index does not

[2] Manifest variable Index on between level is constructed by the same 13-item sum of scores as before but for the sample of HCU providers.

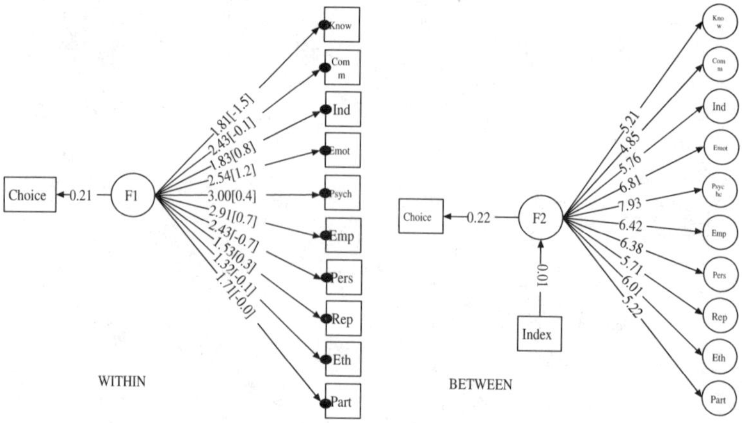

Fig. 2. ML-IRT model (bullets represent the random intercept)

covariate with general HCU climate. However, on both individual and cluster level the relationship bond has significant relation with the choice of HCU.

To sum up, we can conclude that HCU level characteristics explain the perception of relational bonding but the dimensions of relationships have a little influence on the choice of the HCU's provider and seem to be an autotelic value for the customers. In the explanation of HCU choice unidimensional IRT measurement model, besides its lower reliability, has a greater predictive validity in comparison to classical test theory measurement model.

References

BEN-SIRA, Z. (1999): Evaluation of Medical Treatment and Competence Development of a Model of the Function of the Physicians Affective Behavior. *Social Science and Medicine, 16, 132–145.*

BERRY, L.L. (2000): Relationship Marketing of Services Growing Interest,Emerging Perspectives. In: J.N. Sheth and A. Parvatiyar (Eds.): *Handbook of Relationship Marketing.* Sage, London.

FOX, J.P. (2005): Multilevel IRT Using Dichotomous and Polytomous Response Data. *Britisch Journal of Mathematical and Statistical Psychology, 58, 145–172.*

JARVIS, C.B., MACKENZIE, S.B. and PODSAKOFF, P.M. (2003): A Critical Review of Construct Indicators and Measurement Model Misspecification. *Journal of Consumer Research, 11, 30, 1–22.*

MUTHÉN, B. (1994): Multilevel Covariance Structure Analysis. *Sociological Methods and Research, 22, 376–398.*

RUDAWSKA, I. (2006): *Economization of Patient-provider Relationship in Healthcare.* University of Szczecin Press, Szczecin.

WILSON, D.T. (1995): An Integrated Model of Buyer-Seller Relationships, *Journal of Academy of Marketing Science, 23, Fall, 17–29.*

Data Mining in Higher Education

Karoline Schönbrunn and Andreas Hilbert

Technische Universität Dresden, Fakultät Wirtschaftswissenschaften, Professur für Wirtschaftsinformatik, insb. Informationssysteme im Dienstleistungsbereich, D-01062 Dresden; {schoenbrunn, hilbert}@wiid.wiwi.tu-dresden.de

Abstract. The aim of this paper is the critical discussion of different data mining methods in the context of the demand-oriented development of bachelor and master study courses at german universities. The initial point of the investigation was the question, to what extent the knowledge concerning the selection of the so-called "Fachkernkombinationen" (major fields of study) at the Fakultät Wirtschaftswissenschaften of the Technische Universität Dresden, could be used to provide new important and therefore demand-oriented impulses for the development of new bachelor and master courses. In order to identify these entrainment combinations it is obvious to examine the combinations of the major fields of study by means of different data mining methods. Special attention applies to the association analysis which is classical used within the ranges trade (basket analysis) or e-business (web content and web Usage mining) – an application in the higher education management is missing until now.

1 Motivation

German students become more demanding. Thus the students will change thereby the german university landscape. Motives of high school graduates of choosing their study places changed in the last years. An unpublished exploration of first-year students of the university-information-system (HIS) found out that 75% of the high school graduates decide to choose their university due to their place of residence and/or "hotel mummy". Good equipment of the university is an important criterion for 51% of high school graduates, just like for 52% the reputation of the university. For 82% of the high school graduates it is above all important that the courses offered corresponds to the specialized interests. High school graduates use more and more frequently the information from university rankings to choose an university. Therefore it is not amazing that faculties with good ranking results register more students in the following term in relation to the previous year. However the university management and education politicians do not consider this competition trend, because they associate competition rather with the areas of research and professors and not with the students. But good universities are characterized by outstanding researchers and by excellent students. First approaches to commit potential students to the universities are already used by several universities e.g. in the form of workshops for mathematics pupils (University of Munich) or also by "children universities" (Technical University of Dresden). Not only german students but also foreign students select the german universities on basis of the results of university rankings, e.g. it is not allowed for the Goethe Institute and the German academic exchange service to recommend an university, but refer to the university rankings, which are in the meantime although available in English. Crucial for the universities are the increased incomes of study fees, the achievement-dependant allocation of funds and the selection of the best students, which are interested in efficient learning and also at a efficient final degree. The today quoted "burden of students" becomes a source of income for the universities and the students, who have to pay for their studies will more exactly pay attention for which university they spend their money, like the system in the USA (Spiewak (2005)).

2 Student relationship management

New public management deals with the change from the national administrative organisation to an efficient administration by introduction of economical efficiency criteria to public institutions (like e.g. rankings, global households, etc.) (Brggemeier (2004)). An important component of the higher education management is the marketing for an university. This can be understood as the management of relations of the university and its audience and covers establishment, maintenance and reinforcement of relations with "customers", other partners and social groups (Zanger (2005)). An advertisement of an university in a newspaper is an example of university marketing. In this case the social target group and/or the customers are the students. By means of Customer Relationship Management (CRM) customers are acquired, bounded to the enterprise or regained. Thereby several steps have to be gone through, and different strategies have to be used (Neckel and Knobloch (2005)). In contrast to CRM, which focuses on the customers, the higher education management moves the student into focus. The so-called Student Relationship Management (SRM) deals with a holistic, systematic care of the "business relationship" between university and student, whereby particularly the service quality becomes a more and more interesting point (Göpfrich (2002)). The article refers to this definition, since there is an improvement of service quality for students based on the results of the data mining analysis. Figure 1 explains the correspondence of the three steps of CRM in SRM, whereby in this article the step of the operational SRM is discussed.

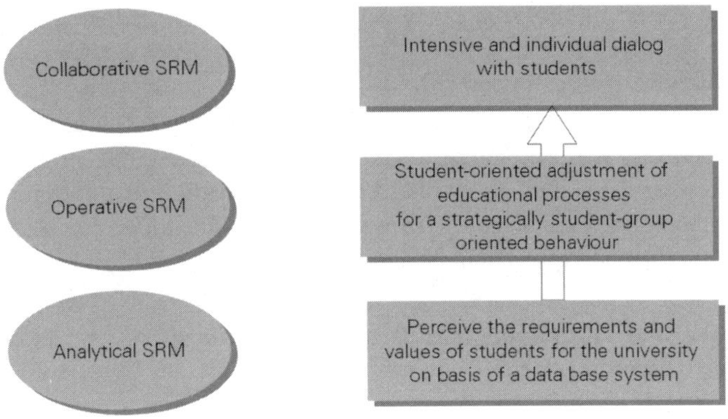

Fig. 1. Student Relationship Management (based on Töpfer (2004))

3 Data mining and its range of applications in higher education management

Data mining is an integral part of knowledge discovery in databases (KDD), which is the overall process of converting raw data into useful information (Tan et al. (2006)). This process is represented in the Figure 2 and will be explained briefly. The data mining process consists of five steps. At first, the

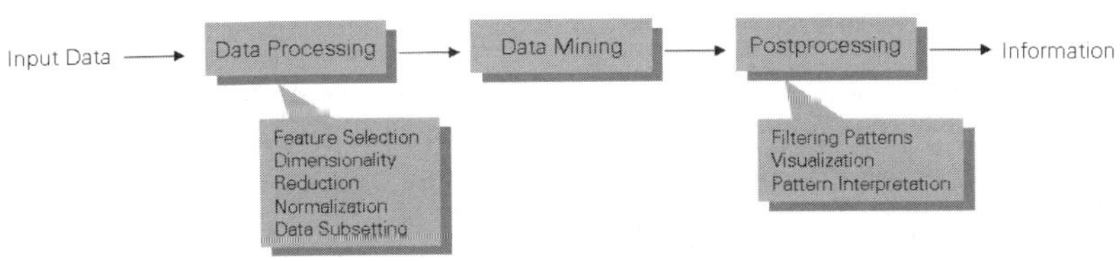

Fig. 2. Process of data mining (Tan et al. (2006))

existing data are prepared (e.g. by means of dimension reduction), then the application of the individual data mining methods take place and afterwards the post processing of the data (e.g. the visualization of the found information) and the extraction of the found information follows.

In the presented article the data mining methods *association analysis* and *link analysis* are used. Association rules describe correlations between conjoined occurring objects. It may concern e.g. products of a supermarket, which customers bought together. Additionally to the information of the objects the rule contains information about the frequency of the conjoined occurring objects, which is denoted as *support*. The Support for the rule $A \rightarrow B$ is computed as $sup(A \rightarrow B) = p(A \cup B)$. Further, it contains information about the strength of the correlation, the so-called *confidence* of a rule – computed as $conf(A \rightarrow B) = \frac{sup(A \rightarrow B)}{sup(A)} = p(B|A)$. An association rule consists of a quantity of items in the rule body and a quantity of items in the rule head (Bollinger (1996)). The link analysis visualizes relations between units in a complex system with a link graph as output.

Particularly in the USA already real and hypothetical (case) studies were accomplished in the range of higher education management. In Table 1, possible questions are specified in the field of higher education management which can be analysed and answered with data mining methods. These are opposed with the appropriate questions of the economy.

Table 1. Possible data mining questions in higher education with its equivalent in the economy (Luan (2004))

Private sector questions	Higher Education equivalents
Who are my profitable customers?	Which students are taking the most "credit hours"?
Who are my repeat Web site visitors?	Which students are most likely to return for more classes?
Who are my loyal customers?	Which are the persisters at my college/university?
Who is likely to increase his/her purchases?	Which alumni are likely to make larger donations?
Which customers are likely to defect to competitiors?	What type of courses will attract more students?

Student typologies are analysed in the context of "similar analysis" in the USA. Students were identified and clustered by means of the data mining cluster methods TwoStep and K-means. Thus university lecturer and the university management receive better insight of needs of the different groups of students. Further forecasts enabled the university to accurately identify good transfer students by means of neural nets, C5.0 and C&RT. Equally there can be made forecasting about the donation behaviour of alumni's due to mass mailings, particularly with the consideration of outliers (e.g. unexpectedly high donations of an alumni). In summary, educational institutions can be used the yielded results of the data mining analysis to better allocate resources and staff, proactively manage student outcomes and improve the effectiveness of alumni development (Luan (2004)).

4 Case study in the faculty management

In the following analysis the frequently selected combinations of major fields of study of the Fakultät Wirtschaftswissenschaften are to be analysed. In addition, the association analysis is transferred to the range of the higher education management. While the basket analysis examines which products are frequently bought together by the customers, the presented article examines the frequently selected major fields of study by the students. Afterwards the yielded results can be used in a further step for e.g. adjustments of the time table corresponding to student desires, avoiding overlaps (event management) and for the development of new bachelor and master studies as basis of information.

The data were evaluated in the time period from 01.02.2001 to 04.11.2005 at the Technischen Universität Dresden. Collectively the data of 1698 students of the Fakultät Wirtschaftswissenschaften were examined, who take one of the study courses of Betriebswirtschaftslehre (BWL), Volkswirtschaftslehre (VWL), Wirtschaftsingenieurwesen (WIng), Wirtschaftsinformatik (WInf), Wirtschaftspädagogik (WiPäd) or Internationales Management (IN). The variables in detail contain the ID of the student, the selected study course, the selected major fields of study, the mark and the date of the closing of the major field of study. A first analysis resulted in the conclusion that the consideration of all students of the complete faculty is not meaningful, since the examination regulations of the individual study courses are too different, so that the results are biased. An interpretation of these results is difficult and/or not possible in this case. Therefore, in a further step the individual study courses were analysed. In this article the results of the study course Wirtschaftsinformatik are exemplary represented, which contains the data of 285 students. The analysis was performed with the SAS Enterprise Miner© 4.3.

Variables included in the analysis were the ID of the students and the major fields of study. In the pre-processing step first the study courses BWL, VWL, WIng, WiPäd an IM were filtered out. Further, only students with a mark between 1.0 and 4.0 in the major field of study were included in the analysis. Minimum confidence for the generation of the rules was 30% and for the support 5%. For a better visualization only rules, which consist of two elements and whose occurance was count more than at least 35 times were used. The results of the association analysis are shown in Figure 3. The legend contains the corresponding values for confidence and support, whereby the symbol represents the confidence, the shades represents the support. The size of the symbol in relation to the size of the other symbols explains the *lift*, which is also a measure of the importance of a rule and – computed as $lift(A \rightarrow B) = \frac{conf(A \rightarrow B)}{sup(B)} = \frac{sup(A \cup B)}{sup(A)\ sup(B)}$. The lift gives information over the change about the distribution of certain objects of a subset to the distribution of the population. Exemplarily one association rule is interpreted and its implications for the development of a time table are represented. The interpretation of the association rule of the major subject of study *Informationssysteme im Dienstleistungsbereich* $\rightarrow$ *Kommunikationswirtschaft* is as follows:

- 17.19% (support) of the considered students selected both *Informationssysteme im Dienstleistungsbereich* and *Kommunikationswirtschaft.*

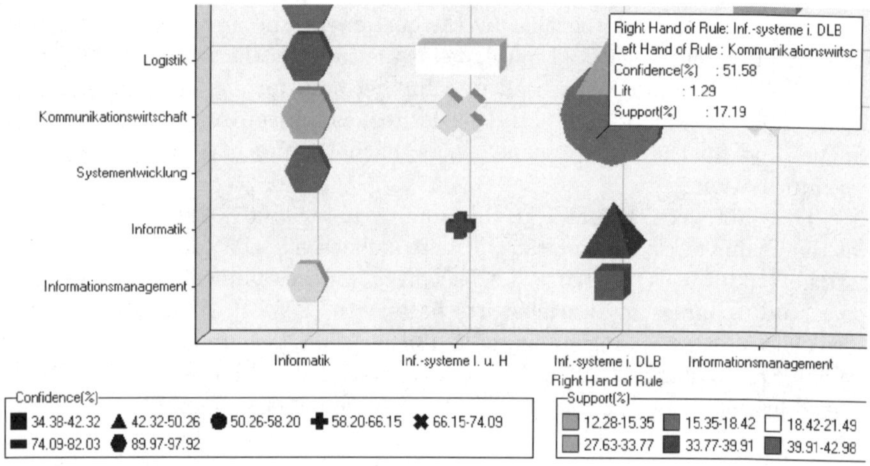

Fig. 3. Detail of the results of the association analysis

- In 51.58% (confidence) of the cases, in which *Informationssysteme im Dienstleistungsbereich* was selected, also *Kommunikationswirtschaft* was selected.
- Students, who selected *Informationssysteme im Dienstleistungsbereich*, selected 1.29 times (lift value) more frequently *Kommunikationswirtschaft* than all other students.

Consequently, attention for the development of the time table could be paid that courses of these two course studies do not overlap and students have the possibility to enrol for both course studies in the same semester and not to wait for one or two semesters.

In addition to the association analysis a link analysis was accomplished to visualize the relations between the individual major subjects of study. However, the problem exists, that students can enrol for some of their major subjects of study at other faculties in Dresden or in other countries. This miscellaneous courses are called "Ersatzfächer" (replacement courses) and should not be analysed in this article. Variables included in the analysis were the ID of the students and the major fields of study. Figure 4 visualizes the link graph of the link analysis. Each node represents a major subject of study. Individual nodes are alphabetically ordered. The size of the major subject of study represents the frequency with which the students enrol for a major subject of study. The major subject of study *Informationsmanagement* was enroled by e.g. 188 students, while *Industrielles Management* was enroled by only 18 students. The lines between the nodes represent the relations between the major subjects of study and the width gives information about the frequency of the relationship. It is indicating that the four professorships belonging to the study course Wirtschaftsinformatik are most frequently connected. For instance, the major subjects of study *Informationsmanagement* and *Infor-*

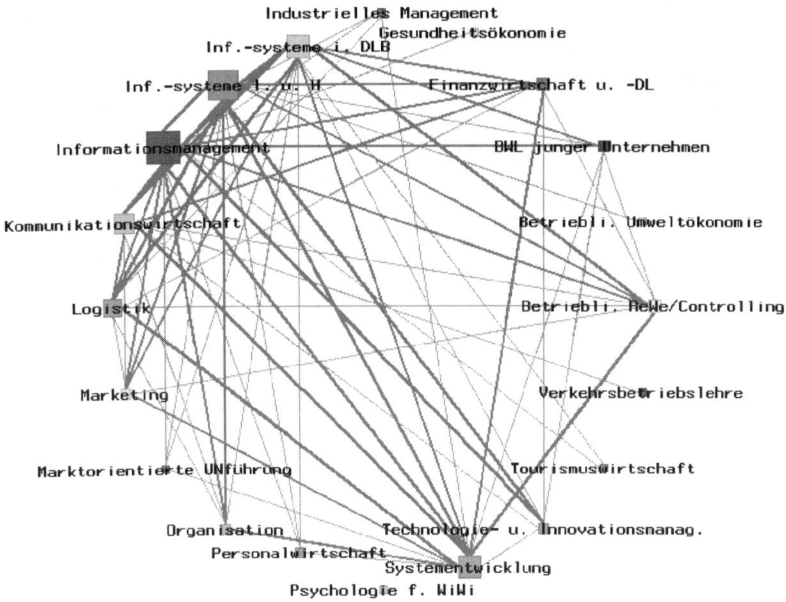

Fig. 4. Link graph as result of the link analysis

mationssysteme in Industrie und Handel show 120 connections in contrast to *Informationsmanagement* and *Personalwirtschaft* with only 16 connections. The link graph can exemplify as basis of information e.g. for the development of new bachelor and master courses, since it represents the favoured combinations of major subjects of study of the students and thus conclusions on the contents of interest of the students are possible.

5 Result and prospect

The presented article provides first approaches for the use of data mining methods in the german higher education management and possibilities for the use of the yielded results particularly in the faculty management. The accomplished analysis supports the service quality for the students, mentioned by the Student Relationship Management, and contributes thus to the connection of the students to the university. The universities can determine the needs of the students and constitute these knowledge aimed, in order to intensify the connection to the students. Altogether the increase of the connection of the students to the university can lead to increasing student quantities in the future and thus also to increasing incomes for the university. By the adjustments of the event management to the needs of the students further sinking periods of study are possible.

In this context, the data security is problematic. For further analysis and methods data are needed, which are available but may be not used (e.g. data of enrolment). According to §4 para. 2 Federal Law for Data Protection (in the version of 2003) (Gola and Schomerus (2005)) data may only be used for the purpose they were evaluated for. A similar law situation concerning the data security is present in America. Indeed the data are evaluated in the context of survey portals, as e.g. NSSE (national survey of student engagement). Therefore, a data warehouse with data of 100 colleges has been generated in the last years, which contains large data sets as in data warehouses in the economy. On this basis the data mining questions raised in chapter 3 can be answered.

References

BOLLINGER, T. (1996): Assoziationsregeln – Analyse eines Data Mining Verfahrens. *Informatik-Spektrum, 19, 257–261.*

BRÜGGEMEIER, M. (2004): Public Management. In: A. Hanft (Ed.): *Grundbegriffe des Hochschulmanagements.* Universitätsverlag Webler, Bielefeld.

GÖPFRICH, H. (2002): SRM Student Relationship Management – Web–unterstützte Kundenorientierung im Bildungsbereich. In: B. Britzelmaier, S. Geberl and S. Weinmann (Eds.): *Der Mensch im Netz–Ubiquitous Computing*, 4. Lichtensteiner Wirtschaftsinformatik-Symposium, FH Lichtenstein, B.G. Teubner, Stuttgart/Leipzig/Wiesbaden.

GOLA, P. and SCHOMERUS, R. (2005): *BDSG: Bundesdatenschutzgesetz*, Kommentar. Beck, München.

LUAN, J. (2004): *Data Mining in Higher Education.* SPSS White Paper, URL: 290/http://193.99.40.183/upload/1122641492_Data.

NECKEL, P. and KNOBLOCH, B. (2005): *Customer Relationship Analytics – Praktische Anwendung des Data Mining im CRM.* dpunkt.verlag, Heidelberg.

SPIEWAK, M. (2005): Studenten erproben die Macht. *Die Zeit, 21, 19.05.2005.*

TAN, P.-N., STEINBACH, M. and KUMAR, V. (2006): *Introduction in Data Mining.* Addison Wesley, Boston.

TÖPFER, A. (2004): Vision und Realität von CRM-Projekten. In: H. Hippner and K.D. Wilde (Eds.): *Management von CRM–Projekten*, Gabler, Wiesbaden.

ZANGER, C. (2005): *Universitätsmarketing–Chancen eines ganzheitlichen Konzeptes am Beispiel der TU Chemnitz.* Vortrag 'Von der Öffentlichkeitsarbeit zum Hochschulmarketing', TU Dresden, 10.06.2005.

Attribute Aware Anonymous Recommender Systems

Manuel Stritt, Karen H.L. Tso and Lars Schmidt-Thieme

Computer-based New Media Group, Department of Computer Science,
Albert-Ludwigs-Universität Freiburg, D-79110 Freiburg, Germany;
{stritt, tso, lst}@informatik.uni-freiburg.de

Abstract. Anonymous recommender systems are the electronic pendant to vendors, who ask the customers a few questions and subsequently recommend products based on the answers. In this article we will propose attribute aware classifier-based approaches for such a system and compare it to classifier-based approaches that only make use of the product IDs and to an existing real-life knowledge-based system. We will show that the attribute-based model is very robust against noise and provides good results in a learning over time experiment.

1 Introduction

Recommender systems (RS) are used by online commercial sites, e.g. amazon.com and ebay.com (giftfinder), to help users to find products that fit their preferences. They use user profiles that contain preference indicators for specific products or types of products to predict the interest a customer may have in product offerings and recommend the customer those which match the customers's likings. Preferences can either be specified directly by customers, i.e., by rating products, or indirectly indicated by their behavior, e.g., search keywords, products viewed in detail, products put in the market basket or products purchased as well as frequencies and durations of such events.

In general, there are two types of RS (Stritt et al. (2005)):
(i) **RS with user identification** that require a user to login (or rely on other unsafe user identification mechanisms such as cookies) and therefore can accumulate preference indicators, e.g., purchase histories over time. Whenever customers log in the system, their past profiles are looked up and used for recommendation. New users have to provide some initial information, e.g., some product ratings, before they can get relevant recommendations. This is sometimes called the **cold start problem** (Rashid et al. (2002)). (ii) **Anonymous recommender systems** that do not require user identification and therefore do not have any initial information about a customer. As in the context of e-commerce, direct ratings and usage-based preference indicators

are too time consuming to collect for single use, other mechanisms have to be used to elicit preference information. Usually, a short questionnaire that is customized to the product category and customer needs is presented up-front. This questionnaire sometimes is called **task specification**.

While RS with identification often can collect referenced indicators that are closely related with specific products, product types or attributes, the main problem of anonymous recommender systems is that the task specification has to be sufficient broad and generic to be filled-in easily by customers, but at the same time specific enough to be useful for recommendations.

In most commercial systems such as the Mentasys Sales Assistant (see http://www.mentasys.de), this relation between customer needs and products is modeled initially explicitly by means of a complex conceptual model, that is adapted later on by usage information. To create such a conceptual model, it requires domain experts and methods for eliciting their knowledge. It also bears many problems in its own, nevertheless, it is time consuming, expensive, and has to be done for each product category.

Alternatively, one could try to learn such preference models automatically with machine learning methods as reported in (Stritt et al. (2005)). These models are based only on the product IDs. In this article we will introduce classifier based models that take the attributes of the products in account, too. For this, we will make the following contributions: (i) we will propose a classification model setup for learning anonymous recommender systems and provide evaluation methods (section 3), (ii) we will introduce a classifier-based model that makes use of the product attributes (section 4) and (iii) we will provide additional empirical evidence for system behavior over time (section 5).

2 Related work

There are in general four recommendation approaches: collaborative filtering, content-based, hybrid and knowledge-based (Huang et al. (2004)).

Collaborative Filtering (Goldberg et al. (1992)), the most commonly-used technique is the attend to find users with the same preferences and to recommend objects to a user that other users with the same preference liked. It uses the simple nearest neighbor methods and does not make use of object attributes. This method has been quite successful in terms of recommendation quality. Hence, due to their simplicity and good quality, collaborative filtering is the prevalent method in practice.

Content-based Filtering (CBF) stems from Information Retrieval (IR), computes comparison of rated items of a single user and the item in the repository. Item attributes information is used for this technique. The performance of using solely CBF have shown to be rather poor. Yet, attributes usually contain valuable information that could improve the performance of recommender systems.

Hybrid collaborative/content-based Filtering combines both CF and CBF techniques. Some hybrid recommender systems are described in Balabanović and Shoham (1997), Melville et al. (2002), Burke (2003) and Ziegler et al. (2004).

Knowledge-based uses knowledge from both the users and the products /items to generate recommendations.

The first three techniques are mostly suited for persistent recommendations, whereas the knowledge-based technique is commonly used for task-based ephemeral recommendation.

The prediction task for various recommendation approaches can be handled using different methods. Commonly used methods are neighborhood formation, association rule mining, machine learning techniques, ...etc. For instance, one RS based on stochastic processes is described in Geyer-Schulz (2003). In general, it can be done in two different ways in the research literature: (i) using heuristic correlation measures and (ii) using learning methods to train a classification model that predicts further ratings or rated items. In most cases, classification models have shown to be suitable for prediction tasks when used with products or users attributes information (Basilico and Hofmann (2004), Basu et al. (1998)). Thus, using classification models would be an appropriate approach for handling knowledge-based RS. A marketing approach of a classification scheme of buying patterns is described in Bunn (1993).

In Stritt et al. (2005) we introduced classifier-based models for anonymous RS that only take the product IDs into account. In contrast to this paper we will now describe models that make use of the product attributes, too. Attribute-based models have already been shown to be useful on data with varying characteristics as described in Tso and Schmidt-Thieme (2006).

3 Framework

The anonymous RS framework makes use of the following entities:

- Set of answers A $= \{a_1, a_2, ..., a_l\}$
- Set of products I $= \{i_1, i_2, ..., i_n\}$
- Product ranklists $R \in I^n$
- Set of profiles $U \in (P(A))^m$, where $P(A)$ is the set of all possible answer combinations.
- Sessions $S \in P(A)\text{x}I$

With this framework an anonymous RS can be modeled as a classification task (Stritt et al. (2005)). The learning table can be gathered from successful[1]

[1] The success-criterion should represent that a user likes a recommended product. In our scenario we define a success as product view (e.g. the user clicked on the recommended product).

(U,I) instances from an existing system or might be generated from a domain expert. In our scenario we have made use of the data gathered from an existing real-life system from an online digital camera shop provided by MENTASYS GmbH that we will call the status-quo system. The models proposed in this article are trained and evaluated on about 20000 instances based on data provided from the status-quo system.

Evaluation

For an in-vitro evaluation, the list of ranked products proposed by the classifier-based model are evaluated on the data of the status-quo system (Stritt et al. (2005)). We define that a hit (i.e. a user likes the proposed product) succeeds if an entry (u,i) appears in the test data. A hit can be seen as an entry in only one session or in all sessions containing the same profile.

An ideal case would be to propose the viewed product on the first position of the ranklist, which is better than listing it at the bottom of the ranklist (which is again better than not proposing it at all). Therefore, the rank positions are assigned with different weights. Breese et al. proposed the so-called breese-score (Breese et al. (1998)) that weights the rank positions with exponential decay. For evaluating different anonymous RS, we use a score that is based on this breese-score but also takes the session s and profile u in account:

$$RankScore(s, u, r) = \frac{hitcount(s, u, r)}{2^{(r-1)/(\alpha-1)}}$$

We set the parameter α to 5 as in Breese et al. (1998). The function *hitcount* depends on the the session s, the profile u and the rank r. This can be calculated in different ways, that lead to different measures:

- **multicount (MC)**: number of product impressions in $TSess$ of product listed at Ranklist(r)
- **singlecount (SC)**: 1 if $MC > 0$, else 0

Where $TSess$:

- **local**: $TSess = s$
- **global**: $TSess \subseteq S$; $\forall s_i \in TSess : \exists(u_j, i_j)|u_j = u$

The rankscore for the complete ranklist R is given by the sum over all ranks:

$$RankScore(s, u, R) = \sum_{i=1}^{|R|} RankScore(s, u, r_i)$$

To get an expressive comparable score, the scores can be calculated as percentage to the maximal rankscore obtained from the optimal ranklist.

$$RankScore = \frac{RankScore_{list}}{RankScore_{max}} * 100$$

The optimal ranklist consists of the products in *TSess* ordered descending by the frequencies. In this article we confine to the global, multicount scores as this one is the most expressive score for customer preferences (Tso and Schmidt-Thieme (2006)).

4 Attribute-based models

In Stritt et al. (2005) we introduced ID-based models for anonymous RS. These classifier-based models have been seen as a U → I training that leads to a A → R assignment. Another possibility is to make use of the product attributes (e.g. price, weight, height) in the hope that they provide more information than just the product-IDs. With this strategy, the model should be able to learn preferences in sense of what attributes customers like, what is closer to the recommendation task of a human salesman.

The idea is to use one classifier for each attribute that propagates one attribute given a profile. This step is followed by a second step that assigns the set of attributes a product-ID distribution. This second step can be seen as a classification problem, too. Figure 1 shows the work flow of this strategy.

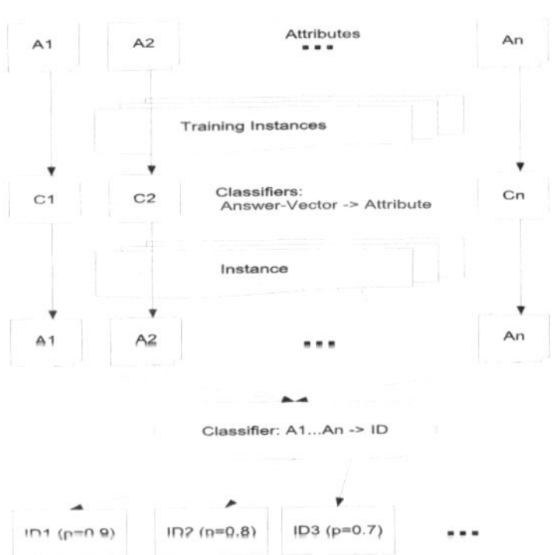

Fig. 1. Attribute-based model.

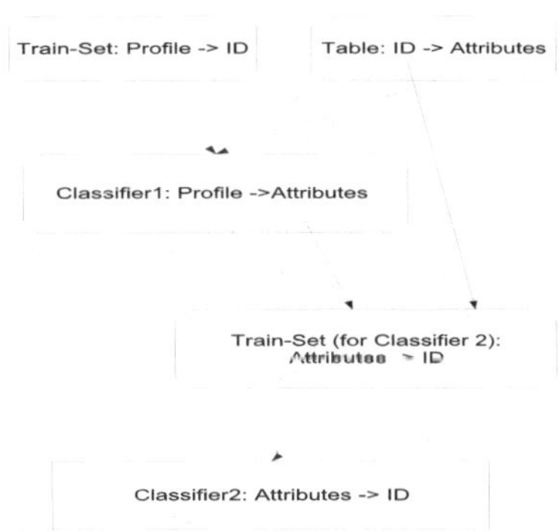

Fig. 2. Classifier: attributes → product-ID.

The training for the classifier of the second step is based on the products of the training table for the first step. The attributes proposed from the first step then are combined with the (original) product-IDs, which leads to the training table for the second classifier that is able to predict a product-ID given a set of attributes. Figure 2 shows an overview of the classifier.

Experiments have been made with using only one attribute, the name of the picture of the product, which is a unique identifier for the product (Attr(bilds)). Further experiments have been done with 5 expressive attributes

(Attr(5)) as well as 15 attributes (Attr(15)) of the products are made as well as experiments with nearly all attributes (Attr(80)) and with nearly all attributes but unique identifier attributes (Attr(80 -bilds)). Figure 3 shows the results for global multicount scores in comparison to the status-quo system and the ID-based model using a NB-Tree classifier (Kohavi (1996)), as well as the status-quo system optimized by a NB-Tree classifier (NB-Tree, sqpp).

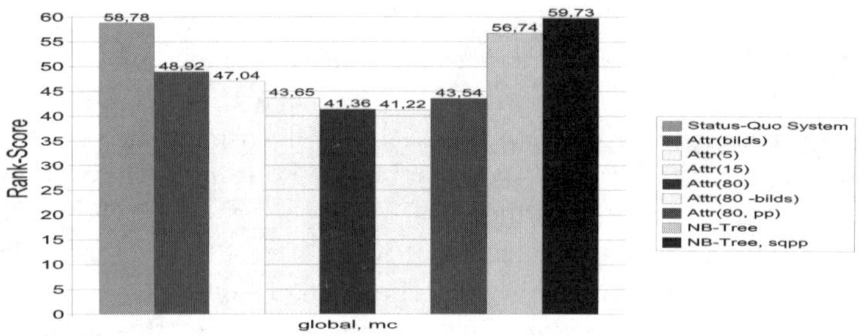

Fig. 3. Results of attribute-based models.

As we can see from Figure 3, the attribute-based results (Attr(x)) are not competitive compared to the ID-based system (NB-Tree). We think that the bad performance of such attribute-based models is due to the fact that the evaluation method only takes the product-IDs into account and does not care about the attributes. Thus, the model would only get a good score if the model proposes exactly the products the status-quo system proposed, even if the attribute-based model proposes products that, regarding the attributes, fits the preferences of the customers better. However, in learning over time (see 5) the results are competitive to the ID-based system and seem to be more robust in sense of a lower variance.

5 Learning over time

In all experiments so far, all sessions have been randomized before splitting into training and testing datasets. In real life, only data from the past can be taken into account for training because no data from the future is available. To simulate such a scenario, the whole data is divided into 10 segments ordered ascending by time. Each time segment contains data of about two weeks and is evaluated using only the time segments in the past for training. This means segment 1 is evaluated using segment 0 as training data and segment 2 is evaluated using segment 0 and 1 as training data and so on.

Figure 4 shows the results from learning over time by comparing the attribute-based model to the ID-based model using a NB-Tree classifier. This shows that the attribute-based model is competitive to the ID-based model in learning over time. The advantage of the attribute-based model is the lower

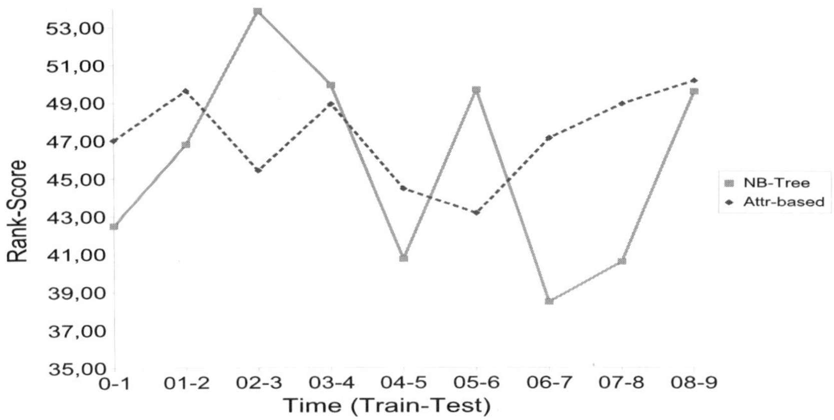

Fig. 4. Learning over time for an attribute-based model in comparison to an ID-based model using a NB-Tree classifier. (Scores from global, multicount.) Mean (attr/id): 47.19 / 45.79. Variance (attr/id): 5.95 / 28.62. (XY-Z on the x-axis means that segment Z was evaluated using segments X to Y as training data.)

variance. This is an evidence that the attribute-based system is more robust and is able to compensate noisy data.

6 Conclusion

Anonymous recommender systems are able to help users to find the products of their needs. The assignment from latent variables (the user preferences) to products is normally designed by experts knowledge. These knowledge-based systems are very successful but it takes a lot of effort to build these systems and to foster them. In this article, we proposed classifier-based models that make use of the product attributes.

In our first experiment, the attribute-based models were not competitive to the classifier-based system (section 4). One of the reasons for this could be due to the evaluation method that is based on "hard" IDs that don't have to stand for the real user preferences. A better way would be to define a metric based on the attributes and then compare the user preferences to the recommendations based on this metric (and not just hit or no hit).

When learning over time, the attribute-based model gives a good performance and it has lower variance. The learning over time experiment is very important because it only takes data from the past into account which is closer to a real life scenario. In this scenario the product set of the training and test data differs more than in a scenario where training and test datasets are taken out of a randomized data set.

Overall, attribute-based models appeared to be a very robust method for recommender systems and especially when noisy data are presented. Further work has to be done to evaluate these models in an online experiment to counterpoise the disadvantages of the in-vitro evaluation.

References

BALABANOVIĆ, M. and SHOHAM, Y. (1997): Fab: Content-based, Collaborative Recommendation. *Communications of the ACM, 40, 66–72.*

BASILICO, J. and HOFMANN, T. (2004): Unifying Collaborative and Content-based Filtering. *ICML 04: Proceedings of the Twenty-first International Conference on Machine Learning, ACM Press, 9.*

BASU, C., HIRSH, H. and COHEN, W. (1998): Recommendation as Classification: Using Social and Content-based Information in Recommendation. *AAAI 98/IAAI98: Proceedings of the Fifteenth National/Tenth Conference on Artificial Intelligence/Innovative Applications of Artificial Intelligence, 714-720.*

BREESE, J.S., HECKERMAN, D. and KADIE, C. (1998): Empirical Analysis of Predictive Algorithms for Collaborative Filtering. *14th Conference on Uncertainty in Artificial Intelligence(UAI-98), 43-52.*

BUNN, M.D. (1993): Taxonomy of Buying Decision Approaches. *Journal of Marketing, 57, 38-56.*

BURKE, R. (2003): Hybrid Recommender Systems: Survey and Experiments. *User Modeling and User-Adapted Interaction, 12, 4, 331-370.*

GEYER-SCHULZ, A., NEUMANN, A. and THEDE, A. (2003): An Architecture for Behaviorbased Library Recommender Systems. *Information Technology and Libraries, 22, 4, 165-174.*

GOLDBERG, D., NICHOLS, D., OKI, B.M. and TERRY, D. (1992): Using Collaborative Filtering to Weave an Information Tapestry. *Communications of the ACM, 35, 12, 61-70.*

HUANG, Z., CHUNG, W. and CHEN, H. (2004): A Graph Model for E-commerce Recommender Systems. *J. Am. Soc. Inf. Sci. Technol., 55, 3, 259-274.*

KOHAVI, R. (1996): Scaling up the Accuracy of Naive-Bayes Classifiers: A Decision-tree Hybrid. *Proceedings of the Second International Conference on Knowledge Discovery and Data Mining, 202-207.*

MELVILLE, P., MOONEY, R. and NAGARAJAN, R. (2002): Content-boosted Collaborative Filtering. *Eighteenth National Conference on Artificial Intelligence (AAAI-2002), 187-192.*

RASHID, A., ALBERT, I., COSELEY, D., LAM, S., MCNEE, S., KONSTAN, J. and RIEDEL, J. (2002): Getting to Know You: Learning New User Preferences in Recommender Systems. *Proc. IUI, 127–134.*

STRITT, M., TSO, K., SCHMIDT-THIEME, L. and SCHWARZ, D. (2005): Anonymous Recommender Systems. *OGAI Journal, 4-11.*

TSO, K. and SCHMIDT-THIEME, L. (2006): Evaluation of Attribute-aware Recommender System Algorithms on Data with Varying Characteristics. *Proceedings of the Tenth Pacific-Asia Conference on Knowledge Discovery and Data Mining (PAKDD 2006), 831–840.*

ZIEGLER, C.-N., SCHMIDT-THIEME, L. and LAUSEN, G. (2004): Exploiting Semantic Product Descriptions for Recommender Systems. *Proceedings of the 2nd ACM SIGIR Semantic Web and Information Retrieval Workshop, 25–29.*

Part VII

Banking and Finance

On the Notions and Properties of Risk and Risk Aversion in the Time Optimal Approach to Decision Making

Martin Bouzaima and Thomas Burkhardt

Lehrstuhl für Finanzierung, Finanzdienstleistungen und eFinance, Universität Koblenz-Landau, Campus Koblenz, Universitätsstr. 1, D-56070 Koblenz, Germany; {bouzaima, tburkha}@uni-koblenz.de

Abstract. This research proposes and discusses proper notions of risk and risk aversion for risks in the dimension of time, which are suitable for the analysis of time optimal decision making according to Burkhardt. The time optimal approach assumes a decision maker with a given goal, e.g. a given future wealth level, that he would like to reach as early as possible. To reach the goal, he can choose from a set of mutually exclusive actions, e.g. risky investments, for which the respective probability distributions of the goal reaching times are known. Our notions of risk and risk aversion are new and derived based on a rational model of choice. They yield intuitively appealing results which are exemplified by an application to an insurance problem. Furthermore, we investigate the choice implications of positive, zero, and negative risk aversion by means of a new St. Petersburg Game. The results indicate that in the time optimal approach nonnegative risk aversion would generally result in counterintuitive choices, whereas negative risk aversion shows the potential to imply plausible choices.

1 Introduction

The aim of this paper is to develop notions and properties of risk and risk aversion for risks in the dimension of time. Hence, we go beyond the existing time preference literature that does not allow for stochastic waiting times (see Frederick et al. (2002) for a review). We assume that the subject aims to achieve a certain goal and that he has complete knowledge about the probability distribution of waiting times resulting from each of his possible actions. Section 2 gives a brief introduction to our constitutive choice theoretical framework. Section 3 sketches the practical relevance of our analysis. Section 4 introduces the new representation of risk preferences over waiting times by utility functions and the derivation of measures of risk aversion thereupon. Section 5 illustrates the intuitive appeal of the new measures by applying them to an insurance problem and section 6 exemplifies how testable behavioral implications can be derived from them. Section 7 introduces a new St. Petersburg Game based on waiting times. In this extreme choice situation, negative risk aversion has the potential to imply plausible choices, but nonnegative has not. This is a striking result, especially since the preceding sections show that our understanding of risk-preferences seems quite reasonable. Section 8 concludes.

2 Time Optimal Decision Making (TODM)

The subsequent analysis builds on a very general choice theoretical framework that we call Time Optimal Decision Making (TODM). The underlying idea of this approach is to shift the focus of attention from the dimension of value to the dimension of time, which is often neglected in the classic theory.

In TODM a subject tries to achieve a given result in the dimension of value, e.g. a certain consumption good or a certain aspired wealth level. It is designed for the application to choice situations where uncertainty occurs exclusively in the dimension of time. In short, the subject knows precisely what he gets, but he does not know when he gets it. The choice task is therefore to choose from a set of mutually exclusive actions that imply known distributions of waiting times. These choice situations are stylized and real world applications will usually feature risk in both dimensions, the dimension of value and the dimension of time. Hence, TODM is no substitute for the classic approach. It is more suitable for some choice situations and the classic approach for others.

In TODM preferences are represented by utility functions, $u_x(t)$. These utility functions have the waiting time t as their argument. The subscript x indicates the result in the dimension of value, such as the consumption good, that the waiting time t refers to. Thus, the utility function u evaluates consumption of a given result x with respect to the temporal distance and the uncertainty of the waiting time until x becomes available from the point of view of the present. This does not mean that subjects generate utility from consumption of time. They rather generate utility from consumption as in the classic sense. In practical applications risk preferences over waiting times may be dependent on x. However, analysis of this dependency is ultimately an empirical task and not subject of this conceptual study.

A further finding of TODM is the existence of a von Neumann/Morgenstern (v.N.M.) Expected Utility representation of preferences over uncertain waiting times (Burkhardt (2006)). It is derived from a set of axioms that characterizes rational choice. Given this approach, a subject is rational if it behaves according to

$$\max E[u_x(\tilde{t})] = \sum_{i=1}^{n} p_i u_x(t_i), \tag{1}$$

where $\tilde{t}$ is a random variable describing waiting times and p_i is the probability of waiting time t_i. The estimation of goal reaching time distributions is a separate, application-specific issue. See Burkhardt and Haasis (2006) for an empirical analysis in the context of DAX stock index investments.

Finally, the subsequent analysis will make use of a monotonicity assumption "earlier is better". It replaces the familiar "more is better" monotonicity assumption from the classic approach and implies that $u_x(t)$ is decreasing in t. We do not claim universal validity of this assumption. Earlier is not necessarily better in every context, however it seems reasonable in many choice situations relevant to TODM.

3 Practical relevance to financial applications

Among financial applications, TODM and concepts of risk and risk aversion derived thereupon are of particular importance to models of Time Optimal Portfolio Selection (TOPS), as originated by Burkhardt (1999, 2000a, 2000b). TOPS assumes that a subject, starting with an initial portfolio value, aspires to reach a certain target portfolio value as soon as possible. To reach this goal the subject chooses from a set of mutually exclusive investment alternatives. Each investment alternative implies a known probability distribution of goal reaching times. Therewith, risks are modelled in (and only in) the dimension of time. In TOPS identification of efficient or optimal portfolios relies on the assumption of some decision criteria. These decision criteria refer to aspects of the probability distribution of target reaching times. Naturally, any attempt to put such decision criteria on a rational foundation requires a sufficient understanding of risk and risk aversion in the dimension of time.

4 Measures of risk aversion in the dimension of time

A subject receives a given outcome x after an uncertain waiting time, which is described by a random variable $\tilde{t}$.

Definition 1. *A subject is {risk averse/ risk neutral/ risk prone} iff for every lottery $\tilde{t}_i$ his preference ordering is $\tilde{t}_i \{\prec / \sim / \succ\} E[\tilde{t}_i]$.*

Our approach to model time preferences using utility functions allows for a qualitative characterization of risk preferences based on the shape of these functions. It is easy to show that the well known result from the classic approach, that a rational subject is {risk averse/risk neutral/risk prone} iff his utility function is {concave/linear/convex} can be directly transfered to the time optimal approach. But because of $u'_x(t) < 0$ the utility functions have an unfamiliar shape.

Beyond the classification into these three qualitatively distinct classes, we can obtain measures based on the *degree* of concavity or convexity of the utility function. The second derivative of $u_x(t)$ with respect to t is therefore a natural candidate for a measure of risk aversion. However, since we make use

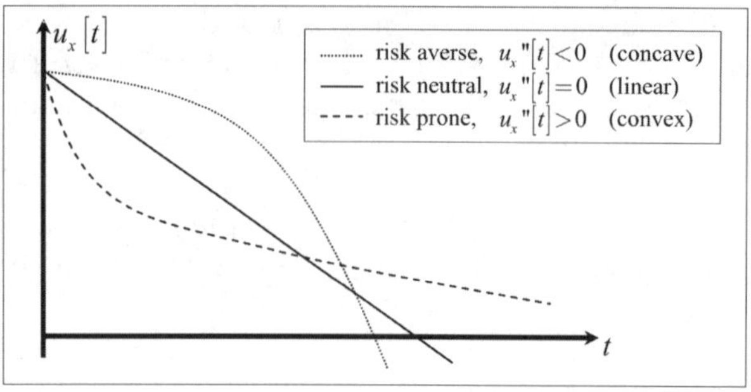

Fig. 1. Risk preferences over waiting times and the curvature of the utility function

of the existence of a v.N.M. expected utility representation of preferences, the utility function is only defined up to positive affine transformations. Proper measures of risk aversion therefore have to be invariant with respect to these transformations. The second derivative of $u_x(t)$ alone does not have the desired property, but Arrow-Pratt type measures of absolute, $r_A(t)$, and relative risk aversion, $r_R(t)$, do:

$$r_A(t) = \mathbf{m}+\frac{u''_x(t)}{u'_x(t)}; \qquad r_R(t) = \mathbf{m}+\frac{u''_x(t)}{u'_x(t)} \cdot t \tag{2}$$

The quotients are not multiplied with -1, as is common for the corresponding measures in the classic approach. Because $u'_x(t) < 0$, a larger quotient indicates a higher degree of risk aversion and no change in sign is required. See Pratt (1964) for the original treatment of these measures in the classic case.

5 Intuitive appeal of the new measures: Demand for insurance as a touchstone example

The economic appeal of the new measures can be tested by applying them to touchstone examples. Touchstones mark important requirements, which should be fulfilled by a theory to be reasonable or useful. The well-functioning of the new measures will be demonstrated by an application to an insurance example. We have a predefined intuitive understanding how a higher degree of risk aversion (risk preference) affects choices in *ceteris paribus* comparisons. We test if a higher degree of risk aversion, as measured by the new measures of risk aversion, has these intuitively presupposed choice implications.

The intuitive understanding of risk aversion is that a more risk averse subject u should be willing to pay a higher risk premium to eliminate a given risk than a less risk averse subject v. A "more risk averse than"-relation is modelled by a variation in the degree of concavity of the utility function (see Mas-Colell et al. (1995, pp. 191) and Wilhelm (1992, pp. 14) for the classic analog of lemma 1 and theorem 1):

Lemma 1. *$u_x(.)$, $v_x(.)$ and $G(.)$ are assumed to be twice differentiable. Then u_x has a higher degree of risk aversion (risk preference) than v_x $\forall t$ in the Arrow-Pratt sense, if and only if a concave (convex) function G exists with $u_x(t) = G(v_x(t))$ $\forall t$.*

In our analysis, insurance premiums π are paid in the "currency" of time units. This means a "more risk averse" subject should be willing to add a larger extra waiting time to eliminate a given waiting time risk than a "less risk averse" subject. A "more risk prone" subject will ask for a higher waiting time reduction than a "less risk prone" subject to trade in a given waiting time risk. Using Lemma 1 we can prove:

Theorem 1. *Assume $u_x(t)$ to be more {risk averse/ risk prone} than $v_x(t)$ $\forall t$ in the Arrow-Pratt sense. Then $\pi_u(\tilde{t})\,\{>/<\}\,\pi_v(\tilde{t})$.*

Proof. From G {concave/ convex} and Jensen's inequality it follows $u_x(E[\tilde{t}]+\pi_u(\tilde{t})) = E[u_x(\tilde{t})] = E[G(v_x(\tilde{t}))]\ \{</>\}\ G(E[v_x(\tilde{t})]) = G(v_x(E[\tilde{t}] + \pi_v(\tilde{t}))) = u_x(E[\tilde{t}]+\pi_v(\tilde{t}))$. Since $u_x'(t) < 0$, it follows that $\pi_u(\tilde{t})\,\{>/<\}\,\pi_v(\tilde{t})$. ■

6 Selected classes of risk preferences: The case of constant absolute risk aversion

The preceding sections sketched the new idea to model time preferences using utility functions and to develop and test measures of risk aversion over waiting times based on the curvature of these functions. Building on that we can analyze the properties of various classes of risk aversion. Derived behavioral implications may be used to analyze the empirical validity of these measures. The subsequent analysis concentrates on constant absolute risk aversion as an example. Once again, we turn to a stylized insurance problem:

Theorem 2. *A risk avers/ risk prone subject with a constant parameter $r_A(t)$ and $u_x'(t) < 0$ is endowed with a fixed goal reaching time $\bar{t}$, that is, a risk free investment. Now a pure risk $\widetilde{\Delta t}$ with $E[\widetilde{\Delta t}] = 0$ is added. Then the maximum additional waiting time $\pi(\widetilde{\Delta t})$ that a risk-averter is willing to add in order to eliminate the risk is independent from $\bar{t}$. Similarly, the minimum waiting time reduction $\pi(\widetilde{\Delta t})$ a risk-preferer asks for in order to trade in the risk is independent from $\bar{t}$.*

Proof. $r_A(t) = \frac{u_x''(t)}{u_x'(t)} = a \Rightarrow u_x(t) = C_1 - \frac{C_2}{a}e^{at}$ for $a \neq 0$ and $u(t) = C_1 - C_2 t$ for $a = 0$; $C_2 > 0$ because $u_x'(t) < 0$. We set $C_1 = 0$ and $C_2 = 1$ without loss of generality. From $E[u_x(\bar{t}+\widetilde{\Delta t})] = u_x(\bar{t}+\pi(\widetilde{\Delta t}))$ it follows that $E[e^{a(\bar{t}+\widetilde{\Delta t})}] = e^{a(\bar{t}+\pi(\widetilde{\Delta t}))}$ for $a \neq 0$ and $E[\bar{t} + \widetilde{\Delta t}] = \bar{t} + \pi(\widetilde{\Delta t})$ for $a = 0$. In both cases, $\bar{t}$ cancels out. ■

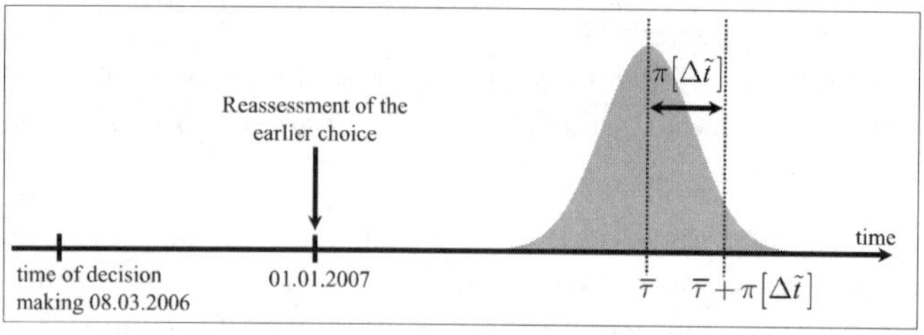

Fig. 2. Max. insurance premium and choice reassessment of the const. $r_A(t)$ subject

The choice implications of constant $r_A(t)$ are illustrated in Figure 2. A subject receives some result x at a certain calendar date $\bar{\tau}$. Now some pure risk is added to the choice situation, inducing the subject's willingness to add an extra waiting time up to π for the elimination of this risk.

Then we obtain two insights:

1. The subject's maximum risk premium in time-units is *independent* of the temporal distance between choice and outcome.

2. There will be no preference reversals. If the constant r_A subject has the opportunity to reassess his decision at some later point in time, he will stick to his choices. Neither would he regret that he purchased the insurance, nor would he regret that he did not use an opportunity to buy an insurance for a certain premium.

We get a different result for a decreasing r_A subject. This subject is willing to pay ever higher risk premiums as time passes. This compares nicely with the phenomenon of preference reversals in the classic analysis of time preferences for the deterministic case (see Frederick et al. (2002)).

7 Insights from a new St. Petersburg Game

St. Petersburg Games define extreme choice situations that highlight the counter-intuitive implications of risk neutral and risk prone choice behavior in classic choice situations. The prominence of this game in the theory of choice particularly stems from Bernoulli's (1738) solution to the paradox that exclusive consideration of expected payoffs is not reasonable. In this section, we introduce a new St. Petersburg Game that allows some new insights into risk preferences over waiting times. The subject receives a given result, such as a given amount of money or a certain consumption good. The outcome of the new St. Petersburg Game defines *when* he receives the result. A fair coin is tossed. If it takes n tosses until heads shows up for the first time, the player receives the result after 2^n time periods. Now we assume that the subject is endowed with a certain finite waiting time: He will receive the given result after this fixed waiting time. We ask: What is the smallest waiting time that the subject is willing to give up to get a lottery ticket for participation in the new St. Petersburg Game in return?

Table 1. A new St. Petersburg Game

Random draw	H	TH	TTH	...	$T...TTH$	...
Probability	$\frac{1}{2}$	$\left(\frac{1}{2}\right)^2$	$\left(\frac{1}{2}\right)^3$	...	$\left(\frac{1}{2}\right)^n$	...
Waiting time (in periods)	2	2^2	2^3	...	2^n	...

Theorem 3. *Neither a risk neutral nor a risk averse subject (always referring to risk preferences in the dimension of time) would trade in a fixed waiting time $t < \infty$ for participation in the new St. Petersburg Game.*

Proof. According to section 4 risk neutral preferences are represented by linear utility functions. Using $u_{lin,x}(t) = a - b \cdot t$, $b > 0$, we obtain:

$$E[u_{lin,x}(\tilde{t})] = \sum_{i=1}^{\infty} \left(\frac{1}{2}\right)^i \cdot \left(a - b2^i\right) \to -\infty \tag{3}$$

Hence, risk neutral subjects will always prefer to keep the fixed finite waiting time they are endowed with.

For every concave utility function, $u_{cv,x}(t)$, we can find a linear utility function, $u_{lin,x}(t)$, with $b > 0$, such that $u_{lin,x}(t) \geq u_{cv,x}(t)$ $\forall t \in \mathbb{R}_+$. Hence, $E[u_{cv,x}(\tilde{t})] \to -\infty$ and risk averse subjects never join the game. ■

Yet, not putting any finite waiting time endowment at stake, however large it is, is counter-intuitive. Assume a waiting time period, as introduced in Table 1, is one month. Furthermore, assume that the subject is endowed with a waiting time of 25 years. Then the probability that participation in the game shortens the waiting time is $p[t < 25\text{years}] = 99.609375\%$ and the expected waiting time in that case is $E[\tilde{t} \mid t < 25\text{years}] \approx 8.03$ months. Choice situations in which it is reasonable not to participate in the new St. Petersburg Game might exist. However, in most choice situations subjects are likely to be willing to take more risk than the risk neutral player.

Theorem 4. *Convex utility functions, $u_{cx,x}(t)$, which imply finite certainty equivalents, t_{CE}, for the new waiting time based St. Petersburg Game, exist.*

Proof. $u_{cx,x}(t) = e^{-t} \Rightarrow E[u_{cx,x}(\tilde{t})] = \sum_{i=1}^{\infty} \left(\frac{1}{2}\right)^i e^{-2^i} \approx 0.072$
$\Rightarrow t_{CE} \approx 2.63$ ■

This proof shows, by way of example, that risk prone subjects may show a more reasonable behavior. For the particular exponential function chosen in this example, the subject is willing to give up an endowment of roughly 2.63 months or more for participation in the new St. Petersburg Game and refuses to put any shorter waiting time at stake. However, the certainty equivalent is not finite for every convex utility function. Similarly, in the classic St. Petersburg Game concavity of the utility function is no sufficient condition for the existence of a finite certainty equivalent.

8 Conclusion

This paper introduced a new approach to analyze risks in the dimension of time. Building on our TODM framework, risk preferences over waiting times were linked to the curvature of suitable utility functions. Our new measures derived thereupon showed very reasonable properties. In this light the insights from our new St. Petersburg Game are all the more striking. Only negative risk aversion (read "risk-loving" behavior) showed the potential to imply plausible choices. Hence, we found some evidence that positive risk preference plays an important role in the dimension of time.

References

BURKHARDT, T. (1999): Time Optimal Portfolio Selection: Mean-variance-efficient Sets for Arithmetic and Geometric Brownian Price Processes. In: R. Decker and W. Gaul (Eds.): *Classification and Information Processing at the Turn of The Millennium.* Springer, Heidelberg, 304–311.

BURKHARDT, T. (2000a): *Portfolioselektion bei unsicherer Anlagedauer.* Habilitationsschrift, Technische Universität Freiberg.

BURKHARDT, T. (2000b): Wachstumsoptimierte Portfolioselektion auf der Grundlage von Zielerreichungszeiten. *OR Spektrum, 22, 203–237.*

BURKHARDT, T. (2006): A Model of Rational Choice among Distributions of Goal Reaching Times. In this volume.

BURKHARDT, T. and HAASIS, M. (2006): On Goal Reaching Time Distributions Estimated from DAX Stock Index Investments. In this volume.

FREDERICK, S., LOEWENSTEIN, G. and O'DONOGHUE, T. (2002): Time Discounting and Time Preference: A Critical Review. *Journal of Economic Literature, 40, 351–401.*

MAS-COLELL, A., WHINSTON, M.D. and GREEN, J.R. (1995): *Microeconomic Theory.* Oxford University Press, Oxford.

PRATT, J.W. (1964): Risk Aversion in the Small and in the Large. *Econometrica, 32, 122–136.*

WILHELM, J. (1992): Risikoaversion, Arbeitspapier. Universität Passau.

A Model of Rational Choice Among Distributions of Goal Reaching Times

Thomas Burkhardt

Lehrstuhl für Finanzierung, Finanzdienstleistungen und eFinance, Universität Koblenz-Landau, Campus Koblenz, Universitätsstr. 1, D-56070 Koblenz, Germany; tburkha@uni-koblenz.de

Abstract. This research note develops a theory of rational choice among distributions of goal reaching times. A motivation of the choice problem considered here is the time optimal approach to portfolio selection by Burkhardt, which considers an investor who is interested to reach a predefined level of wealth and whose preferences can be defined on the feasible probability distributions of the time at which this goal is reached for the first time. Here a more general choice problem is considered, called time optimal decision making. The decision maker is faced with a set of mutually exclusive actions, each of which provides a known distribution of goal reaching times. It is shown that the axiomatic approach of rational choice of von Neumann/Morgenstern can be applied to reach again an expected utility representation for the preferences under consideration. This result not only provides a rational foundation for time optimal decision making, and particularly time optimal portfolio selection, but also for the analysis of time preferences in a stochastic setting, an approach which is completely new to the literature to the best of the author's knowledge. Prime areas of application are decision analysis, portfolio selection, the analysis of saving plans and the development of new financial products.

1 Introduction

Why would we be interested in goal reaching times, and, in addition to that, probability distributions of these quantities? The underlying idea is quite simple: If we have to make a decision to reach a goal, we are quite naturally interested *when* we will reach it! That is, we are interested in the goal reaching time, defined as the time span until or the point in time at which we reach our goal. In an uncertain world, the actions we decide upon will hardly yield certain results, which means that we do not know the goal reaching time with certainty. At best, we might be able to attribute probabilities to the possible goal reaching times for all the alternative actions we consider. In this case, we face a decision under risk, and any choice is a choice from a set of distributions of goal reaching times, which characterizes the available actions.

Section 2 outlines and extents the concept of Time Optimal Decision Making (TODM) as recently developed by Burkhardt (Burkhardt (1999, 2000a, 2000b)) as the proper framework to model this kind of decisions. Section 3 relates the rationality concept of von Neumann/Morgenstern to TODM, and shows that the expected utility principle transfers to TODM without any restrictions. Section 4 undertakes some first steps into the investigation of the resulting utility function, based on the reasonable assumption *earlier is better*. Based on that, Section 5 develops a Time Optimal Stochastic Dominance framework, which is illustrated using the normal distribution. Section 6 concludes.

2 Time optimal decision making

The *time optimal* approach to decision making assumes that the decision maker aspires a well defined material goal. Prime examples are to realize an aspired consumption or to reach a certain wealth level. Objectives like these are typical for all kinds of saving processes, and are characteristic for a multitude of applications. Nevertheless, they are fundamentally different from the objectives that have so far been treated in the economic and financial literature. If future prospects are risky, the time optimal approach models the decision situation so that the decision maker optimizes over the set of feasible goal reaching time distributions for a given goal value. (The time optimal approach covers the more general situations of uncertainty as well, but this is not the subject of the subsequent discussion. Furthermore, the notion of optimization as it is meant here is not restricted to the case of unboundedly rational models. The time optimal approach may well be used in models of bounded rationality.) Consequently, *risk* is characterized *in the dimension of time* — an approach which is completely new to the literature. (There is a rather extensive literature on *time preference* in a *deterministic* setting, which analyzes decisions regarding the valuation of different values at different times with a major focus on discount functions, see Frederick et al. (2002) for a survey. The time optimal approach of decision making discussed here provides a framework for the analysis of time preference in a *stochastic* setting.) As opposed to the time optimal approach, the classic approach assumes that the decision maker has a certain time horizon, and cares about the outcomes of his action at the respective point in time. Therefore, the classic approach may well be characterized as *value optimal*. It models the decision situation so that the decision maker optimizes over the set of feasible value distributions for a given (goal) time, which marks the end of his time horizon. As a result, *risk* is characterized *in the dimension of value*. (This short and abridging sketch of classic decision models is restricted to single period models. But the fundamental focussing on the value dimension remains true even in multiperiod or continuous time models. See Burkhardt (2000b) for a detailed discussion.)

The time optimal approach can be regarded as *dual* to the classic approach. A little reflection shows that real-world goals, particularly financial goals, are characterized in both dimensions, value *and* time, whereas their description is usually quite fuzzy in both dimensions. Classic decisions models abstract from the fuzziness in the dimension of time, time optimal models from the fuzziness in the dimension of value. Both approaches complement each other, by focussing on one of the two dimensions, which is then used to characterize risks. Of particular interest are applications in portfolio selection. They make the aforementioned duality very clear. A given stochastic process $\tilde{v}(t)$ describing the value dynamics of a portfolio defines the distribution of the portfolio value $\tilde{v}$ for any given future point in time t as well as the distribution of the goal reaching time $\tilde{t}$ for any given goal value v. This may be symbolized by

$$\text{value optimal: } (\tilde{v}|t) \leftrightarrow (\tilde{t}|v) \text{ :time optimal} \tag{1}$$

Risk can be described in the dimension of value or time, depending on the chosen perspective, value or time optimal. Both approaches have in common that a decision under risk is effectively a choice from a set of probability distributions, which is determined by the set of available actions. They are distinguished only by the way in which preferences are modelled: Classic models are based on the the distribution of value for a given time, time optimal models on the distribution of goal reaching time for a given goal value. Those are the inputs for the preference model, which are *predefined by the chosen perspective.*

3 Rationality of time optimal decision making

Whether a decision is considered as rational or not, depends on the underlying concept of rationality. One of the most important concepts from the normative point of view is the one originated by von Neumann and Morgenstern (1947), which allows to derive the expected utility principle (EUP) from a small set of intuitively appealing axioms. The EUP is still the fundament on which much of economic theory is build. The classic approach is well in line with it, and it is certainly important to analyze if this also holds for TODM. This is the objective of the following sections. That is, the current investigation will be restricted to this concept, which by no means intends to negate that other concepts might also be of interest.

The first thing is to note that the EUP is a very general one. The underlying axioms require a certain structure regarding the preferences of the decision maker, which are defined on a set of probability distributions. Nothing is contained concerning the events on which those distributions are defined, so they could be almost any kind of good. The next and obvious thing is that an important attribute of any good is the point in time at which it will be available. That is, the EUP allows for the *formal* comparison of risky prospects which might yield different goods at different points in time. From this point of view, the comparison of such prospects defined on different results at the same point in time and on the same result at different points in time are merely special cases, which correspond to the classic and time optimal approaches to decision making. Less clear, and subsequently to investigate, is the question if the EUP for TODM is as sound as it is in the classic case from an economic point of view. (The subsequent investigation is restricted to this comparative question. Positive or more advanced normative issues will not be addressed here.) This is not immediately obvious. At all times, applications of the EUP referred more or less explicitly to the classic, value optimal case, and the underlying axioms were motivated accordingly. Already von Neumann/Morgenstern narrowed their discussion. They assumed "that the aim of all participants in the economic system ... is money, or equivalently a single monetary commodity" and "that difficulties can be obviated by locating all 'events' ... at one and the same, standardized moment", von Neumann and Morgenstern (1947, pp. 8, 19). The value optimal thinking is deeply rooted in financial economics. Without any necessity, some authors formulate even the essential axioms explicitly and thereby exclusively for the value optimal case. (A prominent example is Sinn (1989, p. 80).)

Actually, the EUP can be derived, at least for simple distributions, based on *three Axioms*, which describe preferences using a binary relation $\succsim$ on the set $\mathcal{P}$ of probability measures on a finite set Z of outcomes: 1. *Ordering*: $\succsim$ is a preference relation on $\mathcal{P}$, 2. *Independence* of irrelevant alternatives: For all $p, q, r \in \mathcal{P}$ and $a \in (0,1]$, $p \succ q$ implies $ap + (1-a)r \succ aq + (1-a)r$. 3. *Continuity (or Archimedian Axiom)*: For all $p, q, r \in \mathcal{P}$, if $p \succ q \succ r$, then there exist $a, b \in (0,1)$ such that $ap + (1-a)r \succ q \succ bp + (1-b)r$.

A little more reflection shows that these axioms are equally plausible for the time optimal case as for the classic case from an economic point of view. That is, we can identify the set Z with a set of possible goal reaching times and $\mathcal{P}$ with the goal reaching time distributions defined on them to get the EUP following standard arguments. (Several alternative, but essentially equivalent sets of axioms have been discussed in the literature. The exposition given here follows the well known book by Huang and Litzenberger (1988).)

Theorem 1 (EUP for TODM based on simple distributions). *A preference relation on $\mathcal{P}$ satisfies the Axioms 1–3 if and only if there exists a function $u : Z \to \mathbb{R}$ such that*

$$p \succ q \iff \sum_{t \in Z} u(t)p(t) \succ \sum_{t \in Z} u(t)q(t). \tag{2}$$

Moreover, u is unique up to a positive affine transformation.

The EU-representation according to theorem 1 remains true under more general conditions, which allow for $t \in \mathbb{R}$ and unbounded utility functions, which is important in some applications, particularly such which call for continuous time models. (To overcome the limitation to simple distributions one basically needs to define $\mathcal{P}$ on a Borel algebra on a proper set of goal reaching times Z and to add a few structural axioms which do not change the economic intuition. See e.g. Fishburn (1982, Theorem 4, p. 28).)

The utility function u used in the EU-representation is defined on *times*, which might easily lead to misunderstand that the idea would be to attribute "utility" to time. This is not the case! Utility is assumed to be derived from the availability of the goal value, that is, ultimately from consumption in exactly the same way as in the classic case. But the notion "utility function" in the context of the EU-representation might indeed be more misleading in the time optimal than in the classic case, as deriving "utility" from value seems to make sense, but from time does not. It might be clearer in both cases to consider u simply as a weighting function used in a particular preference representation model, that is EU, which it is, instead of the the somewhat confusing term "utility" – this in mind, we can safely stick to this established notion.

4 Properties of the utility function

What do we know about the utility function? Very little. The EUP does not provide information on even the most basic properties. Very important is the question of monotonicity. In classic models, it is quite common to assume non-saturation, or *more is better*, which results in a strictly increasing classic utility function $u_t(v)$. In time optimal models, it seems quite natural to assume that *earlier is better*, which results in a strictly decreasing utility function $u_v(t)$ (Obviously, these assumptions, while reasonable in many important applications, cannot be generally valid. They are in some sense dual between the two approaches. To stress this duality, the respective utility functions have been indexed by the a priori fixed quantity, t for the classic, v for the time optimal case.) Subsequently, we additionally assume that the utility functions are twice differentiable. Consequently, we have for the first derivatives $u_t'(v) > 0$ and $u_v'(t) < 0$.

The second derivative is important to describe the risk attitude. Risk prone, neutral or averse behavior is characterized by a positive, zero, or negative second derivative. Recent work has shown that these interpretations also hold for TODM (Bouzaima and Burkhardt (2006)). For classic models, there is ample evidence that risk averse behavior is the normal case, that is $u_t''(v) < 0$. For time optimal models, this is still a largely open question. Anyhow, there is at least some evidence that suggests that risk prone behavior is not uncommon or might even be the normal case for TODM, $u_v''(t) > 0$. There are several lines of argument which support such an assertion: 1. The literature on time preference shows that discount functions have the asserted shape. Although they are *not* equivalent to the utility function in TODM, they should be related. 2. A St. Petersburg game adopted to TODM indicates that only risk prone behavior leads to decisions that seem reasonable (Bouzaima and Burkhardt (2006)). 3. If a risk free investment is available, it is possible to map some distributions of value at a given time one to one in distributions of goal reaching times for a given value by usual present value calculations. Then, risk averse decisions in the classic sense correspond to risk prone decisions in the time optimal case. 4. Some first, so far unpublished experimental evidence indicates that the predominant number of participants show indeed risk prone behavior.

This is rather surprising, and certainly more research is necessary to clarify this issue. For the time being, there is at least enough evidence for risk prone behavior in TODM that it seems to be worthwhile to study some of its consequences. We will do just that for stochastic dominance in the next section.

Before we do so, a comment on the application of TODM is in order. Both, the determination or better estimation of the distributions of the goal reaching times as well as of the utility functions is clearly dependent on the application. Burkhardt and Haasis (2006) give an example for the estimation of the distribution of goal reaching times for stock index investments. Utility functions can best be determined by experimental methods, as mentioned above. The conceptual results presented here provide the foundation for such empirical work, which is on the way, but largely not yet published.

5 Towards time optimal stochastic dominance

Stochastic dominance (SD) relations provide a partial ordering of risky prospects based on EU with respect to certain classes of utility functions. In the classic case, most commonly used are first and second order SD relations, SD_1 and SD_2, which are based on the sets of utility functions $U_1^{\mathrm{SD}} = \{u_t | u_t'(v) > 0\}$ and $U_2^{\mathrm{SD}} = \{u_t | u_t'(v) > 0 \wedge u_t''(v) < 0\}$, respectively. Higher order SD relations are defined similarly, each order adding a restriction on one more derivative, with *alternating signs*. The concept allows to derive theorems which provide relatively simple rules to check if a risky prospect $f(v)$ dominates another risky prospect $g(v)$ for all utility functions in the respective set — a prime example is the famous MV-principle. We would certainly like to have corresponding rules to ease TODM. A review of the underlying proofs for the classic case shows that they make use of the alternating signs of the derivatives. It is unclear if we could derive any useful results without that assumption. This observation, together with our previous discussion concerning $u_v''(t)$, motivates the following definition of Time Optimal Stochastic Dominance (TOSD). It is somewhat "experimental", as it will probably be useful only if indeed risk prone behavior is sufficiently common in TODM, and is formulated here just for the first two orders. The extensions are obvious.

Definition 1 (Time optimal stochastic dominance). *Let* $U_1^{TOSD} = \{u_v | u_v'(t) < 0\}$ *and* $U_2^{TOSD} = \{u_v | u_v'(t) < 0 \wedge u_v''(t) > 0\}$. *Then we say that* $f(t)$ *stochastically dominates* $g(t)$ *in the time optimal sense of order* i,

f $TOSD_i$ g, iff $E_f[u_v(t)] \geq E_g[u_v(t)]\ \forall\ u_v \in U_i^{TOSD}$, with strict inequality for at least one u.

This definition allows to state the basic rules for TOSD in a way that shows a nice duality compared to classic SD:

Theorem 2 (Basic TOSD rules). *It holds*

$$f\ TOSD_1\ g \Leftrightarrow f(t) \geq g(t)\ \forall t \quad and \tag{3}$$

$$f\ TOSD_2\ g \Leftrightarrow \int_0^t g(\tau) - f(\tau) d\tau \leq 0\ \forall t, \tag{4}$$

with strict inequality for at least one t. Furthermore, a necessary condition for $TOSD_1$ is $E_f[t] < E_g[t]$, and likewise for $TOSD_2$ it is $E_f[t] \leq E_g[t]$.

This theorem shows that in TOSD the corresponding classic rules hold with $>$ and $\geq$ reversed to $<$ and $\leq$, *if* risk prone behavior is assumed. A nice illustration for this duality is the following time optimal MV-rule, which results for normally distributed goal reaching times.

Theorem 3 (MV-rule for TOSD with normal distribution). *Let $f = N(\mu_1, \sigma_1)$ and $g = N(\mu_2, \sigma_2)$ be two normal distributions of goal reaching times with the parameters as given in the arguments. Then*

$$f\ TOSD_1\ g \Leftrightarrow \mu_1 < \mu_2 \wedge \sigma_1^2 = \sigma_2^2 \quad and \tag{5}$$

$$f\ TOSD_2\ g \Leftrightarrow \mu_1 \leq \mu_2 \wedge \sigma_1^2 \geq \sigma_2^2 \tag{6}$$

Again, the well known relations of the classic rule are reversed. This result is primarily useful as an illustration. We will hardly encounter normally distributed goal reaching times. Times are nonnegative, and relevant distributions tend to be right skewed with heavier tails. A particularly important example which results for Brownian motion to a given goal value is the inverse gaussian distribution, see Burkhardt (2000b) for a detailed discussion.

6 Conclusion

This paper discussed and extended the time optimal approach to decision making. A rational model of choice among distributions of goal reaching times, which is at the core of the time optimal approach, has been based on the axiomatic concept of rationality according to von Neumann/Morgenstern. We showed that the expected utility principle is applicable to time optimal decision making without restrictions. Furthermore, we investigated the dual relations between the classic and the time optimal approach. This duality is far reaching. We saw that the classic approach focuses on the uncertainty of the value for a given point in time, whereas the time optimal approach focuses on the uncertainty of the goal reaching time for a given goal (value). Exploiting this duality, we developed same properties of the von Neumann/Morgenstern type utility function for the time optimal approach by assuming that *earlier is better*, which is the dual assumption to *more is better* in the classic approach. This led to a negative first derivative. Theoretical as well as first experimental evidence has been given to substantiate the assertion that in time optimal decision making risk attitudes might commonly be characterized by a positive second derivative, that is, risk prone behavior. Based on these assumptions, some first notions and theorems towards stochastic dominance relations for time optimal decisions have been developed, again showing a nice duality to the classic approach. The results give help to pave the road for further developments in time optimal decision making. They prove that time optimal decisions models can be based on proper rationality concepts, with immediate applications particularly in finance.

References

BOUZAIMA, M. and BURKHARDT, T. (2006): On the Notions and Properties of Risk and Risk Aversion in the Time Optimal Approach to Decision Making. In this volume.

BURKHARDT, T. (1999): Time Optimal Portfolio Selection: Mean-variance-efficient Sets for Arithmetic and Geometric Brownian Price Processes. In: R. Decker and W. Gaul (Eds.): *Classification and Information Processing at the Turn of The Millennium.* Springer, Heidelberg, 304–311.

BURKHARDT, T. (2000a): Wachstumsoptimierte Portfolioselektion auf der Grundlage von Zielerreichungszeiten. *OR Spektrum, 22, 203–237.*

BURKHARDT, T. (2000b): *Portfolioselektion bei unsicherer Anlagedauer.* Habilitationsschrift, Technische Universität Freiberg.

BURKHARDT, T. and HAASIS, M. (2006): On Goal Reaching Time Distributions Estimated from DAX Stock Index Investments. In this volume.

FISHBURN, P.C. (1982): *The Foundations of Expected Utility.* Reidel Publishing, Dordrecht.

FREDERICK, S., LOEWENSTEIN, G. and O'DONOGHUE, T. (2002): Time Discounting and Time Preference: A Critical Review. *Journal of Economic Literature, 40, 351–401.*

HUANG, C.-F. and LITZENBERGER, R.H. (1988): *Foundations of Financial Economics.* Elsevier, New York.

LEVY, H. (1998): *Stochastic Dominance: Investment Decision Making under Uncertainty.* Kluwer Acad. Publ., Boston, Dordrecht, London.

SINN, H.-W. (1989): *Economic Decisions Under Uncertainty.* 2nd ed., Physica, Heidelberg.

VON NEUMANN, J. and MORGENSTERN, O. (1947): *Theory of Games and Economic Behavior.* Princeton Univ. Press, Princeton.

On Goal Reaching Time Distributions Estimated from DAX Stock Index Investments

Thomas Burkhardt and Michael Haasis

Lehrstuhl für Finanzierung, Finanzdienstleistungen und eFinance, Universität Koblenz-Landau, Campus Koblenz, Universitätsstr. 1, D-56070 Koblenz, Germany; {tburkha, haasis}@uni-koblenz.de

Abstract. This research paper analyzes the distributional properties of stock index time series data from a new perspective, that is, time optimal decision making building on the conceptual foundation of the time optimal approach to portfolio selection introduced by Burkhardt. In this approach, the investor's goal is to reach a predefined level of wealth as soon as possible. We investigate the empirical properties of the goal reaching times for DAX stock index investments for various levels of aspired wealth, compare the observed properties to those expected by the Inverse Gaussian distributional model, investigate the use of overlapping instead of independent goal reaching times, and highlight some methodological issues involved in the empirical analysis. The results are of immediate interest to investors.

1 Motivation

In the Time Optimal approach to portfolio selection by Burkhardt (Burkhardt (1999), (2000a), (2000b)), an investor wants to achieve a certain desired level of wealth above his current level of wealth as soon as possible by making a risky investment. The goal value Z is the relative positive wealth gain he wants to achieve:

$$Z = \frac{\text{desired wealth level}}{\text{current wealth level}} - 1 \quad (\text{in } \%), \qquad z = \log(1 + Z) \tag{1}$$

The stochastic properties of the available assets define the distribution of the time at which the goal is reached for the first time, so choosing an investment strategy effectively means to make a choice among the set of feasible distributions of goal reaching times. From a mathematical point of view, the goal reaching times can be viewed as First Passage Times (FPT) to a given boundary. It can be shown (Burkhardt (2000b, pp. 79).) that standard Brownian asset price processes result in goal reaching times distributed according to the Inverse Gaussian (IG) distribution. In this paper, we analyze the empirical goal reaching times for a stock index time series and compare the results with those expected from the IG reference model. We encounter three primary methodological problems. First, we have to use overlapping, correlated observations, because the length of the time series leads to an insufficient number of independent observations for economically relevant goal values Z. Second, also because of the limited time series, we encounter truncation in the data. Third, the reference model is not discrete, whereas the available time series is.

The Inverse Gaussian reference model is described in Section 2, the analyzed stock index time series is described in Section 3, in Section 4 the methodological issues involved in the empirical analysis are highlighted and in Section 5 the empirical properties of the goal reaching times for the DAX stock index investments are compared to the properties expected by the Inverse Gaussian model for various levels of relative aspired wealth gains.

2 Inverse Gaussian reference model

The dynamics of the (logarithm of) index prices using a Brownian Motion process is:

$$d(\log[P_t]) = dX_t = \mu dt + \sigma dB_t \tag{2}$$

P_t is the price of the stock index, X_t is the logarithm of the price and B_t is the standard Brownian motion. μ is the drift and σ the diffusion parameter. The corresponding IG probability density function (pdf) is defined as

$$g(\tau|\beta,\lambda) = \sqrt{\frac{\lambda}{2\pi\tau^3}} \cdot \exp\left[-\frac{\lambda\,(\tau-\beta)^2}{2\beta^2\tau}\right], \quad \tau > 0 \tag{3}$$

The expected value of the goal reaching time (FPT) is $E[\tau] = \beta$ and the variance is $V[\tau] = \frac{\beta^3}{\lambda}$. The parameters β and λ of the IG pdf are

$$\beta = \frac{z}{\mu}, \qquad \lambda = \frac{z^2}{\sigma^2}. \tag{4}$$

According to Chhikara and Folks (1989) the Inverse Gaussian cumulative distribution function (cdf) can be expressed as a function of the standard normal cdf:

$$G(\tau|\beta,\lambda) = \Phi\left[\sqrt{\frac{\lambda}{\tau}}\left(\frac{\tau}{\beta}-1\right)\right] + \exp\left[\frac{2\lambda}{\beta}\right]\cdot\Phi\left[-\sqrt{\frac{\lambda}{\tau}}\left(1+\frac{\tau}{\beta}\right)\right], \quad \tau > 0 \tag{5}$$

where $\Phi[.]$ denotes the standard normal distribution function.

3 Data

In this paper we analyze certain distributional properties of the Deutscher Aktienindex (DAX). We use the time series from 09/28/1959 to 01/31/2006. Our data sources are http://www.globalfindata.com and http://finance.yahoo.com. Table 1 gives an overview of some properties of the selected stock index time series. The DAX is a performance index (total return). Furthermore, the price

Table 1. Selected empirical properties of the DAX stock index

Index	Start date	Number of obs.	Trading days / year	μ p.a. in %	95% CI μ LB	95% CI μ UB	σ (ann.) in %	95% CI σ LB	95% CI σ UB
DAX	1959.74	11,592	250.14	5.881	0.483	11.279	18.75	18.51	18.99

index time series is in nominal prices and therefore reflects only nominal – and not real – wealth changes. All prices are closing prices, because high and low prices were only available for a fraction of the observation period. For the calculation of the decimal dates the closing times are estimated at 6.00 p.m. For the calculation of the FPTs we use elapsed time, not the number of trading days.

4 Methodology

4.1 Methodological issues

In the reference model, a continuous time series is assumed. All stock index time series closing prices are discrete. This could lead to small differences between the model and the empirically estimated parameters of the IG distribution. This is likely to be the case for very small values of Z. Due to the limited length of the stock index time series we might encounter edge effects, which might lead to a bias in the parameter estimations. This problem will be pronounced if the the expected goal reaching time is large relative to the length of the stock index time series.

If (starting at a day Γ_i) the goal value Z is not reached within the given time series, no FPT will be noted. (The calculation of the goal reaching times is explained in detail in Subsection 4.2.) Including FPTs for which the goal value was not reached would lead to censored data with consequences for the parameter estimation. However, in this paper we concentrate on the more simple case.

4.2 Independent and overlapping observations

The calculation of the goal reaching time in the independent case is as follows. Starting at day 1, the day Γ_1 at which the goal is reached for the first time is noted. The goal reaching time τ_1 is the time from day 1 to day Γ_1 ($\Gamma_1 = 1+\tau_1$). The second goal reaching time τ_2 is calculated starting at day Γ_1, the third starting at day Γ_2. The expected number of observations can be calculated with $E[n_i] =$ length of data series (in years) $\cdot \mu/z$. As can easily be seen, the number of expected observations drops sharply for larger values of Z. The properties of the estimators and their confidence intervals are known.

We suggest an alternative approach to calculate goal reaching times, based on overlapping observations. To indicate the independent and the overlapping case, we use the indices i and o respectively. Starting at day 1, the time interval until the goal is reached for the first time is denoted τ_1. The second goal reaching time (τ_2) is calculated starting at day 2, the third starting at day 3 etc. Now the observations are overlapping and not independent, since the goal reaching times are calculated starting every day. The expected number of observations is $E[n_o] =$ length of data series $\cdot$ number of trading days/year $-$ $(\frac{\mu}{z})$. The expected number of observations for economically significant values of Z (i.e. $Z > 20\%$) is much larger than in the independent case. In this case, the properties of the estimators and the attainable accuracy of measurement are unknown. For the IG model the accuracy of measurement can be estimated by simulating a Geometric Brownian Motion process.

4.3 Parameter estimation

Chhikara and Folks (1989) give the following parameter estimators for β and λ:

$$\hat{\beta} = \frac{1}{n}\sum_{i=1}^{n}\tau_i, \qquad \hat{\lambda} = \frac{n-1}{\sum_{i=1}^{n}\left(\frac{1}{\tau_i} - \frac{1}{\hat{\beta}}\right)} \tag{6}$$

For independent observations the estimator for β is unbiased and the estimator for λ is UMVUE. Using (4) we are able to calculate the implied parameters of the Geometric Brownian Motion process with $\hat{\beta}$ and $\hat{\lambda}$.

The use of $\hat{\lambda}$ for overlapping observations might be problematic as it has been derived for independent observations, but no other estimator is available.

4.4 Accuracy of measurement

For independent observations theoretical $(1-\alpha)$ percent confidence intervals for the parameters of the Inverse Gaussian pdf and the implied parameters of the Geometric Brownian Motion process can be calculated as shown in Wasan (1969, p. 95), and Kamstra and Milevsky (2005, pp. 238). For the value of λ:

$$\hat{\lambda}/\left(n-1\right)\cdot\chi^2_{\alpha/2} \leq \lambda \leq \hat{\lambda}/\left(n-1\right)\cdot\chi^2_{1-\alpha/2} \tag{7}$$

where $\chi^2_{\alpha/2}$ denotes the value from the χ^2 distribution, with $n-1$ degrees of freedom. Using (4) a confidence interval can be obtained for the value of the implied diffusion coefficient σ:

$$z \cdot \left[\sqrt{\hat{\lambda}/(n-1)} \cdot \chi^2_{1-\alpha/2}\right]^{-1} \leq \sigma \leq z \cdot \left[\sqrt{\hat{\lambda}/(n-1)} \cdot \chi^2_{\alpha/2}\right]^{-1} \tag{8}$$

The confidence interval for β is:

$$\hat{\beta} \cdot \left[1 + \sqrt{\hat{\beta}/\left(n\hat{\lambda}\right)} \cdot t_{1-\alpha/2}\right]^{-1} \leq \beta \leq \hat{\beta} \cdot \left[1 - \sqrt{\hat{\beta}/\left(n\hat{\lambda}\right)} \cdot t_{1-\alpha/2}\right]^{-1}, \tag{9}$$

where $t_{1-\alpha/2}$ denotes the value from the student t distribution with n degrees of freedom, provided that $\sqrt{\hat{\beta}/(n\hat{\lambda})} \cdot t^2_{1-\alpha/2} < 1$. Otherwise:

$$\hat{\beta} \cdot \left[1 + \sqrt{\hat{\beta}/(n\hat{\lambda})} \cdot t^2_{1-\alpha/2}\right]^{-1} \leq \beta \leq \infty \tag{10}$$

As above, using (4) the CI for the implied drift coefficient μ can be obtained:

$$z\left[1 - \sqrt{\hat{\beta}/(n\hat{\lambda})} \cdot t_{1-\alpha/2}\right] \cdot \hat{\beta}^{-1} \leq \mu \leq z\left[1 + \sqrt{\hat{\beta}/(n\hat{\lambda})} \cdot t_{1-\alpha/2}\right] \cdot \hat{\beta}^{-1} \tag{11}$$

provided that $\sqrt{\hat{\beta}/(n\hat{\lambda})} \cdot t^2_{1-\alpha/2} < 1$. Otherwise:

$$0 \leq \mu \leq z \cdot \left[1 + \sqrt{\hat{\beta}/(n\hat{\lambda})} \cdot t_{1-\alpha/2}\right] \cdot \hat{\beta}^{-1} \tag{12}$$

For overlapping observations a theory for confidence intervals is not yet available. Hence, we conduct simulations for a substantiated analysis of the accuracy of estimates. The simulations also yield a reference parameter for the parameter estimates. Therefore, we can compare the empirical data with the results of the simulations. In the simulations, all distances between dates are equal, which in reality, they are not. However, the effect on the FPTs is very limited.

4.5 Density estimation: Histogram

Histograms are a particularly simple form of density estimation (Silverman (1986)) which avoids many of the difficulties of other estimation methods in connection with overlapping data. The bin width h of an histogram is $h = \frac{(b-a)}{m}$, where m is the number of bins, a is the observation with the smallest and b the observation with the largest value. For a given bin width the number of bins can be calculated accordingly. The widely used Sturges' rule $m = 1 + \log_2 n$ understates the number of bins according to Scott (2001), especially for large data sets. Therefore, we use the following algorithm (for details see Scott (2001, pp. 12095)) to determine the optimal bin width:

$$CV(h) = \frac{2}{(n-1)h} - \frac{n+1}{(n-1)n^2 h} \sum_{k=1}^{m} \nu_k^2 \tag{13}$$

where ν_k is the number of observations in bin k. The criterion value is calculated for different bin widths (and bin starting points) for one data set of FPTs. The minimum value for $CV(h)$ leads to the optimal bin width. Although the algorithm is not designed for correlated observations, we use this algorithm, because no better theory is available (Unfortunately, the criterion value is also quite noisy). After obtaining the optimal bin width of the histogram, we simulate a Standard Brownian Motion process to estimate CI for each bin of the histogram. RMSEs are used as estimation error measures. The RMSE is calculated from the empirical histogram and one of two suitably parameterized IG distributions. One set of parameters is estimated from the empirical FPTs and the other is calculated from the estimated parameters of the log returns.

5 Results

Our first finding is, that in c.p. comparisons the RMSEs of histograms are substantially higher in the independent case (i) than in the overlapping case (o). E.g., for the DAX time series, $Z = 25\%$ and 20 bins, we obtain the following RMSEs: 0.03165 (i) and 0.01159 (o), calculated using the IG parameterized to fit the empirically observed FPTs, and 0.05345 (i) and 0.01178 (o), calculated using the IG corresponding to the parameters of the log returns.

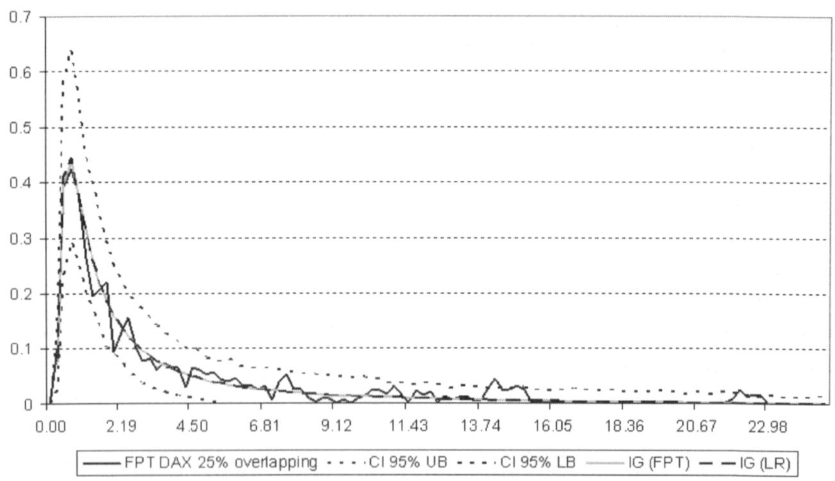

Fig. 1. Histogram of FPT (o) of the DAX with simulated 95% CI, $Z = 25\%$

Our next finding is, that the histogram calculated using the overlapping approach fits well with the predictions of the reference model. This is illustrated in Figure 1. The number of bins is here 101, the number of simulation runs 1000. The empirical histogram (black line; for improved readability we did not plot the bins, but only their midpoints and connected them with a line) lies well within the 95% confidence interval (dotted lines) and very close to the IG distribution, irrespective whether calculated using the FPT (light grey line) or the log return (dashed line) approach. The confidence intervals were calculated for every bin.

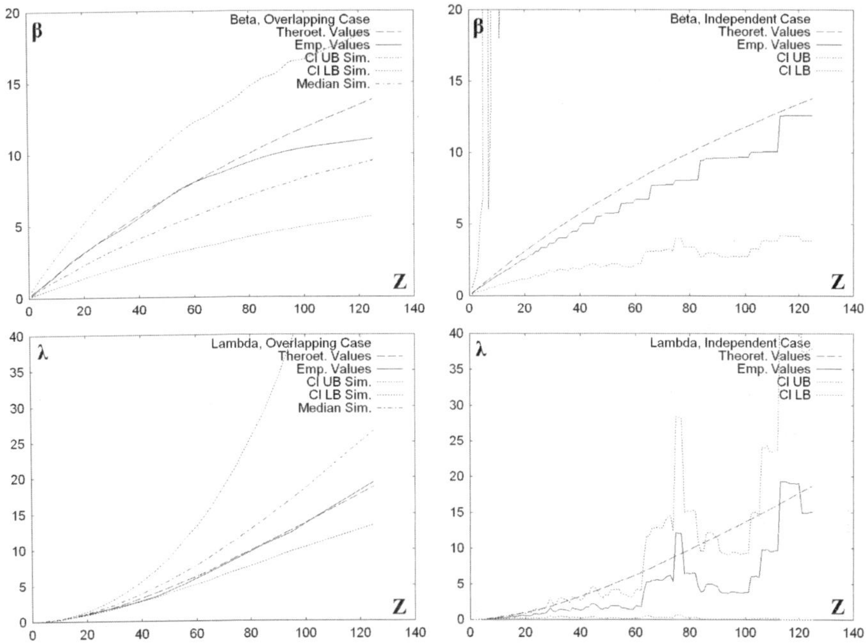

Fig. 2. $\hat{\beta}$ and $\hat{\lambda}$ (in years) and their simulated and theoretical confidence intervals for diff. values of Z in the (o) and (i) case

Finally, we obtain the conclusion that the Geometric Brownian Motion model yields reasonable results for the analyzed goal values within the limits of the measurement accuracy resulting from the methodology introduced in Section 4. Here we have to distinguish between the overlapping (left hand side) and the independent approach (right hand side of Figure 2). In the latter one, the number of observations for economically relevant goal values is too limited (In the independent case for the DAX we have only 13 observations for $Z = 25\%$, 7 for 50%, 4 for 100% and 2 for 200%.) and therefore leads to substantial estimation errors. For values of $Z > 10\%$ the upper bound of the confidence interval becomes very large and infinite for some intervals. Furthermore, the confidence interval estimation for λ is erratic and of limited use.

In the overlapping case, we obtain a sufficient number of observations for parameter estimation. In Figure 2 we plotted the simulated medians (dashed lines). They are the references for the empirically obtained estimations in the (o) case, because they take the dependence and the truncation of the observations into account. The simulated confidence interval for the value of β in the (o) case is clearly smaller than the theoretical one in the (i) case. For small goal values the independent and the overlapping case show almost identical results. However, the estimation error for the overlapping case is smaller. Parameter estimations using FPT observations and log returns yield similar results for the analyzed goal values.

6 Conclusion

As we have shown in this paper, using overlapping FPT observations for parameter estimation is an alternative to overcome the very limited number of observations in the independent case. The IG model and the simulation fit the data well. Within the scope of the achievable accuracy of measurement, we therefore conclude, that the overlapping approach is comparatively advantageous.

References

BURKHARDT, T. (1999): Time Optimal Portfolio Selection: Mean-variance-efficient Sets for Arithmetic and Geometric Brownian Price Processes. In: R. Decker and W. Gaul (Eds.): *Classification and Information Processing at the Turn of The Millenium.* Springer, Heidelberg, 304–311.

BURKHARDT, T. (2000a): Wachstumsorientierte Portfolioselektion auf der Grundlage von Zielerreichungszeiten. *OR Spektrum, 22, 203–237.*

BURKHARDT, T. (2000b) *Portfolioselektion bei unsicherer Anlagedauer.* Habilitationsschrift, Technische Universität Freiberg.

CHHIKARA, R.S. and FOLKS, J.L. (1989): *The Inverse Gaussian Distribution – Statistical Theory, Methodology and Applications.* Dekker, New York.

KAMSTRA, M. and MILEVSKY, M.A. (2005): Waiting for Returns: Using Space-Time Duality to Calibrate Financial Diffusions. *Quantitative Finance 5, 3, 237–244.*

SCOTT, D.M. (2001): Probability Density Estimation. In: N.J. Smelser and P.B. Baltes (Eds.): *International Encyclopedia of the Social and Behavioral Sciences.* Pergamon, Oxford.

SILVERMAN, B.W. (1986): *Density Estimation for Statistics and Data Analysis.* Chapman and Hall, London.

WASAN, M.T. (1969): *First Passage Time Distribution of Brownian Motion with Positive Drift (Inverse Gaussian Distribution).* Queen's Paper in Pure and Applied Mathematics, No. 19. Queen's University, Kingston.

Credit Risk of Collaterals: Examining the Systematic Linkage between Insolvencies and Physical Assets in Germany

Marc Gürtler, Dirk Heithecker and Sven Olboeter

Department of Finance, TU Braunschweig, D-38106 Braunschweig;
{marc.guertler, d.heithecker, s.olboeter}@tu-braunschweig.de

Abstract. According to the new capital adequacy framework (Basel II) the Basel Committee on Banking Supervision (BCBS) strongly advices banks to investigate the relationship between default rates and values of collaterals of secured loan portfolios. This is caused by the fact that the values of collaterals are expected to decline with rising defaults. However, the literature on modelling and examining this effect is rather rare. Therefore, we present a framework based on the Internal Ratings Based (IRB) approach of Basel II in order to examine such dependencies using standard econometric tests. We apply it to insolvency rates and empirical data for physical assets.

1 Introduction

During the last decade many effort has been put into developing credit portfolio models which has been enforced by the BCBS with implementing Basel II (BCBS (1999, 2005a)). Especially for estimating and predicting default rates of portfolios, analytical models like the IRB-model of Basel II as well as simulation and numerical based approaches (e.g., CreditPortfolioView, CreditRisk+, CreditPortfolioManager, and CreditMetrics) have been designed with great success (see Koyluoglu and Hickman (1998), Hamerle and Rösch (2004)).

One of the last challenges in credit risk quantification is the appropriate estimation of the recovery rate (RR) of the amount outstanding (exposure) of a defaulted debtor. This measure, or its corresponding loss (rate) given default (LGD = 1 minus RR), is often treated as a constant or independent parameter in credit risk models. However, this assumption would lead to misspecification of portfolio credit risk, since empirical evidence of a linkage between default rates and recovery rates grows (e.g., Frye (2000a, 2000b), Pykhtin (2003)). Recently, the BCBS became aware of this fact and claimed, that banks shall

take such a relationship into account when estimating the LGD in the IRB-approach. Precisely, for regulatory capital quantification the LGD should be greater than its expected value if the recovery rates or - in case of secured loans - the values of the collaterals are likely to decrease when default rates are high (BCBS (1999), paragraph 468 and BCBS (2005b)).
We present a model-based approach to investigate the dependency between physical assets, that may serves as collaterals, and default rates on a systematic level. Therefore, in the following Section 2 theoretical issues are discussed. In Section 3 and 4 we exemplary implement our approach on empirical data. We sum up our results in Section 5.

2 Analysing collateral credit cycle risk

2.1 The model outline

In the IRB-approach of Basel II defaults are generated from a two-state-one-factor model (see Vasicek (2002), Gordy (2003), BCBS (2005c), or Gürtler and Heithecker (2006)). Concretely, the discrete time process of the "normalized" return of each obligor $i \in \{1, \ldots, I\}$ in period t is represented by the following one-factor model

$$\tilde{a}_{i,t} = \sqrt{\rho_A} \cdot \tilde{x}_t + \sqrt{1-\rho_A} \cdot \tilde{\varepsilon}_{i,t} \tag{1}$$

in which $\tilde{x}_t$ and $\tilde{\varepsilon}_{i,t}$ are independent standard normally distributed variables. Therefore, $\tilde{x}_t$ serves as the common shared, systematic factor that can be interpreted as the overall economic conditions of all obligors whereas $\tilde{\varepsilon}_{i,t}$ represents the individual, idiosyncratic risk of each borrower i. The parameter ρ_A marks the asset correlation between two borrowers i and j.
A borrower defaults if its return $\tilde{a}_{i,t}$ falls short of an exogenously given threshold b_A. Assuming that all borrowers can be grouped into homogeneous risk buckets like industry sectors or rating grades this variable is equal for all obligors. This threshold can be interfered from the (expected) default probability PD_A of the bucket via

$$b_A = N^{-1}(PD_A) \tag{2}$$

that can be determined from empirical data. Here and for the following equations $N(\cdot)$ marks the cumulative standard normal distribution and $N^{-1}(\cdot)$ its inverse. Conditional on a realization of the common random factor $\tilde{x}_t$ the (conditional) probability of default in period t becomes

$$p_{A,t}|\tilde{x}_t = N(\tilde{b}_{A,t}) \text{ with } \tilde{b}_{A,t} := b_A|\tilde{x}_t = \frac{b_A - \sqrt{\rho_A} \cdot \tilde{x}_t}{\sqrt{1-\rho_A}}. \tag{3}$$

Thus, the conditional default threshold $\tilde{b}_{A,t}$ is normal distributed, i.e.

$$\tilde{b}_{A,t} \sim n\left(\frac{b_A}{\sqrt{1-\rho_A}}, \frac{\rho_A}{1-\rho_A}\right). \tag{4}$$

In this context the term $n(\mu, \sigma)$ denotes the normal distribution with expected value μ and standard deviation σ.
With the number I of obligors shrinking to infinity the observed default rate of the portfolio converges to the conditional probability of default $\tilde{p}_{A,t} \rightarrow N(\tilde{b}_{A,t})$ (see Gordy (2003) for a detailed derivation). Obviously, with positive realizations of $\tilde{x}_t$ the default rate declines, but with negative realizations it rises.
Additionally, collateral assets of the value $\tilde{C}_{j,t}$ with $j \in \{1, \ldots, J\}$ are held by the bank in the credit portfolio against potential losses. For simplification, these collaterals are assumed to be represented by an index price $\tilde{C}_t$, i.e., we assume that the values of the collaterals follow a single index model. We model $\tilde{C}_t$ as

$$ln\, \tilde{C}_t = ln\, C_{t-1} + \mu_c + \sigma_c \cdot \tilde{c}_t \text{ where } \tilde{c}_t \sim n(0,1). \tag{5}$$

(Since we only are interested in systematic risk, only the index is modeled.) Thus, when the default threshold $\tilde{b}_{A,t}$ and the standardized return $\tilde{c}_t$ are negatively correlated the number of defaults in the portfolio will increase while at the same time the value of the collateral assets is expected to fall. However, with respect to the recovery process we are more interested in the relationship between the lagged default threshold $\tilde{b}_{A,t-1}$ and the standardized return $\tilde{c}_t$. This is caused by the fact that a bank will make use of the collateral in the workout process after a possible event of default. If

$$\rho_{bC} = corr(\tilde{b}_{A,t-1}, \tilde{c}_t) < 0, \tag{6}$$

the bank will face declining (expected) collateral values in the workout process while the number of defaulted loans has increased. We call this "systematic collateral credit cycle risk" and it serves as an indicator for downturn effects of the LGD like it is mentioned by the BCBS.

2.2 Analysing the dependency between collaterals and default rates

According to our model we should analyse the empirical correlation between the realized collateral return c_t and the realized default threshold $b_{A,t-1}$ with t = 1,..., T from empirical data series. However, two problems occur: (i) banks often fail to present long data series of credit default data and (ii) default rates or prices of physical assets (like automobiles, machines, goods) are often autoregressive (AR), e.g. follow an autoregressive process of the order of one:

$$\tilde{b}_{A,t} = \alpha_{A1} \cdot \tilde{b}_{A,t-1} + \tilde{\varepsilon}_{A,t}\ , \quad \tilde{c}_t = \alpha_{C1} \cdot \tilde{c}_{t-1} + \tilde{\varepsilon}_{C,t} \tag{7}$$
$$\text{in which} \quad \tilde{\varepsilon}_{A,t}, \tilde{\varepsilon}_{C,t} \sim N(\mu_{A/C}, \sigma_{A/C}).$$

For simplification we only use an AR(1) process. Of course, higher order terms are possible and used in the empirical analysis as well. Worth mentioning that one have to ensure that the empirical data series still is stationary, i.e.

$|\alpha_{A1/C1}| < 1$.
Thus, directly measuring correlation between c_t and $b_{A,t-1}$ might lead to spurious results. Therefore, we suggest two proceedings in order to have more statistical reliable analysis on collateral credit cycle risk. Firstly, we conduct the test for Granger (non-)causality (Hamilton (1994), Greene (2003)). Such a causality is interfered when lagged variables of the default threshold (e.g. $b_{A,t-1}$) have explanatory power in a vector autoregression (VAR) on the (leading) return of the collateral c_t. Precisely, we calibrate the VAR model of the order of one, i.e.

$$\begin{pmatrix} \tilde{b}_t \\ \tilde{c}_t \end{pmatrix} = \begin{pmatrix} \phi_{bb} \; \phi_{bc} \\ \phi_{cb} \; \phi_{cc} \end{pmatrix} \cdot \begin{pmatrix} \tilde{b}_{t-1} \\ \tilde{c}_{t-1} \end{pmatrix} + \begin{pmatrix} \tilde{\varepsilon}_{A,t} \\ \tilde{\varepsilon}_{C,t} \end{pmatrix} \tag{8}$$

on empirical data and test the hypothesis $\phi_{cb} = 0$. If this hypothesis can not be rejected changes in $\tilde{b}_{A,t-1}$ would not cause (Granger) changes in $\tilde{c}_t$ and downturn effects of the collateral might not be present.
Secondly, we test the correlation

$$\rho_{bC}^{(\varepsilon)} = corr(\tilde{\varepsilon}_{A,t-1}, \tilde{\varepsilon}_{C,t}) \tag{9}$$

between the residuals of the AR processes due to Equation 6. If the hypothesis $\rho_{bC}^{(\varepsilon)} = 0$ is rejected, there might be a dependency between $\tilde{b}_{A,t-1}$ and $\tilde{c}_t$. Thus, if one of those two hypotheses does not hold, a systematic linkage between both variables may be considered in LGD estimation. However a downturn effect only is present if $\tilde{c}_t$ declines if $\tilde{b}_{A,t-1}$ rises, that leads to $\rho_{bC}^{(\varepsilon)} < 0$. Otherwise there might be an "upturn" effect - the collateral value is to increase with default rate climbing. An easy way to test this behavior is the use of impulse-response functions (Hamilton (1994), Greene (2003)).

3 The data

3.1 Credit cycle from insolvencies in Germany

In order to identify the default threshold we use a database of monthly insolvency rates from 2000 until 2003 from the Federal Statistical Office of Germany (FSO), that incorporates on average 2,000,000 enterprises and therefore provides good information on a "credit cycle" in Germany. Here we present results from data series of entire Germany and three industry sectors due to NACE-code:

(a) Entire Germany
(b) Manufactoring
(c) Construction
(d) Wholesale and Retail Trade

By the use of Equation 3 we get the default thresholds from the insolvency rates. They show a strong (linear) time trend for all four series, that can be verified by an ADF-test and a KPSS-test, both with and without time trend (for the Augmented Dickey-Fuller (ADF) test see e.g. Hamilton (1994), for the test from Kwiatkowski, Phillips, Schmidt, and Shin (KPSS) see Franke et al. (2004)). To be consistent with the model, that does not incorporate a time trend, see Equation 3, the time trend has been removed. However, the (detrended) default threshold remains highly autocorrelated at least for series (c) and (d). It is stationary for all series. The results of the autoregression as well as the ADF-test and KPSS-test of the detrended default threshold $b_t^{detr.}$ are shown in Table 1. Additionally we carried out several other tests like test for normality and seasonal effects as well. We do not report the results here since they do not change the main findings.

Table 1. AR regression and test of stationarity of the default threshold

<table>
<tr><td>series</td><td colspan="4">AR(1)/AR(2)-Regression[I]</td><td colspan="2">ADF-test[II]</td><td colspan="2">KPSS-test[III]</td></tr>
<tr><td>$b_t^{detr.}$</td><td>α_{A1}/α_{A2}</td><td>p-val.</td><td>R^2</td><td>DW</td><td>t-Stat.</td><td>p-val.</td><td>t-stat.</td><td>p-val.</td></tr>
<tr><td>(a)</td><td>0.12</td><td>0.40</td><td>0.02</td><td>2.07</td><td>-5.91</td><td>0.00</td><td>0.13</td><td>> 0.1</td></tr>
<tr><td>(b)</td><td>0.15</td><td>0.32</td><td>0.02</td><td>2.01</td><td>-5.85</td><td>0.00</td><td>0.10</td><td>> 0.1</td></tr>
<tr><td>(c)</td><td>0.45</td><td>0.00</td><td>0.22</td><td>2.16</td><td>-4.31</td><td>0.00</td><td>0.15</td><td>> 0.1</td></tr>
<tr><td rowspan="2">(d)</td><td>0.17</td><td>0.26</td><td rowspan="2">0.04</td><td rowspan="2">1.93</td><td rowspan="2">-6.33</td><td rowspan="2">0.00</td><td rowspan="2">0.10</td><td rowspan="2">> 0.1</td></tr>
<tr><td>0.26</td><td>0.09</td></tr>
</table>

[I] We report the coefficient of the AR(1)-regr. for (a) to (c) and AR(2)-regr. for (d) here. Low p-values indicate, that the hypothesis of $\alpha_{A1/A2}$=0 (no AR) could not be maintained. DW stands for Durban-Watson coefficient.

[II] Low p-values indicate that the hypothesis of non-stationarity shall be rejected for levels of significance greater than p.

[III] High p-values indicate that the hypothesis of stationarity shall not be rejected even for high levels of significance (lower than p).

3.2 Systematic risk of physical assets in Germany

In our analysis we take time series of several indices from the FSO and Eurotax Schwacke GmbH to obtain price changes of physical assets from 2000 to 2003 on a monthly basis. We take into account the indices for: These indices

(A) Machines (Wholesale)
(B) Consumption Goods
(C) Intermediate Outputs
(D) Exported Goods (Worldwide)
(E) Motor Vehicles
(F) Automobiles
(G) Audi A3
(H) Ford Focus
(I) Opel Astra
(J) VW Golf

represent price changes of the market of typical physical assets like machines, goods and automobiles on a highly aggregated level, of course.

In Table 2 we show the autoregression as well as the ADF-test and KPSS-test of the logarithmic returns (see Equation 4) for all ten indices. Thus, all of them can be considered as stationary and at least the series (A), (B), (C) and (D) are autocorrelated.

Table 2. AR regression and test of stationarity of the asset indices

series	AR(1)/AR(2)-Regression[I]				ADF-test[II]		KPSS-test[III]	
$\tilde{c}_t$	$\alpha_{A1/A2}$	p-val.	R^2	DW	t-Stat.	p-val.	t-stat.	p-val.
(A)	0.48	0.00	0.19	1.9	-4,29	0.00	0.18	> 0.1
	-0.26	0.10						
(B)	0.31	0.04	0.12	2.07	-5.40	0.00	0.26	> 0.1
	-0.26	0.08						
(C)	0.52	0.00	0.28	1.88	-3.82	0.01	0.15	> 0.1
(D)	0.59	0.00	0.21	2.02	-4.54	0.00	0.14	> 0.1
	-0.28	0.07						
(E)	0.06	0.68	0.00	1.99	-5.89	0.00	0.14	> 0.1
(F)	0.06	0.69	0.00	1.97	-6.32	0.00	0.08	> 0.1
(G)	-0.05	0.75	0.00	1.98	-7.10	0.00	0.10	> 0.1
(H)	-0.04	0.56	0.00	1.70	-5.14	0.00	0.13	> 0.1
(I)	-0.03	0.71	0.00	2.15	-1.32	0.00	0.20	> 0.1
(J)	0.08	0.54	0.00	2.10	-6.94	0.00	0.08	> 0.1

[I] We report the coefficient of the AR(1)-regr. for (C) and (E) to (I) and AR(2)-regr. for (A), (B) and (D) here. Low p-values indicate that the hypothesis of $\alpha_{A1/A2} = 0$ (no AR) could not be maintained. DW stands for Durban-Watson coefficient.

[II] Low p-values indicate that the hypothesis of non-stationarity shall be rejected for levels of significance greater than p.

[III] High p-values indicate that the hypothesis of stationarity shall not be rejected even for high levels of significance (lower than p).

4 Identifying collateral credit cycle risk

In order to examine the dependency between physical assets and the default threshold we carried out the test for Granger (non-)causality and the test for (Pearson-Bravais) correlation between the residuals of the autoregression. The results are presented in Table 3. For the VAR model we also implemented the typical tests of the residuals like tests for autocorrelation and for normality. We do not show the results. We report levels of significance up to a 50% level since in (regulatory) credit risk management dependencies should be considered in a conservative manner and not with respect to its explanatory power (like desired in a regression analysis). As expected from theory the results of the Granger (non-)causality test mostly agree with the test for

Table 3. Statistical tests on systematic linkage between index returns and default thresholds

	Granger (Non-)Causality Test, p-val^{I,III}				Correlation of the Resids $\rho_{bC}^{(\varepsilon),II}$			
	(a)	(b)	(c)	(d)	(a)	(b)	(c)	(d)
(A)	0.06	0.08	0.01	0.19	-0.36*	-0.33*	-0.42*	-0.27*
(B)	0.16	0.12	0.01	0.34	-0.26*	-0.29*	-0.43*	-0.19**
(C)	0.20	0.13	0.09	0.28	-0.37*	-0.39*	-0.28*	-0.29*
(D)	0.15	0.03	0.11	0.19	-0.42*	-0.48*	-0.33*	-0.37*
(E)	0.08	0.12	0.51	0.10	0.12***	0.09	0.07	0.06
(F)	0.48	0.57	0.71	0.73	-0.3*	-0.25*	-0.13***	-0.28*
(G)	0.79	0.72	0.51	0.91	-0.04	-0.06	0.09	-0.02
(H)	0.00	0.00	0.25	0.00	0.52*	0.51*	0.19**	0.55*
(I)	0.13	0.05	0.76	0.11	0.23**	0.3*	-0.01	0.24**
(J)	0.22	0.26	0.78	0.14	-0.19**	-0.17**	-0.03	-0.23**

^{I} We report p-values here. Low p-values indicate that the hypothesis of non (Granger) causality shall be rejected for levels of significance greater than p.

^{II} We report the empirical observed correlations here. Further we present the p-values on a 10%(*), 30%(**) and 50%(***) level. Low p-values indicate that the hypothesis of no correlation shall be rejected for levels of significance greater than p.

^{III} If the hypotheses on non (Granger) causality as well as no correlation are both rejected on a 10% (30%) level, results are underlined (double underlined).

correlation. Concretely, the results of both tests of (no) dependency between physical assets and the default threshold inhabit either both low or both high levels of significance.

We observe that for machines and goods (series (A) to (D)) both test are highly significant up to a 10% level and correlations are negative. Thus, we expect that the price indices go down when default rates turn up. For the automobile market the relationship is less strong - and positive for the Ford Focus and Opel Astra at a stretch.

5 Conclusion

Correlation between default rates and recovery rates (in the event of default) is one of the main drivers for credit risk of defaulted loans. For collateralised loans, this dependency is mainly caused by a systematic linkage between the value of the collateral and the default rate of the portfolio. We call this "(systematic) credit cycle risk", and it serves as an indicator that the loss rate rises with the default rate at the same time. In this paper we have presented an econometric framework that might help to get a deeper understanding of credit risk on secured loan portfolios. We exemplary implemented our approach on insolvency quotes and several indices of physical assets, that exhibit significant credit cycle risk. We encourage banks to use similar econometric driven frameworks in order to support analyses on identification of LGD-downturn effects like they are specified in the regulatory requirements for Basel II.

References

BASEL COMMITTEE ON BANKING SUPERVISION (1999): A New Capital Adequacy Framework. *Bank for International Settlements.*

BASEL COMMITTEE ON BANKING SUPERVISION (2005a): International Convergence of Capital Measurement and Capital Standards. *Bank for International Settlements.*

BASEL COMMITTEE ON BANKING SUPERVISION (2005b): Guidance on Paragraph 468 of the Framework Document. *Bank for International Settlement.*

BASEL COMMITTEE ON BANKING SUPERVISION (2005c): An Explanatory Note on the Basel II IRB Risk Weight Functions. *Bank for International Settlement.*

FRANKE, J., HÄRDLE, W. and HAFNER, C.M. (2004): *Statistics of Financial Markets.* Springer, Berlin.

FRYE, J. (2000a): Collateral Damage. *Risk, 13, 4, 91–94.*

FRYE, J. (2000b): Depressing Recoveries. *Risk, 13, 11, 108–111.*

GORDY, M. (2003): A Risk-Factor Model Foundation for Ratings-Based Bank Capital Rules. *Journal of Financial Intermediation, 12, 199–232.*

GREENE, W.H. (2003): *Econometric Analysis.* Prentice Hall, New Jersey.

GÜRTLER, M. and HEITHECKER, D. (2006): Modellkonsistente Bestimmung des LGD im IRB-Ansatz von Basel II. *Zeitschrift für betriebswirtschaftliche Forschung, 58, 554-587.*

HAMERLE, A. and RÖSCH, D. (2004): Parameterizing Credit Risk Models. *Working Paper.* www.defaultrisk.com.

HAMILTON, J.D. (1994): *Time Series Analysis.* Prentice Hall, New Jersey.

KOYLUOGLU, H.U. and HICKMAN, A. (1998): Reconcilable Differences. *Risk, 11, 10, 56–62.*

PYKHTIN, M. (2003): Unexpected Recovery. *Risk, 16, 8, 74–78.*

VASICEK, O.A. (2002): Loan Portfolio Value. *Risk, 15, 12, 160–162.*

WAGATHA, M. (2004): *Kointegrationskonzepte für die Kreditrisikomodellierung.* DUV, Wiesbaden.

Foreign Exchange Trading with Support Vector Machines

Christian Ullrich[1], Detlef Seese[2] and Stephan Chalup[3]

[1] AIFB, University of Karlsruhe, D-76128 Karlsruhe, Germany; christian.ullrich@bmw.de
[2] AIFB, University of Karlsruhe, D-76128 Karlsruhe, Germany; seese@aifb.uni-karlsruhe.de
[3] School of Electrical Engineering & Computer Science, University of Newcastle, Callaghan, NSW 2308, Australia; chalup@cs.newcastle.edu.au

Abstract. This paper analyzes and examines the general ability of Support Vector Machine (SVM) models to correctly predict and trade daily EUR exchange rate directions. Seven models with varying kernel functions are considered. Each SVM model is benchmarked against traditional forecasting techniques in order to ascertain its potential value as out-of-sample forecasting and quantitative trading tool. It is found that hyperbolic SVMs perform well in terms of forecasting accuracy and trading results via a simulated strategy. This supports the idea that SVMs are promising learning systems for coping with nonlinear classification tasks in the field of financial time series applications.

1 Introduction

Support Vector Machines (SVMs) have proven to be a principled and very powerful supervised learning system that since its introduction (Cortes and Vapnik (1995)) has outperformed many systems in a variety of applications, such as text categorization (Joachims (1998)), image processing (Quinlan et al. (2004)), and bioinformatic problems (Brown et al. (1999)). Subsequent applications in time series prediction (Müller et al. (1999)) indicate the potential that SVMs have with respect to economics and finance. In predicting Australian foreign exchange rates, Kamruzzaman and Sarker (2003b) showed that a moving average-trained SVM has advantages over an Artificial Neural Network (ANN) based model, which was shown to have advantages over ARIMA models (2003a). Furthermore, Kamruzzaman et al. (2003) had a closer look at SVM regression and investigated how it performs with different standard kernel functions. It was found that Gaussian Radial Basis Function (RBF) and polynomial kernels appear to be a better choice in forecasting the Australian foreign exchange market than linear or spline kernels. Although Gaussian kernels are adequate measures of similarity when the representation dimension of the space remains small, they fail to reach their goal in high dimensional spaces (Francois et al. (2005)). We will examine the general ability of SVMs to correctly classify daily EUR/GBP, EUR/JPY and EUR/USD exchange rate directions. It is more useful for traders and risk managers to predict exchange rate fluctuations than their levels. To predict that the level of the EUR/USD, for instance, is close to the level today is trivial. On the contrary, to determine if the market will rise or fall is much more complex and interesting. Since SVM performance mostly depends on choosing the right kernel, we empirically verify the use of customized p-Gaussians by comparing them with a range of standard kernels. The remainder is organized as follows: Section 2 outlines the procedure for obtaining an explanatory input dataset. Section 3 formulates the SVM as applied to exchange rate forecasting and presents the kernels used. Section 4 describes the benchmarks and trading metrics used for model evaluation. Section 5 gives the empirical results. The conclusion, as well as brief directions for future research, are given in Section 6.

2 Data selection

The obvious place to start selecting data, along with EUR/GBP, EUR/JPY and EUR/USD is with other leading traded exchange rates. Also selected were related financial market data, such as stock market price indices, 3-month interest rates, 10-year government bond yields and spreads, the prices of Brent Crude oil, silver, gold and platinum, several assorted metals being traded on the London Metal Exchange, and agricultural commodities. Macroeconomic variables play a minor role and were disregarded. All data is obtained from Bloomberg and spans a time period from 1 January 1997 to 31 December 2004, totaling 2349 trading days. The data is divided into two periods. The first period (1738 observations) is used for model estimation and is classified in-sample. The second period (350 observations) is reserved for out-of-sample forecasting and evaluation. Missing observations on bank holidays were filled by linear interpolation. The explanatory viability of each variable has been evaluated by removing input variables that do not contribute significantly to model performance. For this purpose, Granger Causality tests (Granger (1969)) with lagged values up to k=20 were performed on stationary I(1) candidate variables. We find that EUR/GBP is Granger-caused by 11 variables:

- EUR/USD, JPY/USD and EUR/CHF exchange rates
- IBEX, MIB30, CAC and DJST stock market indices
- the prices of platinum and nickel
- 10-year Australian and Japanese government bond yields

We identify 10 variables that significantly Granger-cause EUR/JPY:

- EUR/CHF exchange rate

- IBEX stock market index
- the price of silver
- Australian 3-month interest rate
- Australian, German, Japanese, Swiss and US government bond yields
- UK bond spreads

For EUR/USD, the tests yield 7 significant explanatory variables:

- AUD/USD exchange rate
- SPX stock market index and
- the prices of copper, tin, zinc, coffee and cocoa

3 SVM classification model and kernels

3.1 SVM classification model

We will focus on the task of predicting the rise ("+1") or fall ("-1") of daily EUR/GBP, EUR/JPY and EUR/USD exchange rates. We apply the C-Support Vector Classification (C-SVC) algorithm as described in Boser et al. (1992) and Vapnik (1998), and implemented in R packages "e1071" (Chang and Lin (2001)) and "kernlab" (Karatzoglou et al. (2004)): Given training vectors $x_i \in R^n (i = 1, 2, \ldots, l)$, in two classes, and a vector $y \in R^l$ such that $y_i \in \{+1, -1\}$, C-SVC solves the following problem:

$$min_{w,b,\xi} \frac{1}{2} w^T w + C \sum_{i=1}^{l} \xi_i \tag{1}$$

$$\begin{aligned} y_i \left(w^T \phi(x_i) + b \right) &\geq 1 - \xi_i \\ \xi_i \geq 0, i &= 1, 2, \ldots, l \end{aligned}$$

The dual representation is given by

$$min_\alpha \frac{1}{2} \alpha^T Q \alpha - e^T \alpha \tag{2}$$

$$\begin{aligned} 0 \leq \alpha_i \leq C, i = 1, 2, \ldots, l \\ y^T \alpha = 0 \end{aligned}$$

where e is the vector of all ones, C is the upper bound, Q is a lxl positive semidefinite matrix and $Q_{ij} \equiv y_i y_j K(x_i, x_j)$. $K(x_i, x_j) \equiv \phi(x_i)^T \phi(x_j)$ is the kernel, which maps training vectors x_i into a higher dimensional, inner product, feature space by the function ϕ. The decision function is

$$f(x) = sign \left(\sum_{i=1}^{l} y_i y_j K(x_i, x) + b \right) \tag{3}$$

Training a SVM requires the solution of a very large quadratic programming optimization problem (QP) which is solved by using the Sequential Minimization Optimization (SMO) algorithm (Platt (1998)). SMO decomposes a large QP into a series of smaller QP problems which can be solved analytically. Time consuming numerical QP optimization as an inner loop can be avoided.

3.2 Kernel functions

How to find out which kernel is optimal for a given learning task is a rather unexplored problem. Under this circumstance, we compare a range of kernels with regards to their effects on SVM performance. Standard kernels chosen include the following:

- Linear: $k(x, x') = \langle x, x' \rangle$
- Polynomial: $k(x, x') = (scale \cdot \langle x, x' \rangle + offset)^{degree}$
- Laplace: $k(x, x') = exp(-\sigma \|x - x'\|)$
- Gaussian radial basis: $k(x, x') = exp(-\sigma \|x - x'\|^2)$
- Hyperbolic: $k(x, x') = tanh(scale \cdot \langle x, x' \rangle + offset)$
- Bessel: $k(x, x') = \frac{Bessel^n_{v+1}(\sigma\|x-x'\|)}{(\|x-x'\|^{-n(v+1)})}$

Also, the use of customized p-Gaussian kernels $K(x_i, x_j) = exp\left(-d(x_i, x)^p/\sigma^p\right)$ with parameters p and σ is verified. The Euclidean distance between data points is defined by $d(x_i, x) = (\sum_{i=1}^{n} |x_i - x|^2)^{1/2}$. Compared to RBF-kernels, p-Gaussians include a supplementary degree of freedom in order to better adapt to the distribution of data in high-dimensional spaces. p and σ depend on the specific input set for each exchange rate return time series and are calculated as proposed in (Francois et al. (2005)):

$$p = \frac{ln\left(\frac{ln(0.05)}{ln(0.95)}\right)}{ln\left(\frac{dF}{dN}\right)}; \sigma = \frac{d_F}{(-ln(0.05))^{1/p}} = \frac{d_N}{(-ln(0.95))^{1/p}} \tag{4}$$

In the case of EUR/USD, for example, we are considering 1737 8-dimensional objects. We calculate 1737x1737 distances and compute the 5% (d_N) and 95% (d_F) percentiles in that distribution. In order to avoid the known problem of overfitting, we determine robust estimates for C and $scale$ (σ) for each kernel through 20-fold cross validation.

4 Benchmarks and evaluation method

Letting y_t represent the exchange rate at time t, we forecast the variable

$$sign(\Delta y_{t+h}) = sign(y_{t+1} - y_t) \tag{5}$$

where $h = 1$ for a one-period forecast with daily data. The naïve model ($sign(\hat{y}_{t+1}) = sign(y_t)$) and univariate ARMA($p$, q) models are used as benchmarks. ARMA(p, q) models with p autoregressive (AR) terms and q moving averages (MA) are given by

$$y_t = c + \alpha_1 y_{t-1} + \alpha_2 y_{t-2} + \ldots + \alpha_p y_{t-p} + \varepsilon_t + \beta_1 \varepsilon_{t-1} + \ldots + \beta_q \varepsilon_{t-q} \quad (6)$$

where $\varepsilon_t \sim$ i.i.d. $(0, \sigma^2)$. Simple models, that were estimated according to Box and Jenkins (1976), provide the best testing results while preserving generalization ability for forecasting (s-step-ahead predictions for $s \leq q$ are given in parentheses):

- $c = -3.58E - 05, \beta_1 = -0.0535$, and $\beta_3 = -0.0559$ ($\hat{y}_{t+s} = \hat{c} + \hat{\beta}_s \varepsilon_t + \hat{\beta}_{s+2} \varepsilon_{t-2}$) for the EUR/GBP series
- $c = -7.84E - 05$ and $\beta_1 = 0.0288$ ($\hat{y}_{t+s} = \hat{c} + \hat{\beta}_s \varepsilon_t$) for the EUR/JPY series
- $c = -8.32E - 05, \alpha_1 = -0.5840$ and $\beta_1 = 0.5192$ ($\hat{y}_{t+s} - \hat{c} = \hat{\alpha}_1(\hat{y}_{T+s-1} - \hat{c}) + \hat{\beta}_s \varepsilon_t$) for the EUR/USD series

Out-of-sample forecasts are evaluated statistically via confusion matrices and practically via trading simulations. The reason for this twofold evaluation procedure is that trading decisions driven by a model with a small statistical error may not be as profitable as those driven by a model that is selected using financial criteria. In case of the latter, return predictions $\hat{y}_{t+1}$ are first translated into positions. Next, a decision framework is established that tells when the underlying asset is bought or sold depending on the level of the price forecast. We define a single threshold τ, which is set to $\tau = 0$ and use the following mechanism:

$$I_t = \begin{cases} 1 & if \quad \hat{y}_t < y_{t-1} - \tau \\ -1 & if \quad \hat{y}_t > y_{t-1} + \tau \\ 0 & if \quad \text{otherwise} \end{cases}, \text{ with } I_t = \begin{cases} 1 & if \quad \text{the position is long} \\ -1 & if \quad \text{the position is short} \\ 0 & if \quad \text{the position is neutral} \end{cases} \quad (7)$$

The gain or loss π_t on the position at time t is $\pi_t = I_{t-1}(y_t - y_{t-1})$. Since financial goals are user-specific, we examine the models' performances across nine Profit and Loss (P&L) related measures:

- Cumulated P&L: $PL_T^C = \sum_{t=1}^T \pi_t$
- Sharpe ratio: $SR = \frac{PL_T^A}{\sigma_T^A}$, with annualized P&L $PL_T^A = 252\frac{1}{T}\sum_{t=1}^T \pi_t$, and annualized volatility $\sigma_T^A = \sqrt{252}\sqrt{\frac{1}{T-1}\sum_{t=1}^T (\pi_t - \bar{\pi})^2}$
- Maximum daily profit: $Max(\pi_1, \pi_2, \ldots, \pi_T)$
- Maximum daily loss: $Min(\pi_1, \pi_2, \ldots, \pi_T)$
- Maximum drawdown: $MD = Min(PL_t^C - Max_{i=1,2,\ldots,t}(PL_i^C))$
- Value-at-Risk with 95% confidence: $VaR = \mu - Q(\pi, 0.05)$ with $\mu = 0$
- Net PL_T^C: $NPL_T^C = \sum_{t=1}^T (\pi_t - I_t \cdot TC)$, where $I_t = \begin{cases} 1 & \text{if} \quad \pi_{t-1} \cdot \pi_t < 0 \\ 0 & \text{else} \end{cases}$

- Average gain/loss ratio: $\frac{AG}{AL} = \frac{(\text{ Sum of all } \pi_t>0)/\#\text{up}}{(\text{ Sum of all } \pi_t<0)/\#\text{down}}$
- Trader's advantage: $TA = 0.5 \left(1 + \left(\frac{(WT \cdot AG)+(LT \cdot AL)}{\sqrt{(WT \cdot AG^2)+(LT \cdot AL^2)}}\right)\right)$ with WT:= number of winning trades, LT:= number of losing trades, AG:= average gain in up periods, and AL:= average loss in down periods

Accounting for transaction costs (TC) is important for assessing trading performance in realistic ways. An average cost of 3 pips (0.0003) per trade, for a tradable amount of typically 10 to 20 million EUR is considered a reasonable guess and incorporated in NPL_T^C. A model is operationally superior compared to another if it exhibits a larger number of superior performance measures.

5 Empirical results

Accuracy rates for the out-of-sample period are depicted in bar charts as shown in Figure 1. Figures 2 through 4 give results of the trading simulation as described in Section 4. Dominant strategies are represented by the maximum value(s) in each row and are written in bold. We observe the following:

- Statistically, both the naïve and the linear model are beaten by SVM with a suitable kernel choice. The SVM approach is statistically justified.
- Hyperbolic SVMs deliver superior performance for out-of-sample prediction across all currency pairs. In the case of EUR/GBP, the Laplace and hyperbolic SVM perform equally well. In the cases of EUR/JPY and EUR/USD, hyperbolic kernels outperform the other models more clearly. This makes hyperbolic kernels promising candidates to map all sorts of financial market return data into high dimensional feature spaces.
- Operational evaluation results confirm statistical ones in the case of EUR/GBP. The hyperbolic and Laplace SVM give the best results along with the RBF-SVM. For EUR/JPY and EUR/USD, statistical superiority of hyperbolic SVMs cannot be confirmed. Operational evaluation techniques not only measure the number of correctly predicted exchange rate ups and downs but also include the magnitude of returns. Consequently, if local extremes can be exploited, forecasting methods with less statistical performance may yield higher profits than methods with greater statistical performance. In the case of EUR/USD, the trader would have been better off applying a p-Gaussian SVM to maximize profit. In regards to EUR/JPY, no single model is able to outperform the naïve strategy. The hyperbolic SVM, however, dominates two performance measures.
- p-Gaussian SVMs perform reasonably well in predicting EUR/GBP and EUR/USD return directions. For these two currency pairs, p-Gaussian data representations lead to better generalization than Gaussians due to an additional degree of freedom p.

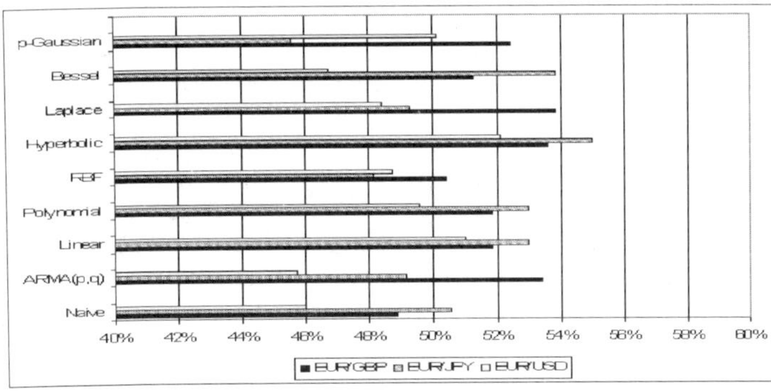

Fig. 1. Classification performance for EUR/GBP, EUR/JPY and EUR/USD

Table 1. Operational performance for EUR/GBP, EUR/JPY and EUR/USD

EUR/GBP	Naive	MA(1,3)	Linear	Polynomial	RBF	Hyperbolic	Laplace	Bessel	p-Gaussian
Cumulative P&L	-0.00750	-0.00953	-0.09360	-0.09360	-0.03896	**0.10360**	0.01546	-0.04114	0.05958
Sharpe ratio	-0.07966	-0.10112	-0.99367	-0.99367	-0.41354	**1.09938**	0.16407	-0.43671	0.63235
Maximum daily profit	0.01492	0.01492	**0.01684**	**0.01684**	**0.01684**	0.01492	**0.01684**	0.01385	0.01232
Maximum daily loss	-0.01684	-0.01684	-0.01492	-0.01492	**-0.01385**	-0.01684	**-0.01385**	-0.01684	-0.01684
Maximum drawdown	-0.03811	-0.03811	-0.03619	-0.03619	**-0.03496**	-0.03811	-0.03512	-0.03564	-0.03811
VaR (alpha = 0.05)	-0.00695	-0.00734	-0.00752	-0.00752	-0.00728	-0.00698	**-0.00691**	-0.00744	-0.00694
Net Cumulative P&L	-0.06120	-0.01013	-0.12750	-0.12750	-0.09026	**0.05590**	-0.01964	-0.09214	0.01428
Avg gain/loss ratio	**1.05178**	0.85038	0.80370	0.80370	0.91714	1.03981	0.89932	0.88235	1.01891
Trader's Advantage	0.00000	**1.00000**	0.53003	0.53003	0.48716	0.48144	0.58986	0.39350	0.43507
EUR/JPY	Naive	MA(1)	Linear	Polynomial	RBF	Hyperbolic	Laplace	Bessel	p-Gaussian
Cumulative P&L	**0.05441**	-0.11333	-0.09477	-0.09477	-0.21907	-0.13867	-0.28671	-0.31145	-0.24980
Sharpe ratio	**0.38680**	-0.80435	-0.67432	-0.67432	-1.55679	-0.98622	-2.03603	-2.21115	-1.77460
Maximum daily profit	**0.02187**	0.02187	0.02068	0.02068	0.02068	0.02174	0.02068	0.02068	0.02050
Maximum daily loss	**-0.02050**	-0.02174	-0.02187	-0.02187	-0.02187	-0.02187	-0.02187	-0.02187	-0.02187
Maximum drawdown	-0.08535	-0.08659	-0.06479	-0.06479	-0.08672	**-0.06197**	-0.08672	-0.06479	-0.08672
VaR (alpha = 0.05)	**-0.01003**	-0.01144	-0.01092	-0.01092	-0.01111	-0.01081	-0.01127	-0.01145	-0.01130
Net cumulative P&L	**0.00281**	-0.11363	-0.15267	-0.15267	-0.27607	-0.19837	-0.34461	-0.36185	-0.30260
Avg gain/loss ratio	**1.04111**	0.92829	0.89996	0.89996	0.88278	0.86458	0.83323	0.83752	0.82177
Trader's advantage	0.00000	0.00000	0.43005	0.43005	0.43247	**0.43647**	0.41154	0.40350	0.40139
EUR/USD	Naive	ARMA(1,1)	Linear	Polynomial	RBF	Hyperbolic	Laplace	Bessel	p-Gaussian
Cumulative P&L	-0.18070	-0.22255	-0.13269	-0.13269	0.00027	0.04797	-0.10055	-0.16166	**0.10182**
Sharpe ratio	-1.23452	-1.52256	-0.90434	-0.90434	-0.06296	0.32520	-0.68505	-1.10372	**0.68905**
Maximum daily profit	0.01962	0.01962	0.01667	0.01667	0.01962	**0.01962**	0.01889	0.01869	0.01889
Maximum daily loss	-0.01889	-0.01889	-0.01962	-0.01962	**-0.01869**	-0.01889	-0.01962	-0.01962	-0.01962
Maximum drawdown	0.04172	**-0.04112**	-0.04484	-0.04484	-0.04391	-0.04410	-0.04484	-0.04484	-0.04484
VaR (alpha = 0.05)	-0.01247	-0.01179	-0.01260	-0.01260	-0.01176	**-0.01085**	-0.01183	-0.01165	-0.01116
Net cumulative P&L	-0.23680	-0.22345	-0.17429	-0.17429	-0.05967	-0.00003	-0.14525	-0.21056	**0.05112**
Avg gain/loss ratio	0.94708	0.93486	0.88117	0.88117	1.03610	0.00209	0.94573	0.94569	**1.10874**
Trader's advantage	0.00000	0.31863	**0.62531**	**0.62531**	0.56826	0.55311	0.58379	0.42194	0.49915

6 Conclusion

The results support the general idea that SVMs are promising learning systems for coping with nonlinear classification and regression tasks in financial time series applications. Future research will likely focus on improvements of SVM models, such as examination of other kernels, adjustment of kernel parameters and development of data mining and optimization techniques for selecting the appropriate kernel. In light of this research, it would also be interesting to see if the dominance of hyperbolic SVMs can be confirmed in further empirical investigations on financial market return prediction.

References

BOSER, B.E., GUYON, I.M. and VAPNIK, V.N. (1992): A Training Algorithm for Optimal Margin Classifiers. In: D. Haussler (Ed.): *Proceedings of the 5th Annual ACM Workshop on Computational Learning Theory.* ACM Press, 144–152.

BOX, G.E.P. and JENKINS, G.M. (1976): *Time Series Analysis: Forecasting and Control.* Holden-Day, San Francisco.

BROWN, M., GRUNDY, W., LIN, D., CRISTIANINI, N., SUGNET, C., FUREY, T., ARES, M. and HAUSSLER, D. (1999): Knowledge-based Analysis of Microarray Gene Expression Data Using Support Vector Machines. *Technical Report.* University of California, Santa Cruz.

CHANG, C.C. and LIN, C.J. (2001): LIBSVM: A Library for Support Vector Machines (Version 2.31). *Technical Report.* Department of Computer Sciences and Information Engineering, National Taiwan University, Taipei, Taiwan.

CORTES, C. and VAPNIK, V. (1995): Support Vector Network. *Machine Learning, 20, 273–297.*

FRANCOIS, D., WERTZ, V. and VERLEYSEN, M. (2005): About the Locality of Kernels in High-dimensional Spaces. *ASMDA 2005 - International Symposium on Applied Stochastic Models and Data Analysis.* Brest, France, 238–245.

GRANGER, C.W.J. (1969): Investigating Causal Relations by Econometric Models and Cross-spectral Methods. *Econometrica, 37, 424–438.*

JOACHIMS, T. (1998): Text Categorization with Support Vector Machines. *Proceedings of European Conference on Machine Learning (ECML).*

KAMRUZZAMAN, J. and SARKER, R.A. (2003a): Forecasting of Currency Exchange Rate: A Case Study. *Proceedings of the IEEE International Conference on Neural Networks & Signal Processing (ICNNSP).* Nanjing.

KAMRUZZAMAN, J. and SARKER, R.A. (2003b): Application of Support Vector Machine to Forex Monitoring. *Third International Conference On Hybrid Intelligent Systems (HIS).* Melbourne.

KAMRUZZAMAN, J., SARKER, R.A. and AHMAD, I. (2003): SVM Based Models for Predicting Foreign Currency Exchange Rates. *Proceedings of the Third IEEE International Conference on Data Mining (ICDM).*

KARATZOGLOU, A., HORNIK, K., SMOLA, A. and ZEILEIS, A. (2004): Kernlab - An S4 Package for Kernel Methods in R. *Journal of Statistical Software, 11, 9.*

KUAN, C.M. and LIU, T. (1995): Forecasting Exchange Rates Using Feedforward and Recurrent Neural Networks. *Journal of Applied Econometrics, 10, 347–364.*

MÜLLER, K.R., SMOLA, A., RÄTSCH, G., SCHÖLKOPF, B., KOHLMORGEN, J. and VAPNIK, V. (1999): Using Support Vector Machines for Time Series Prediction. In: B. Schölkopf , C.J.C. Burges and A.J. Smola (Eds.): *Advances in Kernel Methods.* MIT Press, 242–253.

PLATT, J.C. (1998): Sequential Minimal Optimization: A Fast Algorithm for Training Support Vector Machines. *Technical Report.* Microsoft Research.

QUINLAN, M.J., CHALUP, S.K. and MIDDLETON, R.H. (2004): Application of SVMs for Colour Classification and Collision Detection with AIBO Robots. *Advances in Neural Information Processing Systems, 16, 635–642.*

SCHÖLKOPF, B. (2001): The Kernel Trick for Distances. *Technical Report, MSR 2000-51.* Microsoft Research, Redmond, WA.

VAPNIK, V. (1998): *Statistical Learning Theory.* Wiley, New York.

The Influence of Specific Information on the Credit Risk Level

Miroslaw Wójciak[1] and Aleksandra Wójcicka-Krenz[2]

[1] University of Economics, ul. 1 Maja 50, 40-287 Katowice, Poland; mwojciak@ae.katowice.pl
[2] University of Economics, Al. Niepodleglosci 10, 60-967 Poznan, Poland; aleksandra.wojcicka@ae.poznan.pl

Abstract. The paper presents the influence of specific information on the credit risk level. The effect of some particular information can be expected but in some cases it can be truly surprising. It is due to the exact content and history of previous information and the general standing of the company.

1 Introduction

The credit risk is one of the most important types of financial risk. This risk can be defined as *"a possibility of loss arising from the failure of a counterparty to make a contractual payment"* (Jajuga (2004)). Proper estimation of the credit risk level reduces the threat of bankruptcy not only of a particular company but also of all cooperating companies. The researches all around the world seek for a model that indicates the probability of default (PD) in the best way.

The proper evaluation of PD is the major problem of all credit risk models. Some models estimate probability of default (e.g. Moody's KMV model), some of them calculate it on the basis of historical data and in some cases it is simply impossible to obtain PD (discrimination analysis).

The main aim of this paper is to compare a traditional concept and a new approach of credit risk modelling taking into consideration the specific information published by the quoted companies. Authors want to find connections between certain kinds of information and a credit risk level. The level depends on the investors' reaction reflected in stock prices.

The research was based on financial data from balance sheets of construction industry companies quoted on the Warsaw Stock Exchange (WSE) in 2000-2005 (24 quarters).

2 Models used to evaluate the credit risk

The paper uses two methods of credit risk evaluation including dynamic multidimensional comparative analysis (DMCA) and the credit risk evaluation method based on the option pricing model - Moody's KMV model (MKMV). The former is a traditional way of credit risk evaluation, the latter - a new approach. Both models are general (i.e. can be used in reference to any company), descriptive (i.e. focus only on the level of credit risk and its results), and dynamic (i.e. the data used for their construction comes from from different time periods). In case of DMCA, the data is taken quarterly whereas in MKMV - daily and quarterly. Each model takes different risk measures. Classification into various credit risk levels is the result of DMCA: in MKMV - probability of default. Calculating the market assets value and its volatility (MKMV model) and financial ratios (DMCA) the level of appropriate variables from the previous period is taken. This approach is closer to reality, as the level of liabilities published in a balance sheet is known only from the previous (not current) period.[1]

A combination of two DMCA techniques was used: linear ordering and the Hellwig's pattern method (best value) suggested in Hellwig (1969). Firstly, financial ratios that describe various areas of the analysis (liquidity, profitability, efficiency and debt) are normalised. This means that each nominant is changed into a stimulant and all variables are standardised as follows:

$$z_{ij}^t = \frac{x_{ij}^t - \overline{X}_j}{S(X_j)} \tag{1}$$

where:
t - time of data, $\overline{X}_j$ - arithmetic average of j variable of all objects in all periods, $S(X_j)$ - standard deviation of variable X_j calculated as explained above.

Subsequently, for each separate group of variables a synthetic measure (Q_i) was calculated in comparison to the best value (pattern). Because the variables are standardised, some object-period could be below zero and the nature of variables is not homogeneous. However, in the pattern method the above does not matter because the very nature of variables is taken into consideration when the pattern is set and negative values of normative variables have no influence while using the Euclidean distance. The pattern is based on the formula:

$$\begin{cases} z_{0j} = P90(x_{ij}^t), \; when \; x_{ij} \in S, \\ z_{0j} = P10(x_{ij}^t) \;\; when \; x_{ij} \in D, \end{cases} \tag{2}$$

where S is a stimulus and D is an anti-stimulus.

The pattern uses a 90 (P90) and 10 (P10) percentile because some ratios have untypical observations that could give misleading classification. Those

[1] The delay in publishing (about 4 weeks) was not taken into consideration.

ratios which exceeded the appropriate levels of percentiles where replaced with the levels of P90 or P10.

The next step is to find - for each company - the distance c_{i0}^t from the pattern. The Euclidean distance is used as a distance measure:

$$c_{i0}^t = \sqrt{\sum_{j=1}^m (z_{ij}^t - z_{0j})^2}, \ i = 1, 2, ..., n; t = 1, 2, ..., r; \tag{3}$$

where:
z_{ij}^t - standardised levels of characteristics of i object in time t, z_{0j} - levels of characteristics of the pattern from formula (3), i - number of the object.

Obtained distances allow for establishing synthetic (dynamic) measures of credit risk $SMCR_{ig}^t$:

$$SMCR_{ig}^t = \frac{c_{ig0}^t}{c_{g0}}, \tag{4}$$

where:

$$c_0 = \overline{c}_0 + 2s_0, \tag{5}$$

$$\overline{c}_0 = \frac{1}{n}\frac{1}{r}\sum_{i=1}^n \sum_{t=1}^r c_{i0}^t \text{ and } s_0 = \sqrt{\frac{1}{n}\frac{1}{r}\sum_{i=1}^n \sum_{t=1}^r (c_{i0}^r - \overline{c}_0)^2} \tag{6}$$

SMCR was built on the basis of the following ratios: current ratio, quick ratio, amount due to turnover(4)[2], amount due to liabilities ratio, gross turnover profitability ratio(4), net turnover profitability ratio(4), debt ratio, self-financing ratio. The choice of the above ratios that divide companies into companies of good and bad standing in the best way results from previous research carried out by the authors. These ratios were high for stimula in case of companies with low credit risk level and low for companies with high credit risk level. Such an interdependency is the opposite in case of anti-stimula. SMCR is constructed in such a way that it does not exceed 0. The probability that it will be higher than 1 is insignificant (in case values reached 1 they would be attributed 1). A measure value close to 0 is preferable (increase means financial problems). Appropriate values of ratios of all companies in the construction industry in 2000-2005 are assumed to be development patterns (not only of those companies subject to the analysis).

The other method of credit risk evaluation is based on assets volatility (MKMV). On the basis of the Black-Scholes-Merton option pricing model one can estimate equity and debt value. This is important because when a company is liquidated a bondholder receives some return when the value of equity is higher than 0, which means that the company value (A) is higher than its liabilities (D). Otherwise, a creditor does not get the return because the market value of equity is 0. This means that creditor's return is similar to income of call option writer on the assets of a company taking the loan.

Assuming that assets value changes can be described by Brownian standard geometric motions it is possible to calculate the probability of default of

[2] (4) in the indicator means that it is calculated on the basis of data of four quarters.

any debtor. The probability that the assets value of a company for any fixed time horizon[3] (T) will drop below the critical value (A_{def}) according to the equation (Saunders (1999)) reads:

$$PD_T = P\left[\varepsilon \leq -\frac{ln\left(\frac{A_0}{A_{def}}\right) + \left(\mu - \frac{\sigma_A^2}{2}\right)T}{\sigma_A\sqrt{T}}\right], \tag{7}$$

where:
A - company's assets value, A_{def} - company's critical assets value below which the company cannot pay back its debts[4], T - time of credit, r - risk-free interest rate, σ_A - company's assets value volatility, μ - average return rate of company's assets.

In the equation (7) A_{def}, D, T, r, μ are directly observable. Company's market assets value (A) and its volatility (σ_A) are not directly observable and must be estimated. To calculate them one can use the following relation (Hull (2003)):

$$E = AN(d_1) - De^{-rT}N(d_2), \tag{8}$$

$$\sigma_E = N(d_1)\sigma_A A, \tag{9}$$

where:

$$d_1 = \frac{ln\left(\frac{A}{D}\right) + (r + 0,5\sigma_A^2)T}{\sigma_A\sqrt{T}}, \quad d_2 = \frac{ln\left(\frac{A}{D}\right) + (r - 0,5\sigma_A^2)T}{\sigma_A\sqrt{T}},$$

E - company's market equity value, D - nominal debt value, σ_E- equity volatility, $N(d_i)$ - normal cumulative distribution function for argument d_i, where $i = 1, 2$.

Having equations (8) and (9) one can calculate A and σ_A in several iterations. The equations show that when the debt increases the debt ratio increases as well and company's assets value volatility decreases. Growth of debt ratio will make probability of default. Moreover, company's market assets value volatility will negatively influence PD.

3 Comparison of SMCR and MKMV results with reference to specific information

The probability of default levels of analyzed companies were assigned to SMCR that is divided into intervals of 0.1. But even those are also divided into two groups: levels of PD for companies that face and do not face serious problems (due to published specific information). The problems analyzed

[3] PD is mostly estimated in a year horizon.

[4] According to the model authors A_{def} consists of [short-term liabilities + 0.5 long-term liabilities].

are "real" problems, which means that a company is taken into consideration when it declares its default, or another company applies to declare its debtor bankrupt or banks refuse to credit this company anymore. On the other hand, a company stops being perceived as a "trouble" one when there is a treaty agreed/signed and implemented or when the legal actions were discontinued. The resolutions of a treaty have to be implemented because signing does not always guarantee that such resolutions will be put into practice.

The calculations are carried out for daily and quarterly data. The results obtained are very similar but they are slightly better in case of daily data. This can be caused by the fact that there are more daily observations and due to that the results seem to be more reliable.

Differences in average PD levels between "good" (GC) and "bad" (BC) companies can be observed in Table 2. One can also notice that only companies with a good standing have levels of SMCR below 0.6 and similarly their average PD is low (below 3%). When the SMCR level is exactly 0.6 the difference between the level of average PD for GC and BD is not big - 7% GC and 12% BC. The higher SMCR gets, the bigger the distance between "good" and "bad" companies' average PD is. When SMCR is 0.9 then GC's PD is about 9.9% and BC's - more than 32%. When percentiles are analysed (1st- P10 and 9th- P90) the results are similar.

Table 1. Comparison of SMCR and PD.

SMCR	Problems	average PD_T	no of obs.	st. deviat.	P10	P90
0.1	No	0.0002	366	0.0003	0.0000001	0.0007
0.1	Yes			no observation		
0.2	No	0.0039	1782	0.0093	0.0000001	0.0136
0.2	Yes			no observation		
0.3	No	0.0065	2116	0.0159	0.000002	0.0157
0.3	Yes			no observation		
0.4	No	0.0243	2839	0.0283	0.0000021	0.0661
0.4	Yes			no observation		
0.5	No	0.0295	6559	0.0576	0.0000021	0.1071
0.5	Yes			no observation		
0.6	No	0.0713	7045	0.1062	0.00001	0.2138
0.6	Yes	0.1151	489	0.0705	0.0306	0.2157
0.7	No	0.1530	4008	0.1871	0.00002	0.4631
0.7	Yes	0.2577	793	0.2265	0.0401	0.5959
0.8	No	0.1635	1020	0.2319	0.0025	0.5763
0.8	Yes	0.3562	1375	0.2809	0.0646	0.7439
0.9	No	0.0995	369	0.1064	0.00002	0.2232
0.9	Yes	0.3284	1284	0.1980	0.0937	0.6598
1.0	No	0.0876	113	0.1128	0.0001	0.2808
1.0	Yes	0.5216	1074	0.2936	0.1264	1.0000

One of the companies that is characterised by PD significant drop after restructuring treaties were signed (17.03.04) is Elektromontexp. However, in fact the main drop could be observed by the end of 2004 when the treaties resolutions were implemented. The same situation can be noticed in Energomontaz Pólnoc. Having signed and implemented agreements PD dropped considerably. Pemug is another example. The situation is not exactly the same though. Having implemented the agreement, PD remained quite high because of considerable public liabilities. PD drop can be noticed only after implementation of the resolutions.

The specific information analysed included application to be declared bankrupt (ADB), declared bankrupt (DB), negotiation call (NC), sale of significant assets (SSA), significant contract (SC), restructuring agreement (RA), restructuring process (RP), restructuring agreement accepted (RAA).

In Table 3 the levels of credit risk before (B) and after (A) RA (restructuring process) specific information are presented. As expected this information lowers the credit risk level - both simple change and relative change are negative. RA is a sign of positive actions taken by company authorities.

Table 2. Credit risk levels before and after restructuring agreement (RA) information.

	average PD	change	relative change	no of obs.	st.deviat.
B	49.2%	-11.4%	-23.1%	33	9.3%
A	37.8%	-11.4%	-23.1%	31	1.8%
B	36.7%	-1.7%	-4.7%	19	4.8%
A	34.9%	-1.7%	-4.7%	45	2.0%
	average	-6.55%	-13.93%		

Table 4 shows how many of the companies that face and that do not face real problems had a negative change. There are different possibilities to take this into consideration: ADB - information appeared for the first time in a quarter, ADB(2) - it appeared for the second time in the same quarter, ADB of a dependent company (ADB dep.comp.) - when it appeared for the first time in a quarter for the dependent company and when it appeared for the second time. Figures in the third column inform how many observations were available in a particular category and how many of them had a negative change. General influence of this information is an increase of credit risk level and therefore not many companies experienced a negative change.

A similar situation is observed when DB (declared bankrupt) information is analysed. Generally, this information increases the credit risk level although few companies have a negative change as Table 4 shows.

Quite surprisingly, information on a significant contract (SC) increases the credit risk level although it would seem that it should provide the opposite result (Table 5). Similarly the number in a bracket indicates how many times

Table 3. Changes in credit risk level caused by ADB information.

problems	variable	no of observation	change	relative change
No	ADB	0/0	—	—
No	ADB(2)	0/0	—	—
No	ADB dep.comp.	3/13	0.0146	9.4%
No	ADB dep.comp.(2)	0/2	0.0078	52.1%
Yes	ADB	3/13	0.0762	22.1%
Yes	ADB(2)	2/4	0.0228	8.2%
Yes	ADB dep.comp.	1/2	-0.0787	-34.2%
Yes	ADB dep.comp.(2)	1/1	-0.0176	-5.9%

Table 4. Changes in credit risk level caused by DB information.

problems	variable	no of observation	change	relative change
No	DB	0/0	—	—
No	DB dep.comp.	6/10	0.0019	7.1%
Yes	DB	2/4	0.0221	0.8%
Yes	DB dep.comp.	2/3	0.0088	10.5%
Together	DB	2/4	0.0221	0.8%
Together	DB dep.comp.	8/13	0.00348	7.9%

Table 5. Changes in credit risk level caused by SC information.

problems	variable	no of observation	change	relative change
No	SC	123/291	0.0010	—
No	SC(2)	73/167	0.0017	38.4%
No	SC(3)	55/103	-0.0004	12.6%
No	SC(4)	25/53	0.0021	29.6%
No	SC(5)	15/34	0.0020	61.2%
Yes	SC	29/45	-0.0242	-7.0%
Yes	SC(2)	15/23	-0.0048	-2.7%
Yes	SC(3)	8/14	0.0027	1.7%
Yes	SC(4)	8/10	-0.0052	-5.9%
Yes	SC(5)	1/4	0.0030	3.0%
Yes	SC dep.comp.	4/5	-0.0419	-12.7%

the information appeared during one quarter. It means that investors do not consider a big contract as positive information. Such a situation might be due to the fact that sometimes the standing of really significant contracts are not likely to be met by a company involved and, therefore, the contract in question may even generate some losses.

Sometimes, the same information results in both growth and drop of the credit risk level. It obviously depends on the fact whether the company had problems before or not (SSA, NC, NC dep.comp. are examples of such cases). Some information (like RAA) does not appear at all if a company has a good standing. Although the PD level changes might seem insignificant, the relative changes are usually more than +/- 5% (for NC even 42%).

4 Summary

This paper presents the influence of specific information on the credit risk level. There were 8 main pieces of information taken into consideration. Particular information was supposed to bring some expected effect, in case of other information the results were surprising. ADB information (application to be declared bankrupt) of the company in question has a considerable negative influence on the credit risk level, whereas the same information on a dependent company had almost no influence. DB information (declared bankrupt) of a dependent company had considerable influence on the credit risk level of the main company. Generally, the extent of influence in case of particular specific information cannot be stated because it heavily depends on the exact content of the information, the history of previous information and the general standing of the company. Further research should focus on other, less common, pieces of information and on pieces of information which appear in companies of good financial standing.

References

CAUETTE, J., ALTMAN, E. and NARAYANAN, P. (1998): *Managing Credit Risk - The Next Great Financial Challenge.* John Wiley & Sons, New York.

DEVENTER, D.R., IMAI, K. and MESLER, M. (2005): *Advanced Financial Risk Management.* John Wiley & Sons, Singapore.

HELLWIG, Z. (1968): Application of Taxonomic Method to Typological Division of Countries Considering Their Development, Resources and Structure of Qualified Workers. in: *Przeglad Statystyczny. R. XV. zeszyt 4.*

HULL, J.C. (2003): *Options. Futures and Other Derivatives.* Prentice Hall, New Jersey.

JAJUGA, K. (2001): Statistical Methods in Credit Risk Analysis. In: *Taksonomia 8 - Klasyfikacja i analiza danych - teoria i zastosowania. AE Wrocaw, 224-232.*

JAJUGA, K. (2004): Systematisation of Credit Risk Models. In: D. Appenzeller (Ed.) Upadlosc przedsiebiorstw w Polsce w latach 1990-2003. Teoria i praktyka. *Zeszyty Naukowe AE Poznan nr 49. AE Poznan, 119-126.*

SAUNDERS, A. (1999): *Credit Risk Measurements.* John Wiley & Sons, New York.

Part VIII

Bio- and Health Sciences

Enhancing Bluejay with Scalability, Genome Comparison and Microarray Visualization

Anguo Dong[1], Andrei L. Turinsky[1], Andrew C. Ah-Seng[1], Morgan Taschuk[1], Paul M.K. Gordon[1], Katharina Hochauer[1], Sabrina Fröls[2], Jung Soh[1] and Christoph W. Sensen[1]

[1] Sun Center of Excellence for Visual Genomics, University of Calgary, 3330 Hospital Drive NW, Calgary, Alberta T2N 4N1, Canada; `csensen@ucalgary.ca`

[2] University of Bergen, Department of Biology, Jahnebakken 5, Bergen 5020, Norway

Abstract. The Bluejay genome browser (Browser for Linear Units in Java™) is a flexible visualization environment for biological sequences, which is capable of producing high-quality graphical outputs (`http://bluejay.ucalgary.ca`). We have recently added functionalities to Bluejay to realize the true potential of 2D bioinformatics visualization. We describe the three major new functionalities that will be of added value to the user: (i) exploration of large genomes using level-of-detail management; (ii) comparative visualization of multiple genomes; (iii) visualization of microarray data in a genomic context.

1 Introduction

Bluejay is a Java-based integrated 2D computational environment for the exploration of genomic data. Turinsky et al. (2004) and Gordon and Sensen (1999) describe its key components and features with an emphasis on seamless integration of open standards As Bluejay is a continuously evolving system, we recently augmented its functionalities in three key areas to realize maximal visualization capabilities in 2D bioinformatics.

(i) massive amounts of bioinformatics data can be handled by a semantic level-of-detail management approach. The user can explore a large genome by quickly loading its skeleton, which is generated on-the-fly, and then by visualizing in greater detail only the regions of interest.

(ii) multiple genomes can be visualized in the same display for side-by-side comparison. This allows the comparative investigation of genome structures (synteny) and prediction of gene functions for a number of related organisms.

(iii) Bluejay has been augmented through the integration of TIGR MeV. The system can now be used to explore gene expression results and gene function simultaneously. This enables the user to explore a visual representation of both microarray and genomic data in the same context.

2 Scalable visualization

Analysis and visualization of massive genomics data has been a prominent topic of bioinformatics research in the last several years (Hunt et al. (2001), Chanda and Caldwell (2003), Maurer (2004)). Some of the existing data management and visualization approaches can be classified into three categories as in Reiser et al. (2002): (i) public databanks such as GenBank (Benson et al. (2005)) or EMBL (Guy et al. (2006)). (ii) data query systems like the Sequence Retrieval System (`srs.ebi.ac.uk`) or KEGG (`www.genome.jp/kegg`) (iii) locally installable database solutions like MySQL (`www.mysql.com`) or Oracle (`www.oracle.com`). All three approaches, although excellent in organizing the data, suffer from limitations, in particular in the area of flexible visualization of results.

Users of public databanks and query systems (the first two groups) are often forced to rely on slow, potentially insecure Web connections. The transmission time of visual models increases rapidly as the complexity of the graphics increases. Locally installed database systems (of the third kind mentioned above) allow more focused and customized explorations, but their installation and configuration can be a tedious process that typically requires sophisticated knowledge of computer technology. Besides these options, a large number of downloadable tools exist, but most of these are typically limited to the handling of individual small data files (Brown et al. (2005), Sun and Davuluri (2004)). Our goal is to develop techniques that allow locally installable tools handle very large local data files without resorting to the need for the end user to maintain a sophisticated local database.

Semantic zooming (Loraine and Helt (2002)) is a core technique for our implementation of scalability in visual bioinformatics. This strategy enables Bluejay to run on a memory-constrained client while loading an arbitrarily large genome or multiple genomes simultaneously. A reduced data set, called a skeleton, is first loaded and more details are retrieved later, only if the user zooms into a region of the genome. A data skeletonization mechanism is used to reduce a very large data file into a much smaller, manageable version. Bluejay processes data mostly in the eXtensible Markup Language (XML) format, where pieces of data are embedded within a nested hierarchy of markup tags. Fig. 1 describes the data flow during the skeletonization process. If an XML file containing a large amount of data is loaded into Bluejay, the system performs a data skeletonization on-the-fly, using an eXtensible Stylesheet Language Transformation (XSLT, `www.w3.org/TR/xslt`), which transforms one XML to another as dictated by the XSL file. Browser side caching is also used, to further reduce the data transmission time.

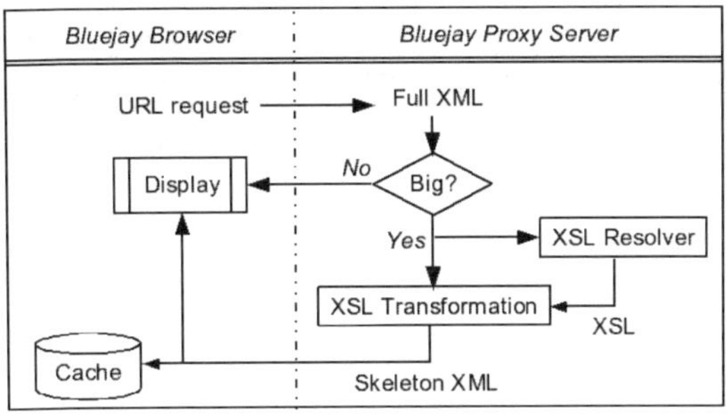

Fig. 1. XSL transformation for XML data skeletonization.

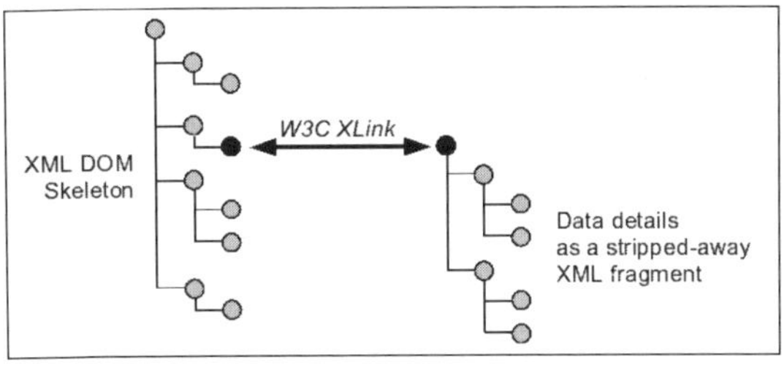

Fig. 2. DOM tree fragment pruning.

The internal representation of XML data is done as a Document Object Model (DOM, `www.w3.org/DOM`), which is a hierarchy of nodes representing nested markups of an XML document. Fig. 2 shows a conceptual skeleton DOM tree. Skeletonization and restoration of data details from the original data involves efficient pruning and inserting of DOM tree fragments. During the skeletonization process, XLink-compliant hyperlinks are inserted into the skeleton in place of the removed data fragments (`www.w3.org/TR/xlink`). The XLinks point to the position of the fragments in the original document, identified by the XPointer standard (`www.w3.org/TR/xptr`). For example, genes may be located by their GenBank ID attribute. The example XLink attributes below show how to specify the role or reason for the link as well as the link location and behavior.

```
xlink:to="fullXMLAnnotationURL#UniqueID" xlink:show="embed"
xlink:actuate="onRequest" xlink:role="URIForRoleOfLink"
```

Semantic retrieval of data entails level-of-detail management and tiling computation. These key operations depend critically on the data structures used behind the scenes. Such a data structure must adequately represent the hierarchy of semantic levels, allow efficient querying of such hierarchy, and facilitate easy addition to, or removal from a specific location in the structure. The DOM trees used in Bluejay naturally satisfy those requirements.

Fig. 3 shows a snapshot of semantic zooming in action. When the user zooms in on the loaded skeleton, Bluejay calculates the region of interest

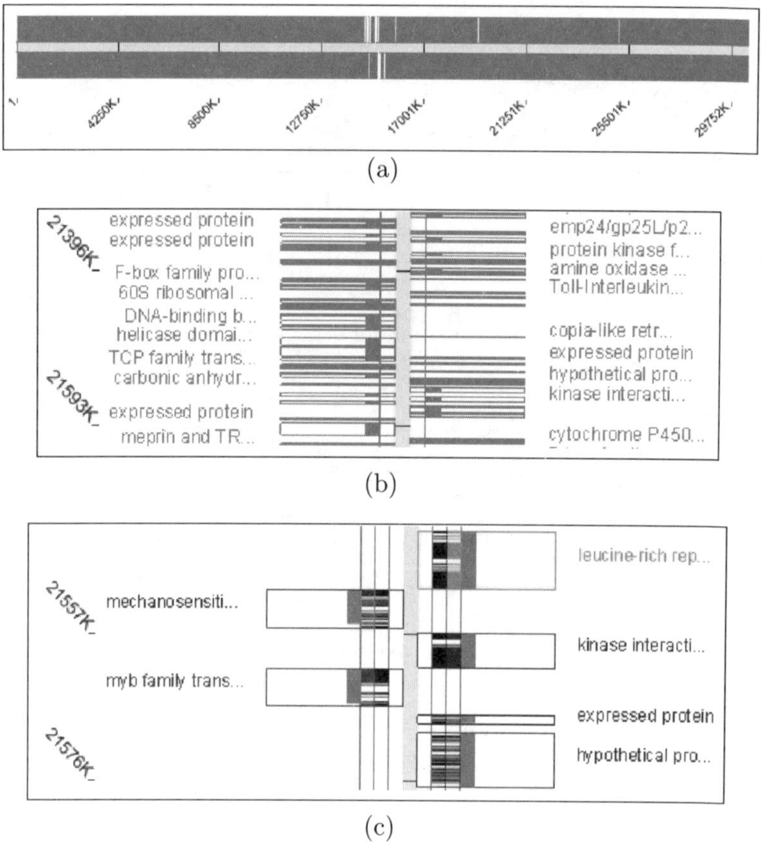

Fig. 3. Tiling in Bluejay: (a) a skeleton of *Arabidopsis thaliana*; (b) zooming brings up transcriptional unit names; (c) further zooming retrieves features.

on the genomic sequence. Only the node data visible in the current "tile" is retrieved. As the user zooms in further, the depth of details increases while the tiles become smaller at the same time. The internal mechanism activates the XLinks based on role and determines the node visibility based on the current display scale. Despite potentially overwhelming volumes of data served to Bluejay, this strategy can be used with a reasonable amount of computational resources, such as a personal computer, at any given time. In the example of Fig. 3, the original full XML file size was 73 Mbytes, but the skeleton holds only 4.7 Mbytes of data, resulting in a data reduction rate of 94%.

3 Multiple genome comparison

Comparative genomics is a promising approach to studying organism diversity, evolution, and gene functions (Wei et al. (2002)). It reveals, at the genome level, the commonalities that exist among different species. Genome comparison is a reliable way to identify genes and predict their functions and interactions. The aim of multiple whole genome comparison in Bluejay is to provide the user with a consistent and intuitive visualization capability for the comparison of genomes.

Several tools have been developed for comparing and analyzing multiple genomes. The BLAST family of programs (Altschul et al. (1990), Altschul et al. (1997)) are widely used for searching protein and nucleotide databases to identify sequence similarities, using local alignment between a query sequence and each of the sequences in a database. MUMmer (Delcher et al. (1999)) is a fast comparison tool that can rapidly align two large nucleotide sequences using a suffix-tree based algorithm. However, these are essentially search or alignment tools that do not offer a rich interface for the comparison of multiple whole genomes. In addition, both the interface and the features provided by those tools are not very user-friendly. For instance, MUMmer is written in Perl and can only be executed on specific operating systems. Some tools require the user to remember a set of commands, which some inexperienced users might feel difficult to use.

For the genome comparison in Bluejay, a *master sequence* is defined as the sequence that the user wants other sequences to be compared with. A *slave sequence* is defined as a sequence that is compared to the master sequence. A *genome pool* is a pool of sequences, which contains the master sequence and all the slave sequences. Bluejay allows the user to select only one master sequence at a time to function as the reference sequence, but there can be as many slave sequences as desired. The default master sequence is the current sequence loaded within Bluejay. Bluejay automatically enters the genome comparison mode if the user loads an additional sequence by using built-in bookmarks or reading from a local file. Fig. 4 shows an example of comparing two sequences in Bluejay.

The lines between the particular genomes link common genes based on their Gene Ontology classification (GO, `www.geneontology.org`). Connecting all common genes belonging to the same classification will clutter the view with too many lines. This problem is alleviated in Bluejay by using the simple strategy of connecting each gene only to the closest one of its kind. For each gene, the closest gene with the same Gene Ontology classification is found among all classified genes, starting through a comparison, which begins with the closest genome. Fig. 5 shows an example of three slave sequences.

The need to rotate sequences often arises, because the manipulation can improve the visualization of the comparison result for circular genomes. Bluejay can rotate all the sequences together or any sequence separately by a user-defined angle. The individual rotation capability is very useful for the alignment and analysis of genome contents in multiple genomes. The lines linking common genes are dynamically recalculated and redrawn when the user rotates a sequence. Sometimes the user might want to unload one or more slave sequences because they are no longer needed for comparison. By

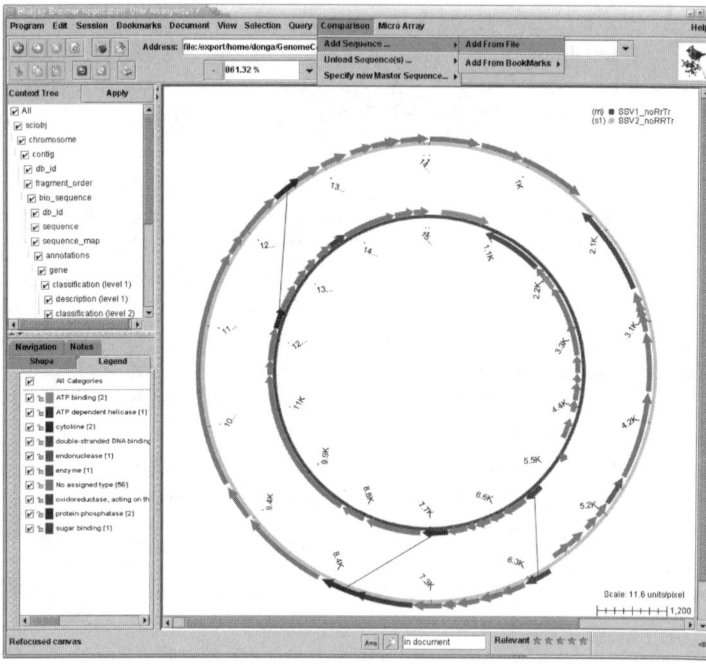

(a)

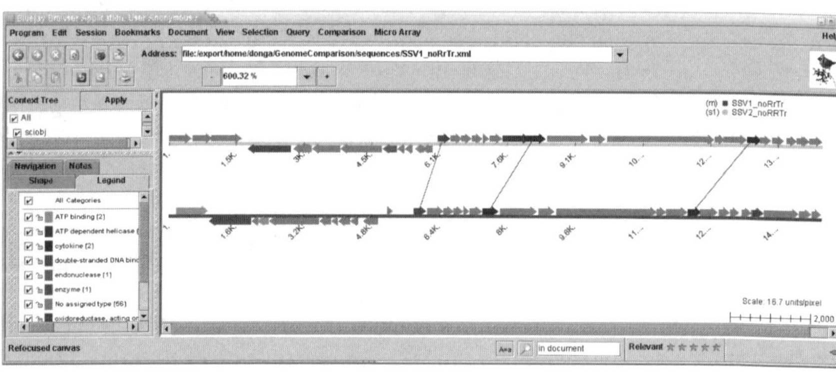

(b)

Fig. 4. Comparing two sequences having three common genes: (a) circular view; (b) linear view.

unloading all slave sequences, the user returns to a non-comparative mode, where only the master genome is displayed. Similar to the sequence rotation mode, the linkage between common genes is dynamically updated after an unloading operation takes place. The genome pool is also updated to reflect the sequence removal. Fig. 6 shows that the linkages are correctly updated following a rotation or unloading operation on a single sequence.

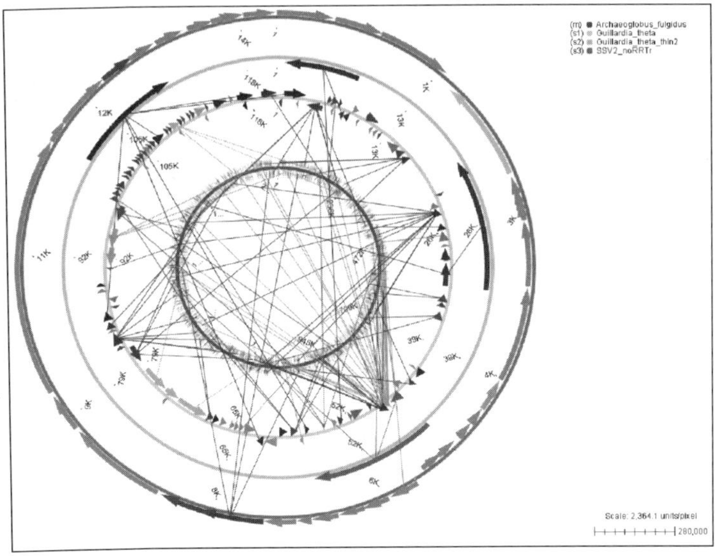

Fig. 5. *Archaeoglobus fulgidus* (innermost) compared with three slave sequences.

4 Microarray data visualization

Microarrays enable the study of expression levels of many genes simultaneously, therefore providing a snapshot of the transcriptional processes within a cell. This has proven to be a valuable tool for the study of gene function in different biological processes and under various conditions (Schena et al. (1995), Schena et al. (1996), Baranzini and Hauser (2002)).

As large datasets are generated from microarray experiments, there is an increasing need for programs, which can generate views of both genomic function and gene expression results in a meaningful and comprehensive manner and perform analysis using the data. We have expanded Bluejay to incorporate TIGR MeV (`www.tigr.org/mev.html`), a Java-based program developed by The Institute for Genomic Research (TIGR) as part of its TM4 (Saeed et al. (2003)). microarray analysis package. TIGR MeV now serves as a module within the Bluejay packagae, which provides access to expression values and gene expression results.

The interaction between TIGR MeV and Bluejay is shown in Fig. 7. TIGR MeV formats and internally stores the expression values from microarray data, along with any annotations. Bluejay reads this formatted dataset and builds hash tables based on the relationships that exist within the data. The hash tables provide fast access to the expression data when a gene name or an identifier like a GenBank number is known.

TIGR MeV can employ several clustering algorithms (i.e., hierarchical clustering, k-means clustering, and self-organizing map) for analysis of microarray data, to identify and separate genes that appear to have similar pat-

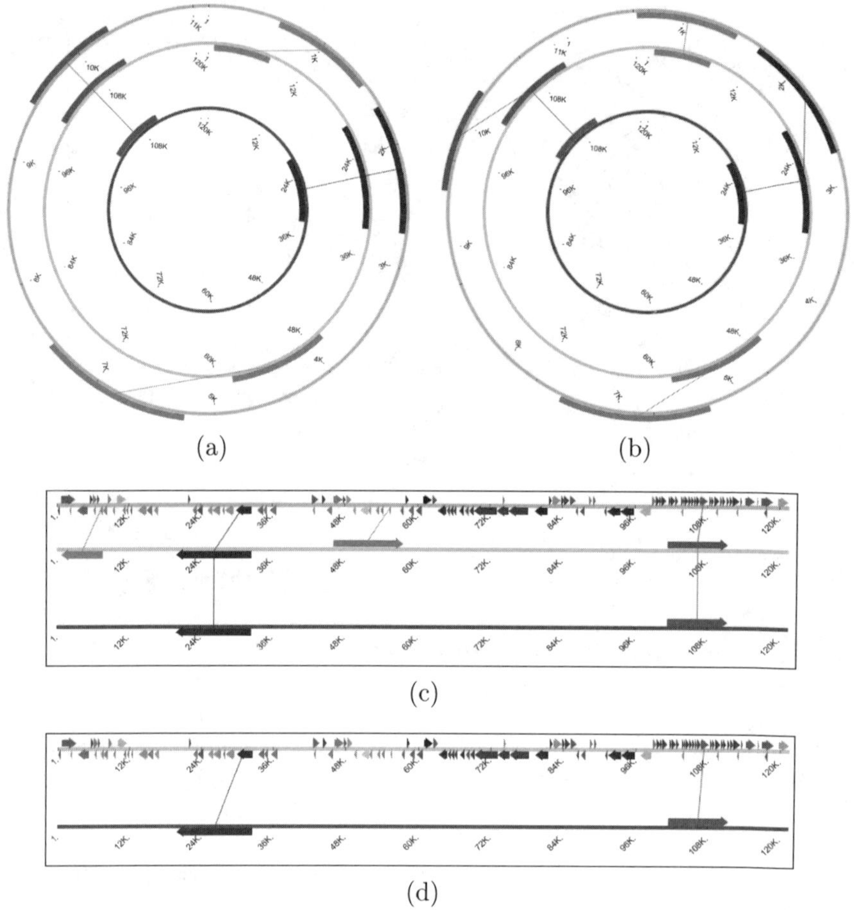

Fig. 6. Rotate and unload operations: (a) before rotating; (b) after rotating the outer slave sequence 30° counterclockwise; (c) before unloading; (d) after unloading the inner slave sequence.

terns of expression under various conditions. Cluster-specific data (i.e., cluster sizes and expression levels) are stored in a legend-type interface, where every item represents a piece of information about the clusters.

We will use a dataset containing expression data from an experiment using *Sulfolobus solfataricus* PH1 (lacS mutant) lysogen infected with *Sulfolobus spindle-shaped virus 1* (SSV1) wild type to demonstrate the microarray analysis functionality of Bluejay. Expression values were calculated using $E_V = log_2(E/C)$, where E is the intensity of the gene in the experiment, and C is the intensity of the same gene in the control.

Bluejay supports both viewing of a single experiment and of multiple experiments. Multiple experiment data can be derived from: (i) different experi-

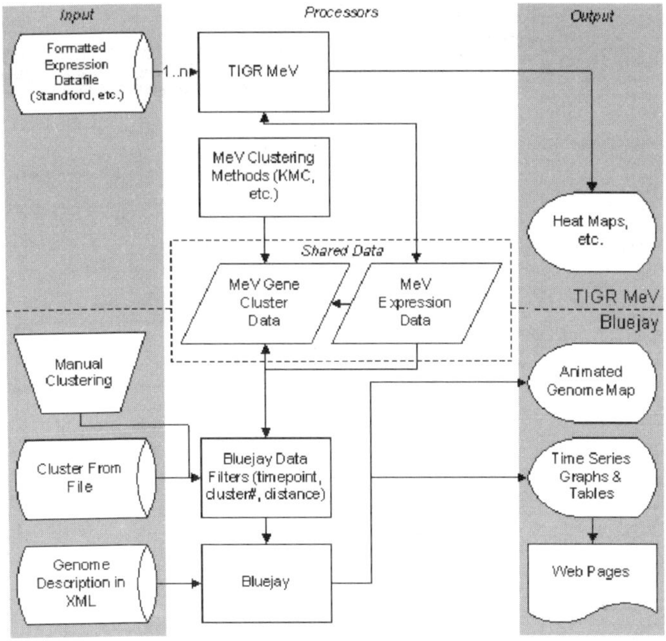

Fig. 7. Interaction between Bluejay and TIGR MeV.

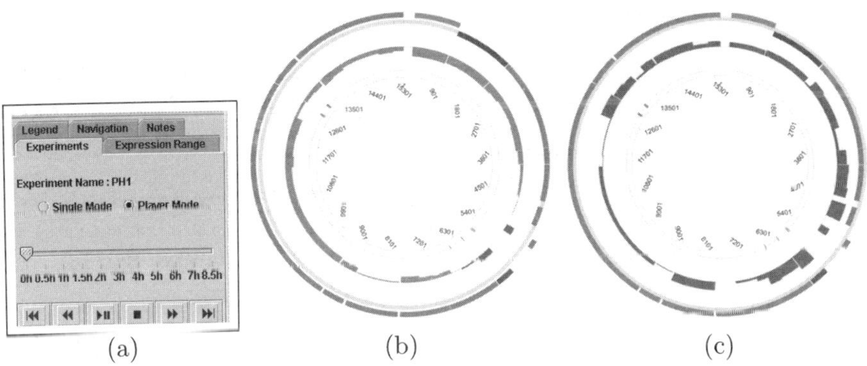

Fig. 8. Navigating multiple experiments: (a) either single mode or player mode can be used; (b) chromosomal wheel at 4h after infection with SSV1 wild type on *Sulfolobus solfataricus* PH1; (c) after 7h.

ments applied to the same organism; or (ii) a time series where data is collected at different time points during the course of the experiment. When data from multiple experiments is loaded, two navigation modes are supported, as shown in Fig. 8. The single mode allows navigating from one time point in the experiment to another by using the timeline slider. The player mode displays a video player interface, where multiple experiments can be displayed like a

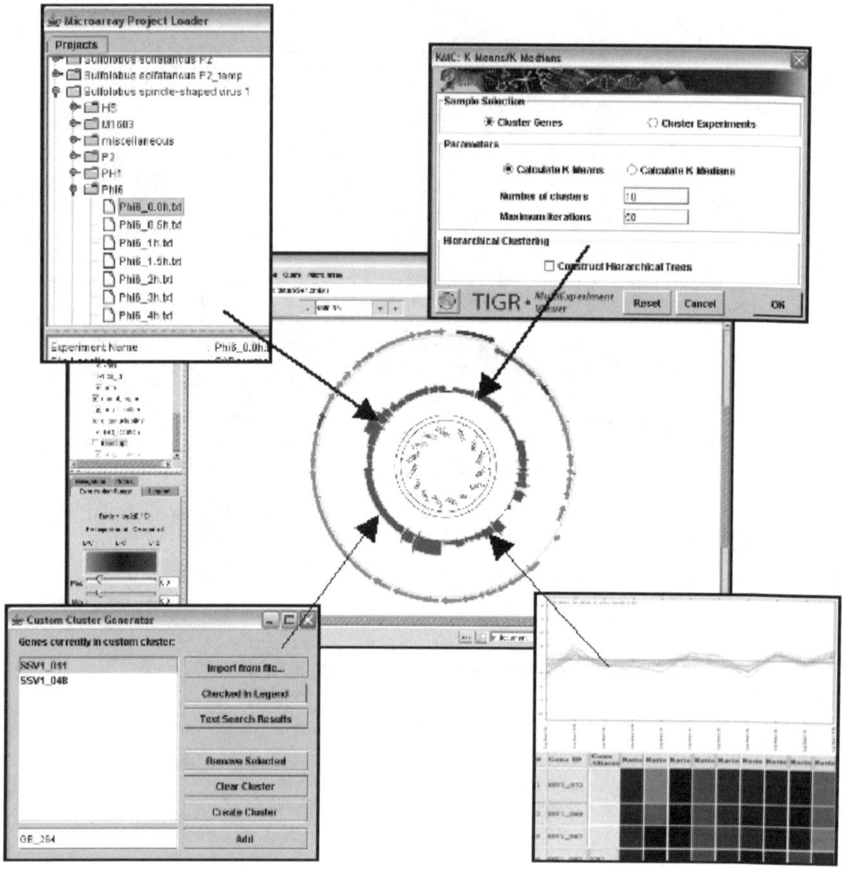

Fig. 9. Integration of microarray analysis tools into Bluejay: Using TIGR MeV, Bluejay can: load microarray data (top left); cluster it (top right); create its own custom clusters (bottom left); generate cluster specific lists (bottom right).

movie. Bluejay allows viewing a user-selected portion of the total number of genes by dynamically adjusting the top and bottom expression level thresholds. The genes with expression levels that are lower than the thresholds in magnitude are ghosted so that the user can focus on highly up-regulated or down-regulated genes.

Bluejay takes all the cluster analysis results generated by TIGR MeV and displays them on the whole genome, as shown in Fig. 9. To distinguish different clusters, Bluejay assigns each cluster its own color and places the color swatch along with the cluster name in the legend. To view only the genes from a specific cluster, one can select that cluster in the analysis window so that the genes falling outside of that cluster are ghosted. Through the use of TIGR MeV analysis and the visualization of expression data by Bluejay, we were able to identify not only the clusters of genes within the microarray dataset, but also their spatial locations on the genome. Further analysis often uncovers if these genes were co-expressed genes; or coded for proteins in the same functional category; or have the function of housekeeping genes. Although a simple visual inspection is usually sufficient and much of this method could be overlooked for smaller genomes with just a few dozen genes such as SSV1, larger genomes with thousands of genomes, such as the *Sulfolobus solfataricus* genome with almost 3000 genes, require the capabilities of viewing the genome as a whole. Bluejay facilitates the analysis of genes not only through the visual inspection of graphs and tabulated results, but also through the study of genes in a genomic context with the corresponding microarray expression values and/or analysis information linked to the gene function.

5 Conclusion

We have described recent enhancements to Bluejay that bring this system closer to the limit of 2D bioinformatics visualization capabilities. Scalable visualization, which can cope with very large datasets, is realized using data skeletonization and just-in-time retrieval of details. Bluejay can now visualize multiple genome sequences for comparison, and provides several display features to facilitate comparative analysis. Bluejay also offers integration with a comprehensive package for the analysis of microarray data, through the seamless integration with TIGR MeV. We expect that these newly added functionalities will make Bluejay an even more versatile and practical visualization tool for the bioinformatics community.

Acknowledgements

We would like to thank Krzysztof Borowski and Emily Xu for their valuable contributions to the enhancement of Bluejay. This work was supported by Genome Canada/Genome Alberta through Integrated and Distributed Bioinformatics Platform for Genome Canada, as well as by the Alberta Science and Research Authority, Western Economic Diversification, National Science and Engineering Research Council, and the Canada Foundation for Innovation. Christoph W. Sensen is the iCORE/Sun Microsystems Industrial Chair for Applied Bioinformatics.

More information on Bluejay and freely downloadable systems in form of a signed applet; Java Web Start; or a local stand-alone application are available at `http://bluejay.ucalgary.ca`.

References

ALTSCHUL, S.F. et al. (1997): Gapped Blast and Psi-Blast: A New Generation of Protein Database Search Programs. *Nucleic Acids Research, 25, 3389–3402.*

ALTSCHUL, S.F., GISH, W., MILLER, W., MYERS, E.W. and LIPMAN, D.J. (1990): Basic Local Alignment Search Tool. *Journal of Molecular Biology, 215, 403–410.*

BARANZINI, S.E. and HAUSER, S.L. (2002): Large-Scale Gene-Expression Studies and the Challenge of Multiple Sclerosis. *Genome Biology, 3, 10, reviews1027.1–1027.5.*

BENSON, D.A., KARSCH-MIZRACHI, I., LIPMAN, D.J., OSTELL, J. and WHEELER, D.L. (2005): GenBank. *Nucleic Acids Research, 33, D34–D38.*

BROWN, C.T., XIE, Y., DAVIDSON, E.H. and CAMERON, R.A. (2005): Paircomp, FamilyRelationsII and Cartwheel: Tools for Interspecific Sequence Comparison. *BMC Bioinformatics, 6, 1, 70.*

CHANDA, S.K. and CALDWELL, J.S. (2003): Fulfilling the Promise: Drug Discovery in the Post-Genomic Era. *Drug Discovery Today, 8, 4, 168–174.*

DELCHER, A.L. et al. (1999): Alignment of Whole Genomes. *Nucleic Acids Research, 27, 11, 2369–2376.*

GORDON, P. and SENSEN C.W. (1999): Bluejay: A Browser for Linear Units in Java. *Proc. 13th Annual International Symposium on High Performance Computing Systems and Applications, 183–194.*

GUY, C. et al. (2006): EMBL Nucleotide Sequence Database: Developments in 2005. *Nucleic Acids Research, 34, D10–D15.*

HUNT, E., ATKINSON, M.P. and IRVING, R.W. (2001): A Database Index to Large Biological Sequences. *International Journal of Very Large Databases, 7, 3, 139–148.*

LORAINE, A.E. and HELT, G.A. (2002): Visualizing the Genome: Techniques for Presenting Human Genome Data and Annotations. *BMC Bioinformatics, 3, 1, 19.*

MAURER. M.H. (2004): The Path to Enlightenment: Making Sense of Genomic and Proteomic Information. *Genomics Proteomics Bioinformatics, 2, 2, 123–131.*

REISER, L., MUELLER, L.A. and RHEE, S.Y. (2002): Surviving in a Sea of Data: A Survey of Plant Genome Data Resources and Issues in Building Data Management Systems. *Plant Molecular Biology, 48, 1-2, 59–74.*

SAEED, A.I. et al. (2003): TM4: A Free, Open-Source System for Microarray Data Management and Analysis. *Biotechniques, 34, 2, 374–378.*

SCHENA, M. et al. (1996): Parallel Human Genome Analysis: Microarray-Based Expression Monitoring of 1000 Genes. *Proc. National Academy of Sciences USA, 93, 20, 10614–10619.*

SCHENA, M., SHALON, D., DAVIS, R.W. and BROWN, P.O. (1995): Quantitative Monitoring of Gene Expression Patterns with a Complementary DNA Microarray. *Science, 270, 5235, 368–369, 371.*

SUN, H. and DAVULURI, R.V. (2004): Java-Based Application Framework for Visualization of Gene Regulatory Region Annotations. *Bioinformatics, 20, 5, 727–734.*

TURINSKY, A.L., AH-SENG, A.C., GORDON, P.M.K., STROMER, J.N., TASCHUK, M.L., XU, E.W. and SENSEN, C.W. (2005): Bioinformatics Visualization and Integration with Open Standards: The Bluejay Genomic Browser. *In Silico Biology, 5, 2, 187–198.*

WEI, L., LIU, Y., DUBCHAK, I., SHON, J. and PARK, J. (2002): Comparative Genomics Approaches to Study Organism Similarities and Differences. *Journal of Biomedical Informatics, 35, 2, 142–150.*

Discovering Biomarkers for Myocardial Infarction from SELDI-TOF Spectra

Christian Höner zu Siederdissen[1], Susanne Ragg[2] and Sven Rahmann[1]

[1] Algorithms and Statistics for Systems Biology Group, Genome Informatics, Department of Technology, Bielefeld University, Germany; {choener,rahmann}@cebitec.uni-bielefeld.de

[2] School of Medicine, Indiana University, Indianapolis, USA, and Riley Hospital for Children, Indianapolis, USA

Abstract. We describe a three-step procedure to separate patients with myocardial infarction from a control group based on SELDI-TOF mass spectra. The procedure returns features ("biomarkers") that are strongly present in one of the two groups. These features should allow future subjects to be classified as at-risk of myocardial infarction. The algorithm uses morphological operations to reduce noise in the input data as well as for performing baseline correction. In contrast to previous approaches on SELDI-TOF spectra, we avoid black-box machine learning procedures and use only features (protein masses) that are easy to interpret.

1 Introduction

Myocardial infarctions cause an estimated 500 000 deaths in the U.S. per year and severely disrupt the lives of many other patients. The risk of imminent myocardial infarction can in principle be diagnosed early by analyzing the blood serum of (future) patients for the presence of certain characteristic biomolecules ("biomarkers"). While some of these are known in the medical literature, several more may exist. The goal of this study is to discover characteristic masses of such biomarkers (by SELDI-TOF mass spectrometry) without attempting to directly identify the molecules in question.

SELDI-TOF (Surface enhanced laser desorption/ionization, time-of-flight) mass spectrometry uses special surfaces on which the blood serum samples are applied. The surfaces can be chemically or biochemically modified to isolate specific (classes of) proteins. After a washing cycle the samples are placed into an evacuated chamber. A pulsed laser beam evaporates part of the sample, thereby ionizing the previously bound proteins. The ionized proteins are accelerated towards a detector by an electro-magnetic field. The time between the ionization of a protein until it hits the detector is a one-to-one function of

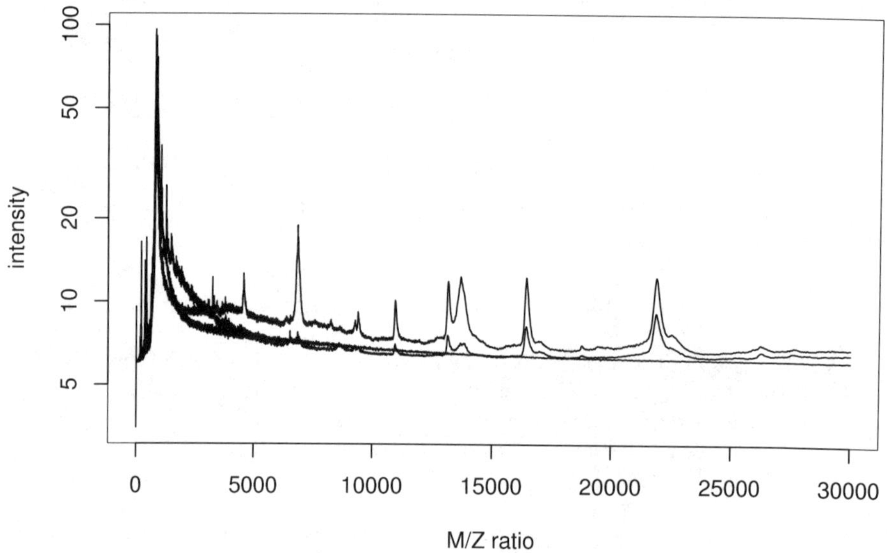

Fig. 1. Three types of spectra: One spectrum from the patient and one from the control group, supplemented by one of the 9 spectra with low signal to noise ratio. These 9 spectra can be easily identified as the raw data shows none of the clearly visible peaks that can be seen in the other spectra (with the exception of the always visible peak at about 840 m/z).

the mass over charge (m/z) ratio of the protein, so the times-of-flight can be directly converted into m/z-values. In this way, the abundance of proteins with a any given m/z-value is determined. A detailed explanation of the approach can be found in (Issaq et al. (2002)). The system used here is a Ciphergen PBS IIc mass spectrometer optimized for measuring masses between 1000 and 15000 Dalton (1 Da approximately corresponds to the mass of one proton).

The complete dataset consists of 64 individual spectra from blood serum samples. The set is divided into the patients group with myocardial infarction and a sex- and age-matched control group. The 64 spectra belong to 32 people (16 patients + 16 control), as each of the blood samples was measured twice. Visual inspection reveals that 9 of the spectra seem to have a low signal to noise ratio (Fig. 1).

For each of the spectra, the relative abundances of 33437 individual m/z ratios were recorded. The Ciphergen software provides automatically detected peaks in the spectra that were discarded in favor of those generated by our own algorithm.

Our goal is to select a small subset of the mass peaks that may correspond to potential biomarkers. A small set (instead of the complete set) of peaks makes it possible to verify the biological relevance of the peaks by wet-lab experiments. A potential biomarker is determined by finding a peak for a given m/z ratio in only one of the two groups or having most of the peaks in one of the groups.

To generate a list of potential biomarkers, several steps have to be taken. (1) We use operations from the field of mathematical morphology to perform baseline correction for the spectra as well as noise-reduction. (2) Raw peaks (spanning several m/z values) are then transformed into point masses. Buckets collect peaks around specific m/z values and are scored, taking into account absolute and relative differences of the number of peaks found in the patient and in the control group. (3) Finally, we select a promising subset of relevant m/z values as potential biomarkers.

2 Baseline correction and noise removal

Morphological operations. For baseline correction and noise removal, we use operations from mathematical morphology, which we introduce below, following Serra (2006), Roerdink (2000) and in particular Maragos (2005), but specializing the concepts to the SELDI-TOF spectra.

Definition 1 (dilation, erosion, opening, closing, tophat). *Let $X, Y \subset \mathbb{R}^n$, let f and g be real-valued functions defined over X and Y, respectively. The following lines define the dilation $\delta_g(f)$, erosion $\varepsilon_g(f)$, opening $\alpha_g(f)$, closing $\phi_g(f)$, and the "top-hat"* $\mathrm{tophat}_g(f)$ *of f by g, respectively:*

$$\begin{aligned}
\delta_g(f)(x) &:= (f \oplus g)(x) := \sup_{y \in Y} f(x-y) + g(y), \\
\varepsilon_g(f)(x) &:= (f \ominus g)(x) := \inf_{y \in Y} f(x+y) - g(y), \\
\alpha_g(f) &:= (f \ominus g) \oplus g, \\
\phi_g(f) &:= (f \oplus g) \ominus g, \\
\mathrm{tophat}_g(f) &:= f - \alpha_g(f).
\end{aligned}$$

The important operations for the following discussion are the closing and the tophat-operation that subtracts from f its g-opening. For us, f will represent the data, and g will be a simple flat structuring element that takes the value of 1 in a neighborhood of zero (and is zero otherwise).

Baseline correction and noise removal. Baseline correction has to be performed to be able to clearly detect the individual peaks. Noise removal removes sinks with small width, thereby removing the small peaks surrounding the sink as well. It is assumed that those sinks and peaks contain no relevant information. Both operations can be performed efficiently using morphological operations. Sauve and Speed (2004) point out that the structuring element for baseline correction has to be chosen carefully. For our data, flat structuring elements for baseline correction as well as noise reduction have proven sufficient.

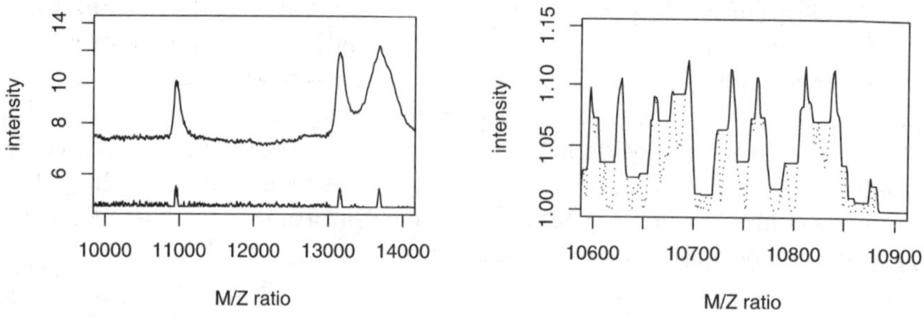

Fig. 2. Left: Baseline correction using the tophat-operator and a flat structuring element of width 51. The lower of the two spectra shows the result of performing the operation. Note that the baseline correction spectrum has artificially been raised 5 units for the figure. **Right:** Noise removal using the tophat-operator and a flat structuring element of width 5. The dotted line is part of the original spectrum. Fast, local changes in intensity are eliminated.

In this case, the required operations can be simplified to minimum and maximum operations operating in small windows on the data. For high-throughput data as is found often in mass spectrometry, this has the additional benefit that the operations are fast.

For baseline correction, the structuring element spans over 25 measurements to the left as well as the right. Applying this structuring element using the tophat operator for grey-value data results in a spectrum that is baseline corrected. The correction is local to small windows over the data thereby ensuring that the baseline is correct independent of local raw baseline.

Noise reduction on the other hand requires a much smaller structuring element that, in our case, spans 5 m/z units. It is applied using the closing operation and fills small sinks that are less than 5 m/z units wide. The closing operation is followed by a baseline correction step.

If both steps, noise removal and baseline correction, are performed, the baseline correction step should be the second. The effects are shown in Fig. 2.

3 Peak extraction, bucketing, and scoring

Peak detection and extraction. The de-noised and baseline corrected spectra are still not ideal to work with: The type of peak we need is a point mass peak. For each peak with contiguous non-zero masses from m/z-position m to $m + k$ and corresponding intensities $I_m, \ldots, I_{m+k}$ the total intensity is given by $\mathcal{I} = \sum_{m \leq j \leq m+k} I_j$. The m/z-position is given by position $m + l$ where $\sum_{m \leq j \leq m+l} I_j \approx \mathcal{I}/2$ (see Fig. 3 left). The calculated position is correct as

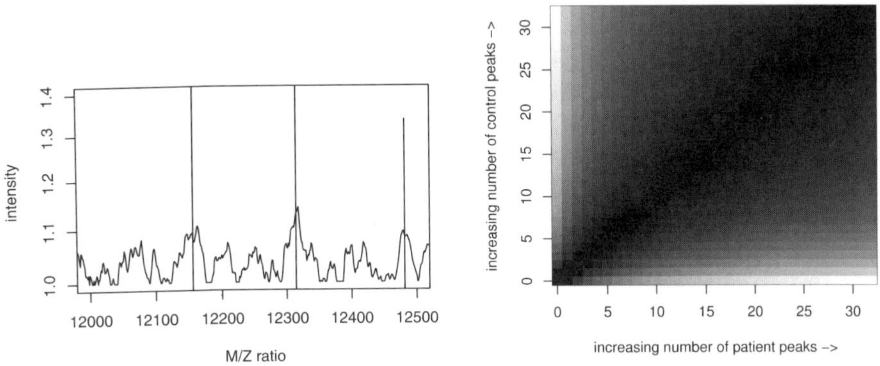

Fig. 3. Left: Peak extraction replaces raw peaks with point mass peaks (vertical lines). For most of the peaks with only one local maximum, the point mass peaks have about the same m/z value as the local maximum. **Right:** Relevance given to a bucket in relation to the number of peaks found in the patient and the control group. The lighter the coloring the higher the score. Setting the parameters α, β and γ to 1 yields the following scores for example: with 5 peaks in the patient group and 0 peaks in the control group, a relevance score of 4; with 12 peaks in the patient group and 2 peaks in the control group the score is 3.

long as the peaks are not heavily skewed to one side. For the spectra used, the point mass peaks were near or at the position of the maximum of the corresponding peak, suggesting that the method is adequate for the data.

Bucketing. We use buckets spanning 5 m/z units to select the relevant peaks and to be able to compare peaks that are not perfectly aligned. To capture the complete neighborhood for one peak, we use buckets that overlap by half their width, resulting in a two-fold coverage of the m/z-axis. This technique is simple and efficient, but can only be successful if only few peaks that do not represent the same potential biomarker fall into one bucket. If this is not the case, a more thorough alignment procedure has to be used.

Relevance scoring. The selection of potential biomarkers is based on a peak (bucket) score function that takes into account the absolute number of peaks found for each of the two classes (patient/control) as well as their relative difference. First, for each bucket, the number of peaks for the patient as well as the control group is counted. Then the i-th bucket can be given a relevance score

$$r_i := \max\left\{0, \frac{|c_i - p_i| - \alpha}{\beta \times \min(c_i, p_i) + \gamma}\right\},$$

where c_i and p_i are the number of peaks found for the control or patient group for the i-th bucket and α, β and γ are parameters. The relevance function is illustrated in Fig. 3 right.

The parameter $\alpha \in \mathbb{N}$, the minimal difference of peaks found in the two groups, will be set close to 0 if the total number of peaks found per bucket is low to capture more peaks and can be set higher if buckets with a higher difference between the number of peaks in the two classes are available. In general, a larger α means discarding buckets with lower relevance. The weight of the smaller group is controlled using the parameter $\beta \in \mathbb{R}$, and normally $0 \leq \beta \leq 2$ is chosen. Setting β controls the influence of the relative difference between the two groups. Normally $\beta = 1$, which prefers buckets where one group is clearly dominant and the other either non-existent or having only very few peaks. The last parameter, $\gamma \in \mathbb{N}^+$, is a constant offset, normally set to 1 to prevent a divide by zero in case of one of the two groups having 0 peaks for a specific bucket.

The parameters were set to $\alpha = 1$, $\beta = 1$ and $\gamma = 1$. Small changes in the choice of the parameters did not lead to large or abrupt changes in the classification results. An appropriate choice of parameters is required for the algorithm, but the correct choice can be made by studies of the behavior of the peak score function and results on the training data.

Bucket selection. In general, it can be assumed that buckets with high relevance scores are potential biomarkers. The selection of the best candidates is based on this assumption. Dependent on the requirements several possibilities exist to reduce the space of candidate biomarkers. While it is easy to select every potential biomarker with a relevance higher than a given threshold k, we have chosen a different criterion: We select the n most relevant buckets for each group (patient and control), plus any additional buckets that achieve the same relevance score as the n-th one (score ties occur frequently).

4 Classification

Classification procedure. After selecting buckets as features of classification, new samples can be classified by calculating two scores: (1) the patient score and (2) the control score. The patient score is calculated by

$$S_{\text{patient}} = \sum_{i \in \text{ patient peaks}} [i \text{ present}] \cdot r_i,$$

where the "patient peaks" are those selected for the patient group, the indicator function $[i \text{ present}]$ takes the value 1 if the corresponding peak is present in the test sample, and 0 otherwise, and r_i is the corresponding relevance score. The score for the control group is defined analogously.

The classification itself is done by fitting a maximum margin line in the 2-dimensional (patient score, control score) plane that separates the two groups. In our tests a perfectly separating line could always be found.

Results. The results presented here were obtained using 8-fold cross validation. The 64 spectra were divided into 8 equal-sized groups using 7 for the training set, used to select the potential biomarkers, and 1 for testing of the

Table 1. Classification results using 8-fold cross-validation. 'bb' is the number of best scores used for selection of the buckets. '#patient' and '#control' gives the number of different buckets found per class on average. 'correct' gives the absolute number of correctly classified spectra summed over the individual cross-validation results.

bb	#patient	#control	correct
3	4.3	4	49/64
4	9.7	7.7	52/64
5	13.5	10.7	57/64
all	1767	1810	61/64

algorithm. We chose 8-fold cross-validation instead of, e.g., 5- or 10-fold cross-validation simply because the size of the dataset immediately suggests this number.

As described in the introduction, 9 spectra out of 64 have a low signal to noise ratio and may possibly be of little or no value. Nonetheless they were included, first because the original set of spectra is small enough, and second because if small peaks are the best candidates for biomarkers then they might be found in the suboptimal spectra as well.

For each of the 8 possible choices of training and test set, the n best buckets were chosen as explained above, with $3 \leq n \leq 5$. Tab. 1 gives the sum of correctly classified spectra depending on the number of buckets used. As the results show, even using only a moderate amount of peaks to classify with, the misclassification rate reaches about 11%. Compared to about 14% 'bad' data, the classification rate appears to reach the best possible.

If the number n of buckets is increased further (cf. the last row in Tab. 1, where all buckets are used), the correct classification rate rises to about 95%. This improvement in classification rate dilutes the value of the selection of possible biomarkers, as now hundreds of potential biomarkers have to be tested in a laboratory.

5 Discussion

Using a simple and easily interpretable method to classify data sets of SELDI-TOF mass spectra, we are able to identify a patient group and a control group with high success rates. The strength of the results gained lies in the possibility to test the small number of potential biomarkers in a laboratory and therefore to check their biological relevance. The high classification rates in a set of spectra with sub-perfect data suggests that the method could be of value for other data sets as well.

On the other hand, we were only able to test the algorithms on one and - as mentioned - small set of spectra. Even though we selected the test step so that the risk of overfitting is small, testing the validity of the results using new data remains an open issue to be resolved in the wet-lab.

Whether new patients can be successfully classified with the peaks that we identified depends of course on the biological relevance of the molecules with these masses. The risk of confounding is high, given the small dataset that was available to us. The value of our approach lies more in the identification of biomarker candidates than in a robust classification for future patients. We believe that the simplicity and direct interpretability of our method – recall that we are essentially automating a procedure that could in principle be performed by a trained eye: the identification of peaks dominating in one group – provide a good chance for successful findings in the lab.

References

ISSAQ, H.J., VEENSTRA, T.D., CONRADS, T.P. and FELSCHOW, D. (2002): The SELDI-TOF MS Approach to Proteomics: Protein Profiling and Biomarker Identification. *Biochemical and Biophysical Research Communications, 292, 587–592.*

MARAGOS, P. (2005): Morphological Filtering for Image Enhancement and Feature Detection. In: A. Bovik (Ed.): *The Image and Video Processing Handbook, 2nd Edition.* Elsevier Academic Press, 135–156.

ROERDINK, J.B.T.M. (2000): Group Morphology. *Pattern Recognition, 33, 877–895.*

SAUVE, A.C. and SPEED, T.P. (2004): Normalization, Baseline Correction and Alignment of High-throughput Mass Spectrometry Data. In: *Proceedings of Gensips 2004.* www.stat.berkeley.edu/~terry/Group/publications/Final2Gensips2004Sauve.pdf.

SERRA, J. (2006): Courses on Mathematical Morphology. http://cmm.ensmp.fr/~serra/cours/index.htm.

Joint Analysis of In-situ Hybridization and Gene Expression Data

Lennart Opitz[1], Alexander Schliep[2] and Stefan Posch[1]

[1] Institut für Informatik, Martin-Luther-Universiät Halle-Wittenberg, D-06120 Halle, Germany; {opitz, posch}@informatik.uni-halle.de
[2] Department Computational Molecular Biology, Max Planck Institute for Molecular Genetics, Ihnestraße 63–73, D-14195 Berlin, Germany; schliep@molgen.mpg.de

Abstract. To understand transcriptional regulation during development a detailed analysis of gene expression is needed. In-situ hybridization experiments measure the spatial distribution of mRNA-molecules and thus complement DNA-microarray experiments. This is of very high biological relevance, as co-location is a necessary condition for possible molecular interactions.

We use publicly available in-situ data from embryonal development of Drosophila and derive a co-location index for pairs of genes. Our image processing pipeline for in-situ images provides a simpler alternative for the image processing part at comparable performance compared to published prior work. We formulate a mixture model which can use the pair-wise co-location indices as constraints in a mixture estimation on gene expression time-courses.

1 Introduction

The cellular processes constituting life as we know it are controlled by highly complex interaction mechanisms, where the most important form of control is transcriptional regulation. That is the control of the amount of proteins which are produced for a gene in the genome. As quantifying protein levels is experimentally difficult, the intermediate product, messenger RNA (mRNA) which is transcribed from a gene and gets translated to a protein, has received a lot of attention. DNA-microarrays are an experimental technique based on hybridization reactions to quantify levels of mRNA-levels for thousands of genes simultaneously. However, these experiments give a view of transcriptional regulation averaged over many cells or tissues. To understand development of organisms and the necessary differentiation of cells with the same genome, it is necessary to obtain a finer grained picture of gene expression.

In-situ hybridization experiments measure the abundance and spatial distribution of specific mRNA-molecules in organisms through staining cells proportionally to mRNA-concentration. Although the experimental technique is quite expensive as experiments have to be repeated for each gene reasonably large amounts of data exist. For example, the Berkeley Drosophila Genome Project (BDGP, http://www.fruitfly.org) provides a database of images for expression patterns during embryonal development. There are problems with data quality due to the experimental errors and the imaging process, however the data provides a unique opportunity to augment gene expression time-courses over embryonal development with co-location information. This is of very high biological relevance, as co-location is a necessary condition for possible interaction.

We introduce an image processing pipeline for processing in-situ hybridization data and a simple co-location index, which performs as well as published results (Peng et al. (2004)) even though it is substantially simpler (for more details see Opitz (2005)). We also formulate a statistical mixture model which allows the use of the co-location data in the form of pair-wise constraints in a mixture estimation on gene expression time-courses. This will provide a self-contained framework for joint analysis of in-situ hybridization and gene expression data.

2 Method

2.1 Image processing pipeline

The majority of hybridization images in the BDGP database contain the projection of exactly one centered embryo. However, there is a substantial portion of images with multiple touching or partially projected embryos. To exploit as much data as possible, the goal of image preprocessing is to locate and extract exactly one complete embryo from each image, even for touching embryos. Subsequently this embryo is registered to standardized orientation and size to allow for comparison of different expression patterns. Figure 1 shows the steps of the image processing pipeline for one example image.

To distinguish between embryo and non-embryo pixels we employ a similar approach as Peng et al. (2004) estimating the local variance of grey level intensities for each pixel in a 3×3 neighborhood. As noted in Peng et al. (2004) the background is much more homogeneous and as a consequence it suffices to apply a fixed predefined threshold for segmentation using variance

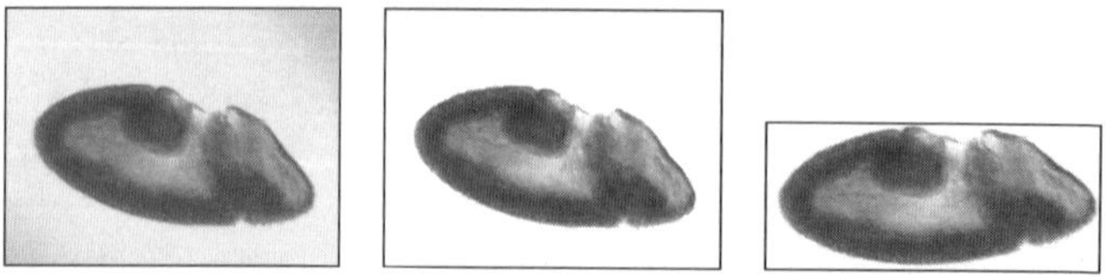

(a) original (b) extracted (c) registrated

Fig. 1. Example for steps of the image processing pipeline.

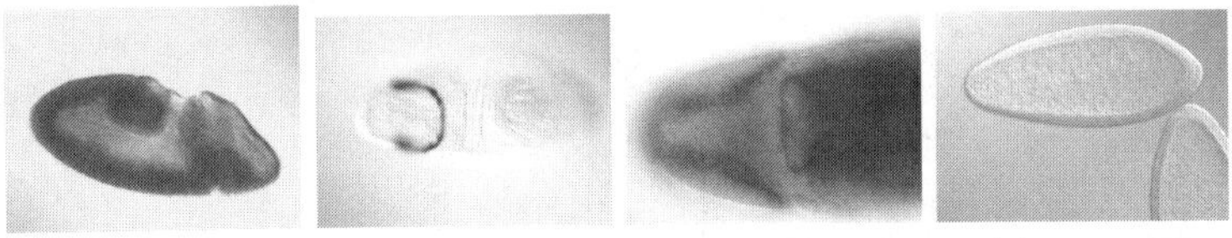

(a) type I (b) type II (c) type III (d) type IV

Fig. 2. The four categories of embryo images.

estimates. To eliminate small holes and erroneous embryo regions a sequence of morphological closing and opening using a circular mask of radius 4 is applied (see e.g. Gonzalez (1991)). Finally, the largest connected component is extracted and remaining holes are filled.

The resulting region may be the projection of a single complete or partial embryo or the projection of a set of multiple touching embryos. For further processing we define four types of regions: (I) one complete high quality embryo, (II) one complete blurred embryo, (III) one partially projected embryo, and (IV) a set of touching embryos (see Figure 2). Embryos of type II and III images are to be eliminated from further processing as they do not allow reliable co-location comparison. For type IV regions the aim of the processing pipeline is to separate the individual embryos and to extract one complete high quality embryo. This classification is realized by a series of simple filters. First, as a measure of ellipticity we compute the deviation of the region extracted from an elliptical. Second, the compactness is computed as the ratio of the squared circumference and the area of the region. Type I regions are defined by a linear separating line in the resulting two-dimensional feature space which is trained by a SVM using a set of 100 training images randomly selected from the BDGP database. To select type II images a threshold on the ellipticity is applied to the remaining images. Type III regions are identified using the area of the region and the number of pixels coincident with its bounding box. The threshold for both filters are determined from the same set of 100 images. To separate multiple touching embryos a new approach has been developed. First, a hypothesis of the coarse location of the centroid of the central embryo is derived using simple heuristic. Using a set of concentric rays emerging from this centroid the contour points of the complete embryo region are computed (see Figure 3). If the distance of two neighboring contour points exceeds the mean of distances by more the 20% it is considered a cut point between different embryos. The set of all detected cut points is used to finally separate the central embryo from the remaining ones. To eliminate invalid embryos separated by this method, an additional set of 50 type IV images is used to train a SVM using again ellipticity and compactness as features.

The final step of preprocessing is to register the embryos extracted to a standardized orientation and size. The embryo is rotated to horizontally align the principal axis. Subsequently the bounding box is scaled to a standard size (658×279 pixels for our experiments). After this registration there is

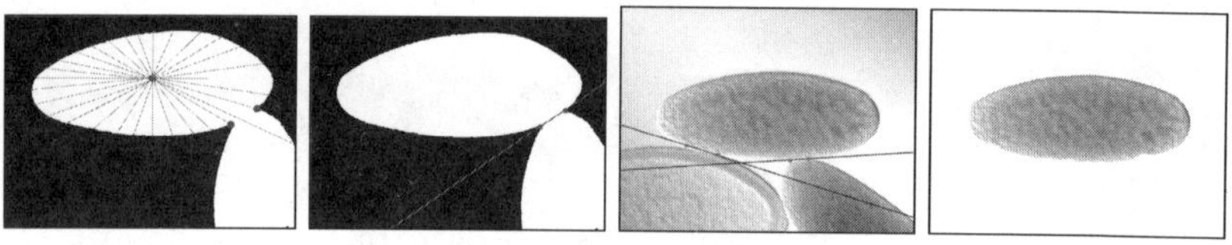

(a) rays (b) cut points (c) original (d) result

Fig. 3. Principle and example for separation of multiple embryos. (a) detection of contour points using a set of concentric rays; (b) detected cut points and separating line; (c) example image; (d) extracted embryo.

still an ambiguity in orientation, which may correspond to dorsal vs ventral and anterior vs posterior position. A part of the BDGP images is taken in a lateral position, giving again rise to four orientations (an example is given in Figure 4). We do not make an attempt for normalization at this step. Rather when comparing a pair of embryos for co-location we take all four orientations into account and use the best result as similarity score.

2.2 Co-location index

To compare in-situ hybridization patterns between genes and/or developmental stages, we developed a simple co-location index for a pair of registered embryos which is directly based on the intensity levels of staining. We prefer this approach to binarization of intensities (cf. Kumar et al. (2002)) or quantization into a set of discrete staining levels (cf. Peng (2004)). Binarization seems too coarse an approximation and disregards completely valuable information on the abundance of mRNA. On the other hand, there is little evidence for a fixed number of homogeneously distributed staining levels across a wide range of genes and developmental stages. As an alternative we propose to use the correlation coefficient of two registered embryo images where intensities of corresponding pixels for two registered embryos are considered as paired data. This score takes both the spatial distribution and the strength of hybridization into account. Using this correlation coefficient, we achieve invariance of the co-location with respect to linear scaling of intensities. This is of importance, as images are acquired under different illumination conditions and this invariance eliminates the need of image normalization. As a consequence, also expression patterns which differ by a uniform scaling of intensities not due to differing illumination are scored as very similar (see Section 3 for an example). Depending on the application this may or may not be desired. In the latter

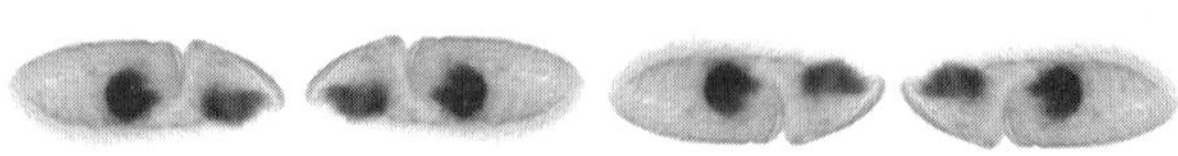

Fig. 4. Four possible orientations of an embryo.

case one may use (unnormalized) cross-correlation that is the scalar product of the expression patterns as a substitute for the correlation coefficient with prior normalization of illumination differences.

2.3 Joint clustering

Mixture models (McLachlan et al. (2000)) are the method of choice for clustering gene expression time-courses; see Bar-Joseph (2004) for a recent review. We extend a framework (Schliep et al. (2005)) using linear Hidden Markov Models as components to allow the joint analysis of gene expression time-courses and co-location information obtained from in-situ experiments. Instead of unsupervised learning we use a partially supervised approach, where constraints between genes are taken from the in-situ data. The pairwise constraints are used in the EM-algorithm with extensions proposed by Lu et al. (2005) and Lange et al. (2005).

Let the real-valued N-by-T matrix $X = \{x_i\}_{i=1}^N$ denote the N gene expression time-courses of length T. A mixture model is a convex combination of K component models; note that here the choice of component model is not important. We write $[x_i|\Theta] = \sum_{k=1}^K \alpha_k [x_i|\theta_k]$, where the non-negative α_j sum to one and the θ_j denote the parameters of the components. Then $\Theta = (\alpha_1, ..., \alpha_K, \theta_1, ..., \theta_K)$ is the set of parameters. Following (Lu et al. (2005)) we assume that—recall the complete data likelihood $[X,Y|\Theta] = [X|Y,\Theta][Y|\Theta]$—there is further dependence on pair-wise constraints $W^+, W^- \in R^{N \times N}$, yielding $[Y|\Theta, W^+, W^-] \propto [Y|\Theta][W^+, W^-|Y, \Theta]$. A positive W_{ij}^+ indicates that genes i and j should be accounted for by the same component, and a positive W_{ij}^- that they should not. Furthermore, we assume that

$$[W|\Theta, Y] = \frac{1}{Z} \exp\left(\sum_i \sum_{j \neq i} -\lambda^+ W_{ij}^+ 1\{y_j \neq y_i\} - \lambda^- W_{ij}^- 1\{y_j = y_i\} \right),$$

where λ^+ and λ^- are global weights of the constraints. The estimation problem can be solved using Gibbs-sampling or mean field approximations (Lu et. al (2005), Lange et al. (2005)). In consequence, when we set entries of W^+, W^- to zero except for strong correlations or anti-correlations, we obtain clusterings in which clusters contain co-located genes with similar expression time-courses. Results will be reported elsewhere.

3 Results

We tested the image processing pipeline using an independent set of 300 randomly chosen images from the BDGP which were manually labeled according to the types introduced in Subsection 2.1. Table 1 shows the proportion of types for manual as well as automatic labeling using the image processing pipeline proposed. About 87% of the images are suited for comparison where

Table 1. Distribution of image types for manual and automatic classification (left); Classification accuracy where the positive class are images of type I and IV (right).

Type	I	II	III	IV
manual	70.7	5.7	7.7	16.0
automatic	71.7	4.3	10.3	13.7

	True	False	Σ
Positive	80,7%	1%	81,7%
Negative	3%	15,3%	18,3%
Σ	83,7%	16,3%	100%

only 71% would be used by approaches like Peng et al. (2004), Kumar et al. (2002). Of the sets of touching embryos 73% could be separated correctly and as a consequence a total of 82% images were rendered as usable with only 1% false positive. The second important issue is the quality of the embryos extracted. Five persons were asked to assess the accuracy of the embryo contours into one of the categories (see Figure 5 for examples). With 69.4% of all images judged as good and 24% as average, the method proves well suited to register and extract embryos.

To evaluate the comparison of expression pattern with the co-location index we first used the same data set of 11 images annotated as "posterior endoderm anlage" as in Peng et al. (2004). Figure 6 shows the ranking comparing *Acf1* as query image to the remaining ten images. These results are as expected, both with regard to detailed annotation from ImaGO (compare Peng et al. (2004)) and with regard to visual impression. We note that our ranks using the co-location index deviates from the results for a correlation coefficient given in Peng (2004) which may be due to different extraction of embryos or quantization of intensities. The rankings obtained between the three methods are very similar, although the co-location index is computationally a much simpler method. Note, that the ranking of *CG5525* ahead of *Slbp* with the co-location index is due to the invariance with respect to illumination changes, see Subsection 2.2.

For a more comprehensive test we extracted ten further groups of images for developmental stage 7-8 with identical annotation. For each of the annotation terms used, we randomly selected ten images from BDGP for the set of genes returned by an ImaGO query. For the ten queries ImaGO return an average of 38.6 genes for each annotation term combination (intersection of 2 different terms) which in turn resulted in an average of 179.5 images in BDGP. We now compare each of the 100 images in turn against the complete

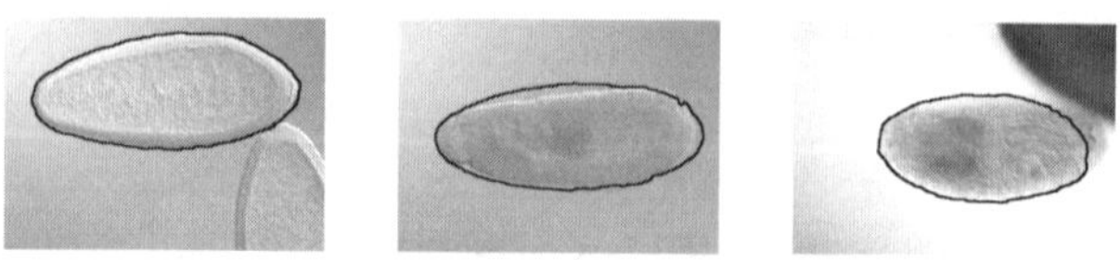

(a) good (b) average (c) bad

Fig. 5. Examples of the three categories to asses the accuracy of embryo contours.

Query	Rank	co-location index	Local GMM	Hybrid
Acf1	1	pont 0.69	pont 0.32	pont 0.102
	2	mam 0.51	mam 0.30	mam 0.051
	3	RhoGAP71E 0.44	RhoGAP71E 0.24	Dcp-1 0.042
	4	Dcp-1 0.42	Dcp-1 0.22	RhoGAP71E 0.038
	5	CG5525 0.37	Slbp 0.20	Slbp 0.026
	6	Slbp 0.31	CG5525 0.18	cl 0.019
	7	cl 0.31	cl 0.17	CG5525 0.018
	8	CG6051 0.29	CG6051 0.12	CG6051 0.008
	9	CG33099 0.18	GATAe 0.07	CG33099 0.002
	10	GATAe -0.21	CG33099 0.04	GATAe 0.001

Fig. 6. Comparing the expression pattern of *Acf1* as query against ten other genes annotated as "posterior endoderm anlage" from ImaGO. Columns 3 to 5 give the ranking using our co-location index compared to the two favored methods of Peng et al. (2004). For each image the name of the gene and the score is given. Ranks and scores for the two matching methods of Peng et al. (2004) are take from the reference, for display we use our imageprocessing method.

BDGP dataset for developmental stage 7-8 comprising a total of 2893 images for 1768 different genes and derive the ranking using the co-location index. For the remaining nine images from the same group as the query we determine the distribution of the resulting ranks shown in Figure 7. As before the expectation is, that these nine remaining images should show up at the top of the ranking list as they share the same annotation with the query. However we can not expect to rank them exactly at the first top positions. First

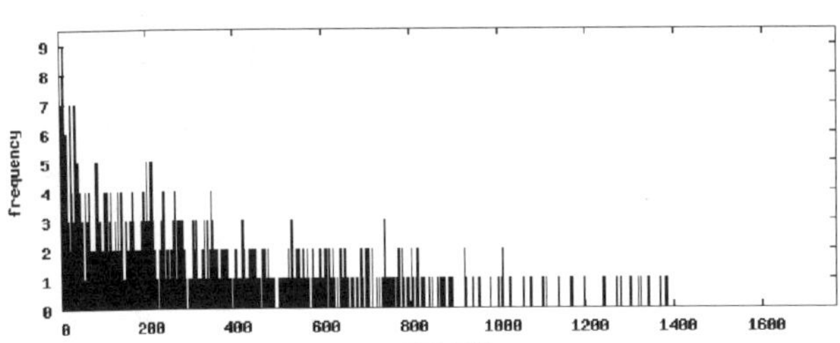

Fig. 7. Distribution of ranks from the 10 genes to each other for 10 different groups.

there are in general more images for genes identically annotated, as we choose only ten images for each group for reasons of computational efforts. Second there are ambiguities with respect to orientation (see Subsection 2.1) and also experimental and annotation inaccuracies. The histogram conforms with our expectation as the nine images are in most cases ranked ($> 67\%$) in the first third of a query result.

4 Conclusions

We presented a reliable yet simple image processing pipeline which allows to compute pair-wise co-location indices for genes from in-situ hybridization images. Furthermore, we formulate a mixture approach to use this co-location data as constraints for clustering gene expression time-courses, potentially leading to more relevant clusters of functionally related, interacting genes. This will be evaluated for the Drosophila development in further studies.

In our image processing pipeline we perform rigid transformations of the embryos with anisotropic scaling. Due to the variations in embryo shape, elastic, non-rigid transformations (Neumann et al. (1999)) might increase robustness of the co-location index. Furthermore, detection of embryo orientation, dorsal/ventral versus lateral position, is a relevant problem which needs to be addressed. Extensions to three-dimensional data as well as more intricate clustering formulations are further exciting questions to pursue.

References

BAR-JOSEPH, Z. (2004): Analyzing Time Series Gene Expression Data. *Bioinformatics, 20, 16, 2493–2503*

GONZALES, R. and WINTZ, P. (1991): *Digital Image Processing.* Addison-Wesley.

KUMAR, S., JAYARAMAN, K., PANCHANATHAN, S., GURUNATHAN, R., MARTI-SUBIRANA, A. and NEWFIELD, S. (2002): BEST - A Novel Computational Approach for Comparing Gene Expression Patterns from Early Stages of Drosophila Melanogaster Development. *Genetics, 169, 2037–2047.*

LANGE, T., LAW, M.H., JAIN, A.K. and BUHMANN, J.M. (2005): Learning with Constrained and Unlabeled Data. *IEEE Computer Society Conference on Computer Vision and Pattern Recognition, 1, 731–738.*

LU, Z. and LEEN, T. (2005): Semi-supervised Learning with Penalized Probabilistic Clustering. *NIPS 17, 849–856.*

MCLACHLAN, G. and PEEL, D. (2000): *Finite Mixture Models.* Wiley, New-York.

NEUMANN, S., POSCH, S. and SAGERER, G. (1999): Towards Evaluation of Docking Hypothesis Using Elastic Matching. *Proceedings of the GCB, 220.*

OPITZ, L. (2005): *Analyse von Bildern der mRNA-in Situ-Hybridisierung.* Master thesis, Institut für Informatik, Universität Halle-Wittenberg.

PENG, H. and MYERS, E.W. (2004): Comparing in situ mRNA Expression Patterns of Drosophila Embryos. *RECOMB'04, 157–166.*

SCHLIEP, A., COSTA, I.G., STEINHOFF, C. and SCHÖNHUTH (2005): Analyzing Gene Expression Time-courses. *IEEE/ACM Transactions on Computational Biology and Bioinformatics, 2, 3, 179–193.*

Unsupervised Decision Trees Structured by Gene Ontology (GO-UDTs) for the Interpretation of Microarray Data

Henning Redestig[1], Florian Sohler[2], Ralf Zimmer[2] and Joachim Selbig[3]

[1] Max Planck Institute for Molecular Plant Physiology, Am Mühlenberg 1, D-14476 Golm, Germany; redestig@mpimp-golm.mpg.de
[2] Institute for Informatics, Ludwig-Maximilians-University, Amalienstraße 17, D-80333 Munich, Germany; {sohler, zimmer}@bio.ifi.lmu.de
[3] University of Potsdam, Am Neuen Palais 10, D-14469 Potsdam, Germany; selbig@mpimp-golm.mpg.de

Abstract. Unsupervised data mining of microarray gene expression data is a standard approach for finding relevant groups of genes as well as samples. Clustering of samples is important for finding e.g. disease subtypes or related treatments. Unfortunately, most sample-wise clustering methods do not facilitate the biological interpretation of the results. We propose a novel approach for microarray sample-wise clustering that computes dendrograms with Gene Ontology terms annotated to each node. These dendrograms resemble decision trees with simple rules which can help to find biologically meaningful differences between the sample groups. We have applied our method to a gene expression data set from a study of prostate cancer. The original clustering which contains clinically relevant features is well reproduced, but in addition our unsupervised decision tree rules give hints for a biological explanation of the clusters.

1 Introduction

Microarray experiments provides information about expression levels of thousands of genes simultaneously. Though the quality and reproducibility of results from famous experiments have been disputed (Michielis and Koscielny (2005)), microarrays are still considered as one of the most effective experimental techniques for the understanding of biological processes on the molecular level.

Clustering of genes or samples is a standard procedure performed for the analysis of gene expression data. While gene-wise clustering can bring functionally related genes together, clustering of samples can provide insight into disease subtypes or patient groups. This is of particular importance when subgroups can be linked to clinically relevant features such as recurrence or severity of disease.

Many different classical clustering algorithms have been applied for sample-wise clustering of microarray data (e.g. Golub et al. (1999)). These methods are usually applied on all measured genes or those satisfying some filtering criteria in order to obtain a grouping of the samples.

The resulting groups or dendrograms computed by classical algorithms do not provide any hints as to in which biological processes the samples were particularly different between clusters — an obvious question to ask from a biological point of view. This issue is instead commonly addressed in a two-step approach by first identifying the sought sample groups via clustering and then the functional gene classes enriched with significantly differentially expressed (DE) genes using the appropriate hypothesis tests.

Using a standard clustering approach and looking for DE genes associated with the result, are two independent concepts. The clustering uses mixed information from a large and heterogeneous set of biological processes; identification of DE genes on the other hand does not take into account that local and global differences between sample groups are not necessarily the same.

Here we introduce a new clustering concept that integrates information about predefined functional gene classes, in this case a mapping of genes to Gene Ontology (GO) (The Gene Ontology Consortium (2000)), directly in the clustering procedure. This is done by first identifying interesting gene classes and then, by making direct use of only these, computing a one-step sample-wise clustering.

Similar to our approach, Lottaz and Spang (2005) introduced a supervised classification approach that also takes advantage of functional gene classes. Our goal is instead to use such information in clustering methods.

Using gene classes based on biological processes, we conceptualize a new clustering method and call it GO unsupervised decision trees (GO-UDTs), as it results in decision tree-like structures. On a publicly available expression dataset from a prostate cancer study, our method exhibits similar clusters as were shown in the original publication by Lapointe et al. (2004); in addition it provides valuable indications for a biological interpretation.

1.1 Unsupervised Decision Trees (UDTs)

Our goal is to remedy one major drawback of current clustering methods – their lack of interpretability. In order to reach that goal, we borrow from the method called decision trees known from classification theory and adapt them to our clustering problem.

Just like classical decision trees do UDTs search the feature space and focus on only a few interesting features at a time. This makes the result directly interpretable, a desirable property that sometimes lets them appear advantageous even when other methods would yield qualitatively better performance. Algorithms for building UDTs have been developed previously and used for clustering (Basak (2005), Karakos (2005)), but have to our knowledge not been applied to biological data. UDTs are constructed by splitting the data into subsets based on information from one single feature at a time. However, instead of using labeled training data to determine the best split, UDTs make use of an objective function that measures the quality of the resulting clustering from each corresponding feature.

1.2 GO-UDTs

In order to make UDTs applicable to the clustering of samples from gene expression data, we define an objective function using features based on single functional gene classes. The intuition behind this function is to score the quality of a split by measuring the quality of the separation of the resulting groups. The GO-UDT algorithm computes a dendrogram in a top-down manner, in each step dividing the samples into subsets that exhibit natural partitionings in at least some gene classes. Given a subset of the samples, the following steps are performed to determine the gene classes that imply the optimal split at a tree node.

1. For each gene class, compute a single feature by
   - selecting all genes belonging to that class,
   - summarizing the data matrix built up with the genes from that gene class with their first principal component (PC).
2. Cluster samples according to each gene class using the computed feature.
3. Score gene classes according to an objective function which measures the quality of the separation of the resulting clusters.
4. Select a set of high scoring gene classes that imply a similar clustering.
5. Re-partition resulting sample groups until stopping criteria is fulfilled.

2 Methods and data

The utilized expression matrix $X_{p,q}$ contains print-tip locss normalized (Yang et al. (2002)) expression estimates of ~24,000 genes from 41 normal prostate specimens, 62 primary prostate tumors and nine lymph node metastases.

In the original publication, the authors used hierarchical clustering to identify sample subgroups. The first identified subgroup mainly consists of samples from healthy patients. Tumor subgroup I was identified as the clinically least aggressive whereas subgroups II and III contain samples from more ill-prognosed tumors and eight of the lymph node metastases. These were the subgroups we expected to rediscover with the addition of attaching labels of relevant biological processes at each node in the resulting tree.

Prior to applying PCA, each gene was mean-centered and scaled to unit-variance to allow discovery of relatively low-variant but still informative gene classes.

Gene annotations for sorting genes to different GO defined biological processes were obtained from ErmineJ (Pavlidis (2005)). Both direct annotations and those inferred from the GO graph were used. Only gene classes with more than nine and less than 300 genes were considered.

2.1 Scoring and clustering

In order to decide the sample clustering and its quality given the expression estimates from a single gene class, c_i, we define the following score. Let $G_{c_i} = \{g_1, g_2, \ldots, g_n\}$ be the set of all sample points, projected onto the first PC of gene class c_i. The parameters μ and σ for two Gaussian models are then fitted to G_{c_i} using both a single Gaussian probability density function (PDF) $p_1 = g(\mu, \sigma)$, and a mixture of two Gaussian PDFs denoted by $p_2 = q g_1(\mu_1, \sigma_1) + (1-q) g_2(\mu_2, \sigma_2)$ where q is a mixture parameter chosen so that $0 < q < 1$. For the fitting of the mixture model, the Mclust algorithm from the *MCLUST* package (Fraley and Raftery (2002)) was utilized. The goodness score is defined as the average log-likelihood ratio of the data:

$$S(G_{c_i}) = \frac{1}{n} \sum_{i=1}^{n} \log \frac{p_2(x_i)}{p_1(x_i)}. \tag{1}$$

2.2 Selecting a good split and deciding when to stop

Obviously, one gene class will have the best score and could be the one finally chosen to split the data, but since small changes in the clustering vector will have large impact further down the tree, a higher amount of stability on the chosen gene class was desired. Therefore a heuristic was applied, choosing a gene class, only if the implied split is supported by other high scoring classes as well.

More precisely, a split for the gene class c_1 can only be made if there are at least A different gene classes among the T highest scoring classes that imply similar clusters to a minimum threshold of S, where all of these gene classes must satisfy a maximal dependence level $d(c_1, \cdot) < D$. If $C_j^{(i)}$ is a vector naming the samples that were ordered to cluster j by the ith clustering then the similarity of two clusterings $C^{(1)}$ and $C^{(2)}$ and the dependency of gene classes c_1 and c_2 are defined as follows:

$$s(C^{(1)}, C^{(2)}) = \frac{1}{2} \left(\frac{|C_1^{(1)} \cap C_1^{(2)}|}{|C_1^{(1)} \cup C_1^{(2)}|} + \frac{|C_2^{(1)} \cap C_2^{(2)}|}{|C_2^{(2)} \cup C_2^{(2)}|} \right); \qquad d(c_1, c_2) = \frac{|c_1 \cap c_2|}{|c_1|},$$

where $|C_k^{(i)} \cap C_k^{(j)}|$ denotes the amount of samples ordered to cluster k by both clustering i and j. Still, the top scoring gene class fulfilling these conditions is not necessarily chosen; instead, we require the A supporting gene classes to have minimum average rank.

For visualization, the supporting terms were also added to the node, as they could be biologically relevant as well. A maximum of five terms were added to each node.

If no A supporting gene classes can be found, the expansion of the tree is discontinued. Additionally the minimum cluster size was set to five. The other parameters were set to $A = 2$, $D = 0.5$, $S = 0.75$ and $T = 30$.

3 Result and discussion

In the publication of Lapointe et al. (2004) three subgroups of cancer samples were identified which are linked to relevant clinical features. Although the discovered grouping is not necessarily the best possible from a biological point of view, we will consider it as a basis for comparison with our results.

A GO-UDT of the expression data was generated using the proposed goodness measure; the result is shown in Figure 1. The subgroups identified in the original publication are reconstructed well — only five tumors are grouped with the non-tumors.

The cell division related gene class *reproduction* was found to exhibit a strong bimodality and separate non-tumor from tumor samples. In accordance with the findings reported in the original publication, also *lipid metabolism* show this behavior. Lapointe et al. also found metabolic activity to be higher in subgroup III by looking at key metabolism genes. This is also indicated in the tree by the gene class *monosaccharide metabolism.*

The second node separates thirteen subgroup III members to a pure leaf with six of the nine lymph node metastases. This is done using the gene classes *dephosphorylation* and *actin filament based process.* A member of the latter class is actinin-4 which is a biomarker for cancer invasion and metastatic capability (Honda et al. (2005)).

Finally, subgroup II is separated from I well using *sensory perception* and the more relevant term *fatty acid metabolism* (Rossi et al. (2003)). The resulting leaf containing subgroup II is also enriched with advanced grade tumors ($p = 0.04$, Fisher's exact test).

The non-tumor samples were also separated into a series of subgroups. Even if these subgroups did correspond to some relevant variable, e.g. age, that subtree was left out as this is only speculative. Furthermore, GO-UDTs represent a kind of clustering technique and as such it can find groupings in the data regardless if they are meaningful or not. Subgroups I, II and III however were shown in the original publication to be truly stable.

Looking at the data projected on to the first principal component using only genes from *reproduction*, two clear clusters are visible separating tumor from non-tumor samples. This motivates the simplified method of only considering the first PC — the second PC provides little extra information for that split, see Figure 2(a).

3.1 Over-representation of DE genes

As comparison to more well-known ways of testing functional classes for importance, we performed an over-representation analysis (ORA). First, DE

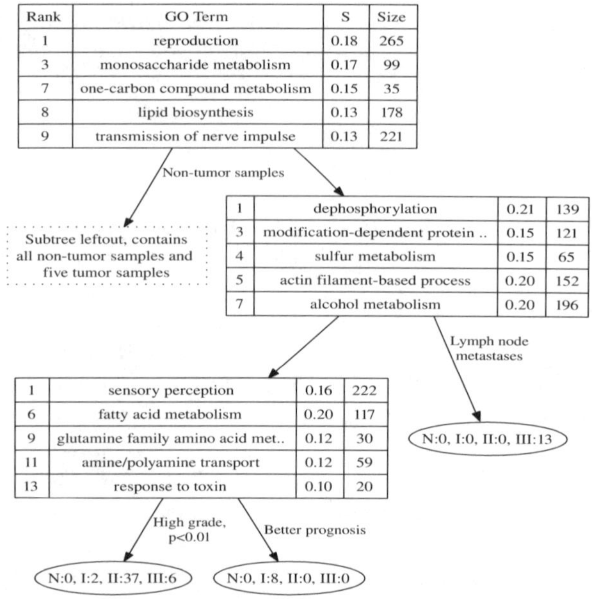

Fig. 1. A GO-UDT of the Lapointe prostate cancer dataset. GO terms are ranked (column *Rank*) according to the score of their induced split (column S) and then a set of GO terms with similar splits and minimum average rank is selected at each node. The column *Size* denotes the number of genes in each gene class. The root node indicates that genes associated with reproduction and metabolism are good at separating the non-tumor from tumor samples. *Dephosphorylation* separates thirteen of the nineteen samples from subgroup II from the others, and *sensory perception* and *fatty acid metabolism* gather nearly all samples from subgroup II to a mixed leaf enriched with advanced grade tumors.

genes between tumor and non-tumor/metastasis samples were identified using the Limma package (Smyth (2005)) to resemble the first split in the generated tree. On significance level $q = 0.05$ (Storey and Tibshirani (2003)), the classes with most enriched with DE genes were identified with Fisher's exact test. The best class, *nerve ensheathment* had ten out of twelve genes identified as DE, $p = 0.0005$. This class was not highly scored by the GO-UDT, the reason is seen in Figure 2(b) showing the unimodal score plot for *nerve ensheathment*. GO-UDTs find *different* gene classes than the more classical two-step approach of first clustering samples and then finding differences. This does not speak against ORA, it just indicates that other measures also should be considered.

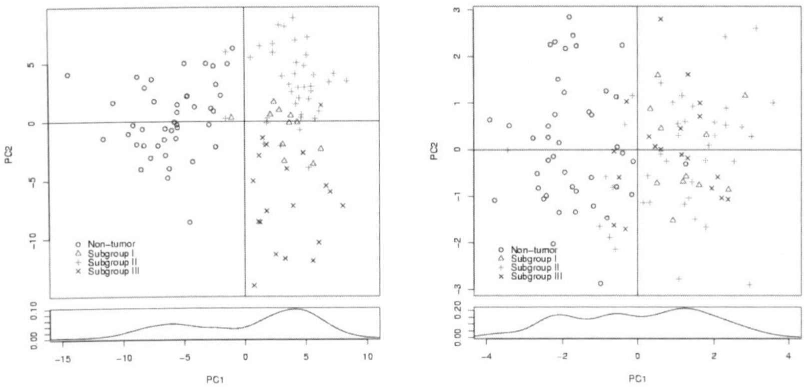

(a) Gene class: *reproduction* (b) Gene class: *nerve ensheathment*

Fig. 2. Shown here are the PCA score plots from the best gene class in the root node of the computed tree and the gene class containing the largest ratio of differentially expressed (DE) genes. The density plots above PC1 shows the distribution of samples on this PC. The gene class *reproduction* shows pronounced bimodality on the first PC and produce strong clusters partitioning non-tumor from tumor samples. The gene class *nerve ensheathment* on the other hand is not bimodal on its first PC despite containing a significantly high ratio of DE genes. Any clustering attempt on the data projected onto this PC would result in highly heterogeneous clusters.

4 Conclusion and future work

Existing methods for sample-wise clustering provide no biological explanation for the observed results. Clustering via GO-UDTs on the other hand is a novel approach for constricting clustering to be done on a one-feature-at-a-time basis in order to increase the interpretability of the final output. In this study, we investigated if GO-UDTs are useful for sample-wise clustering of microarray experiments as a method for discovering the reasons for divergence between subgroups of prostate cancer patients.

Our main objective for the GO-UDT approach was to increase interpretability. We were not looking for better clusters but for a better biological explanation for them. Our results shows that the identified gene classes indeed give hints toward biologically relevant differences and point to possible refined analysis and further experiments.

Introducing the strategy of summarizing expression measurements into condensed features based on their functional ontology, it is tempting to speculate that it also could be used in supervised approaches as well. We see this in combination with refinement of the proposed technique as interesting topics for future work.

References

BASAK, J. and KRISHNAPURAM, R. (2005): Interpretable Hierarchical Clustering by Constructing and Unsupervised Decision Tree. *IEEE Transactions on Knowledge and Data Engineering, 17, 121–132.*

FRALEY, C. and RAFTERY, A.E. (2002): MCLUST: Software for Model-based Clustering, Density Estimation and Discriminat Analysis, and Density Estimation. *J Am. Stat. Ass., 97, 611–631.*

GOLUB, T.R., SLONIM, D.K., TAMAYO, P., HUARD, C., GAASENBEEK, M., MESIROV, J.P., COLLER, H., LOH, M.L., DOWNING, J.R., CALIGIURI, M.A., BLOOMFIELD, C.D. and LANDER, E.S. (1999): Molecular Classification of Cancer: Class Discovery and Class Prediction by Gene Expression Monitoring. *Science, 286, 531–537.*

HONDA, K., YAMADA, T., HAYASHIDA, Y., IDOGAWA, M., SATO, S., HASEGAWA, F., INO, Y., ONO, M. and HIROHASHI, S. (2005): Actinin-4 Increases Cell Motility and Promotes Lymph Node Metastasis of Colorectal Cancer. *Gastroenterology, 128, 51–62.*

KARAKOS, D., KHUDANPUR, S., EISNER, J. and PRIEBE, C.E. (2005): Unsupervised Classification via Decision Trees: An Information-theoretic Perspective. *In Proceedings of the 2005 IEEE Int. Conf. on Acoustics, Speech and Signal Processing, IEEE.*

LAPOINTE, J., LI, C., HIGGINS, J.P., RIJN, M.V.D., BLAIR, E., MONTGOMERY, K., FERRARI, M., EGEVAD, L., RAYFORD, W., BERGERHEIM, U., EKMAN, P., DEMARZO, A., TIBSHIRANI, R., BOTSTEIN, D., BROWN, P., BROOKS, J. and POLLACK, J. (2004): Gene Expression Profiling Identifies Clinically Relevant Subtypes of Prostate Cancer. *PNAS, 101, 811–816.*

LOTTAZ, C. and SPANG, R. (2005): Molecular Decomposition of Complex Clinical Phenotypes Using Biologically Structured Analysis of Microarray Data. *Bioinformatics, 21, 1971–1978.*

MICHIELIS, S., KOSCIELNY, S. and HILL, C. (2005): Prediction of Cancer Outcome with Microarrays: A Multiple Random Validation Study. *The Lancet, 365*

PAVLIDIS, P. (2005): ErmineJ – Gene Ontology Analysis for Microarray Data, v2.0.4. *http://microarray.genomecenter.columbia.edu/ermineJ.*

ROSSI, S., GRANER, E., FEBBO, P., WEINSTEIN, L., BHATTACHARYA, N., ONODY, T., BUBLEY, G., BALK, S. and LODA, M. (2003): Fatty Acid Synthase Expression Defines Distinct Molecular Signatures in Prostate Cancer. *Mol. Cancer Res., 1, 707–715.*

SMYTH, G.K. (2005): Limma: Linear Models for Microarray Data. In: R. Gentleman, V. Carey, S. Dudoit, R. Irizarry and W. Huber (Eds.): *Bioinformatics and Computational Biology Solutions using R and Bioconductor.* Springer, New York, 397–420.

STOREY, J.D. and TIBSHIRANI, R. (2003): Statistical Significance for Genomewide Studies. *PNAS, 100, 9440–9445.*

THE GENE ONTOLOGY CONSORTIUM. (2000): Gene Ontology: Tool for the Unification of Biology. *Nature Genetics, 25, 25–29.*

YANG, Y.H., DUDOIT, S., LUU, P., LIN, D.M., PENG, V., NGAI, J. and SPEED, T.P. (2002): Normalization for cDNA Microarray Data: A Robust Composite Method Addressing Single and Multiple Slide Systematic Variation. *Nucleic Acids Research, 30.*

Part IX

Linguistics and Text Analysis

Clustering of Polysemic Words

Laurent Cicurel[1], Stephan Bloehdorn[2] and Philipp Cimiano[2]

[1] iSOCO S.A., ES-28006 Madrid, Spain; `lcicurel@isoco.com`
[2] Institute AIFB, University of Karlsruhe, D-76128 Karlsruhe, Germany; `{bloehdorn, cimiano}@aifb.uni-karlsruhe.de`

Abstract. In this paper, we propose an approach for constructing clusters of related terms that may be used for deriving formal conceptual structures in a later stage. In contrast to previous approaches in this direction, we explicitly take into account the fact that words can have different, possibly even unrelated, meanings. To account for such ambiguities in word meaning, we consider two alternative soft clustering techniques, namely Overlapping Pole-Based Clustering (PoBOC) and Clustering by Committees (CBC). These soft clustering algorithms are used to detect different contexts of the clustered words, resulting in possibly more than one cluster membership per word. We report on initial experiments conducted on textual data from the tourism domain. [1]

1 Introduction

Since many years, conceptual structures have played a major role in the construction of knowledge management applications. Instantiations of these include highly formalized ontologies, less formal taxonomic structures and, even less formal, groups of descriptors having intuitively similar interpretations. While ontologies (cf. Staab and Studer (2003)) show their full potential in knowledge based systems and reasoning engines, taxonomic structures have been mostly applied as structured controlled vocabularies in the context of library science as well as background knowledge in information retrieval, text mining and natural language processing (see e.g. Bloehdorn et al. (2005)). Plain term clusters have recently attracted attention as a means for structuring descriptors in social tagging systems.

Though all these conceptual structures can provide potential benefits for an increasing number of applications, their construction requires a costly modeling activity, a problem typically referred to as the *knowledge acquisition*

[1] **Acknowledgements:** This work was supported by the European Commission under contract IST-2003-506826 SEKT and the by the German Federal Ministry of Education, Science, Research and Technology in the project SmartWeb.

bottleneck. Recent work in *ontology learning* (cf. Mdche and Staab (2001), Buitelaar et al. (2005)) has started to address this problem by developing methods for the automatic construction of conceptual structures. This is typically done in an unsupervised manner on the basis of text corpora relevant for the domain of interest. A major focus of these approaches has been the usage of term clustering techniques (see e.g. Grefenstette (1994), Faure and Nedellec (1998), Gamallo et al. (2005) and Cimiano et al. (2005)). However, a common weak point of these approaches is that they rarely take into account that words are *ambiguous*, i.e. they can have several – possibly grossly unrelated – meanings. Thus, in most approaches the assignment of words to clusters is assumed to be functional. An exception to this is certainly the work of Pantel and Lin (2003), which also provides the basis for our investigations in the sense that we use *soft clustering algorithms* which can assign words to different clusters, therefore accounting for their different contextual meanings. We restrict our attention on a flat clustering of terms, i.e. we do not aim at constructing a hierarchical structure between the term clusters - the work reported in this paper is thus meant to be a first step in ontology learning. Our contribution lies in the analysis of two different algorithm with respect to their ability to account for several meanings of words.

The remainder of this paper is structured as follows. After a quick review of the Distributional Hypothesis, Section 2 describes the feature representation of terms employed in our approach. In Section 3, we describe the two soft clustering algorithms used in our experiments, namely *Overlapping Pole-Based Clustering (PoBOC)* and *Clustering by Committees (CBC)*. In Section 4, we outline an approach for evaluating clusters of (ambiguous) words with membership in multiple clusters using WordNet as the corresponding gold standard. In Section 5, we report on results of an initial evaluation experiment on a tourism-related corpus consisting of texts obtainedfrom the 'Lonely Planet' website. We conclude in Section 6.

2 Term representation

In this section, we give a short overview of the representation of terms in vector space of syntactic dependencies that will be used to apply the soft clustering techniques described in the next section. Hereby, we adopt the approach described in Cimiano et al. (2005), which is motivated by the *Distributional Hypothesis* (Harris (1968)). The Distributional Hypothesis claims that terms are semantically similar to the extent to which they share similar syntactic contexts. This means that, if two words occur in similar contexts, they are assumed to have a similar meaning. A syntactic context could be, for example, a verb for which the term in question appears as subject or object.

For this purpose, for each term in question, we extract syntactic surface dependencies from the corpus. These surface dependencies are extracted by matching texts tagged with part–of–speech information against a library of patterns encoded as regular expressions. Note that our approach is related to the Generalized Vector Space Model (Wong et al. (1985)) but uses syntactic features instead of plain occurrences of words in documents. In our approach, first the corpus is part-of-speech tagged[2]. The part-of-speech tagger assigns the appropriate syntactic category to every token in the text. Features are then extracted by matching regular expressions defined over tokens and part-of-speech tags which denote syntactic dependencies between a verb and its subject, an adjective and the modified noun and the like. In what follows, we list the syntactic expressions we use and give examples of object–attribute pairs extracted. We employ predicate notation $a(o)$, where a is the attribute and o the object:

- adjective modifiers: e.g. *a nice city* → nice(city)
- prepositional phrase modifiers: e.g. *a city near the river* → near–river(city) and city–near(river), respectively
- possessive modifiers: e.g. *the city's center* → has–center(city)
- noun phrases in subject or object position: e.g. *the city offers an exciting nightlife* → offer–subj (city) and offer–obj(nightlife)
- prepositional phrases following a verb: e.g. *the river flows through the city* → flows–through(city)
- copula constructs: e.g. *a flamingo is a bird* → is–bird(flamingo)
- verb phrases with the verb *to have*: e.g. *every country has a capital* → has–capital(country)

The plain feature extraction gives each feature the same importance. However, it is certainly the case that not all features for a noun are equally representative. Thus, we replace the simple appearance count of a word with a feature by their pointwise mutual information value (cf. Church and Hanks (1990)). The feature vector space is high-dimensional and sparse. To increase the statistical properties, we have thus pruned features and words in our experiments, i.e. we considerd only those words which have at least a given number of features and the features that describe at least a given number of words.

3 Soft clustering algorithms

Clustering words which are potentially ambiguous into semantically homogeneous groups requires two main properties: (i) the clustering algorithm must allow clusters to overlap, i.e. a word can belong to one or more clusters and (ii) it needs to automatically determine an appropriate number of clusters. In our experiments, we have employed two soft-clustering algorithms, namely CBC, an algorithm which was developed by Pantel and Lin (2003), and PoBOC,

[2] In our experiments, we have used TreeTagger (`http://www.ims.uni-stuttgart.de/projekte/corplex/TreeTagger/`).

developed by Cleziou et al. (2004). Note that we define as a *soft-clustering* algorithm every clustering algorithm for which the assignment of objects to clusters is non-functional. For soft-clustering algorithms the clusters thus typically overlap.

Overlapping Pole-Based Clustering: PoBOC

PoBOC is an overlapping cluster algorithm developed by Cleuziou et al. (2004). The similarity measure between two words we have used is the cosine of the angle between their vectors. The main idea of PoBOC is to find so called *poles* at a first step. Poles are very homogeneous clusters which are as far as possible from each other. The elements in the poles can be seen as *monosemous* words, i.e. words which have only one sense or meaning. After this first phase, several words remain unassigned though. These unassigned words are words which, as they do not form part of any pole, potentially feature several meanings and thus could end up in several clusters. After this pole construction phase, the remaining words are assigned to one or many poles.

Clustering by Committee: CBC

CBC was developed by Pantel and Lin (2003). CBC shares its two main phases with PoBOC: first it finds base clusters (poles in PoBOC and *committees* in CBC) and assigns the monosemous words to these committees. The committee construction is a recursive process which applies a hierarchical clustering algorithm on the k most similar neighbors in vector space of each word. Relying on an intra-cluster evaluation, only the best cluster is selected. The committees computed in this way are then filtered using the intra-cluster score and a threshold θ_1 that makes sure that the committees are far enough from each other. The process is then recursively applied to a *residue list* consisting of those words which are far enough from the committees with respect to a threshold θ_2 and thus can not be assigned to any committee.

In the assignment phase, every word is assigned to its most similar committee. Further, it is also assigned to the next most similar committee provided that the similarity is above some threshold θ_3 and that the similarity to the committees the word has been already assigned to is below some threshold θ_4. Hereby, the trick is that the non-zero features which are common to the word and the new committee to which it is assigned are set to '0' in the word vector, thus making sure that the word is always assigned to 'orthogonal' senses at later steps. Overall, CBC thus requires four parameters to be set, in contrast to PoBOC, which is parameter-free. As for PoBOC, we apply the cosine measure to assess the similarity between two terms.

4 Evaluation methodology

In order to evaluate how well the produced clusters correspond to the actual different senses of the word, we compare the clusters with WordNet (cf. Miller (1995)) as a *gold standard*. WordNet is a lexical ontology organized in interconnected synonymous groups of terms called *synsets*. The intuition behind our evaluation methodology is to compare the derived clusterings to the set of synsets defined in WordNet. Obviously, in the ideal case, the produced clusters would be one-by-one copies of the synsets. The procedure to evaluate the term clusters is as follows: first we assign each cluster to one or more synsets (depending on which of the two evaluation modes described below we consider) and then we approximate the semantic similarity between clusters and synsets by comparing the overlap between the words in the cluster and the words in the synsets using a vector-space model[3]. This is approach described in more detail below.

First method: One-to-one Association

This first method consists of assigning each cluster to exactly one synset, i.e. to the most similar one. Hereby the similarity is calculated as follows: for both the cluster and the synset, binary vectors are constructed which are then compared relying on the cosine similarity. The dimension of these vectors corresponds to the union of the words appearing in the cluster or the synset and the value of a dimension of the word/synset vector is 0 in case the corresponding word does not occur in the word/synset, 1 if it does occur. The score of a cluster is then calculated as its similarity with respect to the synset it has been assigned to. Intuitively, a high score means that the cluster resembles very much a synset that is actually defined in WordNet while a low score indicates that even the most similar synset achieves only a small similarity value indicating that the cluster comprises many different synsets. We define the score of the overall clustering as the average of the individual cluster scores.

Second method: One-to-several Association

The previous approach has the advantage of being simple and efficient to compute, but neglects the fact that WordNet is organized hierarchically. According to the above evaluation method, a clustering with clusters being composed of several synsets might still be considered a good one as long as the contributing synsets are semantically close within WordNet. The second evaluation method

[3] It is important to mention that this allows to assess the 'precision' of our clustering, but not the 'recall' , i.e. how many of the actual senses of a word we actually are able to account for. In any case, an evaluation in terms of recall-inspired metrics is quite problematic as not all the senses of a given word contained in WordNet are relevant for all domains.

is thus similar in principle to the first one but takes into account the above intuition by the fact that clusters can be assigned to one or more synsets. Consequently, synset vectors are built possibly taking into account the words of more than one synset. As above, the overall score is the average score of all clusters. The procedure to accomplish the assignment of a cluster to several synsets is as follows: first the cluster is assigned to its most similar synset as explained above. Further, it is iteratively also assigned to the next most similar synset provided that the score does not decrease and that the new synset is not to dissimilar from the original synset. Hereby, similarity between synsets is calculated relying on similarity measure introduced by Lin (1998), which is defined as:

$$sim_{lin} = \frac{2 * IC(lcs(syn_1, syn_2))}{IC(syn_1) + IC(syn_2)}.$$

Hereby lcs denotes the least common subsumer, i.e. the most specific common hypernym of the compared synsets syn_1 and syn_2 and $IC(syn)$ denotes the *Information Content* of a synset given by $IC(syn) = -\log(P(syn))$, where $P(syn)$ is the probability of encountering the synset estimated based on corpus frequency counts.

5 Experimental results

As corpus for our experiments, we use a collection of approximately 1,000 texts downloaded from the LonelyPlanet website describing tourist destinations. The corpus is thus small, consisting of around 523,780 tokens. From this corpus, we extract 10,935 nouns with 19,218 features. Restricting on the nouns that have at least two features and the features that describe at least two nouns, we have an input to the clustering algorithm of 3,769 nouns with 5,041 features. The number of clusters as well as the average cluster size produced by each algorithm, PoBOC and CBC with the parameters $k = 10$, $\theta_1 = 0.35$, $\theta_2 = 0.30$, $\theta_3 = 0.01$, and $\theta_4 = 0.2$, is summarized in Table 1. Further, these numbers are compared to the average synset size in WordNet. We observe that PoBOC outputs bigger (and therefore less) clusters; the standard derivation of its clusters is also higher.

Table 1. Basic clustering statistics.

	No. clusters	Avg (stddev) cluster/synset size
PoBOC	1162	2.90(±2.10)
CBC	2010	1.68(±0.87)
WordNet	7897	2.14(±1.55)

Since this study is mainly concerned ambiguous words, it is interesting to get an idea of the average number of meanings of the clustered words. Considering the number of synsets a word belongs to as the number of meanings, a word has in average 3.42 meanings. In the PoBOC clusters, 92.9% of the terms have more than one meaning, and only 84.5% in the CBC cluster set. In average, 71.4% of the terms in a cluster are ambiguous in the PoBOC cluster set, whereas 70.1% are amiguous in the CBC cluster set. It is also possible to count the number of clusters a word belongs to and use this value as the number of meanings of this word. Surprisingly, PoBOC and CBC obtain the same average score of 1.25 meanings, though they do not arrange the words the same way and do not detect the same words as ambiguous.

Table 2. Average evaluation scores and number of non-associated clusters.

	Average cluster similarity	
	One-to-one	One-to-Several (0.5)
PoBOC	0.645	0.648
Not associated	1 cluster	1 cluster/6716 synsets
CBC	0.750	0.752
Not associated	14 clusters	14 clusters/5879 synsets

The results of our evaluation in terms of average similarity as described in the previous section are presented in Table 2 for both evaluation modes. In the *One-To-Several*-association mode, we set 0.5 as similarity threshold which needs to be exceeded in order for a synset to be added to the evaluation set. In both modes, CBC obtains higher scores than PoBOC. There are at least two reasons to explain these results. First, CBC has a special mechanism designed to cluster words, namely the suppression of the word features in common with the committee, which avoids that words are assigned to distinct but very similar committees. The second reason is that CBC typically creates far more (2,010 versus 1,162) and consequently smaller clusters (average size of 1.68 versus 2.90) than PoBOC resulting in a slight bias in favor of CBC. This bias is due to the fact that the average size of the clusters produced by CBC is closer to the average synset size in WordNet.

We can thus conclude that the clusters produced by CBC correspond better to senses contained in WordNet than those produced by PoBOC. The disadvantage of CBC is certainly that a considerable number of parameters needs to be fixed or tuned, thus making its usage not as straightforward compared to PoBoC.

6 Conclusion and future work

We have analyzed two soft-clustering algorithms, CBC and PoBOC, with respect to the task of capturing the different corpus-specific senses or meanings of a word based on the Distributional Hypothesis and a corresponding representation of terms as vectors of syntactic features. We have further proposed an approach for the evaluation of this and related approaches and reported on experimental results for a dataset for the tourism domain. Our results indicate that clustering using CBC is more adapted for our purposes although we pointed out that PoBOC has the advantage of having little parameterization, which makes it easier to use.

In future work, we aim at using alternative soft clustering approaches, possibly combining CBC and PoBOC. As both algorithms share the same phases, this seems definitely reasonable. Further, it also seems promising to examine a bootstrapping approach in which words are first assigned to clusters corresponding to their different meanings and then the different contexts provided by the clusters are used for disambiguation, yielding sense-specific features.

References

BLOEHDORN, S., CIMIANO, P. and HOTHO A. (2006): Learning Ontologies to Improve Text Clustering and Classification. *Proceedings of the 29th Annual Conference of the German Classification Society (GfKl 2005).* Springer, Berlin.

BUITELAAR, P., CIMIANO, P. and MAGNINI, B. (Eds.) (2005): Ontology Learning from Text: Methods, Evaluation and Applications. IOS Press.

CHURCH, K. and HANKS, P. (1990): Word Association Norms, Mutual Information, and Lexicography. *Computational Linguistics, 16, 22-29.*

CLEUZIOU, G., MARTIN L. and VRAIN, C. (2004): PoBOC: an Overlapping Clustering Algorithm. Application to Rule-Based Classification and Textual Data. *Proceedings of the 16th European Conference on Artificial Intelligence (ECAI-04).*

CIMIANO, P., HOTHO, A. and STAAB, S. (2005): Learning Concept Hierarchies from Text Corpora Using Formal Concept Analysis. *Journal of Artificial Intelligence Research (JAIR), 24, 305–339.*

FAURE, D. and NEDELLEC, C. (1998): A Corpus-based Conceptual Clustering Method for Verb Frames and Ontology. *Proceedings of the LREC Workshop on Adapting Lexical and Corpus Resources to Sublanguages and Applications.*

GAMALLO, P., AGUSTINI, A. and LOPES, G.P. (2005): Clustering Syntactic Positions with Similar Semantic Requirements. *Computational Linguistics, 21, 107–145.*

GREFENSTETTE, G. (1994): *Explorations in Automatic Thesaurus Construction.* Kluwer.

HARRIS, Z. (1968): Mathematical Structures of Language. Wiley.

LIN, D. (1998): Automatic Retrieval and Clustering of Similar Words. *Proceedings of the 36th Annual Meeting of the Association for Computational Linguistics and 17th International Conference on Computational Linguistics (COLING-ACL).*

MDCHE A. and STAAB S. (2001): Ontology Learning for the Semantic Web. *IEEE Intelligent Systems, 16, 72–79.*

MILLER, G.A. (1995): WordNet: A Lexical Database for English. *Communications of the ACM, 38, 39–41.*

PANTEL, P. and LIN, D. (2002): Discovering Word Senses from Text. *Proceedings of the ACM Conference on Knowledge Discovery and Data Mining (KDD-02).*

STAAB, S. and STUDER, R. (Eds.) (2003): *Handbook on Ontologies.* Springer, Berlin.

WONG, S.K.M., ZIARKO, W. and WONG, Patrick C.N. (1985): Generalized Vector Spaces Model in Information Retrieval. *Proceedings of the 8th Annual International ACM SIGIR Conference on Research and Development in Information Retrieval (SIGIR 1985).*

Classifying German Questions According to Ontology-Based Answer Types

Adriana Davidescu, Andrea Heyl, Stefan Kazalski, Irene Cramer and Dietrich Klakow

Spoken Language Systems, Saarland University, D-66123 Saarbrücken, Germany; Dietrich.Klakow@lsv.uni-saarland.de

Abstract. In this paper we describe the evaluation of three machine learning algorithms that assign ontology based answer types to questions in a question-answering task. We used shallow and syntactical features to classify about 1400 German questions with a Decision Tree, a k-nearest Neighbor, and a Naïve Bayes algorithm.

1 Introduction

Although information retrieval techniques have proven to be successful in locating relevant documents, users often prefer to get concise answers to their information need. Question answering systems therefore allow the user to ask natural language questions and provide answers instead of a list of documents. Today, most question answering systems do extensive analyses of the questions to deduce features of the answer. One core task of the question analysis consists of the classification according to an ontology-based answer type. This type represents the most important link between question and answer and normally helps the system to determine what type of answer the user is requesting. Many question analysis components rely on hand-coded rules. However, this strategy is often time-consuming and inflexible: little changes in data and classification may require to make parts of the work over. Our aim therefore was to implement and compare different machine learning methods to classify questions according to a given answer type ontology. This has several advantages: Firstly, it is the most flexible way of building a question analysis component since we are thus able to experiment with different machine learning techniques to find the best fitting algorithm. It furthermore allows us to easily adapt our system as soon as we collect more, different, or complementary data (see Section 2 for more information about the obvious imbalance of answer types considered in the ontology and our question corpus). In addition, it enables us to integrate linguistic knowledge as well as statistic information into one component (see Section 2 for more information about our features). As data material we mainly used about 500 German questions collected in the SmartWeb project (Wahlster (2004)). (This work was partially funded by the BMBF project SmartWeb under contract 01IMD01M.) We trained and tested our classifiers on about 800 additional German questions that we collected with a Web-based experiment (Cramer et al. (2006)). Based on these approximately 1400 questions and on the SmartWeb ontology (Sonntag et al. (2006)) we derived an answer type classification consisting of about 50 hierarchically organized classes. (The SmartWeb ontology is focused on multimodal dialog-based human-computer interaction, therefore several aspects had to be adapted.) Unlike the question answering tracks known form TREC (Voorhees (2001)) we did not restrict ourselves to a small set of factoid answer types (cp. Table 2).

To the best of our knowledge, to date there exists no other attempt comparing various classification methods for German questions. Even for English (in spite of its data richness) there are only few attempts to systematically contrast several feature sets and classifiers. In 2002 Li et al. (2002) studied a hierarchical classification algorithm on the TREC 10 questions. They were able to show that their approach achieves good results compared to heuristic rules. However, their answer type ontology only consisted of a few answer types. Day et al. (2005) integrated a knowledge-based and a machine learning approach for a Chinese question set taking about 60 answer types into account. Most similar to our work are the studies by Zhang et al. (2003) and Li et al. (2005), which both compare several machine learning techniques (inter alia: Support Vector Machine, Decision Tree, k-nearest Neighbor, and AdaBoost) for an English question set. They both make exclusively use of shallow features like bag-of-words and bag-of-ngrams, which appear to be astonishingly appropriate compared to the well established use of rules and rule-like features in traditional question analysis components. Like Li et al. (2002) they only considered a few answer types. In contrast to these approaches we intended to find out how well a machine learning approach would be able to perform given the difficulty of the task in German: different form e.g. English German questions provide a lot of grammatical challenges (among other things complex syntax and morphology). It was therefore questionable whether shallow techniques would be adequate at all. In addition, we intended to acquire more knowledge about the size and properties of a German question corpus necessary to train the various components of a question answering system.

The rest of this paper is organized as follows: In Section 2 we describe our data, the features that we (automatically) annotated, and the answer types. Section 3 gives a very short introduction to the three classifiers and presents the design of our experiments. Section 4 discusses the evaluation and our results. Finally, we draw conclusions and summarize our work in Section 5.

2 Data, answer types and features

We used two question collections. About 2000 questions were complied in the SmartWeb project to foster the development and evaluation of the open-domain question-answering system. Some of the questions were elliptic or

Table 1. Elliptic and anaphoric questions

Q_{1a}:	Definiere die freie Enthalpie. Define the free enthalpy
Q_{1b}:	Wie wird sie noch genannt? How is it also called?
Q_{2a}:	Wann veröffentlichte Milan Füst seine ersten Gedichte? When did Milan Fust publish his first poems?
Q_{2b}:	Und worin? And where?

anaphoric and therefore did not fit in our open-domain approach. Table 1 gives an example of such inappropriate questions. We therefore excluded those questions. The remaining corpus consisted of about 500 questions. We additionally collected about 1400 questions with a Wikipedia-based tool (Cramer et al. (2006)). We merged both collections and manually classified all questions according to the given answer types. Table 2 shows a summary of the most frequent question words occurring in our corpus.

Table 2. Summary of the question types in our collection

question word	percentage	question word	percentage
wann (= when)	10.9%	wie/wie* (= how)	18.0%
warum (= why)	1.4%	wo (= where)	9.3%
was (= what)	20.0%	wo* (= whereby etc.)	3.8%
welch* (= which)	8.9%	not first word	9.1%
wer (= who)	11.8%	without	6.8%

The majority of the questions (about 35%) belong to the inquiry types *concept completion* and *quantification*–here: Location, Person, Date, and Number.Count – which is well reflected by the ontology. The *concept completion* and *quantification* are the most basic and simple types in question-answering. However, there are a lot of questions (roughly 30%) that belong to the inquiry types *comparison*, *definition*, and *request*–here: Definition and Explanation. These types are much more complex and up to date research challenges. We tested several feature sets separately. Most of the features are used in English question-answering systems as well.

Although, there are some that we added considering the rich German morphology and the fact that we aim at building an open domain and Web based question-answering system. As feature sets we used:

- **Collocations** are n-grams of lemmas that normally co-occur. In the first instance we selected them manually. Later we decided not to trust our intuition and calculated the standard deviation for all bigrams that either include a question word or occur at the beginning of a question. We chose those bigrams that had a high standard deviation.
- **Trigger-words** are mainly question words such as "wann" (= when), "wo" (=where). These words are the most obvious features in question-answering.
- **Punctuation marks** consider whether there is a question mark or an exclamation mark at the end of the question. As trigger-words the punctuation mark is an obvious feature although it is normally of minor importance and little robustness.
- **Question length** is counted in word tokens. We found that there is sometimes a though weak relation between specificity of a question and the length of its answer.
- **Named entities** were annotated with LingPipe developed by the Alias-i, Inc., which assigns labels to proper nouns such as person names, locations, and organizations. Unfortunately, the named entity recognition in short sentences and questions shows a lack of robustness. Nevertheless, named entities play a part in the sentence patterns and are supposed to be an important hint for certain answer types.
- **Lemmas** are annotated using the TreeTagger developed at the Institute for Computational Linguistics of the University of Stuttgart. Lemmas were interesting features for a medium size German question corpus since they are able to condense several words to one.
- **POS-tags** are also annotated using the TreeTagger (see above), it uses the Stuttgart-Tuebingen Tagset. We hope to capture certain syntactic structures of the questions with this - kind of shallow - feature.
- **Bag-of-words** is a shallow method often used in information retrieval. It considers documents (here: questions) as a set of words disregarding any syntactic or semantic relation.
- **Sentence patterns** are constructed manually on the basis of about 500 questions annotated with POS-tags, lemmas, and named entities. We identified key words/verbs and encoded the verb arguments as a set of possible POS-tags and named entities with regular expressions.

3 Three classifiers

We implemented three classifiers for our question analysis task: a Decision Tree, a Naïve Bayes, and a k-Nearest Neighbor algorithm. Decision Trees and k-Nearest Neighbor algorithms, respectively, are well known as very robust with only little training data available. We decided to contrast these two with an algorithm that is supposed to be both simple and scalable: the Naïve

Bayes. We give a very rough overview (see Duda et al. (2000) and Mitchell et al. (1997)) of all three in the following sections.

3.1 Decision tree

A Decision Tree over the features $F_1, F_2, \ldots, F_n$ with discrete values and the classes C is a tree where:

- every node is labeled with one of the features $F_1, F_2, \ldots, F_n$.
- every leaf is labeled with a possible class.
- every node with the label F_i has as many outgoing edges as there are possible values for the feature F_i.
- every outgoing edge for F_i is labeled with a possible value for F_i.

In every classification step the algorithm decides about the next feature according to its information gain. Splitting stops when either all examples are correctly classified or the information gain remains under a certain threshold. In addition, there are several more general design decisions: e.g. optimization according to the overall complexity of the tree or pruning after completion of the algorithm. We decided against pruning because of data sparseness and used the information gain as criterion for splitting.

3.2 k-Nearest Neighbor (kNN)

The key idea of this algorithm is to predict the class of an yet unlabeled example by computing the dominant class of its k nearest neighbors. It works as follows:

- select a suitable value for k;
- compute the distances between the feature vector a of an unlabled example and the vectors v_i of all training data by means of Euclidean distance

$$D_i(a, v_i) = \sqrt{\sum_j (a[j] - v_i[j])^2} \tag{1}$$

- find the k nearest neighbors among the training examples;
- predict the label based on the most frequent one among the k nearest neighbors.

To determine an appropriate k-value we conducted several experiments with features on the sentence level such as collocations and named entities. According to the best results in our experiments the value of k was established to 12.

3.3 Naïve Bayes

A Naïve Bayes classifier is a probabilistic classifier, which determines the most probable hypothesis h from a finite set H of hypotheses. The probability model for this classifier is a conditional one, it can be derived using Bayes' theorem. The joint model of Bayes' theorem and the strong independence assumption is expressed as

$$P(h|a_1, ..., a_n) = \frac{P(h) \prod P(a_i|h)}{P(a)}. \tag{2}$$

The classifier calculates the probabilities with the joint model for all classes and chooses the most probable one as result. Following the recommendations in Duda et al. (2000), probabilities for unseen features were smoothed using Lidstone's law. We estimated the parameter on the basis of test runs with features on the sentence level and set it to 0.15.

4 Experiments and results

To evaluate our classifiers and feature sets, we used holdout cross-validation: The questions were randomly split up into 10 sub-samples for training and test. We evaluated accuracy and also considered precision and recall.
We conducted various experiments with our three classifiers and also compared the results with the accuracy of the corresponding classifiers in the WEKA toolkit. Interestingly, Naïve Bayes slightly outperforms the Decision Tree and the kNN classifier – as Table 3 shows – in almost all cases. However, considering all combinations the algorithms do not differ very much. We found that there is no significant difference in performance between WEKA and our classifiers.

To evaluate the various features, we merged them to eleven sets (nine are shown in Table 3) which we examined separately. Our baseline consists of the handpicked trigger words, the question length, and punctuation mark. As Table 3 reveals, the baseline system already performs – with an accuracy of about 45 % – reasonably well. We then stepwise added one or more features. The results are shown in Table 3. The shallow features such as bag-of-words and statistically collected collocations (as opposed to the intuitive ones) make the most important contribution to enhance the performance. Contrary to the accepted opinion named entities and intuitive collocations do not help that much. They even seem to corrupt the performance, as Table 3 shows. Probably, this is due to the fact, that the statistical collocations and bag-of-words already cover information beyond named entities and intuitive collocations. That is to say, these shallow features cover the hand coded and additionally extend the information included in the features further. Comparing the accuracy calculated on the basis of the small test set (500 questions) with the bigger one (1370 questions) clearly shows that all classifiers achieve a generalization over the given data set.

Table 3. Accuracy of our three classifiers

Feature sets	Decision Tree		Naïve Bayes		kNN	
	500	1370	500	1370	500	1370
Baseline	0.37	0.30	0.512	0.48	0.47	0.50
Baseline, bag-of-words	0.45	0.56	0.59	0.61	0.48	0.53
Baseline, named entities	0.48	0.32	0.51	0.50	**0.62**	0.51
Baseline, statistical collocations	**0.54**	0.42	0.63	0.61	0.60	**0.63**
Baseline, statistical collocations, named entities	0.48	0.45	0.63	0.60	0.60	0.58
Baseline, statistical collocations, bag-of-words	0.52	0.59	**0.65**	**0.65**	0.57	0.58
Baseline, statistical collocations, bag-of-words, named entities	0.44	0.50	0.62	0.62	0.53	0.54
Baseline, intuitive collocations, bag-of-words	0.51	**0.60**	0.63	0.64	0.57	0.50
Baseline, intuitive collocations, bag-of-words, named entities	0.44	**0.60**	0.61	0.64	0.52	0.52

5 Discussion

Even though we had relatively little training data the results were sufficiently accurate for the use in a German question-answering system. However, accuracy is still low for small classes, as they are for classes without specific question words. Nevertheless, we feel confident that small classes can either be handled by re-training with a larger amount of data or by integrating the concept of sub-classes/super-classes in our experiments (e.g. location is the super-class of location.mountain). In any case, as data collection continues during the course of the project (SmartWeb) we expect that even for under-represented classes accuracy will increase further. Although some easily predictable classes as e.g. "wer" (=who) almost always ask for a person, there are classes that do not have a specific question word: *Nenne die Mannschaft, in der Beckenbauer als letztes gespielt hat* (*Name Beckenbauer's last soccer team as active player*). We regard the abstract answer types (*Was ist der Unterschied zwischen Weizen- und Roggenpflanzen? What is the difference between wheat and rye plants?*) as another important challenge. Those two question categories are the most difficult ones for our algorithms. In our opinion, there are mainly two possibilities to solve this problem: These questions hopefully either match a sentence pattern or in case our corpus sufficiently grows in the near future may be catched by means of the bag-of-words strategy. We also plan to further explore these classes to distinguish between the fundamental types that need to be correctly handled by our system and noise. During the course of our experiments we continued to manually construct more sentence patterns matching the questions in our corpus. We did not consider them, yet, to avoid overfitting. However, the patterns may improve the performance –

especially for the small classes. Sentence patterns are part of many question-answering systems for English. Although they still have to prove usefulness for German data, what we had to leave for future investigation. We also intend to further improve our answer type ontology. We found that the questions (and answers) in our corpus often do not entirely meet its concepts. In addition, there are questions that the answer type ontology does not cover at all. Especially the abstract types are under-represented. We plan to integrate the results of these experiments into the question analysis component of a German question-answering system. We hope to thus improve the performance of the hole system. While there is still room for improvement, we think – considering the complexity of the task – the achieved performance is surprisingly good.

References

CRAMER, I., LEIDNER, J.L. and KLAKOW, D. (2006): Building an Evaluation Corpus for German Question Answering by Harvesting Wikipedia. *Proceedings of The 5th International Conference on Language Resources and Evaluation, Genoa, Italy.*

DAY, M.-Y., LEE, C.-W., WU, S.-H., ONG, C.-S. and HSU, W.-L. (2005): An Integrated Knowledge-based and Machine Learning Approach for Chinese Question Classification. *IEEE International Conference on Natural Language Processing and Knowledge Engineering.*

DUDA, R.O., HART, P.E. and STORK, G. (2000): *Pattern Classification.* Wiley and Sons, New York.

LI, X., HUANG, X.-J. and WU, L.-D. (2005): Question Classification using Multiple Classifiers. *Proceedings of the 5th Workshop on Asian Language Resources and First Symposium on Asian Language Resources Network.*

LI, X. and ROTH, D. (2002): Learning Question Classifiers. *COLING'02.*

MITCHELL, T.M. (1997): *Machine Learning.* McGraw Hill, Boston.

SONNTAG, D. and ROMANELLI, M. (2006): A Multimodal Result Ontology for Integrated Semantic Web Dialogue Applications. *Proceedings of the 5th international conference on Language Resources and Evaluation, Genoa, Italy.*

VOORHEES, E. (2001): Overview of the TREC 2001 Question Answering Track. *Proceedings of the 10th Text Retrieval Conference, NIST, Gaithersburg, USA.*

WAHLSTER, W. (2004): SmartWeb: Mobile Applications of the Semantic Web. In: P. Adam, M. Reichert (Eds.): *INFORMATIK 2004 - Informatik verbindet, Band 1. Beiträge der 24. Jahrestagung der Gesellschaft fr Informatik e.V. (GI)*, Ulm.

WITTEN, I.H. and FRANK, E. (2000): *Data Mining - Practical Machine Learning Tools and Techniques with Java Implementations.* Morgan Kaufmann Publishers.

ZHANG, D. and LEE, W.S. (2003): Question Classification Using Support Vector Machines. *Proceedings of the 26th Annual International ACM SIGIR Conference on Research and Development in Informaion Retrieval*, Toronto, Canada.

The Relationship of Word Length and Sentence Length: The Inter-Textual Perspective

Peter Grzybek[1], Ernst Stadlober[2] and Emmerich Kelih[1]

[1] University Graz, Department for Slavic Studies, A-8010 Graz, Austria; {peter.grzybek, emmerich.kelih}@uni-graz.at

[2] Graz University of Technology, Institute Statistics, A-8010 Graz, Austria; e.stadlober@tugraz.at

Abstract. The present study concentrates on the relation between sentence length (SL) and word length (WL) as a possible factor in text classification. The dependence of WL and SL is discussed in terms of general system theory and synergetics; the results achieved thus are relevant not only for linguistic studies of text classification, but for the study of other complex systems, as well.

1 Synergetics and the Menzerath-Altmann Law

In a number of previous studies (Grzybek et al. (2005), Kelih et al. (2005), Antić et al. (2006), Kelih et al. (2006)), the impact of word length (WL) and sentence length (SL) for purposes of text classification has been analyzed in detail. It has been shown that both factors play an important role in text classification. Based on this work, the present study focuses less on WL and SL as separate linguistic phenomena in their own right, rather than on the relation between them and implied text typological specifics.

In quantitative and synergetic linguistics, the relations between linguistic units of different levels usually are treated in the framework of the Menzerath-Altmann Law. This law has its origin in the work of the German phonetician Paul Menzerath (1883–1954) who, on the basis of his phonetic and lexicological studies, arrived at the conclusion that the larger a given whole, the smaller its parts. Later, German linguist Gabriel Altmann theoretically explored this concept in detail and provided a mathematical approach which allows for a generalization and an exact formulation of the observations reported above. In this general framework, the relevance of Menzerath's findings is no longer restricted to the lower levels of language; rather, the Menzerath-Altmann Law (MAL) is considered to be a general structural principal of language and, in fact, has become a central concept in the synergetic approach to language and other complex systems in culture and nature (cf. Altmann and Schwibbe (1989), Cramer (2005, p. 663)).

In its most general form, the *MAL* (cf. Altmann (1980, p. 1)) sounds as follows: "The longer a language construct the shorter its components (constituents)." In our case, if 'sentence' is the construct (i.e. the independent variable), then 'words' can be considered to be its components (i.e., the dependent variable). As to the mathematical formulation of the *MAL*, Altmann (1980), in his seminal "Prolegemona on Menzerath's Law", suggested formula (1a) to be the most general form:

$$y = Ax^b e^{-cx} . \tag{1a}$$

In this context, Altmann also presented two special cases of equation (1a), namely, equation (1b) for $c = 0$, and equation (1c) for $b = 0$; whereas equation (1a) is the most general form, equation (1b) has turned out to be the most commonly used "standard form" for linguistic purposes:

$$y = Ax^b , \tag{1b}$$

$$y = Ae^{-cx} . \tag{1c}$$

In a subsequent study on the relation between *SL* (x, measured as number of words per sentence) and *WL* (y, measured as the number of syllables per word) – which is of immediate relevance in our context – Altmann (1983, p. 31) pointed out that the *MLA* as described above holds true only as long as one is concerned with the direct constituents of a given construct, that is to say when dealing with the relation between units from immediately neighboring linguistic levels. In this case, an increase of the units of one level is related with a decrease of the units of the neighboring level. Therefore, in its direct form, the *MAL* might fail to grasp the relation between *SL* and *WL*, when we are not concerned with the word as the direct constituent of the sentence. In fact, an intermediate level is likely to come into play – such as phrases or clauses as the direct constituents of the sentence. In this case, words might well be the direct constituents of clauses or phrases, but they would only be indirect constituents of a sentence. Consequently, an increase in *SL* should result in an increase in *WL*, too. Corresponding observations must therefore not be misinterpreted in terms of a counterproof of the *MAL*.

In fact, the most prominent example of such an indirect relation is the one between *SL* and *WL*. This specific relation has been termed Arens Law by Altmann (1983), and it has hence become well known as Altmann-Arens Law, respectively. The relevant study goes back to Altmann's (1983) re-analysis of Hans Arens' (1965) book *Verborgene Ordnung*. In this book, Arens analyzed mean *WL* $\bar{y}_i$ and mean *SL* $\bar{x}_i$ of 117 German literary prose texts from 52 different authors. As a result, Arens observed an obvious increase of mean *WL* with an increase of mean *SL*. Whereas Arens assumed a more or less linear increase, Altmann (1983) went a different way, interpreting the observed relation in terms of Menzerath's Law. Consequently, the increase in WL with increasing SL should not be linear; rather it should follow Menzerath's Law.

Given that the "standard case" (1b) has often been sufficient to describe the relation between SL and clause length (i.e., $z = Ax^b$), as well as the one between clause length and WL (i.e., $y = A'z^{b'}$), Altmann (1983, p. 32) argued in favor of using this special case, consequently obtaining $y = A''x^{b''}$, corresponding to (1b). The only difference to be expected for the relation between directly and indirectly related units of different levels is that, in case of directly neighboring units, parameters b and b' should be negative (due to the prognosed decline); in case of indirectly related units, with intermediate levels, $b'' = b \cdot b'$ will become positive.

Re-analyzing Arens' data by fitting equation (1b), Altmann found the results to be highly significant, thus allegedly corroborating his assumptions on the Menzerathian relation between SL and WL.[1] However, taking an unbiased look at Altmann's (1983) re-analysis of Arens' data, doubt may arise as to his positive interpretation. Reason for doubt is provided by Grotjahn's (1992) discussion of methodological weaknesses of the F-test when testing linguistic data; its major flaw is its sensibility in case of large data material, thus tending to indicate significant results with in increase of sample size. Grotjahn therefore arrived at the conclusion that in linguistics, the calculation of the determination coefficient R^2 should be favored, instead of F-tests; in fact, this has hence become the common procedure, a value of $R^2 > 0.85$ usually being interpreted to be an index of a good fit. Surprisingly enough, Grzybek and Stadlober (2006), in their recent re-analysis of Arens' data, found that in this case we are concerned with a rather poor value of $R^2 = 0.70$. Thus, notwithstanding the fact that the result for the power model (1b) is definitely better than the one for the linear model (with $R^2 = 0.58$), it is far from being convincing, consequently shedding doubt on the adequacy of Altmann's Menzerathian interpretation. Therefore, this achievement asks for a general and systematic re-analysis of the WL and SL problem.

A first step in this direction has been undertaken by Grzybek and Stadlober (2006). Given the relatively weak empirical evidence, they point out a number of general problems which are relevant for a better understanding of the MAL. In addition to the test theoretical problems just mentioned, a crucial question seems to be if Arens' Law may in fact be interpreted in terms of inter-textually relevant regularity: according to its basic idea, the scope of Menzerath's Law has been to describe the relation between the constituting components of a given construct; consequently, the MAL originally was designed in terms of an **intra-textual** law, relevant for the internal structure

[1] In detail, an F-test was calculated to test the goodness of fit; with parameter values $a = 1.2183$ and $b = 0.1089$ the result was $\hat{F}_{1,115} = 241.40$ $(p < 0.001)$.

of a given text sample.[2] Arens' data, however, are of a different kind, implying **inter-textual** relations, based on the calculation of the mean lengths of words ($\bar{x}_i$) and sentences ($\bar{y}_i$) for each of the 117 text samples, resulting in two vectors of arithmetic means ($\bar{\mathbf{x}}$ and $\bar{\mathbf{y}}$).

Yet, over the last decades, both perspectives have not been clearly kept apart; strictly speaking, Arens' Law may be interpreted to be a logical derivation of the MAL, on an intra-textual perspective, due to the intervention of intermediate levels. From an inter-textual perspective, however, Arens' Law is not necessarily a logical consequence of the MAL; rather, it has the very same status of a strong hypothesis as has the MAL itself (an interpretation which does not generally rule out its validity on this level). Still, the poor result obtained asks for an explanation. In their re-analysis of Arens' data, Grzybek and Stadlober (2006) found out that one crucial problem is a merely statistical one: as long as there are not enough data points, variation is extremely large – consequently, pooling turned out to be a highly efficient procedure, resulting in a determination coefficient of $R^2 > 0.90$, the exact value depending on the concrete pooling procedure chosen.

In addition to this conclusion – which may seem to be trivial at first sight, since it is not surprising that pooling yields better results –, Grzybek and Stadlober (2006) found out something even more important: namely, that pooling alone is not the decisive factor. Rather, simply adding more data following a naive "The more the better" principle, may even worsen the result obtained. This tendency became very clear when relevant data from Fucks' (1955) study were added to Arens's data. These data are based on two different text types: literary texts and scholarly texts. Combining Arens' and Fucks data, all literary texts displayed a more or less similar tendency, but the scholarly texts were characterized differently; thus, despite the larger number of analyzed texts, the overall results became even worse ($R^2 = 0.22$, cf. Figure 1).

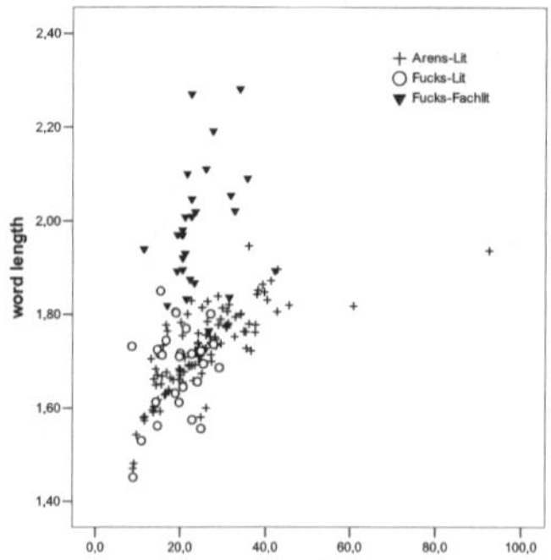

Fig. 1. WL and SL (Arens'/Fucks' data)

The conclusion by Grzybek and Stadlober (2006) was that data homogeneity must be obeyed more than this has been done hitherto in related studies. This means that text typological implications must be carefully controlled, since texts of different types are likely to follow different rules. This interpretation would be in line with the isolated WL and SL studies reported above, in which these two characteristics were shown to be efficient factors in discriminating text or discourse types. However, due to the small sample size of Fucks's data ($N = 27$ per text type), Grzybek and Stadlober (2006) could not test their assumption. The purpose of the present study therefore is (i) to reproduce Arens' study

[2] We need not discuss the notion of 'text' here; for the sake of simplification we tolerate that a 'text' may be represented by homogeneous material, as well as by a mixed corpus, or by dictionary material, etc.

on a broader material basis, using Russian material, and (ii) to control the factor of text typology in these studies.

2 The inter-textual perspective

In a first step, WL and SL of Russian literary prose texts shall be analyzed, in analogy to Arens' procedure. For this purpose, and in order to exclude any author-specific factors, we concentrate on Lev N. Tolstoj's novel *Anna Karenina*, which consists of 239 chapters in eight books. The mean values are calculated for each chapter separately, so we get 239 data points. As a result, we see that there is only a weak relation between the means of WL and SL: for the linear model, we obtain a low $R^2 = 0.15$ with parameter values $a = 2.08$ and $b = 0.01$, for the power model (1b) we have $R^2 = 0.18$ with $a = 1.80$ and $b = 0.08$. Figure 2 illustrates these findings.

A closer inspection of Figure 2 shows that, with regard to SL, the bulk (90%) of data points is within the relatively small interval $10 \leq x_i \leq 22$. Therefore, it seems reasonable to assume that, despite the large amount of data, the texts do not display enough variance for Arens's Law to become relevant. This might be interpreted in terms of an author-specific characteristic; consequently, for the sake of comparison, the inclusion of texts from different authors would be necessary, similar to Arens' study. Although there seems to be increasing evidence not to consider WL and SL an author-specific trait, this option has to be checked in future. It seems more reasonable, to see this observation in line with the studies mentioned above, showing that WL and SL are specific to particular discourse types – in this case, further text types would have to be added.

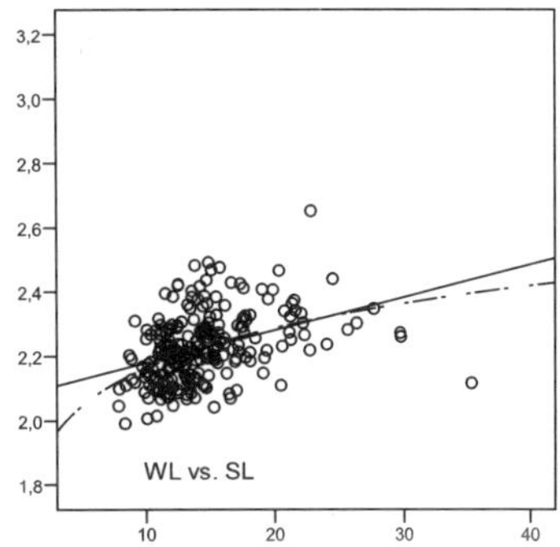

Fig. 2. Anna Karenina

In order to test the effect of this extension, further texts from five different text types were additionally included. In order to exclude undesired effects, and to base the analyses on a text-typologically balanced corpus (including ca. 30 texts per text type), only the first book of *Anna Karenina* with its 34 chapters remained in the corpus. This results in a corpus of 199 Russian texts from six different text types; the first three columns of Table 1 illustrate the composition of the text corpus.

Calculating mean WL and SL for each individual text, we obtain a two-dimensional vector consisting of 199 data points. Before analyzing these texts as a complete corpus, the texts shall first be analyzed controlling their attribution to one of the six text types. Table 1 contains the detailed fitting results both for the linear and the power models. The results are equally poor

Table 1. Corpus of texts: linear model and power model (1b)

Text type	Author	Number	a	b	R^2	a	b	R^2
Private Letter	A.P. Čechov	30	2.27	−0.003	.09	2.36	−0.022	.02
Drama	A.P. Čechov	44	2.08	−0.001	.02	2.12	−0.012	.01
Short Story	A.P. Cechov	31	2.06	0.027	.20	1.88	0.092	.19
Novel	L.N. Tolstoj	34	2.04	0.011	.26	1.77	0.082	.27
Comment	(various)	30	2.67	0.005	.02	2.56	0.028	.02
Scientific	(various)	30	2.65	0.006	.01	2.44	0.042	.01
Total		199	1.90	0.039	.49	1.57	0.169	.47

in either case: for the linear model, values of $R^2 = 0.20$ and $R^2 = 0.26$ are obtained for the short stories and the novel texts, respectively; for the remaining text types, the fitting results are even worse. For the power model, the values are similar (with $R^2 = 0.19$ and $R^2 = 0.27$ for the short stories and the novel texts). Figure 3 represents the results for the separately analyzed text types: there is only a weak relation between WL and SL, the degree varying between the text types.

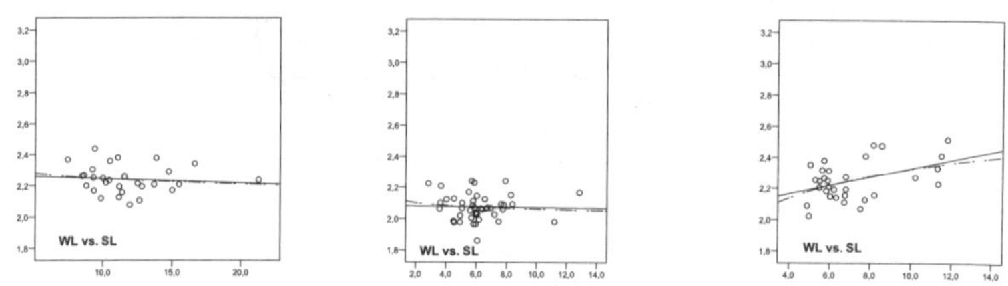

(a) Private letter (b) Drama (c) Short story

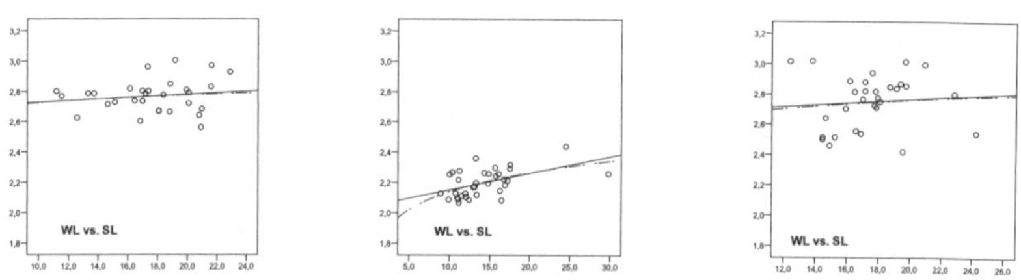

(d) Novel (e) Comment (f) Scientific

Fig. 3. Dependence of WL on SL in six text types

The results are slightly better, though still far from being convincing, if one analyzes all texts simultaneously, without genre distinction, by way of a

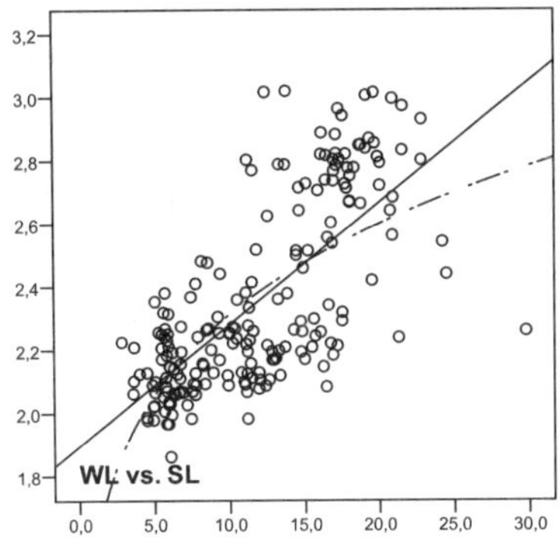

Fig. 4. Dependence of WL on SL in the whole text corpus

corpus analysis: under this condition, the results are $R^2 = 0.49$ and $R^2 = 0.47$, respectively, for the linear and power model (cf. Figure 4).

3 Conclusion

In summary, on the basis of 199 Russian texts, there seems to be no strong relationship between WL and SL. It must be emphasized once more, however, that this result is related to the **inter-textual perspective**, only. Obviously, the individual text types are rather limited with regard to the variation of WL and/or SL; it is a mere matter of logics to arrive at the conclusion that, if there is no variation of WL and/or SL, there can be no remarkable correlation. This seemingly negative finding should be interpreted in accordance with the results obtained previously, arguing in favor of both WL and SL to be good discriminating factors for the distinction of particular discourse types. This is completely in line with the previous observation that these two factors can serve for discriminating purposes. this would not be the case unless WL and/or SL were not relatively specific for a given text or discourse type – and this discriminating power could not be efficient unless WL and/or SL were not characteristic for a given group of texts.

The most important finding of the present study is that, at least for the Russian texts analyzed, there is only a weak relationship between the means of WL and SL. As to an explanation of this result, one should not wrongly search for specifics of the Russian language; rather, the reason seems to be the application of the Arens-Altmann Law on an inter-textual level, as has been initially done by Arens, and subsequently by Altmann. Consequently, in this respect, the conclusions made by Arens (1965) and Altmann (1983) cannot be generally accepted; from an inter-textual perspective, it does not seem to be justified to speak of a law-like regularity as to the WL and SL relation.

Yet, one should not generally discard the claims implied in the two laws mentioned: in Altmann's original interpretation, both Menzerath and Arens Laws were conceived in an **intra-textual perspective**. Hence it will be necessary to study the relevance of these laws from an intra-textual point of view as well, before one can arrive at any serious conclusions as to the validity of these laws. Exploring this has to be the task of another study, in which again, text-typological aspects must be adequately controlled.

References

ALTMANN, G. (1980): Prolegomena to Menzerath's Law. In: *Glottometrika 2.* Brockmeyer, Bochum, 1–10.

ALTMANN, G. (1983): H. Arens' "Verborgene Ordnung" und das Menzerathsche Gesetz. In: M. Faust et al. (Eds.), *Allgemeine Sprachwissenschaft, Sprachtypologie und Textlinguistik.* Narr, Tübingen, 31–39.

ALTMANN, G. and SCHWIBBE, M.H. (1989): *Das Menzerathsche Gesetz in informationsverarbeitenden Systemen.* Olms, Hildesheim.

ANTIĆ, G., STADLOBER, E., GRZYBEK, P. and KELIH, E. (2006): Word Length and Frequency Distributions. In: M. Spiliopoulou et al. (Eds.): *From Data and Information Analysis to Knowledge Engineering.* Springer, Berlin, 310–317.

ARENS, H. (1965) *Verborgene Ordnung. Die Beziehungen zwischen Satzlnge und Wortlnge in deutscher Erzhlprosa vom Barock bis heute.* Pädagogischer Verlag Schwann, Düsseldorf.

CRAMER, I.M. (2005): Das Menzerathsche Gesetz. In: R. Köhler, G. Altmann, Piotrowski and G. Raimund (Eds.): *Quantitative Linguistics. An International Handbook.* de Gruyter, Berlin, 659–688.

FUCKS, W. (1955): Unterschied des Prosastils von Dichtern und Schriftstellern. Ein Beispiel mathematischer Stilanalyse. In: *Sprachforum*, 1, 234–241.

GROTJAHN, R. (1992): Evaluating the Adequacy of Regression Models: Some Potential Pitfalls. In: *Glottometrika 13.* Brockmeyer, Bochum, 121–172.

GRZYBEK, P. and STADLOBER, E. (2006): Do We Have Problems with Arens' Law? The Sentence-Word Relation Revisited. In: P. Grzybek and R. Köhler (Eds.): *Exact Methods in the Study of Language and Text.* de Gruyter, Berlin, (In print).

GRZYBEK, P., STADLOBER, E., KELIH, E. and ANTIĆ, G. (2005): Quantitative Text Typology: The Impact of Word Length. In: C. Weihs and W. Gaul (Eds.): *Classification – The Ubiquitous Challenge.* Springer, Berlin, 53–64.

KELIH, E., ANTIĆ, G., GRZYBEK, P. and STADLOBER, E. (2005): Classification of Author and/or Genre? The Impact of Word Length. In: C. Weihs and W. Gaul (Eds.): *Classification – The Ubiquitous Challenge.* Springer, Berlin, 498–505.

KELIH, E., GRZYBEK, P., ANTIĆ, G. and STADLOBER, E. (2006): Quantitative Text Typology: The Impact of Sentence Length. In: M. Spiliopoulou et al. (Eds.): *From Data and Information Analysis to Knowledge Engineering.* Springer, Berlin, 382–389.

Comparing the Stability of Different Clustering Results of Dialect Data

Edgar Haimerl[1] and Hans-Joachim Mucha[2]

[1] Institut für Romanistik, Universität Salzburg, Akademiestraße 24, A-5020 Salzburg, Austria; info@hedco.de
[2] Weierstraß-Institut für Angewandte Analysis und Stochastik, D-10117 Berlin, Germany; mucha@wias-berlin.de

Abstract. Mucha and Haimerl (2005) proposed an algorithm to determine the stability of clusters found in hierarchical cluster analysis (HCA) and to calculate the rate of recovery by which an element can be reassigned to the same cluster in successive classifications of bootstrap samples. As proof of the concept this algorithm was applied to quantitative linguistics data. These investigations used only HCA algorithms. This paper will take a broader look at the stability of clustering results, and it will take different cluster algorithms into account; e.g. we compare the stability values of partitions from HCA with results from partitioning algorithms. To ease the comparison, the same data set - from dialect research of Northern Italy, as in Mucha and Haimerl (2005) - will be used here.

1 The starting point

1.1 The data under investigation

In order to make the results from different cluster algorithms comparable, the algorithms are always applied to the same set of data. The similarity matrix goes back to a nominal data matrix of classified maps from the ALD atlas of Northern Italy[1], which presents the dialect responses as deep phonetic transcription. The classification of the transcription was done automatically or manually based on different linguistic criteria (see Haimerl (1998), Bauer (2003)). The resulting nominal data matrix with 3386 classified maps and 218 locations each was the starting point for the calculation of the ALD-DM similarity matrix using RIW[2] as similarity measure. The ALD-DM distance matrix is easily calculated as 100 - ALD-DM-similarity-value.

[1] The data are published as atlas maps in Goebl et al. (1998).

[2] The relative identity value (RIW) is similar to the simple matching coefficient but rules out missing values.

1.2 Prior achievements

Hierarchical clustering algorithms have been successfully applied to linguistic dialect data in the last decades (e.g. Goebl (1984)). The downside of the manual procedure of this time was that the generation of dendrograms and the corresponding maps was very time-consuming. Thus it was not possible to experiment with different partitions or to select the most suitable segmentation from a set of output maps. The implementation of interactive dendrograms in VDM[3] in 2000 helped to overcome this downside. The results of HCA calculations are presented as dendrograms and colors can be interactively assigned to branches of the dendrogram. These changes of the dendrograms colors are instantly reflected in a polygon map where every location corresponds to one leaf in the dendrogram. The downside of this convenient method to investigate and spatialize the HCA results is that the maps, while visually convincing, do not show how stable and reproducible the partitions are. Procedures are needed to examine the stability of the partition with respect to the optimal number of clusters, the stability of each cluster and the stability of the assignment of a single element to a cluster. Only with these stability values in mind does an in depth linguistic interpretations of the clustering results make sense. If we have not been able to prove that neighboring groups have a strict and reproducible demarcation line between them, there is no statistical basis for further investigations at that point.

2 The stability of hierarchical clustering results

2.1 The procedure

The objective is to investigate the stability of HCA on three levels:

1. Determine the number of clusters that result in the most stable partition.
2. Calculate the rate of recovery for every cluster in a partition.
3. Calculate the reliability of cluster membership for every cluster element.

The algorithm proposed and applied in Mucha and Haimerl (2005) uses the bootstrap resample technique. For every arbitrarily chosen sample, a hierarchical cluster analysis is calculated. Applying the adjusted Rand's measure to compare the partitions pairwise yields values that represent the stability of the partitions from 2 to N groups[4]. Furthermore the comparison of several

[3] VDM (Visual DialectoMetry) is an implementation of dialectometric methods and visualization procedures that among others calculates HCA, draws dendrograms and corresponding maps. For further details see `http://ald.sbg.ac.at/DM` (accessed April 22nd 2006).

[4] The Rand's index of two partitions ranges from 0 to 1. It is based on counts of pairs of objects with respect to their class membership. When the partitions agree perfectly, the Rand index is 1. The adjusted Rand's index avoids the bias towards number and size of clusters; see Hubert and Arabie (1985) or Mucha (2004).

partitions based on bootstrap samples allows for calculating the stability of the clusters as the rate of recovery of subset $\mathcal{E}$ by subset $\mathcal{F}$:

$$\gamma(\mathcal{E}, \mathcal{F}) = \frac{|\mathcal{E} \cap \mathcal{F}|}{|\mathcal{E}|}. \tag{1}$$

The rate of recovery ranges from 0 for disjoined sets to 1 only if $\mathcal{E} \subseteq \mathcal{F}$. As measure for the stability of the assignment of an element to a group we propose the reliability of cluster membership which counts the most frequent assignment to a group (k) relative to the total number of simulations (n):

$$m = \frac{k}{n}. \tag{2}$$

2.2 The results and their visualization

The implementation of this algorithm is called Hierarchical Cluster Analysis with Validation (HCAValid). The HCAValid algorithm with Ward distance measure was applied to the ALD-DM distance matrix taking partitions from 2 to 10 clusters into account. It found that the partition with 7 groups is the most stable solution and has an average adjusted Rand's index of 0.79.

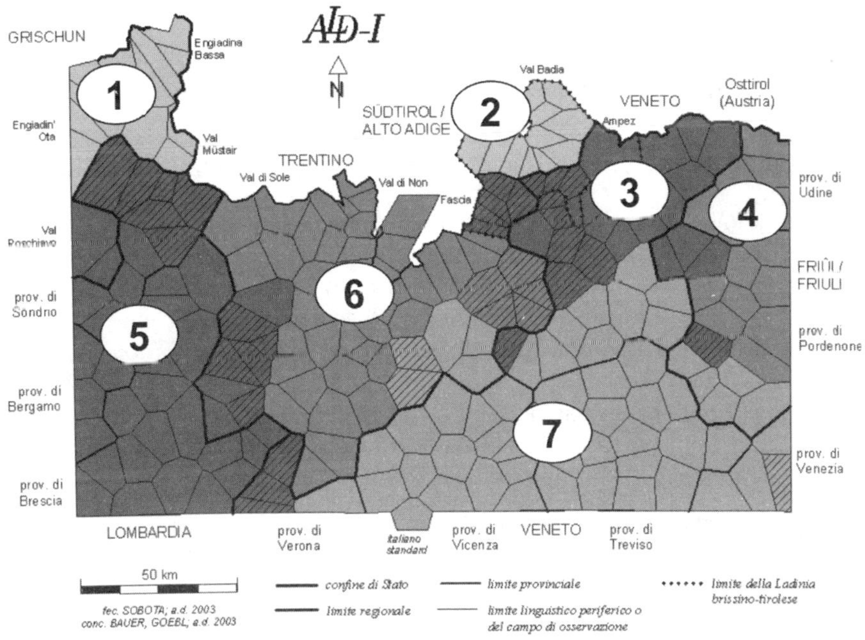

Fig. 1. Result from HCAValid - seven-cluster partition with unstable locations in hatching pattern

In order to make the cluster results clear we visualize the clustering as polygon maps. This gives us the ability to

- let elements in the clusters coalesce to areas on the polygon map
- distinguish clusters by color
- find stability on the element level: bright colors signal a reliability of cluster member-ship higher than 0.99.

Figure 1 shows a Thiessen polygon map on which the groups found by HCAValid are represented by areas which are fully covered by polygons in the same grey scale. We define unstable objects as those objects with a reliability of cluster membership smaller than 0.9953. The map in Figure 1 marks these unstable objects with hatching patterns. HCAValid locates three groups with outstanding stability. They have no unstable objects. In two clusters, not one single object has been misclassified in all cluster analysis of bootstrap samples. These are Grisons (1), the northern Dolomitic Ladin valleys (2) and Friuli (4).

3 The stability of partitioning clustering results

Hierarchical cluster algorithms start with joining those elements to groups that are closest to each other and later add more and more elements to these groups. Contrary to this procedure partitioning algorithms start with a number of arbitrarily chosen cluster centers and the researcher must input the number of clusters. They repeat the assignment of elements to the clusters until the objective to minimize the total intra-cluster variance is met.

3.1 Partitioning clustering using bootstrap resample technique

A well known downside of partitioning cluster algorithms is that they rely on an "initial clustering that is obtained by the successive selection of representative objects" (Kaufman and Rousseeuw (1990, p. 102)) for all groups. Applying the bootstrap resample technique to partitioning clustering algorithms overcomes this downside by calculating multiple partitions of subsets and choosing the best partition.

Calculating the adjusted Rand's values and the rates of recovery makes the results from partitioning cluster algorithms comparable with the results from hierarchical cluster analysis. For both of them

- the adjusted Rand's measure is used for pairwise comparison of the partitions
- the adjusted Rand's value is used to detect the best number of clusters
- the rate of recovery of clusters and the reliability of cluster membership of elements helps to distinguish stable from non stable clusters and elements.

An implementation of a partitioning algorithm is TIHEXM (Späth (1980)). The TIHEXM-Sim module is based on TIHEXM and enhances it with bootstrap sample technique and calculates various stability values, e.g. adjusted Rand's values, rates of recovery, reliabilities of cluster membership and Jaccard coefficients. Applying TIHEXM-Sim to the ALD-DM distance matrix yields the best adjusted Rand's value of 0.9 for the seven-cluster partition (Figure 3).

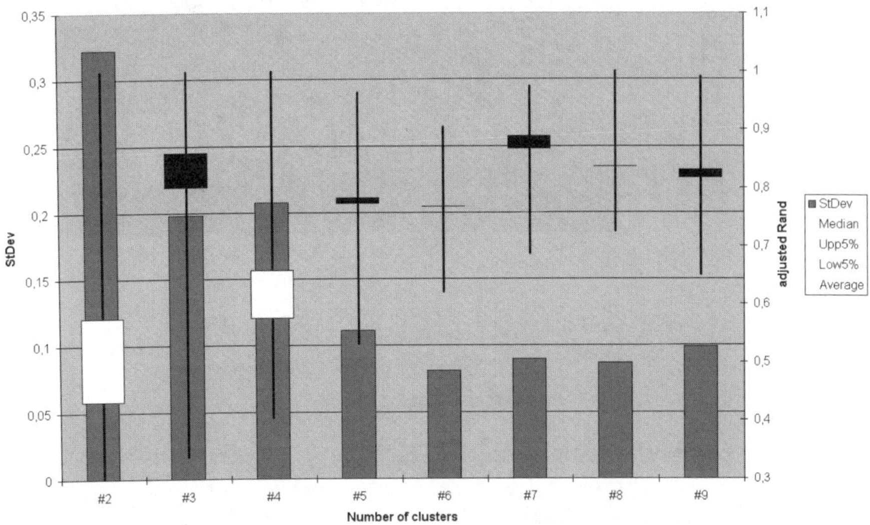

Fig. 2. Adjusted Rand's values from 50 simulations of k-means in TIHEXM-Sim

4 Visualizations of the results

4.1 The three-cluster partition

The three-cluster partition is not qualified as the most stable one: though the Rand's value for three clusters is pretty good, there is a high standard deviation. The three-cluster partition is close to the traditional dialectal classification according to Ascoli (1873) and similar to the two-cluster partition from k-means or PAM (Kaufman and Rousseeuw (1990, pp. 68)). The three areas - Grisons, the Northern Dolomitic Ladin valleys and Friuli - are joined in one group (2) and stand in opposition to Northern Italian which is split into a Eastern (3) and Western (1) group (Figure 3). The Northern Italian group is rather unstable, due to many unstable locations, whereas Grisons and the two Northern Dolomitic Ladin valleys only contain locations with reliabilities of cluster membership better than 0.99.

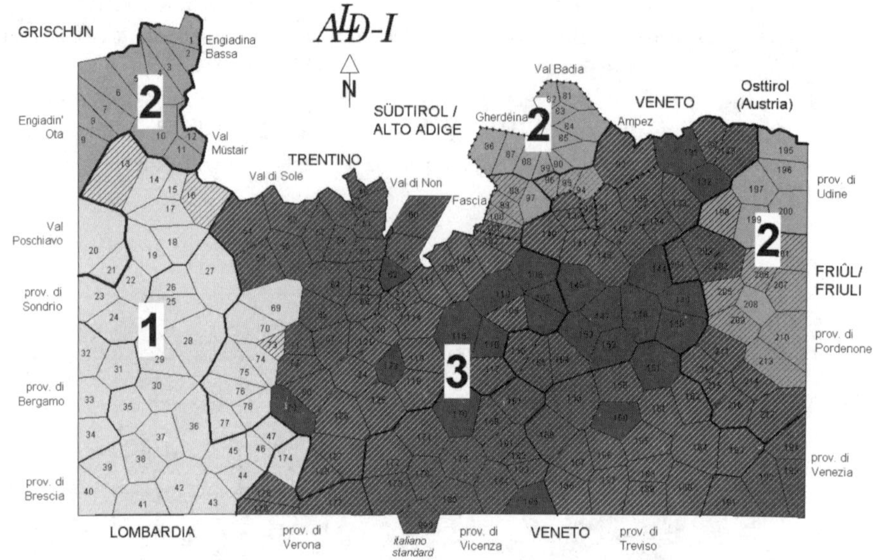

Fig. 3. Three-cluster partition by TIHEXM-Sim

4.2 The seven-cluster partition

Figure 4 presents the result of the seven-cluster partition with TIHEXM-Sim as a polygon map. Interestingly the Grisons area (1) forms the most stable group with only stable locations. The Fiuli area (2) can also be considered as very stable (only one location with a reliability of cluster membership of less than 0.99) but the Dolomitic Ladin cluster (7) has the less stability of all clusters. The instability of the Dolomitic Ladin valleys has its root in unstable locations in the southern valleys which are under strong influence of Northeastern Italian from Veneto (6).

5 Comparing the results from hierarchical and partitioning clustering

In order to make the cluster results clear and HCAValid results comparable with TIHEXM-Sim partitioning, the clustering results of both methods are visualized as polygon maps. There are striking similarities between the result from hierarchical clustering (see Figure 1) and partitioning clustering (see Figure 4). The areas generated by the HCAValid classification coincide in most locations with the areas generated by TIHEXM-Sim. Though the reliabilities of cluster membership of elements are higher in hierarchical clustering than in partitioning clustering, the rate of recovery of the clusters is still very similar, as the following table proves. For most areas the rate of

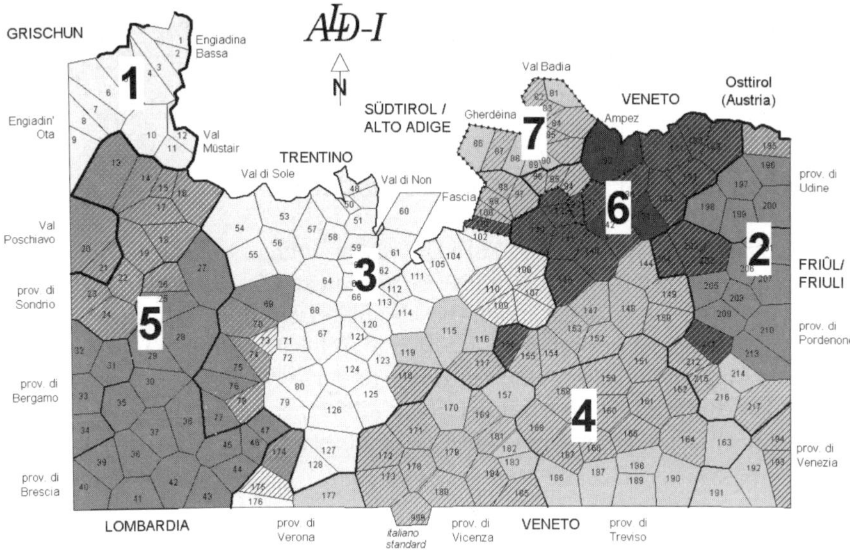

Fig. 4. Seven-cluster partition by TIHEXM-Sim

recovery is higher in the partitioning result than in the hierarchical result. The most striking difference is the rate of recovery of the Dolomitic Ladin

Table 1. Comparison HCAValid versus TihExm

Area	HCAValid	TihExm
Grisons	1.00	1.00
Dolomitic Ladin valleys	**0.99**	**0.80**
Friuli	0.99	0.99
Veneto-south	0.93	0.95
Trentino	0.91	0.95
Lombardy	0.89	0.91
Veneto-north	0.8	0.86
Mean	0.92	0.91

valleys, which is significantly lower for the partitioning result. The reason for this low stability might be the tendency of the k-means distance definition to generate groups of similar magnitude. HCA does not tend to integrate the southern Dolomitic Ladin valleys into a bigger "all Dolomitic" group. But the bootstrap sample technique shows that the partitioning analysis is not good in isolating small groups. Another significant difference is that all partitioning methods (k-means, PAM, TIHEXM-Sim) joined Grisons, the Dolomitic Ladin valleys and Friuli in the two or three-cluster solution. This makes sense from a linguistic point of view and reflects a major tendency in the data. The reference point map of standard Italian (location # 999) as well as multidimensional scaling with two or seven dimensions also shows this opposition. With hierarchical cluster analysis only complete linkage grouped these three areas together.

6 Objectives for further investigations

The exciting thing about the ALD-DM data from a statistical viewpoint is that small groups with high stability coexist next to large groups with lower stability. We suspect that a model-based clustering approach might yield results with even higher stability values that reflect the inherent structure of the ALD data more closely. But in order to verify this expectation, an algorithm like MCLUST (Fraley and Raftery (2002)) must be modified to generate results that are comparable with the stability values from HCAValid and TIHEXM-Sim.

References

ASCOLI, G.I. (1873): Saggi ladini. *Archivio glottologico italiano, 1, 1–556.*

BAUER, R. (2003): Dolomitenladinische Ähnlichkeitsprofile aus dem Gadertal; ein Werkstattbericht zur dialektometrischen Analyse des ALD-I. In: *Ladinia XXVI-XXVII (2002-2003)*, 209–250.

FRALEY, C. and RAFTERY, A.E. (2002): MCLUST: Software for Model-Based Clustering, Density Estimation and Discriminant Analysis. *Technical Report 415*, Department of Statistics, University of Washington.

GOEBL, H. (1984): *Dialektometrische Studien anhand italoromanischer, rätoromanischer und galloromanischer Sprachmaterialien aus AIS und ALF.* Bd.1-3. Max Niemeyer, Tübingen.

GOEBL, H., BAUER, R. and HAIMERL, E. (1998): *Atlante linguistico del ladino dolomitico e dei dialetti limitrofi 1a parte.* Dr. Ludwig Reichert Verlag, Wiesbaden.

HAIMERL, E. (1998): A Database Application for the Generation of Phonetic Atlas Maps. In: J. Nerbonne (Ed.): *Linguistic Databases.* CSLI, Stanford, 103–116.

HUBERT, L.J. and ARABIE, P. (1985): Comparing Partitions. *Journal of Classification, 2, 193–218.*

KAUFMAN, L. and ROUSSEEUW, P.J. (1990): *Finding Groups in Data.* Wiley, New York.

MUCHA, H.-J. (2004): Automatic Validation of Hierarchical Clustering. In: J. Antoch (Ed.): *Proceedings in Computational Statistics, COMPSTAT 2004, 16th Symposium.* Physica, Heidelberg, 1535–1542.

MUCHA, H.-J. and HAIMERL, E. (2005): Automatic Validation of Hierarchical Cluster Analysis with Application in Dialectometry. In: C. Weihs and W. Gaul (Eds.): *Classification - The Ubiquitous Challenge*, Springer, Berlin, 513–520.

SPÄTH, H. (1980): *Cluster Analysis Algorithms.* Ellis Horwood Limited, Chichester.

Part-of-Speech Discovery by Clustering Contextual Features

Reinhard Rapp

Universitat Rovira i Virgili, GRLMC, Tarragona, Spain; reinhard.rapp@urv.cat

Abstract. An unsupervised method for part-of-speech discovery is presented whose aim is to induce a system of word classes by looking at the distributional properties of words in raw text. Our assumption is that the word pair consisting of the left and right neighbors of a particular token is characteristic of the part of speech to be selected at this position. Based on this observation, we cluster all such word pairs according to the patterns of their middle words. This gives us centroid vectors that are useful for the induction of a system of word classes and for the correct classification of ambiguous words.

1 Introduction

The aim of part-of-speech induction is to discover a system of word classes that is in agreement with human intuition, and then to automatically assign all possible parts of speech to a given ambiguous or unambiguous word type. One of the pioneering studies concerning this as yet unsolved problem is Schütze (1993), where a statistical and a neural network approach are combined. The method relies on a matrix of global co-occurrence vectors, i.e. on vectors that have been derived from an entire corpus. Since many words are ambiguous and can belong to more than one part of speech, this means that these vectors reflect the sum of the contextual behavior of a word's underlying parts of speech, i.e. they are mixtures of its different kinds of syntactical behavior. The difficult problem of separating the mixtures is addressed by the neural network. In subsequent work, the neural network is ambandoned in favor of combining several global vectors (Schütze (1995)). More recent studies try to solve the problem of ambiguity by combining distributional with morphological information (Clark (2003), Freitag (2004)), or by demixing the global vectors by means of vector projection (Rapp (2006)).

In this paper we rise the question if it is really necessary to use an approach based on mixtures or if there is some way to avoid the mixing beforehand. For this purpose, we suggest to look at local contexts instead of global co occurrence vectors. As can be seen from human performance, in almost all cases the local context of an ambiguous word is sufficient to disambiguate its part of speech. The aim of this study is to show how this observation can be successfully exploited for part-of-speech induction.

The core assumption underlying our approach is that words of a particular part of speech often have the same left and right neighbors, i.e. a pair of such neighbors can be considered to be characteristic of a part of speech. For example, a noun may be surrounded by the pair "the ... is" a verb by the pair "he ... the", and an adjective by the pair "the ... thing". For ease of reference, in the remainder of this paper we call these local contexts *neighbor pairs*. The idea is now to cluster the neighbor pairs on the basis of the middle words they occur with. This way neighbor pairs typical of the same part of speech are grouped together. For classification, a word is assigned to that cluster where its neighbor pairs are found. If its neighbor pairs are spread over several clusters, the word can be assumed to be ambiguous. This way ambiguity resolution follows naturally from the methodology, without any sophisticated demixing.

Although this is not in the focus of the paper, let us mention that the method is not only suitable for part-of-speech induction, but also for part-of-speech tagging. If a new word token is to be tagged, we simply compute the distance of its neighbor pair to each of the cluster centroids, and assign it to the part of speech implied by the closest cluster.

2 Approach

Our algorithm comprises the following steps:

- select neighbor pairs with high corpus frequency
- count co-occurrence frequencies of neighbor pairs and middle words; create matrix (rows = pairs, columns = middle words), normalize columns
- K-means clustering of rows (clusters = discovered parts of speech)
- sum up column values that belong to the same cluster; interpret sums as likelihood of a certain word occurring as a particular part of speech

Let us illustrate this by looking at the example in Table 1. The rows are the neighbor pairs and the columns are middle words as observed in a corpus. Most words in our example are syntactically unambiguous. Only *link* can be either a noun or a verb and therefore shows the co-occurrence patterns of both. Apart from the particular choice of features, what distinguishes our approach from others is that we do not cluster the words (columns) which would be the more straightforward thing to do. Instead we cluster the neighbor pairs (rows). Clustering the columns would be fine for unambiguous words, but has the drawback that (with common algorithms) ambiguous words are only assigned to the cluster relating to their dominant part of speech. This means that no disambiguation takes place at this stage. Although we previously showed that disambiguation can be conducted in a subsequent step which involves vector projection (Rapp (2006)), this way of demixing is difficult and error prone.

Table 1. Matrix of neighbor pairs and their corresponding middle words.

	car	cup	drink	link	quick	seek	tall	thin
a ... has	•	•		•				
a ... is	•	•		•				
a ... boy					•		•	•
a ... girl					•		•	•
the ... has	•	•		•				
the ... is	•	•		•				
the ... boy					•		•	•
the ... girl					•		•	•
to ... a			•	•		•		
to ... the			•	•		•		
you ... a			•	•		•		
you ... the			•	•		•		

Table 2. Clusters of neighbor pairs.

	car	cup	drink	link	quick	seek	tall	thin
a ... has, a ... is, the ... has, the ... is	•	•		•				
a ... boy, a ... girl, the ... boy, the ... girl					•		•	•
to ... a, to ... the, you ... a, you ... the			•	•		•		

The problem of demixing can be avoided by clustering the rows. This may lead, for example, to the condensed representation shown in Table 2. The neighbor pairs have been grouped in such a way that the clusters correspond to classes that can be linguistically interpreted as nouns, adjectives, and verbs. As desired, all unambiguous words have been assigned to only a single cluster, and only the ambiguous word *link* has been assigned to two clusters.

Although it is not obvious from our example, there is a drawback with this approach: By avoiding the ambiguity problem for words we introduce it for the neighbor pairs, i.e. ambiguities concerning neighbor pairs are not resolved. An example is the neighbor pair "then ... comes", where the middle word can either be a personal pronoun like *he* or a proper noun like *John*. However, we believe that this is a problem that for several reasons is of less importance: Firstly, we are not explicitly interested in the ambiguities of neighbor pairs. Secondly, the ambiguities of neighbor pairs seem less frequent and less systematic than those of words (an example is the very common noun - verb ambiguity in English), and therefore chances of misclustering are lower. Thirdly, this problem can be reduced by considering longer contexts (e.g. ± 2 words) which tend to be far less ambiguous. That is, by choosing an appropriate context width the best tradeoff between data sparseness and ambiguity reduction can be chosen. In preliminary experiments where we achieved the most promising results using a context width of ± 1 one word.

3 Implementation

Our computations are based on the British National Corpus which is a balanced sample of written and spoken English comprising about 100 million words. As the number of word types and neighbor pairs is prohibitively high in a corpus of this size, we restricted ourselves to a selected vocabulary, as detailed in section 4. From all neighbor pairs we chose the top 2000 which had the highest co-occurrence frequency with the union of all words in the vocabulary. Note, however, that neighbor pairs containing punctuation marks or special characters had been eliminated beforehand, as Schütze (1995) pointed out that punctuation marks are rather unspecific and give only little information regarding the preceding or succeeding word. Since punctuation marks have high corpus frequencies, without this heuristic many of the top 2000 neighbor pairs would include them, and results would be adversely affected.

By searching through the full corpus, we constructed a matrix as exemplified in Table 1. However, as a large corpus may contain errors and idiosyncrasies, the matrix cells were not filled with binary yes/no decisions, but with the frequency of a word type occurring as the middle word of the respective neighbor pair. Note that we used raw co-occurrence frequencies and did not apply any association measure. However, to account for the potentially large variation in word frequency and to give an equal chance to each word in the subsequent computations, the matrix columns were normalized. This implies that the rows can not be normalized, which is no problem since our vector similarity measure does not require normalization.

As our method for grouping the rows we used K-means clustering as provided by Matlab 7 with the cosine coefficient as our similarity measure. The cosine coefficient computes the cosine of the angle between two vectors, that is it only takes the directions of the vectors into account but not their lengths. The clustering algorithm was started using random initialization. Since random initialization can occasionally lead to poor results as K-means clustering can easily get stuck in local optima, we replicated the clustering process 50 times and automatically selected the best result which is the one with the smallest sum of point-to-centroid distances. In order to be able to easily compare the clustering results with expectation, the number of clusters was specified to correspond to the number of expected word classes. That is, for this particular purpose we considered it an advantage that K-means clustering requires the number of clusters to be predetermined. Of course, for other applications the number of clusters may be unknown. In this case the results for a range of cluster counts could be compared using silhouette plots, whereby the best result is indicated by the largest overall silhouette value. Alternatively, a clustering method could be applied that automatically determines the number of clusters.

After the clustering is complete, to obtain their centroids in analogy to Table 2 the column vectors for each cluster are summed up. The centroid values for each word can now be interpreted as evidence of this word belonging to the class described by the respective cluster. For example, if we obtained three clusters corresponding to nouns, verbs, and adjectives, and if the corresponding centroid values e.g. for the word *link* would be 0.7, 0.3, and 0.0, this could be interpreted such that in 70% of its corpus occurrences *link* has the function of a noun, in 30% of the cases it appears as a verb, and that it never occurs as an adjective. Note that the centroid values for a particular word will always add up to 1 since, as mentioned above, the column vectors had been normalized beforehand.

Another useful application of the centroid vectors is that they allow us to judge the quality of our neighbor pairs with respect to their selectivity regarding the particular word class. If the row vector of a neighbor pair is very similar to the centroid of its cluster, then it can be assumed that this neighbor pair only accepts middle words of the correct class, whereas neighbor pairs with lower similarity to the centroid are probably less selective, i.e. they occasionally allow for words from other clusters. As a measure of a neighbor pair's selectivity we use the cosine distance which has a range between zero and one. Hereby a value close to zero indicates a high similarity (low distance) to the centroid and consequently high selectivity, whereas a value close to one means low similarity to the centroid and low selectivity.

4 Results

Word classification in general, especially for function words, has been under dispute among linguists for centuries, and a definitive solution is still not agreed upon. Given these difficulties as well as severe computational limitations, asking our system to process an unrestricted vocabulary would be rather ambitious. Also, evaluating the outcome of the system would not be trivial in the general case. Therefore, to make the task easier, we selected a sample vocabulary of 50 words where the number of expected word classes is limited and where the results can be easily judged by any speaker of English. The list of words is included in Table 3 (columns 1 and 5). They are common words that were selected from the mid frequency range (around 5000) of the British National Corpus.

The other columns of Table 3 show the centroid values corresponding to each word after the procedure described in the previous section had been conducted, that is, the 2000 most frequent neighbor pairs of the 50 words were clustered into three groups. For clarity, all values have been multiplied by 1000 and rounded. Therefore, apart from rounding errors, the values for each word add up to 1000.

Instead of naming each cluster by a number or by specifying the corresponding long list of neighbor pairs (as shown in Table 2), in Table 3 we manually chose linguistically motivated names, namely *noun*, *verb*, and *adjective*. Note, however, that this has only been done to make the interpretation of the results easier. Obviously, the selection of such meaningful names can not be the output of an unsupervised algorithm.

If we look at Table 3, we find that some words, such as *encourage*, *imagine*, and *option*, have one value close to 1000, with the other two values in the one digit range. This is a typical pattern for unambiguous words that belong to only a single word class. However, perhaps unexpectedly, the majority of words has values in the upper two digit or three digit range in two or even three columns. This means that according to our system most words seem to be ambiguous in one or another way. For example, the word *brief*, although in the majority of cases clearly an adjective in the sense of *short*, can occasionally also occur as a noun (in the sense of *document*) or a verb (in the sense of *to instruct somebody*). In other cases, the occurrences of different parts of speech are more balanced; an example is the verb *to strike* versus the noun *the strike*.

Table 3. List of 50 words and their values (scaled by 1000) from each of the three cluster centroids.

	Noun	Verb	Adj.		Noun	Verb	Adj.
accident	978	8	15	lunch	741	198	60
belief	972	17	11	maintain	4	993	3
birth	968	15	18	occur	15	973	13
breath	946	21	33	option	984	10	7
brief	132	50	819	pleasure	931	16	54
broad	59	7	934	protect	4	995	1
busy	22	22	956	prove	5	989	6
catch	71	920	9	quick	47	14	938
critical	51	13	936	rain	881	64	56
cup	957	23	21	reform	756	221	23
dangerous	37	29	934	rural	66	13	921
discuss	3	991	5	screen	842	126	32
drop	334	643	24	seek	8	955	37
drug	944	10	46	serve	20	958	22
empty	48	187	765	slow	43	141	816
encourage	7	990	3	spring	792	130	78
establish	2	995	2	strike	544	424	32
expensive	55	14	931	suit	200	789	11
familiar	42	17	941	surprise	818	141	41
finance	483	473	44	tape	868	109	23
grow	15	973	12	thank	14	983	3
imagine	4	993	4	thin	32	58	912
introduction	989	0	11	tiny	27	1	971
link	667	311	23	widely	9	4	988
lovely	41	7	952	wild	220	6	774

According to our judgment, the values for all words seem roughly plausible. Only the values for *rain* as a noun versus a verb seemed on first glance counterintuitive, as we had expected the noun and verb readings to be at similar levels, which is not the case. Note, however, that for semantic reasons the uninflected verb form *rain* (as opposed to *rains*) can normally only occur as an infinitive, whereas base forms of most other verbs can also occur as first and second person singular and as first, second and third person plural. This means that these possibilities can not contribute to the appearance of *rain* as a verb, which explains its low observed frequency.

Having successfully assigned words to parts of speech, let us now consider the question of which neighbor pairs are especially selective regarding a particular class. As described in the previous section, the answer to this question can be obtained by looking at the cosine distances between each neighbor pair and the centroid of its cluster. Table 4 lists the top 10 neighbor pairs for each cluster. As can be seen, for most pairs it is syntactically not possible to insert an inappropriate middle word. This means that our hypothesis that the cluster centroid stands for high selectivity has been empirically confirmed.

Table 4. Top 10 most selective neighbor pairs for nouns, verbs, and adjectives.

	Cluster 1: Nouns (609 items)		Cluster 2: Verbs (723 items)		Cluster 3: Adjectives (668 items)	
Rank	Distance	Neighbor pair	Distance	Neighbor pair	Distance	Neighbor pair
1	0.269	the ... she	0.161	could ... their	0.145	as ... as
2	0.273	the ... he	0.197	to ... the	0.168	being ... and
3	0.290	the ... in	0.215	to ... some	0.180	so ... as
4	0.296	of ... and	0.226	to ... his	0.228	a ... one
5	0.327	the ... was	0.228	they ... the	0.231	are ... in
6	0.331	that ... is	0.231	not ... the	0.249	be ... and
7	0.338	of ... as	0.239	not ... a	0.252	very ... one
8	0.343	of ... that	0.242	must ... the	0.253	so ... that
9	0.344	the ... and	0.245	to ... both	0.257	is ... and
10	0.349	of ... or	0.248	to ... my	0.276	is ... but

5 Summary, conclusions and future work

A novel statistical approach has been presented which clusters contextual features (neighbor pairs) as observed in a large text corpus and derives syntactically oriented word classes from the clusters. In addition, for each word a probability of its occurrence as a member of one of these classes is computed. This is a feature that is usually not provided by other systems, but which could make a difference for possible applications such as lexicography, part-of-speech tagging, syntax parsing, or machine translation. Although further evaluation is required, qualitative inspection showed that the computed probabilities seem to at least roughly agree with human intuition. At the time of writing, we are not aware of any other work leading to comparable quantitative results.

On the plus side, the algorithm can be extended from part-of-speech induction to part-of-speech tagging in a straightforward manner, as the contextual features required for tagging are explicitly assigned to parts of speech. It can also be easily turned from unsupervised to supervised learning by simply providing some sample words for each class and by collecting appropriate features from a corpus. Note that neighbor pairs sharing not even a single middle word (such as "a ... is" and "an ... is") can nevertheless end up in the same cluster, as other features (e.g. "the ... is") can function as mediators.

On the negative side, the feature vectors to be clustered tend to be rather sparse, which may lead to difficulties in finding the best clusters, especially if many word classes are to be distinguished. Also, it has not been shown that the features considered are well suited not only for content but also for function words. As function words are usually surrounded by content words which tend to be too rare to produce reliable patterns of neighbor pairs, it is possible that the method described in Rapp (2006), which by clustering words instead of features is in some sense complementary to the approach described here, may be better suited for clustering function words.

Remaining questions include the following: Can singular value decomposition (to be in effect only temporarily for clustering) reduce the problem of data sparseness? Can 2-dimensional clustering (i.e. the stepwise reduction of both rows and columns of the matrix) possibly lead to better clusters? Should other features be explored? For example longer contexts or the four concatenated context vectors as proposed in Schütze (1995), namely the left and right context vector of the token under consideration, the right context vector of the previous word and the left context vector of the next word? Can the approach be extended to word sense induction by looking at longer distance equivalents to our neighbor pairs and their middle words (which would probably be pairs of strongly associated words and the products of their co-occurrence vectors)? All these are strands of research that we look forward to explore.

References

CLARK, A. (2003): Combining Distributional and Morphological Information for Part of Speech Induction. *Proceedings of 10th EACL*, Budapest, 59–66.

FREITAG, D. (2004): Toward Unsupervised Whole-corpus Tagging. *Proceedings of 20th COLING*, Geneva, 357–363.

RAPP, R. (2006): Part-of-speech Induction by Singular Value Decomposition and Hierarchical Clustering. In: M. Spiliopoulou, R. Kruse, C. Borgelt, A. Nürnberger and W. Gaul (Eds.) *From Data and Information Analysis to Knowledge Engineering. Proceedings of the 29th Annual Conference of the GfKl, Magdeburg 2005.* Springer, Berlin, 422–429.

SCHÜTZE, H. (1993): Part-of-speech Induction from Scratch. *Proceedings of 31st ACL, Columbus*, Ohio, 251–258.

SCHÜTZE, H. (1995): Distributional Part-of-speech Tagging. *Proceedings of 7th EACL, Dublin*, Ireland, 141–148.

Part X

Statistical Musicology and Sound Classification

A Probabilistic Framework for Audio-Based Tonal Key and Chord Recognition

Benoit Catteau[1], Jean-Pierre Martens[1] and Marc Leman[2]

[1] ELIS - Electronics & Information Systems, Ghent University, Gent, Belgium; Benoit.Catteau@elis.UGent.be
[2] IPEM - Department of Musicology, Ghent University, Gent, Belgium; Marc.Leman@UGent.be

Abstract. A unified probabilistic framework for audio-based chord and tonal key recognition is described and evaluated. The proposed framework embodies an acoustic observation likelihood model and key & chord transition models. It is shown how to conceive these models and how to use music theory to link key/chord transition probabilities to perceptual similarities between keys/chords. The advantage of a theory based model is that it does not require any training, and consequently, that its performance is not affected by the quality of the available training data.

1 Introduction

Tonal key and chord recognition from audio are important steps towards the construction of a mid-level representation of Western tonal music for e.g. Music Information Retrieval (MIR) applications.

A straightforward approach to key recognition (e.g., Pauws (2004), Leman (2000)) is to represent the acoustic observations and the keys by chroma vectors and chroma profiles respectively, and to use an ad hoc distance measure to assess how well the observations match a suggested key profile. Well-known profiles are the Krumhansl and Kessler (1982) and Temperley (1999) profiles, and a popular distance measure is the cosine distance.

The classical approach to chord recognition is one of key detection before chord recognition. Recently however, Shenoy and Wang (2005) proposed a 3-step algorithm performing chord detection first, key detection then, and finally, chord enhancement on the basis of high-level knowledge and key information.

Our point of departure is that tonal key and chord recognition should preferably be accomplished simultaneously on the basis of a unified probabilistic framework. We will propose a segment-based framework that extends e.g. the frame-based HMM framework for chord detection proposed by Bello and Pickens (2005).

In the subsequent sections we provide a general outline (Section 2) and a detailed description (Sections 3 and 4) of our approach, as well as an experimental evaluation (Section 5) of our current implementation of this approach.

2 General outline of the approach

Before introducing our probabilistic framework, we want to recall some basics about the links between notes, chords and keys in Western tonal music.

The pitch of a periodic sound is usually mapped to a pitch class (a chroma) collecting all the pitches that are in an octave relation to each other. Chroma's are represented on a log-frequency scale of 1 octave long, and this chromatic scale is divided into 12 equal intervals, the borders of which are labeled as notes: A, As, B, .. , Gs.

A tonal key is represented by 7 eligible notes selected from the set of 12. Characteristics of a key are its tonic (the note with the lowest chroma) and the mode (major, minor harmonic, ..) that was used to select the 7 notes starting from the tonic.

A chord refers to a stack of three (=triad) or more notes sounding together during some time. It can be represented by a 12-bit binary chromatic vector with ones on the chord note positions and zeroes on the remaining positions. This vector leads to a unique chord label as soon as the key is available.

Having explained the links between keys, chords and notes, we can now present our probabilistic framework. We suppose that an acoustic front-end has converted the audio into a sequence of N events which are presumed to represent individual chords. Each event is characterized by an acoustic observation vector $\mathbf{x}_n$, and the whole observation sequence is denoted as $\mathbf{X} = \{\mathbf{x}_1, .., \mathbf{x}_N\}$. The aim is now to assign key labels k_n and chord chroma vectors $\mathbf{c}_n$ to these events. More precisely, we seek for the sequence pair $(\hat{\mathbf{K}}, \hat{\mathbf{C}})$ that maximizes the posterior probability $P(\mathbf{K}, \mathbf{C}|\mathbf{X})$. By applying Bayes' law, and by noting that the prior probability $P(\mathbf{X})$ is independent of $(\mathbf{K},\mathbf{C})$, one comes to the conclusion that the problem can also be formulated as

$$\hat{\mathbf{K}}, \hat{\mathbf{C}} = \arg\max_{\mathbf{K},\mathbf{C}} P(\mathbf{K}, \mathbf{C}, \mathbf{X}) = \arg\max_{\mathbf{K},\mathbf{C}} P(\mathbf{K}, \mathbf{C})\, P(\mathbf{X}|\mathbf{K}, \mathbf{C}) \tag{1}$$

By sticking to two key modes, namely the major and minor harmonic mode, and by only examining the 4 most important triads (major, minor, augmented and diminished) per tonic we achieve that only 48 chord vectors and 24 keys per event have to be tested. If we can then assume that the acoustic likelihood $P(\mathbf{X}|\mathbf{K}, \mathbf{C})$ can be factorized as

$$P(\mathbf{X}|\mathbf{K}, \mathbf{C}) = \prod_{n=1}^{N} P(\mathbf{x}_n|k_n, \mathbf{c}_n) \tag{2}$$

and if $P(\mathbf{K}, \mathbf{C})$ can be modeled by the following bigram music model

$$P(\mathbf{K}, \mathbf{C}) = \prod_{n=1}^{N} P(k_n, \mathbf{c}_n | k_{n-1}, \mathbf{c}_{n-1}) \quad (3)$$

then it is straightforward to show that the problem can be reformulated as

$$\hat{\mathbf{K}}, \hat{\mathbf{C}} = \arg\max_{\mathbf{K},\mathbf{C}} \prod_{n=1}^{N} P(k_n|k_{n-1}, \mathbf{c}_{n-1}) P(\mathbf{c}_n|k_{n-1}, \mathbf{c}_{n-1}, k_n) P(\mathbf{x}_n|k_n, \mathbf{c}_n) \quad (4)$$

The solution can be found by means of a Dynamic Programming search.

In the subsequent two sections we describe the front-end that was used to construct the acoustic observations and the models that were developed to compute the probabilities involved.

3 The acoustic front-end

The objective of the acoustic front-end is to segment the audio into chord and rest intervals and to create a chroma vector for each chord interval.

3.1 Frame-by-frame analysis

The front-end first performs a frame-by-frame short-time power spectrum (STPS) analysis. The frames are 150 ms long and two subsequent frames overlap by 130 ms. The frames are Hamming windowed and the STPS is computed in 1024 points equidistantly spaced on a linear frequency scale.

The STPS is then mapped to a log-frequency spectrum comprising 84 samples: 7 octaves (between the MIDI scores C1 and C8) and 12 samples per octave. By convolving this spectrum with a Hamming window of 1 octave wide, one obtains a so-called background spectrum. Subtracting this from the original spectrum leads to an enhanced log-frequency spectrum. By means of sub-harmonic summation (Terhardt et al. (1982)), the latter is converted to a sub-harmonic sum spectrum $T(i)$, $i = 0, ..83$ which is finally folded into one octave to yield the components of the chroma vector $\mathbf{x}$ of the analyzed frame:

$$x_m = \sum_{j=0}^{6} T_n(12j + m), \qquad m = 0, .., 11 \quad (5)$$

3.2 Segmentation

The chroma vectors of the individual frames are used to perform a segmentation of the audio signal. A frame can either be appended to a previously started event or it can be assigned to a new event. The latter happens if the absolute value of the correlation between consecutive chroma vectors drops below a certain threshold.

On the basis of its mean frame energy each event is labeled as chord or rest and for each chord, a chroma vector is computed by first taking the mean chroma vector over its frames, and by then normalizing this mean vector so as to achieve that its elements sum up to 1.

4 Modeling the probabilities

For solving Equation 4, one needs good models for the observation likelihoods $P(\mathbf{x}_n|k_n, \mathbf{c}_n)$, the key transition probabilities $P(k_n|k_{n-1}, \mathbf{c}_{n-1})$ and the chord transition probabilities $P(\mathbf{c}_n|k_{n-1}, \mathbf{c}_{n-1}, k_n)$.

4.1 Modeling the observation likelihoods

The observation likelihood expresses how well the observations support a proposed chord hypothesis. Although they sum up to one, we assume weak dependencies among the vector components and propose to use the following model:

$$P(\mathbf{x}_n|k_n, \mathbf{c}_n) = \prod_{m=0}^{11} P(x_{nm}|c_{nm}), \qquad \sum_{m=0}^{11} x_{nm} = 1 \tag{6}$$

In its most simple form this model requires two statistical distributions: $P(x|c = 1)$ and $P(x|c = 0)$ (x and c denote individual notes here). We have chosen for

$$P(x|0) = G_o\,(e^{-\frac{x^2}{2\sigma^2}} + P_o) \qquad x \in (0, 1) \tag{7}$$

$$P(x|1) = G_1\,(e^{-\frac{(x-X)^2}{2\sigma^2}} + P_o) \qquad x \in (0, X) \tag{8}$$

$$= G_1\,(1 + P_o) \qquad x \in (X, 1) \tag{9}$$

(see Figure 1) with G_o and G_1 being normalization factors. Offset P_o must preserve some evidence in case an expected large x_{nm} is missing or an unexpected large x_{nm} (e.g. caused by an odd harmonic of the pitch) is present. In our experiments X and σ were kept fixed to 0.33 and 0.13 respectively (these values seem to explain the observation statistics).

4.2 Modeling the key transition probabilities

Normally it would take a large chord and key annotated music corpus to determine appropriate key and chord transition probabilities. However, we argue that (1) transitions between similar keys/chords are more likely to occur

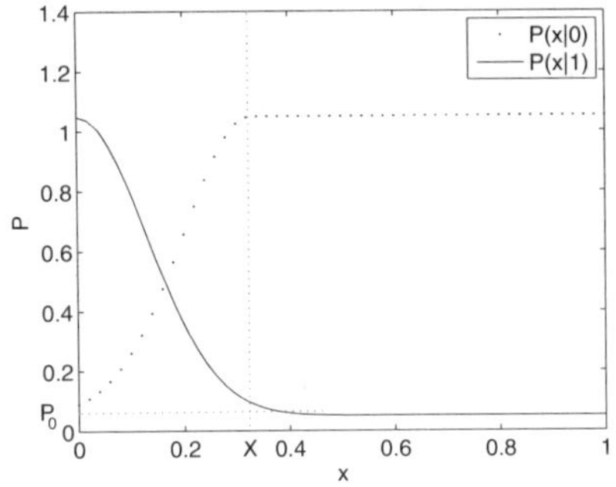

Fig. 1. Distributions (without normalization factors) to model the observation likelihoods of x given that the note chroma vector contains a $c = 1$ or $c = 0$.

than transitions between less similar keys/chords, and (2) chords comprising the key tonic or fifth are more likely to appear than others. We therefore propose to retrieve the requested probabilities from music theory and to avoid the need for a labeled training database.

Lerdahl (2001) has proposed a three-dimensional representation of the tonal space and a scheme for quantizing the perceptual differences between chords as well as keys. Lerdahl distinguishes five note levels, namely the chromatic, diatonic, triadic, fifth and tonic levels and he accumulates the differences observed at all these levels in a distance metric.

If we can assume that in the case of a key modulation the probability of k_n is dominated by the distance $d(k_n, k_{n-1})$ emerging from Lerdahl's theory, then we can propose the following model:

$$P(k_n|k_{n-1}, \mathbf{c}_{n-1}) = P_{os} \qquad k_n = k_{n-1} \tag{10}$$

$$= \beta_s \, e^{-\frac{d(k_n, k_{n-1})}{d_s}} \qquad k_n \neq k_{n-1} \tag{11}$$

with β_s being a normalization factor and $d_s = 15$, the mean distance between keys. By changing P_{os} we can control the chance of hypothesizing a key modulation.

4.3 Modeling the chord transition probabilities

For computing these probabilities we rely on the distances between diatonic chords (= chords solely composed of notes that fit into the key) as they follow from Lerdahl's theory, and on the tonicity of the chord. Reserving some probability mass for transitions to non-diatonic chords we obtain

$$P(\mathbf{c}_n|\mathbf{c}_{n-1}, k_n, k_{n-1}) = P_{oc} \qquad \mathbf{c}_n = \text{non-diatonic in } k_n \tag{12}$$

$$= \beta_c \, e^{-\frac{d(\mathbf{c}_n, \mathbf{c}_{n-1})}{d_c}} \, g(\mathbf{c}_n, k_n) \quad \mathbf{c}_n = \text{diatonic in } k_n \tag{13}$$

as a model. β_c is a normalizaton factor, $d_c = 6$ (the mean distance between chord vectors) and $g(\mathbf{c}_n, k_n)$ is a model that favors chords comprising the key tonic ($g = 1.5$) or fifth ($g = 1.25$) over others ($g = 1$). By changing P_{oc} we can control the chance of hypothesizing a non-diatonic chord.

5 Experimental results

For parameter tuning and system evaluation we have used four databases.
Cadences. A set of 144 files: 3 classical cadences times 24 keys (12 major and 12 minor keys) times 2 synthesis methods (Shepard tones and MIDI-to-wave).
Modulations. A set of 20 files: 10 chord sequences of length 9 (copied from Krumhansl and Kessler (1982)) times 2 synthesis methods. All sequences start in C major or C minor and on music theoretical grounds a unique key can be assigned to each chord. Eight sequences show a key modulation at position 5, the other two do not, but they explore chords on various degrees.
Real audio. A set of 10 polyphonic audio fragments (60 seconds) from 10 different songs (see Table 1). Each fragment was chord and key labeled.
MIREX. A set of 96 MIDI-to-wave synthesized fragments: compiled as a training database for the systems participating in the MIREX-2005 key detection contest. Each fragment was supplied with one key label. In case of modulation it is supposed to represent the dominant key for that fragment.

Table 1. The test songs and their key.

	Artist	Title	Key
1	CCR	Proud Mary	D Major
2	CCR	Who'll stop the rain	G Major
3	CCR	Bad moon rising	D Major
4	America	Horse with no name	E Minor
5	Dolly Parton	Jolene	Cs Minor
6	Toto Cutugno	L'Italiano	A Minor
7	Iggy Pop	The passenger	A Minor
8	Marco Borsato	Dromen zijn bedrog	C Minor
9	Live	I Alone	Gb Major → Eb Major
10	Ian McCulloch	Sliding	C Major

5.1 Free parameter tuning

In order to tune the free parameters (P_o, P_{os}, P_{oc}) we worked on all the cadences and modulation sequences and one song from the real audio database. Since P_{os} and P_{oc} were anticipated to be the most critical parameters we explored them first in combination with $P_o = 0.1$. There is a reasonably large area in the (P_{os}, P_{oc})-plane where the performances on all the tuning data are good and stable ($0.3 < P_{os} < 0.5$ and $0 \leq P_{oc} < 0.2$). We have chosen for $P_{os} = 0.4$ and $P_{oc} = 0.15$ to get a fair chance of selecting key modulations and non-diatonic chords when present in the audio. For these values we got 100%, 96.7% and 92.1% of correct key labels for the cadences, the modulation sequences and the song. The corresponding correct chord label percentages were 100%, 93.8% and 73.7%. Changing P_o did not cause any further improvement.

5.2 System evaluation

Real audio. For real audio we have measured the percentages of deleted reference chords (D), inserted chords (I), frames with the correct key label (C_k) and frames with the correct chord label (C_c). We obtained $D = 4.3\%, I = 82\%, C_k = 51.2\%$ and $C_c = 75.7\%$. An illustration of the reference and computed labels for song 1 is shown on Figure 2.

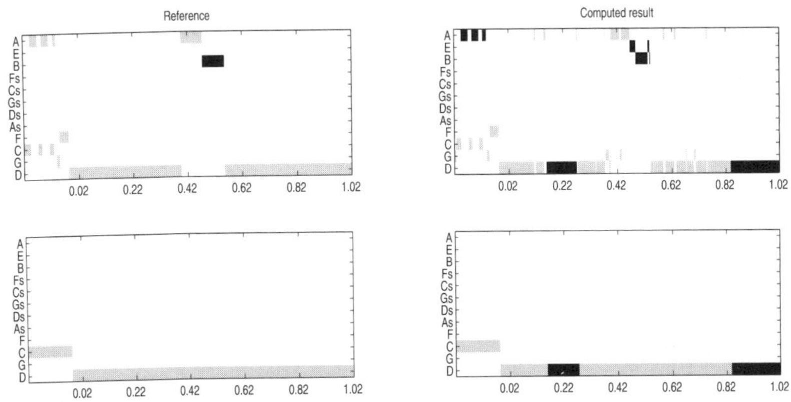

Fig. 2. Annotated (left) and computed (right) chords (top) and keys (bottom) for song 1. The grey zones refer to major and the black ones to minor labels.

A first observation is that our system produces a lot of chord insertions. This must be investigated in more detail, but possibly the annotator discarded some of the short chord changes.

A second observation is that the key accuracy is rather low. However, a closer analysis showed that more than 60% of the key errors were confusions between a minor and its relative major. Another 15% were confusions between keys whose tonics differ by a fifth. By applying a weighted error measure as recommended by MIREX (weights of 0.3 for minor to relative major, 0.5 for a tonic difference of a fifth, and 1 otherwise) we obtain a key accuracy of 75.5%.

Our chord recognition results seem to be very good. Without chord enhancement on the basis of high-level musical knowledge (this knowledge can also be applied on our system outputs) Shenoy and Wang (2005) report a chord accuracy of 48%. Although there are differences in the data set, the assumptions made by the system (e.g. fixed key) and the evaluation procedure, we believe that the above figure supports our claim that simultaneous chord and key labeling can outperform a cascaded approach.

MIREX data. Since we did not participate in the MIREX contest, we only had access to the MIREX training set and not to the evaluation set. However since we did not perform any parameter tuning on this set, we believe that the results of our system on the MIREX training set are representative of thoses we would be able to attain on the MIREX evaluation set.

Using the recommended MIREX evaluation approach we obtained a key accuracy of 83%. The best result reported in the MIREX contest İzmirli (2005) was 89.5%. We hope that by further refining our models we will soon be able to bridge the gap with that performance.

6 Summary and conclusion

We have proposed a segment-based probabilistic framework for the simultaneous recognition of chords and keys. The framework incorporates a novel observation likelihood model and key & chord transition models that were not trained but derived from the tonal space theory of Lerdahl.

Our system was evaluated on real audio fragments and on MIDI-to-wave synthesized chord sequences (MIREX-2005 contest data). Apparently, real audio is hard to process correctly, but nevertheless our system does appear to outperform its counterparts in advanced chord labeling systems that have recently been developed by others. The key labeling results for the MIREX data are also very good and already close to the best results previously reported for these data.

References

BELLO, J.P. and PICKENS, J. (2005): A Robust Mid-level Representation for Harmonic Content in Music Signals. In: *Procs 6th Int. Conference on Music Information Retrieval (ISMIR 2005)*. London, 304–311.

İZMIRLI, Ö. (2005): Tonal Similarity from Audio Using a Template Based Attractor Model. In: *Procs 6th Int. Conference on Music Information Retrieval (ISMIR 2005)*. London, 540–545.

KRUMHANSL, C. and KESSLER, E. (1982): Tracing the Dynamic Changes in Perceived Tonal Organization in a Spatial Representation of Musical Keys. *Psychological Review, 89, 334–368.*

LEMAN, M. (2000): An Auditory Model of the Role of Short-term Memory in Probe-tone Ratings. *Music Perception, 17, 435-464.*

LERDAHL, F. (2001): *Tonal Pitch Space.* Oxford University Press, New York.

PAUWS, S. (2004): Musical Key Extraction from Audio. In: *Proc. of the 5th Int. Conference on Music Information Retrieval (ISMIR 2004)*. Barcelona, 96–99.

SHENOY, A. and WANG, Y. (2005): Key, Chord, and Rhythm Tracking of Popular Music Recordings. *Computer Music Journal, 29, 3, 75–86.*

TEMPERLEY, D. (1999): What's Key for Key? The Krumhansl-Schmuckler Key-Finding Algorithm Reconsidered. *Music Perception, 17, 1, 65–100.*

TERHARDT, E., STOLL, G. and SEEWANN, M. (1982): Algorithm for Extraction of Pitch and Pitch Salience for Complex Tonal Signals. *Soc. Am., 71, 679–688.*

Using MCMC as a Stochastic Optimization Procedure for Monophonic and Polyphonic Sound

Katrin Sommer and Claus Weihs

Lehrstuhl für Computergestützte Statistik, Universität Dortmund,
D-44221 Dortmund; sommer@statistik.uni-dortmund.de

Abstract. Based on a model of Davy and Godsill (2002) we describe a general model for time series from monophonic and polyphonic musical sound to estimate the pitch. The model is a hierarchical Bayes Model which will be estimated with MCMC methods. For parameter estimation an MCMC based stochastic optimization is introduced. A comparative study illustrates usefullness of the MCMC algorithm.

1 Introduction

In this research paper we describe a general model for the estimation of pitch from monophonic and polyphonic musical time series data, based on a model from Davy and Godsill (2002). We will first discuss this hierarchical Bayes Model for the monophonic case. For parameter estimation an MCMC based stochastic optimization is used. Next we extend our model to the case of polyphonic muscial time series. In a study, different optimization procedures will be compared using real audio data from the McGill University Master Samples (Opolko and Wapnick (1987)). For monophonic samples, the estimation is almost perfect. We compare and contrast our model with two other models based on a heuristic estimator and a frequency domain procedure. Finally first results of the work with polyphonic sounds will be illustrated.

2 Harmonic model

In this section a harmonic model is introduced and its components are illustrated. The whole model which is based on the model from Davy and Godsill (2002) has the following structure:

$$y_t = \sum_{h=1}^{H} \sum_{i=0}^{I} \phi_{t,i} \left[a_{h,i} \cos(2\pi h t f_0 / f_s) + b_{h,i} \sin(2\pi h t f_0 / f_s) \right] + \varepsilon_t .$$

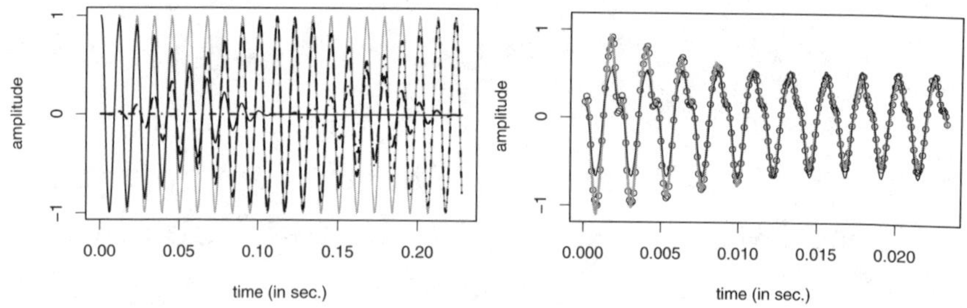

Fig. 1. Illustration of the modelling with basis functions. A cosine function summed up by 3 basis functions (left), modelling time-variant amplitudes of a real tone (right).

In this model one tone y_t is composed out of harmonics from H partial tones. The tone is normalized to [-1,1]. The number of observations is T, $t \in \{0, \ldots, T-1\}$. The first partial is the fundamental frequency f_0, the other $H-1$ partials are called overtones. The amplitudes of each partial tone are $a_{h,i}$ and $b_{h,i}$. Finally, f_s is the sampling rate and ε_t is the model error. The amplitudes of all partial tones are time-variant. They are modelled with so-called basis functions to avoid higher complexity.

In our model the basis functions $\phi_{t,i}$ are Hanning windows with 50% overlap. Hanning windows are shifted stretched squared cosine functions. The i-th basis function is defined as

$$\phi_{t,i} := \cos^2\left[\pi(t - i\Delta)/(2\Delta)\right] \mathbf{1}_{[(i-1)\Delta,(i+1)\Delta]}(t),$$
$$\Delta = (T-1)/I, \quad T = \text{no. of observations},$$

where $\mathbf{1}$ is the indicator function. In principle a basis function can be any non-oscillating function (Davy and Godsill (2002)).

In Figure 1, left, a cosine function (grey dotted line) is shown which can be modelled with three basis functions (black). The sum of the basis functions in one timepoint is 1. Figure 1, right, illustrates the difference of modelling a tone with oder without time-variant amplitudes. The points are the real tone. The assumption of constant amplitudes over time cannot depict the higher amplitudes at the beginning of the tone (black line). Modelling with time-variant amplitudes (grey dotted line) leads to better results.

3 Hierarchical Bayes-Model

In this section a hierarchical Bayes-Model is introduced for the parameters in the pitch model.

The likelihood of the tone, assuming normal distribution, can be written as

$$p(y \mid \theta, f_0, H, \sigma_\varepsilon^2) = \frac{1}{(2\pi\sigma_\varepsilon^2)^{T/2}} \exp\left[-\frac{1}{2\sigma_\varepsilon^2}(y - D\theta)^T(y - D\theta)\right],$$

where D is a matrix with sine and cosine entries multiplied by the values of the basis functions

$$D_{t+1,2(Hi+h)-1} = \phi_{t,i} \cos(hf_0/f_s t), \quad D_{t+1,2(Hi+h)} = \phi_{t,i} \sin(hf_0/f_s t).$$

For each time point and for each basis function the number of entries is 2 times the number of partial tones. So the matrix has the dimension $T \times 2H(I+1)$.

The priors in the Bayes-Model have a hierarchical structure

$$p(\theta, f_0, H, \sigma_\varepsilon^2) = p(\theta \mid f_0, H, \sigma_\varepsilon^2) p(f_0 \mid H) p(H) p(\sigma_\varepsilon^2).$$

The amplitudes $a_{h,i}$ and $b_{h,i}$ are combined in one amplitude-vector θ:

$$\theta_{2(Hi+m)-1} = a_{h,i}, \quad \theta_{2(Hi+m)} = b_{h,i}.$$

The following parameters determine the model: fundamental frequency f_0, number of partials H, parameter vector θ, and the predefined number of basis functions $I+1$. For these parameters we assume priors and hyperparameters which are chosen uninformative in order to obtain a flexible model for different instruments (Sommer and Weihs (2006)).

4 Stochastic optimization

The basic idea used here for parameter estimation, is to maximize the likelihood by stochastic search for the best coefficients in a given region with a given probability distribution. In standard MCMC methods (Gilks et al. (1996)) the distributions are fully generated, which leads to a heavy computational burden. As a short cut, we used an optimal model fit criterion instead. This means, that every 50 MCMC iterations we check whether linear regression of the last 50 residuals against the iteration number delivers a slope significant at a previously specified level. If this is not the case, we stop iterating. At most 2 000 iteratiosn are peformed.

Figure 2 shows the decreasing error with an increasing number of iterations. Between iteration number 350 and iteration number 400 the slope is no more significant at a level of 10 %.

5 Extension to polyphonic time series data

In this section the monophonic model will be extended to a polyphonic model

$$y_t = \sum_{k=1}^{K} \sum_{h=1}^{H} \sum_{i=0}^{I} \Phi_{t,i} \left\{ a_{k,h,i} \cos\left(2\pi h t f_0/f_s\right) + b_{k,h,i} \sin\left(2\pi h t f_0/f_s\right) \right\} + \varepsilon_t.$$

In this model an additional parameter K is added to represent the number of tones. For our purposes it is assumed that K is known. In future research the number of tones will be determined by a new MCMC-step in the MCMC-algorithm.

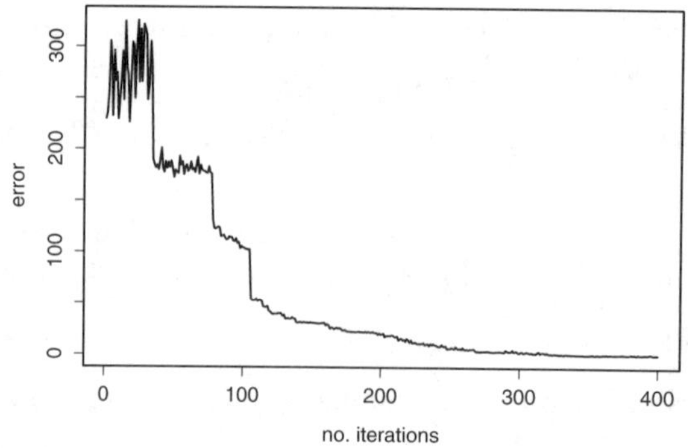

Fig. 2. Progression of model error

6 Comparative study

6.1 Data

In a comparative study we attempt to find optimal levels for some of the unknown parameters of the hierarchical Bayes model and the estimation procedure. The data used are real audio recordings of different musical instruments. We chose 5 instruments (flute, electric guitar, piano, trumpet and violin) each with 5 notes (220, 262, 440, 523 and 880 Hz) from the McGill University Master Samples (Opolko and Wapnick (1987)).

The instruments were chosen out of two groups, bowed instruments and wind instruments. There are three ways a bowed instrument can be played. They can be picked, bowed or stroke. We chose one instrument for each way. The two wind instruments can be distinguished as a woodwind instrument and a brass instrument.

The choice of tones was restricted by the availability of the data from the McGill database and the different ranges of the instruments.

For each tone $T = 512$ data points are sampled at 11 025 Hz. The sample size is a tradeoff between the computational burden and the quality of the estimate. For longer time series we use a sequence of overlapping intervals. Since the absolute overall loudness of the different recordings is not relevant, the time series data is normalized to the interval $[-1, 1]$.

6.2 Monophonic data

In the study we estimated all the unknown parameters of the hierarchical Bayes model except the number of basis functions which was fixed to 1 to 5, where one basis function implies constant amplitudes over time.

Some unknown parameters in the estimation algorithm are optimized in our study. The main aim was to find a good value for the stopping criterion. This is equivalent to finding a significance level for the linear regression which leads to good results and whether non-significance should be allowed more than once before stopping in order to avoid local maxima. The significance level was varied from 0.05 to 0.55 in steps of 0.05. Stopping criterion of reaching non-significance once, twice or three times where used.

Further more, 3 Markov chains with different frequency starting points are simulated. The frequency starting points are 175 Hz, 1230 Hz and the result ff_{Heur} of a Heuristic fundamental frequency estimation (Ligges et al. (2002)):

$$ff_{\text{Heur}} = h + \frac{s-h}{2}\sqrt{ds/dh},$$

where h is the peaking Fourier frequency, s is the peaking neighbor frequency. The corresponding density values are dh and ds. The chain started at ff_{Heur} is simulated with a Burn In of 200 iterations.

Overall, the design of the study leads to a full factorial design (5 instruments * 5 notes * 11 levels * 3 stops * 5 basis functions) with 4 125 experiments, each applied to 3 chains. In Table 1 there is an overview of all components of the design. Instruments, notes and starting points define the environment of the simulation. For the three parameters level, stops and number of basis functions we looked for the best level. Now the results of the study

Table 1. Factors and correspondending levels of the full factorial design

	factor	level
data	instruments	flute, electric guitar, piano, trumpet, violin
	notes	220, 262, 440, 523, 880 Hz
chains	starting points	175 Hz, 1230 Hz and ff_{Heur}
optimization	level	0.05, 0.1, ..., 0.55
	stops	1, 2, 3
	no. of basis functions	1, 2, ..., 5

are discussed. The deviation from the real tone is measured in cents. 100 cents are equivalent one halftone. Many estimates lie outside [-50,50] cents, the interval where one can assign the estimation to the correct tone. The number of incorrectly estimated tones decreases if the stopping criterion is met more than once. The Heuristic as starting point of the chain leads to best results of the three chains. Combining all chains by choosing the chain with minimal error leads to very good results. There are only few estimates outside the bound of [-50,50] cents. Most values are enclosed by the interval [-25,25] cents. Because of the combination of the chains, the number of times the stopping criterion is met is irrelevant.

There are almost no differences in the results for the different levels, if one chooses the chain and the number of basis functions which minimize the error. Most estimated values are lie within an interval of $[-25, 25]$ cents. Only three frequencies are incorrectly estimated. They are outside the interval of $[-50, 50]$ cents, by a significance level of 0.05 and accordingly 0.4. Therefore we deem a level of 0.1 as appropriate.

Using only one basis function, meaning constant amplitudes over time, leads to the highest number of incorrectly estimated fundamental frequencies. The modulation with time-variant amplitudes generally lead to far better results with an optimum for 3 basis functions. In Table 2 the number of deviations from the fundamental frequencies bigger than 50 cents are shown.

Table 2. Number of deviations larger than 50 cents

	no. basis functions				
stop	1	2	3	4	5
1	21	3	0	8	2
2	11	1	1	2	1
3	8	2	1	0	1

Now the results of the study are compared with two other optimization procedures to judge the MCMC algorithm. These two procedures are the Heuristic procedure described above and a frequency domain procedure, which is optimized with a Nelder-Mead optimization (Nelder and Mead (1965)). 25 experiments (5 instruments * 5 notes) are estimated with these two methods.

In the model there are included only the first three partials whereas the amplitude of the first partial is set to 1. There are no basis functions in the model, constant amplitudes over the time are assumed. Starting points of the optimization are

$$ff = ff_{\text{Heur}} + \{2, 0, -2\}\text{Hz} \qquad B_2 = 0.5 \qquad B_3 = 0.5$$

The stopping criterion is the default stopping criterion of the R function `optim()` (R Core (2005)), but the maximum number of iterations is set to 5 000 (Weihs and Ligges (2006)).

For the MCMC-algorithm a significance level of 0.1 and "meet non-significance once" is chosen as the stopping criterion. In Figure 3 one can see that the MCMC optimization yields perfect results within the interval of $[-25, 25]$ cents. The Nelder-Mead optimization and the Heuristic both result in one estimation outside $[-50, 50]$ cents. Considering the median leads to the finding, that the frequency domain procedure estimates the median perfect while the MCMC optimization slightly overestimates and the Heuristic underestimates the true pitch.

6.3 Polyphonic data

As database for the study with polyphonic data the McGill audio data has been used again. We chose the same 5 instruments, but other pitches (262,

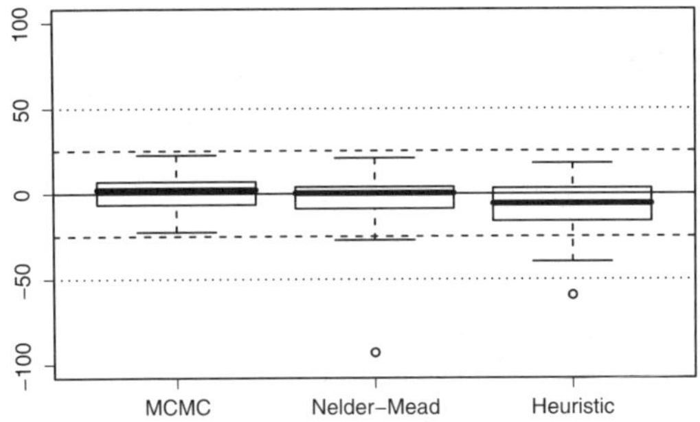

Fig. 3. Boxplots of deviations of estimated fundamental frequency of the optimization procedures

Table 3. 1 if both tones were correctly identified, 0 otherwise. The left hand table requires the exact tone to be estimated, the right table also counts multiples of the tone as correct.

	instrument				
tone	flu	guit	pian	trum	viol
c4	1	1	1	1	1
e4	0	1	0	0	1
g4	0	0	0	0	0
a4	1	1	1	0	0
c5	1	1	1	1	1

	instrument				
tone	flu	guit	pian	trum	viol
c4	1	1	1	1	1
e4	0	1	0	1	1
g4	1	1	1	1	1
a4	1	1	1	0	0
c5	1	1	1	1	1

330, 392, 440 and 523 Hz). The corresponding tones lie within the same octave. The first tone is chosen to be 262 Hz (c4) played by the piano. This tone is combined with each tone of each instrument of the above list. Altogether there are 25 experiments.

The amplitudes are modelled with three basis functions. Starting point of one chain is the Heuristic ff_{Heur}. "Meet the level of non-significance at a significance level of 0.1" was chosen as stopping criterion. The pitches are tracked over 10 time-intervals of size $T = 512$ with 50% overlap.

The results of the study with the polyphonic data are not as promising as those from the study with the monophonic datasets since many tones are not estimated correctly. The left side of Table 3 shows 1 if both tones were correctly estimated and 0 otherwise. In 15 of the 25 experiments both notes were estimated correctly. Counting multiples of the tones increases the number of correct estimates to 21 (see the right hand side of Table 3). It is obvious, that our model has problems with tones in the range of c4 – g4. Note that c4 and g4 are the first and second overtone of c3. Also, the trumpet is problematic, in 3 of 5 cases the notes were not identified correctly. Repeating the 25 experiment with 30 and 50 time intervals leads to a slight improvement of 16 and 23 correctly estimated sets of tones in both cases, instead of 15 and 21.

7 Conclusion

In this paper a pitch tracking model for monophonic sound has been introduced and extended to a polyphonic model. The unknown parameters have been estimated with an MCMC algorithm as a stochastic optimization procedure. Our studies have shown optimal results for monophonic sound, but worse results for polyphonic sound, which could possibly be improved by parameter tuning. Further aims of our work are the modelling of 3 and more tones and an additional MCMC step for finding the correct number K of tones by the MCMC algorithm.

Acknowledgements

This work has been supported by the Graduiertenkolleg "Statistical Modelling" of the German Research Foundation (DFG). We thank Uwe Ligges for his kind support.

References

DAVY, M. and GODSILL, S.J. (2002): Bayesian Harmonic Models for Musical Pitch Estimation and Analysis. Technical Report 431, Cambridge University Engineering Department.

GILKS, W.R., RICHARDSON, S. and SPIEGELHALTER, D.J. (1996): *Markov Chain Monte Carlo in Practice*, Chapman & Hall.

LIGGES, U., WEIHS, C. and HASSE-BECKER, P. (2002): Detection of Locally Stationary Segments in Time Series. In: W. Härdle and B. Rönz (Eds.): *COMPSTAT 2002 - Proceedings in Computational Statistics - 15th Symposium held in Berlin, Germany.* Physica, Heidelberg, 285–290.

NELDER J.A. and MEAD R. (1965): A Simplex Method for Function Minimization. *The Computer Journal, 7, 308–313.*

OPOLKO, F. and WAPNICK, J. (1987): McGill University Master Samples [Compact disc]: Montreal, Quebec: McGill University.

R Development Core Team (2006): R: A Language and Environment for Statistical Computing. R Foundation of Statistical Computing, Vienna, Austria. `http://www.r-project.org`

SOMMER, K. and WEIHS, C. (2006): Using MCMC as a Stochastic Optimization Procedure for Music Time Series. In: V. Batagelj, H.H. Bock, A. Ferligoj and A. Ziberna (Eds.): *Data Science and Classification*, Springer, Heidelberg, 307–314.

WEIHS, C. and LIGGES, U. (2006): Parameter Optimization in Automatic Transcription of Music. In: M. Spiliopoulou, R. Kruse, A. Nürnberger, C. Borgelt and W. Gaul (Eds.): *From Data and Information Analysis to Knowledge Engineering.* Springer, Berlin, 740–747.

Vowel Classification by a Neurophysiologically Parameterized Auditory Model

Gero Szepannek[1], Tamás Harczos[2], Frank Klefenz[3], András Katai[3], Patrick Schikowski[3] and Claus Weihs[1]

[1] Fachbereich Statistik, Universität Dortmund, D-44221 Dortmund; szepannek@statistik.uni-dortmund.de

[2] Péter Pázmány Catholic University, H-1083 Budapest, Hungary

[3] Fraunhofer Institut für Digitale Medientechnologie (IDMT), D-98693 Ilmenau

Abstract. A meaningful feature extraction is a very important challenge indispensable to allow good classification results. In Automatic Speech Recognition human performance is still superior to technical solutions. In this paper a feature extraction for sound data is presented that is perceptually motivated by the signal processing of the human auditory system. The physiological mechanisms of signal transduction in the human ear and its neural representation are described. The generated pulse spiking trains of the inner hair cells are connected to a feed forward timing artificial Hubel-Wiesel network, which is a structured computational map for higher cognitive functions as e.g. vowel recognition. According to the theory of Greenberg a signal triggers a set of delay trajectories. In the paper this is shown for classification of different vowels from several speakers.

1 Introduction

Speech recognition is a wide research area. In this paper, a method for audio signal processing is presented. In Section 2 the time-varying audio signal is first transformed by a neurophysiologically parameterized auditory model into a two-dimensional spatiotemporal neurotransmitter vesicle release distribution. A Hubel-Wiesel timing neural network is then applied to this image to detect the emerging vesicle release patterns in Section 3. In Sections 4 and 5 a method is described to extract features from this representation that can serve for vowel recognition using Penalized Discriminant Analysis.

2 Modelling the auditory nerve response

Physiology of the ear

Signal processing in the ear consists in a chain of consecutive steps that are described in Szepannek et al. (2005). Figure 1[1] shows the human ear, consisting of the outer ear, the middle ear and the inner ear. The outer ear (pinna) and the middle ear (ear canal) work like resonators amplyfing frequencies in the range of $0.2 - 5\ kHz$. This leads to the well-known curves of the range of audibility. Via the stapes, hammer and anvil, the incoming sound waves are transmitted to the inner ear (cochlea) at the oval window into propagating fluid waves along the basilar membrane [BM].

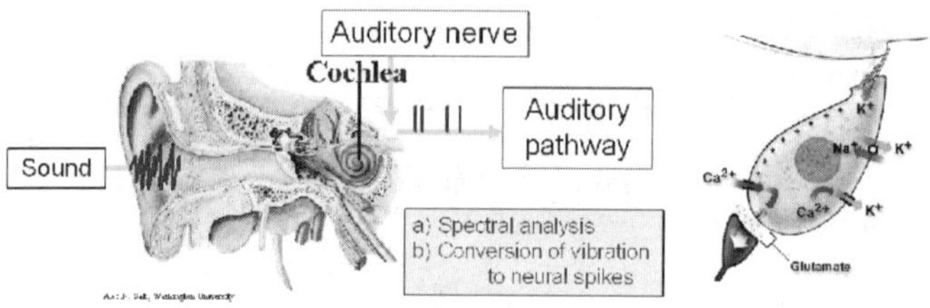

Fig. 1. Processing in the ear, schematic overview (left), inner hair cell (right).

Physiology of the inner ear

Along the human basilar membrane three rows of outer hair cells and one row of inner hair cells [IHCs] are located. The outer hair cells are assumed to enhance perception, while the inner hair cells are transducers of the mechanical waves into electrical potentials that generate action potentials (APs or spikes) at the auditory nerve fibres [ANFs]. Due to different properties of the BM along the cochlea its movement at any position is dominated by a small range of frequencies. This phenomenon is called tonotopic bandpass-filtering.

Transduction of sound waves into electrical nerve potentials

The signal-transmission in the inner hair cells is illustrated in the right part of Figure 1. On top of the hair cell, three rows of Stereocilia-hairs follow the movement of the basilar membrane at the location of the IHC. Stereocilia deflection results in opening and closing of $[K^+]$-Ion-channels. Influx of $[K^+]$ leads to depolarisation (or inversely hyper-polarisation) of the IHC resulting in half-way rectification of the bandpass-filtered sound wave. As a function of the IHC-membrane potential $[Ca^{++}]$-Ions enter the IHC and evoke the release of neurotransmittter at its presynaptic end.

Modelling the auditory nerve response

To model all these consecutive processing steps, results of neurophysiologic studies have to be incorporated. A review is given in Szepannek et al. (2005). A common state of the art inner hair cell model that represents the is developed by Sumner et al. (2002). This model is extended by the work of Baumgarte (2002) who accurately modelled masking phenomenons by BM-bandpass filtering. Both models are coupled at the Fraunhofer IDMT (Ilmenau). The diffusion of the neurotransmitter across the synaptic cleft causes postsynap-

[1] Taken with permission from http://oto.wustl.edu/cochlea/ and www.cochlee.org

tic depolarization of the auditory nerve fibre and can be modelled by the well-known Hodgkin-Huxley equations or simple Integrate-And-Fire neurons [unpublished work 2006 at Fraunhofer IDMT]. If the postsynaptic membrane potential crosses some threshold, an action potential is generated. After firing, the ANF underlies some refractory period where the threshold for firing is largely increased and thus firing is less probable.

Of course, the neural response to an incoming sound signal is not deterministic. Random fluctuations are measured at any step of the signal processing. In the model of Sumner et al. there is noise modelled in binomially distributed release of neurotransmitter vesicles in a system of three (replenishing) vesicle-pools (with release rate driven by the $[Ca^{++}]$ at the presynaptic end). Additionally some white noise is modelled in the deflection of the stereocilia hair bundles due to thermal deflections according to Ehrenberger et al. (1999).

The neural representation consists of the firing times called spikes (abscissa) of the 251 ANFs which are tonotopically ordered (ordinate) according to their different center frequencies [CFs], see Figure 2.

For the CFs of neighbouring ANFs holds $CF_{[i]} - CF_{[i-1]} = 0.1\,bark$. We calculated the response of an ANF is by averaging 50 repetitive simulations since the information available to the brain results from 3500 not only 251 IHCs each being connected to 8-10 ANFs.

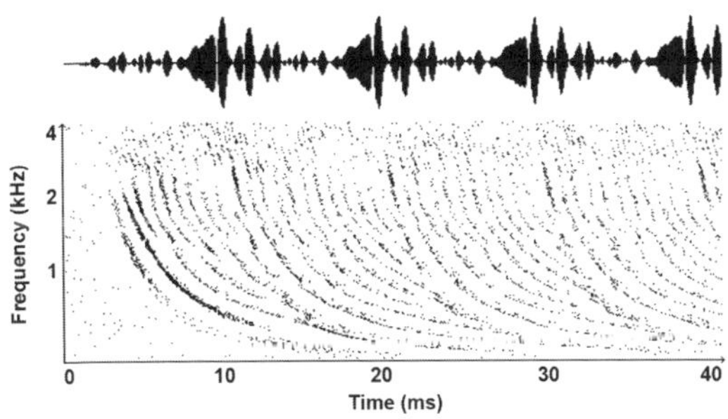

Fig. 2. Waveform (top) and vesicle release trajectories (bottom) of vowel /aa/.

Motivation: Delay trajectories

The application of the Hubel-Wiesel network as it will be described later to the output of an auditory model is motivated by the fact that a sound might be represented by regular shapes in an intermediate representation. Here, the auditory nerve patterns serve as input for the network. They appear to be bundles of different curvature for quasi stationary signals like vowels.

According to the auditory image study from Greenberg et al. (1997) when speaking of complex sounds like e.g. speech, then the resulting curves do have different shape. We concentrate on these emerging neural spike data sets. We will try to detect the resulting curves composed of spikes by fitting appropriate curves to the spike firing times along the cochlea, and then see whether a sound can be classified by the generated sequence of curve parameters. The curve parameters are the time of occurrence and their specific curvature. The Hubel-Wiesel delay network is then applied to the auditory image of vowel sounds intonated by speakers of different sex and dialect.

3 Hubel-Wiesel neural network

The core of the system is a feed-forward timing artificial Hubel-Wiesel network [HWN], which is extensively described in (Brückmann et al. (2004)). This neural net is able to learn almost any set of different slopes or a set of sinusoids of different frequencies. It has been also shown to be capable of self-learning. Harczos et al. (2006) coupled the network with the neural ANF spike output.

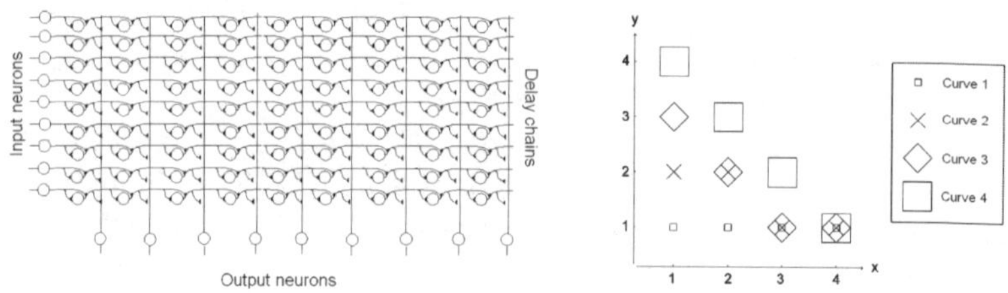

Fig. 3. Structure of a Hubel-Wiesel net (left) curvatures of a 4x4 network (right).

Please note that the curves created according to the method defined above will be inverted before use, i.e., the first curve being looked for in the auditory image will be a straight vertical line.
Greenberg pointed out that the motion of the BM proceeds in an orderly fashion from the base to the point of maximum displacement, beyond which it damps out relatively quickly. The transit of the travelling wave is extremely fast at the base, but slowing dramatically for peak displacements at the apex of the cochlea (Greenberg (1997), Shamma (1985)). He showed, furthermore, that the delay trajectories can be efficiently modelled by the simple equation:

$$d_a = nf_i^{-1} + k \tag{1}$$

where the cochlear delay d_a can be calculated from the given frequency f_i and delay constant k. Based on this statement Harczos et al. found the following curve-equation for the digital system:

$$f_\nu(j) = \frac{f_{min}\nu(n_p - 1 - j)}{(j + f_{min})(n_p - 1)} \tag{2}$$

where n_p is the size of the (quadratic) network (measured in pixels). ν and j, $(\nu, j \in 0, \ldots, n_p - 1)$ denote the index of the current curve (ν) and the index of the current pixel being calculated j. f_{min} sets the average curvature.

The extraction of the curves is performed by an HWN in a parallel way. Parallel operation here means a line-wise instead of a pixel-wise approach. Our auditory model has 251 ANFs each corresponding to a specific centre frequency. Since speech processing does not require the whole spectral information, a spectral crop can be applied to decrease the number of neurons to be processed. 25 bottom neurons as well as 85 top neurons are cropped, yielding a network size of $h = 141$ f_{min} is chosen to be 30. The ANF output is also time-scaled: 6 consecutive data on each neuron are averaged and build one input data, reducing the sampling rate to 7350 Hz. Harczos et al. found out that these parameters were a good compromise for both sexes.
Once the input sound file has been transformed into an auditory image [AI], the HWN will be configured, i.e., the curves corresponding to a given f_{min} will be taught. Next, the cropped AI will be fed into the network. In each step, the image will be shifted by one column that the network will transform. Each step generates an output array of 141 elements. The output image would give clear information about "when" and "what curvature" were in the AI.

4 The vowel data

The vowel data to be analyzed is taken from the well-known TIMIT corpus of continous speech (NIST (1995)). This data base contains English speech of females and males from 8 different dialectic regions of the USA. Four different vowel types are taken into the analysis: /aa/, /ae/, /iy/ and /uw/. These vowels are chosen according to Ali (2002) since they represent the four extremes of vowel pronounciation and tongue position.
A data set is constructed for model evaluation consisting of totally 128 vowels (= 32*4) that are randomly chosen from the data base restricted to equally represent both sexes and each of 8 dialectic regions of the USA. The models are tested using (8 fold) cross validation. It should be noted that there is relatively few training data for modelling.
Let $X_i(t)$ be the output of the Hubel-Wiesel-Network of neuron i, $i = 1, \ldots, 141$ at time t. The output is of varying length (since pronounciation time of the vowels varies). To be able to implement any of the common classification methods we have to extract variables from the output of the Hubel-Wiesel network. The idea for feature extraction is as follows: the Hubel-Wiesel network synchronizes the spikes of the auditory nerve fibres. For a period that can be considered as stationary (here $20ms$ in the center of the sounds) we perform a Fast Fourier Transformation of any neuron i. The variables X_f used for classification are the maximal amplitudes of the fourier frequencies f over all neurons which represent shapes of different delay lines:

$$X_f := \max_i FFT(f, i) \qquad (3)$$

These variables can be interpreted as *intensity of the perceived formant frequencies of the signal* since the formant extraction here is performed after the auditory processing.

5 Vowel classification using penalized discriminant analysis

Several different classification methods are implemented for the vowel classification task on the given variables. Among them *Discriminant Analysis* as well as *Nearest Neighbour, Trees, Naive Bayes, SVMs* and *Neural Networks*. The best performance is obtained using *Penalized Discriminant Analysis (PDA)* which is briefly introduced below. In *PDA* a $K-$class classification problem is transformed into a regression problem with criterion

$$ASR(\{\Theta_l, \beta_l\}_{l=1}^{L}) = 1/N \sum_{l=1}^{L} \sum_{i=1}^{N} \left[\left(\Theta_l(g_i) - h^T(x_i)\beta_l\right)^2 - \lambda \beta_l^T \Omega \beta_l \right] \qquad (4)$$

to be minimized. g_i is the class label of object $i = 1, \ldots, N$ and $\Theta_l(g_i)$ are optimal scores for the classes that have to be predicted by the functional $h(x_i)^T \beta$. This is a projection of the classes into $\mathbb{R}^L$, $L \leq K - 1$ dimensions. For $h(x)$ being the identity (also used here) and $\lambda = 0$ Hastie et al. (1995) showed equality of the β_l to the coefficents in Linear Discriminant Analysis (up to some constant). The reformulation of the problem allows an expansion of classical (Linear) Discriminant Analysis to non-linearity, using nonlinear spline basis functions in $h(x)$. The penalty term Ω can be included in the model to enforce smoothness of model coefficients to avoid the problem of multicollinearity of the predictors.

The requirement of smoothness of the parameters is meaningful, if strong collinearity is present in the predictor variables. This is given here due to the implicite order within the set of variables. In such case, the variance of the estimated β_l can drastically increase. A standard way to tackle this problem is Ridge-Regression (i.e. taking Ω to be the identity matrix, see e.g. Schmidt and Trenkler (1998)) as it is also done here. The magnitude of λ indicates the degree of penalization opposing smoothness of the model coefficients to accuracy of the fit. λ can be equivalently formulated as degrees of freedom (df, defined as $tr(\Omega)$) indicating how many independent components may be hidden within the model's coefficients (see Hastie et al. (1995)).

Figure 4 shows the model coefficients for different dfs. The best result is obtained for 28 dfs requiring smooth but not too smooth coefficients. Despite hopeful results on the training data there is confusion between /uw/ and the other vowels on the cross-validated data as can be seen in Table 1. The other 3 vowels are recognized quite well. A classification model on only three vowel classes (/aa/, /ae/ and /iy/) shows a cross-validated recognition rate

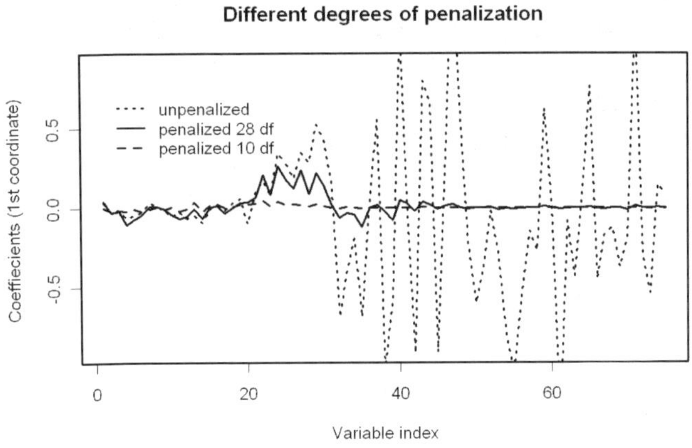

Fig. 4. PDA models with different degrees of freedom.

Table 1. Confusion matrices on training (left) and cv (right) data.

true	/aa/	/ae/	/iy/	/uw/	/aa/	/ae/	/iy/	/uw/
/aa/	29	3	0	0	26	5	0	1
/ae/	0	31	0	1	3	24	0	5
/iy/	0	0	27	5	0	3	21	11
/uw/	0	0	6	26	0	3	10	19

of 85.1% and demonstrates the general ability of the features to recognize vowels, independently from sex or the speaker's dialect. The benefit of such auditory/Hubel-Wiesel modelling may result from the incorporation of perceptive phenomena like masking and adaption.

6 Summary

In this paper an approach is presented to encode information as it is done in the human auditory system. A representation of sound at the auditory nerve fibres can be simulated. An artificial Hubel-Wiesel network is introduced for further processing. It is shown that this representation can be used for automatic vowel recognition, especially due to its temporal composition by means of the *perceptual intensity of the formants.*
Penalized Discriminant Analysis has been found to perform best for vowel recognition on the generated features, but not all vowels can be identified correctly by the model. Further research may include search for additional information contained in the shape of the delay trajectories. Furthermore, a more accurate scaling of the Hubel-Wiesel network's output values should be investigated, since Shamma (1985) stated that *"the amplitude of the cochlear*

travelling wave rapidly decreases after reaching the point of maximal excitation", i.e. the possible length of a travelling wave is limited by its frequency-position along the cochlea. The system's behaviour under noisy conditions has also to be investigated since this is today's biggest impact auf ASR systems compared human speech recognition as well as its performance on a larger speech data base and the extension to consonant recognition. Finally, the obtained classifiers may serve to improve existing HMM-based speech recognition systems (see e.g. Fink (2003)).

References

ALI, A., VAN DER SPIEGEL, J. and MUELLER, P. (2002): Robust Auditory-Based Speech Recognition Using the Average Localized Synchrony-Detection. *Proc. IEEE Transactions on Speech and Audio Processing, 10, 5, 279–292.*

BAUMGARTE, F. (2000): Ein psychophysiologisches Gehörmodell zur Nachbildung von Wahrnehmungsschwellen für die Audiocodierung. PhD Thesis, University of Hannover, Germany.

BRÜCKMANN, A., KLEFENZ, F. and WÜNSCHE, A. (2004): A Neural Net for 2d-slope and Sinusoidal Shape Detection. *Int. Scient. J. of Computing, 3, 1, 21–26.*

EHRENBERGER, K., FELIX, D. and SVOZIL, K. (1999): Stochastic Resonance in Cochlear Signal Transduction. *Acta Otolaryngol, 119, 166 – 170.*

FINK, G. (2003): *Mustererkennung mit Markov-Modellen: Theorie - Praxis - Anwendungsgebiete.* Teubner, Stuttgart.

GREENBERG, S., POEPPEL, D. and ROBERTS, T. (1997): A Space-Time Theory of Pitch and Timbre Based on Cortical Expansion of the Cochlear Traveling Wave delay. In: *Proceedings of the XIth Int Symposium on Hearing, Grantham, 1997.* Springer, Berlin, 203–210.

HARCZOS, T., KLEFENZ, F. and KATAI, A. (2006): A Neurobiologically Inspired Vowel Recognizer Using Hough-Transform - A Novel Approach to Auditory Image Processing. In: *Proc. of Visapp. 2006, 1, 251–256.*

HASTIE, T., BUJA, A. and TIBSHIRANI, R. (1995): Penalized Discriminant Analysis, *Annals of Statistics, 23, 73-102.*

NIST [National Institute of Standards and Technology] (1995): TIMIT Acoustic-Phonetic Contimous Speech Corpus, *Disc 1-1.1, NIST order No. PB91-505065.*

SCHMIDT, K. and TRENKLER, G. (1998): *Moderne Matrix-Algebra.* Springer, Heidelberg.

SHAMMA, S. (1985): Speech Processing in the Auditory System I: The Representation of Speech Sounds in the Responses of the Auditory Nerve, *Journ. Acoust. Soc. Am., 78, 5, 1612-1621.*

SUMNER, C., O'MARD, L., LOPEZ-POVEDA, E. and MEDDIS, R. (2002): A Revised Model of the Inner-hair Cell and Auditory Nerve Complex. *Journal of the Acoustical Society of America, 111, 2178–2189.*

SZEPANNEK, G., KLEFENZ, F. and WEIHS, C. (2005): Schallanalyse - Neuronale Repräsentation des Hörvorgangs als Basis, *Informatik Spektrum, 28, 5, 389–395.*

Part XI

Archaeology

Uncovering the Internal Structure of the Roman Brick and Tile Making in Frankfurt-Nied by Cluster Validation

Jens Dolata[1], Hans-Joachim Mucha[2] and Hans-Georg Bartel[3]

[1] Archaeolocical Service for Rhineland-Palatinate: Landesamt für Denkmalpflege Rheinland-Pfalz, Abt. Archäologische Denkmalpflege, Amt Mainz, Große Langgasse 29, D-55116 Mainz; dolata@ziegelforschung.de

[2] Weierstraß-Institut für Angewandte Analysis und Stochastik (WIAS), Mohrenstraße 39, D-10117 Berlin; mucha@wias-berlin.de

[3] Institut für Chemie, Humboldt-Universität zu Berlin, Brook-Taylor-Straße 2, D-12489 Berlin

Abstract. During the past few years, a complex model of history and relations of Roman brick and tile production in south-west Germany has been developed by archaeologists. However, open questions remain concerning the brickyard of Frankfurt-Nied. From the statistical point of view the set of bricks and tiles of this location is divided into two clusters. These clusters can be confirmed by cluster validation. As a result of these validations, archaeologists can now modify and consolidate their ideas about the internal structures of Roman brick and tile making in Frankfurt-Nied.

1 Archaeological material and question

About 1000 Roman stamped bricks and tiles from the Upper Rhine area have been investigated by chemical analysis until now. The combination of archaeological (especially epigraphic) information and the results of mineralogy and chemistry allowed archaeologists to develop a complex model of brick and tile making in Roman times. In south-west Germany, a few large brickyards existed, of which the operating authority was the Roman army. The period of running is from the middle of the first century A.D. until the end of the fourth century. One of the important areas for brick-production is located at the lower Main-bank at Frankfurt-Nied. The running of this site has its coherence with the construction of buildings of the Roman frontier area in Upper Germany, the so-called *Obergermanischer Limes* (*Limes Germaniae Superioris*). The beginning of the brick and tile production in Frankfurt-Nied is to be set in the years 83/85 and the main output is that of the *legio XXII Primigenia* in the beginning of the second century A.D. More than 500 different brickstamps of this military-unit only used in the brickyard of Frankfurt-Nied can be distinguished. Archaeologists are most interested in the chronology of the production-marks, which are found on the building-material. Here we will take a closer look at the internal structures of the Roman brickyard in Frankfurt-Nied by statistical investigation of more than hundred samples. This work is based on the chemical composition of bricks and tiles only.

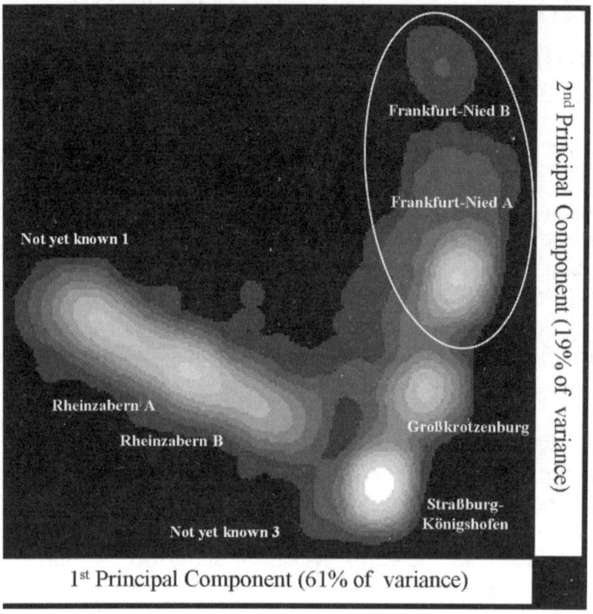

Fig. 1. Several cuts of the mountains of bivariate density and clusters

Today, we have a long history of chemical and statistical analysis of Roman brick and tile making in *Germania Superior*, see for example, Dolata (2000), Bartel et al. (2002), Dolata et al. (2003), and Mucha et al. (2005). As a result the following locations of brickyards are identified: Frankfurt-Nied, Groß-Krotzenburg, Rheinzabern, Straßburg-Königshofen, Worms and two with respect to their provenience not yet known ones. In Figure 1, several cuts of the mountains of the bivariate density estimation are shown. This estimation is based on the first two principal components of 613 objects that are described by 19 chemical elements; see Dolata (2000) and Section 2 for details. The bivariate density looks like a main data body with two arms. The right arm consists of three compact clusters. In opposition, the ridge at the left hand side shows neither clear peaks nor clear separation into clusters. Here let us focus on local statistical analysis of Frankfurt-Nied that is located at the upper right hand side area and that is additionally marked by an ellipse.

From the archaeological point of view, bricks and tiles of Frankfurt-Nied stem from one brickyard only. This opinion is mainly based on the brick-stamps that were found on the objects. But, as can be seen in the principal component density plot (Figure 1), the samples assigned to Frankfurt-Nied seem to be divided into two clusters. Therefore, in this paper, the cluster analysis of the relevant 137 samples of Frankfurt-Nied with a subsequent cluster validation will be presented. The aim is finding homogeneous clusters. The solution will be verified by decision making about the number of clusters and by assessing both the stability of individual clusters and the reliability of assignment of each object to its cluster.

2 Preparation of the data

The statistical analysis of the 137 objects is based on the contents of 9 oxides and 10 chemical trace elements that are measured with X-ray fluorescence analysis in quite different scales. Thus, it was necessary to transform the data by preprocessing the original data matrix by dividing each column (variable) by its arithmetic mean. This quite simple transformation has the useful property that the relative variability in the original variables become the variability of the transformed ones (Underhill and Peisach (1985)). As a consequence of this transformation, the original variables become comparable with each other. The arithmetic mean of each new variable is equal to 1. The variables preserve their different original variability and therefore have different influence on the cluster analysis.

3 The cluster analysis method

There is a well-known partitional clustering technique for minimizing the within-class sum of squares criterion based on pairwise distances (Späth (1985)). Concretely, the squared Euclidean distance was applied. Starting with a random partition of the set of objects into K clusters, the method works by exchanging observations between these clusters in order to minimize the well-known within-class sum of squares criterion (for details, see Bartel et al. (2003)). Here we investigated partitions into $K = 2, 3, ..., 6$ clusters.

4 Cluster validation

An automatic validation of the clustering results is applied here. It is based on the resampling technique without replacement (Mucha (2004)). Here, the result of clustering the set of all 137 objects is compared with 250 cluster analyses of random selected subsets. Concretely, every time about three-fourths of the 137 objects is selected randomly in order to establish the 250 subsets. The first step of the cluster validation consists of a decision making about the appropriate number of clusters. A partition gives the appropriate number of clusters K if it is very stable, thus, if it can be reproduced to a maximum degree by clustering the random subsets (Jain and Dubes (1988)). For instance, the adjusted Rand index measures the correspondence between partitions (Rand (1971), Hubert and Arabie (1985)).

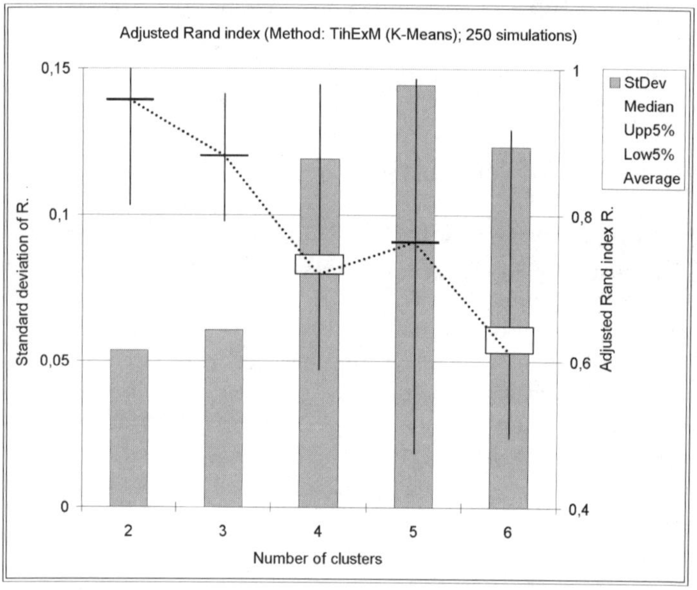

Fig. 2. Summary statistics of the adjusted Rand measure

Table 1. Summary statistics of the adjusted Rand values with respect to different number of clusters

Number of clusters	Standard Deviation	Median	Upp5%	Low5%	Average	Maximum	Minimum
2	0.054	**0.956**	1	0.813	0.957	1	0.733
3	0.061	**0.88**	0.965	0.791	0.881	1	0.563
4	0.119	**0.719**	0.978	0.589	0.745	1	0.444
5	0.144	**0.764**	0.986	0.474	0.761	1	0.385
6	0.123	**0.612**	0.917	0.495	0.648	0.993	0.459

Both the Figure 2 and the corresponding Table 1 show the simulation results. The two cluster solution is the most likely one with 106 objects in the very compact cluster 1 and only 31 objects in the widespread cluster 2, respectively (see also Figure 3). The partition into two clusters is the most stable one with the highest median of the 250 adjusted Rand indexes (=0.956, see also the scale at the right hand side and the corresponding line that connects the medians for different number of clusters in Figure 2). Moreover, the 250 adjusted Rand values have the smallest standard deviation (see the scale at the left hand side and the corresponding bars in Figure 2). The resulting two clusters will be investigated in detail later on.

In the next step of cluster validation, the stability of each individual cluster is assessed based on measures of similarity between sets, e.g., the asymmetric measure of cluster agreement or the symmetric Jaccard measure. Here we do not consider special properties like compactness and isolation. It should be mentioned that it makes sense to investigate the (often quite different) specific stability of clusters of the same clustering on the same data. Often it can be observed that the clusters have a quite different stability. Some of them are very stable. Thus, they can be reproduced and confirmed to a high degree. To define stability with respect to the individual clusters, measures of correspondence between a cluster $\mathcal{E}$ and a cluster $\mathcal{F}$ like

$$\tau(\mathcal{E},\mathcal{F}) = \frac{|\mathcal{E}\cap\mathcal{F}|}{|\mathcal{E}\cup\mathcal{F}|}, \quad \gamma(\mathcal{E},\mathcal{F}) = \frac{|\mathcal{E}\cap\mathcal{F}|}{|\mathcal{E}|}, \quad \eta(\mathcal{E},\mathcal{F}) = \frac{|\mathcal{E}\cap\mathcal{F}|}{|\mathcal{E}|+|\mathcal{F}|} \tag{1}$$

have to be defined. ($\mathcal{E}$ and $\mathcal{F}$ are nonempty subsets of some finite set.) The Jaccard coefficient τ as well as the Dice coefficient η are symmetric and they attain their minimum 0 only for disjoint sets and their maximum 1 only for equal ones. The asymmetric measure γ assesses the rate of recovery of subset $\mathcal{E}$ by the subset $\mathcal{F}$. It attains its minimum 0 only for disjoint sets and its maximum 1 only if $\mathcal{E}\subseteq\mathcal{F}$ holds. Obviously, it is necessary $\tau\leq\gamma$.

Table 2. Assessment of the stability of clusters

Statistic	Cluster	Jaccard	Rate of Recovery	Dice
Median	1	**0.987**	**1**	**0.993**
	2	**0.958**	**1**	**0.979**
Average	1	**0.988**	**0.993**	**0.994**
	2	**0.958**	**0.982**	**0.978**
Standard deviation	1	0.016	0.014	0.008
	2	0.052	0.035	0.029

Table 2 shows the assessment of the stability of the individual clusters of the two clusters solution by three different measures. Setting a cutoff level of 0.95 for the median or average, all the three measures assess the individual clusters as stable. Thus, the two cluster solution can be affirmed here. For instance, the mean recovery rate says that there are two most stable clusters with $\gamma = 0.993$ and $\gamma = 0.982$, respectively. Further, the Jaccard measure also verifies the stability of both clusters: The corresponding median values are $\tau = 0.987$ and $\tau = 0.958$, respectively. This is significant concerning a cutoff value of 5% to consider a cluster as stable.

In the third and most detailed level of validation, the reliability of the cluster membership of each individual observation will be assessed. Figure 3 shows the cluster membership of each object. Additionally, instable objects

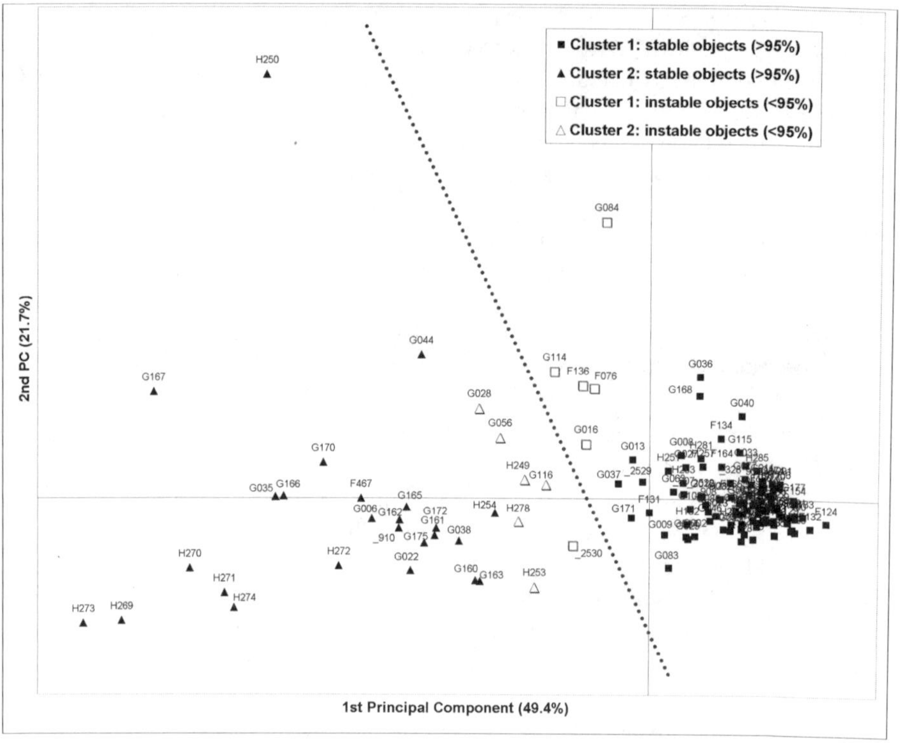

Fig. 3. Cluster analysis results (local clustering and validation by bootstrapping)

Table 3. Univariate statistics of oxides of each cluster of the two cluster solution

Statistic	Cluster	SiO_2	TiO_2	Al_2O_3	Fe_2O_3	MnO	MgO	CaO	Na_2O	K_2O
Average	1	72.76	1.667	17.14	**4.288**	**0.0228**	0.926	0.633	0.2212	2.235
	2	71.28	1.819	16.28	**6.43**	**0.0803**	0.888	0.734	0.2973	2.002
Standard	1	2.45	0.267	1.85	**0.722**	**0.009**	0.159	0.301	0.0675	0.325
deviation	2	3.54	0.407	2.15	**1.53**	**0.0243**	0.281	0.556	0.1273	0.447
Minima	1	64.26	0.894	11.96	2.721	0.012	0.584	0.184	0.114	1.141
	2	64.45	1.071	11.34	4.746	0.0496	0.501	0.304	0.133	1.421
Maxima	1	81.1	2.39	25.1	7.404	0.0571	1.521	2.715	0.459	3.634
	2	77.83	2.523	20.16	9.974	0.136	1.676	3.463	0.689	2.964

are marked by light symbols. The most instable objects are G114 and G116 with a rate of reliability of 0.53 and 0.63, respectively. In detail, that means that object G114 was classified with nearly the same frequency into cluster 1 or into cluster 2. As a result of these validations, the archaeologists can now modify and consolidate their ideas. The archaeologists consider additional information about the objects. Thus, the main aim of cluster analysis is to give assistance for the experts. The corresponding graphical and numerical

Table 4. Univariate statistics of trace elements of each cluster

Statistic	Cluster	V	Cr	Ni	Zn	Rb	Sr	Y	Zr	Nb	Ba
Average	1	2.2	95.3	**148.8**	**43.1**	**43.9**	123.7	134.4	34.2	284.6	38.2
	2	2	92.7	**219.9**	**78.4**	**79.2**	96.7	134.7	34.4	314.6	42.9
Standard	1	0.3	16.6	**23.5**	**10.8**	**16.5**	15.8	24.5	3.6	29.6	8
deviation	2	0.4	15.9	**58.3**	**22.7**	**13.7**	23.6	21.1	4.3	23.9	10.8
Minima	1	1.1	61	90	24	15	78	67	27	217	16
	2	1.4	66	129	42	54	69	88	28	265	21
Maxima	1	3.6	147	208	82	106	175	263	49	346	65
	2	3	119	323	123	124	154	175	44	363	61

outputs are helpful for the experts. The chemical attributes contribute in quite different amount to the difference between the two clusters; see Table 3 and Table 4. For instance, cluster 2 has very high values in the oxides Fe_2O_3 and MnO and in the trace elements Ni, Zn and Rb.

5 Archaeological result and interpretation

Based on the investigation of different types of brick-stamps from various military-units, it is possible to interpret the results obtained by statistical examination: The objects assigned to Frankfurt-Nied can be separated into two clusters. In the small cluster, 6 samples are situated with brick-stamps of the *cohors I Asturum*, and 12 samples are contained with brick-stamps of the *legio XXI Rapax*. This result is of archaeological importance: Brick-stamps of the auxiliary-unit *cohors I Aturum* are very rare. The examples analyzed here have been found in the brickyard of Frankfurt-Nied and in the settlement of Frankfurt-Heddernheim. Archaeologists do not yet know when soldiers of this unit have been sent to the brickyard for making bricks and what was the historical context. Brick-stamps of a second military-unit are found together in the statistical cluster: stamps of the *legio XXI Rapax*. The chronological assignment of bricks and tiles from this legion is very precise. The *legio XXI Rapax* has been transferred under the emperor *Domitianus* in 83 A.D. from *Bonna* in *Germania Inferior* to *Mogontiacum* in *Germania Superior*. Already in 89/90 the unit has been withdrawn from *Germania Superior* as a consequence of the rebellion of *Saturninus*. There is known one stamp-type from the brickyard of Rheinzabern and only two from the brickyard of Frankfurt-Nied. Because of the shifting of brick-production from the Rhine to Main in 83/85 A.D. it is possible to date the 12 samples separated in the cluster in a very small period of time (between 83/85 and 89/90). We consider the stamps of *cohors I Asturum* to be pressed in bricks and tiles of similar clay, perhaps in the same time. In the future, this hypothesis has to be proofed by investigation of additional objects. In connection with this other open questions on the archaeological model of Roman brick and tile production in the Upper Rhine area can be answered.

6 Conclusions

From the statistical point of view, the set of bricks and tiles from Frankfurt-Nied is divided into two clusters. Both clusters can be confirmed or reproduced to a high degree by resampling techniques. As a result of these validations accompanied by helpful information about the stability of each cluster and each single observation, archaeologists have been updated their ideas about the internal structures of the Roman brick and tile making in this location. Cluster 2 has 31 observations only and is not homogeneous. Thus, in the future, much more objects of the corresponding *cohors I Asturum* and *legio XXI Rapax* should be prepaired for a statistical investigation.

References

BARTEL, H.-G., MUCHA, H.-J. and DOLATA, J. (2002): Automatische Klassifikation in der Archäometrie: Berliner und Mainzer Arbeiten zu oberrheinischen Ziegeleien in römischer Zeit. *Berliner Beiträge zur Archäometrie, 19, 31–62.*

BARTEL, H.-G., MUCHA, H.-J. and DOLATA, J. (2003): Über eine Modifikation eines graphentheoretisch basierten partitionierenden Verfahrens der Clusteranalyse. *Match, 48, 209–223.*

DOLATA, J. (2000): *Römische Ziegelstempel aus Mainz und dem nördlichen Obergermanien.* Dissertation, Johann Wolfgang Goethe-Universität, Frankfurt.

DOLATA, J., MUCHA, H.-J. and BARTEL, H.-G. (2003): Archäologische und mathematisch-statistische Neuordnung der Orte römischer Baukeramikherstellung im nördlichen Obergermanien. *Xantener Berichte, 13, 381–409.*

HUBERT, L.J. and ARABIE, P. (1985): Comparing Partitions. *Journal of Classification, 2, 193–218.*

JAIN, A.K. and DUBES, R.C. (1988): *Algorithms for Clustering Data.* Prentice Hall, New Jersey.

MUCHA, H.-J. (2004): Automatic Validation of Hierarchical Clustering. In: J. Antoch (Ed.): *Proceedings in Computational Statistics, COMPSTAT 2004, 16th Symposium.* Physica, Heidelberg, 1535–1542.

MUCHA, H.-J., BARTEL, H.-G. and DOLATA, J. (2005): Model-based Cluster Analysis of Roman Bricks and Tiles from Worms and Rheinzabern. In: C. Weihs and W. Gaul, W. (Eds.): *Classification - The Ubiquitous Challenge*, Springer, Berlin, 317–324.

RAND, W.M. (1971): Objective Criteria for the Evaluation of Clustering Methods. *Journal of the American Statistical Association, 66, 846–850.*

SPÄTH, H. (1985): *Cluster Dissection and Analysis.* Ellis Horwood, Chichester.

UNDERHILL, L.G. and PEISACH, M. (1985): Correspondence Analysis and Its Application in Multielement Trace Analysis. *J. Trace and Microprobe Techniques, 3, 1-2, 41–65.*

Where Did I See You Before... A Holistic Method to Compare and Find Archaeological Artifacts

Vincent Mom

DPP (Digital Preservation Projects) Dordrecht, The Netherlands; v.mom@wxs.nl

Abstract. This paper describes the Secanto (Section Analysis Tool) computer program designed to find look-alikes of archaeological objects by comparing their shapes (sections, profiles). The current database contains the low resolution images of about 1000 profiles of handmade Iron Age ceramic vessels from The Netherlands and Northern Germany, taken from 14 'classic' publications. A point-and-click data entry screen enables the user to enter her/his own profile and within 2 minutes the best look-alikes (best, according to a calculated similarity parameter) are retrieved from the database. The images, essentially treated as two-dimensional information carriers, are directly compared by measuring their surface curvatures. The differences between these curvatures are expressed in a similarity parameter, which can also be interpreted as a 'distance between'. The method looks very promising, also for other types of artifacts like stone tools and coins.

1 Introduction

Archaeologists spend much time comparing artifacts. The individual qualities of these artifacts as such of course have a descriptive value, but the archaeological story does not come alive until these properties are being compared with those of other artifacts in (dis)similar circumstances. Therefore comparing processes provide the foundations of the archaeological discourse and are henceforth of utmost importance.

Now these archaeological artifacts, in this process of comparing, can have two formats so to speak: the artifact can be 'real', a tangible object obtained from e.g. an excavation or a (museum) collection, or the artifact is not physically present but available as an image (a drawing, a picture) and/or a textual description, usually as part of a publication, and the process of comparing artifacts, irrespective of the format in which they manifest themselves is often not a simple, trivial matter. Of course, the question asked is leading in this matter. 'Which object is the heaviest', 'Are these objects made of the same material?'

and 'Are these objects the same?' span a whole universe of complications. If the question only considers properties which can be simply measured then the comparison may be evenly simple ('these vessels have the same height', 'both these vessels are made of bronze') although the level of precision may start playing a role (e.g. 'This bronze contains X % arsenic while the other contains Y %...'). And especially when the question of 'sameness' arises ('Are these objects the same?') then the comparison process is of utmost importance.

In this contribution I will describe the usage of a computer program, Secanto (Section Analysis Tool) that compares shapes of objects and which currently primarily is used to retrieve objects from a database, based on the similarity of the object shapes. First I will aim some arrows at the usual way that shapes are being compared by archaeologists and which practice inspired me to develop this computer program. Then I will describe shortly the working of Secanto. Finally I will give an overview of the experiments that are currently ongoing with Secanto.

2 How archaeologists compare shapes

Archaeology is, for an important part, a craft. At excavations for instance the digging is a disciplined activity, moreover so as the object under study is more or less destroyed and errors made during the excavation process usually cannot be restored. Training and supervised practicing form part of the archaeologists' curriculum. The same counts for the 'detailed determination' of the finds. The first time a student archaeologist opens a box with the finds of an excavation, everything looks new. But having done it a number of times, and especially when the finds are more or less from the same area and time period, a certain measure of accustomisation takes place: the raw materials the objects are made from (of which there is usually only a limited number) are readily recognised and as certain materials often are used for certain functionalities only, it might not be so difficult to see whether a piece of ceramics comes from a plate, an amphora or a drinking vessel.

But archaeologists want more. It is not enough to say that the object is a coin of bronze, but more details ('a bronze denarius with the portrait of the emperor Hadrian') may attach a time stamp to the object and, what is more, to the layer the object is found in. And archaeologists need these time stamps to be able to lift the value of their finds above the simple level of 'nice goodies' and tell their stories of what happened in those times long gone. This means that we are not satisfied with the statement 'this is a handmade vessel from the Iron Age' but that we want more: we want to know whether this vessel 'is known', 'is of a certain type' etc.

Partly, the archaeologist uses his/her past experience to do these detailed determinations and the more experience an archaeologist has, the more he/she will be inclined to do so. But this, of course, is a dangerous approach: there will be a tendency to fit an object into the personal reference frame, with an emphasis on corresponding properties and ignoring differences. Secondly, if an object is 'unknown', one will try to find information about the object in literature. And here also quite subjective tendencies creep in. In the first place, some publications are more en vogue than others and the order in which different publications will be scanned depends on the research group that one is a member of. A second issue is the lack of an objective quantifier to indicate the difference in shape between two objects. The height of an object can be measured, and there is nothing subjective about a statement like 'A is higher than B'. But a statement like 'A looks more like B than like C' is hardly ever underpinned by objective, measured properties. And then there is the (often implicit) up- and downgrading of certain properties: sometimes the shape of the rim of the vessel is leading while in other cases the height/diameter ratio determines the type of the vessel. But hardly seldom there is an explicit algorithm available that determines a. whether an object belongs in a certain typology and b. if so, to which type it belongs. Finally the method of paging through books until a good lookalike presents itself is frustrated by the lack of aforementioned quantifier in combination with the subjective 'this is a good fit' notion of the individual researcher: some people will be quickly satisfied in this respect, while others will hardly ever find a good lookalike. This makes that archaeological science has its foundations on rather subjective, paradigmatic 'facts'.

In the now following sections I will zoom in on this 'objective quantifier' and the application of such a quantifier in the Secanto retrieval system.

3 Comparing shapes

The shape of a vessel (and other categories of artifacts) can be described in two ways:

1. One can define a number of parameters of which the values are determined by measurement and/or visual inspection. One may call this a reductionistic approach as the underlying idea is that the object is sufficiently described by these parameters and the object is 'reduced' to these parameter values.

2. One can approach a shape as a uniform mathematical entity, defined by (relative) co-ordinates which may be expressed as x, y and z values, spline coefficients, Fourier transform coefficients etc.

The parametric approach may be well suited to describe objects, but that does not automatically imply that these parameter sets also provide the best method to compare the objects. And when considering shapes, parametric descriptions are hardly ever superior to drawings or images, at least to the human eye. Other problems are the (often strong) correlations that exist between the parameters, and the fact that sometimes the values of certain parameters are not available because the object is incomplete. The way to solve the correlation problem by mathematical methods results in new sets of parameters which are often difficult to relate 'visually' to the original set of parameters. Finally there are the obvious problems of the names of the parameters (width or diameter?), not to speak of terminology (plate, dish or saucer?) and multilingual aspects.

In the second approach, however, the whole issue of describing the object using parameters may be skipped: the shapes are essentially treated as integral two-dimensional information carriers and the images are *directly* compared ("holistic approach") by measuring the differences between their surface curvatures or surface overlap. These differences are expressed in a dissimilarity parameter, which value can also directly be interpreted as a "distance between" the objects.

4 The Sliced method

Shennan and Wilcock (1975) applied a relative simple numeric method to compare vessels: the Sliced method. The objective was to derive automatically a typology for Bell Beaker pottery from Central Germany, but an even simpler method, comparing height/width ratios using Principal Component Analysis, gave comparable results. Summarised, automatically generated typologies in general never really got a very strong foothold in archaeology.

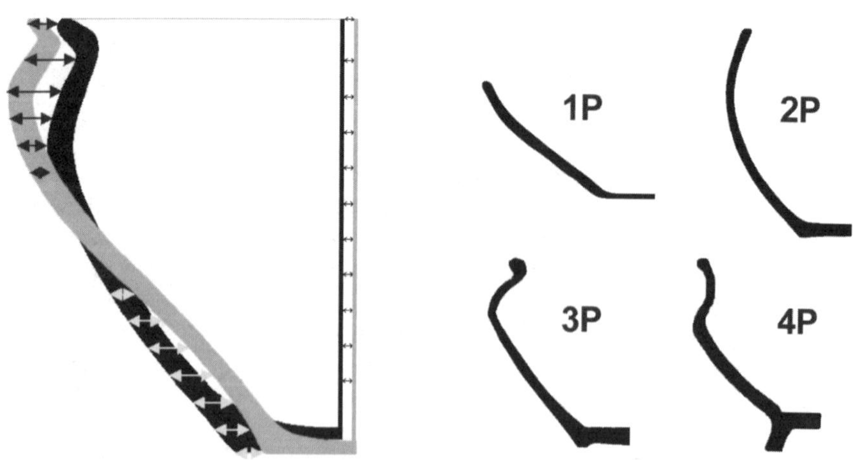

Fig. 1. a) The Sliced method b) Four global shape categories

However, IT environments have changed dramatically since the main frame/punched card were the standard. The current vehicle is Internet and the availability of data has increased enormously. And the Sliced method, maybe not completely fit to generate typologies, still provides a simple, good method to compare the shapes of objects. So if we concentrate on shape alone, and ignore all other properties that may determine whether an object belongs to a certain type or not, then we can use this method as a key to retrieve lookalike vessels from (large) databases.

The Sliced method, as the name suggests, divides the object in (equidistant) slices, and the dissimilarity is calculated as the sum of the differences between the points on the surfaces of the two vessels, squared (see Fig. 1a). To incorporate not only the shape of the outer vessel walls but also the overall vessel width, the distances between the central axes (squared) are added as well. One shape is shifted over the other until an optimal (smallest) sum of squares value is obtained.

5 Iron Age handmade pottery

The idea of 'Type', at least when dealing with artifacts, is a typical (sic) industrial notion. Roman pottery like the terra sigillata from Arrezo (Central Italy) and Le Graufesenque (Southern France) was produced by the millions according to rather 'modern' manufacturing practices like instruction lists and bills-of-materials in order to be able to make reproducable wares. So speaking of a type 'Dragendorff 25' or 'Ritterling 18' will not give any rise to dispute about the appearance of such Roman ceramics.

But handmade Iron Age pottery comes from a completely different manufacturing mode. In contrast to the aforementioned Roman ceramics that was transported all over the Roman empire, this local pottery may be best characterized as artisanal backyard production. No throwing wheel was used and decoration, if present, usually was limited to fingerprints and coarse brushes on the surface. Its usage was in practical purposes (cooking, storage) although also remains of cremations were buried in such vessels.

The variety of shapes is large. On a global, functional level, one may distinguish four shape categories, based on the number of partitions of the vessel curve: 1 partition is a plate or open bowl, 2 partitions a closed bowl without rim, 3 partitions a vessel with a rim (the majority of the shapes is of this category) and shapes with a pronounced bottom shape have 4 partitions (see Fig. 1b). Several typologies do exist, but none of these has in general been adopted by the archaeological community because none of them is well applicable to the material from other excavations than the ones they are based upon. Therefore Iron Age pottery is a good candidate to test automated, more objective ways of comparing and retrieving lookalikes from literature.

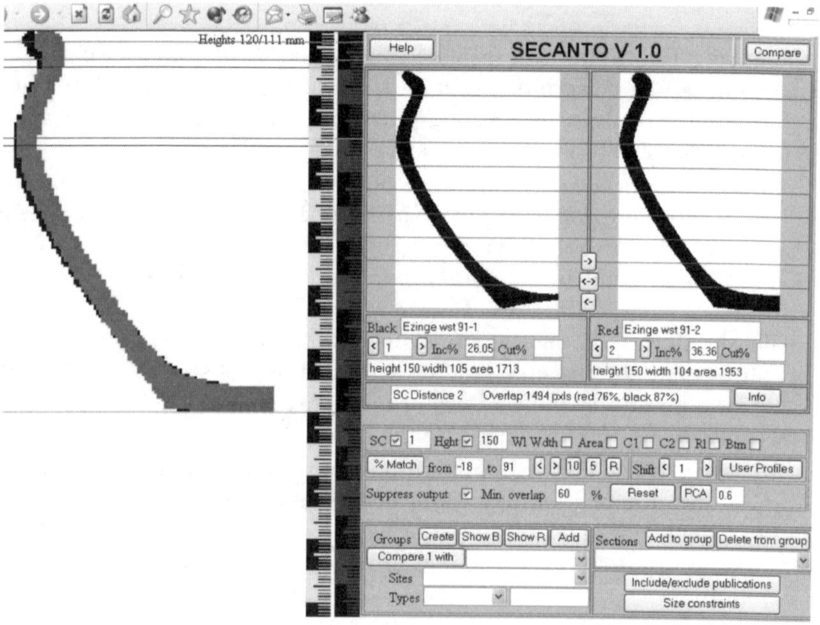

Fig. 2. Secanto user interface. Profiles of two Iron Age vessels

6 How Secanto works

The Secanto system consists of HTML pages with embedded JavaScript for the calculations. The underlying database currently consists of about 1000 profiles (low-resolution .GIF files) of handmade vessels taken from a number of well known books about Iron Age excavations and Roman Period excavations in the Netherlands and near surroundings: Abbink (1999), Bloemers (1978), Diederik (2002), Van Es (1965, 1968), Van Heeringen and Van Trierum (1981), Van Heeringen (1987, 1989), Reichmann (1979), Taayke (1987) and Tol et al. (2000). The core of the system is the comparison engine. Fig. 2 shows a screenshot of the userinterface with at the lefthand side the optimal fit. Currently two methods have been implemented: the Sliced method (see before) and a calculation based on maximising the overlap of the profiles. The latter is not suited for profiles of vessels because of the elongated shapes of the profiles, but looks promising for more bulky objects like stone axes. The vessel profiles are distributed over about 30 shape groups. These groups contain vessels that 'look alike' according to the Sliced method. The creation of these shape groups was an interactive process, using the original typologies as a starting point and adding/removing vessels based on calculated dissimilarity parameters (for more details Mom (2006)). The user uploads a profile file from his own PC to the server, which is then compared to the 'best representatives' of each shape group (see Fig. 3). Next, a refinement step follows within the selected shape group and an ordered list of 'best fits' is returned to the user (see

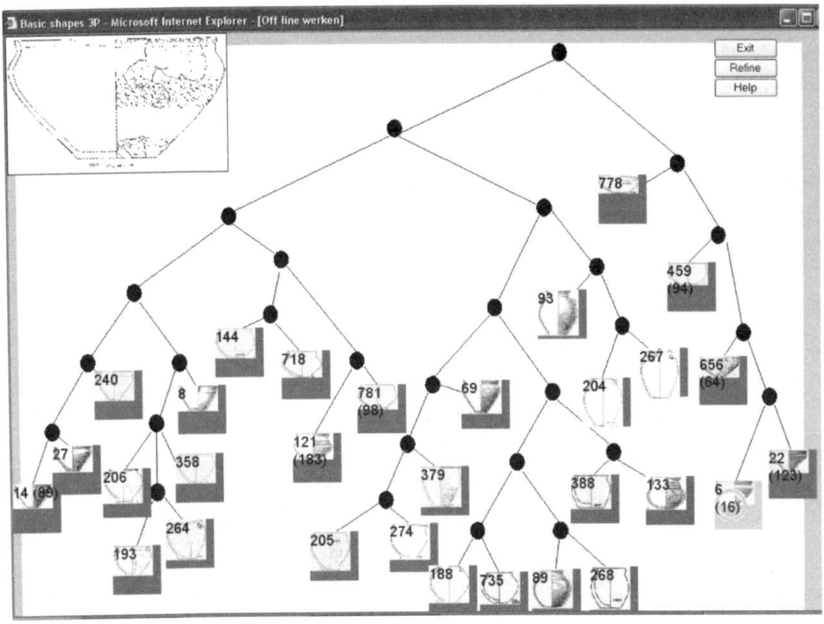

Fig. 3. Search results. In the upper left corner the uploaded file. Shape group 6 (lower right corner) gives the best match: a 'distance' of 16.

Fig. 4). The whole process takes about two minutes but has not yet been optimized with respect to speed. The user/archaeologist is, of course, expected

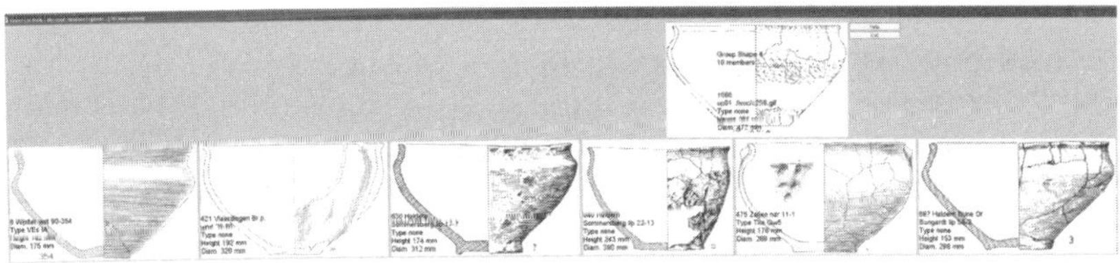

Fig. 4. Vessel cat walk. Above the uploaded file, below vessels from shape group 6, displayed from left to right according to dissimilarity.

to scrutinise the results and to refer to the original publications for background information, and the method should be regarded as an quick and reliable *first step* only, not as an automated determination system.

7 Other types of artifacts

The above text describes the currently largest Secanto database with handmade Iron Age vessels, but the technique of course works for all types of vessels that possess rotational symmetry along the vertical axis. The Netherlands National Service for Archaeological Heritage applied the algorithm to their database with images of medieval glass and the results are comparable with the results for the Iron Age vessels. It is tempting to try other types of artifacts but the filling of relevant databases is a time consuming activity and it appears that every new field entails specific new technical challenges. The majority of coins, for example, have a circular shape, but a polar transformation can remedy this feature (see Fig. 5). But the extraction of a 'profile' is not a trivial matter and other comparison methods probably are more fruitful than the Sliced method.

Fig. 5. A coin, and it's polar transformation. Images courtesy of the Netherlands National Service for Archaeological Heritage.

Apart from the ceramic vessels, the usability of the Sliced method is currently being investigated for two types of stone tools: Mesolithic stone axes and Neolithic arrow points. It must be said that the shape of these stone tools is, by most archaeologists, not seen as a primary property of the tool which determines to which type a certain tool belongs: the technology applied is seen as the determining factor. Partly this has to do with the fact that the raw material from which these tools are made, usually flint stone, is not so easily shaped as clay, and an error during the making of the tool cannot be repaired. Nevertheless, within a group of stone tools made according to the same technology, a comparison of shapes may provide additional information to categorize these tools. A complicating factor is the fact that stone tools do not have rotational symmetry, so using 2 dimensional projections for the comparison ignores differences which may be quite significant. On the other hand, using two, or even three projections complicates the method and also brings up the question of the relative importance of the two (or three) calculated differences between these projections. In Fig. 6 a two-dimensional representation of the calculated dissimilarity matrix for a set of 23 arrow points is shown. A simple 'rattle and shake' algorithm (Monte Carlo combined with Deepest

Descent) was used to match the two-dimensional matrix with the real dissimilarity matrix, and the sum of squares and the calculated stress were used to observe the goodness of fit, see, e.g., Everitt and Rabe-Hesketh (1997) for some background on these topics.

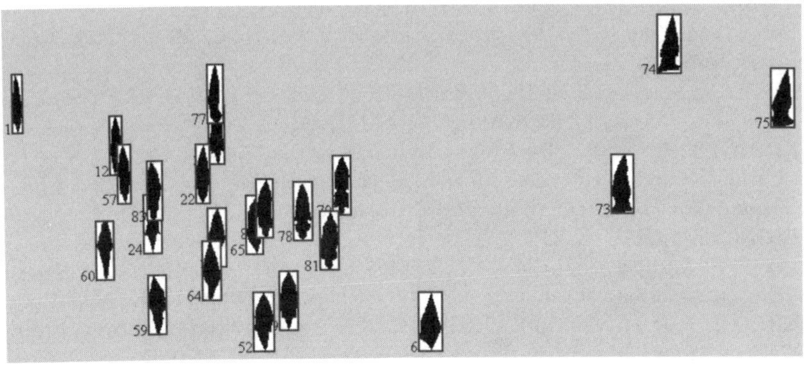

Fig. 6. A set of arrow points. The calculated dissimilarity matrix is the basis for this representation.

8 Summary

With the increasing availability of archaeological data sets the need for simple retrieving mechanisms based on *graphical* comparison increases accordingly as keywords can only describe the objects to a limited extent. Also, objectivity, repeatability and efficientcy are important issues. The Slicing method gives quite good results for handmade vessels, a category of artifacts that is infamous for its large variety of shapes as there was no industrial straitjacket to limit its number. For stone tools the usability might be somewhat less, but there also the usage of automated tools may initially help the archaeologist when she looks at that piece of flint and wonders 'Where did I see you before...'.

References

ABBINK, A.A. (1999): *Make It and Break It: The Cycles of Pottery. A Study of the Technology, Form, Function, and Use of Pottery from the Settlements at Uitgeest - Groot Dorregeest and Schagen - Muggenburg 1, Roman Period, North Holland, the Netherlands.* PhD Thesis, Leiden: Faculty of Archaeology, Leiden University, The Netherlands.

BLOEMERS, J.H.F. (1978): *Rijswijk (Z.H.) 'De Bult' Eine siedlung der Cananefaten.* ROB, Amersfoort.

DIEDERIK, F. (2002): *'Schervengericht': een onderzoek naar inheems aardewerk uit de late derde en de vierde eeuw in de Kop van Noord-Holland.* Archeologische Werkgemeenschap voor Nederland (AWN), Amsterdam.

ES, W.A. van (1965): Wijster, A Native Village Beyond the Imperial Frontier 150 - 425 A.D. In *Palaeohistoria Acta et Commvnicationes Institvti Bio-Archaeologici Vniversitatis Groninganae VOL. XI.* J.B. Wolters, Groningen.

ES, W.A. van (1968): *Paddepoel, Excavations of Frustrated Terps, 200 B.C. - 250 A.D.* ROB, Amersfoort.

EVERITT, B.S. and RABE-HESKETH, S. (1997): *The Analysis of Proximity Data. Kendall's Library of Statistics 4.* Arnold, London.

HEERINGEN, R.M. van (1987): The Iron Age in the Western Netherlands, II: Site Catalogue and Pottery Description. *Berichten van de Rijksdienst voor het Oudheidkundig Bodemonderzoek 37.* ROB, Amersfoort.

HEERINGEN, R.M. van (1989): The Iron Age in the Western Netherlands, III / IV / V: Site Catalogue, Pottery Description and Synthesis. *Berichten van de Rijksdienst voor het Oudheidkundig Bodemonderzoek 39.* ROB, Amersfoort.

HEERINGEN, R.M. van and TRIERUM, M.C. van (1981): The Iron Age in the Western Netherlands, I: Introduction and Method of Pottery Description. *Berichten van de Rijksdienst voor het Oudheidkundig Bodemonderzoek 39.* ROB, Amersfoort.

MOM, V. (2006): SECANTO, the Section Analysis Tool. *Computer Applications and Quantitative Methods in Archaeology. Proceedings of the 33nd Conference, Tomar, April 2005, forthcoming.*

REICHMANN, C. (1979): *Zur Besiedlungsgeschichte des Lippemündungsgebietes während der jüngeren vorrömischen Eisenzeit und der ältesten römischen Kaiserzeit: Ein Beitrag zur archäologischen Interpretation schriftlicher Überlieferung.* Dambeck, Wesel.

SHENNAN, S.J. and WILCOCK, J.D. (1975): Shape and Style Variation in Central German Bell Beakers. *Science and Archaeology, 15, 17-31.*

TAAYKE, E. (1987): Die einheimische Keramik der nordlichen Niederlande, 600 v.Chr. bis 300 n. Chr. *Berichten van de Rijksdienst voor het Oudheidkundig Bodemonderzoek 42.* ROB, Amersfoort.

TOL, A. et al. (2000): *Twee urnenvelden in Limburg : een verslag verslag van opgravingen te Roermond en Sittard, 1997-1998.* Vossiuspers AUP, Amsterdam.

Keywords

Author Index

Printing: Krips bv, Meppel
Binding: Stürtz, Würzburg